A BRIDGE TO
LINEAR ALGEBRA

A BRIDGE TO LINEAR ALGEBRA

Dragu Atanasiu
University of Borås, Sweden

Piotr Mikusiński
University of Central Florida, USA

World Scientific

NEW JERSEY · LONDON · SINGAPORE · BEIJING · SHANGHAI · HONG KONG · TAIPEI · CHENNAI · TOKYO

Published by

World Scientific Publishing Co. Pte. Ltd.

5 Toh Tuck Link, Singapore 596224

USA office: 27 Warren Street, Suite 401-402, Hackensack, NJ 07601

UK office: 57 Shelton Street, Covent Garden, London WC2H 9HE

Library of Congress Cataloging-in-Publication Data

Names: Atanasiu, Dragu, author. | Mikusiński, Piotr, author.

Title: A bridge to linear algebra / by Dragu Atanasiu (University of Borås, Sweden),
    Piotr Mikusiński (University of Central Florida, USA).

Description: New Jersey : World Scientific, 2019. | Includes index.

Identifiers: LCCN 2018061427| ISBN 9789811200229 (hardcover : alk. paper) |
    ISBN 9789811201462 (pbk. : alk. paper)

Subjects: LCSH: Algebras, Linear--Textbooks. | Algebra--Textbooks.

Classification: LCC QA184.2 .A83 2019 | DDC 512/.5--dc23

LC record available at https://lccn.loc.gov/2018061427

British Library Cataloguing-in-Publication Data

A catalogue record for this book is available from the British Library.

For any available supplementary material, please visit
https://www.worldscientific.com/worldscibooks/10.1142/11276#t=suppl

We dedicate this book to our wives,

Delia and Grażyna

Contents

Preface   **ix**

1   Basic ideas of linear algebra   **1**
  1.1 2×2 matrices   1
  1.2   Inverse matrices   10
  1.3   Determinants   28
  1.4   Diagonalization of 2×2 matrices   39

2   Matrices   **55**
  2.1   General matrices   55
  2.2   Gaussian elimination   71
  2.3   The inverse of a matrix   101

3   The vector space $\mathbb{R}^2$ **131**
  3.1   Vectors in $\mathbb{R}^2$ 132
  3.2   The dot product and the projection on a vector line in $\mathbb{R}^2$ 143
  3.3   Symmetric 2×2 matrices   162

4   The vector space $\mathbb{R}^3$ **179**
  4.1   Vectors in $\mathbb{R}^3$ 180
  4.2   Projections in $\mathbb{R}^3$ 200

5   Determinants and bases in $\mathbb{R}^3$ **233**
  5.1   The cross product   233
  5.2   Calculating inverses and determinants of 3×3 matrices   250
  5.3   Linear dependence of three vectors in $\mathbb{R}^3$ 264
  5.4   The dimension of a vector subspace of $\mathbb{R}^3$ 283

6   Singular value decomposition of 3×2 matrices   **291**

7   Diagonalization of 3×3 matrices   **307**
  7.1   Eigenvalues and eigenvectors of 3×3 matrices   307
  7.2   Symmetric 3×3 matrices   331

8   Applications to geometry   **355**
   8.1   Lines in $\mathbb{R}^2$ 355
   8.2   Lines and planes in $\mathbb{R}^3$ 370

9   Rotations   **391**
   9.1   Rotations in $\mathbb{R}^2$ 391
   9.2   Quadratic forms 400
   9.3   Rotations in $\mathbb{R}^3$ 414
   9.4   Cross product and the right-hand rule 420

10 Problems in plane geometry   **429**
   10.1 Lines and circles 429
   10.2 Triangles 433
   10.3 Geometry and trigonometry 443
   10.4 Geometry problems from the International Mathematical Olympiads . 446

11 Problems for a computer algebra system   **457**

12 Answers to selected exercises   **459**

Bibliography   **491**

Index   **493**

Preface

As teachers in the classroom, we have noticed that some hardworking students have trouble finding their feet when tackling linear algebra for the first time. One of us has been teaching students in Sweden for many years using a more accessible method which became the eventual foundation and inspiration for this book.

Why do we need yet another linear algebra text?

To provide introductory level mathematics students greater opportunities for success in both grasping, practicing, and internalizing the foundation tools of Linear Algebra. We present these tools in concrete examples prior to being presented with higher level complex concepts, properties and operations.

TO STUDENTS:

This book is intended to be read, with or without help from an instructor, as an introduction to the general theory presented in a standard linear algebra course. Students are encouraged to read it before or parallel with a standard linear algebra textbook as a study guide, practice book, or reference source for whatever and whenever they have problems understanding the general theory. This book can also be recommended as a student aid and its material assigned by an instructor as a reference source for students needing some coaching, clarification, or PRACTICE!

It is our goal to provide a "lifesaver" for students drowning in a standard linear algebra course. When students get confused, lost, or stuck with a general result, they can find a particular case of that result in this book done with all the details and consequently easy to read. Then the general result will make much more sense.

We welcome students to use this guide to become more comfortable, confident, and successful in understanding the concepts and tools of linear algebra.

GOOD LUCK!

TO INSTRUCTORS:

Let's face it, many students experience difficulties when they learn linear algebra for the first time. For example, they struggle to understand concepts like linear independence and bases. In order to help students we propose the following pedagogical approach: We present in depth all major topics of a standard course in linear algebra in the context of $\mathbb{R}^2$ and $\mathbb{R}^3$, including linear independence, bases, dimension, change of basis, rank theorem, rank nullity theorem, orthogonality, projections, determinant, eigenvalues and eigenvectors, diagonalization, spectral decomposition, rotations, quadratic forms. We also give an elementary and very detailed presentation of the singular value decomposition of a 3×2 matrix.

Students gain understanding of these ideas by studying concrete cases and solving relatively simple but nontrivial problems, where the essential ideas are not lost in computational complexities of higher dimensions.

The only part where we do not restrict ourselves to particular cases is when we present the algebra of matrices, Gauss elimination, and inverse matrices.

There are many repetitions in order to facilitate understanding of the presented ideas. For example, the dimension is first defined for $\mathbb{R}^2$ and vector subspaces of dimension 2 in $\mathbb{R}^3$ and then for $\mathbb{R}^3$. QR factorization is first presented for 2×2 matrices, then 3×2 matrices, and finally for 3×3 matrices.

Our approach uses more geometry than most books on linear algebra. In our opinion, this is a very natural presentation of linear algebra. Using concepts of linear algebra we obtain powerful tools to solve plane geometry problems. At the same time, geometry offers a way to use and understand linear algebra. This book proves that there in no conflict between analytic geometry and linear algebra, as it was presented in older books.

When writing this book we were influenced by the recommendations of the Linear Algebra Curriculum Study Group.

Now a few words about the content of the book.

Chapter 1 presents most of the basic ideas of this book in the context of 2×2 matrices. We attempted to make this chapter more dynamic, introducing from the beginning elementary matrices, inverse of a matrix, determinant, LU decomposition, eigenvalues and eigenvectors, and in this way hoping that students would find it more attractive and that it will stimulate curiosity of students about the content of the rest of the book.

Chapter 2 is about the algebra of general matrices, Gauss elimination, and inverse matrices. This chapter is less abstract and easier to understand.

Chapters 3, 4, and 5 form the kernel of this book. Here we present vectors in $\mathbb{R}^2$ and $\mathbb{R}^3$, linear independence, bases, dimension and orthogonality. We can say that Chapter 3 is about the vector space $\mathbb{R}^2$, Chapter 4 about the vector subspaces of dimension 2 of $\mathbb{R}^3$ and Chapter 5 is about the vector space $\mathbb{R}^3$.

Some applications are also discussed. In Chapter 3 we present QR factorization for 2×2 matrices and in Chapter 4 we present the least square method for 3×2 matrices and QR factorization for matrices 3×2 matrices. In Chapter 5 we discuss practical methods for calculating determinants of 3×3 matrices.

Chapter 6 is a short chapter about singular value decomposition for 3×2 matrices. The meaning of this chapter is to give more opportunities to use matrices. It will also help students understand the singular value decomposition in the general case.

Chapter 7 is about diagonalization in $\mathbb{R}^3$. We include complete calculations for many determinants and solve numerous systems of equations. At the end of the chapter we present 3×3 symmetric matrices and QR factorization for 3×3 matrices.

Chapter 8 gives a presentation of classical analytic geometry compatible with the concepts of linear algebra.

Chapter 9 is about rotations in $\mathbb{R}^2$ and $\mathbb{R}^3$. Quadratic forms in $\mathbb{R}^2$ are also discussed here because our presentation makes use of rotations.

Chapter 10 contains, for readers interested in geometry, completely solved problems in plane geometry. Among them are four problems given at International Mathematical Olympiads. In our solutions we use concepts and tools from linear algebra, including vectors, norm, linear independence, and rotations.

Because the use of technology is also important for students, we give some examples using Maple in an appendix at the end of the book. This part is not emphasized, since practically all examples and exercises in the book are designed for "paper and pencil" calculations. We believe that the experience of working through these examples improves understanding of the presented material.

In several places of this book we refer to the book *Core Topics in Linear Algebra*, which presents the standard topics of an introductory course in linear algebra. These two books can be used in parallel, with *A Bridge to Linear Algebra* providing a wealth of examples for the ideas discussed in *Core Topics in Linear Algebra*. On the other hand, the books are written so that they can be used independently. When the reader is directed to the book *Core Topics in Linear Algebra*, actually any standard book for an introductory course in linear algebra can be used.

ACKOWLEDGEMENTS:

We would like to thank Joseph Brennan from the University of Central Florida for fruitful discussions that influenced the final version of the book. We acknowledge the effort and time spent of our colleagues from the University of Borås, Anders Bengtsson, Martin Bohlén, Anders Mattsson, and Magnus Lundin, who critiqued portions of the earlier versions of the manuscript. We also benefitted from the comments of the reviewers. We are indebted to the students from the University of Borås who were the inspiration for writing this book. We would like to thank Delia Dumitrescu for drawing the hand needed for the right-hand rule and designing the figures for the problems from the International Mathematical Olympiads. We are grateful to the World Scientific Publishing team, including Rochelle Kronzek-Miller, Lai Fun Kwong, Rok Ting Tan, Yolande Koh, and Uthrapathy Janarthanan, for their support and assistance. Finally, we would like to express our gratitude the TeX-LaTeX Stack Exchange community for helping us on several occasions with LaTeX questions.

Chapter 1

Basic ideas of linear algebra

1.1 2×2 **matrices**

The introduction of matrices is one of the great ideas of linear algebra. Matrices were invented to solve some mathematical problems, like systems of linear equations, in a shorter, more transparent and more elegant way. In this chapter we describe some operations on matrices. The purpose of this chapter is to provide motivation and an opportunity for the reader to work with matrices. The ideas introduced here will be generalized and discussed in a more systematic way in the following chapters.

Solving linear equations is one of the basic problems of mathematics. Linear equations are also among the most common models for real life problems. The simplest linear equation is

$$ax = b, \tag{1.1}$$

where a and b are known real numbers and x is the unknown quantity. The equation has a unique solution if and only if $a \neq 0$. The solution is $x = \frac{b}{a}$.

Now we consider the system of equations:

$$\begin{cases} ax + by = e \\ cx + dy = f \end{cases}, \tag{1.2}$$

where a, b, c, d, e, and f are known real numbers and x and y are to be determined. This looks much more complicated than the equation $ax = b$. Linear algebra gives us tools that allow us to treat (1.2), and in fact many other more complicated problems, as a special case of the basic equation

$$Ax = \mathbf{b}, \tag{1.3}$$

where A, $\mathbf{x}$, and $\mathbf{b}$ are no longer numbers, but many similarities between this equation and (1.1) remain. If we think of $\mathbf{x}$ as the solution of (1.2), then it should be represented by both x and y. We will use the notation

$$\mathbf{x} = \begin{bmatrix} x \\ y \end{bmatrix}$$

and call **x** a 2 × 1 ***matrix*** or a 2 × 1 ***vector***. The geometric interpretation of vectors will be discussed in Chapter 3. At this time we think of $\begin{bmatrix} x \\ y \end{bmatrix}$ as a way of representing a solution of the system (1.2). Similarly, we write $\mathbf{b} = \begin{bmatrix} e \\ f \end{bmatrix}$.

By definition

$$\begin{bmatrix} a_1 \\ b_1 \end{bmatrix} = \begin{bmatrix} a_2 \\ b_2 \end{bmatrix} \quad \text{if and only if} \quad a_1 = a_2 \text{ and } b_1 = b_2.$$

If we go back to the system (1.2) we quickly realize that A has to contain the information about all coefficients, that is, a, b, c, and d. To capture this information we will write

$$A = \begin{bmatrix} a & b \\ c & d \end{bmatrix}.$$

Such an array is called a 2 × 2 *matrix*.

We also have by definition

$$\begin{bmatrix} a_1 & b_1 \\ c_1 & d_1 \end{bmatrix} = \begin{bmatrix} a_2 & b_2 \\ c_2 & d_2 \end{bmatrix}$$

if and only if

$$a_1 = a_2, \ b_1 = b_2, \ c_1 = c_2, \ \text{and } d_1 = d_2.$$

Consequently, $\begin{bmatrix} 1 & 2 \\ 3 & 4 \end{bmatrix} \neq \begin{bmatrix} 1 & 2 \\ 4 & 3 \end{bmatrix}$.

Now (1.2) can be written as $A\mathbf{x} = \mathbf{b}$ or

$$\begin{bmatrix} a & b \\ c & d \end{bmatrix} \begin{bmatrix} x \\ y \end{bmatrix} = \begin{bmatrix} e \\ f \end{bmatrix}$$

if we define

$$\begin{bmatrix} a & b \\ c & d \end{bmatrix} \begin{bmatrix} x \\ y \end{bmatrix} = \begin{bmatrix} ax + by \\ cx + dy \end{bmatrix}. \tag{1.4}$$

Definition 1.1.1. The vector $\begin{bmatrix} ax + by \\ cx + dy \end{bmatrix}$ is called the ***product*** of the matrix $\begin{bmatrix} a & b \\ c & d \end{bmatrix}$ and the vector $\begin{bmatrix} x \\ y \end{bmatrix}$.

Example 1.1.2. The system

$$\begin{cases} x + 3y = 6 \\ 2x + \ y = 1 \end{cases}$$

can be written as

$$\begin{bmatrix} 1 & 3 \\ 2 & 1 \end{bmatrix} \begin{bmatrix} x \\ y \end{bmatrix} = \begin{bmatrix} 6 \\ 1 \end{bmatrix},$$

where

$$\begin{bmatrix} 1 & 3 \\ 2 & 1 \end{bmatrix} \begin{bmatrix} x \\ y \end{bmatrix} = \begin{bmatrix} x+3y \\ 2x+y \end{bmatrix}.$$

By a 1×2 matrix we mean a row $\begin{bmatrix} a_1 & a_2 \end{bmatrix}$ of two real numbers a_1 and a_2. As in the case of other matrices, we write $\begin{bmatrix} a_1 & a_2 \end{bmatrix} = \begin{bmatrix} b_1 & b_2 \end{bmatrix}$ if and only if $a_1 = b_1$ and $a_2 = b_2$.

The operation in (1.4) can be viewed as the result of combining two simpler operations. To this end we define the product of a 1×2 matrix $\begin{bmatrix} a_1 & a_2 \end{bmatrix}$ by a 2×1 matrix $\begin{bmatrix} b_1 \\ b_2 \end{bmatrix}$:

$$\begin{bmatrix} a_1 & a_2 \end{bmatrix} \begin{bmatrix} b_1 \\ b_2 \end{bmatrix} = a_1 b_1 + a_2 b_2. \qquad (1.5)$$

Example 1.1.3.

$$\begin{bmatrix} 5 & 4 \end{bmatrix} \begin{bmatrix} 2 \\ -6 \end{bmatrix} = 5 \cdot 2 + 4 \cdot (-6) = -14.$$

Using the operation defined in (1.5), the operation introduced in (1.4) can be written as

$$\begin{bmatrix} a_1 & a_2 \\ a_3 & a_4 \end{bmatrix} \begin{bmatrix} b_1 \\ b_2 \end{bmatrix} = \begin{bmatrix} \begin{bmatrix} a_1 & a_2 \end{bmatrix} \begin{bmatrix} b_1 \\ b_2 \end{bmatrix} \\ \begin{bmatrix} a_3 & a_4 \end{bmatrix} \begin{bmatrix} b_1 \\ b_2 \end{bmatrix} \end{bmatrix} = \begin{bmatrix} a_1 b_1 + a_2 b_2 \\ a_3 b_1 + a_4 b_2 \end{bmatrix}.$$

This might look like a more complicated expression than (1.4), but it is actually a convenient way of interpreting the product of a 2×2 matrix and a 2×1 matrix and it will serve us well in more complicated situations considered later.

Example 1.1.4. We want to calculate

$$\begin{bmatrix} 4 & -9 \\ 3 & 2 \end{bmatrix} \begin{bmatrix} 6 \\ 1 \end{bmatrix}.$$

Since

$$\begin{bmatrix} 4 & -9 \end{bmatrix} \begin{bmatrix} 6 \\ 1 \end{bmatrix} = 15 \quad \text{and} \quad \begin{bmatrix} 3 & 2 \end{bmatrix} \begin{bmatrix} 6 \\ 1 \end{bmatrix} = 20,$$

we obtain

$$\begin{bmatrix} 4 & -9 \\ 3 & 2 \end{bmatrix} \begin{bmatrix} 6 \\ 1 \end{bmatrix} = \begin{bmatrix} 15 \\ 20 \end{bmatrix}.$$

Using the operation defined in (1.5), we can also define the product of a 1×2 matrix $[a_1 \ a_2]$ and a 2×2 matrix $\begin{bmatrix} b_1 & b_3 \\ b_2 & b_4 \end{bmatrix}$:

$$[a_1 \ a_2] \begin{bmatrix} b_1 & b_3 \\ b_2 & b_4 \end{bmatrix} = \left[[a_1 \ a_2] \begin{bmatrix} b_1 \\ b_2 \end{bmatrix} \ \ [a_1 \ a_2] \begin{bmatrix} b_3 \\ b_4 \end{bmatrix} \right] = [a_1 b_1 + a_2 b_2 \ \ a_1 b_3 + a_2 b_4].$$

Example 1.1.5. To calculate

$$[7 \ -1] \begin{bmatrix} 2 & 4 \\ 8 & -2 \end{bmatrix}$$

we first find

$$[7 \ -1] \begin{bmatrix} 2 \\ 8 \end{bmatrix} = 6 \quad \text{and} \quad [7 \ -1] \begin{bmatrix} 4 \\ -2 \end{bmatrix} = 30.$$

Hence

$$[7 \ -1] \begin{bmatrix} 2 & 4 \\ 8 & -2 \end{bmatrix} = [6 \ 30].$$

Finally we define the product of two 2×2 matrices, again using the operation defined in (1.5):

$$\begin{bmatrix} a_1 & a_2 \\ a_3 & a_4 \end{bmatrix} \begin{bmatrix} b_1 & b_2 \\ b_3 & b_4 \end{bmatrix} = \begin{bmatrix} [a_1 \ a_2] \begin{bmatrix} b_1 \\ b_3 \end{bmatrix} & [a_1 \ a_2] \begin{bmatrix} b_2 \\ b_4 \end{bmatrix} \\ [a_3 \ a_4] \begin{bmatrix} b_1 \\ b_3 \end{bmatrix} & [a_3 \ a_4] \begin{bmatrix} b_2 \\ b_4 \end{bmatrix} \end{bmatrix}.$$

Note that the product of two 2×2 matrices can be equivalently expressed in one of the following three ways:

$$\begin{bmatrix} a_1 & a_2 \\ a_3 & a_4 \end{bmatrix} \begin{bmatrix} b_1 & b_2 \\ b_3 & b_4 \end{bmatrix} = \begin{bmatrix} [a_1 \ a_2] \begin{bmatrix} b_1 & b_2 \\ b_3 & b_4 \end{bmatrix} \\ [a_3 \ a_4] \begin{bmatrix} b_1 & b_2 \\ b_3 & b_4 \end{bmatrix} \end{bmatrix}$$

$$= \left[\begin{bmatrix} a_1 & a_2 \\ a_3 & a_4 \end{bmatrix} \begin{bmatrix} b_1 \\ b_3 \end{bmatrix} \ \ \begin{bmatrix} a_1 & a_2 \\ a_3 & a_4 \end{bmatrix} \begin{bmatrix} b_2 \\ b_4 \end{bmatrix} \right]$$

$$= \begin{bmatrix} a_1 b_1 + a_2 b_3 & a_1 b_2 + a_2 b_4 \\ a_3 b_1 + a_4 b_3 & a_3 b_2 + a_4 b_4 \end{bmatrix}.$$

Example 1.1.6. We wish to calculate the product

$$\begin{bmatrix} 1 & 5 \\ 3 & 2 \end{bmatrix} \begin{bmatrix} 4 & -3 \\ -1 & 6 \end{bmatrix}.$$

We have

$$[1 \ 5]\begin{bmatrix} 4 \\ -1 \end{bmatrix} = -1, \quad [1 \ 5]\begin{bmatrix} -3 \\ 6 \end{bmatrix} = 27$$

and

$$[3 \ 2]\begin{bmatrix} 4 \\ -1 \end{bmatrix} = 10, \quad [3 \ 2]\begin{bmatrix} -3 \\ 6 \end{bmatrix} = 3.$$

This means that

$$\begin{bmatrix} 1 & 5 \\ 3 & 2 \end{bmatrix}\begin{bmatrix} 4 & -3 \\ -1 & 6 \end{bmatrix} = \begin{bmatrix} -1 & 27 \\ 10 & 3 \end{bmatrix}.$$

It is important to remember that the product of matrices is not commutative, that is, the result usually depends on the order of matrices.

Example 1.1.7. For the product

$$\begin{bmatrix} 4 & -3 \\ -1 & 6 \end{bmatrix}\begin{bmatrix} 1 & 5 \\ 3 & 2 \end{bmatrix}$$

we calculate

$$[4 \ -3]\begin{bmatrix} 1 \\ 3 \end{bmatrix} = -5, \quad [4 \ -3]\begin{bmatrix} 5 \\ 2 \end{bmatrix} = 14$$

and

$$[-1 \ 6]\begin{bmatrix} 1 \\ 3 \end{bmatrix} = 17, \quad [-1 \ 6]\begin{bmatrix} 5 \\ 2 \end{bmatrix} = 7.$$

This means that

$$\begin{bmatrix} 4 & -3 \\ -1 & 6 \end{bmatrix}\begin{bmatrix} 1 & 5 \\ 3 & 2 \end{bmatrix} = \begin{bmatrix} -5 & 14 \\ 17 & 7 \end{bmatrix},$$

while in the previous example we found that

$$\begin{bmatrix} 1 & 5 \\ 3 & 2 \end{bmatrix}\begin{bmatrix} 4 & -3 \\ -1 & 6 \end{bmatrix} = \begin{bmatrix} -1 & 27 \\ 10 & 3 \end{bmatrix}.$$

The results are completely different.

Now we consider products of three matrices. There are two ways we can calculate a product of three matrices, as the next example illustrates.

Example 1.1.8. Show that

$$\left([2 \ 3]\begin{bmatrix} -1 & 2 \\ 1 & 1 \end{bmatrix}\right)\begin{bmatrix} -8 \\ 2 \end{bmatrix} = [2 \ 3]\left(\begin{bmatrix} -1 & 2 \\ 1 & 1 \end{bmatrix}\begin{bmatrix} -8 \\ 2 \end{bmatrix}\right).$$

Solution. First we calculate the product

$$\left([2 \ 3] \begin{bmatrix} -1 & 2 \\ 1 & 1 \end{bmatrix}\right) \begin{bmatrix} -8 \\ 2 \end{bmatrix}.$$

We find

$$[2 \ 3] \begin{bmatrix} -1 & 2 \\ 1 & 1 \end{bmatrix} = [1 \ 7]$$

and then

$$[1 \ 7] \begin{bmatrix} -8 \\ 2 \end{bmatrix} = 6.$$

Now we calculate

$$[2 \ 3] \left(\begin{bmatrix} -1 & 2 \\ 1 & 1 \end{bmatrix} \begin{bmatrix} -8 \\ 2 \end{bmatrix}\right).$$

We find

$$\begin{bmatrix} -1 & 2 \\ 1 & 1 \end{bmatrix} \begin{bmatrix} -8 \\ 2 \end{bmatrix} = \begin{bmatrix} 12 \\ -6 \end{bmatrix}$$

and then

$$[2 \ 3] \begin{bmatrix} 12 \\ -6 \end{bmatrix} = 6.$$

□

In the above example we get the same result regardless of the way the product is calculated. This is always true as the next theorem shows.

Theorem 1.1.9. *For any numbers $a_1, a_2, b_1, b_2, b_3, b_4, c_1, c_2$ we have*

$$\left([a_1 \ a_2] \begin{bmatrix} b_1 & b_2 \\ b_3 & b_4 \end{bmatrix}\right) \begin{bmatrix} c_1 \\ c_2 \end{bmatrix} = [a_1 \ a_2] \left(\begin{bmatrix} b_1 & b_2 \\ b_3 & b_4 \end{bmatrix} \begin{bmatrix} c_1 \\ c_2 \end{bmatrix}\right).$$

Proof. The equality can be verified by simply calculating the products on both sides and comparing the results. On the left-hand side we have

$$[a_1 \ a_2] \begin{bmatrix} b_1 & b_2 \\ b_3 & b_4 \end{bmatrix} = [a_1 b_1 + a_2 b_3 \quad a_1 b_2 + a_2 b_4]$$

and

$$[a_1 b_1 + a_2 b_3 \quad a_1 b_2 + a_2 b_4] \begin{bmatrix} c_1 \\ c_2 \end{bmatrix} = a_1 b_1 c_1 + a_2 b_3 c_1 + a_1 b_2 c_2 + a_2 b_4 c_2,$$

so

$$\left([a_1 \ a_2] \begin{bmatrix} b_1 & b_2 \\ b_3 & b_4 \end{bmatrix}\right) \begin{bmatrix} c_1 \\ c_2 \end{bmatrix} = a_1 b_1 c_1 + a_2 b_3 c_1 + a_1 b_2 c_2 + a_2 b_4 c_2.$$

We obtain the same result if we calculate

$$\begin{bmatrix} a_1 & a_2 \end{bmatrix} \left(\begin{bmatrix} b_1 & b_2 \\ b_3 & b_4 \end{bmatrix} \begin{bmatrix} c_1 \\ c_2 \end{bmatrix} \right).$$

The calculations are left as an exercise. □

The result in the above lemma is an example of associativity of matrix multiplication. It is an important property of matrix multiplication and it allows us to write the product

$$\begin{bmatrix} a_1 & a_2 \end{bmatrix} \begin{bmatrix} b_1 & b_2 \\ b_3 & b_4 \end{bmatrix} \begin{bmatrix} c_1 \\ c_2 \end{bmatrix}$$

without parentheses. In the next theorem we prove the associativity property for the product of three 2 × 2 matrices.

Theorem 1.1.10. *For any numbers* $a_1, a_2, a_3, a_4, b_1, b_2, b_3, b_4, c_1, c_2, c_3, c_4$ *we have*

$$\left(\begin{bmatrix} a_1 & a_2 \\ a_3 & a_4 \end{bmatrix} \begin{bmatrix} b_1 & b_2 \\ b_3 & b_4 \end{bmatrix} \right) \begin{bmatrix} c_1 & c_2 \\ c_3 & c_4 \end{bmatrix} = \begin{bmatrix} a_1 & a_2 \\ a_3 & a_4 \end{bmatrix} \left(\begin{bmatrix} b_1 & b_2 \\ b_3 & b_4 \end{bmatrix} \begin{bmatrix} c_1 & c_2 \\ c_3 & c_4 \end{bmatrix} \right).$$

Proof. The equality can be verified by calculating the products on both sides and comparing the results. However, such approach would lead to rather tedious calculations. We can significantly simplify our proof by employing Theorem 1.1.9.

First we observe that

$$\begin{bmatrix} a_1 & a_2 \\ a_3 & a_4 \end{bmatrix} \begin{bmatrix} b_1 & b_2 \\ b_3 & b_4 \end{bmatrix} = \begin{bmatrix} \begin{bmatrix} a_1 & a_2 \end{bmatrix} \begin{bmatrix} b_1 & b_2 \\ b_3 & b_4 \end{bmatrix} \\ \begin{bmatrix} a_3 & a_4 \end{bmatrix} \begin{bmatrix} b_1 & b_2 \\ b_3 & b_4 \end{bmatrix} \end{bmatrix}$$

and consequently

$$\left(\begin{bmatrix} a_1 & a_2 \\ a_3 & a_4 \end{bmatrix} \begin{bmatrix} b_1 & b_2 \\ b_3 & b_4 \end{bmatrix} \right) \begin{bmatrix} c_1 & c_2 \\ c_3 & c_4 \end{bmatrix} = \begin{bmatrix} \left(\begin{bmatrix} a_1 & a_2 \end{bmatrix} \begin{bmatrix} b_1 & b_2 \\ b_3 & b_4 \end{bmatrix} \right) \begin{bmatrix} c_1 \\ c_3 \end{bmatrix} & \left(\begin{bmatrix} a_1 & a_2 \end{bmatrix} \begin{bmatrix} b_1 & b_2 \\ b_3 & b_4 \end{bmatrix} \right) \begin{bmatrix} c_2 \\ c_4 \end{bmatrix} \\ \left(\begin{bmatrix} a_3 & a_4 \end{bmatrix} \begin{bmatrix} b_1 & b_2 \\ b_3 & b_4 \end{bmatrix} \right) \begin{bmatrix} c_1 \\ c_3 \end{bmatrix} & \left(\begin{bmatrix} a_3 & a_4 \end{bmatrix} \begin{bmatrix} b_1 & b_2 \\ b_3 & b_4 \end{bmatrix} \right) \begin{bmatrix} c_2 \\ c_4 \end{bmatrix} \end{bmatrix}.$$

Similarly,

$$\begin{bmatrix} a_1 & a_2 \\ a_3 & a_4 \end{bmatrix} \left(\begin{bmatrix} b_1 & b_2 \\ b_3 & b_4 \end{bmatrix} \begin{bmatrix} c_1 & c_2 \\ c_3 & c_4 \end{bmatrix} \right) = \begin{bmatrix} \begin{bmatrix} a_1 & a_2 \end{bmatrix} \left(\begin{bmatrix} b_1 & b_2 \\ b_3 & b_4 \end{bmatrix} \begin{bmatrix} c_1 \\ c_3 \end{bmatrix} \right) & \begin{bmatrix} a_1 & a_2 \end{bmatrix} \left(\begin{bmatrix} b_1 & b_2 \\ b_3 & b_4 \end{bmatrix} \begin{bmatrix} c_2 \\ c_4 \end{bmatrix} \right) \\ \begin{bmatrix} a_3 & a_4 \end{bmatrix} \left(\begin{bmatrix} b_1 & b_2 \\ b_3 & b_4 \end{bmatrix} \begin{bmatrix} c_1 \\ c_3 \end{bmatrix} \right) & \begin{bmatrix} a_3 & a_4 \end{bmatrix} \left(\begin{bmatrix} b_1 & b_2 \\ b_3 & b_4 \end{bmatrix} \begin{bmatrix} c_2 \\ c_4 \end{bmatrix} \right) \end{bmatrix}.$$

According to Theorem 1.1.9, the two matrices on the right-hand side are equal. □

We can see that matrix multiplication shares some properties with number multiplication, like associativity, but there are also some significant differences. For example matrix multiplication is not commutative. The number one plays a very special role in number multiplication, namely, $1 \cdot a = a \cdot 1 = a$ for any real number a. It turns out that there is a matrix that plays the same role in matrix multiplication.

Theorem 1.1.11. *For any numbers a, b, c, d we have*

$$\begin{bmatrix} 1 & 0 \\ 0 & 1 \end{bmatrix}\begin{bmatrix} a & b \\ c & d \end{bmatrix} = \begin{bmatrix} a & b \\ c & d \end{bmatrix}\begin{bmatrix} 1 & 0 \\ 0 & 1 \end{bmatrix} = \begin{bmatrix} a & b \\ c & d \end{bmatrix}.$$

Proof. The equalities can be verified by direct calculations.   □

Besides the matrix multiplication we will use addition of matrices of the same size. To add two matrices we simply add the corresponding entries of the matrices:

$$\begin{bmatrix} a_1 \\ a_2 \end{bmatrix} + \begin{bmatrix} b_1 \\ b_2 \end{bmatrix} = \begin{bmatrix} a_1 + b_1 \\ a_2 + b_2 \end{bmatrix}$$

$$\begin{bmatrix} a_1 & a_2 \end{bmatrix} + \begin{bmatrix} b_1 & b_2 \end{bmatrix} = \begin{bmatrix} a_1 + b_1 & a_2 + b_2 \end{bmatrix}$$

$$\begin{bmatrix} a_1 & a_2 \\ a_3 & a_4 \end{bmatrix} + \begin{bmatrix} b_1 & b_2 \\ b_3 & b_4 \end{bmatrix} = \begin{bmatrix} a_1 + b_1 & a_2 + b_2 \\ a_3 + b_3 & a_4 + b_4 \end{bmatrix}$$

We will also multiply matrices by real numbers. To multiply a matrix by a real number t we multiply every entry of that matrix by t:

$$t\begin{bmatrix} a_1 \\ a_2 \end{bmatrix} = \begin{bmatrix} t a_1 \\ t a_2 \end{bmatrix}$$

$$t\begin{bmatrix} a_1 & a_2 \end{bmatrix} = \begin{bmatrix} t a_1 & t a_2 \end{bmatrix}$$

$$t\begin{bmatrix} a_1 & a_2 \\ a_3 & a_4 \end{bmatrix} = \begin{bmatrix} t a_1 & t a_2 \\ t a_3 & t a_4 \end{bmatrix}$$

1.1.1 Exercises

Find the products of the given matrices.

1. $[3 \ -2] \begin{bmatrix} 4 \\ 5 \end{bmatrix}$

2. $[5 \ 3] \begin{bmatrix} 2 \\ -1 \end{bmatrix}$

3. $[2 \ -3] \begin{bmatrix} 5 & 2 \\ 3 & -4 \end{bmatrix}$

4. $[1 \ 7] \begin{bmatrix} 2 & -2 \\ 4 & 3 \end{bmatrix}$

5. $\begin{bmatrix} 7 & -1 \\ 2 & 4 \end{bmatrix} \begin{bmatrix} 1 \\ -5 \end{bmatrix}$

6. $\begin{bmatrix} 3 & 5 \\ -2 & 8 \end{bmatrix} \begin{bmatrix} 2 \\ 3 \end{bmatrix}$

7. $\begin{bmatrix} 4 & 1 \\ 5 & -1 \end{bmatrix} \begin{bmatrix} 2 & 1 \\ 5 & 9 \end{bmatrix}$

8. $\begin{bmatrix} 1 & 1 \\ -7 & 3 \end{bmatrix} \begin{bmatrix} 3 & -1 \\ 4 & 1 \end{bmatrix}$

9. $\begin{bmatrix} 2 & 1 \\ 5 & 9 \end{bmatrix} \begin{bmatrix} 4 & 1 \\ 5 & -1 \end{bmatrix}$

10. $\begin{bmatrix} 3 & -1 \\ 4 & 1 \end{bmatrix} \begin{bmatrix} 1 & 1 \\ -7 & 3 \end{bmatrix}$

11. $\begin{bmatrix} 7 & -2 \\ 5 & 3 \end{bmatrix} \begin{bmatrix} p & q \\ r & s \end{bmatrix}$

12. $\begin{bmatrix} 3 & 4 \\ 8 & 1 \end{bmatrix} \begin{bmatrix} p & q \\ r & s \end{bmatrix}$

13. $\begin{bmatrix} p & q \\ r & s \end{bmatrix} \begin{bmatrix} 7 & -2 \\ 5 & 3 \end{bmatrix}$

14. $\begin{bmatrix} p & q \\ r & s \end{bmatrix} \begin{bmatrix} 3 & 4 \\ 8 & 1 \end{bmatrix}$

15. Show by direct calculations that

$$[a_1 \ a_2] \left(\begin{bmatrix} b_1 & b_2 \\ b_3 & b_4 \end{bmatrix} \begin{bmatrix} c_1 \\ c_2 \end{bmatrix} \right) = a_1 b_1 c_1 + a_2 b_3 c_1 + a_1 b_2 c_2 + a_2 b_4 c_2$$

16. Show that the product

$$\begin{bmatrix} a_1 & a_2 \\ a_3 & a_4 \end{bmatrix} \left(\begin{bmatrix} b_1 & b_2 \\ b_3 & b_4 \end{bmatrix} \begin{bmatrix} c_1 & c_2 \\ c_3 & c_4 \end{bmatrix} \right)$$

can be written in the form

$$\begin{bmatrix} [a_1 \ a_2] \left(\begin{bmatrix} b_1 & b_2 \\ b_3 & b_4 \end{bmatrix} \begin{bmatrix} c_1 \\ c_3 \end{bmatrix} \right) & [a_1 \ a_2] \left(\begin{bmatrix} b_1 & b_2 \\ b_3 & b_4 \end{bmatrix} \begin{bmatrix} c_2 \\ c_4 \end{bmatrix} \right) \\ [a_3 \ a_4] \left(\begin{bmatrix} b_1 & b_2 \\ b_3 & b_4 \end{bmatrix} \begin{bmatrix} c_1 \\ c_3 \end{bmatrix} \right) & [a_3 \ a_4] \left(\begin{bmatrix} b_1 & b_2 \\ b_3 & b_4 \end{bmatrix} \begin{bmatrix} c_2 \\ c_4 \end{bmatrix} \right) \end{bmatrix}.$$

17. Show that

$$\begin{bmatrix} 1 & 0 \\ 0 & 1 \end{bmatrix} \begin{bmatrix} a & b \\ c & d \end{bmatrix} = \begin{bmatrix} a & b \\ c & d \end{bmatrix} \begin{bmatrix} 1 & 0 \\ 0 & 1 \end{bmatrix} = \begin{bmatrix} a & b \\ c & d \end{bmatrix}.$$

10. Show that

$$\left(s \begin{bmatrix} a_1 & a_2 \\ a_3 & a_4 \end{bmatrix} \right) \begin{bmatrix} b_1 & b_2 \\ b_3 & b_4 \end{bmatrix} = \begin{bmatrix} a_1 & a_2 \\ a_3 & a_4 \end{bmatrix} \left(s \begin{bmatrix} b_1 & b_2 \\ b_3 & b_4 \end{bmatrix} \right) = s \left(\begin{bmatrix} a_1 & a_2 \\ a_3 & a_4 \end{bmatrix} \begin{bmatrix} b_1 & b_2 \\ b_3 & b_4 \end{bmatrix} \right)$$

19. Show that

$$\left(s \begin{bmatrix} a_1 & a_2 \\ a_3 & a_4 \end{bmatrix} \right) \begin{bmatrix} b_1 \\ b_2 \end{bmatrix} = \begin{bmatrix} a_1 & a_2 \\ a_3 & a_4 \end{bmatrix} \left(s \begin{bmatrix} b_1 \\ b_2 \end{bmatrix} \right) = s \left(\begin{bmatrix} a_1 & a_2 \\ a_3 & a_4 \end{bmatrix} \begin{bmatrix} b_1 \\ b_2 \end{bmatrix} \right).$$

20. Show that if A is a 2×2 matrix and B and C are 2×1 vectors, then

$$A(B + C) = AB + AC.$$

21. Show that if A, B, and C are 2×2 matrices, then

$$A(B + C) = AB + AC.$$

22. Show that if A and B are 2×2 matrices and C is a 2×1 vector, then

$$(A + B)C = AC + BC.$$

23. Show that if A, B, and C are 2×2 matrices, then

$$(A + B)C = AC + BC.$$

1.2   Inverse matrices

When solving a linear equation $ax = b$, with $a \neq 0$, we multiply both sides of the equation by $\frac{1}{a}$ to obtain the solution $x = \frac{b}{a}$. We are now going to describe a generalization of this idea to matrix equations of the form $A\mathbf{x} = \mathbf{b}$.

Definition 1.2.1. If

$$\begin{bmatrix} \alpha & \beta \\ \gamma & \delta \end{bmatrix} \begin{bmatrix} a & b \\ c & d \end{bmatrix} = \begin{bmatrix} a & b \\ c & d \end{bmatrix} \begin{bmatrix} \alpha & \beta \\ \gamma & \delta \end{bmatrix} = \begin{bmatrix} 1 & 0 \\ 0 & 1 \end{bmatrix}, \tag{1.6}$$

then the matrix $\begin{bmatrix} \alpha & \beta \\ \gamma & \delta \end{bmatrix}$ is called an ***inverse of the matrix*** $\begin{bmatrix} a & b \\ c & d \end{bmatrix}$. A matrix that has an inverse is called an ***invertible*** matrix.

Note that, if $\begin{bmatrix} \alpha & \beta \\ \gamma & \delta \end{bmatrix}$ is an inverse matrix of $\begin{bmatrix} a & b \\ c & d \end{bmatrix}$, then $\begin{bmatrix} a & b \\ c & d \end{bmatrix}$ is an inverse matrix of $\begin{bmatrix} \alpha & \beta \\ \gamma & \delta \end{bmatrix}$. If (1.6) holds, we can say that the matrices $\begin{bmatrix} \alpha & \beta \\ \gamma & \delta \end{bmatrix}$ and $\begin{bmatrix} a & b \\ c & d \end{bmatrix}$ are inverses of each other.

Example 1.2.2. Since

$$\begin{bmatrix} 2 & 0 \\ 0 & \frac{1}{7} \end{bmatrix} \begin{bmatrix} \frac{1}{2} & 0 \\ 0 & 7 \end{bmatrix} = \begin{bmatrix} \frac{1}{2} & 0 \\ 0 & 7 \end{bmatrix} \begin{bmatrix} 2 & 0 \\ 0 & \frac{1}{7} \end{bmatrix} = \begin{bmatrix} 1 & 0 \\ 0 & 1 \end{bmatrix},$$

the matrices

$$\begin{bmatrix} 2 & 0 \\ 0 & \frac{1}{7} \end{bmatrix} \quad \text{and} \quad \begin{bmatrix} \frac{1}{2} & 0 \\ 0 & 7 \end{bmatrix}$$

are inverses of each other.

Example 1.2.3. Since

$$\begin{bmatrix} 1 & 5 \\ 0 & 1 \end{bmatrix} \begin{bmatrix} 1 & -5 \\ 0 & 1 \end{bmatrix} = \begin{bmatrix} 1 & -5 \\ 0 & 1 \end{bmatrix} \begin{bmatrix} 1 & 5 \\ 0 & 1 \end{bmatrix} = \begin{bmatrix} 1 & 0 \\ 0 & 1 \end{bmatrix},$$

the matrices

$$\begin{bmatrix} 1 & 5 \\ 0 & 1 \end{bmatrix} \quad \text{and} \quad \begin{bmatrix} 1 & -5 \\ 0 & 1 \end{bmatrix}$$

are inverses of each other.

Example 1.2.4. Since

$$\begin{bmatrix} 6 & 8 \\ 2 & 3 \end{bmatrix} \begin{bmatrix} \frac{3}{2} & -4 \\ -1 & 3 \end{bmatrix} = \begin{bmatrix} \frac{3}{2} & -4 \\ -1 & 3 \end{bmatrix} \begin{bmatrix} 6 & 8 \\ 2 & 3 \end{bmatrix} = \begin{bmatrix} 1 & 0 \\ 0 & 1 \end{bmatrix},$$

the matrices

$$\begin{bmatrix} 6 & 8 \\ 2 & 3 \end{bmatrix} \quad \text{and} \quad \begin{bmatrix} \frac{3}{2} & -4 \\ -1 & 3 \end{bmatrix}$$

are inverses of each other.

Theorem 1.2.5. *If a matrix has an inverse, then that inverse is unique.*

Proof. We need to show that, if

$$\begin{bmatrix} \alpha & \beta \\ \gamma & \delta \end{bmatrix} \begin{bmatrix} a & b \\ c & d \end{bmatrix} = \begin{bmatrix} a & b \\ c & d \end{bmatrix} \begin{bmatrix} \alpha & \beta \\ \gamma & \delta \end{bmatrix} = \begin{bmatrix} 1 & 0 \\ 0 & 1 \end{bmatrix},$$

and

$$\begin{bmatrix} s & t \\ u & v \end{bmatrix} \begin{bmatrix} a & b \\ c & d \end{bmatrix} = \begin{bmatrix} a & b \\ c & d \end{bmatrix} \begin{bmatrix} s & t \\ u & v \end{bmatrix} = \begin{bmatrix} 1 & 0 \\ 0 & 1 \end{bmatrix},$$

then

$$\begin{bmatrix} \alpha & \beta \\ \gamma & \delta \end{bmatrix} = \begin{bmatrix} s & t \\ u & v \end{bmatrix}.$$

Indeed, from the above assumptions and Theorem 1.1.11, we have

$$\begin{bmatrix} \alpha & \beta \\ \gamma & \delta \end{bmatrix} = \begin{bmatrix} \alpha & \beta \\ \gamma & \delta \end{bmatrix}\begin{bmatrix} 1 & 0 \\ 0 & 1 \end{bmatrix}$$

$$= \begin{bmatrix} \alpha & \beta \\ \gamma & \delta \end{bmatrix}\left(\begin{bmatrix} a & b \\ c & d \end{bmatrix}\begin{bmatrix} s & t \\ u & v \end{bmatrix}\right)$$

$$= \left(\begin{bmatrix} \alpha & \beta \\ \gamma & \delta \end{bmatrix}\begin{bmatrix} a & b \\ c & d \end{bmatrix}\right)\begin{bmatrix} s & t \\ u & v \end{bmatrix}$$

$$= \begin{bmatrix} 1 & 0 \\ 0 & 1 \end{bmatrix}\begin{bmatrix} s & t \\ u & v \end{bmatrix}$$

$$= \begin{bmatrix} s & t \\ u & v \end{bmatrix}.$$

□

The inverse of a matrix $\begin{bmatrix} a & b \\ c & d \end{bmatrix}$ will be denoted $\begin{bmatrix} a & b \\ c & d \end{bmatrix}^{-1}$. With the aid of inverse matrices we can easily solve matrix equations.

Theorem 1.2.6. *If the matrix* $\begin{bmatrix} a & b \\ c & d \end{bmatrix}$ *is invertible, then the equation*

$$\begin{bmatrix} a & b \\ c & d \end{bmatrix}\begin{bmatrix} x \\ y \end{bmatrix} = \begin{bmatrix} e \\ f \end{bmatrix} \tag{1.7}$$

has an unique solution which is

$$\begin{bmatrix} x \\ y \end{bmatrix} = \begin{bmatrix} a & b \\ c & d \end{bmatrix}^{-1}\begin{bmatrix} e \\ f \end{bmatrix}. \tag{1.8}$$

Proof. First we show that the numbers x and y defined by (1.8) satisfy equation (1.7). Indeed, from

$$\begin{bmatrix} x \\ y \end{bmatrix} = \begin{bmatrix} a & b \\ c & d \end{bmatrix}^{-1}\begin{bmatrix} e \\ f \end{bmatrix}$$

we obtain

$$\begin{bmatrix} a & b \\ c & d \end{bmatrix}\begin{bmatrix} x \\ y \end{bmatrix} = \begin{bmatrix} a & b \\ c & d \end{bmatrix}\left(\begin{bmatrix} a & b \\ c & d \end{bmatrix}^{-1}\begin{bmatrix} e \\ f \end{bmatrix}\right) = \left(\begin{bmatrix} a & b \\ c & d \end{bmatrix}\begin{bmatrix} a & b \\ c & d \end{bmatrix}^{-1}\right)\begin{bmatrix} e \\ f \end{bmatrix} = \begin{bmatrix} 1 & 0 \\ 0 & 1 \end{bmatrix}\begin{bmatrix} e \\ f \end{bmatrix} = \begin{bmatrix} e \\ f \end{bmatrix}.$$

Now suppose that we have

$$\begin{bmatrix} a & b \\ c & d \end{bmatrix}\begin{bmatrix} x \\ y \end{bmatrix} = \begin{bmatrix} e \\ f \end{bmatrix}.$$

Then

$$\begin{bmatrix} x \\ y \end{bmatrix} = \begin{bmatrix} 1 & 0 \\ 0 & 1 \end{bmatrix}\begin{bmatrix} x \\ y \end{bmatrix} = \left(\begin{bmatrix} a & b \\ c & d \end{bmatrix}^{-1}\begin{bmatrix} a & b \\ c & d \end{bmatrix} \right)\begin{bmatrix} x \\ y \end{bmatrix} = \begin{bmatrix} a & b \\ c & d \end{bmatrix}^{-1}\left(\begin{bmatrix} a & b \\ c & d \end{bmatrix}\begin{bmatrix} x \\ y \end{bmatrix} \right) = \begin{bmatrix} a & b \\ c & d \end{bmatrix}^{-1}\begin{bmatrix} e \\ f \end{bmatrix}.$$

□

Example 1.2.7. Solve the system of equations

$$\begin{cases} 6x + 8y = 3 \\ 2x + 3y = 2 \end{cases}.$$

Solution. The above system can be written as a matrix equation

$$\begin{bmatrix} 6 & 8 \\ 2 & 3 \end{bmatrix}\begin{bmatrix} x \\ y \end{bmatrix} = \begin{bmatrix} 3 \\ 2 \end{bmatrix}.$$

In Example 1.2.4 we found that

$$\begin{bmatrix} 6 & 8 \\ 2 & 3 \end{bmatrix}^{-1} = \begin{bmatrix} \frac{3}{2} & -4 \\ -1 & 3 \end{bmatrix}.$$

Consequently, the unique solution is

$$\begin{bmatrix} x \\ y \end{bmatrix} = \begin{bmatrix} \frac{3}{2} & -4 \\ -1 & 3 \end{bmatrix}\begin{bmatrix} 3 \\ 2 \end{bmatrix} = \begin{bmatrix} -\frac{7}{2} \\ 3 \end{bmatrix},$$

that is, $x = -\frac{7}{2}$ and $y = 3$. □

Note that once we know that $\begin{bmatrix} 6 & 8 \\ 2 & 3 \end{bmatrix}^{-1} = \begin{bmatrix} \frac{3}{2} & -4 \\ -1 & 3 \end{bmatrix}$, all we need to solve the system

$$\begin{cases} 6x + 8y = 1 \\ 2x + 3y = 5 \end{cases},$$

is to calculate the product $\begin{bmatrix} \frac{3}{2} & -4 \\ -1 & 3 \end{bmatrix}\begin{bmatrix} 1 \\ 5 \end{bmatrix} = \begin{bmatrix} -\frac{37}{2} \\ 14 \end{bmatrix}$. The solution is $x = -\frac{37}{2}$ and $y = 14$.

This indicates that being able to decide if a matrix is invertible and finding the inverse of an invertible matrix is important. We will consider different ways these problems can be solved. The first one uses elementary matrices.

Elementary matrices

> **Definition 1.2.8.** A 2×2 matrix is called an ***elementary matrix*** if it has the
> form of one of the following matrices
>
> $$\begin{bmatrix} 0 & 1 \\ 1 & 0 \end{bmatrix}, \quad \begin{bmatrix} s & 0 \\ 0 & 1 \end{bmatrix}, \quad \begin{bmatrix} 1 & 0 \\ 0 & s \end{bmatrix}, \quad \begin{bmatrix} 1 & 0 \\ t & 1 \end{bmatrix}, \quad \text{and} \quad \begin{bmatrix} 1 & t \\ 0 & 1 \end{bmatrix},$$
>
> where s and t are arbitrary numbers with $s \neq 0$.

The product of an elementary matrix and an arbitrary matrix behaves in a predictable fashion:

> $$\begin{bmatrix} 0 & 1 \\ 1 & 0 \end{bmatrix}\begin{bmatrix} a & b \\ c & d \end{bmatrix} = \begin{bmatrix} c & d \\ a & b \end{bmatrix}$$
>
> $$\begin{bmatrix} s & 0 \\ 0 & 1 \end{bmatrix}\begin{bmatrix} a & b \\ c & d \end{bmatrix} = \begin{bmatrix} sa & sb \\ c & d \end{bmatrix}$$
>
> $$\begin{bmatrix} 1 & 0 \\ 0 & s \end{bmatrix}\begin{bmatrix} a & b \\ c & d \end{bmatrix} = \begin{bmatrix} a & b \\ sc & sd \end{bmatrix}$$
>
> $$\begin{bmatrix} 1 & 0 \\ t & 1 \end{bmatrix}\begin{bmatrix} a & b \\ c & d \end{bmatrix} = \begin{bmatrix} a & b \\ c+ta & d+tb \end{bmatrix}$$
>
> $$\begin{bmatrix} 1 & t \\ 0 & 1 \end{bmatrix}\begin{bmatrix} a & b \\ c & d \end{bmatrix} = \begin{bmatrix} a+tc & b+td \\ c & d \end{bmatrix}$$

We will use elementary matrices to find inverse matrices. The process will usually require several multiplications by elementary matrices. To be able to do it quickly and correctly you should know the above identities very well.

Example 1.2.9.

$$\begin{bmatrix} 0 & 1 \\ 1 & 0 \end{bmatrix}\begin{bmatrix} 3 & 1 \\ 2 & 7 \end{bmatrix} = \begin{bmatrix} 2 & 7 \\ 3 & 1 \end{bmatrix}$$

$$\begin{bmatrix} 5 & 0 \\ 0 & 1 \end{bmatrix}\begin{bmatrix} 3 & 1 \\ 2 & 7 \end{bmatrix} = \begin{bmatrix} 15 & 5 \\ 2 & 7 \end{bmatrix}$$

$$\begin{bmatrix} 1 & 0 \\ 0 & 5 \end{bmatrix}\begin{bmatrix} 3 & 1 \\ 2 & 7 \end{bmatrix} = \begin{bmatrix} 3 & 1 \\ 10 & 35 \end{bmatrix}$$

$$\begin{bmatrix} 1 & 0 \\ 4 & 1 \end{bmatrix} \begin{bmatrix} 3 & 1 \\ 2 & 7 \end{bmatrix} = \begin{bmatrix} 3 & 1 \\ 14 & 11 \end{bmatrix}$$

$$\begin{bmatrix} 1 & 4 \\ 0 & 1 \end{bmatrix} \begin{bmatrix} 3 & 1 \\ 2 & 7 \end{bmatrix} = \begin{bmatrix} 11 & 29 \\ 2 & 7 \end{bmatrix}$$

Example 1.2.10. Calculate the product

$$\begin{bmatrix} 1 & 3 \\ 0 & 1 \end{bmatrix} \begin{bmatrix} 0 & 1 \\ 1 & 0 \end{bmatrix} \begin{bmatrix} 2 & 0 \\ 0 & 1 \end{bmatrix} \begin{bmatrix} 1 & 0 \\ 5 & 1 \end{bmatrix} \begin{bmatrix} 4 & 1 \\ 2 & 3 \end{bmatrix}.$$

Solution.

$$\begin{bmatrix} 1 & 3 \\ 0 & 1 \end{bmatrix} \begin{bmatrix} 0 & 1 \\ 1 & 0 \end{bmatrix} \begin{bmatrix} 2 & 0 \\ 0 & 1 \end{bmatrix} \begin{bmatrix} 1 & 0 \\ 5 & 1 \end{bmatrix} \begin{bmatrix} 4 & 1 \\ 2 & 3 \end{bmatrix} = \begin{bmatrix} 1 & 3 \\ 0 & 1 \end{bmatrix} \begin{bmatrix} 0 & 1 \\ 1 & 0 \end{bmatrix} \begin{bmatrix} 2 & 0 \\ 0 & 1 \end{bmatrix} \begin{bmatrix} 4 & 1 \\ 22 & 8 \end{bmatrix}$$

$$= \begin{bmatrix} 1 & 3 \\ 0 & 1 \end{bmatrix} \begin{bmatrix} 0 & 1 \\ 1 & 0 \end{bmatrix} \begin{bmatrix} 8 & 2 \\ 22 & 8 \end{bmatrix}$$

$$= \begin{bmatrix} 1 & 3 \\ 0 & 1 \end{bmatrix} \begin{bmatrix} 22 & 8 \\ 8 & 2 \end{bmatrix}$$

$$= \begin{bmatrix} 46 & 14 \\ 8 & 2 \end{bmatrix}$$

□

Elementary matrices are invertible and their inverses are elementary matrices:

Theorem 1.2.11.

$$\begin{bmatrix} 0 & 1 \\ 1 & 0 \end{bmatrix}^{-1} = \begin{bmatrix} 0 & 1 \\ 1 & 0 \end{bmatrix}$$

$$\begin{bmatrix} s & 0 \\ 0 & 1 \end{bmatrix}^{-1} = \begin{bmatrix} \frac{1}{s} & 0 \\ 0 & 1 \end{bmatrix}$$

$$\begin{bmatrix} 1 & 0 \\ 0 & s \end{bmatrix}^{-1} = \begin{bmatrix} 1 & 0 \\ 0 & \frac{1}{s} \end{bmatrix}$$

$$\begin{bmatrix} 1 & 0 \\ t & 1 \end{bmatrix}^{-1} = \begin{bmatrix} 1 & 0 \\ -t & 1 \end{bmatrix}$$

$$\begin{bmatrix} 1 & t \\ 0 & 1 \end{bmatrix}^{-1} = \begin{bmatrix} 1 & -t \\ 0 & 1 \end{bmatrix}$$

Proof. We have

$$\begin{bmatrix} 0 & 1 \\ 1 & 0 \end{bmatrix}\begin{bmatrix} 0 & 1 \\ 1 & 0 \end{bmatrix} = \begin{bmatrix} 1 & 0 \\ 0 & 1 \end{bmatrix}$$

$$\begin{bmatrix} s & 0 \\ 0 & 1 \end{bmatrix}\begin{bmatrix} \frac{1}{s} & 0 \\ 0 & 1 \end{bmatrix} = \begin{bmatrix} \frac{1}{s} & 0 \\ 0 & 1 \end{bmatrix}\begin{bmatrix} s & 0 \\ 0 & 1 \end{bmatrix} = \begin{bmatrix} 1 & 0 \\ 0 & 1 \end{bmatrix}$$

$$\begin{bmatrix} 1 & 0 \\ 0 & s \end{bmatrix}\begin{bmatrix} 1 & 0 \\ 0 & \frac{1}{s} \end{bmatrix} = \begin{bmatrix} 1 & 0 \\ 0 & \frac{1}{s} \end{bmatrix}\begin{bmatrix} 1 & 0 \\ 0 & s \end{bmatrix} = \begin{bmatrix} 1 & 0 \\ 0 & 1 \end{bmatrix}$$

$$\begin{bmatrix} 1 & 0 \\ t & 1 \end{bmatrix}\begin{bmatrix} 1 & 0 \\ -t & 1 \end{bmatrix} = \begin{bmatrix} 1 & 0 \\ -t & 1 \end{bmatrix}\begin{bmatrix} 1 & 0 \\ t & 1 \end{bmatrix} = \begin{bmatrix} 1 & 0 \\ 0 & 1 \end{bmatrix}$$

$$\begin{bmatrix} 1 & t \\ 0 & 1 \end{bmatrix}\begin{bmatrix} 1 & -t \\ 0 & 1 \end{bmatrix} = \begin{bmatrix} 1 & -t \\ 0 & 1 \end{bmatrix}\begin{bmatrix} 1 & t \\ 0 & 1 \end{bmatrix} = \begin{bmatrix} 1 & 0 \\ 0 & 1 \end{bmatrix}$$

$\square$

If the matrices $\begin{bmatrix} a & b \\ c & d \end{bmatrix}$ and $\begin{bmatrix} \alpha & \beta \\ \gamma & \delta \end{bmatrix}$ are invertible then the product $\begin{bmatrix} a & b \\ c & d \end{bmatrix}\begin{bmatrix} \alpha & \beta \\ \gamma & \delta \end{bmatrix}$ is an invertible matrix and we have

$$\left(\begin{bmatrix} a & b \\ c & d \end{bmatrix}\begin{bmatrix} \alpha & \beta \\ \gamma & \delta \end{bmatrix}\right)^{-1} = \begin{bmatrix} \alpha & \beta \\ \gamma & \delta \end{bmatrix}^{-1}\begin{bmatrix} a & b \\ c & d \end{bmatrix}^{-1}.$$

Indeed, we have

$$\left(\begin{bmatrix} \alpha & \beta \\ \gamma & \delta \end{bmatrix}^{-1}\begin{bmatrix} a & b \\ c & d \end{bmatrix}^{-1}\right)\left(\begin{bmatrix} a & b \\ c & d \end{bmatrix}\begin{bmatrix} \alpha & \beta \\ \gamma & \delta \end{bmatrix}\right) = \begin{bmatrix} \alpha & \beta \\ \gamma & \delta \end{bmatrix}^{-1}\left(\begin{bmatrix} a & b \\ c & d \end{bmatrix}^{-1}\begin{bmatrix} a & b \\ c & d \end{bmatrix}\right)\begin{bmatrix} \alpha & \beta \\ \gamma & \delta \end{bmatrix}$$

$$= \begin{bmatrix} \alpha & \beta \\ \gamma & \delta \end{bmatrix}^{-1}\begin{bmatrix} 1 & 0 \\ 0 & 1 \end{bmatrix}\begin{bmatrix} \alpha & \beta \\ \gamma & \delta \end{bmatrix}$$

$$= \begin{bmatrix} \alpha & \beta \\ \gamma & \delta \end{bmatrix}^{-1}\begin{bmatrix} \alpha & \beta \\ \gamma & \delta \end{bmatrix}$$

$$= \begin{bmatrix} 1 & 0 \\ 0 & 1 \end{bmatrix}.$$

A similar argument shows that

$$\left(\begin{bmatrix} a & b \\ c & d \end{bmatrix}\begin{bmatrix} \alpha & \beta \\ \gamma & \delta \end{bmatrix}\right)\left(\begin{bmatrix} \alpha & \beta \\ \gamma & \delta \end{bmatrix}^{-1}\begin{bmatrix} a & b \\ c & d \end{bmatrix}^{-1}\right) = \begin{bmatrix} 1 & 0 \\ 0 & 1 \end{bmatrix}.$$

This property is not limited to two matrices.

Theorem 1.2.12. *The product $A_1 \cdots A_m$ of invertible 2×2 matrices is invertible and we have*

$$(A_1 \cdots A_m)^{-1} = A_m^{-1} \cdots A_1^{-1}.$$

Proof. The proof follows from the argument for two matrices presented above. □

From Theorem 1.2.12 and the fact that elementary matrices are invertible it follows that every matrix that is a product of elementary matrices is invertible and the inverse is a product of elementary matrices.

Example 1.2.13. Write the matrix

$$\left(\begin{bmatrix} 1 & 3 \\ 0 & 1 \end{bmatrix} \begin{bmatrix} 0 & 1 \\ 1 & 0 \end{bmatrix} \begin{bmatrix} 2 & 0 \\ 0 & 1 \end{bmatrix} \begin{bmatrix} 1 & 0 \\ 5 & 1 \end{bmatrix} \right)^{-1}$$

as a product of elementary matrices.

Solution.

$$\left(\begin{bmatrix} 1 & 3 \\ 0 & 1 \end{bmatrix} \begin{bmatrix} 0 & 1 \\ 1 & 0 \end{bmatrix} \begin{bmatrix} 2 & 0 \\ 0 & 1 \end{bmatrix} \begin{bmatrix} 1 & 0 \\ 5 & 1 \end{bmatrix} \right)^{-1} = \begin{bmatrix} 1 & 0 \\ 5 & 1 \end{bmatrix}^{-1} \begin{bmatrix} 2 & 0 \\ 0 & 1 \end{bmatrix}^{-1} \begin{bmatrix} 0 & 1 \\ 1 & 0 \end{bmatrix}^{-1} \begin{bmatrix} 1 & 3 \\ 0 & 1 \end{bmatrix}^{-1}$$

$$= \begin{bmatrix} 1 & 0 \\ -5 & 1 \end{bmatrix} \begin{bmatrix} \frac{1}{2} & 0 \\ 0 & 1 \end{bmatrix} \begin{bmatrix} 0 & 1 \\ 1 & 0 \end{bmatrix} \begin{bmatrix} 1 & -3 \\ 0 & 1 \end{bmatrix}$$

□

The following is the first theorem that addresses the question of invertibility of 2×2 matrices.

Theorem 1.2.14. *For an arbitrary* 2×2 *matrix* $\begin{bmatrix} a & b \\ c & d \end{bmatrix}$ *the following conditions are equivalent:*

(i) *The matrix* $\begin{bmatrix} a & b \\ c & d \end{bmatrix}$ *is invertible;*

(ii) *The only solution of the equation* $\begin{bmatrix} a & b \\ c & d \end{bmatrix} \begin{bmatrix} x \\ y \end{bmatrix} = \begin{bmatrix} 0 \\ 0 \end{bmatrix}$ *is the trivial solution* $x = 0$ *and* $y = 0$;

(iii) *There are elementary matrices* $E_1, \dots, E_m$ *such that*

$$E_1 \cdots E_m \begin{bmatrix} a & b \\ c & d \end{bmatrix} = \begin{bmatrix} 1 & 0 \\ 0 & 1 \end{bmatrix}.$$

Proof. If the matrix $\begin{bmatrix} a & b \\ c & d \end{bmatrix}$ is invertible and we have

$$\begin{bmatrix} a & b \\ c & d \end{bmatrix} \begin{bmatrix} x \\ y \end{bmatrix} = \begin{bmatrix} 0 \\ 0 \end{bmatrix},$$

then

$$\begin{bmatrix} x \\ y \end{bmatrix} = \begin{bmatrix} a & b \\ c & d \end{bmatrix}^{-1} \begin{bmatrix} 0 \\ 0 \end{bmatrix} = \begin{bmatrix} 0 \\ 0 \end{bmatrix}.$$

This proves that (i) implies (ii).

Now assume that (ii) holds. If $a = 0$ and $c = 0$, then

$$\begin{bmatrix} a & b \\ c & d \end{bmatrix} \begin{bmatrix} 1 \\ 0 \end{bmatrix} = \begin{bmatrix} 0 \\ 0 \end{bmatrix},$$

contradicting (ii). Thus (ii) implies that $a \neq 0$ or $c \neq 0$.

If $a \neq 0$, then

$$\begin{bmatrix} \frac{1}{a} & 0 \\ 0 & 1 \end{bmatrix} \begin{bmatrix} a & b \\ c & d \end{bmatrix} = \begin{bmatrix} 1 & \frac{b}{a} \\ c & d \end{bmatrix}$$

and

$$\begin{bmatrix} 1 & 0 \\ -c & 1 \end{bmatrix} \begin{bmatrix} 1 & \frac{b}{a} \\ c & d \end{bmatrix} = \begin{bmatrix} 1 & \frac{b}{a} \\ 0 & d - c\frac{b}{a} \end{bmatrix} = \begin{bmatrix} 1 & \frac{b}{a} \\ 0 & \frac{ad-bc}{a} \end{bmatrix}.$$

We must have $ad - bc \neq 0$ because, if $ad - bc = 0$, then we would have

$$\begin{bmatrix} a & b \\ c & d \end{bmatrix} \begin{bmatrix} -\frac{b}{a} \\ 1 \end{bmatrix} = \begin{bmatrix} 0 \\ 0 \end{bmatrix},$$

contradicting (ii). Since

$$\begin{bmatrix} 1 & 0 \\ 0 & \frac{a}{ad-bc} \end{bmatrix} \begin{bmatrix} 1 & \frac{b}{a} \\ 0 & \frac{ad-bc}{a} \end{bmatrix} = \begin{bmatrix} 1 & \frac{b}{a} \\ 0 & 1 \end{bmatrix}$$

and

$$\begin{bmatrix} 1 & -\frac{b}{a} \\ 0 & 1 \end{bmatrix} \begin{bmatrix} 1 & \frac{b}{a} \\ 0 & 1 \end{bmatrix} = \begin{bmatrix} 1 & 0 \\ 0 & 1 \end{bmatrix},$$

we have

$$\begin{bmatrix} 1 & -\frac{b}{a} \\ 0 & 1 \end{bmatrix} \begin{bmatrix} 1 & 0 \\ 0 & \frac{a}{ad-bc} \end{bmatrix} \begin{bmatrix} 1 & 0 \\ -c & 1 \end{bmatrix} \begin{bmatrix} \frac{1}{a} & 0 \\ 0 & 1 \end{bmatrix} \begin{bmatrix} a & b \\ c & d \end{bmatrix} = \begin{bmatrix} 1 & 0 \\ 0 & 1 \end{bmatrix}. \tag{1.9}$$

This proves that (ii) implies (iii) when $a \neq 0$.

If $a = 0$, then $c \neq 0$. In this case we use the equality

$$\begin{bmatrix} 0 & 1 \\ 1 & 0 \end{bmatrix} \begin{bmatrix} a & b \\ c & d \end{bmatrix} = \begin{bmatrix} c & d \\ a & b \end{bmatrix},$$

and apply the proof given in the case $a \neq 0$ to the matrix $\begin{bmatrix} c & d \\ a & b \end{bmatrix}$. Thus (ii) implies (iii).

Now we show that (iii) implies (i). If

$$E_1 \cdots E_m \begin{bmatrix} a & b \\ c & d \end{bmatrix} = \begin{bmatrix} 1 & 0 \\ 0 & 1 \end{bmatrix},$$

then

$$\begin{bmatrix} a & b \\ c & d \end{bmatrix} = (E_1 \cdots E_m)^{-1} E_1 \cdots E_m \begin{bmatrix} a & b \\ c & d \end{bmatrix} = (E_1 \cdots E_m)^{-1} \begin{bmatrix} 1 & 0 \\ 0 & 1 \end{bmatrix} = (E_1 \cdots E_m)^{-1}$$

which gives us

$$\begin{bmatrix} a & b \\ c & d \end{bmatrix} = (E_1 \cdots E_m)^{-1} = E_m^{-1} \cdots E_1^{-1}.$$

Now is easy to see that

$$E_1 \cdots E_m \begin{bmatrix} a & b \\ c & d \end{bmatrix} = E_1 \cdots E_m E_m^{-1} \cdots E_1^{-1} = \begin{bmatrix} 1 & 0 \\ 0 & 1 \end{bmatrix}.$$

and

$$\begin{bmatrix} a & b \\ c & d \end{bmatrix} E_1 \cdots E_m = E_m^{-1} \cdots E_1^{-1} E_1 \cdots E_m = \begin{bmatrix} 1 & 0 \\ 0 & 1 \end{bmatrix}.$$

This means that the matrix $\begin{bmatrix} a & b \\ c & d \end{bmatrix}$ is invertible and

$$\begin{bmatrix} a & b \\ c & d \end{bmatrix}^{-1} = E_1 \cdots E_m.$$

□

The above theorem suggests a method for calculating the inverse of an arbitrary invertible 2 × 2 matrix.

Corollary 1.2.15. *If*

$$E_1 \cdots E_m \begin{bmatrix} a & b \\ c & d \end{bmatrix} = \begin{bmatrix} 1 & 0 \\ 0 & 1 \end{bmatrix},$$

where $E_1, \ldots, E_m$ are elementary matrices, then

$$\begin{bmatrix} a & b \\ c & d \end{bmatrix}^{-1} = E_1 \cdots E_m \quad and \quad \begin{bmatrix} a & b \\ c & d \end{bmatrix} = E_m^{-1} \cdots E_1^{-1}.$$

Example 1.2.16. Write the matrix $\begin{bmatrix} 2 & 3 \\ 5 & 4 \end{bmatrix}$ and its inverse as products of elementary matrices and find the inverse of the matrix.

Solution. Since

$$\begin{bmatrix} \frac{1}{2} & 0 \\ 0 & 1 \end{bmatrix} \begin{bmatrix} 2 & 3 \\ 5 & 4 \end{bmatrix} = \begin{bmatrix} 1 & \frac{3}{2} \\ 5 & 4 \end{bmatrix},$$

$$\begin{bmatrix} 1 & 0 \\ -5 & 1 \end{bmatrix}\begin{bmatrix} 1 & \frac{3}{2} \\ 5 & 4 \end{bmatrix} = \begin{bmatrix} 1 & \frac{3}{2} \\ 0 & -\frac{7}{2} \end{bmatrix},$$

$$\begin{bmatrix} 1 & 0 \\ 0 & -\frac{2}{7} \end{bmatrix}\begin{bmatrix} 1 & \frac{3}{2} \\ 0 & -\frac{7}{2} \end{bmatrix} = \begin{bmatrix} 1 & \frac{3}{2} \\ 0 & 1 \end{bmatrix},$$

$$\begin{bmatrix} 1 & -\frac{3}{2} \\ 0 & 1 \end{bmatrix}\begin{bmatrix} 1 & \frac{3}{2} \\ 0 & 1 \end{bmatrix} = \begin{bmatrix} 1 & 0 \\ 0 & 1 \end{bmatrix},$$

the inverse of the matrix $\begin{bmatrix} 2 & 3 \\ 5 & 4 \end{bmatrix}$ is

$$\begin{aligned} \begin{bmatrix} 2 & 3 \\ 5 & 4 \end{bmatrix}^{-1} &= \begin{bmatrix} 1 & -\frac{3}{2} \\ 0 & 1 \end{bmatrix}\begin{bmatrix} 1 & 0 \\ 0 & -\frac{2}{7} \end{bmatrix}\begin{bmatrix} 1 & 0 \\ -5 & 1 \end{bmatrix}\begin{bmatrix} \frac{1}{2} & 0 \\ 0 & 1 \end{bmatrix} \\ &= \begin{bmatrix} 1 & -\frac{3}{2} \\ 0 & 1 \end{bmatrix}\begin{bmatrix} 1 & 0 \\ 0 & -\frac{2}{7} \end{bmatrix}\begin{bmatrix} \frac{1}{2} & 0 \\ -\frac{5}{2} & 1 \end{bmatrix} \\ &= \begin{bmatrix} 1 & -\frac{3}{2} \\ 0 & 1 \end{bmatrix}\begin{bmatrix} \frac{1}{2} & 0 \\ \frac{5}{7} & -\frac{2}{7} \end{bmatrix} \\ &= \begin{bmatrix} \frac{1}{2} - \frac{15}{14} & \left(-\frac{3}{2}\right)\left(-\frac{2}{7}\right) \\ \frac{5}{7} & -\frac{2}{7} \end{bmatrix} \\ &= \begin{bmatrix} -\frac{4}{7} & \frac{3}{7} \\ \frac{5}{7} & -\frac{2}{7} \end{bmatrix} \end{aligned}$$

and

$$\begin{aligned} \begin{bmatrix} 2 & 3 \\ 5 & 4 \end{bmatrix} &= \left(\begin{bmatrix} 1 & -\frac{3}{2} \\ 0 & 1 \end{bmatrix}\begin{bmatrix} 1 & 0 \\ 0 & -\frac{2}{7} \end{bmatrix}\begin{bmatrix} 1 & 0 \\ -5 & 1 \end{bmatrix}\begin{bmatrix} \frac{1}{2} & 0 \\ 0 & 1 \end{bmatrix}\right)^{-1} \\ &= \begin{bmatrix} \frac{1}{2} & 0 \\ 0 & 1 \end{bmatrix}^{-1}\begin{bmatrix} 1 & 0 \\ -5 & 1 \end{bmatrix}^{-1}\begin{bmatrix} 1 & 0 \\ 0 & -\frac{2}{7} \end{bmatrix}^{-1}\begin{bmatrix} 1 & -\frac{3}{2} \\ 0 & 1 \end{bmatrix}^{-1} \\ &= \begin{bmatrix} 2 & 0 \\ 0 & 1 \end{bmatrix}\begin{bmatrix} 1 & 0 \\ 5 & 1 \end{bmatrix}\begin{bmatrix} 1 & 0 \\ 0 & -\frac{7}{2} \end{bmatrix}\begin{bmatrix} 1 & \frac{3}{2} \\ 0 & 1 \end{bmatrix}. \end{aligned}$$

□

In the next example we use what we learned in this section to solve a system of linear equations.

Example 1.2.17. Solve the system of equations

$$\begin{cases} 2x + 3y = 1 \\ 5x + y = 2 \end{cases}$$

using elementary matrices.

Solution. The system can be written as a matrix equation:

$$\begin{bmatrix} 2 & 3 \\ 5 & 1 \end{bmatrix} \begin{bmatrix} x \\ y \end{bmatrix} = \begin{bmatrix} 1 \\ 2 \end{bmatrix}.$$

Now we solve the equation by multiplying both sides of the equation by appropriately chosen elementary matrices:

$$\begin{bmatrix} \frac{1}{2} & 0 \\ 0 & 1 \end{bmatrix} \begin{bmatrix} 2 & 3 \\ 5 & 1 \end{bmatrix} \begin{bmatrix} x \\ y \end{bmatrix} = \begin{bmatrix} 1 & \frac{3}{2} \\ 5 & 1 \end{bmatrix} \begin{bmatrix} x \\ y \end{bmatrix} = \begin{bmatrix} \frac{1}{2} & 0 \\ 0 & 1 \end{bmatrix} \begin{bmatrix} 1 \\ 2 \end{bmatrix} = \begin{bmatrix} \frac{1}{2} \\ 2 \end{bmatrix},$$

$$\begin{bmatrix} 1 & 0 \\ -5 & 1 \end{bmatrix} \begin{bmatrix} 1 & \frac{3}{2} \\ 5 & 1 \end{bmatrix} \begin{bmatrix} x \\ y \end{bmatrix} = \begin{bmatrix} 1 & \frac{3}{2} \\ 0 & -\frac{13}{2} \end{bmatrix} \begin{bmatrix} x \\ y \end{bmatrix} = \begin{bmatrix} 1 & 0 \\ -5 & 1 \end{bmatrix} \begin{bmatrix} \frac{1}{2} \\ 2 \end{bmatrix} = \begin{bmatrix} \frac{1}{2} \\ -\frac{1}{2} \end{bmatrix},$$

$$\begin{bmatrix} 1 & 0 \\ 0 & -\frac{2}{13} \end{bmatrix} \begin{bmatrix} 1 & \frac{3}{2} \\ 0 & -\frac{13}{2} \end{bmatrix} \begin{bmatrix} x \\ y \end{bmatrix} = \begin{bmatrix} 1 & \frac{3}{2} \\ 0 & 1 \end{bmatrix} \begin{bmatrix} x \\ y \end{bmatrix} = \begin{bmatrix} 1 & 0 \\ 0 & -\frac{2}{13} \end{bmatrix} \begin{bmatrix} \frac{1}{2} \\ -\frac{1}{2} \end{bmatrix} = \begin{bmatrix} \frac{1}{2} \\ \frac{1}{13} \end{bmatrix},$$

$$\begin{bmatrix} 1 & -\frac{3}{2} \\ 0 & 1 \end{bmatrix} \begin{bmatrix} 1 & \frac{3}{2} \\ 0 & 1 \end{bmatrix} \begin{bmatrix} x \\ y \end{bmatrix} = \begin{bmatrix} 1 & 0 \\ 0 & 1 \end{bmatrix} \begin{bmatrix} x \\ y \end{bmatrix} = \begin{bmatrix} x \\ y \end{bmatrix} = \begin{bmatrix} 1 & -\frac{3}{2} \\ 0 & 1 \end{bmatrix} \begin{bmatrix} \frac{1}{2} \\ \frac{1}{13} \end{bmatrix} = \begin{bmatrix} \frac{5}{13} \\ \frac{1}{13} \end{bmatrix}.$$

The solution is $x = \frac{5}{13}$ and $y = \frac{1}{13}$. □

In the definition of an invertible 2×2 matrix we require that

$$\begin{bmatrix} \alpha & \beta \\ \gamma & \delta \end{bmatrix} \begin{bmatrix} a & b \\ c & d \end{bmatrix} = \begin{bmatrix} a & b \\ c & d \end{bmatrix} \begin{bmatrix} \alpha & \beta \\ \gamma & \delta \end{bmatrix} = \begin{bmatrix} 1 & 0 \\ 0 & 1 \end{bmatrix}.$$

This seems to imply that it is necessary to verify that both equalities

$$\begin{bmatrix} \alpha & \beta \\ \gamma & \delta \end{bmatrix} \begin{bmatrix} a & b \\ c & d \end{bmatrix} = \begin{bmatrix} 1 & 0 \\ 0 & 1 \end{bmatrix} \quad \text{and} \quad \begin{bmatrix} a & b \\ c & d \end{bmatrix} \begin{bmatrix} \alpha & \beta \\ \gamma & \delta \end{bmatrix} = \begin{bmatrix} 1 & 0 \\ 0 & 1 \end{bmatrix}$$

hold. Actually, as the next theorem shows, if we verify one of these equalities, then the other one follows.

Theorem 1.2.18. *For an arbitrary* 2×2 *matrix* $\begin{bmatrix} a & b \\ c & d \end{bmatrix}$ *the following condi-tions are equivalent:*

(i) *The matrix* $\begin{bmatrix} a & b \\ c & d \end{bmatrix}$ *is invertible;*

(ii) *There is a matrix* $\begin{bmatrix} \alpha & \beta \\ \gamma & \delta \end{bmatrix}$ *such that* $\begin{bmatrix} \alpha & \beta \\ \gamma & \delta \end{bmatrix}\begin{bmatrix} a & b \\ c & d \end{bmatrix} = \begin{bmatrix} 1 & 0 \\ 0 & 1 \end{bmatrix};$

(iii) *There is a matrix* $\begin{bmatrix} \alpha & \beta \\ \gamma & \delta \end{bmatrix}$ *such that* $\begin{bmatrix} a & b \\ c & d \end{bmatrix}\begin{bmatrix} \alpha & \beta \\ \gamma & \delta \end{bmatrix} = \begin{bmatrix} 1 & 0 \\ 0 & 1 \end{bmatrix}.$

Proof. Clearly (i) implies (ii) and (iii).

If (ii) holds, then the equality

$$\begin{bmatrix} a & b \\ c & d \end{bmatrix}\begin{bmatrix} x \\ y \end{bmatrix} = \begin{bmatrix} 0 \\ 0 \end{bmatrix}$$

gives us

$$\begin{bmatrix} x \\ y \end{bmatrix} = \begin{bmatrix} 1 & 0 \\ 0 & 1 \end{bmatrix}\begin{bmatrix} x \\ y \end{bmatrix} = \begin{bmatrix} \alpha & \beta \\ \gamma & \delta \end{bmatrix}\begin{bmatrix} a & b \\ c & d \end{bmatrix}\begin{bmatrix} x \\ y \end{bmatrix} = \begin{bmatrix} \alpha & \beta \\ \gamma & \delta \end{bmatrix}\begin{bmatrix} 0 \\ 0 \end{bmatrix} = \begin{bmatrix} 0 \\ 0 \end{bmatrix}.$$

This implies, by Theorem 1.2.14, that the matrix $\begin{bmatrix} a & b \\ c & d \end{bmatrix}$ is invertible. This shows that (ii) implies (i).

Similarly, if (iii) holds, then the equality

$$\begin{bmatrix} \alpha & \beta \\ \gamma & \delta \end{bmatrix}\begin{bmatrix} x \\ y \end{bmatrix} = \begin{bmatrix} 0 \\ 0 \end{bmatrix}$$

gives us

$$\begin{bmatrix} x \\ y \end{bmatrix} = \begin{bmatrix} 1 & 0 \\ 0 & 1 \end{bmatrix}\begin{bmatrix} x \\ y \end{bmatrix} = \begin{bmatrix} a & b \\ c & d \end{bmatrix}\begin{bmatrix} \alpha & \beta \\ \gamma & \delta \end{bmatrix}\begin{bmatrix} x \\ y \end{bmatrix} = \begin{bmatrix} a & b \\ c & d \end{bmatrix}\begin{bmatrix} 0 \\ 0 \end{bmatrix} = \begin{bmatrix} 0 \\ 0 \end{bmatrix}.$$

This implies, by Theorem 1.2.14, that the matrix $\begin{bmatrix} \alpha & \beta \\ \gamma & \delta \end{bmatrix}$ is invertible. Now, the equal-ity

$$\begin{bmatrix} a & b \\ c & d \end{bmatrix}\begin{bmatrix} \alpha & \beta \\ \gamma & \delta \end{bmatrix} = \begin{bmatrix} 1 & 0 \\ 0 & 1 \end{bmatrix}$$

implies that

$$\begin{bmatrix} a & b \\ c & d \end{bmatrix} = \begin{bmatrix} \alpha & \beta \\ \gamma & \delta \end{bmatrix}^{-1}.$$

This means that

$$\begin{bmatrix} a & b \\ c & d \end{bmatrix}\begin{bmatrix} \alpha & \beta \\ \gamma & \delta \end{bmatrix} = \begin{bmatrix} \alpha & \beta \\ \gamma & \delta \end{bmatrix}\begin{bmatrix} a & b \\ c & d \end{bmatrix} = \begin{bmatrix} 1 & 0 \\ 0 & 1 \end{bmatrix}.$$

Consequently, the matrix $\begin{bmatrix} a & b \\ c & d \end{bmatrix}$ is invertible, completing the proof that (iii) implies (i). $\qquad\qquad\qquad\qquad\qquad\qquad\qquad\qquad\qquad\qquad\qquad\qquad$ □

LU-decomposition of 2×2 matrices

Now we present a representation of 2×2 matrices in a form that is useful in applications.

Definition 1.2.19. A matrix of the form $\begin{bmatrix} a & 0 \\ b & c \end{bmatrix}$ is called a *lower triangular matrix*. A matrix of the form $\begin{bmatrix} a & b \\ 0 & c \end{bmatrix}$ is called *lower triangular matrix*.

Lower triangular and upper triangular matrices are used in the so-called LU-decomposition of matrices.

Definition 1.2.20. By an *LU-decomposition* (or an *LU-factorization*) of a 2×2 matrix A we mean the representation of A in the form

$$A = LU$$

where U is an upper triangular matrix and L is a lower triangular matrix with every entry on the main diagonal equal 1.

An LU-decomposition of a 2×2 matrix will have the form

$$\begin{bmatrix} a_1 & b_1 \\ a_2 & b_2 \end{bmatrix} = \begin{bmatrix} 1 & 0 \\ l & 1 \end{bmatrix} \begin{bmatrix} u_1 & u_2 \\ 0 & u_3 \end{bmatrix}.$$

When finding an LU-decomposition of a 2×2 matrix it is useful to note that

$$\begin{bmatrix} a_1 & b_1 \\ a_2 & b_2 \end{bmatrix} = \begin{bmatrix} 1 & 0 \\ l & 1 \end{bmatrix} \begin{bmatrix} u_1 & u_2 \\ 0 & u_3 \end{bmatrix} \quad \text{if and only if} \quad \begin{bmatrix} 1 & 0 \\ -l & 1 \end{bmatrix} \begin{bmatrix} a_1 & b_1 \\ a_2 & b_2 \end{bmatrix} = \begin{bmatrix} u_1 & u_2 \\ 0 & u_3 \end{bmatrix},$$

because $\begin{bmatrix} 1 & 0 \\ l & 1 \end{bmatrix}^{-1} = \begin{bmatrix} 1 & 0 \\ -l & 1 \end{bmatrix}.$

Example 1.2.21. Find an LU-decomposition of the matrix $\begin{bmatrix} 2 & 7 \\ 5 & 3 \end{bmatrix}$.

Solution. Since

$$\begin{bmatrix} 1 & 0 \\ -\frac{5}{2} & 1 \end{bmatrix} \begin{bmatrix} 2 & 7 \\ 5 & 3 \end{bmatrix} = \begin{bmatrix} 2 & 7 \\ 0 & -\frac{29}{2} \end{bmatrix},$$

we have

$$\begin{bmatrix} 2 & 7 \\ 5 & 3 \end{bmatrix} = \begin{bmatrix} 1 & 0 \\ \frac{5}{2} & 1 \end{bmatrix} \begin{bmatrix} 2 & 7 \\ 0 & -\frac{29}{2} \end{bmatrix}.$$

□

Not every 2 × 2 matrix has an LU-decomposition.

Example 1.2.22. Show that the matrix $\begin{bmatrix} 0 & 1 \\ 4 & 3 \end{bmatrix}$ has no LU-decomposition.

Solution. Suppose, to the contrary, that the matrix has an LU-decomposition, that is, there are numbers l, u_1, u_2, and u_3 such that

$$\begin{bmatrix} 0 & 1 \\ 4 & 3 \end{bmatrix} = \begin{bmatrix} 1 & 0 \\ l & 1 \end{bmatrix} \begin{bmatrix} u_1 & u_2 \\ 0 & u_3 \end{bmatrix}.$$

Since

$$\begin{bmatrix} 1 & 0 \\ l & 1 \end{bmatrix} \begin{bmatrix} u_1 & u_2 \\ 0 & u_3 \end{bmatrix} = \begin{bmatrix} u_1 & u_2 \\ lu_1 & lu_2 + u_3 \end{bmatrix},$$

we must have $u_1 = 0$. But then

$$\begin{bmatrix} 0 & 1 \\ 4 & 3 \end{bmatrix} = \begin{bmatrix} 0 & u_2 \\ 0 & lu_2 + u_3 \end{bmatrix},$$

which is not true.

□

Example 1.2.23. Let $A = \begin{bmatrix} p & q \\ r & s \end{bmatrix}$. Assuming that $p \neq 0$, find an LU-decomposition of A.

Solution. Since

$$\begin{bmatrix} 1 & 0 \\ -\frac{r}{p} & 1 \end{bmatrix} \begin{bmatrix} p & q \\ r & s \end{bmatrix} = \begin{bmatrix} p & q \\ 0 & s - \frac{qr}{p} \end{bmatrix},$$

we have

$$\begin{bmatrix} p & q \\ r & s \end{bmatrix} = \begin{bmatrix} 1 & 0 \\ \frac{r}{p} & 1 \end{bmatrix} \begin{bmatrix} p & q \\ 0 & s - \frac{qr}{p} \end{bmatrix}.$$

□

The above shows that a matrix $\begin{bmatrix} p & q \\ r & s \end{bmatrix}$ has an LU-decomposition as long as $p \neq 0$.

The next example illustrates how we can use LU-decomposition to solve systems of linear equations.

Example 1.2.24. Use LU-decomposition to solve the system

$$\begin{cases} 2x_1 + x_2 = b_1 \\ 5x_1 + 2x_2 = b_2 \end{cases}.$$

Solution. The system can be written as a matrix equation

$$\begin{bmatrix} 2 & 1 \\ 5 & 2 \end{bmatrix} \begin{bmatrix} x_1 \\ x_2 \end{bmatrix} = \begin{bmatrix} b_1 \\ b_2 \end{bmatrix}. \tag{1.10}$$

Since

$$\begin{bmatrix} 1 & 0 \\ -\frac{5}{2} & 1 \end{bmatrix} \begin{bmatrix} 2 & 1 \\ 5 & 2 \end{bmatrix} = \begin{bmatrix} 2 & 1 \\ 0 & -\frac{1}{2} \end{bmatrix},$$

we have

$$\begin{bmatrix} 2 & 1 \\ 5 & 2 \end{bmatrix} = \begin{bmatrix} 1 & 0 \\ \frac{5}{2} & 1 \end{bmatrix} \begin{bmatrix} 2 & 1 \\ 0 & -\frac{1}{2} \end{bmatrix}.$$

Consequently, the equation (1.10) is equivalent to

$$\begin{bmatrix} 1 & 0 \\ \frac{5}{2} & 1 \end{bmatrix} \begin{bmatrix} 2 & 1 \\ 0 & -\frac{1}{2} \end{bmatrix} \begin{bmatrix} x_1 \\ x_2 \end{bmatrix} = \begin{bmatrix} b_1 \\ b_2 \end{bmatrix}.$$

First we let

$$\begin{bmatrix} 2 & 1 \\ 0 & -\frac{1}{2} \end{bmatrix} \begin{bmatrix} x_1 \\ x_2 \end{bmatrix} = \begin{bmatrix} y_1 \\ y_2 \end{bmatrix} \tag{1.11}$$

and solve first the system

$$\begin{bmatrix} 1 & 0 \\ \frac{5}{2} & 1 \end{bmatrix} \begin{bmatrix} y_1 \\ y_2 \end{bmatrix} = \begin{bmatrix} b_1 \\ b_2 \end{bmatrix}.$$

Since $y_1 = b_1$ and $\frac{5}{2} y_1 + y_2 = b_2$, we get $y_1 = b_1$ and $y_2 = b_2 - \frac{5}{2} b_1$. (This step is called *forward substitution*.)

Now the matrix equation (1.11) becomes

$$\begin{bmatrix} 2 & 1 \\ 0 & -\frac{1}{2} \end{bmatrix} \begin{bmatrix} x_1 \\ x_2 \end{bmatrix} = \begin{bmatrix} b_1 \\ b_2 - \frac{5}{2} b_1 \end{bmatrix}$$

or

$$\begin{cases} 2x_1 + x_2 = b_1 \\ -\frac{1}{2} x_2 = b_2 - \frac{5}{2} b_1 \end{cases}.$$

We first get (This step is called *back substitution* and will be discussed later in this book in connection with Gauss elimination)

$$x_2 = 2b_2 + 5b_1$$

and then

$$x_1 = \frac{1}{2}b_1 - \frac{1}{2}x_2 = -2b_1 + b_2.$$

□

1.2.1 Exercises

Calculate the following products.

1. $\begin{bmatrix} 0 & 1 \\ 1 & 0 \end{bmatrix}\begin{bmatrix} 3 & 8 \\ 4 & 5 \end{bmatrix}$

2. $\begin{bmatrix} 9 & 0 \\ 0 & 1 \end{bmatrix}\begin{bmatrix} 3 & 8 \\ 4 & 5 \end{bmatrix}$

3. $\begin{bmatrix} 1 & 0 \\ 3 & 1 \end{bmatrix}\begin{bmatrix} 3 & 8 \\ 4 & 5 \end{bmatrix}$

4. $\begin{bmatrix} 1 & -2 \\ 0 & 1 \end{bmatrix}\begin{bmatrix} 3 & 8 \\ 4 & 5 \end{bmatrix}$

5. $\begin{bmatrix} 1 & 3 \\ 0 & 1 \end{bmatrix}\begin{bmatrix} 4 & 0 \\ 0 & 1 \end{bmatrix}\begin{bmatrix} 1 & 0 \\ 5 & 1 \end{bmatrix}\begin{bmatrix} 1 & -2 \\ 0 & 1 \end{bmatrix}$

6. $\begin{bmatrix} 1 & 0 \\ 0 & 2 \end{bmatrix}\begin{bmatrix} 1 & -1 \\ 0 & 1 \end{bmatrix}\begin{bmatrix} 1 & 0 \\ 7 & 1 \end{bmatrix}\begin{bmatrix} 3 & 0 \\ 0 & 1 \end{bmatrix}$

7. $\begin{bmatrix} 0 & 1 \\ 1 & 0 \end{bmatrix}\begin{bmatrix} 2 & 0 \\ 0 & 1 \end{bmatrix}\begin{bmatrix} 1 & 0 \\ 5 & 1 \end{bmatrix}\begin{bmatrix} 1 & 2 \\ -1 & 1 \end{bmatrix}$

8. $\begin{bmatrix} 1 & 0 \\ 3 & 1 \end{bmatrix}\begin{bmatrix} 2 & 0 \\ 0 & 1 \end{bmatrix}\begin{bmatrix} 1 & 5 \\ 0 & 1 \end{bmatrix}\begin{bmatrix} 0 & 1 \\ 1 & 0 \end{bmatrix}$

9. $\begin{bmatrix} 1 & 7 \\ 0 & 1 \end{bmatrix}\begin{bmatrix} 0 & 1 \\ 1 & 0 \end{bmatrix}\begin{bmatrix} 1 & 0 \\ 1 & 1 \end{bmatrix}\begin{bmatrix} w & x \\ y & z \end{bmatrix}$

10. $\begin{bmatrix} 1 & 0 \\ 4 & 1 \end{bmatrix}\begin{bmatrix} 3 & 0 \\ 0 & 1 \end{bmatrix}\begin{bmatrix} 1 & 2 \\ 0 & 1 \end{bmatrix}\begin{bmatrix} w & x \\ y & z \end{bmatrix}$

11. $\begin{bmatrix} 5 & 0 \\ 0 & 1 \end{bmatrix}\begin{bmatrix} 1 & 3 \\ 0 & 1 \end{bmatrix}\begin{bmatrix} 1 & 0 \\ -2 & 1 \end{bmatrix}\begin{bmatrix} w & x \\ y & z \end{bmatrix}$

12. $\begin{bmatrix} 1 & -4 \\ 0 & 1 \end{bmatrix}\begin{bmatrix} 0 & 1 \\ 1 & 0 \end{bmatrix}\begin{bmatrix} 1 & 0 \\ 0 & 7 \end{bmatrix}\begin{bmatrix} w & x \\ y & z \end{bmatrix}$

Write the inverse of the given matrix as a product of elementary matrices and find the inverse.

13. $\begin{bmatrix} 9 & 0 \\ 0 & 1 \end{bmatrix}\begin{bmatrix} 1 & 8 \\ 0 & 1 \end{bmatrix}$

14. $\begin{bmatrix} 5 & 0 \\ 0 & 1 \end{bmatrix}\begin{bmatrix} 1 & -4 \\ 0 & 1 \end{bmatrix}\begin{bmatrix} 1 & 0 \\ 0 & 7 \end{bmatrix}$

15. $\begin{bmatrix} 1 & 0 \\ 0 & 2 \end{bmatrix}\begin{bmatrix} 1 & -1 \\ 0 & 1 \end{bmatrix}\begin{bmatrix} 1 & 0 \\ 7 & 1 \end{bmatrix}\begin{bmatrix} 3 & 0 \\ 0 & 1 \end{bmatrix}$

16. $\begin{bmatrix} 1 & 0 \\ -3 & 1 \end{bmatrix}\begin{bmatrix} 0 & 1 \\ 1 & 0 \end{bmatrix}\begin{bmatrix} 4 & 1 \\ 0 & 1 \end{bmatrix}\begin{bmatrix} 1 & 4 \\ 0 & 1 \end{bmatrix}$

Find a matrix P satisfying the given equation for some a, b, and c.

17. $P\begin{bmatrix} 2 & 4 \\ 3 & 1 \end{bmatrix} = \begin{bmatrix} a & 0 \\ b & c \end{bmatrix}$

18. $P\begin{bmatrix} 1 & 3 \\ 5 & 8 \end{bmatrix} = \begin{bmatrix} a & b \\ 0 & c \end{bmatrix}$

19. $P\begin{bmatrix} 3 & 2 \\ 2 & 5 \end{bmatrix} = \begin{bmatrix} a & 0 \\ b & 1 \end{bmatrix}$

20. $P\begin{bmatrix} 3 & 2 \\ 2 & 3 \end{bmatrix} = \begin{bmatrix} 1 & b \\ 0 & c \end{bmatrix}$

Use elementary matrices to calculate the inverse of the given matrix.

21. $\begin{bmatrix} 1 & 4 \\ 5 & 22 \end{bmatrix}$ 22. $\begin{bmatrix} 2 & 3 \\ 4 & 1 \end{bmatrix}$ 23. $\begin{bmatrix} 0 & 8 \\ 3 & -12 \end{bmatrix}$ 24. $\begin{bmatrix} 4 & 5 \\ 3 & 7 \end{bmatrix}$

Write the given matrix as a product of elementary matrices.

25. $\begin{bmatrix} 36 & 5 \\ 7 & 1 \end{bmatrix}$ 27. $\begin{bmatrix} 3 & -2 \\ 15 & -8 \end{bmatrix}$ 29. $\begin{bmatrix} 2 & 4 \\ 2 & 8 \end{bmatrix}$

26. $\begin{bmatrix} 4 & 0 \\ -1 & 1 \end{bmatrix}$ 28. $\begin{bmatrix} 5 & 4 \\ 2 & 1 \end{bmatrix}$ 30. $\begin{bmatrix} 1 & 2 \\ 3 & 4 \end{bmatrix}$

31. Show that the matrix $\begin{bmatrix} a & b \\ 0 & 0 \end{bmatrix}$ is not invertible.

32. Show that the matrix $\begin{bmatrix} a & 0 \\ c & 0 \end{bmatrix}$ is not invertible.

33. Show that if the matrix A is invertible and the matrix B is not invertible, then the matrix AB is not invertible.

34. Show that if the matrix A is not invertible and the matrix B is invertible, then the matrix AB is not invertible.

35. Show that, if the matrix $A = \begin{bmatrix} a & b \\ c & d \end{bmatrix}$ is not invertible and $a \neq 0$ or $c \neq 0$, then

$\begin{bmatrix} a & b \\ c & d \end{bmatrix} = \begin{bmatrix} a & ka \\ c & kc \end{bmatrix}$ for some real number k.

36. Using elementary matrices to show that the matrix $\begin{bmatrix} a & b \\ c & d \end{bmatrix}$ is not invertible if and only if one of the following conditions occurs:

(a) $\begin{bmatrix} a & b \\ c & d \end{bmatrix} = \begin{bmatrix} a & ka \\ c & kc \end{bmatrix}$ for some real number k,

(b) $\begin{bmatrix} a & b \\ c & d \end{bmatrix} = \begin{bmatrix} 0 & b \\ 0 & d \end{bmatrix}$.

Find a number a such that the given matrix A is not invertible and then determine a product of elementary matrices $\begin{bmatrix} \alpha & \beta \\ \gamma & \delta \end{bmatrix}$ such that $\begin{bmatrix} \alpha & \beta \\ \gamma & \delta \end{bmatrix} A = \begin{bmatrix} 1 & k \\ 0 & 0 \end{bmatrix}$.

37. $\begin{bmatrix} 2 & 8 \\ 5 & a \end{bmatrix}$ 38. $\begin{bmatrix} 1 & a \\ 5 & 3 \end{bmatrix}$ 39. $\begin{bmatrix} a & 2 \\ 7 & 3 \end{bmatrix}$ 40. $\begin{bmatrix} 3 & 2 \\ 8 & a \end{bmatrix}$

Solve the given system of equations using elementary matrices.

41. $\begin{cases} 2x + y = 7 \\ 3x - 2y = 3 \end{cases}$ 42. $\begin{cases} 3x + 2y = 1 \\ 5x + 4y = 0 \end{cases}$

43. $\begin{cases} 4x + 3y = u \\ 2x + y = v \end{cases}$ 　　　　　　44. $\begin{cases} 2x + y = u \\ x + 4y = v \end{cases}$

Determine the LU-decomposition of the given matrix.

45. $\begin{bmatrix} 5 & 7 \\ 3 & 8 \end{bmatrix}$ 　　　　　　48. $\begin{bmatrix} 4 & 1 \\ 1 & 1 \end{bmatrix}$

46. $\begin{bmatrix} 7 & 1 \\ 2 & 2 \end{bmatrix}$ 　　　　　　49. $\begin{bmatrix} p & p \\ q & p \end{bmatrix}, p \neq 0$

47. $\begin{bmatrix} 3 & 2 \\ 5 & 3 \end{bmatrix}$

50. Suppose that the matrices A, B and C are invertible. Show that the matrix ABC is invertible and we have

$$(ABC)^{-1} = C^{-1}B^{-1}A^{-1}.$$

51. Suppose that the matrices A, B, C, and D are invertible. Show that the matrix $ABCD$ is invertible and we have

$$(ABCD)^{-1} = D^{-1}C^{-1}B^{-1}A^{-1}.$$

1.3　Determinants

Theorem 1.2.14 characterizes invertible 2×2 matrices. In the proof of that theorem the number $ad - bc$ seems to play a significant role.

Definition 1.3.1. The number $ad - bc$ is called the **determinant** of the matrix $\begin{bmatrix} a & b \\ c & d \end{bmatrix}$ and is denoted by $\det \begin{bmatrix} a & b \\ c & d \end{bmatrix}$, that is,

$$\det \begin{bmatrix} a & b \\ c & d \end{bmatrix} = ad - bc.$$

Example 1.3.2.
$$\det \begin{bmatrix} 6 & 3 \\ 13 & 4 \end{bmatrix} = 6 \cdot 4 - 13 \cdot 3 = 24 - 39 = -15$$

If $\mathbf{a} = \begin{bmatrix} a_1 \\ a_2 \end{bmatrix}$ and $\mathbf{b} = \begin{bmatrix} b_1 \\ b_2 \end{bmatrix}$, then the matrix $\begin{bmatrix} a_1 & b_1 \\ a_2 & b_2 \end{bmatrix}$ can be written as $[\mathbf{a} \ \mathbf{b}]$. We will use the notation

$$\det[\mathbf{a} \ \mathbf{b}] = \det \begin{bmatrix} a_1 & b_1 \\ a_2 & b_2 \end{bmatrix} = a_1 b_2 - a_2 b_1.$$

In the following theorem we list some useful properties of determinants.

Theorem 1.3.3. *For any* $\mathbf{a}, \mathbf{b}, \mathbf{c}$ *in* $\mathbb{R}^2$ *and* t *in* $\mathbb{R}$ *we have*

(a) $\det \begin{bmatrix} \mathbf{a} & \mathbf{a} \end{bmatrix} = 0$;

(b) $\det \begin{bmatrix} \mathbf{a} & \mathbf{b} \end{bmatrix} = -\det \begin{bmatrix} \mathbf{b} & \mathbf{a} \end{bmatrix}$;

(c) $t \det \begin{bmatrix} \mathbf{a} & \mathbf{b} \end{bmatrix} = \det \begin{bmatrix} t\mathbf{a} & \mathbf{b} \end{bmatrix}) = \det \begin{bmatrix} \mathbf{a} & t\mathbf{b} \end{bmatrix}$;

(d) $\det \begin{bmatrix} \mathbf{a} + \mathbf{c} & \mathbf{b} \end{bmatrix} = \det \begin{bmatrix} \mathbf{a} & \mathbf{b} \end{bmatrix} + \det \begin{bmatrix} \mathbf{c} & \mathbf{b} \end{bmatrix}$;

(e) $\det \begin{bmatrix} \mathbf{a} & \mathbf{b} + \mathbf{c} \end{bmatrix} = \det \begin{bmatrix} \mathbf{a} & \mathbf{b} \end{bmatrix} + \det \begin{bmatrix} \mathbf{a} & \mathbf{c} \end{bmatrix}$;

(f) $\det \begin{bmatrix} \mathbf{a} + t\mathbf{b} & \mathbf{b} \end{bmatrix} = \det \begin{bmatrix} \mathbf{a} & \mathbf{b} \end{bmatrix}$;

(g) $\det \begin{bmatrix} \mathbf{a} & \mathbf{b} + t\mathbf{a} \end{bmatrix} = \det \begin{bmatrix} \mathbf{a} & \mathbf{b} \end{bmatrix}$.

Moreover, for any real numbers a_1, a_2, b_1 *and* b_2, *we have*

(h) $\det \begin{bmatrix} a_1 + t a_2 & b_1 + t b_2 \\ a_2 & b_2 \end{bmatrix} = \det \begin{bmatrix} a_1 & b_1 \\ a_2 & b_2 \end{bmatrix}$;

(i) $\det \begin{bmatrix} a_1 & b_1 \\ a_2 + t a_1 & b_2 + t b_1 \end{bmatrix} = \det \begin{bmatrix} a_1 & b_1 \\ a_2 & b_2 \end{bmatrix}$;

(j) $\det \begin{bmatrix} a_1 & a_2 \\ b_1 & b_2 \end{bmatrix} = \det \begin{bmatrix} a_1 & b_1 \\ a_2 & b_2 \end{bmatrix}$.

Proof. These identities follow easily from the definition of the determinant. We leave the proofs as exercises. $\qquad \square$

The following theorem describes an essential property of determinants. The property is not just convenient for calculations, but it has important theoretical consequences. The proof of Theorem 1.3.5 is a good example of that.

Theorem 1.3.4. *For arbitrary numbers* $a, b, c, d, \alpha, \beta, \gamma, \delta$ *we have*

$$\det \left(\begin{bmatrix} a & b \\ c & d \end{bmatrix} \begin{bmatrix} \alpha & \beta \\ \gamma & \delta \end{bmatrix} \right) = \det \begin{bmatrix} a & b \\ c & d \end{bmatrix} \det \begin{bmatrix} \alpha & \beta \\ \gamma & \delta \end{bmatrix}.$$

Proof. Since

$$\begin{bmatrix} a & b \\ c & d \end{bmatrix} \begin{bmatrix} \alpha & \beta \\ \gamma & \delta \end{bmatrix} = \begin{bmatrix} a\alpha + b\gamma & a\beta + b\delta \\ c\alpha + d\gamma & c\beta + d\delta \end{bmatrix},$$

we have

$$\det\left(\begin{bmatrix} a & b \\ c & d \end{bmatrix}\begin{bmatrix} \alpha & \beta \\ \gamma & \delta \end{bmatrix}\right) = \det\begin{bmatrix} a\alpha + b\gamma & a\beta + b\delta \\ c\alpha + d\gamma & c\beta + d\delta \end{bmatrix}$$

$$= (a\alpha + b\gamma)(c\beta + d\delta) - (a\beta + b\delta)(c\alpha + d\gamma)$$
$$= ad(\alpha\delta - \beta\gamma) + bc(\beta\gamma - \alpha\delta)$$
$$= (ad - bc)(\alpha\delta - \beta\gamma)$$
$$= \det\begin{bmatrix} a & b \\ c & d \end{bmatrix}\det\begin{bmatrix} \alpha & \beta \\ \gamma & \delta \end{bmatrix}.$$

$\square$

Theorem 1.3.5. *If the matrix $\begin{bmatrix} a & b \\ c & d \end{bmatrix}$ is invertible, then $\det\begin{bmatrix} a & b \\ c & d \end{bmatrix} \neq 0$.*

Proof. If the matrix $\begin{bmatrix} a & b \\ c & d \end{bmatrix}$ is invertible, then

$$\begin{bmatrix} a & b \\ c & d \end{bmatrix}\begin{bmatrix} a & b \\ c & d \end{bmatrix}^{-1} = \begin{bmatrix} 1 & 0 \\ 0 & 1 \end{bmatrix}.$$

From Theorem 1.3.4 we have

$$\det\begin{bmatrix} a & b \\ c & d \end{bmatrix}\det\left(\begin{bmatrix} a & b \\ c & d \end{bmatrix}^{-1}\right) = \det\left(\begin{bmatrix} a & b \\ c & d \end{bmatrix}\begin{bmatrix} a & b \\ c & d \end{bmatrix}^{-1}\right) = \det\begin{bmatrix} 1 & 0 \\ 0 & 1 \end{bmatrix} = 1.$$

Hence $\det\begin{bmatrix} a & b \\ c & d \end{bmatrix} \neq 0$.
$\square$

The proof of Theorem 1.2.14 suggests a method for finding the inverse of an invertible matrix. Consider an invertible matrix $\begin{bmatrix} a & b \\ c & d \end{bmatrix}$. If $a \neq 0$, then we have

$$\begin{bmatrix} 1 & -\frac{b}{a} \\ 0 & 1 \end{bmatrix}\begin{bmatrix} 1 & 0 \\ 0 & \frac{a}{ad-bc} \end{bmatrix}\begin{bmatrix} 1 & 0 \\ -c & 1 \end{bmatrix}\begin{bmatrix} \frac{1}{a} & 0 \\ 0 & 1 \end{bmatrix}\begin{bmatrix} a & b \\ c & d \end{bmatrix} = \begin{bmatrix} 1 & 0 \\ 0 & 1 \end{bmatrix},$$

and consequently we must have

$$\begin{bmatrix} 1 & -\frac{b}{a} \\ 0 & 1 \end{bmatrix}\begin{bmatrix} 1 & 0 \\ 0 & \frac{a}{ad-bc} \end{bmatrix}\begin{bmatrix} 1 & 0 \\ -c & 1 \end{bmatrix}\begin{bmatrix} \frac{1}{a} & 0 \\ 0 & 1 \end{bmatrix} = \begin{bmatrix} a & b \\ c & d \end{bmatrix}^{-1}.$$

To find the inverse of the matrix $\begin{bmatrix} a & b \\ c & d \end{bmatrix}$ we calculate product of the elementary matrices on the left-hand side:

$$\begin{bmatrix} a & b \\ c & d \end{bmatrix}^{-1} = \begin{bmatrix} 1 & -\frac{b}{a} \\ 0 & 1 \end{bmatrix}\begin{bmatrix} 1 & 0 \\ 0 & \frac{a}{ad-bc} \end{bmatrix}\begin{bmatrix} 1 & 0 \\ -c & 1 \end{bmatrix}\begin{bmatrix} \frac{1}{a} & 0 \\ 0 & 1 \end{bmatrix} = \begin{bmatrix} 1 & -\frac{b}{a} \\ 0 & 1 \end{bmatrix}\begin{bmatrix} 1 & 0 \\ 0 & \frac{a}{ad-bc} \end{bmatrix}\begin{bmatrix} \frac{1}{a} & 0 \\ -\frac{c}{a} & 1 \end{bmatrix}$$

$$= \begin{bmatrix} 1 & -\frac{b}{a} \\ 0 & 1 \end{bmatrix} \begin{bmatrix} \frac{1}{a} & 0 \\ -\frac{c}{ad-bc} & \frac{a}{ad-bc} \end{bmatrix}$$

$$= \begin{bmatrix} \frac{d}{ad-bc} & \frac{-b}{ad-bc} \\ \frac{-c}{ad-bc} & \frac{a}{ad-bc} \end{bmatrix}.$$

It turns out that the obtained matrix is the inverse of a invertible matrix $\begin{bmatrix} a & b \\ c & d \end{bmatrix}$ also in the case when $a = 0$.

Theorem 1.3.6. *If* $\det \begin{bmatrix} a & b \\ c & d \end{bmatrix} = ad - bc \neq 0$, *then the matrix* $\begin{bmatrix} a & b \\ c & d \end{bmatrix}$ *is invertible and*

$$\begin{bmatrix} a & b \\ c & d \end{bmatrix}^{-1} = \begin{bmatrix} \frac{d}{ad-bc} & \frac{-b}{ad-bc} \\ \frac{-c}{ad-bc} & \frac{a}{ad-bc} \end{bmatrix} = \frac{1}{ad-bc} \begin{bmatrix} d & -b \\ -c & a \end{bmatrix}.$$

Proof. It is easy to verify that

$$\begin{bmatrix} a & b \\ c & d \end{bmatrix} \begin{bmatrix} \frac{d}{ad-bc} & \frac{-b}{ad-bc} \\ \frac{-c}{ad-bc} & \frac{a}{ad-bc} \end{bmatrix} = \begin{bmatrix} \frac{d}{ad-bc} & \frac{-b}{ad-bc} \\ \frac{-c}{ad-bc} & \frac{a}{ad-bc} \end{bmatrix} \begin{bmatrix} a & b \\ c & d \end{bmatrix} = \begin{bmatrix} 1 & 0 \\ 0 & 1 \end{bmatrix}.$$

□

The formula in the above theorem allows us to solve more efficiently problems that require finding the inverse of a matrix.

Example 1.3.7. Solve the system

$$\begin{cases} x + 3y = 6 \\ 2x + y = 1 \end{cases}.$$

Solution. Since the system is equivalent to the equation

$$\begin{bmatrix} 1 & 3 \\ 2 & 1 \end{bmatrix} \begin{bmatrix} x \\ y \end{bmatrix} = \begin{bmatrix} 6 \\ 1 \end{bmatrix}$$

and

$$\det \begin{bmatrix} 1 & 3 \\ 2 & 1 \end{bmatrix} = 1 \cdot 1 - 3 \cdot 2 = -5 \neq 0,$$

we obtain

$$\begin{bmatrix} x \\ y \end{bmatrix} = \begin{bmatrix} 1 & 3 \\ 2 & 1 \end{bmatrix}^{-1} \begin{bmatrix} 6 \\ 1 \end{bmatrix} = \frac{1}{-5} \begin{bmatrix} 1 & -3 \\ -2 & 1 \end{bmatrix} \begin{bmatrix} 6 \\ 1 \end{bmatrix} = -\frac{1}{5} \begin{bmatrix} 3 \\ -11 \end{bmatrix} = \begin{bmatrix} -\frac{3}{5} \\ \frac{11}{5} \end{bmatrix}.$$

In other words, the solution of the system is $x = -\frac{3}{5}$ and $y = \frac{11}{5}$.   □

Example 1.3.8. Find a 2×2 matrix X such that

$$\begin{bmatrix} 4 & 3 \\ 1 & 1 \end{bmatrix} X = \begin{bmatrix} 1 & 2 \\ 1 & 1 \end{bmatrix}.$$

Solution. If

$$\begin{bmatrix} 4 & 3 \\ 1 & 1 \end{bmatrix} X = \begin{bmatrix} 1 & 2 \\ 1 & 1 \end{bmatrix},$$

then

$$X = \begin{bmatrix} 4 & 3 \\ 1 & 1 \end{bmatrix}^{-1} \begin{bmatrix} 1 & 2 \\ 1 & 1 \end{bmatrix}.$$

Since $\begin{bmatrix} 4 & 3 \\ 1 & 1 \end{bmatrix}^{-1} = \begin{bmatrix} 1 & -3 \\ -1 & 4 \end{bmatrix}$, we have

$$X = \begin{bmatrix} 1 & -3 \\ -1 & 4 \end{bmatrix} \begin{bmatrix} 1 & 2 \\ 1 & 1 \end{bmatrix} = \begin{bmatrix} -2 & -1 \\ 3 & 2 \end{bmatrix}$$

□

Theorems 1.3.5 and 1.3.6 yield:

Theorem 1.3.9. *The matrix* $\begin{bmatrix} a & b \\ c & d \end{bmatrix}$ *is invertible if and only if* $\det \begin{bmatrix} a & b \\ c & d \end{bmatrix} \neq 0.$

Cramer's Rule

Theorems 1.2.6 and 1.3.6 lead to the following result that gives the solution of a linear system in a explicit form in terms of determinants. The theorem is known as *Cramer's Rule*.

Theorem 1.3.10. *If* $\det \begin{bmatrix} a & b \\ c & d \end{bmatrix} \neq 0$, *then the system*

$$\begin{cases} ax + by = e \\ cx + dy = f \end{cases}$$

has a unique solution for any real numbers e and f. The solution is

$$x = \frac{\det \begin{bmatrix} e & b \\ f & d \end{bmatrix}}{\det \begin{bmatrix} a & b \\ c & d \end{bmatrix}} \quad and \quad y = \frac{\det \begin{bmatrix} a & e \\ c & f \end{bmatrix}}{\det \begin{bmatrix} a & b \\ c & d \end{bmatrix}}.$$

Proof. Since

$$\begin{bmatrix} x \\ y \end{bmatrix} = \begin{bmatrix} a & b \\ c & d \end{bmatrix}^{-1} \begin{bmatrix} e \\ f \end{bmatrix} = \begin{bmatrix} \frac{d}{ad-bc} & \frac{-b}{ad-bc} \\ \frac{-c}{ad-bc} & \frac{a}{ad-bc} \end{bmatrix} \begin{bmatrix} e \\ f \end{bmatrix} = \begin{bmatrix} \frac{ed-bf}{ad-bc} \\ \frac{af-ce}{ad-bc} \end{bmatrix},$$

we have

$$x = \frac{\det \begin{bmatrix} e & b \\ f & d \end{bmatrix}}{\det \begin{bmatrix} a & b \\ c & d \end{bmatrix}} \quad and \quad y = \frac{\det \begin{bmatrix} a & e \\ c & f \end{bmatrix}}{\det \begin{bmatrix} a & b \\ c & d \end{bmatrix}}.$$

$\square$

Example 1.3.11. In this example we solve the system

$$\begin{cases} 2x + 5y = 4 \\ x + 3y = 3 \end{cases}$$

using Cramer's Rule. Since

$$\det \begin{bmatrix} 2 & 5 \\ 1 & 3 \end{bmatrix} = 2 \cdot 3 - 1 \cdot 5 = 1,$$

$$\det \begin{bmatrix} 4 & 5 \\ 3 & 3 \end{bmatrix} = 4 \cdot 3 - 3 \cdot 5 = -3,$$

and

$$\det \begin{bmatrix} 2 & 4 \\ 1 & 3 \end{bmatrix} = 2 \cdot 3 - 1 \cdot 4 = 2,$$

we have

$$x = \frac{\det \begin{bmatrix} 4 & 5 \\ 3 & 3 \end{bmatrix}}{\det \begin{bmatrix} 2 & 5 \\ 1 & 3 \end{bmatrix}} = -3 \quad \text{and} \quad y = \frac{\det \begin{bmatrix} 2 & 4 \\ 1 & 3 \end{bmatrix}}{\det \begin{bmatrix} 2 & 5 \\ 1 & 3 \end{bmatrix}} = 2.$$

From Theorem 1.3.10 we obtain the following useful result.

Theorem 1.3.12. *Let* $\mathbf{u}$ *and* $\mathbf{v}$ *be vectors in* $\mathbb{R}^2$ *such that* $\det \begin{bmatrix} \mathbf{u} & \mathbf{v} \end{bmatrix} \neq 0$. *Then for any vector* $\mathbf{x}$ *in* $\mathbb{R}^2$ *there exist unique real numbers* α *and* β *such that*

$$\mathbf{x} = \alpha \mathbf{u} + \beta \mathbf{v}.$$

Proof. If $\mathbf{u} = \begin{bmatrix} u_1 \\ u_2 \end{bmatrix}$, $\mathbf{v} = \begin{bmatrix} v_1 \\ v_2 \end{bmatrix}$, and $\mathbf{x} = \begin{bmatrix} x_1 \\ x_2 \end{bmatrix}$, then the equation $\mathbf{x} = \alpha \mathbf{u} + \beta \mathbf{v}$ is equivalent to the system of linear equations

$$\begin{cases} u_1 \alpha + v_1 \beta = x_1 \\ u_2 \alpha + v_2 \beta = x_2 \end{cases}.$$

If $\det \begin{bmatrix} \mathbf{u} & \mathbf{v} \end{bmatrix} = \det \begin{bmatrix} u_1 & v_1 \\ u_2 & v_2 \end{bmatrix} \neq 0$, then the system has a unique solution by Theorem 1.3.10. $\qquad \square$

The next result is a consequence of Theorems 1.2.14 and 1.3.9. It will be used in some arguments in the next section.

Theorem 1.3.13. *The following conditions are equivalent:*

(a) $\det \begin{bmatrix} a & b \\ c & d \end{bmatrix} = 0$;

(b) *The system*

$$\begin{cases} ax + by = 0 \\ cx + dy = 0 \end{cases} \tag{1.12}$$

has a nontrivial solution, that is, a solution different from the trivial solution $x = y = 0$.

We can easily find nontrivial solutions of the stystem (1.12) when $\det \begin{bmatrix} a & b \\ c & d \end{bmatrix} = 0$. Indeed, if $d \neq 0$ or $c \neq 0$, then $x = d$ and $y = -c$ is a nontrivial solution of the system.

Similarly, if $a \neq 0$ or $b \neq 0$, then $x = b$ and $y = -a$ is a nontrivial solution. Finaly, if $a = b = c = d = 0$, then any pair of numbers x and y is a solution of the system.

If $\det \begin{bmatrix} a & b \\ c & d \end{bmatrix} = 0$, then the system

$$\begin{cases} ax + by = e \\ cx + dy = f \end{cases}$$

either has no solutions or infinitely many solutions, as illustrated by the following two examples.

Example 1.3.14. The system

$$\begin{cases} x + y = 1 \\ 2x + 2y = 3 \end{cases}$$

has no solutions. Indeed, if some x and y were a solution of the system, then we would have $x + y = 1$. But then $2x + 2y = 2$, which is not possible in view of the second equation of the system.

Example 1.3.15. We will show that the system

$$\begin{cases} x + y = 1 \\ 2x + 2y = 2 \end{cases}$$

has infinitely many solutions.

It is easy to check that $x = t$ and $y = 1 - t$ is a solution of the system for any real number t.

The Leontief model

The **Leontief model** is a model introduced to describe the economics of a whole country or a large region. It was proposed by Wassily Leontief, who won the Nobel prize in economics in 1973. The following theorem gives us the mathematical foundation of the model in the case of 2×2 matrices.

Theorem 1.3.16. *Let* $C = \begin{bmatrix} a_{11} & a_{12} \\ a_{21} & a_{22} \end{bmatrix}$ *be a matrix with* **nonnegative entries** *and such that* $a_{11} + a_{21} < 1$ *and* $a_{12} + a_{22} < 1$. *Then*

   (a) *The matrix* $\begin{bmatrix} 1 & 0 \\ 0 & 1 \end{bmatrix} - \begin{bmatrix} a_{11} & a_{12} \\ a_{21} & a_{22} \end{bmatrix}$ *is invertible;*

   (b) *All entries of the matrix* $\left(\begin{bmatrix} 1 & 0 \\ 0 & 1 \end{bmatrix} - \begin{bmatrix} a_{11} & a_{12} \\ a_{21} & a_{22} \end{bmatrix} \right)^{-1}$ *are* **nonnegative;**

   (c) *For every vector* $\mathbf{d} = \begin{bmatrix} d_1 \\ d_2 \end{bmatrix}$ *with* **nonnegative entries** *the equation* $\mathbf{x} =$
$C\mathbf{x} + \mathbf{d}$ *has a unique solution* $\mathbf{x} = \begin{bmatrix} x_1 \\ x_2 \end{bmatrix}$ *with* **nonnegative entries.**

Proof. If

$$a_{11} \geq 0, \; a_{21} \geq 0, \; a_{12} \geq 0, \; a_{22} \geq 0$$

and

$$a_{11} + a_{21} < 1 \quad \text{and} \quad a_{12} + a_{22} < 1,$$

then

$$\det\left(\begin{bmatrix} 1 & 0 \\ 0 & 1 \end{bmatrix} - \begin{bmatrix} a_{11} & a_{12} \\ a_{21} & a_{22} \end{bmatrix} \right) = (1 - a_{11})(1 - a_{22}) - a_{12}a_{21} > a_{12}a_{21} - a_{12}a_{21} = 0,$$

and thus the matrix

$$\begin{bmatrix} 1 & 0 \\ 0 & 1 \end{bmatrix} - \begin{bmatrix} a_{11} & a_{12} \\ a_{21} & a_{22} \end{bmatrix}$$

is invertible. Since

$$\left(\begin{bmatrix} 1 & 0 \\ 0 & 1 \end{bmatrix} - \begin{bmatrix} a_{11} & a_{12} \\ a_{21} & a_{22} \end{bmatrix} \right)^{-1} = \frac{1}{(1 - a_{11})(1 - a_{22}) - a_{12}a_{21}} \begin{bmatrix} 1 - a_{22} & a_{12} \\ a_{21} & 1 - a_{11} \end{bmatrix},$$

all entries of the matrix $\left(\begin{bmatrix} 1 & 0 \\ 0 & 1 \end{bmatrix} - \begin{bmatrix} a_{11} & a_{12} \\ a_{21} & a_{22} \end{bmatrix} \right)^{-1}$ are nonnegative. Moreover, if $d_1 \geq 0$ and $d_2 \geq 0$, then the solution of the equation

$$\begin{bmatrix} x_1 \\ x_2 \end{bmatrix} = \begin{bmatrix} a_{11} & a_{12} \\ a_{21} & a_{22} \end{bmatrix} \begin{bmatrix} x_1 \\ x_2 \end{bmatrix} + \begin{bmatrix} d_1 \\ d_2 \end{bmatrix}$$

is

$$\begin{bmatrix} x_1 \\ x_2 \end{bmatrix} = \left(\begin{bmatrix} 1 & 0 \\ 0 & 1 \end{bmatrix} - \begin{bmatrix} a_{11} & a_{12} \\ a_{21} & a_{22} \end{bmatrix} \right)^{-1} \begin{bmatrix} d_1 \\ d_2 \end{bmatrix}.$$

Consequently, $x_1 \geq 0$ and $x_2 \geq 0$.   ☐

Example 1.3.17. Solve the equation $\mathbf{x} = C\mathbf{x} + \mathbf{d}$ for

$$C = \begin{bmatrix} \frac{1}{10} & \frac{2}{5} \\ \frac{3}{10} & \frac{1}{5} \end{bmatrix} \quad \text{and} \quad \mathbf{d} = \begin{bmatrix} 70 \\ 5 \end{bmatrix}.$$

Proof. Since

$$\begin{bmatrix} 1 & 0 \\ 0 & 1 \end{bmatrix} - C = \begin{bmatrix} 1 & 0 \\ 0 & 1 \end{bmatrix} - \begin{bmatrix} \frac{1}{10} & \frac{2}{5} \\ \frac{3}{10} & \frac{1}{5} \end{bmatrix} = \begin{bmatrix} \frac{9}{10} & -\frac{2}{5} \\ -\frac{3}{10} & \frac{4}{5} \end{bmatrix}$$

and

$$\left(\begin{bmatrix} 1 & 0 \\ 0 & 1 \end{bmatrix} - \begin{bmatrix} \frac{1}{10} & \frac{2}{5} \\ \frac{3}{10} & \frac{1}{5} \end{bmatrix} \right)^{-1} = \begin{bmatrix} \frac{4}{3} & \frac{2}{3} \\ \frac{1}{2} & \frac{3}{2} \end{bmatrix},$$

we obtain the solution

$$\begin{bmatrix} \frac{4}{3} & \frac{2}{3} \\ \frac{1}{2} & \frac{3}{2} \end{bmatrix} \begin{bmatrix} 70 \\ 5 \end{bmatrix} = \begin{bmatrix} \frac{290}{3} \\ \frac{85}{2} \end{bmatrix}.$$

□

We illustrate an application of the above theorem in economics with a very simple example. We consider two products, product A and product B. We assume that product A is used in the production of product A and in the production of product B. Similarly, product B is used in the production of product A and in the production of product B. The variables in Theorem 1.3.16 are interpreted as follows:

x_1 = the total output of product A

x_2 = the total output of product B

$a_{11}x_1$ = the amount of product A used in the production of A

$a_{12}x_2$ = the amount of product A used in the production of B

$a_{21}x_1$ = the amount of product B used in the production of A

$a_{22}x_2$ = the amount of product B used in the production of B

d_1 = the remaining amount of product A

d_2 = the remaining amount of product B

We have

$$x_1 = a_{11}x_1 + a_{12}x_2 + d_1 \quad \text{and} \quad x_2 = a_{21}x_1 + a_{22}x_2 + d_2.$$

The matrix $C = \begin{bmatrix} a_{11} & a_{12} \\ a_{21} & a_{22} \end{bmatrix}$ is called the **consumption matrix**, the vector $\mathbf{x} = \begin{bmatrix} x_1 \\ x_2 \end{bmatrix}$ is called the **output vector** or the **production vector**, and the vector $\mathbf{d} = \begin{bmatrix} d_1 \\ d_2 \end{bmatrix}$ is called the **demand vector**. Theorem 1.3.16 guarantees that, under the assumption of the theorem, for any demand vector with nonnegative entries there is a unique output vector with nonnegative entries.

1.3.1 Exercises

Find the determinants of the given matrices.

1. $\begin{bmatrix} 3 & 2 \\ 2 & 3 \end{bmatrix}$

2. $\begin{bmatrix} 4 & -2 \\ 5 & 7 \end{bmatrix}$

3. $\begin{bmatrix} 1 & 5 \\ 3 & 4 \end{bmatrix}$

4. $\begin{bmatrix} 3 & 2 \\ -3 & -2 \end{bmatrix}$

5. $\begin{bmatrix} a & b \\ c & 0 \end{bmatrix}$

6. $\begin{bmatrix} a & b \\ 0 & c \end{bmatrix}$

7. $\begin{bmatrix} a & ta \\ b & tb \end{bmatrix}$

8. $\begin{bmatrix} a & b \\ ta & tb \end{bmatrix}$

Find the inverses of the following matrices when possible.

9. $\begin{bmatrix} 3 & 2 \\ 1 & 5 \end{bmatrix}$

10. $\begin{bmatrix} 1 & -3 \\ 2 & -4 \end{bmatrix}$

11. $\begin{bmatrix} 1 & 2 \\ 3 & 4 \end{bmatrix}$

12. $\begin{bmatrix} 1 & \frac{2}{3} \\ 0 & \frac{5}{3} \end{bmatrix}.$

13. $\begin{bmatrix} a & b \\ b & a \end{bmatrix}$

14. $\begin{bmatrix} a+1 & -1 \\ 1 & a \end{bmatrix}$

15. $\begin{bmatrix} 2 & a \\ -a & 5 \end{bmatrix}$

16. $\begin{bmatrix} a & -2 \\ 2 & a \end{bmatrix}$

17. $\begin{bmatrix} 3 & a \\ 2 & b \end{bmatrix}$

18. $\begin{bmatrix} 4 & 1 \\ a & b \end{bmatrix}$

Use Theorem 1.2.6 to solve the following systems of equations.

19. $\begin{cases} x + 2y = 2 \\ 3x + 4y = 1 \end{cases}$

20. $\begin{cases} 2x - y = 1 \\ x + 2y = -1 \end{cases}$

21. $\begin{cases} 7x - 2y = 0 \\ 5x + 3y = 4 \end{cases}$

22. $\begin{cases} 5x + 4y = 3 \\ 2x + 3y = 0 \end{cases}$

Show that the following identities hold for any real numbers a_1, a_2, b_1, b_2, and t.

23. $\det \begin{bmatrix} a_1 + ta_2 & b_1 + tb_2 \\ a_2 & b_2 \end{bmatrix} = \det \begin{bmatrix} a_1 & b_1 \\ a_2 & b_2 \end{bmatrix}$

24. $\det \begin{bmatrix} a_1 & b_1 \\ a_2 + ta_1 & b_2 + tb_1 \end{bmatrix} = \det \begin{bmatrix} a_1 & b_1 \\ a_2 & b_2 \end{bmatrix}$

Use Theorem 1.3.10 to solve the following systems of equations.

25. $\begin{cases} 3x + 2y = 1 \\ 2x + 3y = 1 \end{cases}$

30. $\begin{cases} (a-1)x - y = 1 \\ x + ay = 0 \end{cases}$

26. $\begin{cases} 5x - 4y = 1 \\ 3x + 5y = 0 \end{cases}$

31. $\begin{cases} 3x + ay = s \\ 2x + 5y = t \end{cases}$

27. $\begin{cases} 7x + 2y = 0 \\ x + 7y = -1 \end{cases}$

32. $\begin{cases} 3x + ay = s \\ x + 2y = t \end{cases}$

28. $\begin{cases} 5x + 3y = 2 \\ 3x + 4y = -1 \end{cases}$

33. $\begin{cases} 2x + ay = s \\ bx + 2y = t \end{cases}$

29. $\begin{cases} (a-4)x - 5y = 1 \\ 2x + ay = 1 \end{cases}$

34. $\begin{cases} 4x + ay = s \\ 5x + by = t \end{cases}$

1.4 Diagonalization of 2 × 2 matrices

Matrices of the form $\begin{bmatrix} \alpha & 0 \\ 0 & \beta \end{bmatrix}$ are called ***diagonal matrices***. They are easy to work with. For example, we have

$$\begin{bmatrix} \alpha & 0 \\ 0 & \beta \end{bmatrix} \begin{bmatrix} x \\ y \end{bmatrix} = \begin{bmatrix} \alpha x \\ \beta y \end{bmatrix}.$$

Multiplying diagonal matrices is equally easy:

$$\begin{bmatrix} \alpha_1 & 0 \\ 0 & \beta_1 \end{bmatrix} \begin{bmatrix} \alpha_2 & 0 \\ 0 & \beta_2 \end{bmatrix} = \begin{bmatrix} \alpha_1 \alpha_2 & 0 \\ 0 & \beta_1 \beta_2 \end{bmatrix} \tag{1.13}$$

In this section we show how these and other properties of diagonal matrices can be useful when dealing with matrices that are not diagonal.

From (1.13) we obtain the following nice and useful property of diagonal matrices

$$\begin{bmatrix} \alpha & 0 \\ 0 & \beta \end{bmatrix} \begin{bmatrix} \alpha & 0 \\ 0 & \beta \end{bmatrix} \begin{bmatrix} \alpha & 0 \\ 0 & \beta \end{bmatrix} = \begin{bmatrix} \alpha & 0 \\ 0 & \beta \end{bmatrix}^3 = \begin{bmatrix} \alpha^3 & 0 \\ 0 & \beta^3 \end{bmatrix}$$

and more generally

$$\underbrace{\begin{bmatrix} \alpha & 0 \\ 0 & \beta \end{bmatrix} \cdots \begin{bmatrix} \alpha & 0 \\ 0 & \beta \end{bmatrix}}_{n \text{ times}} = \begin{bmatrix} \alpha & 0 \\ 0 & \beta \end{bmatrix}^n = \begin{bmatrix} \alpha^n & 0 \\ 0 & \beta^n \end{bmatrix}.$$

This property can be extended to any 2 × 2 matrix that can be written in the form PDP^{-1} where P is an invertible 2×2 matrix and D is a diagonal matrix. Indeed, for such a matrix we have

$$\left(PDP^{-1}\right)^3 = PDP^{-1}PDP^{-1}PDP^{-1}$$
$$= PD(P^{-1}P)D(P^{-1}P)DP^{-1}$$
$$= PD^3P^{-1}.$$

More generally, if n is any natural number, we have

$$\left(PDP^{-1}\right)^n = \underbrace{PDP^{-1}PDP^{-1}\ldots PDP^{-1}}_{n \text{ times}} = PD^nP^{-1},$$

which means that

$$\left(P\begin{bmatrix} \alpha & 0 \\ 0 & \beta \end{bmatrix}P^{-1}\right)^n = P\begin{bmatrix} \alpha^n & 0 \\ 0 & \beta^n \end{bmatrix}P^{-1}$$

for any numbers α and β and any invertible 2×2 matrix P.

It is not clear when a matrix can be written in the form PDP^{-1}. Moreover, even if we know that a matrix has such a representation, it is not obvious how to find P and D. In this section we investigate these questions. We begin by considering an example.

Example 1.4.1. We want to write the matrix

$$A = \begin{bmatrix} 5 & -1 \\ 2 & 2 \end{bmatrix}$$

as a product

$$\begin{bmatrix} 5 & -1 \\ 2 & 2 \end{bmatrix} = \begin{bmatrix} u_1 & v_1 \\ u_2 & v_2 \end{bmatrix}\begin{bmatrix} \alpha & 0 \\ 0 & \beta \end{bmatrix}\begin{bmatrix} u_1 & v_1 \\ u_2 & v_2 \end{bmatrix}^{-1} \tag{1.14}$$

where

$$\begin{bmatrix} u_1 & v_1 \\ u_2 & v_2 \end{bmatrix}$$

is an invertible matrix.

Multiplying from the right both sides of the equation (1.14) by the matrix $\begin{bmatrix} u_1 & v_1 \\ u_2 & v_2 \end{bmatrix}$ we get

$$\begin{bmatrix} 5 & -1 \\ 2 & 2 \end{bmatrix}\begin{bmatrix} u_1 & v_1 \\ u_2 & v_2 \end{bmatrix} = \begin{bmatrix} u_1 & v_1 \\ u_2 & v_2 \end{bmatrix}\begin{bmatrix} \alpha & 0 \\ 0 & \beta \end{bmatrix}\begin{bmatrix} u_1 & v_1 \\ u_2 & v_2 \end{bmatrix}^{-1}\begin{bmatrix} u_1 & v_1 \\ u_2 & v_2 \end{bmatrix}$$

$$= \begin{bmatrix} u_1 & v_1 \\ u_2 & v_2 \end{bmatrix}\begin{bmatrix} \alpha & 0 \\ 0 & \beta \end{bmatrix}\begin{bmatrix} 1 & 0 \\ 0 & 1 \end{bmatrix}$$

$$= \begin{bmatrix} u_1 & v_1 \\ u_2 & v_2 \end{bmatrix}\begin{bmatrix} \alpha & 0 \\ 0 & \beta \end{bmatrix}.$$

This means that equation (1.14) is equivalent to the equation

$$\begin{bmatrix} 5 & -1 \\ 2 & 2 \end{bmatrix}\begin{bmatrix} u_1 & v_1 \\ u_2 & v_2 \end{bmatrix} = \begin{bmatrix} u_1 & v_1 \\ u_2 & v_2 \end{bmatrix}\begin{bmatrix} \alpha & 0 \\ 0 & \beta \end{bmatrix}. \tag{1.15}$$

Since

$$\begin{bmatrix} u_1 & v_1 \\ u_2 & v_2 \end{bmatrix} \begin{bmatrix} \alpha & 0 \\ 0 & \beta \end{bmatrix} = \begin{bmatrix} \alpha u_1 & \beta v_1 \\ \alpha u_2 & \beta v_2 \end{bmatrix}$$

the equation (1.15) can be written as

$$\begin{bmatrix} 5 & -1 \\ 2 & 2 \end{bmatrix} \begin{bmatrix} u_1 & v_1 \\ u_2 & v_2 \end{bmatrix} = \begin{bmatrix} \alpha u_1 & \beta v_1 \\ \alpha u_2 & \beta v_2 \end{bmatrix}. \tag{1.16}$$

Equation (1.16) is equivalent to the following two equations:

$$\begin{bmatrix} 5 & -1 \\ 2 & 2 \end{bmatrix} \begin{bmatrix} u_1 \\ u_2 \end{bmatrix} = \begin{bmatrix} \alpha u_1 \\ \alpha u_2 \end{bmatrix} = \alpha \begin{bmatrix} u_1 \\ u_2 \end{bmatrix} \tag{1.17}$$

and

$$\begin{bmatrix} 5 & -1 \\ 2 & 2 \end{bmatrix} \begin{bmatrix} v_1 \\ v_2 \end{bmatrix} = \begin{bmatrix} \beta v_1 \\ \beta v_2 \end{bmatrix} = \beta \begin{bmatrix} v_1 \\ v_2 \end{bmatrix}. \tag{1.18}$$

So far we have shown that the equation

$$\begin{bmatrix} 5 & -1 \\ 2 & 2 \end{bmatrix} = \begin{bmatrix} u_1 & v_1 \\ u_2 & v_2 \end{bmatrix} \begin{bmatrix} \alpha & 0 \\ 0 & \beta \end{bmatrix} \begin{bmatrix} u_1 & v_1 \\ u_2 & v_2 \end{bmatrix}^{-1}$$

is equivalent to finding real numbers α and β and vectors $\begin{bmatrix} u_1 \\ u_2 \end{bmatrix}$ and $\begin{bmatrix} v_1 \\ v_2 \end{bmatrix}$ such that

$$\begin{bmatrix} 5 & -1 \\ 2 & 2 \end{bmatrix} \begin{bmatrix} u_1 \\ u_2 \end{bmatrix} = \alpha \begin{bmatrix} u_1 \\ u_2 \end{bmatrix} \quad \text{and} \quad \begin{bmatrix} 5 & -1 \\ 2 & 2 \end{bmatrix} \begin{bmatrix} v_1 \\ v_2 \end{bmatrix} = \beta \begin{bmatrix} v_1 \\ v_2 \end{bmatrix}$$

and such that the matrix $\begin{bmatrix} u_1 & v_1 \\ u_2 & v_2 \end{bmatrix}$ is invertible.

The equation (1.17) can be written as

$$\begin{bmatrix} 5-\alpha & -1 \\ 2 & 2-\alpha \end{bmatrix} \begin{bmatrix} u_1 \\ u_2 \end{bmatrix} = \begin{bmatrix} 0 \\ 0 \end{bmatrix}$$

or

$$\begin{cases} (5-\alpha)u_1 - u_2 = 0 \\ 2u_1 + (2-\alpha)u_2 = 0 \end{cases}. \tag{1.19}$$

We are interested in a solution such that $u_1 \neq 0$ or $u_2 \neq 0$. This means that, by Theorem 1.3.13, we must have

$$\det \begin{bmatrix} 5-\alpha & -1 \\ 2 & 2-\alpha \end{bmatrix} = 0.$$

Similarly, equation (1.18) can be written as

$$\begin{bmatrix} 5-\beta & -1 \\ 2 & 2-\beta \end{bmatrix}\begin{bmatrix} v_1 \\ v_2 \end{bmatrix} = \begin{bmatrix} 0 \\ 0 \end{bmatrix}$$

or

$$\begin{cases} (5-\beta)v_1 - v_2 = 0 \\ 2v_1 + (2-\beta)v_2 = 0 \end{cases}. \tag{1.20}$$

Since we are interested in a solution such that $v_1 \neq 0$ or $v_2 \neq 0$, we must have

$$\det \begin{bmatrix} 5-\beta & -1 \\ 2 & 2-\beta \end{bmatrix} = 0.$$

Consequently, both α and β are roots of the equation

$$\det \begin{bmatrix} 5-\lambda & -1 \\ 2 & 2-\lambda \end{bmatrix} = 0.$$

We calculate the determinant and obtain the equation

$$\lambda^2 - 7\lambda + 12 = 0,$$

which has roots $\alpha = 3$ and $\beta = 4$.

With $\alpha = 3$ the system (1.19) becomes

$$\begin{cases} 2u_1 - u_2 = 0 \\ 2u_1 - u_2 = 0 \end{cases}$$

and has a nontrivial solution $\begin{bmatrix} u_1 \\ u_2 \end{bmatrix} = \begin{bmatrix} 1 \\ 2 \end{bmatrix}$. With $\beta = 4$ the system (5.5) becomes

$$\begin{cases} v_1 - v_2 = 0 \\ 2v_1 - 2v_2 = 0 \end{cases}$$

and has a nontrivial solution $\begin{bmatrix} v_1 \\ v_2 \end{bmatrix} = \begin{bmatrix} 1 \\ 1 \end{bmatrix}$.

We note that

$$\det \begin{bmatrix} u_1 & v_1 \\ u_2 & v_2 \end{bmatrix} = \det \begin{bmatrix} 1 & 1 \\ 2 & 1 \end{bmatrix} = -1 \neq 0,$$

so the matrix is invertible and we have

$$\begin{bmatrix} u_1 & v_1 \\ u_2 & v_2 \end{bmatrix}^{-1} = \begin{bmatrix} 1 & 1 \\ 2 & 1 \end{bmatrix}^{-1} = -\begin{bmatrix} 1 & -1 \\ -2 & 1 \end{bmatrix} = \begin{bmatrix} -1 & 1 \\ 2 & -1 \end{bmatrix}.$$

Consequently, the desired representation of the matrix $\begin{bmatrix} 5 & -1 \\ 2 & 2 \end{bmatrix}$ is

$$\begin{bmatrix} 5 & -1 \\ 2 & 2 \end{bmatrix} = \begin{bmatrix} u_1 & v_1 \\ u_2 & v_2 \end{bmatrix} \begin{bmatrix} \alpha & 0 \\ 0 & \beta \end{bmatrix} \begin{bmatrix} u_1 & v_1 \\ u_2 & v_2 \end{bmatrix}^{-1} = \begin{bmatrix} 1 & 1 \\ 2 & 1 \end{bmatrix} \begin{bmatrix} 3 & 0 \\ 0 & 4 \end{bmatrix} \begin{bmatrix} -1 & 1 \\ 2 & -1 \end{bmatrix}.$$

We illustrate the advantage of the obtained representation by calculating $\begin{bmatrix} 5 & -1 \\ 2 & 2 \end{bmatrix}^{77}$, which is practically impossible to do by direct calculations with the matrix $\begin{bmatrix} 5 & -1 \\ 2 & 2 \end{bmatrix}$. On the other hand, since

$$\begin{bmatrix} 5 & -1 \\ 2 & 2 \end{bmatrix} = \begin{bmatrix} 1 & 1 \\ 2 & 1 \end{bmatrix} \begin{bmatrix} 3 & 0 \\ 0 & 4 \end{bmatrix} \begin{bmatrix} -1 & 1 \\ 2 & -1 \end{bmatrix},$$

we have

$$\begin{aligned}
\begin{bmatrix} 5 & -1 \\ 2 & 2 \end{bmatrix}^{77} &= \left(\begin{bmatrix} 1 & 1 \\ 2 & 1 \end{bmatrix} \begin{bmatrix} 3 & 0 \\ 0 & 4 \end{bmatrix} \begin{bmatrix} -1 & 1 \\ 2 & -1 \end{bmatrix} \right)^{77} \\
&= \begin{bmatrix} 1 & 1 \\ 2 & 1 \end{bmatrix} \begin{bmatrix} 3 & 0 \\ 0 & 4 \end{bmatrix}^{77} \begin{bmatrix} -1 & 1 \\ 2 & -1 \end{bmatrix} \\
&= \begin{bmatrix} 1 & 1 \\ 2 & 1 \end{bmatrix} \begin{bmatrix} 3^{77} & 0 \\ 0 & 4^{77} \end{bmatrix} \begin{bmatrix} -1 & 1 \\ 2 & -1 \end{bmatrix} \\
&= \begin{bmatrix} 3^{77} & 4^{77} \\ 2 \cdot 3^{77} & 4^{77} \end{bmatrix} \begin{bmatrix} -1 & 1 \\ 2 & -1 \end{bmatrix} \\
&= \begin{bmatrix} -3^{77} + 2 \cdot 4^{77} & 3^{77} - 4^{77} \\ -2 \cdot 3^{77} + 2 \cdot 4^{77} & 2 \cdot 3^{77} - 4^{77} \end{bmatrix}.
\end{aligned}$$

Now we will show that the method used in the above example leads to a more general result. We observe that finding the roots of the equation

$$\det \begin{bmatrix} 5 - \lambda & -1 \\ 2 & 2 - \lambda \end{bmatrix} = 0$$

was crucial in the presented solution. The roots $\alpha = 3$ and $\beta = 4$ turned out to be the numbers needed for the representation

$$\begin{bmatrix} 5 & -1 \\ 2 & 2 \end{bmatrix} = \begin{bmatrix} u_1 & v_1 \\ u_2 & v_2 \end{bmatrix} \begin{bmatrix} \alpha & 0 \\ 0 & \beta \end{bmatrix} \begin{bmatrix} u_1 & v_1 \\ u_2 & v_2 \end{bmatrix}^{-1}.$$

This observation leads to one of the most important ideas in linear algebra.

Definition 1.4.2. A real number λ is called an **_eigenvalue_** of a 2×2 matrix A if the equation

$$A\begin{bmatrix} x_1 \\ x_2 \end{bmatrix} = \lambda \begin{bmatrix} x_1 \\ x_2 \end{bmatrix}$$

has a nontrivial solution, that is, a solution $\begin{bmatrix} x_1 \\ x_2 \end{bmatrix} \neq \begin{bmatrix} 0 \\ 0 \end{bmatrix}$.

Example 1.4.3. Since

$$\begin{bmatrix} 5 & -1 \\ 2 & 2 \end{bmatrix}\begin{bmatrix} 1 \\ 2 \end{bmatrix} = \begin{bmatrix} 3 \\ 6 \end{bmatrix} = 3\begin{bmatrix} 1 \\ 2 \end{bmatrix}$$

and

$$\begin{bmatrix} 5 & -1 \\ 2 & 2 \end{bmatrix}\begin{bmatrix} 1 \\ 1 \end{bmatrix} = \begin{bmatrix} 4 \\ 4 \end{bmatrix} = 4\begin{bmatrix} 1 \\ 1 \end{bmatrix},$$

$\lambda = 3$ and $\lambda = 4$ are eigenvalues of the matrix $\begin{bmatrix} 5 & -1 \\ 2 & 2 \end{bmatrix}$.

The following theorem gives us a practical way of finding eigenvalues of a 2×2 matrix.

Theorem 1.4.4. *The real number λ is an eigenvalue of the matrix $\begin{bmatrix} a & b \\ c & d \end{bmatrix}$ if*

$$\det\begin{bmatrix} a-\lambda & b \\ c & d-\lambda \end{bmatrix} = 0.$$

Proof. First note that the equation

$$\begin{bmatrix} a & b \\ c & d \end{bmatrix}\begin{bmatrix} x \\ y \end{bmatrix} = \lambda\begin{bmatrix} x \\ y \end{bmatrix}$$

can be written as

$$\begin{bmatrix} a-\lambda & b \\ c & d-\lambda \end{bmatrix}\begin{bmatrix} x \\ y \end{bmatrix} = \begin{bmatrix} 0 \\ 0 \end{bmatrix}.$$

The above equation has a nontrivial solution if

$$\det\begin{bmatrix} a-\lambda & b \\ c & d-\lambda \end{bmatrix} = 0,$$

by Theorem 1.3.13. $\square$

Example 1.4.5. Find the eigenvalues of the matrix $\begin{bmatrix} 1 & 5 \\ 3 & 3 \end{bmatrix}$.

Solution. Since

$$\det \begin{bmatrix} 1-\lambda & 5 \\ 3 & 3-\lambda \end{bmatrix} = (1-\lambda)(3-\lambda) - 15 = \lambda^2 - 4\lambda - 12,$$

we need to solve the quadratic equation

$$\lambda^2 - 4\lambda - 12 = 0.$$

The solutions are $\lambda = 6$ and $\lambda = -2$, which are the eigenvalues of the matrix $\begin{bmatrix} 1 & 5 \\ 3 & 3 \end{bmatrix}$.

□

Example 1.4.6. Since

$$\det \begin{bmatrix} 1-\lambda & 3 \\ 2 & -1-\lambda \end{bmatrix} = \lambda^2 - 7,$$

the eigenvalues of the matrix $A = \begin{bmatrix} 1 & 3 \\ 2 & -1 \end{bmatrix}$ are $\lambda = \sqrt{7}$ and $\lambda = -\sqrt{7}$.

Definition 1.4.7. Let λ be an eigenvalue of a 2 × 2 matrix A. A vector $\begin{bmatrix} x_1 \\ x_2 \end{bmatrix} \neq \begin{bmatrix} 0 \\ 0 \end{bmatrix}$ such that $A \begin{bmatrix} x_1 \\ x_2 \end{bmatrix} = \lambda \begin{bmatrix} x_1 \\ x_2 \end{bmatrix}$ is called an ***eigenvector corresponding to the eigenvalue*** λ.

Example 1.4.8. Since

$$\begin{bmatrix} 1 & 5 \\ 3 & 3 \end{bmatrix} \begin{bmatrix} 1 \\ 1 \end{bmatrix} = \begin{bmatrix} 6 \\ 6 \end{bmatrix} = 6 \begin{bmatrix} 1 \\ 1 \end{bmatrix}$$

and

$$\begin{bmatrix} 1 & 5 \\ 3 & 3 \end{bmatrix} \begin{bmatrix} -5 \\ 3 \end{bmatrix} = \begin{bmatrix} 10 \\ -6 \end{bmatrix} = -2 \begin{bmatrix} -5 \\ 3 \end{bmatrix},$$

the vector $\begin{bmatrix} 1 \\ 1 \end{bmatrix}$ is an eigenvector corresponding to the eigenvalue $\lambda = 6$ and the vector $\begin{bmatrix} -5 \\ 3 \end{bmatrix}$ is an eigenvector corresponding to the eigenvalue $\lambda = -2$,

Note that eigenvectors are not unique. If $\begin{bmatrix} x_1 \\ x_2 \end{bmatrix}$ is an eigenvector of A corresponding to the eigenvalue λ, then for any real number t we have

$$A\left(t\begin{bmatrix} x_1 \\ x_2 \end{bmatrix}\right) = t\left(A\begin{bmatrix} x_1 \\ x_2 \end{bmatrix}\right) = t\left(\lambda\begin{bmatrix} x_1 \\ x_2 \end{bmatrix}\right) = \lambda\left(t\begin{bmatrix} x_1 \\ x_2 \end{bmatrix}\right),$$

so $t\begin{bmatrix} x_1 \\ x_2 \end{bmatrix}$ is also an eigenvector of A corresponding to the eigenvalue λ as long as $t \neq 0$.

Similarly, if $\begin{bmatrix} x_1 \\ x_2 \end{bmatrix}$ and $\begin{bmatrix} y_1 \\ y_2 \end{bmatrix}$ are eigenvectors of A corresponding to the eigenvalue λ, then it is easy to verify that the vector $\begin{bmatrix} x_1 + y_1 \\ x_2 + y_2 \end{bmatrix}$ is also an eigenvector of the matrix A as long as $\begin{bmatrix} x_1 + y_1 \\ x_2 + y_2 \end{bmatrix} \neq \begin{bmatrix} 0 \\ 0 \end{bmatrix}$.

Definition 1.4.9. A matrix A is called **_diagonalizable_** if there is a diagonal matrix D and an invertible matrix P such that

$$A = PDP^{-1}.$$

In other words, a matrix $A = \begin{bmatrix} a_1 & b_1 \\ a_2 & b_2 \end{bmatrix}$ is diagonalizable if there are real numbers α and β and an invertible matrix $\begin{bmatrix} u_1 & v_1 \\ u_2 & v_2 \end{bmatrix}$ such that

$$\begin{bmatrix} a_1 & b_1 \\ a_2 & b_2 \end{bmatrix} = \begin{bmatrix} u_1 & v_1 \\ u_2 & v_2 \end{bmatrix} \begin{bmatrix} \alpha & 0 \\ 0 & \beta \end{bmatrix} \begin{bmatrix} u_1 & v_1 \\ u_2 & v_2 \end{bmatrix}^{-1}.$$

Note that we are not assuming that α and β are different.

Example 1.4.10. In Example 1.4.1 we have shown that

$$\begin{bmatrix} 5 & -1 \\ 2 & 2 \end{bmatrix} = \begin{bmatrix} 1 & 1 \\ 2 & 1 \end{bmatrix} \begin{bmatrix} 3 & 0 \\ 0 & 4 \end{bmatrix} \begin{bmatrix} 1 & 1 \\ 2 & 1 \end{bmatrix}^{-1}.$$

Therefore the matrix

$$A = \begin{bmatrix} 5 & -1 \\ 2 & 2 \end{bmatrix}$$

is diagonalizable.

Not all matrices are diagonalizable.

Example 1.4.11. If possible, diagonalize the matrix $A = \begin{bmatrix} 2 & 0 \\ 3 & 2 \end{bmatrix}$.

Solution. The eigenvalues of A are the roots of the equation

$$\det \begin{bmatrix} 2-\lambda & 0 \\ 3 & 2-\lambda \end{bmatrix} = \lambda^2 - 4\lambda + 4 = 0.$$

We solve this quadratic equation and find that the only eigenvalue is $\lambda = 2$.

The eigenvectors corresponding to the eigenvalue $\lambda = 2$ are the solutions of the equation

$$\begin{bmatrix} 2 & 0 \\ 3 & 2 \end{bmatrix} \begin{bmatrix} x \\ y \end{bmatrix} = 2 \begin{bmatrix} x \\ y \end{bmatrix}$$

that is equivalent to the equation

$$3x = 0.$$

So a vector $\begin{bmatrix} x \\ y \end{bmatrix}$ is an eigenvector of A corresponding to the eigenvalue $\lambda = 2$ if $x = 0$.

Consider two eigenvectors $\begin{bmatrix} 0 \\ a \end{bmatrix}$ and $\begin{bmatrix} 0 \\ b \end{bmatrix}$ corresponding to the eigenvalue $\lambda = 2$.

Since $\det \begin{bmatrix} 0 & 0 \\ a & b \end{bmatrix} = 0$, the matrix $\begin{bmatrix} 0 & 0 \\ a & b \end{bmatrix}$ is not invertible. Because this happens no matter what a and b we use, the matrix A cannot be diagonalized. □

The next theorem describes a property that guarantees diagonalizability of a 2×2 matrix. Note that it is not an "if and only if" statement.

Theorem 1.4.12. *Every* 2 × 2 *matrix with two different real eigenvalues is diagonalizable.*

Proof. Let $A = \begin{bmatrix} a_1 & b_1 \\ a_2 & b_2 \end{bmatrix}$. Assume there exist two different real numbers α and β and nonzero vectors $\begin{bmatrix} u_1 \\ u_2 \end{bmatrix}$ and $\begin{bmatrix} v_1 \\ v_2 \end{bmatrix}$ such that

$$\begin{bmatrix} a_1 & b_1 \\ a_2 & b_2 \end{bmatrix} \begin{bmatrix} u_1 \\ u_2 \end{bmatrix} = \alpha \begin{bmatrix} u_1 \\ u_2 \end{bmatrix} = \begin{bmatrix} \alpha u_1 \\ \alpha u_2 \end{bmatrix} \quad \text{and} \quad \begin{bmatrix} a_1 & b_1 \\ a_2 & b_2 \end{bmatrix} \begin{bmatrix} v_1 \\ v_2 \end{bmatrix} = \beta \begin{bmatrix} v_1 \\ v_2 \end{bmatrix} = \begin{bmatrix} \beta v_1 \\ \beta v_2 \end{bmatrix}. \tag{1.21}$$

We will show that the matrix $\begin{bmatrix} u_1 & v_1 \\ u_2 & v_2 \end{bmatrix}$ is invertible. We use Theorem 1.2.14, that is, we assume that

$$\begin{bmatrix} u_1 & v_1 \\ u_2 & v_2 \end{bmatrix} \begin{bmatrix} x \\ y \end{bmatrix} = \begin{bmatrix} 0 \\ 0 \end{bmatrix} \tag{1.22}$$

and show that $x = y = 0$.

We first note that the equations in (1.21) can be written as a single equation

$$\begin{bmatrix} a_1 & b_1 \\ a_2 & b_2 \end{bmatrix} \begin{bmatrix} u_1 & v_1 \\ u_2 & v_2 \end{bmatrix} = \begin{bmatrix} \alpha u_1 & \beta v_1 \\ \alpha u_2 & \beta v_2 \end{bmatrix} \tag{1.23}$$

which implies

$$\begin{bmatrix} a_1 & b_1 \\ a_2 & b_2 \end{bmatrix} \begin{bmatrix} u_1 & v_1 \\ u_2 & v_2 \end{bmatrix} \begin{bmatrix} x \\ y \end{bmatrix} = \begin{bmatrix} \alpha u_1 & \beta v_1 \\ \alpha u_2 & \beta v_2 \end{bmatrix} \begin{bmatrix} x \\ y \end{bmatrix}. \tag{1.24}$$

From (1.24) and (1.22) we obtain

$$\begin{bmatrix} \alpha u_1 & \beta v_1 \\ \alpha u_2 & \beta v_2 \end{bmatrix} \begin{bmatrix} x \\ y \end{bmatrix} = \begin{bmatrix} 0 \\ 0 \end{bmatrix}$$

which can be written as

$$\begin{cases} \alpha u_1 x + \beta v_1 y = 0 \\ \alpha u_2 x + \beta v_2 y = 0 \end{cases}. \tag{1.25}$$

Equation (1.22) can be written as

$$\begin{cases} u_1 x + v_1 y = 0 \\ u_2 x + v_2 y = 0 \end{cases}.$$

Multiplying these equations by α and subtracting them from the corresponding equations in (1.25) we get

$$\begin{cases} (\beta - \alpha) v_1 y = 0 \\ (\beta - \alpha) v_2 y = 0 \end{cases},$$

which gives us $y = 0$ because $\beta - \alpha \neq 0$ and at least one of the numbers v_1 and v_2 is different from 0.

By modifying the above argument appropriately we obtain $x = 0$. This allows us to conclude, by Theorem 1.2.14, that the matrix $\begin{bmatrix} u_1 & v_1 \\ u_2 & v_2 \end{bmatrix}$ is invertible. Multiplying (1.23) by $\begin{bmatrix} u_1 & v_1 \\ u_2 & v_2 \end{bmatrix}^{-1}$ on the left produces the desired result:

$$\begin{bmatrix} a_1 & b_1 \\ a_2 & b_2 \end{bmatrix} = \begin{bmatrix} u_1 & v_1 \\ u_2 & v_2 \end{bmatrix} \begin{bmatrix} \alpha & 0 \\ 0 & \beta \end{bmatrix} \begin{bmatrix} u_1 & v_1 \\ u_2 & v_2 \end{bmatrix}^{-1}.$$

□

Note that the above proof shows more than just that every 2×2 matrix with two different real eigenvalues is diagonalizable. It gives us a practical method for diagonalizing such a matrix. First we need to find the eigenvalues. If the eigenvalues are two different real numbers α and β, then we need to find an eigenvector $\begin{bmatrix} u_1 \\ u_2 \end{bmatrix}$ corresponding to the eigenvalue α and an eigenvector $\begin{bmatrix} v_1 \\ v_2 \end{bmatrix}$ corresponding to the eigenvalue β. Then the matrix $\begin{bmatrix} u_1 & v_1 \\ u_2 & v_2 \end{bmatrix}$ is invertible and we have

$$\begin{bmatrix} a_1 & b_1 \\ a_2 & b_2 \end{bmatrix} = \begin{bmatrix} u_1 & v_1 \\ u_2 & v_2 \end{bmatrix} \begin{bmatrix} \alpha & 0 \\ 0 & \beta \end{bmatrix} \begin{bmatrix} u_1 & v_1 \\ u_2 & v_2 \end{bmatrix}^{-1}.$$

In the next example we use this method to diagonalize a given 2 × 2 matrix.

Example 1.4.13. If possible, diagonalize the matrix $A = \begin{bmatrix} 5 & 7 \\ 3 & 9 \end{bmatrix}$.

Solution. The eigenvalues of A are the roots of the equation

$$\det \begin{bmatrix} 5-\lambda & 7 \\ 3 & 9-\lambda \end{bmatrix} = \lambda^2 - 14\lambda + 24 = 0.$$

We solve this quadratic equation and find that the eigenvalues are $\lambda = 2$ and $\lambda = 12$.

The eigenvectors corresponding to the eigenvalue $\lambda = 2$ are the solutions of the equation

$$\begin{bmatrix} 5 & 7 \\ 3 & 9 \end{bmatrix} \begin{bmatrix} x \\ y \end{bmatrix} = 2 \begin{bmatrix} x \\ y \end{bmatrix}$$

that is equivalent to the equation

$$\begin{bmatrix} 3 & 7 \\ 3 & 7 \end{bmatrix} \begin{bmatrix} x \\ y \end{bmatrix} = \begin{bmatrix} 0 \\ 0 \end{bmatrix}.$$

So the vector $\begin{bmatrix} x \\ y \end{bmatrix}$ is an eigenvector A corresponding to the eigenvalue $\lambda = 2$ if $3x + 7y = 0$. This means that we can take $\begin{bmatrix} x \\ y \end{bmatrix} = \begin{bmatrix} 7 \\ -3 \end{bmatrix}$ as an eigenvector of A corresponding to $\lambda = 2$.

The eigenvectors corresponding to the eigenvalue $\lambda = 12$ are the solutions of the equation

$$\begin{bmatrix} 5 & 7 \\ 3 & 9 \end{bmatrix} \begin{bmatrix} x \\ y \end{bmatrix} = 12 \begin{bmatrix} x \\ y \end{bmatrix},$$

or, equivalently, of the equation

$$\begin{bmatrix} -7 & 7 \\ 3 & -3 \end{bmatrix} \begin{bmatrix} x \\ y \end{bmatrix} = \begin{bmatrix} 0 \\ 0 \end{bmatrix}.$$

So the vector $\begin{bmatrix} x \\ y \end{bmatrix}$ is an eigenvector A corresponding to the eigenvalue $\lambda = 12$ if $x - y = 0$. This means that $\begin{bmatrix} x \\ y \end{bmatrix} = \begin{bmatrix} 1 \\ 1 \end{bmatrix}$ as an eigenvector of A corresponding to $\lambda = 12$.

Now we have everything we need to diagonalize the matrix A:

$$A = \begin{bmatrix} 7 & 1 \\ -3 & 1 \end{bmatrix} \begin{bmatrix} 2 & 0 \\ 0 & 12 \end{bmatrix} \begin{bmatrix} 7 & 1 \\ -3 & 1 \end{bmatrix}^{-1}.$$

Since $\begin{bmatrix} 7 & 1 \\ -3 & 1 \end{bmatrix}^{-1} = \frac{1}{10}\begin{bmatrix} 1 & -1 \\ 3 & 7 \end{bmatrix}$, we have

$$A = \begin{bmatrix} 7 & 1 \\ -3 & 1 \end{bmatrix}\begin{bmatrix} 2 & 0 \\ 0 & 12 \end{bmatrix}\begin{bmatrix} \frac{1}{10} & -\frac{1}{10} \\ \frac{3}{10} & \frac{7}{10} \end{bmatrix}$$

We could also write

$$A = \begin{bmatrix} 1 & 7 \\ 1 & -3 \end{bmatrix}\begin{bmatrix} 12 & 0 \\ 0 & 2 \end{bmatrix}\begin{bmatrix} 1 & 7 \\ 1 & -3 \end{bmatrix}^{-1} = \begin{bmatrix} 1 & 7 \\ 1 & -3 \end{bmatrix}\begin{bmatrix} 12 & 0 \\ 0 & 2 \end{bmatrix}\begin{bmatrix} \frac{3}{10} & \frac{7}{10} \\ \frac{1}{10} & -\frac{1}{10} \end{bmatrix}.$$

It is important to note that the order of the eigenvectors matches the order of the eigenvalues in the diagonal matrix.   □

Applications

Example 1.4.14. Let p, q, r, and s be real numbers such that $p > 0, q > 0, r > 0$, $s > 0, p + r = 1$ and $q + s = 1$. Show that the matrix

$$A = \begin{bmatrix} p & q \\ r & s \end{bmatrix}$$

has eigenvalues 1 and k, where $-1 < k < 1$.

Solution. We can write $r = 1 - p$ and $q = 1 - s$. The eigenvalues of A are the solutions of the equation

$$\det\begin{bmatrix} p - \lambda & 1 - s \\ 1 - p & s - \lambda \end{bmatrix} = 0.$$

Since

$$\det\begin{bmatrix} p - \lambda & 1 - s \\ 1 - p & s - \lambda \end{bmatrix} = (p - \lambda)(s - \lambda) - (1 - p)(1 - s)$$

$$= \lambda^2 - (p + s)\lambda + p + s - 1$$

$$= (\lambda - 1)(\lambda - p - s + 1),$$

the eigenvalues are 1 and $k = p + s - 1$. Moreover

$$-1 < k = p + s - 1 < 1,$$

because $0 < p < 1$ and $0 < s < 1$.   □

Example 1.4.15. Let p, q, r, and s be real numbers such that $p > 0$, $q > 0$, $r > 0$, $s > 0$, $p + r = 1$ and $q + s = 1$ and let $A = \begin{bmatrix} p & q \\ r & s \end{bmatrix}$. If x_0 and y_0 are arbitrary real numbers and

$$\begin{bmatrix} x_{n+1} \\ y_{n+1} \end{bmatrix} = A \begin{bmatrix} x_n \\ y_n \end{bmatrix} \quad \text{for } n \geq 1,$$

express x_n and y_n in terms of x_0 and y_0 and find the limits $\lim_{n \to \infty} x_n$ and $\lim_{n \to \infty} y_n$.

Solution. The eigenvalues of A are 1 and k with $-1 < k < 1$. Let $\mathbf{u}$ be an eigenvector corresponding to the eigenvalue 1 and $\mathbf{v}$ an eigenvector corresponding to the eigenvalue k. Because $\det \begin{bmatrix} \mathbf{u} & \mathbf{v} \end{bmatrix} \neq 0$, there are real numbers α and β such that

$$\begin{bmatrix} x_0 \\ y_0 \end{bmatrix} = \alpha \mathbf{u} + \beta \mathbf{v}$$

by Theorem 1.3.12. Since

$$\begin{bmatrix} x_n \\ y_n \end{bmatrix} = A^n \begin{bmatrix} x_0 \\ y_0 \end{bmatrix} = \alpha A^n \mathbf{u} + \beta A^n \mathbf{v} = \alpha \mathbf{u} + \beta k^n \mathbf{v},$$

if $\mathbf{u} = \begin{bmatrix} u_1 \\ u_2 \end{bmatrix}$, we get

$$\lim_{n \to \infty} x_n = \alpha u_1 \quad \text{and} \quad \lim_{n \to \infty} y_n = \alpha u_2.$$

We note that, if we denote $P = \begin{bmatrix} \mathbf{u} & \mathbf{v} \end{bmatrix}$ and $D = \begin{bmatrix} 1 & 0 \\ 0 & k \end{bmatrix}$, then

$$\lim_{n \to \infty} \begin{bmatrix} x_n \\ y_n \end{bmatrix} = \lim_{n \to \infty} A^n \begin{bmatrix} x_0 \\ y_0 \end{bmatrix}$$

$$= \lim_{n \to \infty} P D^n P^{-1} \begin{bmatrix} x_0 \\ y_0 \end{bmatrix}$$

$$= \lim_{n \to \infty} P \begin{bmatrix} 1 & 0 \\ 0 & k^n \end{bmatrix} P^{-1} \begin{bmatrix} x_0 \\ y_0 \end{bmatrix}$$

$$= P \begin{bmatrix} 1 & 0 \\ 0 & 0 \end{bmatrix} P^{-1} \begin{bmatrix} x_0 \\ y_0 \end{bmatrix}$$

$$= P \begin{bmatrix} 1 & 0 \\ 0 & 0 \end{bmatrix} P^{-1} P \begin{bmatrix} \alpha \\ \beta \end{bmatrix}$$

$$= P \begin{bmatrix} \alpha \\ 0 \end{bmatrix} = \alpha \mathbf{u}.$$

$\square$

1.4.1 Exercises

Find an eigenvector of the given matrix A corresponding to the given eigenvalue λ.

1. $A = \begin{bmatrix} 1 & 2 \\ 2 & 4 \end{bmatrix}, \lambda = 0$

3. $A = \begin{bmatrix} -2 & 2 \\ 4 & 5 \end{bmatrix}, \lambda = 6$

2. $A = \begin{bmatrix} 2 & 4 \\ 1 & 5 \end{bmatrix}, \lambda = 1$

4. $A = \begin{bmatrix} 1 & 8 \\ 2 & 7 \end{bmatrix}, \lambda = -1$

Find the eigenvalues of the given matrix.

5. $\begin{bmatrix} 5 & 3 \\ 1 & 3 \end{bmatrix}$

9. $\begin{bmatrix} 4 & 3 \\ 4 & 3 \end{bmatrix}$

13. $\begin{bmatrix} a & b \\ 1-a & 1-b \end{bmatrix}$

6. $\begin{bmatrix} 2 & 3 \\ 5 & 4 \end{bmatrix}$

10. $\begin{bmatrix} 2 & 5 \\ -1 & -4 \end{bmatrix}$

14. $\begin{bmatrix} a & b \\ 3-a & 3-b \end{bmatrix}$

7. $\begin{bmatrix} 10 & 4 \\ 6 & 5 \end{bmatrix}$

11. $\begin{bmatrix} 1 & 7 \\ 9 & 3 \end{bmatrix}$

15. $\begin{bmatrix} a+3 & 2a \\ 7a & 14a+3 \end{bmatrix}$

8. $\begin{bmatrix} 3 & 3 \\ 1 & 5 \end{bmatrix}$

12. $\begin{bmatrix} 3 & 5 \\ 12 & 10 \end{bmatrix}$

16. $\begin{bmatrix} a+k & b \\ 2a & 2b+k \end{bmatrix}$

17. Find a matrix A such that $\begin{bmatrix} 4 \\ -1 \end{bmatrix}$ is an eigenvector of A corresponding to the eigenvalue $\lambda = 3$ and $\begin{bmatrix} -1 \\ 1 \end{bmatrix}$ is an eigenvector of A corresponding to the eigenvalue $\lambda = 0$.

18. Find a matrix A such that $\begin{bmatrix} 3 \\ 2 \end{bmatrix}$ is an eigenvector of A corresponding to the eigenvalue $\lambda = 2$ and $\begin{bmatrix} 2 \\ 3 \end{bmatrix}$ is an eigenvector of A corresponding to the eigenvalue $\lambda = -4$.

19. Let s and t be two real numbers. Find a matrix A such that $\begin{bmatrix} 2 \\ 1 \end{bmatrix}$ is an eigenvector of A corresponding to the eigenvalue s and $\begin{bmatrix} 1 \\ 4 \end{bmatrix}$ is an eigenvector of A corresponding to the eigenvalue t.

20. Let s and t be two real numbers. Find a matrix A such that $\begin{bmatrix} 3 \\ 4 \end{bmatrix}$ is an eigenvector of A corresponding to the eigenvalue s and $\begin{bmatrix} 2 \\ 0 \end{bmatrix}$ is an eigenvector of A corresponding to the eigenvalue t.

21. Find real numbers a and b such that the matrix $\begin{bmatrix} a & b \\ 5 & 3 \end{bmatrix}$ has eigenvalues 1 and 2.

22. Assuming a matrix $\begin{bmatrix} a & b \\ c & d \end{bmatrix}$ has only one eigenvalue α, find α.

If possible, diagonalize the given matrix.

23. $A = \begin{bmatrix} 9 & 3 \\ 5 & 7 \end{bmatrix}$

25. $A = \begin{bmatrix} 2 & 4 \\ 3 & 13 \end{bmatrix}$

24. $A = \begin{bmatrix} 2 & 2 \\ 1 & 3 \end{bmatrix}$

26. $A = \begin{bmatrix} 5 & 3 \\ 3 & 13 \end{bmatrix}$

27. For $A = \begin{bmatrix} 2 & 1 \\ 8 & 9 \end{bmatrix}$ find two different invertible matrices P_1 and P_2 and a diagonal matrix D such that $A = P_1 D P_1^{-1}$ and $A = P_2 D P_2^{-1}$.

28. For $A = \begin{bmatrix} 3 & 3 \\ 8 & 13 \end{bmatrix}$ find two different invertible matrices P_1 and P_2 and two different diagonal matrices D_1 and D_2 such that $A = P_1 D P_1^{-1}$ and $A = P_2 D P_2^{-1}$.

Find the given products of matrices.

29. $\begin{bmatrix} 2 & 4 \\ 2 & 9 \end{bmatrix}^n$

30. $\begin{bmatrix} 3 & 2 \\ 4 & 10 \end{bmatrix}^n$

31. Consider sequences (x_n) and (y_n) defined by the following recurrence relations
$$\begin{cases} x_{n+1} = 3x_n + 7y_n \\ y_{n+1} = x_n + 9y_n \end{cases}$$
with $x_0 = y_0 = 1$. Find x_{33} and y_{33}.

32. Consider sequences (x_n) and (y_n) defined by the following recurrence relations
$$\begin{cases} x_{n+1} = 2x_n + 2y_n \\ y_{n+1} = 5x_n + 11y_n \end{cases}$$
with $x_0 = 2$ and $y_0 = 3$. Find x_n and y_n.

33. Consider sequences (x_n) and (y_n) defined by the following recurrence relations
$$\begin{cases} x_{n+1} = \frac{3}{5}x_n + \frac{1}{4}y_n \\ y_{n+1} = \frac{2}{5}x_n + \frac{3}{4}y_n \end{cases}$$
with $x_0 = 2$ and $y_0 = 3$. Find $\lim_{n \to \infty} x_n$ and $\lim_{n \to \infty} y_n$.

34. Consider sequences (x_n) and (y_n) defined by the following recurrence relations

$$\begin{cases} x_{n+1} = \frac{1}{2}x_n + \frac{1}{3}y_n \\ y_{n+1} = \frac{1}{2}x_n + \frac{2}{3}y_n \end{cases}$$

with $x_0 = 5$ and $y_0 = 2$. Find $\lim_{n \to \infty} x_n$ and $\lim_{n \to \infty} y_n$.

Chapter 2

Matrices

2.1 General matrices

In Chapter 1 we considered some of the fundamental ideas of linear algebra in the context of 2×2 matrices. The discussed ideas and methods can be generalized and used to solve problems that require matrices of larger size. In this chapter we study algebraic properties of matrices of arbitrary size.

Basic definitions

Definition 2.1.1. By an $m \times n$ *matrix* we mean an array of numbers

$$A = \begin{bmatrix} a_{11} & a_{12} & \cdots & a_{1n} \\ a_{21} & a_{22} & \cdots & a_{2n} \\ \vdots & \vdots & \ddots & \vdots \\ a_{m1} & a_{m2} & \cdots & a_{mn} \end{bmatrix}.$$

The matrices

$$\begin{bmatrix} a_{i1} & a_{i2} & \cdots & a_{in} \end{bmatrix},$$

where $1 \leq i \leq m$, are called the *rows* of the matrix A. The matrices

$$\begin{bmatrix} a_{1j} \\ a_{2j} \\ \vdots \\ a_{mj} \end{bmatrix},$$

where $1 \leq j \leq n$, are called the *columns* of the matrix A. The numbers a_{ij} are referred to as *entries* of the matrix. An $n \times n$ matrix is called a *square matrix*.

An $m \times n$ matrix has m rows, n columns, and mn entries.

Example 2.1.2. The matrix

$$A = \begin{bmatrix} a_{11} & a_{12} & a_{13} & a_{14} \\ a_{21} & a_{22} & a_{23} & a_{24} \\ a_{31} & a_{32} & a_{33} & a_{34} \end{bmatrix}$$

has three rows

$$\begin{bmatrix} a_{11} & a_{12} & a_{13} & a_{14} \end{bmatrix}, \begin{bmatrix} a_{21} & a_{22} & a_{23} & a_{24} \end{bmatrix}, \begin{bmatrix} a_{31} & a_{32} & a_{33} & a_{34} \end{bmatrix},$$

four columns

$$\begin{bmatrix} a_{11} \\ a_{21} \\ a_{31} \end{bmatrix}, \begin{bmatrix} a_{12} \\ a_{22} \\ a_{32} \end{bmatrix}, \begin{bmatrix} a_{13} \\ a_{23} \\ a_{33} \end{bmatrix}, \begin{bmatrix} a_{14} \\ a_{24} \\ a_{34} \end{bmatrix},$$

and 12 entries.

Definition 2.1.3. *Two matrices are equal* if they have the same size and the corresponding entries are the same. More precisely, two $m \times n$ matrices

$$A = \begin{bmatrix} a_{11} & a_{12} & \cdots & a_{1n} \\ a_{21} & a_{22} & \cdots & a_{2n} \\ \vdots & \vdots & \ddots & \vdots \\ a_{m1} & a_{m2} & \cdots & a_{mn} \end{bmatrix} \quad \text{and} \quad B = \begin{bmatrix} b_{11} & b_{12} & \cdots & b_{1n} \\ b_{21} & b_{22} & \cdots & b_{2n} \\ \vdots & \vdots & \ddots & \vdots \\ b_{m1} & b_{m2} & \cdots & b_{mn} \end{bmatrix}$$

are equal if and only if $a_{ij} = b_{ij}$ for every $1 \le i \le m$ and every $1 \le j \le n$.

Example 2.1.4. The matrices

$$\begin{bmatrix} 1 & -1 & 2 & 5 \\ 0 & \pi & -\frac{1}{2} & 7 \\ \sqrt{2} & 0 & -0.77 & \frac{1}{3} \end{bmatrix} \quad \text{and} \quad \begin{bmatrix} 1 & -1 & 2 & 5 \\ 0 & \pi & -\frac{1}{2} & 7 \\ \sqrt{2} & 0 & 0.77 & \frac{1}{3} \end{bmatrix}$$

are not equal.

Sum of matrices

Definition 2.1.5. If A and B are two $m \times n$ matrices, the **sum** $A + B$ is the matrix whose entry in the i-th row and j-th column is $a_{ij} + b_{ij}$, where a_{ij} is the element of the matrix A which is in the i-th row and j-th column and b_{ij} is the element of the matrix B which is in the i-th row and j-th column:

$$\begin{bmatrix} a_{11} & a_{12} & \cdots & a_{1n} \\ a_{21} & a_{22} & \cdots & a_{2n} \\ \vdots & \vdots & \ddots & \vdots \\ a_{m1} & a_{m2} & \cdots & a_{mn} \end{bmatrix} + \begin{bmatrix} b_{11} & b_{12} & \cdots & b_{1n} \\ b_{21} & b_{22} & \cdots & b_{2n} \\ \vdots & \vdots & \ddots & \vdots \\ b_{m1} & b_{m2} & \cdots & b_{mn} \end{bmatrix} = \begin{bmatrix} a_{11}+b_{11} & a_{12}+b_{12} & \cdots & a_{1n}+b_{1n} \\ a_{21}+b_{21} & a_{22}+b_{22} & \cdots & a_{nj}+b_{2n} \\ \vdots & \vdots & \ddots & \vdots \\ a_{m1}+b_{31} & a_{m2}+b_{m2} & \cdots & a_{mn}+b_{mn} \end{bmatrix}$$

In other words, we add matrices by adding the corresponding entries.

Note that $A + B$ does not make sense if the the matrices A and B are not of the same size.

Example 2.1.6.

$$\begin{bmatrix} a_{11} & a_{12} \\ a_{21} & a_{22} \\ a_{31} & a_{32} \end{bmatrix} + \begin{bmatrix} b_{11} & b_{12} \\ b_{21} & b_{22} \\ b_{31} & b_{32} \end{bmatrix} = \begin{bmatrix} a_{11}+b_{11} & a_{12}+b_{12} \\ a_{21}+b_{21} & a_{22}+b_{22} \\ a_{31}+b_{31} & a_{32}+b_{32} \end{bmatrix}$$

$$\begin{bmatrix} 3 & 4 & -1 \\ 2 & 5 & 1 \end{bmatrix} + \begin{bmatrix} 1 & 0 & 3 \\ 4 & -3 & 7 \end{bmatrix} = \begin{bmatrix} 4 & 4 & 2 \\ 6 & 2 & 8 \end{bmatrix}$$

Definition 2.1.7. The $m \times n$ matrix whose all entries are 0,

$$\begin{bmatrix} 0 & 0 & \cdots & 0 \\ 0 & 0 & \cdots & 0 \\ \vdots & \vdots & \ddots & \vdots \\ 0 & 0 & \cdots & 0 \end{bmatrix},$$

is called the $m \times n$ **zero matrix**. The $m \times n$ zero matrix will be denoted by $\mathbf{0}_{m,n}$ or simply by $\mathbf{0}$ when the dimension of the matrix is clear from the context.

If A, B, and C are three $m \times n$ matrices and $\mathbf{0}$ is the $m \times n$ zero matrix, then

$$A + \mathbf{0} = \mathbf{0} + A = A,$$

$$A + B = B + A,$$

$$(A + B) + C = A + (B + C).$$

The above properties are immediate consequences of the definition of addition of matrices and the corresponding properties of addition of real numbers.

Scalar multiplication

Definition 2.1.8. If A is a $m \times n$ matrix and t is a real number, then tA is the matrix whose entry in the i-th row and j-th column is ta_{ij}, where a_{ij} is the entry of the matrix A in the i-th row and j-th column:

$$t \begin{bmatrix} a_{11} & a_{12} & \cdots & a_{1n} \\ a_{21} & a_{22} & \cdots & a_{2n} \\ \vdots & & \ddots & \vdots \\ a_{m1} & a_{m2} & \cdots & a_{mn} \end{bmatrix} = \begin{bmatrix} ta_{11} & ta_{12} & \cdots & ta_{1n} \\ ta_{21} & ta_{22} & \cdots & ta_{2n} \\ \vdots & & \ddots & \vdots \\ ta_{m1} & ta_{m2} & \cdots & ta_{mn} \end{bmatrix}$$

In other words, to multiply a matrix A by a number t we multiply every entry of A by t. The operation of multiplication of a matrix by a number is referred to as **scalar multiplication**.

Example 2.1.9.

$$t \begin{bmatrix} a_{11} & a_{12} \\ a_{21} & a_{22} \\ a_{31} & a_{32} \end{bmatrix} = \begin{bmatrix} ta_{11} & ta_{12} \\ ta_{21} & ta_{22} \\ ta_{31} & ta_{32} \end{bmatrix}$$

$$5 \begin{bmatrix} 3 & 4 \\ 2 & 5 \end{bmatrix} = \begin{bmatrix} 15 & 20 \\ 10 & 25 \end{bmatrix}$$

> If A and B are $m \times n$ matrices and s and t are real numbers, then
>
> $$(s+t)A = sA + tA,$$
>
> $$(st)A = s(tA),$$
>
> $$t(A+B) = tA + tB.$$

The above properties are immediate consequences of the definitions of scalar multiplication and addition of matrices and the corresponding properties of addition and multiplication of real numbers.

Product of matrices

Products of matrices play a fundamental role in linear algebra. So far we considered the following products:

$$\begin{bmatrix} a_1 & a_2 \end{bmatrix} \begin{bmatrix} b_1 \\ b_2 \end{bmatrix}, \quad \begin{bmatrix} a_1 & a_2 \\ a_3 & a_4 \end{bmatrix} \begin{bmatrix} b_1 \\ b_2 \end{bmatrix}, \quad \begin{bmatrix} a_1 & a_2 \end{bmatrix} \begin{bmatrix} b_1 & b_3 \\ b_2 & b_4 \end{bmatrix}$$

as well as

$$\begin{bmatrix} a_{11} & a_{12} \\ a_{21} & a_{22} \end{bmatrix} \begin{bmatrix} b_{11} & b_{12} \\ b_{21} & b_{22} \end{bmatrix}.$$

The first product can be easily extended to higher dimensions:

$$\begin{bmatrix} a_1 & a_2 & a_3 \end{bmatrix} \begin{bmatrix} b_1 \\ b_2 \\ b_3 \end{bmatrix} = a_1 b_1 + a_2 b_2 + a_3 b_3,$$

$$\begin{bmatrix} a_1 & a_2 & a_3 & a_4 \end{bmatrix} \begin{bmatrix} b_1 \\ b_2 \\ b_3 \\ b_4 \end{bmatrix} = a_1 b_1 + a_2 b_2 + a_3 b_3 + a_4 b_4,$$

$$\begin{bmatrix} a_1 & a_2 & a_3 & a_4 & a_5 \end{bmatrix} \begin{bmatrix} b_1 \\ b_2 \\ b_3 \\ b_4 \\ b_5 \end{bmatrix} = a_1 b_1 + a_2 b_2 + a_3 b_3 + a_4 b_4 + a_5 b_5,$$

and, in general,

$$\begin{bmatrix} a_1 & a_2 & \dots & a_m \end{bmatrix} \begin{bmatrix} b_1 \\ b_2 \\ \vdots \\ b_m \end{bmatrix} = a_1 b_1 + a_2 b_2 + \dots + a_m b_m.$$

Two matrices can be multiplied, if the number of columns of the one on the left is the same as the number of rows of the one on the right. So, if A is a $k \times l$ matrix and B is a $m \times n$ matrix, then the product AB is well-defined if $l = m$.

Definition 2.1.10. The **product** of a $k \times m$ matrix A and an $m \times n$ matrix B is the $k \times n$ matrix AB such that the entry in the i-th row and the j-th column is

$$\begin{bmatrix} a_{i1} & a_{i2} & \cdots & a_{im} \end{bmatrix} \begin{bmatrix} b_{1j} \\ b_{2j} \\ \vdots \\ b_{mj} \end{bmatrix},$$

where $\begin{bmatrix} a_{i1} & a_{i2} & \cdots & a_{im} \end{bmatrix}$ is the i-th row of the matrix A and $\begin{bmatrix} b_{1j} \\ b_{2j} \\ \vdots \\ b_{mj} \end{bmatrix}$ is the j-th column of the matrix B.

For example,

$$\begin{bmatrix} a_1 & a_2 & a_3 & a_4 \\ c_1 & c_2 & c_3 & c_4 \\ f_1 & f_2 & f_3 & f_4 \end{bmatrix} \begin{bmatrix} b_1 & d_1 \\ b_2 & d_2 \\ b_3 & d_3 \\ b_4 & d_4 \end{bmatrix}$$

$$= \begin{bmatrix} \begin{bmatrix} a_1 & a_2 & a_3 & a_4 \end{bmatrix} \begin{bmatrix} b_1 \\ b_2 \\ b_3 \\ b_4 \end{bmatrix} & \begin{bmatrix} a_1 & a_2 & a_3 & a_4 \end{bmatrix} \begin{bmatrix} d_1 \\ d_2 \\ d_3 \\ d_4 \end{bmatrix} \\ \begin{bmatrix} c_1 & c_2 & c_3 & c_4 \end{bmatrix} \begin{bmatrix} b_1 \\ b_2 \\ b_3 \\ b_4 \end{bmatrix} & \begin{bmatrix} c_1 & c_2 & c_3 & c_4 \end{bmatrix} \begin{bmatrix} d_1 \\ d_2 \\ d_3 \\ d_4 \end{bmatrix} \\ \begin{bmatrix} f_1 & f_2 & f_3 & f_4 \end{bmatrix} \begin{bmatrix} b_1 \\ b_2 \\ b_3 \\ b_4 \end{bmatrix} & \begin{bmatrix} f_1 & f_2 & f_3 & f_4 \end{bmatrix} \begin{bmatrix} d_1 \\ d_2 \\ d_3 \\ d_4 \end{bmatrix} \end{bmatrix}$$

$$= \begin{bmatrix} a_1 b_1 + a_2 b_2 + a_3 b_3 + a_4 b_4 & a_1 d_1 + a_2 d_2 + a_3 d_3 + a_4 d_4 \\ c_1 b_1 + c_2 b_2 + c_3 b_3 + c_4 b_4 & c_1 d_1 + c_2 d_2 + c_3 d_3 + c_4 d_4 \\ f_1 b_1 + f_2 b_2 + f_3 b_3 + f_4 b_4 & f_1 d_1 + f_2 d_2 + f_3 d_3 + f_4 d_4 \end{bmatrix}$$

Here are some concrete examples of products of matrices.

Example 2.1.11.

$$\begin{bmatrix} 1 & 3 & 0 & 2 \\ 2 & 0 & 1 & 0 \end{bmatrix} \begin{bmatrix} 4 & 1 & 1 \\ 0 & 2 & 1 \\ 1 & 1 & 6 \\ 0 & 1 & 1 \end{bmatrix} = \begin{bmatrix} 4 & 9 & 6 \\ 9 & 3 & 8 \end{bmatrix}$$

$$\begin{bmatrix} 1 & 2 & 0 \\ 0 & 3 & 1 \\ 1 & -1 & 1 \end{bmatrix} \begin{bmatrix} 2 & 2 & 3 & 6 \\ 3 & 1 & 1 & 6 \\ 2 & 0 & 1 & 0 \end{bmatrix} = \begin{bmatrix} 8 & 4 & 5 & 18 \\ 11 & 3 & 4 & 18 \\ 1 & 1 & 3 & 0 \end{bmatrix}$$

$$\begin{bmatrix} 1 & 3 \\ 2 & 0 \\ 3 & 4 \end{bmatrix} \begin{bmatrix} 2 \\ -1 \end{bmatrix} = \begin{bmatrix} -1 \\ 4 \\ 2 \end{bmatrix}$$

$$\begin{bmatrix} 1 \\ 3 \end{bmatrix} \begin{bmatrix} 2 & 1 & -5 & 4 \end{bmatrix} = \begin{bmatrix} 2 & 1 & -5 & 4 \\ 6 & 3 & -15 & 12 \end{bmatrix}$$

In Theorem 1.1.10 we show that the product of 2×2 matrices is associative, that is, $A(BC) = (AB)C$ for any 2×2 matrices A, B, and C. This property is not limited to 2×2 matrices.

Theorem 2.1.12. *If A is an $m \times n$ matrix, B is an $n \times p$ matrix, and C is a $p \times q$ matrix, then we have*
$$A(BC) = (AB)C.$$

Proof of a particular case. The identity can be obtained by direct calculations. We illustrate the idea of the proof by considering the case when $m = 1$, $n = 3$, $p = 2$, and $q = 1$.

Let

$$A = \begin{bmatrix} a_1 & a_2 & a_3 \end{bmatrix}, \quad B = \begin{bmatrix} b_{11} & b_{12} \\ b_{21} & b_{22} \\ b_{31} & b_{32} \end{bmatrix}, \quad \text{and} \quad C = \begin{bmatrix} c_1 \\ c_2 \end{bmatrix}.$$

Then

$$A(BC) = \begin{bmatrix} a_1 & a_2 & a_3 \end{bmatrix} \left(\begin{bmatrix} b_{11} & b_{12} \\ b_{21} & b_{22} \\ b_{31} & b_{32} \end{bmatrix} \begin{bmatrix} c_1 \\ c_2 \end{bmatrix} \right)$$

$$= \begin{bmatrix} a_1 & a_2 & a_3 \end{bmatrix} \begin{bmatrix} b_{11}c_1 + b_{12}c_2 \\ b_{21}c_1 + b_{22}c_2 \\ b_{31}c_1 + b_{32}c_2 \end{bmatrix}$$

$$= a_1 b_{11} c_1 + a_1 b_{12} c_2 + a_2 b_{21} c_1 + a_2 b_{22} c_2 + a_3 b_{31} c_1 + a_3 b_{32} c_2$$

and

$$(AB)C = \left(\begin{bmatrix} a_1 & a_2 & a_3 \end{bmatrix} \begin{bmatrix} b_{11} & b_{12} \\ b_{21} & b_{22} \\ b_{31} & b_{32} \end{bmatrix} \right) \begin{bmatrix} c_1 \\ c_2 \end{bmatrix}$$

$$= \begin{bmatrix} a_1 b_{11} + a_2 b_{21} + a_3 b_{31} & a_1 b_{12} + a_2 b_{22} + a_3 b_{32} \end{bmatrix} \begin{bmatrix} c_1 \\ c_2 \end{bmatrix}$$

$$= a_1 b_{11} c_1 + a_2 b_{21} c_1 + a_3 b_{31} c_1 + a_1 b_{12} c_2 + a_2 b_{22} c_2 + a_3 b_{32} c_2$$

$$= a_1 b_{11} c_1 + a_1 b_{12} c_2 + a_2 b_{21} c_1 + a_2 b_{22} c_2 + a_3 b_{31} c_1 + a_3 b_{32} c_2.$$

$\square$

Definition 2.1.13. The $n \times n$ matrix with 1's on the main diagonal and 0's everywhere else is called a **unit matrix** or an **identity matrix** and denoted by I_n:

$$I_n = \begin{bmatrix} 1 & 0 & \cdots & 0 \\ 0 & 1 & \cdots & 0 \\ \vdots & & \ddots & \vdots \\ 0 & 0 & \cdots & 1 \end{bmatrix}$$

I_n is called an identity matrix because when we multiply a matrix A by the identity matrix of the appropriate size, then the result is the original matrix A.

Theorem 2.1.14. *Let A be an m × n matrix. Then*

$$I_m A = A \quad and \quad A I_n = A.$$

Proof of a particular case. We verify the result for the matrix

$$A = \begin{bmatrix} a_{11} & a_{12} & a_{13} \\ a_{21} & a_{22} & a_{23} \end{bmatrix}.$$

Indeed, we have

$$I_2 A = \begin{bmatrix} 1 & 0 \\ 0 & 1 \end{bmatrix} \begin{bmatrix} a_{11} & a_{12} & a_{13} \\ a_{21} & a_{22} & a_{23} \end{bmatrix}$$

$$= \begin{bmatrix} \begin{bmatrix} 1 & 0 \end{bmatrix} \begin{bmatrix} a_{11} \\ a_{21} \end{bmatrix} & \begin{bmatrix} 1 & 0 \end{bmatrix} \begin{bmatrix} a_{12} \\ a_{22} \end{bmatrix} & \begin{bmatrix} 1 & 0 \end{bmatrix} \begin{bmatrix} a_{13} \\ a_{23} \end{bmatrix} \\ \begin{bmatrix} 0 & 1 \end{bmatrix} \begin{bmatrix} a_{11} \\ a_{21} \end{bmatrix} & \begin{bmatrix} 0 & 1 \end{bmatrix} \begin{bmatrix} a_{12} \\ a_{22} \end{bmatrix} & \begin{bmatrix} 0 & 1 \end{bmatrix} \begin{bmatrix} a_{13} \\ a_{23} \end{bmatrix} \end{bmatrix}$$

$$= \begin{bmatrix} a_{11} & a_{12} & a_{13} \\ a_{21} & a_{22} & a_{23} \end{bmatrix}$$

and

$$AI_3 = \begin{bmatrix} a_{11} & a_{12} & a_{13} \\ a_{21} & a_{22} & a_{23} \end{bmatrix} \begin{bmatrix} 1 & 0 & 0 \\ 0 & 1 & 0 \\ 0 & 0 & 1 \end{bmatrix}$$

$$= \begin{bmatrix} \begin{bmatrix} a_{11} & a_{12} & a_{13} \end{bmatrix} \begin{bmatrix} 1 \\ 0 \\ 0 \end{bmatrix} & \begin{bmatrix} a_{11} & a_{12} & a_{13} \end{bmatrix} \begin{bmatrix} 0 \\ 1 \\ 0 \end{bmatrix} & \begin{bmatrix} a_{11} & a_{12} & a_{13} \end{bmatrix} \begin{bmatrix} 0 \\ 0 \\ 1 \end{bmatrix} \\ \begin{bmatrix} a_{21} & a_{22} & a_{23} \end{bmatrix} \begin{bmatrix} 1 \\ 0 \\ 0 \end{bmatrix} & \begin{bmatrix} a_{21} & a_{22} & a_{23} \end{bmatrix} \begin{bmatrix} 0 \\ 1 \\ 0 \end{bmatrix} & \begin{bmatrix} a_{21} & a_{22} & a_{23} \end{bmatrix} \begin{bmatrix} 0 \\ 0 \\ 1 \end{bmatrix} \end{bmatrix}$$

$$= \begin{bmatrix} a_{11} & a_{12} & a_{13} \\ a_{21} & a_{22} & a_{23} \end{bmatrix}.$$

□

Theorem 2.1.15. *If A is an $m \times n$ matrix and B and C are $n \times p$ matrices, then*

$$A(B + C) = AB + AC.$$

Proof of a particular case. We illustrate the method of the proof by considering the case when $m = 1$, $n = 3$, and $p = 1$.

$$\begin{bmatrix} a_1 & a_2 & a_3 \end{bmatrix} \left(\begin{bmatrix} b_1 \\ b_2 \\ b_3 \end{bmatrix} + \begin{bmatrix} c_1 \\ c_2 \\ c_3 \end{bmatrix} \right) = \begin{bmatrix} a_1 & a_2 & a_3 \end{bmatrix} \begin{bmatrix} b_1 + c_1 \\ b_2 + b_2 \\ b_3 + c_3 \end{bmatrix}$$

$$= \begin{bmatrix} a_1 & a_2 & a_3 \end{bmatrix} \begin{bmatrix} b_1 \\ b_2 \\ b_3 \end{bmatrix} + \begin{bmatrix} a_1 & a_2 & a_3 \end{bmatrix} \begin{bmatrix} c_1 \\ c_2 \\ c_3 \end{bmatrix}.$$

□

Theorem 2.1.16. *If A and B are $m \times n$ matrices and C is an $n \times p$ matrix, then*

$$(A + B)C = AB + BC.$$

Proof of a particular case. To illustrate the method of the proof we consider the case when $m = 1$, $n = 3$, and $p = 1$.

$$\left(\begin{bmatrix} a_1 & a_2 & a_3 \end{bmatrix} + \begin{bmatrix} b_1 & b_2 & b_3 \end{bmatrix} \right) \begin{bmatrix} c_1 \\ c_2 \\ c_3 \end{bmatrix} = \begin{bmatrix} a_1 + b_1 & a_2 + b_2 & a_3 + b_3 \end{bmatrix} \begin{bmatrix} c_1 \\ c_2 \\ c_3 \end{bmatrix}$$

$$= \begin{bmatrix} a_1 & a_2 & a_3 \end{bmatrix} \begin{bmatrix} c_1 \\ c_2 \\ c_3 \end{bmatrix} + \begin{bmatrix} b_1 & b_2 & b_3 \end{bmatrix} \begin{bmatrix} c_1 \\ c_2 \\ c_3 \end{bmatrix}.$$

$\square$

Theorem 2.1.17. *If A is an $m \times n$ matrix, B is an $n \times p$ matrix, and t is a real number, then*

$$t\,AB = (t\,A)B = A(t\,B).$$

Proof of a particular case. We illustrate the method of the proof by considering the case when $m = 1$, $n = 3$ and $p = 1$.

$$t\left(\begin{bmatrix} a_1 & a_2 & a_3 \end{bmatrix} \begin{bmatrix} b_1 \\ b_2 \\ b_3 \end{bmatrix} \right) = \begin{bmatrix} ta_1 & ta_2 & ta_3 \end{bmatrix} \begin{bmatrix} b_1 \\ b_2 \\ b_3 \end{bmatrix}$$

$$= \begin{bmatrix} a_1 & a_2 & a_3 \end{bmatrix} \begin{bmatrix} tb_1 \\ tb_2 \\ tb_3 \end{bmatrix}$$

$$= ta_1 b_1 + ta_2 b_2 + ta_3 b_3.$$

$\square$

We close this section with a simple theorem that is often quite useful in arguments when it is necessary to show that two matrices are equal.

Theorem 2.1.18. *Let A and B be $m \times n$ matrices. If*

$$A\mathbf{x} = B\mathbf{x}$$

for every vector $\mathbf{x}$ in $\mathbb{R}^n$, then $A = B$.

Proof of a particular case. We verify the result when A and B are 2×3 matrices.
   If the equality

$$\begin{bmatrix} a_{11} & a_{12} & a_{13} \\ a_{21} & a_{22} & a_{23} \end{bmatrix} \begin{bmatrix} x_1 \\ x_2 \\ x_3 \end{bmatrix} = \begin{bmatrix} b_{11} & b_{12} & b_{13} \\ b_{21} & b_{22} & b_{23} \end{bmatrix} \begin{bmatrix} x_1 \\ x_2 \\ x_3 \end{bmatrix}$$

holds for arbitrary vector $\mathbf{x} = \begin{bmatrix} x_1 \\ x_2 \\ x_3 \end{bmatrix}$ in $\mathbb{R}^3$, then it must hold for the vectors $\begin{bmatrix} 1 \\ 0 \\ 0 \end{bmatrix}$, $\begin{bmatrix} 0 \\ 1 \\ 0 \end{bmatrix}$,

and $\begin{bmatrix} 0 \\ 0 \\ 1 \end{bmatrix}$. Consequently, we have

$$\begin{bmatrix} a_{11} & a_{12} & a_{13} \\ a_{21} & a_{22} & a_{23} \end{bmatrix} \begin{bmatrix} 1 \\ 0 \\ 0 \end{bmatrix} = \begin{bmatrix} b_{11} & b_{12} & b_{13} \\ b_{21} & b_{22} & b_{23} \end{bmatrix} \begin{bmatrix} 1 \\ 0 \\ 0 \end{bmatrix},$$

$$\begin{bmatrix} a_{11} & a_{12} & a_{13} \\ a_{21} & a_{22} & a_{23} \end{bmatrix} \begin{bmatrix} 0 \\ 1 \\ 0 \end{bmatrix} = \begin{bmatrix} b_{11} & b_{12} & b_{13} \\ b_{21} & b_{22} & b_{23} \end{bmatrix} \begin{bmatrix} 0 \\ 1 \\ 0 \end{bmatrix},$$

and

$$\begin{bmatrix} a_{11} & a_{12} & a_{13} \\ a_{21} & a_{22} & a_{23} \end{bmatrix} \begin{bmatrix} 0 \\ 0 \\ 1 \end{bmatrix} = \begin{bmatrix} b_{11} & b_{12} & b_{13} \\ b_{21} & b_{22} & b_{23} \end{bmatrix} \begin{bmatrix} 0 \\ 0 \\ 1 \end{bmatrix}.$$

The above equalities can be written together as

$$\begin{bmatrix} a_{11} & a_{12} & a_{13} \\ a_{21} & a_{22} & a_{23} \end{bmatrix} \begin{bmatrix} 1 & 0 & 0 \\ 0 & 1 & 0 \\ 0 & 0 & 1 \end{bmatrix} = \begin{bmatrix} b_{11} & b_{12} & b_{13} \\ b_{21} & b_{22} & b_{23} \end{bmatrix} \begin{bmatrix} 1 & 0 & 0 \\ 0 & 1 & 0 \\ 0 & 0 & 1 \end{bmatrix}.$$

Now we have

$$\begin{bmatrix} a_{11} & a_{12} & a_{13} \\ a_{21} & a_{22} & a_{23} \end{bmatrix}$$

$$= \begin{bmatrix} a_{11} & a_{12} & a_{13} \\ a_{21} & a_{22} & a_{23} \end{bmatrix} \begin{bmatrix} 1 & 0 & 0 \\ 0 & 1 & 0 \\ 0 & 0 & 1 \end{bmatrix} = \begin{bmatrix} b_{11} & b_{12} & b_{13} \\ b_{21} & b_{22} & b_{23} \end{bmatrix} \begin{bmatrix} 1 & 0 & 0 \\ 0 & 1 & 0 \\ 0 & 0 & 1 \end{bmatrix}.$$

$$= \begin{bmatrix} b_{11} & b_{12} & b_{13} \\ b_{21} & b_{22} & b_{23} \end{bmatrix}.$$

But this means that $A = B$. $\qquad\qquad\qquad\qquad\qquad\qquad\qquad\qquad\qquad\square$

Transpose of a matrix

Definition 2.1.19. The *transpose* of a matrix

$$A = \begin{bmatrix} a_{11} & a_{12} & \cdots & a_{1n} \\ a_{21} & a_{22} & \cdots & a_{2n} \\ \vdots & \vdots & \ddots & \vdots \\ a_{m1} & a_{m2} & \cdots & a_{mn} \end{bmatrix}$$

is the matrix denoted by A^T whose rows are the columns of A in the same order, that is

$$A^T = \begin{bmatrix} a_{11} & a_{21} & \cdots & a_{m1} \\ a_{12} & a_{22} & \cdots & a_{m2} \\ \vdots & \vdots & \ddots & \vdots \\ a_{1n} & a_{2n} & \cdots & a_{mn} \end{bmatrix}.$$

The first row of A^T is the same as the first column of A, the second row of A^T is the same as the second column of A, and so on. Note that the columns of A^T are the same as the rows of A. If A is an $m \times n$ matrix, then A^T is an $n \times m$ matrix.

Example 2.1.20. Here are some examples of transposes of matrices of different sizes:

$$\begin{bmatrix} 1 & 3 \\ 2 & 0 \\ 3 & 4 \end{bmatrix}^T = \begin{bmatrix} 1 & 2 & 3 \\ 3 & 0 & 4 \end{bmatrix}$$

$$\begin{bmatrix} 1 & -1 & 0 \\ 2 & 3 & 5 \\ -3 & 4 & -7 \end{bmatrix}^T = \begin{bmatrix} 1 & 2 & -3 \\ -1 & 3 & 4 \\ 0 & 5 & -7 \end{bmatrix}$$

$$\begin{bmatrix} 1 \\ 2 \\ 3 \\ 4 \end{bmatrix}^T = \begin{bmatrix} 1 & 2 & 3 & 4 \end{bmatrix}$$

Note that the products AA^T and $A^T A$ are always defined. If A is an $m \times n$ matrix, then AA^T is a square $m \times m$ matrix and $A^T A$ is a square $n \times n$ matrix.

Example 2.1.21. We have

$$\begin{bmatrix} 1 & 3 \\ 2 & 0 \\ 3 & 4 \end{bmatrix} \begin{bmatrix} 1 & 3 \\ 2 & 0 \\ 3 & 4 \end{bmatrix}^T = \begin{bmatrix} 1 & 3 \\ 2 & 0 \\ 3 & 4 \end{bmatrix} \begin{bmatrix} 1 & 2 & 3 \\ 3 & 0 & 4 \end{bmatrix} = \begin{bmatrix} 10 & 2 & 15 \\ 2 & 4 & 6 \\ 15 & 6 & 25 \end{bmatrix}$$

and

$$\begin{bmatrix} 1 & 3 \\ 2 & 0 \\ 3 & 4 \end{bmatrix}^T \begin{bmatrix} 1 & 3 \\ 2 & 0 \\ 3 & 4 \end{bmatrix} = \begin{bmatrix} 1 & 2 & 3 \\ 3 & 0 & 4 \end{bmatrix} \begin{bmatrix} 1 & 3 \\ 2 & 0 \\ 3 & 4 \end{bmatrix} = \begin{bmatrix} 14 & 15 \\ 15 & 25 \end{bmatrix}$$

Example 2.1.22. We have

$$\begin{bmatrix} 3 \\ 1 \\ 5 \\ -1 \end{bmatrix} \begin{bmatrix} 3 \\ 1 \\ 5 \\ -1 \end{bmatrix}^T = \begin{bmatrix} 3 \\ 1 \\ 5 \\ -1 \end{bmatrix} \begin{bmatrix} 3 & 1 & 5 & -1 \end{bmatrix} = \begin{bmatrix} 9 & 3 & 15 & -3 \\ 3 & 1 & 5 & -1 \\ 15 & 5 & 25 & -5 \\ -3 & -1 & -5 & 1 \end{bmatrix}$$

and

$$\begin{bmatrix} 3 \\ 1 \\ 5 \\ -1 \end{bmatrix}^T \begin{bmatrix} 3 \\ 1 \\ 5 \\ -1 \end{bmatrix} = \begin{bmatrix} 3 & 1 & 5 & -1 \end{bmatrix} \begin{bmatrix} 3 \\ 1 \\ 5 \\ -1 \end{bmatrix} = 3 \cdot 3 + 1 \cdot 1 + 5 \cdot 5 + (-1) \cdot (-1) = 36$$

For the next result we leave the student to verify some particular cases by direct calculations in exercises.

Theorem 2.1.23. *If A is a k × m matrix and B is an m × n matrix, then*

$$(AB)^T = B^T A^T.$$

Note that the order of matrices in the above equality changes. Since B^T is an $n \times m$ matrix and A^T is an $m \times k$, the product $B^T A^T$ makes sense. For example, we have

$$\left(\begin{bmatrix} a_{11} & a_{12} \\ a_{21} & a_{22} \end{bmatrix} \begin{bmatrix} b_{11} & b_{12} \\ b_{21} & b_{22} \end{bmatrix} \right)^T = \begin{bmatrix} b_{11} & b_{12} \\ b_{21} & b_{22} \end{bmatrix}^T \begin{bmatrix} a_{11} & a_{12} \\ a_{21} & a_{22} \end{bmatrix}^T$$

and

$$\left(\begin{bmatrix} a_{11} & a_{12} & a_{13} \\ a_{21} & a_{22} & a_{23} \end{bmatrix} \begin{bmatrix} b_{11} & b_{12} \\ b_{21} & b_{22} \\ b_{31} & b_{32} \end{bmatrix} \right)^T = \begin{bmatrix} b_{11} & b_{12} \\ b_{21} & b_{22} \\ b_{31} & b_{32} \end{bmatrix}^T \begin{bmatrix} a_{11} & a_{12} & a_{13} \\ a_{21} & a_{22} & a_{23} \end{bmatrix}^T.$$

These properties can be verified by direct calculations using the obvious equalities

$$\begin{bmatrix} a_1 & a_2 \end{bmatrix} \begin{bmatrix} b_1 \\ b_2 \end{bmatrix} = \begin{bmatrix} b_1 & b_2 \end{bmatrix} \begin{bmatrix} a_1 \\ a_2 \end{bmatrix}$$

and

$$\begin{bmatrix} a_1 & a_2 & a_3 \end{bmatrix} \begin{bmatrix} b_1 \\ b_2 \\ b_3 \end{bmatrix} = \begin{bmatrix} b_1 & b_2 & b_3 \end{bmatrix} \begin{bmatrix} a_1 \\ a_2 \\ a_3 \end{bmatrix}.$$

Definition 2.1.24. A square matrix A is called **symmetric** if $A^T = A$.

Example 2.1.25. Here are some examples of symmetric matrices:

$$\begin{bmatrix} 1 & -2 \\ -2 & 1 \end{bmatrix}, \quad \begin{bmatrix} 1 & 3 & 0 \\ 3 & -4 & 2 \\ 0 & 2 & -1 \end{bmatrix}, \quad \begin{bmatrix} 1 & 2 & 3 & 4 \\ 2 & 0 & -1 & 6 \\ 3 & -1 & -5 & 0 \\ 4 & 6 & 0 & 7 \end{bmatrix}.$$

Note that both matrices AA^T and $A^T A$ found in Example 2.1.21 are symmetric. It is not difficult to show that for an arbitrary matrix A the matrices AA^T and $A^T A$ are symmetric.

2.1.1   Exercises

Perform the indicated operations on matrices.

1. $\begin{bmatrix} 1 & 2 & -3 & 5 \end{bmatrix} + \begin{bmatrix} 9 & -2 & 4 & 8 \end{bmatrix}$

2. $\begin{bmatrix} 2 \\ 0 \\ -1 \\ 3 \end{bmatrix} + \begin{bmatrix} 9 \\ 1 \\ 1 \\ -3 \end{bmatrix}$

3. $\begin{bmatrix} 1 & 2 & 3 \\ 1 & 0 & -1 \end{bmatrix} + \begin{bmatrix} 7 & 1 & -4 \\ 11 & 2 & -3 \end{bmatrix}$

4. $\begin{bmatrix} 1 & 1 & 2 \\ 0 & 1 & -1 \\ 3 & -2 & 0 \end{bmatrix} + \begin{bmatrix} 3 & 1 & 4 \\ -2 & 5 & -1 \\ -3 & 1 & 2 \end{bmatrix}$

5. $5 \begin{bmatrix} 3 & 2 & 1 \\ 1 & 4 & -1 \end{bmatrix}$

6. $3 \begin{bmatrix} 1 & 2 \\ 5 & -2 \\ 1 & 0 \\ 2 & 7 \end{bmatrix}$

7. $2 \begin{bmatrix} 5 & 1 \\ 1 & 7 \\ 2 & 3 \end{bmatrix} + 3 \begin{bmatrix} 4 & -1 \\ 2 & 8 \\ 0 & -5 \end{bmatrix}$

8. $3 \begin{bmatrix} 2 & 2 \\ 1 & 5 \\ 3 & 0 \end{bmatrix} + 8 \begin{bmatrix} 0 & 1 \\ 1 & -1 \\ -3 & 1 \end{bmatrix} + 2 \begin{bmatrix} 2 & 1 \\ -1 & 0 \\ 1 & -1 \end{bmatrix}$

9. $\begin{bmatrix} 1 \\ 4 \end{bmatrix} \begin{bmatrix} 2 & 5 \end{bmatrix}$

10. $\begin{bmatrix} 5 \\ 2 \end{bmatrix} \begin{bmatrix} 1 & 7 \end{bmatrix}$

11. $\begin{bmatrix} 2 & -1 & 1 & 4 \end{bmatrix} \begin{bmatrix} 1 \\ 3 \\ 2 \\ 5 \end{bmatrix}$

12. $\begin{bmatrix} -1 & 4 & 2 & 4 & 3 \end{bmatrix} \begin{bmatrix} 3 \\ 2 \\ 6 \\ -3 \\ 4 \end{bmatrix}$

13. $\begin{bmatrix} 4 \\ 2 \\ 1 \\ 5 \end{bmatrix} \begin{bmatrix} 2 & 7 & 1 \end{bmatrix}$

14. $\begin{bmatrix} 7 \\ 2 \\ 1 \end{bmatrix} \begin{bmatrix} 2 & 1 & 1 & 2 \end{bmatrix}$

15. $\begin{bmatrix} 1 & 1 & 0 & 1 \\ 1 & 0 & 0 & 1 \end{bmatrix} \begin{bmatrix} 1 & 1 & 1 \\ 1 & 0 & 0 \\ 0 & 1 & 0 \\ 0 & 0 & 1 \end{bmatrix}$

16. $\begin{bmatrix} 1 & -1 & 0 \\ 1 & 2 & 1 \\ 1 & 1 & 1 \\ 1 & 0 & 1 \end{bmatrix} \begin{bmatrix} 1 & 1 \\ 1 & 3 \\ 0 & 1 \end{bmatrix}$

17. $\begin{bmatrix} 1 & 2 & 5 \\ 3 & 2 & 1 \end{bmatrix} \begin{bmatrix} 1 & 1 \\ 0 & 3 \\ 1 & 0 \end{bmatrix} \begin{bmatrix} 2 & 0 \\ 1 & 1 \end{bmatrix}$

18. $\begin{bmatrix} 2 & 0 \\ 1 & 1 \end{bmatrix} \begin{bmatrix} 1 & 2 & 5 \\ 3 & 2 & 1 \end{bmatrix} \begin{bmatrix} 1 & 1 \\ 0 & 3 \\ 1 & 0 \end{bmatrix}$

19. $\begin{bmatrix} 5 \\ 1 \\ 2 \end{bmatrix} \begin{bmatrix} 4 & -1 \end{bmatrix} \begin{bmatrix} 1 & 1 & 0 & 0 \\ 1 & 0 & 1 & 1 \end{bmatrix}$

20. $\begin{bmatrix} 1 \\ 1 \end{bmatrix} \begin{bmatrix} x & y & z \end{bmatrix} \begin{bmatrix} 1 \\ 3 \\ 2 \end{bmatrix}$

Explain why the following products are not defined.

21. $\begin{bmatrix} 1 & 1 & -1 & 1 \\ 0 & 1 & 1 & 1 \end{bmatrix} \begin{bmatrix} 1 & 1 & 1 \\ 1 & 0 & 1 \\ 1 & 1 & 0 \end{bmatrix}$

22. $\begin{bmatrix} 1 & -1 \\ 1 & 1 \\ 1 & 3 \\ 1 & 2 \end{bmatrix} \begin{bmatrix} 1 & 1 \\ 1 & 3 \\ 0 & 1 \end{bmatrix}$

23. Show, without using Theorem 2.1.18, that if

$$\begin{bmatrix} a_{11} & a_{12} \\ a_{21} & a_{22} \end{bmatrix} \begin{bmatrix} x_1 \\ x_2 \end{bmatrix} = \begin{bmatrix} b_{11} & b_{12} \\ b_{21} & b_{22} \end{bmatrix} \begin{bmatrix} x_1 \\ x_2 \end{bmatrix} \quad \text{for every} \quad \begin{bmatrix} x_1 \\ x_2 \end{bmatrix},$$

then $\begin{bmatrix} a_{11} & a_{12} \\ a_{21} & a_{22} \end{bmatrix} = \begin{bmatrix} b_{11} & b_{12} \\ b_{21} & b_{22} \end{bmatrix}.$

24. Show, without using Theorem 2.1.18, that if

$$\begin{bmatrix} a_{11} & a_{12} & a_{13} \\ a_{21} & a_{22} & a_{23} \\ a_{31} & a_{32} & a_{33} \end{bmatrix} \begin{bmatrix} x_1 \\ x_2 \\ x_3 \end{bmatrix} = \begin{bmatrix} b_{11} & b_{12} & b_{13} \\ b_{21} & b_{22} & b_{23} \\ b_{31} & b_{32} & b_{33} \end{bmatrix} \begin{bmatrix} x_1 \\ x_2 \\ x_3 \end{bmatrix} \quad \text{for every} \quad \begin{bmatrix} x_1 \\ x_2 \\ x_3 \end{bmatrix},$$

then $\begin{bmatrix} a_{11} & a_{12} & a_{13} \\ a_{21} & a_{22} & a_{23} \\ a_{31} & a_{32} & a_{33} \end{bmatrix} = \begin{bmatrix} b_{11} & b_{12} & b_{13} \\ b_{21} & b_{22} & b_{23} \\ b_{31} & b_{32} & b_{33} \end{bmatrix}.$

25. Show, without using Theorem 2.1.18, that if

$$\begin{bmatrix} a_{11} & a_{12} & a_{13} & a_{14} \end{bmatrix} \begin{bmatrix} x_1 \\ x_2 \\ x_3 \\ x_4 \end{bmatrix} = \begin{bmatrix} b_{11} & b_{12} & b_{13} & b_{14} \end{bmatrix} \begin{bmatrix} x_1 \\ x_2 \\ x_3 \\ x_4 \end{bmatrix} \quad \text{for every} \quad \begin{bmatrix} x_1 \\ x_2 \\ x_3 \\ x_4 \end{bmatrix},$$

then $\begin{bmatrix} a_{11} & a_{12} & a_{13} & a_{14} \end{bmatrix} = \begin{bmatrix} b_{11} & b_{12} & b_{13} & b_{14} \end{bmatrix}$.

26. Show, without using Theorem 2.1.18, that if

$$\begin{bmatrix} a_{11} & a_{12} & a_{13} & a_{14} \\ a_{21} & a_{22} & a_{23} & a_{24} \end{bmatrix} \begin{bmatrix} x_1 \\ x_2 \\ x_3 \\ x_4 \end{bmatrix} = \begin{bmatrix} b_{11} & b_{12} & b_{13} & b_{14} \\ b_{21} & b_{22} & b_{23} & b_{24} \end{bmatrix} \begin{bmatrix} x_1 \\ x_2 \\ x_3 \\ x_4 \end{bmatrix} \quad \text{for every} \quad \begin{bmatrix} x_1 \\ x_2 \\ x_3 \\ x_4 \end{bmatrix},$$

then $\begin{bmatrix} a_{11} & a_{12} & a_{13} & a_{14} \\ a_{21} & a_{22} & a_{23} & a_{24} \end{bmatrix} = \begin{bmatrix} b_{11} & b_{12} & b_{13} & b_{14} \\ b_{21} & b_{22} & b_{23} & b_{24} \end{bmatrix}$.

Find A^T for the given matrix A.

27. $A = \begin{bmatrix} 1 & 2 & -3 \end{bmatrix}$

29. $A = \begin{bmatrix} 1 & 2 & 5 \\ 3 & 4 & 2 \end{bmatrix}$

28. $A = \begin{bmatrix} 1 & 1 & 1 \\ 1 & 2 & 3 \\ 4 & 5 & 6 \\ 7 & 8 & 9 \end{bmatrix}$

30. $A = \begin{bmatrix} 1 & 1 & 2 & 1 \\ 0 & 1 & -1 & 9 \\ 3 & -2 & 0 & 3 \end{bmatrix}$

31. Show that

$$\left(\begin{bmatrix} a_1 & a_2 \\ a_3 & a_4 \end{bmatrix} \begin{bmatrix} b_1 & b_2 \\ b_3 & b_4 \end{bmatrix} \right)^T = \begin{bmatrix} b_1 & b_2 \\ b_3 & b_4 \end{bmatrix}^T \begin{bmatrix} a_1 & a_2 \\ a_3 & a_4 \end{bmatrix}^T.$$

32. Show that

$$\left(\begin{bmatrix} a_{11} & a_{12} & a_{13} \\ a_{21} & a_{22} & a_{23} \end{bmatrix} \begin{bmatrix} b_{11} & b_{12} \\ b_{21} & b_{22} \\ b_{31} & b_{32} \end{bmatrix} \right)^T = \begin{bmatrix} b_{11} & b_{12} \\ b_{21} & b_{22} \\ b_{31} & b_{32} \end{bmatrix}^T \begin{bmatrix} a_{11} & a_{12} & a_{13} \\ a_{21} & a_{22} & a_{23} \end{bmatrix}^T.$$

33. Let A be a 2×2 matrix. Show that $\det A = \det A^T$.

34. Show that $\begin{bmatrix} a_1 & a_2 \end{bmatrix} \left(\begin{bmatrix} b_{11} & b_{12} & b_{13} \\ b_{21} & b_{22} & b_{23} \end{bmatrix} \begin{bmatrix} c_1 \\ c_2 \\ c_3 \end{bmatrix} \right) = \left(\begin{bmatrix} a_1 & a_2 \end{bmatrix} \begin{bmatrix} b_{11} & b_{12} & b_{13} \\ b_{21} & b_{22} & b_{23} \end{bmatrix} \right) \begin{bmatrix} c_1 \\ c_2 \\ c_3 \end{bmatrix}$.

35. Show that for an arbitrary matrix A the matrix AA^T is symmetric.

2.2 Gaussian elimination

In Chapter 1 we solved a system of linear equations by writing it in the matrix form and then solving it by inverting the matrix. In this chapter we present a different approach. First we note that solving the system

$$\begin{cases} a_1 x + b_1 y + c_1 z = d_1 \\ a_2 x + b_2 y + c_2 z = d_2 \\ a_3 x + b_3 y + c_3 z = d_3 \end{cases}$$

is more difficult and time consuming than solving the system that has the form

$$\begin{cases} a_1 x + b_1 y + c_1 z = d_1 \\ \qquad\quad b_2 y + c_2 z = d_2 \\ \qquad\qquad\quad c_3 z = d_3 \end{cases} \tag{2.1}$$

Then we observe that we will not affect the solution of the system of linear equations if we multiply one of the equations by a number different from 0 or multiply one of the equations by a number and then add the result to another equation. Moreover, if necessary, we can always change the order of equations in the system. It turns out that by manipulating the system as described above we can eventually change it to the form (2.1) or a similar form that makes solving the system very easy. This process is referred to as *Gaussian elimination*. In this chapter we discuss the process of Gaussian elimination in detail and examine different possible outcomes of the process.

Elementary operations

The operations used in the Gaussian elimination process are called elementary operations.

Definition 2.2.1. By *elementary operations* on a system of linear equations we mean the following three operations:

- **Interchange two equations.**

- **Multiply an equation by a nonzero constant.**

- **Multiply an equation by a constant and then add the result to another equation.**

Now we give a number of examples in order to illustrate and clarify the meaning of these operations. In these examples we give a system of linear equations, then we describe the elementary operation that will be applied to the system, and then show the resulting system.

Example 2.2.2.

$$\begin{cases} 2x + 3y - 2z = 1 \\ x + 2y + z = 3 \\ 3x + 4y - 5z = -1 \end{cases}$$

interchange equation 1 and equation 3

$$\begin{cases} 3x + 4y - 5z = -1 \\ x + 2y + z = 3 \\ 2x + 3y - 2z = 1 \end{cases}$$

Example 2.2.3.

$$\begin{cases} 2x + 3y - 2z = 1 \\ x + 2y + z = 3 \\ 3x + 4y - 5z = -1 \end{cases}$$

multiply equation 2 by 7

$$\begin{cases} 2x + 3y - 2z = 1 \\ 7x + 14y + 7z = 21 \\ 3x + 4y - 5z = -1 \end{cases}$$

Example 2.2.4.

$$\begin{cases} 2x + 3y - 2z = 1 \\ x + 2y + z = 3 \\ 3x + 4y - 5z = -1 \end{cases}$$

multiply equation 1 by 4 and then add to equation 3

$$\begin{cases} 2x + 3y - 2z = 1 \\ x + 2y + z = 3 \\ 11x + 16y - 13z = 3 \end{cases}$$

In the next example we use two elementary operations to modify a system. It is usually necessary to use several elementary operations to modify the system to a form that is easy to solve.

Example 2.2.5.

$$\begin{cases} 2x + 3y - 2z = 1 \\ x + 2y + z = 3 \\ 3x + 4y - 5z = -1 \end{cases}$$

multiply equation 2 by -3 and then add to equation 3

$$\begin{cases} 2x + 3y - 2z = 1 \\ x + 2y + z = 3 \\ -2y - 8z = -10 \end{cases}$$

add equation 3 to equation 2

$$\begin{cases} 2x + 3y - 2z = 1 \\ x - 7z = -7 \\ -2y - 8z = -10 \end{cases}$$

It may seem that the second operation, namely **add equation 3 to equation 2**, is not an elementary operation, since it is not listed in the definition at the beginning of this section, but in fact, it is a special case of the third operation, since we could describe it as **multiply equation 3 by 1 and then add to equation 2**.

Now we illustrate how we can use elementary operations to solve a system of linear equations.

Example 2.2.6. Solve the system

$$\begin{cases} x + y + z = -1 \\ 2x + 4y + 3z = 0 \\ 3x + y + 5z = 1 \end{cases}$$

Solution. First we eliminate the x-terms from the second and third equations.
multiply equation 1 by -2 and then add to equation 2

$$\begin{cases} x + y + z = -1 \\ 2y + z = 2 \\ 3x + y + 5z = 1 \end{cases}$$

multiply equation 1 by -3 and then add to equation 3

$$\begin{cases} x + y + z = -1 \\ 2y + z = 2 \\ -2y + 2z = 4 \end{cases}$$

To replace by 1 the number 2 in front of y in the second equation we
multiply equation 2 by $\frac{1}{2}$

$$\begin{cases} x + \ y + \ \ z = -1 \\ \qquad y + \frac{1}{2}z = \ \ 1 \\ \quad -2y + 2z = \ \ 4 \end{cases}$$

Next we eliminate the y-term from the third equation.

multiply equation 2 by 2 and then add to equation 3

$$\begin{cases} x + y + \ \ z = -1 \\ \quad y + \frac{1}{2}z = \ \ 1 \\ \qquad\quad 3z = \ \ 6 \end{cases}$$

To remove the 3 in front of z from the third equation we

multiply equation 3 by $\frac{1}{3}$

$$\begin{cases} x + y + \ \ z = -1 \\ \quad y + \frac{1}{2}z = \ \ 1 \\ \qquad\quad z = \ \ 2 \end{cases}$$

Now we eliminate the z-terms from the first and second equations.

multiply equation 3 by $-\frac{1}{2}$ and then add to equation 2

$$\begin{cases} x + y + z = -1 \\ \quad y \qquad = \ \ 0 \\ \qquad z = \ \ 2 \end{cases}$$

multiply equation 3 by -1 and then add to equation 1

$$\begin{cases} x + y \qquad = -3 \\ \quad y \qquad = \ \ 0 \\ \qquad z = \ \ 2 \end{cases}$$

Finally, we eliminate the y-term from the first equation.

multiply equation 2 by -1 and then add to equation 1

$$\begin{cases} x \qquad\quad = -3 \\ \quad y \qquad = \ \ 0 \\ \qquad z = \ \ 2 \end{cases}$$

The solution of the system is

$$\begin{cases} x = -3 \\ y = 0 \\ z = 2 \end{cases} .$$

$\square$

Example 2.2.7. Solve the system

$$\begin{cases} 2x + 3y - 2z = & 1 \\ x + 2y + \ z = & 3 \\ 3x + 4y - 5z = -1 \end{cases}$$

Solution. Because the second equation looks simpler than the first one we **interchange equation 1 and equation 2**.

$$\begin{cases} x + 2y + \ z = & 3 \\ 2x + 3y - 2z = & 1 \\ 3x + 4y - 5z = -1 \end{cases}$$

Next we eliminate the x-terms from the second and third equations. **multiply equation 1 by -2 and then add to equation 2**

$$\begin{cases} x + 2y + \ z = & 3 \\ -y - 4z = -5 \\ 3x + 4y - 5z = -1 \end{cases}$$

multiply equation 1 by -3 and then add to equation 3

$$\begin{cases} x + 2y + \ z = & 3 \\ -y - 4z = & -5 \\ -2y - 8z = -10 \end{cases}$$

To remove the minus sign in front of y from the second equation we **multiply equation 2 by -1**

$$\begin{cases} x + 2y + \ z = & 3 \\ y + 4z = & 5 \\ -2y - 8z = -10 \end{cases}$$

Now we eliminate the y-terms from the third equation **multiply equation 2 by 2 and then add to equation 3**

$$\begin{cases} x + 2y + \ z = 3 \\ y + 4z = 5 \\ 0 + \ 0 = 0 \end{cases}$$

Finally we eliminate the y-terms from the first equation. **multiply equation 2 by -2 and then add to equation 1**

$$\begin{cases} x \quad - 7z = -7 \\ y + 4z = \ 5 \\ 0 = \ 0 \end{cases}$$

Since the last equation does not contribute anything, the system is equivalent to the system with two equations:

$$\begin{cases} x & -7z = -7 \\ & y + 4z = 5 \end{cases}$$

This system has infinitely many solutions. The solutions can be described in the form

$$\begin{cases} x = -7 + 7z \\ y = 5 - 4z \end{cases},$$

where z is an arbitrary real number.   □

The process of solving systems of linear equations using the Gaussian elimination method can be simplified if we use matrices. We observe that the complete information about a system of linear equations

$$\begin{cases} 2x + 3y - 2z = 1 \\ x + 2y + z = 3 \\ 3x + 4y - 5z = -1 \end{cases}$$

is contained in the matrix

$$\begin{bmatrix} 2 & 3 & -2 & 1 \\ 1 & 2 & 1 & 3 \\ 3 & 4 & -5 & -1 \end{bmatrix}.$$

It suffices to remember that the first column corresponds to x, the second column to y, the third column to z, and that the last column contains the numbers on the other side of the $=$ sign. This matrix is called the **augmented matrix** of the system.

We will solve the system of linear equations from the last example by first converting it to the augmented matrix of the system and then performing elementary operations on rows of the matrix. In this case we call these operations elementary row operations.

Definition 2.2.8. By **elementary row operations** on a matrix we mean the following three operations:

- **Row interchange: Interchange two rows of the matrix.**

- **Row scaling: Multiply a row of the matrix by a nonzero constant.**

- **Row replacement: Multiply a row of the matrix by a constant and then add the result to another a row of the matrix.**

Example 2.2.9. Solve the system

$$\begin{cases} 2x + 3y - 2z = 1 \\ x + 2y + z = 3 \\ 3x + 4y - 5z = -1 \end{cases}.$$

Solution. The augmented matrix of the system is

$$\begin{bmatrix} 2 & 3 & -2 & 1 \\ 1 & 2 & 1 & 3 \\ 3 & 4 & -5 & -1 \end{bmatrix}.$$

interchange rows 1 and 2

$$\begin{bmatrix} 1 & 2 & 1 & 3 \\ 2 & 3 & -2 & 1 \\ 3 & 4 & -5 & -1 \end{bmatrix}$$

multiply row 1 by -2 and then add to row 2

$$\begin{bmatrix} 1 & 2 & 1 & 3 \\ 0 & -1 & -4 & -5 \\ 3 & 4 & -5 & -1 \end{bmatrix}$$

multiply row 1 by -3 and then add to row 3

$$\begin{bmatrix} 1 & 2 & 1 & 3 \\ 0 & -1 & -4 & -5 \\ 0 & -2 & -8 & -10 \end{bmatrix}$$

multiply row 2 by -1

$$\begin{bmatrix} 1 & 2 & 1 & 3 \\ 0 & 1 & 4 & 5 \\ 0 & -2 & -8 & -10 \end{bmatrix}$$

multiply row 2 by 2 and then add to equation 3

$$\begin{bmatrix} 1 & 2 & 1 & 3 \\ 0 & 1 & 4 & 5 \\ 0 & 0 & 0 & 0 \end{bmatrix}$$

multiply row 2 by -2 and then add to equation 1

$$\begin{bmatrix} 1 & 0 & -7 & -7 \\ 0 & 1 & 4 & 5 \\ 0 & 0 & 0 & 0 \end{bmatrix}.$$

The last matrix corresponds to the system

$$\begin{cases} x & -7z = -7 \\ & y + 4z = 5 \\ & 0 = 0 \end{cases}$$

and now we obtain the solutions as in Example 2.2.7.   □

We note that the two operations
                    multiply row 1 by -2 **and then add to row 2**
                    multiply row 1 by -3 **and then add to row 3**
in the above example do not depend on each other in any way and the order of these two operations does not matter. We can combine them together. More precisely, from the matrix

$$\begin{bmatrix} 1 & 2 & 1 & 3 \\ 2 & 3 & -2 & 1 \\ 3 & 4 & -5 & -1 \end{bmatrix}$$

by applying the following operations
                    multiply row 1 by -2 **and then add to row 2**
                    multiply row 1 by -3 **and then add to row 3**
we get the matrix

$$\begin{bmatrix} 1 & 2 & 1 & 3 \\ 0 & -1 & -4 & -5 \\ 0 & -2 & -8 & -10 \end{bmatrix}.$$

The same thing is true for the last two operations in the above example. That is, the operations
                    multiply row 2 by 2 **and then add to equation 3**
                    multiply row 2 by -2 **and then add to equation 1**
could be applied together to the matrix

$$\begin{bmatrix} 1 & 2 & 1 & 3 \\ 0 & 1 & 4 & 5 \\ 0 & -2 & -8 & -10 \end{bmatrix}$$

in order to obtain the matrix

$$\begin{bmatrix} 1 & 0 & -7 & -7 \\ 0 & 1 & 4 & 5 \\ 0 & 0 & 0 & 0 \end{bmatrix}.$$

The Gaussian elimination algorithm

In the previous section we illustrated the Gaussian elimination method with a couple of examples. Now we will discuss it in a more formal way. First we need to introduce some new terminology.

Definition 2.2.10. By a *leading term* in a row of a matrix we mean the first nonzero entry. In other words, a leading term in a row is a nonzero entry in that row such that all entries to the left of it are 0. If a leading term is equal to 1, then we call it a *leading* 1.

Example 2.2.11. Here are some examples of matrices with the leading terms enclosed in boxes.

$$\begin{bmatrix} \boxed{1} & -1 & 0 & 9 \\ 0 & \boxed{2} & 0 & 2 \\ 0 & 0 & \boxed{-1} & 5 \end{bmatrix}, \quad \begin{bmatrix} \boxed{1} & 3 & -2 & 0 & 3 \\ 0 & 0 & 0 & \boxed{1} & 7 \\ 0 & 0 & 0 & 0 & 0 \\ 0 & 0 & 0 & 0 & 0 \end{bmatrix}, \quad \begin{bmatrix} \boxed{3} & 5 & 0 & 2 & 3 & 2 & 0 \\ 0 & 0 & \boxed{-2} & 7 & 2 & 4 & 0 \\ 0 & 0 & 0 & 0 & 0 & 0 & \boxed{3} \\ 0 & 0 & 0 & 0 & 0 & 0 & 0 \end{bmatrix}.$$

Definition 2.2.12. A matrix is in a *reduced row echelon form* (or *Gauss-Jordan form*) if it satisfies all of the following conditions:

(a) All leading terms of the matrix are equal to 1;

(b) In each column of with a leading 1 all other terms are equal to 0;

(c) Each leading 1 is in a column to the right of the leading 1 in the row above it;

(d) Rows whose entries are all zero are below rows with nonzero entries.

Example 2.2.13. Here are examples of matrices in a reduced row echelon form.

$$\begin{bmatrix} 1 & 0 \\ 0 & 1 \end{bmatrix}, \quad \begin{bmatrix} 1 & 0 & 0 & 9 \\ 0 & 1 & 0 & 2 \\ 0 & 0 & 1 & 5 \end{bmatrix}, \quad \begin{bmatrix} 1 & 3 & -2 & 0 & 3 \\ 0 & 0 & 0 & 1 & 7 \\ 0 & 0 & 0 & 0 & 0 \\ 0 & 0 & 0 & 0 & 0 \end{bmatrix}, \quad \begin{bmatrix} 1 & 0 & 4 & 0 & 3 & 1 & 0 & 2 \\ 0 & 1 & 7 & 0 & 2 & 0 & 0 & 2 \\ 0 & 0 & 0 & 1 & 8 & 3 & 0 & 7 \\ 0 & 0 & 0 & 0 & 0 & 0 & 1 & 5 \end{bmatrix}$$

Example 2.2.14. Here are examples of matrices that are not in a reduced row echelon form.

1. The matrix

$$\begin{bmatrix} 1 & 0 & 0 & 9 \\ 0 & 1 & 0 & 2 \\ 0 & 0 & \boxed{2} & 5 \end{bmatrix}$$

is not in a reduced row echelon form because condition (a) is not satisfied. The leading entry in the third row is not 1.

2. The matrix

$$\begin{bmatrix} 1 & 0 & 0 \\ 0 & 1 & \boxed{-1} \\ 0 & 0 & 1 \end{bmatrix}$$

is not in a reduced row echelon form because condition (b) is not satisfied. The third column has a leading 1 and another nonzero term.

3. The matrix

$$\begin{bmatrix} 1 & 3 & 0 & 0 & 3 \\ 0 & 0 & 0 & \boxed{1} & 7 \\ 0 & 0 & \boxed{1} & 0 & 0 \\ 0 & 0 & 0 & 0 & 0 \end{bmatrix}$$

is not in a reduced row echelon form because condition (c) is not satisfied. The leading 1 in the third row is not in a column to the right of the leading 1 in the row above it.

4. The matrix

$$\begin{bmatrix} 1 & 5 & 0 & 2 & 3 & 2 & 0 & 5 \\ 0 & 0 & 1 & 7 & 2 & 4 & 0 & 3 \\ 0 & 0 & 0 & 0 & 0 & 0 & 0 & 0 \\ 0 & 0 & 0 & 0 & 0 & 0 & \boxed{1} & 0 \end{bmatrix}$$

is not in a reduced row echelon form because condition (d) is not satisfied. All terms in the third row are 0's, but the fourth row has a nonzero term.

Definition 2.2.15. Every entry in a matrix, which is in a reduced row echelon form, where a leading 1 is located is called a ***pivot position*** and every column that contains a pivot position is called a ***pivot column***.

Example 2.2.16. All pivot positions in the matrix

$$\begin{bmatrix} \boxed{1} & 5 & 0 & 2 & 3 & 2 & 0 & 5 \\ 0 & 0 & \boxed{1} & 7 & 2 & 4 & 0 & 3 \\ 0 & 0 & 0 & 0 & 0 & 0 & \boxed{1} & 2 \\ 0 & 0 & 0 & 0 & 0 & 0 & 0 & 0 \end{bmatrix}$$

are marked. Columns 1, 3, and 7 are the pivot columns of this matrix.

> **Theorem 2.2.17.** *Every matrix can be reduced using row operations in a reduced row echelon form.*
> *The reduced row echelon form does not depend of the row operations we choose to get this form.*

We do not prove this theorem in the book.

Because the reduced row echelon form is unique we can extend the definition 2.2.15:

> **Definition 2.2.18.** Every entry in a matrix where a leading 1 is located in the reduced row echelon form of the matrix is called a ***pivot position*** and every column that contains a pivot position is called a ***pivot column***.

The general algorithm for obtaining the reduced row echelon form of any matrix is based on three basic ideas used in this process:

- If there is a nonzero term in a column, we can always move it to a desired position in that column by applying an appropriate **row interchange**.

- Any nonzero term can be changed to 1 by an appropriate **scaling**.

- If a column has a term equal to 1, then any other nonzero term in that column can be changed to 0 by an appropriate **row replacement**.

Now we describe the general Gaussian elimination process. Pivot columns and pivot elements play an important role in this algorithm.

> **Step 1** Identify the first (from the left) nonzero column. (This is a pivot column.)
>
> **Step 2** If necessary, move a row with a nonzero entry in the pivot column to the top using an appropriate **row interchange**. (After this operation the entry at the top of the pivot column is in the pivot position.)
>
> **Step 3** If necessary, change the number in the pivot position to 1 using an appropriate **scaling**. (The 1 in the pivot position is a leading 1.)
>
> **Step 4** Replace, if necessary, every term below the leading 1 by 0 using appropriate **row replacements**.

We are not done yet, but we take a break to look at an example.

Example 2.2.19. We consider the matrix

$$\begin{bmatrix} 0 & 0 & 1 & -1 & 0 & 1 & 2 \\ 0 & 0 & 3 & 0 & 3 & -2 & 0 \\ 0 & 2 & 0 & 1 & 3 & -1 & 3 \\ 0 & -1 & 2 & 0 & 1 & 0 & 5 \end{bmatrix}.$$

Step 1 The second column is the first one that has nonzero terms. This is the pivot column.

Step 2 We interchange rows 1 and 4 to get the nonzero term -1 at the top of the pivot column:

$$\begin{bmatrix} 0 & -1 & 2 & 0 & 1 & 0 & 5 \\ 0 & 0 & 3 & 0 & 3 & -2 & 0 \\ 0 & 2 & 0 & 1 & 3 & -1 & 3 \\ 0 & 0 & 1 & -1 & 0 & 1 & 2 \end{bmatrix}.$$

Step 3 We multiply the first row by -1 to get 1 in the pivot position:

$$\begin{bmatrix} 0 & 1 & -2 & 0 & -1 & 0 & -5 \\ 0 & 0 & 3 & 0 & 3 & -2 & 0 \\ 0 & 2 & 0 & 1 & 3 & -1 & 3 \\ 0 & 0 & 1 & -1 & 0 & 1 & 2 \end{bmatrix}.$$

Step 4 We add the first row multiplied by -2 to the third row to replace the 2 in the pivot column by 0:

$$\begin{bmatrix} 0 & 1 & -2 & 0 & -1 & 0 & -5 \\ 0 & 0 & 3 & 0 & 3 & -2 & 0 \\ 0 & 0 & 4 & 1 & 5 & -1 & 13 \\ 0 & 0 & 1 & -1 & 0 & 1 & 2 \end{bmatrix}.$$

Now we continue with the general algorithm.

Step 5 Temporarily ignore the top row of the matrix.

Step 6 Apply steps 1–4 to the smaller matrix that remains. Continue then with step 5 followed by steps 1–4 until there are no nonzero rows left.

We now return to our example.

Steps 5 and 6 We ignore (cover) the top row of the matrix after step 4, that is

$$\begin{bmatrix} 0 & 1 & -2 & 0 & -1 & 0 & -5 \\ 0 & 0 & 3 & 0 & 3 & -2 & 0 \\ 0 & 0 & 4 & 1 & 5 & -1 & 13 \\ 0 & 0 & 1 & -1 & 0 & 1 & 2 \end{bmatrix}$$

and proceed with the algorithm on the submatrix under the top row:
interchange rows 2 and 4
(that is, interchange rows 1 and 3 in the submatrix under the top row)

$$\begin{bmatrix} 0 & 1 & -2 & 0 & -1 & 0 & -5 \\ 0 & 0 & 1 & -1 & 0 & 1 & 2 \\ 0 & 0 & 4 & 1 & 5 & -1 & 13 \\ 0 & 0 & 3 & 0 & 3 & -2 & 0 \end{bmatrix}$$

multiply row 2 by −4 and then add to row 3
(that is, multiply row 1 by −4 and then add to row 3 in the submatrix)
multiply row 2 by −3 and then add to row 4
(that is, multiply row 1 by −3 and then add to row 3 in the submatrix under the top
row)

$$\begin{bmatrix} 0 & 1 & -2 & 0 & -1 & 0 & -5 \\ 0 & 0 & 1 & -1 & 0 & 1 & 2 \\ 0 & 0 & 0 & 5 & 5 & -5 & 5 \\ 0 & 0 & 0 & 3 & 3 & -5 & -6 \end{bmatrix}$$

Now we ignore the top two rows from the obtained matrix, that is,

$$\begin{bmatrix} 0 & 1 & -2 & 0 & -1 & 0 & -5 \\ 0 & 0 & 1 & -1 & 0 & 1 & 2 \\ 0 & 0 & 0 & 5 & 5 & -5 & 5 \\ 0 & 0 & 0 & 3 & 3 & -5 & -6 \end{bmatrix}$$

and proceed with the algorithm on the matrix under these rows:
multiply row 3 by $\frac{1}{5}$
(that is, multiply row 1 by $\frac{1}{5}$ in the new submatrix)

$$\begin{bmatrix} 0 & 1 & -2 & 0 & -1 & 0 & -5 \\ 0 & 0 & 1 & -1 & 0 & 1 & 2 \\ 0 & 0 & 0 & 1 & 1 & -1 & 1 \\ 0 & 0 & 0 & 3 & 3 & -5 & -6 \end{bmatrix}$$

multiply row 3 by −3 and then add to row 4

(that is, multiply row 1 by -3 and then add to row 2 in the new submatrix)

$$\begin{bmatrix} 0 & 1 & -2 & 0 & -1 & 0 & -5 \\ 0 & 0 & 1 & -1 & 0 & 1 & 2 \\ 0 & 0 & 0 & 1 & 1 & -1 & 1 \\ 0 & 0 & 0 & 0 & 0 & -2 & -9 \end{bmatrix}$$

Now we ignore the top three rows from the obtained matrix, that is,

$$\begin{bmatrix} 0 & 1 & -2 & 0 & -1 & 0 & -5 \\ 0 & 0 & 1 & -1 & 0 & 1 & 2 \\ 0 & 0 & 0 & 1 & 1 & -1 & 1 \\ 0 & 0 & 0 & 0 & 0 & -2 & -9 \end{bmatrix}$$

and proceed with the algorithm on the matrix under these rows which is the row 4 in the original matrix:

multiply row 4 by $-\frac{1}{2}$

(that is, multiply the only row of the new submatrix by $-\frac{1}{2}$)

$$\begin{bmatrix} 0 & 1 & -2 & 0 & -1 & 0 & -5 \\ 0 & 1 & 1 & -1 & 0 & 1 & 2 \\ 0 & 0 & 0 & 1 & 1 & -1 & 1 \\ 0 & 0 & 0 & 0 & 0 & 1 & \frac{9}{2} \end{bmatrix}$$

The obtained matrix has four pivot positions:

$$\begin{bmatrix} 0 & \boxed{1} & -2 & 0 & -1 & 0 & -5 \\ 0 & 0 & \boxed{1} & -1 & 0 & 1 & 2 \\ 0 & 0 & 0 & \boxed{1} & 1 & -1 & 1 \\ 0 & 0 & 0 & 0 & 0 & \boxed{1} & \frac{9}{2} \end{bmatrix}$$

Columns 2, 3, 4, and 6 are the pivot columns.

The matrix obtained in the example above is not yet in the reduced row echelon form. Condition (b) is not satisfied: in addition to the leading 1's there are other nonzero terms in columns 3, 4, and 6. One more step is necessary.

Step 7 Use appropriate **row replacements** to replace with 0 all nonzero terms, other than the leading 1's, in all pivot columns starting from the last pivot column to the right and then continuing with next pivot to the left and so on.

Now we are finally ready to finish our example and obtain the reduced row echelon form of the matrix.

Step 7 First we take care of column 6, the last pivot column on the right:

add row 4 to row 3

multiply row 4 by −1 and then add to row 2

$$\begin{bmatrix} 0 & 1 & -2 & 0 & -1 & 0 & -5 \\ 0 & 0 & 1 & -1 & 0 & 0 & -\frac{5}{2} \\ 0 & 0 & 0 & 1 & 1 & 0 & \frac{11}{2} \\ 0 & 0 & 0 & 0 & 0 & 1 & \frac{9}{2} \end{bmatrix}$$

Now the only nonzero term in column 6 is the pivot term. Next we take care of column 4. We skip column 5 because it is not a pivot column.

add row 3 to row 2

$$\begin{bmatrix} 0 & 1 & -2 & 0 & -1 & 0 & -5 \\ 0 & 0 & 1 & 0 & 1 & 0 & 3 \\ 0 & 0 & 0 & 1 & 1 & 0 & \frac{11}{2} \\ 0 & 0 & 0 & 0 & 0 & 1 & \frac{9}{2} \end{bmatrix}$$

Now the only nonzero term in column 4 is the pivot term. The final step is to replace the −2 in column 3 by 0.

multiply row 2 by 2 and then add to row 1

$$\begin{bmatrix} 0 & 1 & 0 & 0 & 1 & 0 & 1 \\ 0 & 0 & 1 & 0 & 1 & 0 & 3 \\ 0 & 0 & 0 & 1 & 1 & 0 & \frac{11}{2} \\ 0 & 0 & 0 & 0 & 0 & 1 & \frac{9}{2} \end{bmatrix}$$

Now the matrix is in the reduced row echelon form.

Here are all the steps of the Gaussian elimination process put together.

Step 1 Identify the first (from the left) nonzero column. (This is the pivot column.)

Step 2 If necessary, move a row with a nonzero term in the pivot column to the top using an appropriate **row interchange**. (The position at the top of the pivot column is now in the pivot position.)

Step 3 If necessary, change the number in the pivot position to 1 using an appropriate **scaling**. (The 1 in the pivot position is a leading 1.)

Step 4 Replace, if necessary, every term below the leading 1 by 0 using appropriate **row replacements**.

Step 5 Temporarily cover (ignore) the top row of the matrix.

Step 6 Apply steps 1–4 to the smaller matrix that remains. Continue then with step 5 followed by steps 1–4 until there are no nonzero rows left.

Step 7 Use appropriate **row replacements** to replace with 0 all nonzero terms, other than the leading 1's, in all pivot columns starting from the last pivot column on the right and working to the left.

While it is important to understand the above algorithm, it would not make much sense to try to memorize it. It is important to understand what the reduced row echelon form is and what row operations are allowed. Following the above algorithm exactly may not be the best strategy. For example, in the matrix

$$\begin{bmatrix} 1 & -1 & 0 & 1 & 2 \\ 2 & -3 & 0 & 5 & 1 \\ -2 & 3 & 3 & -2 & 1 \end{bmatrix}$$

instead of multiplying row 1 by -2 and adding it to row 2 and then multiplying row 1 by 2 and adding it to row 3, it makes more sense to start by adding row 2 to row 3

$$\begin{bmatrix} 1 & -1 & 0 & 1 & 2 \\ 2 & -3 & 0 & 5 & 1 \\ 0 & 0 & 3 & 3 & 2 \end{bmatrix}$$

and then multiply row 1 by -2 and add to row 2

$$\begin{bmatrix} 1 & -1 & 0 & 1 & 2 \\ 0 & -1 & 0 & 3 & -3 \\ 0 & 0 & 3 & 3 & 2 \end{bmatrix}.$$

The algorithm described above is designed in such a way that, if we already have the desired values in a pivot column, then the following steps will not mess them up. For example, in the above matrix the values in the first column are exactly the values we want:

$$\begin{bmatrix} \boxed{1} & -1 & 0 & 1 & 2 \\ \boxed{0} & -1 & 0 & 3 & -3 \\ \boxed{0} & 0 & 3 & 3 & 2 \end{bmatrix}.$$

When continuing the Gaussian elimination process we have to make sure that the first column remains unchanged.

Using the idea presented above we show how to get the reduced row echelon form of the matrix in Example 2.2.19 in a different way. From the original matrix

$$\begin{bmatrix} 0 & 0 & 1 & -1 & 0 & 1 & 2 \\ 0 & 0 & 3 & 0 & 3 & -2 & 0 \\ 0 & 2 & 0 & 1 & 3 & -1 & 3 \\ 0 & -1 & 2 & 0 & 1 & 0 & 5 \end{bmatrix}$$

we obtain the matrix

$$\begin{bmatrix} 0 & 1 & -2 & 0 & -1 & 0 & -5 \\ 0 & 0 & 1 & -1 & 0 & 1 & 2 \\ 0 & 0 & 4 & 1 & 5 & -1 & 13 \\ 0 & 0 & 3 & 0 & 3 & -2 & 0 \end{bmatrix}$$

proceeding as presented in example 2.2.19. Then we replace by 0 the entries in the third column in rows 1, 3, and 4, that is, **above and below** the leading 1 in the second row. Note that this way we do not change columns 1 and 2.

multiply row 2 by 2 and then add to row 1
multiply row 2 by −4 and then add to row 3
multiply row 2 by −3 and then add to row 4

$$\begin{bmatrix} 0 & 1 & 0 & -2 & -1 & 2 & -1 \\ 0 & 0 & 1 & -1 & 0 & 1 & 2 \\ 0 & 0 & 0 & 5 & 5 & -5 & 5 \\ 0 & 0 & 0 & 3 & 3 & -5 & -6 \end{bmatrix}$$

multiply row 3 by $\frac{1}{5}$

$$\begin{bmatrix} 0 & 1 & 0 & -2 & -1 & 2 & -1 \\ 0 & 0 & 1 & -1 & 0 & 1 & 2 \\ 0 & 0 & 0 & 1 & 1 & -1 & 1 \\ 0 & 0 & 0 & 3 & 3 & -5 & -6 \end{bmatrix}$$

Next we replace by 0 the entries in the forth column in rows 1,2 and 4, that is, **above and below** the leading 1 in the third row. Note that this way we do not change columns 1, 2, and 3 because the entries to the left of the leading 1 in the third row are 0.

multiply row 3 by 2 and then add to row 1
multiply row 3 by 1 and then add to row 2
multiply row 3 by −3 and then add to row 4

$$\begin{bmatrix} 0 & 1 & 0 & 0 & 1 & 0 & 1 \\ 0 & 0 & 1 & 0 & 1 & 0 & 3 \\ 0 & 0 & 0 & 1 & 1 & -1 & 1 \\ 0 & 0 & 0 & 0 & 0 & -2 & -9 \end{bmatrix}$$

multiply row 4 by $-\frac{1}{2}$

$$\begin{bmatrix} 0 & 1 & 0 & 0 & 1 & 0 & 1 \\ 0 & 0 & 1 & 0 & 1 & 0 & 3 \\ 0 & 0 & 0 & 1 & 1 & -1 & 1 \\ 0 & 0 & 0 & 0 & 0 & 1 & \frac{9}{2} \end{bmatrix}$$

multiply row 4 by 1 and then add to row 3

$$\begin{bmatrix} 0 & 1 & 0 & 0 & 1 & 0 & 1 \\ 0 & 0 & 1 & 0 & 1 & 0 & 3 \\ 0 & 0 & 0 & 1 & 1 & 0 & \frac{11}{2} \\ 0 & 0 & 0 & 0 & 0 & 1 & \frac{9}{2} \end{bmatrix}$$

The following observation is useful when obtaining 0 above and below a leading 1.

If there is a leading 1 in the row p and column q, then the row replacement

multiply row p by α and then add to row $r \neq p$

does not change the columns 1, 2, ..., $q-1$ because all the entries to the left of the leading 1 in the p-th row are 0.

The above observations lead to the following modified algorithm for the Gaussian elimination process. Note that the steps 1, 2, and 3 have not changed.

Step 1 Identify the first (from the left) nonzero column. (This is the pivot column.)

Step 2 If necessary, move a row with a nonzero entry in the pivot column to the top using an appropriate **row interchange**. (The entry at the top of the pivot column is now in the pivot position.)

Step 3 If necessary, change the entry in the pivot position to 1 using an appropriate **scaling**. (The 1 in the pivot position is a leading 1.)

Step 4 Replace, if necessary, every entry **above and below** the leading 1 by 0 using appropriate **row replacements**.

Step 5 Temporarily cover (ignore) the top row of the matrix.

Step 6 Apply steps 1–3 to the smaller matrix that remains and step 4 to the whole matrix. Continue then with step 5 followed by steps 1–4 until there are no nonzero rows left.

We consider another example to illustrate the differences between the original algorithm and the modified algorithm.

Example 2.2.20. Find the reduced row echelon form of the matrix

$$\begin{bmatrix} 2 & 4 & 7 & 3 & 2 & 1 & 1 \\ 1 & 2 & 1 & 2 & 5 & 0 & 1 \\ 4 & 8 & 9 & 7 & 12 & 1 & 1 \end{bmatrix}$$

Solution. First we follow the original algorithm:

interchange rows 1 and 2

$$\begin{bmatrix} 1 & 2 & 1 & 2 & 5 & 0 & 1 \\ 2 & 4 & 7 & 3 & 2 & 1 & 1 \\ 4 & 8 & 9 & 7 & 12 & 1 & 1 \end{bmatrix}$$

multiply first row by −2 and then add to row 2
multiply first row by −4 and then add to row 3

$$\begin{bmatrix} 1 & 2 & 1 & 2 & 5 & 0 & 1 \\ 0 & 0 & 5 & -1 & -8 & 1 & -1 \\ 0 & 0 & 5 & -1 & -8 & 1 & -3 \end{bmatrix}$$

multiply row 2 by $\frac{1}{5}$

$$\begin{bmatrix} 1 & 2 & 1 & 2 & 5 & 0 & 1 \\ 0 & 0 & 1 & -\frac{1}{5} & -\frac{8}{5} & \frac{1}{5} & -\frac{1}{5} \\ 0 & 0 & 5 & -1 & -8 & 1 & -3 \end{bmatrix}$$

multiply row 2 by -5 and then add to row 3

$$\begin{bmatrix} 1 & 2 & 1 & 2 & 5 & 0 & 1 \\ 0 & 0 & 1 & -\frac{1}{5} & -\frac{8}{5} & \frac{1}{5} & -\frac{1}{5} \\ 0 & 0 & 0 & 0 & 0 & 0 & 2 \end{bmatrix}$$

multiply row 3 by $\frac{1}{2}$

$$\begin{bmatrix} 1 & 2 & 1 & 2 & 5 & 0 & 1 \\ 0 & 0 & 1 & -\frac{1}{5} & -\frac{8}{5} & \frac{1}{5} & -\frac{1}{5} \\ 0 & 0 & 0 & 0 & 0 & 0 & 1 \end{bmatrix}$$

multiply row 3 by $\frac{1}{5}$ and then add to row 2
multiply row 3 by -1 and then add to row 1

$$\begin{bmatrix} 1 & 2 & 1 & 2 & 5 & 0 & 0 \\ 0 & 0 & 1 & -\frac{1}{5} & -\frac{8}{5} & \frac{1}{5} & 0 \\ 0 & 0 & 0 & 0 & 0 & 0 & 1 \end{bmatrix}$$

multiply row 2 by -1 and then add to row 1

$$\begin{bmatrix} 1 & 2 & 0 & \frac{11}{5} & \frac{33}{5} & -\frac{1}{5} & 0 \\ 0 & 0 & 1 & -\frac{1}{5} & -\frac{8}{5} & \frac{1}{5} & 0 \\ 0 & 0 & 0 & 0 & 0 & 0 & 1 \end{bmatrix}$$

This is the reduced row echelon form of our matrix.

Now we apply the modified algorithm. After we obtain, using the original algorithm, the matrix

$$\begin{bmatrix} 1 & 2 & 1 & 2 & 5 & 0 & 1 \\ 0 & 0 & 1 & -\frac{1}{5} & -\frac{8}{5} & \frac{1}{5} & -\frac{1}{5} \\ 0 & 0 & 5 & -1 & -8 & 1 & -3 \end{bmatrix}$$

we replace by 0 the entries in the third column in rows 1 and 3, that is, above and below the leading 1 in the second row.

multiply row 2 by -1 and then add to row 1
multiply row 2 by -5 and then add to row 3

$$\begin{bmatrix} 1 & 2 & 0 & \frac{11}{5} & \frac{33}{5} & -\frac{1}{5} & \frac{6}{5} \\ 0 & 0 & 1 & -\frac{1}{5} & -\frac{8}{5} & \frac{1}{5} & -\frac{1}{5} \\ 0 & 0 & 0 & 0 & 0 & 0 & 2 \end{bmatrix}$$

multiply row 3 by $\frac{1}{2}$

$$\begin{bmatrix} 1 & 2 & 0 & \frac{11}{5} & \frac{33}{5} & -\frac{1}{5} & \frac{6}{5} \\ 0 & 0 & 1 & -\frac{1}{5} & -\frac{8}{5} & \frac{1}{5} & -\frac{1}{5} \\ 0 & 0 & 0 & 0 & 0 & 0 & 1 \end{bmatrix}$$

multiply row 3 by $-\frac{6}{5}$ and then add to row 1
multiply row 3 by $\frac{1}{5}$ and then add to row 2

$$\begin{bmatrix} 1 & 2 & 0 & \frac{11}{5} & \frac{33}{5} & -\frac{1}{5} & 0 \\ 0 & 0 & 1 & -\frac{1}{5} & -\frac{8}{5} & \frac{1}{5} & 0 \\ 0 & 0 & 0 & 0 & 0 & 0 & 1 \end{bmatrix}$$

As expected, this is the reduced row echelon form obtained using the original algorithm. □

Solving systems of linear equations using Gaussian elimination

At the beginning of this section we presented some examples of systems of linear equations that were solved by Gaussian elimination. In the first couple of examples we worked directly with the equations. In the last example in that section we first represented the system by a matrix and then worked with the matrices until we were ready to give the final answer. Those examples were used to motivate the Gaussian elimination algorithm presented in the previous section. In this section we present more examples illustrating the use of Gaussian elimination to solve systems of linear equations.

Example 2.2.21. Solve the system

$$\begin{cases} 2x + 7y = 1 \\ 5x + 3y = 2 \end{cases}.$$

Solution. We represent the system by its augmented matrix

$$\begin{bmatrix} 2 & 7 & 1 \\ 5 & 3 & 2 \end{bmatrix}$$

and then apply the Gaussian elimination algorithm to obtain the reduced row echelon form of the matrix:

multiply row 1 by $\frac{1}{2}$

$$\begin{bmatrix} 1 & \frac{7}{2} & \frac{1}{2} \\ 5 & 3 & 2 \end{bmatrix}$$

multiply row 1 by -5 and then add to row 2

$$\begin{bmatrix} 1 & \frac{7}{2} & \frac{1}{2} \\ 0 & -\frac{29}{2} & -\frac{1}{2} \end{bmatrix}$$

multiply row 2 by $-\frac{2}{29}$

$$\begin{bmatrix} 1 & \frac{7}{2} & \frac{1}{2} \\ 0 & 1 & -\frac{1}{29} \end{bmatrix}$$

multiply row 2 by $-\frac{7}{2}$ and then add to row 1

$$\begin{bmatrix} 1 & 0 & \frac{18}{29} \\ 0 & 1 & -\frac{1}{29} \end{bmatrix}$$

The solution of the system is

$$x = \frac{18}{29} \quad \text{and} \quad y = -\frac{1}{29}.$$

$\square$

Example 2.2.22. Solve the system

$$\begin{cases} 2x + y + 3z + 2w = 1 \\ 3x + 2y + z + 2w = 3 \\ x + 5z + 2w = -1 \end{cases}.$$

Solution. The augmented matrix of the system is

$$\begin{bmatrix} 2 & 1 & 3 & 2 & 1 \\ 3 & 2 & 1 & 2 & 3 \\ 1 & 0 & 5 & 2 & -1 \end{bmatrix}.$$

Now we apply the Gaussian elimination algorithm to obtain the reduced row echelon form of the matrix:

interchange rows 1 and 3

$$\begin{bmatrix} 1 & 0 & 5 & 2 & -1 \\ 3 & 2 & 1 & 2 & 3 \\ 2 & 1 & 3 & 2 & 1 \end{bmatrix}$$

multiply row 1 by -3 and then add to row 2

multiply row 1 by −2 and then add to row 3

$$\begin{bmatrix} 1 & 0 & 5 & 2 & -1 \\ 0 & 2 & -14 & -2 & 6 \\ 0 & 1 & -7 & -2 & 3 \end{bmatrix}$$

interchange rows 2 and 3

$$\begin{bmatrix} 1 & 0 & 5 & 2 & -1 \\ 0 & 1 & -7 & -2 & 3 \\ 0 & 2 & -14 & -2 & 6 \end{bmatrix}$$

multiply row 2 by −2 and then add to row 3

$$\begin{bmatrix} 1 & 0 & 5 & 2 & -1 \\ 0 & 1 & -7 & -2 & 3 \\ 0 & 0 & 0 & 0 & 0 \end{bmatrix}$$

The original system is reduced to

$$\begin{cases} x + & 5z + 2w = -1 \\ & y - 7z - 2w = 3 \end{cases}.$$

The solution of the system is

$$x = -5z - 2w - 1 \quad \text{and} \quad y = 7z + 2w + 3.$$

$\square$

Definition 2.2.23. The variables in a system of equations which correspond to the pivot columns of the augmented matrix are called *pivot variables* or *basic variables*. The other variables are called *free variables*.

In the above example we have two pivot columns, namely columns 1 and 2. The variables corresponding to columns 1 and 2 are x and y. These are the basic variables of our system. The variable z and w corresponding to columns 3 and 4, which are not pivot columns, are the free variable of the system. Note that in the solution of the system the basic variables are expressed in terms of the free variables.

Example 2.2.24. Solve the system

$$\begin{cases} 2x + 2y + z = 0 \\ 3x + y + z = 1 \\ x + 3y + z = 2 \end{cases},$$

Solution. The augmented matrix of the system is

$$\left[\begin{array}{ccc|c} 2 & 2 & 1 & 0 \\ 3 & 1 & 1 & 1 \\ 1 & 3 & 1 & 2 \end{array} \right].$$

Now we apply the Gaussian elimination algorithm to obtain the reduced row echelon form of the matrix.

interchange rows 1 and 3

$$\left[\begin{array}{ccc|c} 1 & 3 & 1 & 2 \\ 3 & 1 & 1 & 1 \\ 2 & 2 & 1 & 0 \end{array} \right]$$

multiply row 1 by -3 and then add to row 2
multiply row 1 by -2 and then add to row 3

$$\left[\begin{array}{ccc|c} 1 & 3 & 1 & 2 \\ 0 & -8 & -2 & -5 \\ 0 & -4 & -1 & -4 \end{array} \right]$$

multiply row 2 by $-\frac{1}{8}$

$$\left[\begin{array}{ccc|c} 1 & 3 & 1 & 2 \\ 0 & 1 & \frac{1}{4} & \frac{5}{8} \\ 0 & -4 & -1 & -4 \end{array} \right]$$

multiply row 2 by 4 and then add to row 3

$$\left[\begin{array}{ccc|c} 1 & 3 & 1 & 2 \\ 0 & 1 & \frac{1}{4} & \frac{5}{8} \\ 0 & 0 & 0 & -\frac{3}{2} \end{array} \right]$$

We have arrived to something that is not possible, because from the last row we get $0 = -\frac{3}{2}$. The system has no solution.   □

Echelon form of a matrix and back substitution

When solving a system of linear equations using Gaussian elimination it is not necessary to reduce the augmented matrix representing a system all the way to the Gauss-Jordan form to solve the system. The solution can be easily obtained "half way" to the Gauss-Jordan form. We proceed with Gaussian elimination until we get the so-called row echelon form of the matrix and then we use a method called **back substitution** to solve the system. In this section we take a closer look at this method.

Example 2.2.25. Solve the system

$$\begin{cases} 2x + 7y = 1 \\ 5x + 3y = 2 \end{cases}$$

Solution. We start by transforming the augmented matrix of the system of equation, that is, the matrix

$$\begin{bmatrix} 2 & 7 & 1 \\ 5 & 3 & 2 \end{bmatrix}.$$

multiply row 1 by $-\frac{5}{2}$ and add to row 2

$$\begin{bmatrix} 2 & 7 & 1 \\ 0 & -\frac{29}{2} & -\frac{1}{2} \end{bmatrix}$$

We have not found the reduced row echelon form of the augmented matrix, but we can easily get the solution of the system from the above matrix. Indeed, when converted back to a system of equations we get

$$\begin{cases} 2x & +7y & = 1 \\ & -\frac{29}{2}y = -\frac{1}{2} \end{cases}.$$

The second equation gives us $y = \frac{1}{29}$. We substitute back the obtained value of y to the first equation and obtain an equation for x:

$$2x + 7 \cdot \frac{1}{29} = 1$$

Solving for x we get $x = \frac{11}{29}$. □

Definition 2.2.26. A matrix is in a *row echelon form* if it satisfies all of the following conditions:

   (a) In any two rows with leading terms the leading term of the row above is to the left of the leading term of the row below;

   (b) In a column with a leading term all entries below the leading term are 0;

   (c) Rows whose entries are all zero are below rows with nonzero entries.

It is clear that all matrices in a reduced row echelon form are in echelon form.

Example 2.2.27. Here are some examples of matrices in a row echelon form, but not in a reduced row echelon form:

$$\begin{bmatrix} 3 & 2 & 0 \\ 0 & 1 & 4 \\ 0 & 0 & 7 \end{bmatrix}, \begin{bmatrix} 4 & 5 \\ 0 & 1 \\ 0 & 0 \end{bmatrix}, \begin{bmatrix} 2 & 3 & 5 & -2 & 3 \\ 0 & 0 & 0 & 3 & 7 \\ 0 & 0 & 0 & 0 & 0 \\ 0 & 0 & 0 & 0 & 0 \end{bmatrix}, \begin{bmatrix} 5 & 3 & 4 & 0 & 3 & 1 & 4 & 2 \\ 0 & 2 & 7 & 1 & 2 & 1 & 0 & 2 \\ 0 & 0 & 0 & 3 & 8 & 3 & 1 & 7 \\ 0 & 0 & 0 & 0 & 0 & 0 & 3 & 5 \end{bmatrix}.$$

Example 2.2.28. Here are some examples of matrices that are not in a row echelon form:

$$\begin{bmatrix} 0 & 1 \\ 1 & 3 \end{bmatrix}, \begin{bmatrix} 3 & 2 & 0 \\ 0 & 1 & 4 \\ 0 & 1 & 7 \end{bmatrix}, \begin{bmatrix} 2 & 8 & 2 & 9 \\ 0 & 0 & 5 & 2 \\ 0 & 1 & 4 & 5 \end{bmatrix}, \begin{bmatrix} 4 & 5 \\ 0 & 0 \\ 0 & 1 \end{bmatrix}.$$

As first indicated in Example 2.2.25, we can solve a system of linear equations by first reducing the augmented matrix of the system to a row echelon form and then solving the reduced system of equations by back substitution. We illustrate this method with some more examples.

Example 2.2.29. Solve the system

$$\begin{cases} 2x + y + z = 2 \\ x + 3y + z = 1 \\ -x + y + z = 3 \end{cases}.$$

Solution. We represent the system by its augmented matrix

$$\begin{bmatrix} 2 & 1 & 1 & 2 \\ 1 & 3 & 1 & 1 \\ -1 & 1 & 1 & 3 \end{bmatrix}$$

and then obtain a row echelon form of the matrix.

interchange rows 1 and 2

$$\begin{bmatrix} 1 & 3 & 1 & 1 \\ 2 & 1 & 1 & 2 \\ -1 & 1 & 1 & 3 \end{bmatrix}$$

multiply row 1 by -2 and then add to row 2
add row 1 to row 3

$$\begin{bmatrix} 1 & 3 & 1 & 1 \\ 0 & -5 & -1 & 0 \\ 0 & 4 & 2 & 4 \end{bmatrix}$$

multiply row 2 by $-\frac{1}{5}$

$$\begin{bmatrix} 1 & 3 & 1 & 1 \\ 0 & 1 & \frac{1}{5} & 0 \\ 0 & 4 & 2 & 4 \end{bmatrix}$$

multiply row 2 by -4 and then add to row 3

$$\begin{bmatrix} 1 & 3 & 1 & 1 \\ 0 & 1 & \frac{1}{5} & 0 \\ 0 & 0 & \frac{6}{5} & 4 \end{bmatrix}$$

multiply row 3 by $\frac{5}{6}$

$$\begin{bmatrix} 1 & 3 & 1 & 1 \\ 0 & 1 & \frac{1}{5} & 0 \\ 0 & 0 & 1 & \frac{10}{3} \end{bmatrix}$$

When converted back to a system of equations we get

$$\begin{cases} x + 3y + z = 1 \\ \quad\quad y + \frac{1}{5}z = 0 \\ \quad\quad\quad\quad z = \frac{10}{3} \end{cases}.$$

Consequently,

$$z = \frac{10}{3},$$

$$y = -\frac{1}{5}z = -\frac{1}{5} \cdot \frac{10}{3} = -\frac{2}{3},$$

and

$$x = 1 - 3y - z = 1 - 3 \cdot \left(-\frac{2}{3}\right) - \frac{10}{3} = -\frac{1}{3}.$$

$\square$

Example 2.2.30. Solve the system

$$\begin{cases} x + y + z = -1 \\ 2x + 4y + 3z = 0 \\ 3x + y + 5z = 1 \end{cases}$$

Solution. First we obtain a row echelon form of the augmented matrix of the system and then converted it back to a system of equations:

$$\begin{cases} x + y + \phantom{\frac{1}{2}}z = -1 \\ y + \frac{1}{2}z = 1 \\ z = 2 \end{cases}$$

So $z = 2$, $y = 0$, and $x = -3$.

$\square$

Definition 2.2.31. Two matrices A and B of the same size are called ***equivalent*** if A can be transformed into B by elementary row operations. If A and B are equivalent, we write

$$A \sim B.$$

Example 2.2.32. The work done in Example 2.2.24 can be summarized as follows:

$$\begin{bmatrix} 2 & 2 & 1 & 0 \\ 3 & 1 & 1 & 1 \\ 1 & 3 & 1 & 2 \end{bmatrix} \sim \begin{bmatrix} 1 & 3 & 1 & 2 \\ 3 & 1 & 1 & 1 \\ 2 & 2 & 1 & 0 \end{bmatrix} \sim \begin{bmatrix} 1 & 3 & 1 & 2 \\ 0 & -8 & -2 & -5 \\ 0 & -4 & -1 & -4 \end{bmatrix}$$

$$\sim \begin{bmatrix} 1 & 3 & 1 & 2 \\ 0 & 1 & \frac{1}{4} & \frac{5}{8} \\ 0 & -4 & -1 & -4 \end{bmatrix} \sim \begin{bmatrix} 1 & 3 & 1 & 2 \\ 0 & 1 & \frac{1}{4} & \frac{5}{8} \\ 0 & 0 & 0 & -\frac{3}{2} \end{bmatrix}$$

In future examples we will give the matrices obtained by elementary row operations without explicitly describing those operations. With sufficient experience it should be clear what operations were used.

Example 2.2.33. Solve the system

$$\begin{cases} x + 2y + z = 1 \\ 2x + 2y + 3z = 4 \\ 3x + 2y + 5z = 7 \end{cases}.$$

Solution. Since

$$\begin{bmatrix} 1 & 2 & 1 & 1 \\ 2 & 2 & 3 & 4 \\ 3 & 2 & 5 & 7 \end{bmatrix} \sim \begin{bmatrix} 1 & 2 & 1 & 1 \\ 0 & -2 & 1 & 2 \\ 0 & 0 & 0 & 0 \end{bmatrix},$$

the system can be reduced to

$$\begin{cases} x + 2y + z = 1 \\ \quad\;\; -2y + z = 2 \end{cases}$$

From the second equation we get $y = \frac{1}{2}z - 1$ and then from the first equation we get $x = -2z + 3$, where z is a free variable. □

2.2.1 Exercises

Find the system whose augmented matrix is the given matrix.

1. $\begin{bmatrix} 2 & 1 & 3 \\ 3 & 2 & 4 \end{bmatrix}$

2. $\begin{bmatrix} 2 & 3 & -7 & -7 & 0 \\ 5 & 1 & 4 & 5 & 2 \\ 3 & 2 & 4 & 4 & 1 \end{bmatrix}$

Find the reduced row echelon form of the following matrices.

3. $\begin{bmatrix} 2 & 7 \\ 8 & 28 \end{bmatrix}$

4. $\begin{bmatrix} 2 & 1 \\ 2 & 1 \end{bmatrix}$

5. $\begin{bmatrix} 1 & 2 & 3 \\ 4 & 0 & 1 \end{bmatrix}$

6. $\begin{bmatrix} 1 & -1 & 3 \\ -1 & 1 & -3 \end{bmatrix}$

7. $\begin{bmatrix} 3 & 1 & 1 \\ 4 & 2 & 1 \\ 5 & 3 & 1 \end{bmatrix}$

8. $\begin{bmatrix} 3 & 0 & 1 \\ 3 & 2 & 1 \\ 1 & 1 & 3 \end{bmatrix}$

9. $\begin{bmatrix} 2 & 1 & 3 \\ 2 & 3 & 1 \\ 2 & 5 & -1 \\ 1 & 1 & 1 \end{bmatrix}$

10. $\begin{bmatrix} 0 & 1 & 2 \\ 3 & -2 & 0 \\ 3 & -1 & 3 \\ 1 & 0 & 5 \end{bmatrix}$

11. $\begin{bmatrix} 2 & 1 & 3 & 1 \\ 1 & 2 & 0 & 1 \\ 1 & 1 & 2 & 1 \end{bmatrix}$

12. $\begin{bmatrix} 2 & 2 & 1 & 1 \\ 1 & 3 & 1 & 1 \\ 3 & 1 & 0 & 1 \end{bmatrix}$

13. $\begin{bmatrix} 2 & -3 & 0 & 5 & 1 \\ -2 & 3 & 3 & -2 & 1 \\ 1 & -1 & 0 & 1 & 2 \end{bmatrix}$

14. $\begin{bmatrix} 2 & 1 & 1 & 1 & 0 & 0 \\ 1 & 1 & 1 & 0 & 1 & 0 \\ 2 & 5 & -1 & 0 & 0 & 1 \end{bmatrix}$

15. $\begin{bmatrix} 1 & 2 & 4 & p \\ 1 & 1 & 1 & q \\ 5 & 7 & 11 & 2p+3q \end{bmatrix}$

16. $\begin{bmatrix} 2 & 1 & 1 & p \\ 3 & 1 & 2 & q \\ 1 & 1 & 2 & p-5q \end{bmatrix}$

17. $\begin{bmatrix} 2 & 1 & 0 & 3 & p \\ 3 & 1 & 1 & 5 & q \\ 5 & 3 & -1 & 8 & 4p+q \end{bmatrix}$

18. $\begin{bmatrix} 1 & 1 & 1 & 2 & p \\ 2 & 4 & 1 & 3 & q \\ 3 & 5 & 2 & 5 & p+q \end{bmatrix}$

20. $\begin{bmatrix} 0 & 1 & 1 & p \\ 1 & 0 & 1 & q \\ 1 & 1 & 0 & r \end{bmatrix}$

19. $\begin{bmatrix} 2 & 1 & 3 & p \\ 1 & 2 & 1 & q \\ 1 & 1 & 2 & r \end{bmatrix}$

21. List all reduced row echelon forms of 4×2 matrices with 1 pivot.

22. List all reduced row echelon forms of 3×3 matrices with 1 pivot.

23. List all reduced row echelon forms of 3×3 matrices with 2 pivots.

24. List all reduced row echelon forms of 2×4 matrices with 2 pivots.

25. List all reduced row echelon forms of 3×4 matrices with 3 pivots.

26. List all reduced row echelon forms of 4×4 matrices with 3 pivots.

Solve the given system of linear equations using Gaussian elimination method.

27. $\begin{cases} 3x + 2y = 1 \\ 4x + 3y = 2 \end{cases}$

28. $\begin{cases} 2x + 3y = 0 \\ x + 2y = 1 \end{cases}$

29. $\begin{cases} 2x + y - z = 1 \\ 3x + 2y + z = 4 \end{cases}$

30. $\begin{cases} x + y + 2z = 2 \\ 2x + 3y + 5z = 7 \end{cases}$

31. $\begin{cases} 2x + y + z = 4 \\ x + 2y + z = 3 \\ x + y + 2z = 0 \end{cases}$

32. $\begin{cases} 3x + y + 2z = 7 \\ 3x - y + z = 5 \\ x + 9y + 5z = 11 \end{cases}$

33. $\begin{cases} x + 3y + 2z = 1 \\ x + y + z = 2 \\ 2x + 4y + 3z = 2 \end{cases}$

34. $\begin{cases} 2x + y = 0 \\ 3x + 2y = 1 \\ 2x + 3y = 1 \end{cases}$

35. $\begin{cases} 3x + y + 2z = p \\ x + 3y + 2z = q \\ x + y + z = r \end{cases}$

36. $\begin{cases} 2x + y + z = p \\ x + y + 2z = q \\ x + 2y + z = r \end{cases}$

37. $\begin{cases} x + y + 3z = p \\ x - y + z = q \\ 3x - y + 5z = r \end{cases}$

38. $\begin{cases} 4x + 3y = p \\ 3x + 4y = q \\ x + y = r \end{cases}$

39. $\begin{cases} 2x + y + z = p \\ x + 2y + z = q \\ 3x + 3y + 2z = r \\ 5x + 4y + 3z = s \end{cases}$

40. $\begin{cases} 3x + 2y + z = p \\ x + 2y + 2z = q \\ x + y + 2z = r \\ x + y + z = s \end{cases}$

41. $\begin{cases} 3x + 5y + 3z + 4w = p \\ x + 2y + 2z + w = q \\ x + y - z + 2w = r \end{cases}$

$$42. \quad \begin{cases} 2x + y + 3z + w = p \\ x + 2y \qquad + w = q \\ x + y + 2z + w = r \end{cases}$$

2.3 The inverse of a matrix

The inverse of a matrix plays a special role in linear algebra. There are two basic questions in connection with invertible matrices: how can we check if a matrix is invertible and how do we find the inverse of an invertible matrix. While up to now we only considered invertibility of 2×2 matrices, in this section we extend our considerations to matrices of arbitrary sizes. We also apply what we learned in the previous section to the problem of finding inverse matrices.

Elementary matrices

Elementary 2×2 matrices were introduced in Section 1.2. We noticed there that elementary matrices can be used to find the inverse of a matrix. In this section we extend those ideas to matrices of a larger size.

> **Definition 2.3.1.** By an ***elementary matrix*** we mean a matrix obtained from an identity matrix by one elementary row operation.

Example 2.3.2. The matrix

$$\begin{bmatrix} 1 & 0 & 0 \\ 0 & 0 & 1 \\ 0 & 1 & 0 \end{bmatrix}$$

is an elementary matrix since it can be obtained from the unit matrix $\begin{bmatrix} 1 & 0 & 0 \\ 0 & 1 & 0 \\ 0 & 0 & 1 \end{bmatrix}$ by

interchanging rows 1 and 2.
    The matrix

$$\begin{bmatrix} 1 & 0 & 0 & 0 \\ 0 & 1 & 0 & 0 \\ 0 & 0 & k & 0 \\ 0 & 0 & 0 & 1 \end{bmatrix}$$

is an elementary matrix since it can be obtained from the unit matrix

$$\begin{bmatrix} 1 & 0 & 0 & 0 \\ 0 & 1 & 0 & 0 \\ 0 & 0 & 1 & 0 \\ 0 & 0 & 0 & 1 \end{bmatrix}$$

by multiplication of row 3 by k.

The matrix

$$\begin{bmatrix} 1 & 0 & 0 & 0 & 0 \\ 0 & 1 & 0 & k & 0 \\ 0 & 0 & 1 & 0 & 0 \\ 0 & 0 & 0 & 1 & 0 \\ 0 & 0 & 0 & 0 & 1 \end{bmatrix}$$

is an elementary matrix since it can be obtained from the unit matrix

$$\begin{bmatrix} 1 & 0 & 0 & 0 & 0 \\ 0 & 1 & 0 & 0 & 0 \\ 0 & 0 & 1 & 0 & 0 \\ 0 & 0 & 0 & 1 & 0 \\ 0 & 0 & 0 & 0 & 1 \end{bmatrix}$$

by multiplying row 4 by k and adding it to row 2.

The following observation is very useful in calculations involving elementary matrices.

> The result of applying an elementary row operation to a matrix is equivalent to multiplication by an elementary matrix obtained from the unit matrix by the same elementary row operation.

Example 2.3.3. Multiplying a 3×4 matrix by the elementary matrix

$$\begin{bmatrix} 1 & 0 & 0 \\ 0 & 0 & 1 \\ 0 & 1 & 0 \end{bmatrix}$$

is equivalent to the elementary row operation of interchanging rows 2 and 3:

$$\begin{bmatrix} 1 & 0 & 0 \\ 0 & 0 & 1 \\ 0 & 1 & 0 \end{bmatrix} \begin{bmatrix} a_1 & b_1 & c_1 & d_1 \\ a_2 & b_2 & c_2 & d_2 \\ a_3 & b_3 & c_3 & d_3 \end{bmatrix} = \begin{bmatrix} a_1 & b_1 & c_1 & d_1 \\ a_3 & b_3 & c_3 & d_3 \\ a_2 & b_2 & c_2 & d_2 \end{bmatrix}.$$

Multiplying a 4×2 matrix by the matrix

$$\begin{bmatrix} 1 & 0 & 0 & 0 \\ 0 & 1 & 0 & 0 \\ 0 & 0 & k & 0 \\ 0 & 0 & 0 & 1 \end{bmatrix}$$

is equivalent to the elementary row operation of multiplication of row 3 by k:

$$\begin{bmatrix} 1 & 0 & 0 & 0 \\ 0 & 1 & 0 & 0 \\ 0 & 0 & k & 0 \\ 0 & 0 & 0 & 1 \end{bmatrix} \begin{bmatrix} a_1 & b_1 \\ a_2 & b_2 \\ a_3 & b_3 \\ a_4 & b_4 \end{bmatrix} = \begin{bmatrix} a_1 & b_1 \\ a_2 & b_2 \\ ka_3 & kb_3 \\ a_4 & b_4 \end{bmatrix}.$$

Multiplying a 5×3 matrix by the matrix

$$\begin{bmatrix} 1 & 0 & 0 & 0 & 0 \\ 0 & 1 & 0 & k & 0 \\ 0 & 0 & 1 & 0 & 0 \\ 0 & 0 & 0 & 1 & 0 \\ 0 & 0 & 0 & 0 & 1 \end{bmatrix}$$

is equivalent to the elementary row operation of multiplying row 4 by k and adding it to row 2:

$$\begin{bmatrix} 1 & 0 & 0 & 0 & 0 \\ 0 & 1 & 0 & k & 0 \\ 0 & 0 & 1 & 0 & 0 \\ 0 & 0 & 0 & 1 & 0 \\ 0 & 0 & 0 & 0 & 1 \end{bmatrix} \begin{bmatrix} a_1 & b_1 & c_1 \\ a_2 & b_2 & c_2 \\ a_3 & b_3 & c_3 \\ a_4 & b_4 & c_4 \\ a_5 & b_5 & c_5 \end{bmatrix} = \begin{bmatrix} a_1 & b_1 & c_1 \\ a_2 + ka_4 & b_2 + kb_4 & c_2 + kc_4 \\ a_3 & b_3 & c_3 \\ a_4 & b_4 & c_4 \\ a_5 & b_5 & c_5 \end{bmatrix}.$$

In the next example the situation is somewhat different from what we considered so far in this section.

Example 2.3.4. Find a matrix P such that

$$P \begin{bmatrix} a_1 & b_1 & c_1 \\ a_2 & b_2 & c_2 \\ a_3 & b_3 & c_3 \end{bmatrix} = \begin{bmatrix} a_1 & b_1 & c_1 \\ a_2 + ja_1 & b_2 + jb_1 & c_2 + jc_1 \\ a_3 + ka_1 & b_3 + kb_1 & c_3 + kc_1 \end{bmatrix}$$

Solution. The matrix on the right is the result of applying two elementary operations: multiply row 1 by j and add to row 2 and then multiply row 1 by k and add to row 3. To represent these two elementary operations we need two elementary matrices:

$$\begin{bmatrix} 1 & 0 & 0 \\ j & 1 & 0 \\ 0 & 0 & 1 \end{bmatrix} \quad \text{and} \quad \begin{bmatrix} 1 & 0 & 0 \\ 0 & 1 & 0 \\ k & 0 & 1 \end{bmatrix}.$$

Thus the matrix P is the product of these two matrices:

$$P = \begin{bmatrix} 1 & 0 & 0 \\ j & 1 & 0 \\ 0 & 0 & 1 \end{bmatrix} \begin{bmatrix} 1 & 0 & 0 \\ 0 & 1 & 0 \\ k & 0 & 1 \end{bmatrix} = \begin{bmatrix} 1 & 0 & 0 \\ j & 1 & 0 \\ k & 0 & 1 \end{bmatrix}.$$

Now we have

$$P \begin{bmatrix} a_1 & b_1 & c_1 \\ a_2 & b_2 & c_2 \\ a_3 & b_3 & c_3 \end{bmatrix} = \begin{bmatrix} 1 & 0 & 0 \\ j & 1 & 0 \\ k & 0 & 1 \end{bmatrix} \begin{bmatrix} a_1 & b_1 & c_1 \\ a_2 & b_2 & c_2 \\ a_3 & b_3 & c_3 \end{bmatrix} = \begin{bmatrix} a_1 & b_1 & c_1 \\ a_2 + ja_1 & b_2 + jb_1 & c_2 + jc_1 \\ a_3 + ka_1 & b_3 + kb_1 & c_3 + kc_1 \end{bmatrix}.$$

Note that the matrix P does not depend on the order of the elementary matrices, that is, we have

$$\begin{bmatrix} 1 & 0 & 0 \\ j & 1 & 0 \\ 0 & 0 & 1 \end{bmatrix} \begin{bmatrix} 1 & 0 & 0 \\ 0 & 1 & 0 \\ k & 0 & 1 \end{bmatrix} = \begin{bmatrix} 1 & 0 & 0 \\ 0 & 1 & 0 \\ k & 0 & 1 \end{bmatrix} \begin{bmatrix} 1 & 0 & 0 \\ j & 1 & 0 \\ 0 & 0 & 1 \end{bmatrix} = \begin{bmatrix} 1 & 0 & 0 \\ j & 1 & 0 \\ k & 0 & 1 \end{bmatrix}.$$

□

The matrix $P = \begin{bmatrix} 1 & 0 & 0 \\ j & 1 & 0 \\ k & 0 & 1 \end{bmatrix}$ in the above example is not an elementary matrix since we need to apply two elementary operations to the 3×3 identity matrix to obtain P. When applying Gaussian elimination to a matrix, we often combine elementary operations that change a single column in the matrix. For example, we write

multiply row 3 by $-\frac{2}{5}$ and then add to row 1

multiply row 3 by $\frac{1}{5}$ and then add to row 2

Note that the combination of these two elementary operations correspond to multiplication by a single matrix, namely, the matrix

$$\begin{bmatrix} 1 & 0 & -\frac{2}{5} \\ 0 & 1 & \frac{1}{5} \\ 0 & 0 & 1 \end{bmatrix}.$$

In the remainder of this chapter we often represent matrices as products of elementary matrices. It is convenient to combine those elementary matrices that correspond to multiplication of the same row and addition to the remaining rows into a single matrix, as in the example above. Since traditionally the name "elementary matrix" is reserved to mean a matrix obtained from an identity matrix by one elementary row operation, we will use the name "simple matrix" to mean a matrix from this larger class of matrices.

Definition 2.3.5. By a *simple matrix* we mean an elementary matrix or a product of elementary matrices that correspond to multiplication of the same row and addition to the remaining rows.

Example 2.3.6. The matrices

$$\begin{bmatrix} 1 & p & 0 \\ 0 & 1 & 0 \\ 0 & q & 1 \end{bmatrix}, \quad \begin{bmatrix} 1 & 0 & 0 & p \\ 0 & 1 & 0 & q \\ 0 & 0 & 1 & r \\ 0 & 0 & 0 & 1 \end{bmatrix}, \quad \begin{bmatrix} 1 & 0 & 0 & 0 & 0 \\ p & 1 & 0 & 0 & 0 \\ q & 0 & 1 & 0 & 0 \\ r & 0 & 0 & 1 & 0 \\ s & 0 & 0 & 0 & 1 \end{bmatrix}$$

are simple matrices for arbitrary real numbers p, q, r, s.

Example 2.3.7. We consider the matrix

$$A = \begin{bmatrix} 2 & 3 & 2 & 5 \\ 1 & 2 & -1 & 3 \\ 1 & 1 & 3 & 2 \end{bmatrix}.$$

Find a matrix P such that the product PA is the reduced row echelon form of the matrix A. Express the matrix P as a product of simple matrices.

Solution. We have

$$\begin{bmatrix} 0 & 1 & 0 \\ 1 & 0 & 0 \\ 0 & 0 & 1 \end{bmatrix} \begin{bmatrix} 2 & 3 & 2 & 5 \\ 1 & 2 & -1 & 3 \\ 1 & 1 & 3 & 2 \end{bmatrix} = \begin{bmatrix} 1 & 2 & -1 & 3 \\ 2 & 3 & 2 & 5 \\ 1 & 1 & 3 & 2 \end{bmatrix},$$

$$\begin{bmatrix} 1 & 0 & 0 \\ -2 & 1 & 0 \\ -1 & 0 & 1 \end{bmatrix} \begin{bmatrix} 1 & 2 & -1 & 3 \\ 2 & 3 & 2 & 5 \\ 1 & 1 & 3 & 2 \end{bmatrix} = \begin{bmatrix} 1 & 2 & -1 & 3 \\ 0 & -1 & 4 & -1 \\ 0 & -1 & 4 & -1 \end{bmatrix},$$

$$\begin{bmatrix} 1 & 0 & 0 \\ 0 & -1 & 0 \\ 0 & 0 & 1 \end{bmatrix} \begin{bmatrix} 1 & 2 & -1 & 3 \\ 0 & -1 & 4 & -1 \\ 0 & -1 & 4 & -1 \end{bmatrix} = \begin{bmatrix} 1 & 2 & -1 & 3 \\ 0 & 1 & -4 & 1 \\ 0 & -1 & 4 & -1 \end{bmatrix},$$

and finally

$$\begin{bmatrix} 1 & -2 & 0 \\ 0 & 1 & 0 \\ 0 & 1 & 1 \end{bmatrix} \begin{bmatrix} 1 & 2 & -1 & 3 \\ 0 & 1 & -4 & 1 \\ 0 & -1 & 4 & -1 \end{bmatrix} = \begin{bmatrix} 1 & 0 & 7 & 1 \\ 0 & 1 & -4 & 1 \\ 0 & 0 & 0 & 0 \end{bmatrix}.$$

This means that we can take

$$P = \begin{bmatrix} 1 & -2 & 0 \\ 0 & 1 & 0 \\ 0 & 1 & 1 \end{bmatrix} \begin{bmatrix} 1 & 0 & 0 \\ 0 & -1 & 0 \\ 0 & 0 & 1 \end{bmatrix} \begin{bmatrix} 1 & 0 & 0 \\ -2 & 1 & 0 \\ -1 & 0 & 1 \end{bmatrix} \begin{bmatrix} 0 & 1 & 0 \\ 1 & 0 & 0 \\ 0 & 0 & 1 \end{bmatrix}.$$

$$= \begin{bmatrix} 1 & -2 & 0 \\ 0 & 1 & 0 \\ 0 & 1 & 1 \end{bmatrix} \begin{bmatrix} 1 & 0 & 0 \\ 0 & -1 & 0 \\ 0 & 0 & 1 \end{bmatrix} \begin{bmatrix} 0 & 1 & 0 \\ 1 & -2 & 0 \\ 0 & -1 & 1 \end{bmatrix}$$

$$= \begin{bmatrix} 1 & -2 & 0 \\ 0 & 1 & 0 \\ 0 & 1 & 1 \end{bmatrix} \begin{bmatrix} 0 & 1 & 0 \\ -1 & 2 & 0 \\ 0 & -1 & 1 \end{bmatrix}$$

$$= \begin{bmatrix} 2 & -3 & 0 \\ -1 & 2 & 0 \\ -1 & 1 & 1 \end{bmatrix}.$$

It is easy to verify that

$$PA = \begin{bmatrix} 2 & -3 & 0 \\ -1 & 2 & 0 \\ -1 & 1 & 1 \end{bmatrix} \begin{bmatrix} 2 & 3 & 2 & 5 \\ 1 & 2 & -1 & 3 \\ 1 & 1 & 3 & 2 \end{bmatrix} = \begin{bmatrix} 1 & 0 & 7 & 1 \\ 0 & 1 & -4 & 1 \\ 0 & 0 & 0 & 0 \end{bmatrix}.$$

□

Invertible matrices

Definition 2.3.8. An $n \times n$ matrix A is **invertible** if there is a matrix B such that
$$AB = BA = I_n,$$
where I_n is the $n \times n$ identity matrix.

Theorem 2.3.9. *If an $n \times n$ matrix A is invertible, then there is a unique matrix B such that $AB = BA = I_n$.*

Proof. We need to show that, if
$$AB = BA = I_n \quad \text{and} \quad AC = CA = I_n,$$
then $B = C$. Indeed, if $BA = I_n$ and $AC = I_n$, then
$$B = BI_n = B(AC) = (BA)C = I_nC = C.$$

□

Definition 2.3.10. Let A be an invertible $n \times n$ matrix. The unique matrix B such that $AB = BA = I_n$ is called the **inverse** of the matrix A and is denoted by A^{-1}.

Now we consider the problem of finding the inverse of an invertible matrix. We start by noting that, since elementary row operations are reversible, elementary matrices are invertible. Moreover, it is easy to find the inverse of an elementary matrix or, more generally, a simple matrix.

Example 2.3.11. Assuming $k \neq 0$, find the inverses of the elementary matrices

$$\begin{bmatrix} k & 0 & 0 \\ 0 & 1 & 0 \\ 0 & 0 & 1 \end{bmatrix}, \quad \begin{bmatrix} 1 & 0 & 0 \\ 0 & k & 0 \\ 0 & 0 & 1 \end{bmatrix}, \quad \text{and} \quad \begin{bmatrix} 1 & 0 & 0 \\ 0 & 1 & 0 \\ 0 & 0 & k \end{bmatrix}.$$

Solution.

$$\begin{bmatrix} k & 0 & 0 \\ 0 & 1 & 0 \\ 0 & 0 & 1 \end{bmatrix}^{-1} = \begin{bmatrix} \frac{1}{k} & 0 & 0 \\ 0 & 1 & 0 \\ 0 & 0 & 1 \end{bmatrix},$$

$$\begin{bmatrix} 1 & 0 & 0 \\ 0 & k & 0 \\ 0 & 0 & 1 \end{bmatrix}^{-1} = \begin{bmatrix} 1 & 0 & 0 \\ 0 & \frac{1}{k} & 0 \\ 0 & 0 & 1 \end{bmatrix},$$

and

$$\begin{bmatrix} 1 & 0 & 0 \\ 0 & 1 & 0 \\ 0 & 0 & k \end{bmatrix}^{-1} = \begin{bmatrix} 1 & 0 & 0 \\ 0 & 1 & 0 \\ 0 & 0 & \frac{1}{k} \end{bmatrix}.$$

□

Example 2.3.12. Find the inverses of the elementary matrices

$$\begin{bmatrix} 0 & 1 & 0 \\ 1 & 0 & 0 \\ 0 & 0 & 1 \end{bmatrix}, \quad \begin{bmatrix} 1 & 0 & 0 \\ 0 & 0 & 1 \\ 0 & 1 & 0 \end{bmatrix} \quad \text{and} \quad \begin{bmatrix} 0 & 0 & 1 \\ 0 & 1 & 0 \\ 1 & 0 & 0 \end{bmatrix}.$$

Solution.

$$\begin{bmatrix} 0 & 1 & 0 \\ 1 & 0 & 0 \\ 0 & 0 & 1 \end{bmatrix}^{-1} = \begin{bmatrix} 0 & 1 & 0 \\ 1 & 0 & 0 \\ 0 & 0 & 1 \end{bmatrix},$$

$$\begin{bmatrix} 1 & 0 & 0 \\ 0 & 0 & 1 \\ 0 & 1 & 0 \end{bmatrix}^{-1} = \begin{bmatrix} 1 & 0 & 0 \\ 0 & 0 & 1 \\ 0 & 1 & 0 \end{bmatrix},$$

and

$$\begin{bmatrix} 0 & 0 & 1 \\ 0 & 1 & 0 \\ 1 & 0 & 0 \end{bmatrix}^{-1} = \begin{bmatrix} 0 & 0 & 1 \\ 0 & 1 & 0 \\ 1 & 0 & 0 \end{bmatrix}.$$

Note that for these elementary matrices the inverse is the same as the original matrix.   □

Example 2.3.13. Find the inverse of the simple matrices

$$\begin{bmatrix} 1 & 0 & 0 \\ a & 1 & 0 \\ b & 0 & 1 \end{bmatrix}, \quad \begin{bmatrix} 1 & c & 0 \\ 0 & 1 & 0 \\ 0 & d & 1 \end{bmatrix} \quad \text{and} \quad \begin{bmatrix} 1 & 0 & e \\ 0 & 1 & f \\ 0 & 0 & 1 \end{bmatrix}.$$

Solution.

$$\begin{bmatrix} 1 & 0 & 0 \\ a & 1 & 0 \\ b & 0 & 1 \end{bmatrix}^{-1} = \begin{bmatrix} 1 & 0 & 0 \\ -a & 1 & 0 \\ -b & 0 & 1 \end{bmatrix},$$

$$\begin{bmatrix} 1 & c & 0 \\ 0 & 1 & 0 \\ 0 & d & 1 \end{bmatrix}^{-1} = \begin{bmatrix} 1 & -c & 0 \\ 0 & 1 & 0 \\ 0 & -d & 1 \end{bmatrix},$$

and

$$\begin{bmatrix} 1 & 0 & e \\ 0 & 1 & f \\ 0 & 0 & 1 \end{bmatrix}^{-1} = \begin{bmatrix} 1 & 0 & -e \\ 0 & 1 & -f \\ 0 & 0 & 1 \end{bmatrix}.$$

□

The inverses found in the above three examples easily generalize to larger matrices.

Example 2.3.14.

$$\begin{bmatrix} 1 & a & 0 & 0 & 0 \\ 0 & 1 & 0 & 0 & 0 \\ 0 & b & 1 & 0 & 0 \\ 0 & c & 0 & 1 & 0 \\ 0 & d & 0 & 0 & 1 \end{bmatrix}^{-1} = \begin{bmatrix} 1 & -a & 0 & 0 & 0 \\ 0 & 1 & 0 & 0 & 0 \\ 0 & -b & 1 & 0 & 0 \\ 0 & -c & 0 & 1 & 0 \\ 0 & -d & 0 & 0 & 1 \end{bmatrix}.$$

In the process of determining the inverse of a matrix the following theorem is often useful.

> **Theorem 2.3.15.** *If $A_1, \ldots, A_m$ are invertible $n \times n$ matrices, then the product matrix $A_1 \cdots A_m$ is invertible and we have*
>
> $$(A_1 \cdots A_m)^{-1} = A_m^{-1} \cdots A_1^{-1}.$$

Note that the order of the matrices in the equality $(A_1 \cdots A_m)^{-1} = A_m^{-1} \cdots A_1^{-1}$ is different on the left-hand side and the right-hand side.

Proof for $m = 3$. We have

$$A_1 A_2 A_3 A_3^{-1} A_2^{-1} A_1^{-1} = A_1 A_2 A_2^{-1} A_1^{-1} = A_1 A_1^{-1} = I_n$$

and

$$A_3^{-1} A_2^{-1} A_1^{-1} A_1 A_2 A_3 = A_3^{-1} A_2^{-1} A_2 A_3 = A_3^{-1} A_3 = I_n.$$

It should be clear why this argument easily generalizes to any number of matrices. □

Example 2.3.16. Calculate the matrix

$$A = \begin{bmatrix} 1 & 0 & 0 & 0 \\ 0 & 0 & 1 & 0 \\ 0 & 1 & 0 & 0 \\ 0 & 0 & 0 & 1 \end{bmatrix} \begin{bmatrix} 1 & 0 & 3 & 0 \\ 0 & 1 & 2 & 0 \\ 0 & 0 & 1 & 0 \\ 0 & 0 & 7 & 1 \end{bmatrix} \begin{bmatrix} 1 & 0 & 0 & 0 \\ 0 & \frac{1}{5} & 0 & 0 \\ 0 & 0 & 1 & 0 \\ 0 & 0 & 0 & 1 \end{bmatrix}$$

and then write its inverse as a product of three simple matrices. Use the result to calculate the inverse of the matrix A.

Solution. First we find that

$$A = \begin{bmatrix} 1 & 0 & 3 & 0 \\ 0 & 0 & 1 & 0 \\ 0 & \frac{1}{5} & 2 & 0 \\ 0 & 0 & 7 & 1 \end{bmatrix}.$$

Now

$$A^{-1} = \left(\begin{bmatrix} 1 & 0 & 0 & 0 \\ 0 & 0 & 1 & 0 \\ 0 & 1 & 0 & 0 \\ 0 & 0 & 0 & 1 \end{bmatrix} \begin{bmatrix} 1 & 0 & 3 & 0 \\ 0 & 1 & 2 & 0 \\ 0 & 0 & 1 & 0 \\ 0 & 0 & 7 & 1 \end{bmatrix} \begin{bmatrix} 1 & 0 & 0 & 0 \\ 0 & \frac{1}{5} & 0 & 0 \\ 0 & 0 & 1 & 0 \\ 0 & 0 & 0 & 1 \end{bmatrix} \right)^{-1}$$

$$= \begin{bmatrix} 1 & 0 & 0 & 0 \\ 0 & \frac{1}{5} & 0 & 0 \\ 0 & 0 & 1 & 0 \\ 0 & 0 & 0 & 1 \end{bmatrix}^{-1} \begin{bmatrix} 1 & 0 & 3 & 0 \\ 0 & 1 & 2 & 0 \\ 0 & 0 & 1 & 0 \\ 0 & 0 & 7 & 1 \end{bmatrix}^{-1} \begin{bmatrix} 1 & 0 & 0 & 0 \\ 0 & 0 & 1 & 0 \\ 0 & 1 & 0 & 0 \\ 0 & 0 & 0 & 1 \end{bmatrix}^{-1}$$

$$
= \begin{bmatrix} 1 & 0 & 0 & 0 \\ 0 & 5 & 0 & 0 \\ 0 & 0 & 1 & 0 \\ 0 & 0 & 0 & 1 \end{bmatrix} \begin{bmatrix} 1 & 0 & -3 & 0 \\ 0 & 1 & -2 & 0 \\ 0 & 0 & 1 & 0 \\ 0 & 0 & -7 & 1 \end{bmatrix} \begin{bmatrix} 1 & 0 & 0 & 0 \\ 0 & 0 & 1 & 0 \\ 0 & 1 & 0 & 0 \\ 0 & 0 & 0 & 1 \end{bmatrix}
$$

$$
= \begin{bmatrix} 1 & 0 & 0 & 0 \\ 0 & 5 & 0 & 0 \\ 0 & 0 & 1 & 0 \\ 0 & 0 & 0 & 1 \end{bmatrix} \begin{bmatrix} 1 & -3 & 0 & 0 \\ 0 & -2 & 1 & 0 \\ 0 & 1 & 0 & 0 \\ 0 & -7 & 0 & 1 \end{bmatrix}
$$

$$
= \begin{bmatrix} 1 & -3 & 0 & 0 \\ 0 & -10 & 5 & 0 \\ 0 & 1 & 0 & 0 \\ 0 & -7 & 0 & 1 \end{bmatrix}.
$$

It is easy to verify that

$$
\begin{bmatrix} 1 & 0 & 3 & 0 \\ 0 & 0 & 1 & 0 \\ 0 & \frac{1}{5} & 2 & 0 \\ 0 & 0 & 7 & 1 \end{bmatrix} \begin{bmatrix} 1 & -3 & 0 & 0 \\ 0 & -10 & 5 & 0 \\ 0 & 1 & 0 & 0 \\ 0 & -7 & 0 & 1 \end{bmatrix} = \begin{bmatrix} 1 & 0 & 0 & 0 \\ 0 & 1 & 0 & 0 \\ 0 & 0 & 1 & 0 \\ 0 & 0 & 0 & 1 \end{bmatrix}
$$

and

$$
\begin{bmatrix} 1 & -3 & 0 & 0 \\ 0 & -10 & 5 & 0 \\ 0 & 1 & 0 & 0 \\ 0 & -7 & 0 & 1 \end{bmatrix} \begin{bmatrix} 1 & 0 & 3 & 0 \\ 0 & 0 & 1 & 0 \\ 0 & \frac{1}{5} & 2 & 0 \\ 0 & 0 & 7 & 1 \end{bmatrix} = \begin{bmatrix} 1 & 0 & 0 & 0 \\ 0 & 1 & 0 & 0 \\ 0 & 0 & 1 & 0 \\ 0 & 0 & 0 & 1 \end{bmatrix}.
$$

□

A characterization of invertible matrices

Now we turn our attention to the problem of determining whether a given matrix is invertible. We start with the following important theorem.

Theorem 2.3.17. *Let A be an $n \times n$ matrix with columns $c_1, \ldots, c_n$, that is, $A = [c_1 \ldots c_n]$. The following conditions are equivalent:*

(a) *The matrix A is invertible;*

(b) *The equation*
$$
x_1 c_1 + \cdots + x_n c_n = 0
$$
has only the trivial solution, that is, the solution $x_1 = \cdots = x_n = 0$;

(c) *The reduced row echelon form of the matrix A is the identity matrix I_n;*

(d) *The matrix A can be written as a product of elementary matrices.*

Proof for n = 4. First we assume that the matrix

$$A = \begin{bmatrix} \mathbf{c}_1 & \mathbf{c}_2 & \mathbf{c}_3 & \mathbf{c}_4 \end{bmatrix} = \begin{bmatrix} a_1 & b_1 & c_1 & d_1 \\ a_2 & b_2 & c_2 & d_2 \\ a_3 & b_3 & c_3 & d_3 \\ a_4 & b_4 & c_4 & d_4 \end{bmatrix}$$

is invertible. The equation

$$x_1 \mathbf{c}_1 + x_2 \mathbf{c}_2 + x_3 \mathbf{c}_3 + x_4 \mathbf{c}_4 = \mathbf{0}$$

can be written as a matrix equation

$$\begin{bmatrix} a_1 & b_1 & c_1 & d_1 \\ a_2 & b_2 & c_2 & d_2 \\ a_3 & b_3 & c_3 & d_3 \\ a_4 & b_4 & c_4 & d_4 \end{bmatrix} \begin{bmatrix} x_1 \\ x_2 \\ x_3 \\ x_4 \end{bmatrix} = \begin{bmatrix} 0 \\ 0 \\ 0 \\ 0 \end{bmatrix}.$$

Since the matrix A is invertible, we have

$$\begin{bmatrix} x_1 \\ x_2 \\ x_3 \\ x_4 \end{bmatrix} = \begin{bmatrix} 1 & 0 & 0 & 0 \\ 0 & 1 & 0 & 0 \\ 0 & 0 & 1 & 0 \\ 0 & 0 & 0 & 1 \end{bmatrix} \begin{bmatrix} x_1 \\ x_2 \\ x_3 \\ x_4 \end{bmatrix}$$

$$= \begin{bmatrix} a_1 & b_1 & c_1 & d_1 \\ a_2 & b_2 & c_2 & d_2 \\ a_3 & b_3 & c_3 & d_3 \\ a_4 & b_4 & c_4 & d_4 \end{bmatrix}^{-1} \begin{bmatrix} a_1 & b_1 & c_1 & d_1 \\ a_2 & b_2 & c_2 & d_2 \\ a_3 & b_3 & c_3 & d_3 \\ a_4 & b_4 & c_4 & d_4 \end{bmatrix} \begin{bmatrix} x_1 \\ x_2 \\ x_3 \\ x_4 \end{bmatrix}$$

$$= \begin{bmatrix} a_1 & b_1 & c_1 & d_1 \\ a_2 & b_2 & c_2 & d_2 \\ a_3 & b_3 & c_3 & d_3 \\ a_4 & b_4 & c_4 & d_4 \end{bmatrix}^{-1} \begin{bmatrix} 0 \\ 0 \\ 0 \\ 0 \end{bmatrix} = \begin{bmatrix} 0 \\ 0 \\ 0 \\ 0 \end{bmatrix}.$$

This shows that (a) implies (b).

Next we assume that the only solution of the equation $x_1 \mathbf{c}_1 + x_2 \mathbf{c}_2 + x_3 \mathbf{c}_3 + x_4 \mathbf{c}_4 = \mathbf{0}$ is $x_1 = x_2 = x_3 = x_4 = 0$. This means that the only solution of the equation

$$\begin{bmatrix} a_1 & b_1 & c_1 & d_1 \\ a_2 & b_2 & c_2 & d_2 \\ a_3 & b_3 & c_3 & d_3 \\ a_4 & b_4 & c_4 & d_4 \end{bmatrix} \begin{bmatrix} x_1 \\ x_2 \\ x_3 \\ x_4 \end{bmatrix} = \begin{bmatrix} 0 \\ 0 \\ 0 \\ 0 \end{bmatrix}$$

is

$$\begin{bmatrix} x_1 \\ x_2 \\ x_3 \\ x_4 \end{bmatrix} = \begin{bmatrix} 0 \\ 0 \\ 0 \\ 0 \end{bmatrix}.$$

We note that the same is true for any matrix obtained from the matrix A by elementary row operations.

If $a_1 = a_2 = a_3 = a_4 = 0$, then

$$
\begin{bmatrix}
a_1 & b_1 & c_1 & d_1 \\
a_2 & b_2 & c_2 & d_2 \\
a_3 & b_3 & c_3 & d_3 \\
a_4 & b_4 & c_4 & d_4
\end{bmatrix}
\begin{bmatrix} 1 \\ 0 \\ 0 \\ 0 \end{bmatrix}
=
\begin{bmatrix}
0 & b_1 & c_1 & d_1 \\
0 & b_2 & c_2 & d_2 \\
0 & b_3 & c_3 & d_3 \\
0 & b_4 & c_4 & d_4
\end{bmatrix}
\begin{bmatrix} 1 \\ 0 \\ 0 \\ 0 \end{bmatrix}
=
\begin{bmatrix} 0 \\ 0 \\ 0 \\ 0 \end{bmatrix}
$$

contrary to our assumption. So at least one of the numbers a_1, a_2, a_3, or a_4 must be different from 0. Without loss of generality, we can assume that $a_1 \neq 0$. Now we multiply the first row of the matrix

$$
\begin{bmatrix}
a_1 & b_1 & c_1 & d_1 \\
a_2 & b_2 & c_2 & d_2 \\
a_3 & b_3 & c_3 & d_3 \\
a_4 & b_4 & c_4 & d_4
\end{bmatrix}
$$

by $\frac{1}{a_1}$ and get

$$
\begin{bmatrix}
1 & \frac{b_1}{a_1} & \frac{c_1}{a_1} & \frac{d_1}{a_1} \\
a_2 & b_2 & c_2 & d_2 \\
a_3 & b_3 & c_3 & d_3 \\
a_4 & b_4 & c_4 & d_4
\end{bmatrix}
$$

and then, by row replacement, we obtain

$$
\begin{bmatrix}
1 & \frac{b_1}{a_1} & \frac{c_1}{a_1} & \frac{d_1}{a_1} \\
0 & b_2 - \frac{a_2}{a_1}b_1 & c_2 - \frac{a_2}{a_1}c_1 & d_2 - \frac{a_2}{a_1}d_1 \\
0 & b_3 - \frac{a_3}{a_1}b_1 & c_3 - \frac{a_3}{a_1}c_1 & d_3 - \frac{a_3}{a_1}d_1 \\
0 & b_4 - \frac{a_4}{a_1}b_1 & c_4 - \frac{a_4}{a_1}c_1 & d_4 - \frac{a_4}{a_1}d_1
\end{bmatrix}
$$

We will write this last matrix as

$$
\begin{bmatrix}
1 & b_1' & c_1' & d_1' \\
0 & b_2' & c_2' & d_2' \\
0 & b_3' & c_3' & d_3' \\
0 & b_4' & c_4' & d_4'
\end{bmatrix}.
$$

If $b_2' = b_3' = b_4' = 0$, then

$$
\begin{bmatrix}
a_1 & b_1 & c_1 & d_1 \\
a_2 & b_2 & c_2 & d_2 \\
a_3 & b_3 & c_3 & d_3 \\
a_4 & b_4 & c_4 & d_4
\end{bmatrix}
\begin{bmatrix} -b_1' \\ 1 \\ 0 \\ 0 \end{bmatrix}
=
\begin{bmatrix}
1 & b_1' & c_1' & d_1' \\
0 & 0 & c_2' & d_2' \\
0 & 0 & c_3' & d_3' \\
0 & 0 & c_4' & d_4'
\end{bmatrix}
\begin{bmatrix} -b_1' \\ 1 \\ 0 \\ 0 \end{bmatrix}
=
\begin{bmatrix} 0 \\ 0 \\ 0 \\ 0 \end{bmatrix}
$$

contrary to our assumption. So at least one of the numbers b_2', b_3', or b_4' must be different from 0. Without loss of generality, we can assume that $b_2' \neq 0$. Now we continue as before with the matrix

$$\begin{bmatrix} 1 & b_1' & c_1' & d_1' \\ 0 & b_2' & c_2' & d_2' \\ 0 & b_3' & c_3' & d_3' \\ 0 & b_4' & c_4' & d_4' \end{bmatrix}$$

and by multiplying the second row by $\frac{1}{b_2'}$ we get

$$\begin{bmatrix} 1 & b_1' & c_1' & d_1' \\ 0 & 1 & \frac{c_2'}{b_2'} & \frac{d_2'}{b_2'} \\ 0 & b_3' & c_3' & d_3' \\ 0 & b_4' & c_4' & d_4' \end{bmatrix}$$

and then by row replacement we get

$$\begin{bmatrix} 1 & 0 & c_1' - \frac{b_1'}{b_2'}c_2' & d_1' - \frac{b_1'}{b_2'}d_2' \\ 0 & 1 & \frac{c_2'}{b_2'} & \frac{d_2'}{b_2'} \\ 0 & 0 & c_3' - \frac{b_3'}{b_2'}c_2' & d_3' - \frac{b_3'}{b_2'}d_2' \\ 0 & 0 & c_4' - \frac{b_4'}{b_2'}c_2' & d_4' - \frac{b_4'}{b_2'}d_2' \end{bmatrix}.$$

We will write this last matrix as

$$\begin{bmatrix} 1 & 0 & c_1'' & d_1'' \\ 0 & 1 & c_2'' & d_2'' \\ 0 & 0 & c_3'' & d_3'' \\ 0 & 0 & c_4'' & d_4'' \end{bmatrix}.$$

If $c_3'' = c_4'' = 0$, then

$$\begin{bmatrix} a_1 & b_1 & c_1 & d_1 \\ a_2 & b_2 & c_2 & d_2 \\ a_3 & b_3 & c_3 & d_3 \\ a_4 & b_4 & c_4 & d_4 \end{bmatrix} \begin{bmatrix} -c_1'' \\ -c_2'' \\ 1 \\ 0 \end{bmatrix} = \begin{bmatrix} 1 & 0 & c_1'' & d_1'' \\ 0 & 1 & c_2'' & d_2'' \\ 0 & 0 & 0 & d_3'' \\ 0 & 0 & 0 & d_4'' \end{bmatrix} \begin{bmatrix} -c_1'' \\ -c_2'' \\ 1 \\ 0 \end{bmatrix} = \begin{bmatrix} 0 \\ 0 \\ 0 \\ 0 \end{bmatrix}$$

contrary to our assumption. So at least one of the numbers c_3'' or c_4'' must be different from 0. Without loss of generality, we can assume that $c_3'' \neq 0$. Now we continue as before with the matrix

$$\begin{bmatrix} 1 & 0 & c_1'' & d_1'' \\ 0 & 1 & c_2'' & d_2'' \\ 0 & 0 & c_3'' & d_3'' \\ 0 & 0 & c_4'' & d_4'' \end{bmatrix}$$

and by multiplying the third row by $\frac{1}{c_3''}$ we get

$$
\begin{bmatrix}
1 & 0 & c_1'' & d_1'' \\
0 & 1 & c_2'' & d_2'' \\
0 & 0 & 1 & \frac{d_3''}{c_3''} \\
0 & 0 & c_4'' & d_4''
\end{bmatrix}
$$

and then by row replacement

$$
\begin{bmatrix}
1 & 0 & 0 & d_1'' - \frac{c_1'' d_3''}{c_3''} \\
0 & 1 & 0 & d_2'' - \frac{c_2'' d_3''}{c_3''} \\
0 & 0 & 1 & \frac{d_3''}{c_3''} \\
0 & 0 & 0 & d_4'' - \frac{c_4'' d_3''}{c_3''}
\end{bmatrix}.
$$

We will write this last matrix as

$$
\begin{bmatrix}
1 & 0 & 0 & d_1''' \\
0 & 1 & 0 & d_2''' \\
0 & 0 & 1 & d_3''' \\
0 & 0 & 0 & d_4'''
\end{bmatrix}.
$$

If $d_4''' = 0$, then

$$
\begin{bmatrix}
a_1 & b_1 & c_1 & d_1 \\
a_2 & b_2 & c_2 & d_2 \\
a_3 & b_3 & c_3 & d_3 \\
a_4 & b_4 & c_4 & d_4
\end{bmatrix}
\begin{bmatrix}
-d_1''' \\
-d_2''' \\
-d_3''' \\
1
\end{bmatrix}
=
\begin{bmatrix}
1 & 0 & 0 & d_1''' \\
0 & 1 & 0 & d_2''' \\
0 & 0 & 1 & d_3''' \\
0 & 0 & 0 & 0
\end{bmatrix}
\begin{bmatrix}
-d_1''' \\
-d_2''' \\
-d_3''' \\
1
\end{bmatrix}
=
\begin{bmatrix}
0 \\
0 \\
0 \\
0
\end{bmatrix}
$$

contrary to our assumption. So we must have $d_4''' \neq 0$.

Now we multiply the last row of the matrix

$$
\begin{bmatrix}
1 & 0 & 0 & d_1''' \\
0 & 1 & 0 & d_2''' \\
0 & 0 & 1 & d_3''' \\
0 & 0 & 0 & d_4'''
\end{bmatrix}
$$

by $\frac{1}{d_4'''}$ and get

$$
\begin{bmatrix}
1 & 0 & 0 & d_1''' \\
0 & 1 & 0 & d_2''' \\
0 & 0 & 1 & d_3''' \\
0 & 0 & 0 & 1
\end{bmatrix}
$$

and then by row replacement

$$\begin{bmatrix} 1 & 0 & 0 & 0 \\ 0 & 1 & 0 & 0 \\ 0 & 0 & 1 & 0 \\ 0 & 0 & 0 & 1 \end{bmatrix}$$

This long argument shows that if $x_1 = x_2 = x_3 = x_4 = 0$ is the only solution of the equation $x_1 \mathbf{c}_1 + x_2 \mathbf{c}_2 + x_3 \mathbf{c}_3 + x_4 \mathbf{c}_4 = \mathbf{0}$, then the reduced row echelon form of the matrix A is the identity matrix I_4. Therefore (b) implies (c).

Now we assume that the reduced row echelon form of the matrix A is the identity matrix I_4. Each step in the Gaussian elimination process corresponds to an elementary matrix E_j. When the matrix A is multiplied by the product $E = E_m E_{m-1} \dots E_2 E_1$ of all those elementary matrices, then we obtain I_4, that is, we have $EA = I_4$. Consequently,

$$A = E^{-1} = (E_m E_{m-1} \dots E_2 E_1)^{-1} = E_1^{-1} E_2^{-1} \dots E_{m-1}^{-1} E_m^{-1}.$$

Therefore (c) implies (d) because the inverse of a elementary matrix is an elementary matrix.

Finally, if the matrix A can be written as a product of elementary matrices, then it is invertible, because elementary matrices are invertible and the product of invertible matrices is an invertible matrix. Therefore (d) implies (a).

All essential ingredients of the general proof are present in the considered case and generalizing the argument to larger matrices is easy, but tedious. □

Example 2.3.18. Show that the matrix $A = \begin{bmatrix} 2 & 1 & 2 \\ 1 & 2 & 1 \\ 1 & 2 & 2 \end{bmatrix}$ is invertible. Write the matrix and its inverse as products of simple matrices and calculate A^{-1}.

Solution. Since

$$\begin{bmatrix} 0 & 1 & 0 \\ 1 & 0 & 0 \\ 0 & 0 & 1 \end{bmatrix} \begin{bmatrix} 2 & 1 & 2 \\ 1 & 2 & 1 \\ 1 & 2 & 2 \end{bmatrix} = \begin{bmatrix} 1 & 2 & 1 \\ 2 & 1 & 2 \\ 1 & 2 & 2 \end{bmatrix},$$

$$\begin{bmatrix} 1 & 0 & 0 \\ -2 & 1 & 0 \\ -1 & 0 & 1 \end{bmatrix} \begin{bmatrix} 1 & 2 & 1 \\ 2 & 1 & 2 \\ 1 & 2 & 2 \end{bmatrix} = \begin{bmatrix} 1 & 2 & 1 \\ 0 & -3 & 0 \\ 0 & 0 & 1 \end{bmatrix},$$

$$\begin{bmatrix} 1 & 0 & 0 \\ 0 & -\frac{1}{3} & 0 \\ 0 & 0 & 1 \end{bmatrix} \begin{bmatrix} 1 & 2 & 1 \\ 0 & -3 & 0 \\ 0 & 0 & 1 \end{bmatrix} = \begin{bmatrix} 1 & 2 & 1 \\ 0 & 1 & 0 \\ 0 & 0 & 1 \end{bmatrix},$$

$$\begin{bmatrix} 1 & -2 & 0 \\ 0 & 1 & 0 \\ 0 & 0 & 1 \end{bmatrix} \begin{bmatrix} 1 & 2 & 1 \\ 0 & 1 & 0 \\ 0 & 0 & 1 \end{bmatrix} = \begin{bmatrix} 1 & 0 & 1 \\ 0 & 1 & 0 \\ 0 & 0 & 1 \end{bmatrix},$$

and finally

$$\begin{bmatrix} 1 & 0 & -1 \\ 0 & 1 & 0 \\ 0 & 0 & 1 \end{bmatrix} \begin{bmatrix} 1 & 0 & 1 \\ 0 & 1 & 0 \\ 0 & 0 & 1 \end{bmatrix} = \begin{bmatrix} 1 & 0 & 0 \\ 0 & 1 & 0 \\ 0 & 0 & 1 \end{bmatrix},$$

we have

$$
\begin{aligned}
A^{-1} &= \begin{bmatrix} 1 & 0 & -1 \\ 0 & 1 & 0 \\ 0 & 0 & 1 \end{bmatrix} \begin{bmatrix} 1 & -2 & 0 \\ 0 & 1 & 0 \\ 0 & 0 & 1 \end{bmatrix} \begin{bmatrix} 1 & 0 & 0 \\ 0 & -\frac{1}{3} & 0 \\ 0 & 0 & 1 \end{bmatrix} \begin{bmatrix} 1 & 0 & 0 \\ -2 & 1 & 0 \\ -1 & 0 & 1 \end{bmatrix} \begin{bmatrix} 0 & 1 & 0 \\ 1 & 0 & 0 \\ 0 & 0 & 1 \end{bmatrix} \\
&= \begin{bmatrix} 1 & 0 & -1 \\ 0 & 1 & 0 \\ 0 & 0 & 1 \end{bmatrix} \begin{bmatrix} 1 & -2 & 0 \\ 0 & 1 & 0 \\ 0 & 0 & 1 \end{bmatrix} \begin{bmatrix} 1 & 0 & 0 \\ 0 & -\frac{1}{3} & 0 \\ 0 & 0 & 1 \end{bmatrix} \begin{bmatrix} 0 & 1 & 0 \\ 1 & -2 & 0 \\ 0 & -1 & 1 \end{bmatrix} \\
&= \begin{bmatrix} 1 & 0 & -1 \\ 0 & 1 & 0 \\ 0 & 0 & 1 \end{bmatrix} \begin{bmatrix} 1 & -2 & 0 \\ 0 & 1 & 0 \\ 0 & 0 & 1 \end{bmatrix} \begin{bmatrix} 0 & 1 & 0 \\ -\frac{1}{3} & \frac{2}{3} & 0 \\ 0 & -1 & 1 \end{bmatrix} \\
&= \begin{bmatrix} 1 & 0 & -1 \\ 0 & 1 & 0 \\ 0 & 0 & 1 \end{bmatrix} \begin{bmatrix} \frac{2}{3} & -\frac{1}{3} & 0 \\ -\frac{1}{3} & \frac{2}{3} & 0 \\ 0 & -1 & 1 \end{bmatrix} \\
&= \begin{bmatrix} \frac{2}{3} & \frac{2}{3} & -1 \\ -\frac{1}{3} & \frac{2}{3} & 0 \\ 0 & -1 & 1 \end{bmatrix}.
\end{aligned}
$$

Now we use the representation of A^{-1} as a product of simple matrices to find a representation of A as a product of simple matrices.

$$
\begin{aligned}
A &= \left(A^{-1} \right)^{-1} \\
&= \left(\begin{bmatrix} 1 & 0 & -1 \\ 0 & 1 & 0 \\ 0 & 0 & 1 \end{bmatrix} \begin{bmatrix} 1 & -2 & 0 \\ 0 & 1 & 0 \\ 0 & 0 & 1 \end{bmatrix} \begin{bmatrix} 1 & 0 & 0 \\ 0 & -\frac{1}{3} & 0 \\ 0 & 0 & 1 \end{bmatrix} \begin{bmatrix} 1 & 0 & 0 \\ -2 & 1 & 0 \\ -1 & 0 & 1 \end{bmatrix} \begin{bmatrix} 0 & 1 & 0 \\ 1 & 0 & 0 \\ 0 & 0 & 1 \end{bmatrix} \right)^{-1} \\
&= \begin{bmatrix} 0 & 1 & 0 \\ 1 & 0 & 0 \\ 0 & 0 & 1 \end{bmatrix}^{-1} \begin{bmatrix} 1 & 0 & 0 \\ -2 & 1 & 0 \\ -1 & 0 & 1 \end{bmatrix}^{-1} \begin{bmatrix} 1 & 0 & 0 \\ 0 & -\frac{1}{3} & 0 \\ 0 & 0 & 1 \end{bmatrix}^{-1} \begin{bmatrix} 1 & -2 & 0 \\ 0 & 1 & 0 \\ 0 & 0 & 1 \end{bmatrix}^{-1} \begin{bmatrix} 1 & 0 & -1 \\ 0 & 1 & 0 \\ 0 & 0 & 1 \end{bmatrix}^{-1} \\
&= \begin{bmatrix} 0 & 1 & 0 \\ 1 & 0 & 0 \\ 0 & 0 & 1 \end{bmatrix} \begin{bmatrix} 1 & 0 & 0 \\ 2 & 1 & 0 \\ 1 & 0 & 1 \end{bmatrix} \begin{bmatrix} 1 & 0 & 0 \\ 0 & -3 & 0 \\ 0 & 0 & 1 \end{bmatrix} \begin{bmatrix} 1 & 2 & 0 \\ 0 & 1 & 0 \\ 0 & 0 & 1 \end{bmatrix} \begin{bmatrix} 1 & 0 & 1 \\ 0 & 1 & 0 \\ 0 & 0 & 1 \end{bmatrix}.
\end{aligned}
$$

$\square$

According to Definition 2.3.8 an $n \times n$ matrix A is invertible if there is a matrix B such that $AB = I_n$ and $BA = I_n$. In the case when $n = 2$ we discovered that it is not necessary to check both conditions. The same is true for square matrices of any size.

Theorem 2.3.19. *If A and B are n × n matrices such that*

$$AB = I_n,$$

then both matrices A and B are invertible and we have $A^{-1} = B$ and $B^{-1} = A$.

Proof. Assume that A and B are $n \times n$ matrices such that

$$AB = I_n.$$

If

$$B\mathbf{x} = \mathbf{0},$$

then

$$AB\mathbf{x} = A\mathbf{0}$$

and thus

$$\mathbf{x} = I_n\mathbf{x} = AB\mathbf{x} = A\mathbf{0} = \mathbf{0}.$$

This shows that the only solution of the equation $B\mathbf{x} = \mathbf{0}$ is the trivial solution $\mathbf{x} = \mathbf{0}$, which implies that B is invertible, by Theorem 2.3.17. Let $B^{-1} = C$. Then

$$BC = CB = I_n$$

and consequently

$$A = AI_n = ABC = I_nC = C,$$

which means that

$$BA = AB = I_n.$$

$\square$

Example 2.3.20. Since

$$\begin{bmatrix} 1 & 1 & 0 \\ 2 & 0 & -1 \\ 0 & -1 & 0 \end{bmatrix} \begin{bmatrix} 1 & 0 & 1 \\ 0 & 0 & -1 \\ 2 & -1 & 2 \end{bmatrix} = I_3,$$

both matrices $\begin{bmatrix} 1 & 1 & 0 \\ 2 & 0 & -1 \\ 0 & -1 & 0 \end{bmatrix}$ and $\begin{bmatrix} 1 & 0 & 1 \\ 0 & 0 & -1 \\ 2 & -1 & 2 \end{bmatrix}$ are invertible and we have

$$\begin{bmatrix} 1 & 1 & 0 \\ 2 & 0 & -1 \\ 0 & -1 & 0 \end{bmatrix}^{-1} = \begin{bmatrix} 1 & 0 & 1 \\ 0 & 0 & -1 \\ 2 & -1 & 2 \end{bmatrix}$$

and
$$\begin{bmatrix} 1 & 0 & 1 \\ 0 & 0 & -1 \\ 2 & -1 & 2 \end{bmatrix}^{-1} = \begin{bmatrix} 1 & 1 & 0 \\ 2 & 0 & -1 \\ 0 & -1 & 0 \end{bmatrix}.$$

A matrix equation

In Theorem 1.2.6 we show how inverse matrices can be used to solve systems of two linear equations with two unknowns. That method works equally well for more general systems.

> **Theorem 2.3.21.** *Let A be an n × n invertible matrix and B an n × p matrix. The equation*
> $$AX = B \qquad\qquad (2.2)$$
> *has a unique solution and that solution is the n × p matrix*
> $$X = A^{-1}B.$$

Proof. If A is invertible, we can multiply the equation $AX = B$ by the matrix A^{-1} and get
$$A^{-1}(AX) = A^{-1}B$$
which simplifies to
$$X = A^{-1}B. \qquad\qquad (2.3)$$

This shows that, if the equation (2.2) has a solution, then it is given by the equation (2.3).

Now we need to check that $X = A^{-1}B$ is a solution. Indeed, we have
$$AX = A(A^{-1}B) = (AA^{-1})B = I_n B = B.$$

$\square$

> **Theorem 2.3.22.** *Let A be an n × n invertible matrix and B an n × p matrix. Then the reduced row echelon form of the n × (n + p) matrix*
> $$\begin{bmatrix} A & B \end{bmatrix}$$
> *is the matrix*
> $$\begin{bmatrix} I_n & A^{-1}B \end{bmatrix}.$$

Proof. According to Theorem 2.3.17 there are elementary matrices $E_1, E_2, \ldots, E_{m-1}, E_m$ such that

$$E_m E_{m-1} \ldots E_2 E_1 A = I_n.$$

Consequently,

$$E_m E_{m-1} \ldots E_2 E_1 \begin{bmatrix} A & B \end{bmatrix} = \begin{bmatrix} E_m E_{m-1} \ldots E_2 E_1 A & E_m E_{m-1} \ldots E_2 E_1 B \end{bmatrix} = \begin{bmatrix} I_n & A^{-1}B \end{bmatrix},$$

because $E_m E_{m-1} \ldots E_2 E_1 = A^{-1}$. $\qquad\square$

The above result applied to the case when $B = I_n$ tells us that the reduced row echelon form of the $n \times 2n$ matrix $\begin{bmatrix} A & I_n \end{bmatrix}$ is the matrix $\begin{bmatrix} I_n & A^{-1} \end{bmatrix}$. This suggests a convenient method for finding the inverse of a matrix, as illustrated in the following example.

Example 2.3.23. Determine if the matrix

$$A = \begin{bmatrix} 2 & 1 & 1 & 1 \\ 1 & 2 & 1 & 1 \\ 1 & 1 & 2 & 1 \\ 1 & 1 & 1 & 2 \end{bmatrix}$$

is invertible. If A is invertible, then find its inverse.

Solution. We find that

$$\begin{bmatrix} 2 & 1 & 1 & 1 & 1 & 0 & 0 & 0 \\ 1 & 2 & 1 & 1 & 0 & 1 & 0 & 0 \\ 1 & 1 & 2 & 1 & 0 & 0 & 1 & 0 \\ 1 & 1 & 1 & 2 & 0 & 0 & 0 & 1 \end{bmatrix} \sim \begin{bmatrix} 1 & 0 & 0 & 0 & \frac{4}{5} & -\frac{1}{5} & -\frac{1}{5} & -\frac{1}{5} \\ 0 & 1 & 0 & 0 & -\frac{1}{5} & \frac{4}{5} & -\frac{1}{5} & -\frac{1}{5} \\ 0 & 0 & 1 & 0 & -\frac{1}{5} & -\frac{1}{5} & \frac{4}{5} & -\frac{1}{5} \\ 0 & 0 & 0 & 1 & -\frac{1}{5} & -\frac{1}{5} & -\frac{1}{5} & \frac{4}{5} \end{bmatrix}.$$

This means that the matrix A is invertible and the inverse is

$$A^{-1} = \begin{bmatrix} \frac{4}{5} & -\frac{1}{5} & -\frac{1}{5} & -\frac{1}{5} \\ -\frac{1}{5} & \frac{4}{5} & -\frac{1}{5} & -\frac{1}{5} \\ -\frac{1}{5} & -\frac{1}{5} & \frac{4}{5} & -\frac{1}{5} \\ -\frac{1}{5} & -\frac{1}{5} & -\frac{1}{5} & \frac{4}{5} \end{bmatrix} = \frac{1}{5} \begin{bmatrix} 4 & -1 & -1 & -1 \\ -1 & 4 & -1 & -1 \\ -1 & -1 & 4 & -1 \\ -1 & -1 & -1 & 4 \end{bmatrix}.$$

$\qquad\square$

Example 2.3.24. Find a matrix

$$\begin{bmatrix} x & p \\ y & q \\ z & r \end{bmatrix}$$

such that

$$\begin{bmatrix} 2 & 1 & 1 \\ 3 & 3 & 1 \\ 1 & 1 & 2 \end{bmatrix} \begin{bmatrix} x & p \\ y & q \\ z & r \end{bmatrix} = \begin{bmatrix} 2 & 1 \\ 1 & 2 \\ 3 & 4 \end{bmatrix}.$$

Solution. Since

$$\begin{bmatrix} 2 & 1 & 1 & 2 & 1 \\ 3 & 3 & 1 & 1 & 2 \\ 1 & 1 & 2 & 3 & 4 \end{bmatrix} \sim \begin{bmatrix} 1 & 0 & 0 & \frac{3}{5} & -1 \\ 0 & 1 & 0 & -\frac{4}{5} & 1 \\ 0 & 0 & 1 & \frac{8}{5} & 2 \end{bmatrix},$$

we have

$$\begin{bmatrix} x & p \\ y & q \\ z & r \end{bmatrix} = \begin{bmatrix} 2 & 1 & 1 \\ 3 & 3 & 1 \\ 1 & 1 & 2 \end{bmatrix}^{-1} \begin{bmatrix} 2 & 1 \\ 1 & 2 \\ 3 & 4 \end{bmatrix} = \begin{bmatrix} \frac{3}{5} & -1 \\ -\frac{4}{5} & 1 \\ \frac{8}{5} & 2 \end{bmatrix},$$

by Theorem 2.3.22.   □

LU-decomposition of 3×3 matrices

In Chapter 1 we introduced LU-decomposition of 2×2 matrices. The idea can be easily generalized to 3×3 matrices.

> **Definition 2.3.25.** By an *LU-decomposition* (or an *LU-factorization*) of a 3×3 matrix A we mean the representation of A in the form
>
> $$A = LU$$
>
> where U is an upper triangular matrix and L is a lower triangular matrix with every entry on the main diagonal equal 1, that is,
>
> $$A = \begin{bmatrix} 1 & 0 & 0 \\ l_1 & 1 & 0 \\ l_2 & l_3 & 1 \end{bmatrix} \begin{bmatrix} u_1 & u_2 & u_3 \\ 0 & u_4 & u_5 \\ 0 & 0 & u_6 \end{bmatrix}.$$

When finding an LU-decomposition of a 3×3 matrix it is useful to note that, if $a_1 \neq 0$, then

$$\begin{bmatrix} 1 & 0 & 0 \\ -\frac{a_2}{a_1} & 1 & 0 \\ -\frac{a_3}{a_1} & 0 & 1 \end{bmatrix} \begin{bmatrix} a_1 & b_1 & c_1 \\ a_2 & b_2 & c_2 \\ a_3 & b_3 & c_3 \end{bmatrix} = \begin{bmatrix} a_1 & b_1 & c_1 \\ 0 & b_2' & c_2' \\ 0 & b_3' & c_3' \end{bmatrix}$$

and, if $b_2' \neq 0$,

$$\begin{bmatrix} 1 & 0 & 0 \\ 0 & 1 & 0 \\ 0 & -\frac{b_3'}{b_2'} & 1 \end{bmatrix} \begin{bmatrix} a_1 & b_1 & c_1 \\ 0 & b_2' & c_2' \\ 0 & b_3' & c_3' \end{bmatrix} = \begin{bmatrix} a_1 & b_1 & c_1 \\ 0 & b_2' & c_2' \\ 0 & 0 & c_3'' \end{bmatrix} = \begin{bmatrix} u_1 & u_2 & u_3 \\ 0 & u_4 & u_5 \\ 0 & 0 & u_6 \end{bmatrix}.$$

Example 2.3.26. Find an LU-decomposition of the matrix $\begin{bmatrix} 1 & 1 & 2 \\ 3 & 1 & 1 \\ 2 & 4 & 3 \end{bmatrix}$.

Solution. Since

$$\begin{bmatrix} 1 & 0 & 0 \\ -3 & 1 & 0 \\ -2 & 0 & 1 \end{bmatrix} \begin{bmatrix} 1 & 1 & 2 \\ 3 & 1 & 1 \\ 2 & 4 & 3 \end{bmatrix} = \begin{bmatrix} 1 & 1 & 2 \\ 0 & -2 & -5 \\ 0 & 2 & -1 \end{bmatrix}$$

and

$$\begin{bmatrix} 1 & 0 & 0 \\ 0 & 1 & 0 \\ 0 & 1 & 1 \end{bmatrix} \begin{bmatrix} 1 & 1 & 2 \\ 0 & -2 & -5 \\ 0 & 2 & -1 \end{bmatrix} = \begin{bmatrix} 1 & 1 & 2 \\ 0 & -2 & -5 \\ 0 & 0 & -6 \end{bmatrix},$$

we have

$$\begin{bmatrix} 1 & 0 & 0 \\ 0 & 1 & 0 \\ 0 & 1 & 1 \end{bmatrix} \begin{bmatrix} 1 & 0 & 0 \\ -3 & 1 & 0 \\ -2 & 0 & 1 \end{bmatrix} \begin{bmatrix} 1 & 1 & 2 \\ 3 & 1 & 1 \\ 2 & 4 & 3 \end{bmatrix} = \begin{bmatrix} 1 & 1 & 2 \\ 0 & -2 & -5 \\ 0 & 0 & -6 \end{bmatrix}.$$

Hence

$$\begin{bmatrix} 1 & 1 & 2 \\ 3 & 1 & 1 \\ 2 & 4 & 3 \end{bmatrix} = \begin{bmatrix} 1 & 0 & 0 \\ 3 & 1 & 0 \\ 2 & 0 & 1 \end{bmatrix} \begin{bmatrix} 1 & 0 & 0 \\ 0 & 1 & 0 \\ 0 & -1 & 1 \end{bmatrix} \begin{bmatrix} 1 & 1 & 2 \\ 0 & -2 & -5 \\ 0 & 0 & -6 \end{bmatrix},$$

because

$$\begin{bmatrix} 1 & 0 & 0 \\ 0 & 1 & 0 \\ 0 & 1 & 1 \end{bmatrix}^{-1} = \begin{bmatrix} 1 & 0 & 0 \\ 0 & 1 & 0 \\ 0 & -1 & 1 \end{bmatrix} \quad \text{and} \quad \begin{bmatrix} 1 & 0 & 0 \\ -3 & 1 & 0 \\ -2 & 0 & 1 \end{bmatrix}^{-1} = \begin{bmatrix} 1 & 0 & 0 \\ 3 & 1 & 0 \\ 2 & 0 & 1 \end{bmatrix}.$$

Since

$$\begin{bmatrix} 1 & 0 & 0 \\ 3 & 1 & 0 \\ 2 & 0 & 1 \end{bmatrix} \begin{bmatrix} 1 & 0 & 0 \\ 0 & 1 & 0 \\ 0 & -1 & 1 \end{bmatrix} = \begin{bmatrix} 1 & 0 & 0 \\ 3 & 1 & 0 \\ 2 & -1 & 1 \end{bmatrix},$$

we obtain the following LU-decomposition of the matrix

$$\begin{bmatrix} 1 & 1 & 2 \\ 3 & 1 & 1 \\ 2 & 4 & 3 \end{bmatrix} = \begin{bmatrix} 1 & 0 & 0 \\ 3 & 1 & 0 \\ 2 & -1 & 1 \end{bmatrix} \begin{bmatrix} 1 & 1 & 2 \\ 0 & -2 & -5 \\ 0 & 0 & -6 \end{bmatrix}.$$

□

Example 2.3.27. Find an LU-decomposition of the matrix $\begin{bmatrix} 2 & 1 & 4 \\ 1 & 1 & 1 \\ 2 & 3 & 1 \end{bmatrix}$.

Solution. Since

$$\begin{bmatrix} 1 & 0 & 0 \\ -\frac{1}{2} & 1 & 0 \\ -1 & 0 & 1 \end{bmatrix} \begin{bmatrix} 2 & 1 & 4 \\ 1 & 1 & 1 \\ 2 & 3 & 1 \end{bmatrix} = \begin{bmatrix} 2 & 1 & 4 \\ 0 & \frac{1}{2} & -1 \\ 0 & 2 & -3 \end{bmatrix}$$

and

$$\begin{bmatrix} 1 & 0 & 0 \\ 0 & 1 & 0 \\ 0 & -4 & 1 \end{bmatrix} \begin{bmatrix} 2 & 1 & 4 \\ 0 & \frac{1}{2} & -1 \\ 0 & 2 & -3 \end{bmatrix} = \begin{bmatrix} 2 & 1 & 4 \\ 0 & \frac{1}{2} & -1 \\ 0 & 0 & 1 \end{bmatrix},$$

we have

$$\begin{bmatrix} 2 & 1 & 4 \\ 1 & 1 & 1 \\ 2 & 3 & 1 \end{bmatrix} = \begin{bmatrix} 1 & 0 & 0 \\ \frac{1}{2} & 1 & 0 \\ 1 & 0 & 1 \end{bmatrix} \begin{bmatrix} 1 & 0 & 0 \\ 0 & 1 & 0 \\ 0 & 4 & 1 \end{bmatrix} \begin{bmatrix} 2 & 1 & 4 \\ 0 & \frac{1}{2} & -1 \\ 0 & 0 & 1 \end{bmatrix}.$$

Consequently,

$$\begin{bmatrix} 2 & 1 & 4 \\ 1 & 1 & 1 \\ 2 & 3 & 1 \end{bmatrix} = \begin{bmatrix} 1 & 0 & 0 \\ \frac{1}{2} & 1 & 0 \\ 1 & 4 & 1 \end{bmatrix} \begin{bmatrix} 2 & 1 & 4 \\ 0 & \frac{1}{2} & -1 \\ 0 & 0 & 1 \end{bmatrix}.$$

$\square$

Example 2.3.28. Find an LU-decomposition of the matrix $\begin{bmatrix} 2 & 1 & 0 \\ 3 & 1 & 0 \\ 2 & 3 & 0 \end{bmatrix}$.

Solution. Since

$$\begin{bmatrix} 1 & 0 & 0 \\ -\frac{3}{2} & 1 & 0 \\ -1 & 0 & 1 \end{bmatrix} \begin{bmatrix} 2 & 1 & 0 \\ 3 & 1 & 0 \\ 2 & 3 & 0 \end{bmatrix} = \begin{bmatrix} 2 & 1 & 0 \\ 0 & -\frac{1}{2} & 0 \\ 0 & 2 & 0 \end{bmatrix}$$

and

$$\begin{bmatrix} 1 & 0 & 0 \\ 0 & 1 & 0 \\ 0 & 4 & 1 \end{bmatrix} \begin{bmatrix} 2 & 1 & 0 \\ 0 & -\frac{1}{2} & 0 \\ 0 & 2 & 0 \end{bmatrix} = \begin{bmatrix} 2 & 1 & 0 \\ 0 & -\frac{1}{2} & 0 \\ 0 & 0 & 0 \end{bmatrix},$$

we obtain

$$\begin{bmatrix} 2 & 1 & 0 \\ 3 & 1 & 0 \\ 2 & 3 & 0 \end{bmatrix} = \begin{bmatrix} 1 & 0 & 0 \\ \frac{3}{2} & 1 & 0 \\ 1 & -4 & 1 \end{bmatrix} \begin{bmatrix} 2 & 1 & 0 \\ 0 & -\frac{1}{2} & 0 \\ 0 & 0 & 0 \end{bmatrix}.$$

$\square$

As in the case of 2×2 matrices, not every 3×3 matrix has an LU-decomposition.

Example 2.3.29. Show that the matrix $A = \begin{bmatrix} 1 & 1 & 2 \\ 1 & 1 & 3 \\ 0 & 1 & 4 \end{bmatrix}$ has no LU-decomposition.

Solution. Suppose to the contrary, that the matrix A has an LU-decomposition, that is there are numbers l_j's and u_k's such that

$$\begin{bmatrix} 1 & 1 & 2 \\ 1 & 1 & 3 \\ 0 & 1 & 4 \end{bmatrix} = \begin{bmatrix} 1 & 0 & 0 \\ l_1 & 1 & 0 \\ l_2 & l_3 & 1 \end{bmatrix} \begin{bmatrix} u_1 & u_2 & u_3 \\ 0 & u_4 & u_5 \\ 0 & 0 & u_6 \end{bmatrix}.$$

Since

$$\begin{bmatrix} 1 & 0 & 0 \\ l_1 & 1 & 0 \\ l_2 & l_3 & 1 \end{bmatrix} \begin{bmatrix} u_1 & u_2 & u_3 \\ 0 & u_4 & u_5 \\ 0 & 0 & u_6 \end{bmatrix} = \begin{bmatrix} u_1 & u_2 & u_3 \\ l_1 u_1 & l_1 u_2 + u_4 & l_1 u_3 + u_5 \\ l_2 u_1 & l_2 u_2 + l_3 u_4 & l_2 u_3 + l_3 u_5 + u_6 \end{bmatrix},$$

we obtain $u_1 = u_2 = 1$, then $l_1 = 1$ and $l_2 = 0$, and then $u_4 = 0$. But this is not possible, because then we would have

$$1 = l_2 u_2 + l_3 u_4 = 0 \cdot 1 + l_3 \cdot 0 = 0.$$

$\square$

The following theorem gives us a condition that guarantees existence of an LU-decomposition of a 3×3 matrix.

Theorem 2.3.30. *If $r \neq ap$, then the matrix*

$$A = \begin{bmatrix} 1 & p & q \\ a & r & s \\ b & t & u \end{bmatrix}$$

has an LU-decomposition of the matrix A.

Proof. Since

$$\begin{bmatrix} 1 & 0 & 0 \\ -a & 1 & 0 \\ -b & 0 & 1 \end{bmatrix} \begin{bmatrix} 1 & p & q \\ a & r & s \\ b & t & u \end{bmatrix} = \begin{bmatrix} 1 & p & q \\ 0 & r-ap & s-aq \\ 0 & t-bp & u-bq \end{bmatrix}$$

and

$$\begin{bmatrix} 1 & 0 & 0 \\ 0 & 1 & 0 \\ 0 & -\frac{t-bp}{r-ap} & 1 \end{bmatrix} \begin{bmatrix} 1 & p & q \\ 0 & r-ap & s-aq \\ 0 & t-bp & u-bq \end{bmatrix} = \begin{bmatrix} 1 & p & q \\ 0 & r-ap & s-aq \\ 0 & 0 & u-bq-\frac{(s-aq)(t-bp)}{r-ap} \end{bmatrix},$$

we have

$$\begin{bmatrix} 1 & p & q \\ a & r & s \\ b & t & u \end{bmatrix} = \begin{bmatrix} 1 & 0 & 0 \\ a & 1 & 0 \\ b & \frac{t-bp}{r-ap} & 1 \end{bmatrix} \begin{bmatrix} 1 & p & q \\ 0 & r-ap & s-aq \\ 0 & 0 & u-bq-\frac{(s-aq)(t-bp)}{r-ap} \end{bmatrix}.$$

☐

In the next example we use LU-decomposition to solve a system of three linear equations with three unknowns.

Example 2.3.31. Use the LU-decomposition

$$\begin{bmatrix} 2 & 1 & 4 \\ 1 & 1 & 1 \\ 2 & 3 & 1 \end{bmatrix} = \begin{bmatrix} 1 & 0 & 0 \\ \frac{1}{2} & 1 & 0 \\ 1 & 4 & 1 \end{bmatrix} \begin{bmatrix} 2 & 1 & 4 \\ 0 & \frac{1}{2} & -1 \\ 0 & 0 & 1 \end{bmatrix}$$

to solve the system

$$\begin{cases} 2x_1 + x_2 + 4x_3 = 2 \\ x_1 + x_2 + x_3 = 1 \\ 2x_1 + 3x_2 + x_3 = 0 \end{cases}.$$

Solution. The system can be written as a matrix equation

$$\begin{bmatrix} 2 & 1 & 4 \\ 1 & 1 & 1 \\ 2 & 3 & 1 \end{bmatrix} \begin{bmatrix} x_1 \\ x_2 \\ x_3 \end{bmatrix} = \begin{bmatrix} 2 \\ 1 \\ 0 \end{bmatrix}$$

or

$$\begin{bmatrix} 1 & 0 & 0 \\ \frac{1}{2} & 1 & 0 \\ 1 & 4 & 1 \end{bmatrix} \begin{bmatrix} 2 & 1 & 4 \\ 0 & \frac{1}{2} & -1 \\ 0 & 0 & 1 \end{bmatrix} \begin{bmatrix} x_1 \\ x_2 \\ x_3 \end{bmatrix} = \begin{bmatrix} 2 \\ 1 \\ 0 \end{bmatrix}$$

We let

$$\begin{bmatrix} 2 & 1 & 4 \\ 0 & \frac{1}{2} & -1 \\ 0 & 0 & 1 \end{bmatrix} \begin{bmatrix} x_1 \\ x_2 \\ x_3 \end{bmatrix} = \begin{bmatrix} y_1 \\ y_2 \\ y_3 \end{bmatrix}$$

and solve the equation

$$\begin{bmatrix} 1 & 0 & 0 \\ \frac{1}{2} & 1 & 0 \\ 1 & 4 & 1 \end{bmatrix} \begin{bmatrix} y_1 \\ y_2 \\ y_3 \end{bmatrix} = \begin{bmatrix} 2 \\ 1 \\ 0 \end{bmatrix}.$$

By forward substitution we get

$$y_1 = 2, \quad y_2 = 0, \quad \text{and} \quad y_3 = -2.$$

Next we solve the equation

$$\begin{bmatrix} 2 & 1 & 4 \\ 0 & \frac{1}{2} & -1 \\ 0 & 0 & 1 \end{bmatrix} \begin{bmatrix} x_1 \\ x_2 \\ x_3 \end{bmatrix} = \begin{bmatrix} 2 \\ 0 \\ -2 \end{bmatrix}.$$

By backward substitution we get

$$x_3 = -2, \quad x_2 = -4, \quad \text{and} \quad x_1 = 7.$$

$\square$

2.3.1 Exercises

Find the given products of matrices.

1. $\begin{bmatrix} 1 & 0 & 0 \\ 0 & 1 & 4 \\ 0 & 0 & 1 \end{bmatrix} \begin{bmatrix} a_1 & b_1 & c_1 & d_1 \\ a_2 & b_2 & c_2 & d_2 \\ a_3 & b_3 & c_3 & d_3 \end{bmatrix}$

2. $\begin{bmatrix} 1 & 0 & 0 \\ 0 & 1 & 0 \\ 0 & 1 & 1 \end{bmatrix} \begin{bmatrix} a_1 & b_1 & c_1 & d_1 \\ a_2 & b_2 & c_2 & d_2 \\ a_3 & b_3 & c_3 & d_3 \end{bmatrix}$

3. $\begin{bmatrix} 1 & 0 & 0 \\ -1 & 1 & 0 \\ 7 & 0 & 1 \end{bmatrix} \begin{bmatrix} a_1 & b_1 & c_1 & d_1 \\ a_2 & b_2 & c_2 & d_2 \\ a_3 & b_3 & c_3 & d_3 \end{bmatrix}$

4. $\begin{bmatrix} 1 & 0 & -4 \\ 0 & 1 & -5 \\ 0 & 0 & 1 \end{bmatrix} \begin{bmatrix} a_1 & b_1 & c_1 & d_1 \\ a_2 & b_2 & c_2 & d_2 \\ a_3 & b_3 & c_3 & d_3 \end{bmatrix}$

5. $\begin{bmatrix} 1 & -a_1 & 0 & 0 & 0 \\ 0 & 1 & 0 & 0 & 0 \\ 0 & -a_3 & 1 & 0 & 0 \\ 0 & -a_2 & 0 & 1 & 0 \\ 0 & -a_5 & 0 & 0 & 1 \end{bmatrix} \begin{bmatrix} a_1 \\ 1 \\ a_3 \\ a_2 \\ a_5 \end{bmatrix}$

6. $\begin{bmatrix} 1 & 0 & 0 & 0 & 0 \\ 0 & 0 & 0 & 1 & 0 \\ 0 & 0 & 1 & 0 & 0 \\ 0 & 1 & 0 & 0 & 0 \\ 0 & 0 & 0 & 0 & 1 \end{bmatrix} \begin{bmatrix} a_1 \\ a_2 \\ a_3 \\ a_4 \\ a_5 \end{bmatrix}$

7. $\begin{bmatrix} 1 & 3 & 0 & 0 \\ 0 & 1 & 0 & 0 \\ 0 & 0 & 1 & 0 \\ 0 & 8 & 0 & 1 \end{bmatrix} \begin{bmatrix} a_1 & b_1 \\ a_2 & b_2 \\ a_3 & b_3 \\ a_4 & b_4 \end{bmatrix}$

8. $\begin{bmatrix} 1 & 0 & -5 & 0 \\ 0 & 1 & 0 & 0 \\ 0 & 0 & 1 & 0 \\ 0 & 0 & 1 & 1 \end{bmatrix} \begin{bmatrix} a_1 & b_1 & c_1 \\ a_2 & b_2 & c_2 \\ a_3 & b_3 & c_3 \\ a_4 & b_4 & c_4 \end{bmatrix}$

9. $\begin{bmatrix} 1 & 2 & 0 \\ 0 & 1 & 0 \\ 0 & 5 & 1 \end{bmatrix} \begin{bmatrix} 1 & 0 & 1 \\ 0 & 1 & -2 \\ 0 & 0 & 1 \end{bmatrix} \begin{bmatrix} 1 & 0 & 0 \\ 3 & 1 & 0 \\ -1 & 0 & 1 \end{bmatrix}$

10. $\begin{bmatrix} 1 & 0 & 0 \\ 0 & 3 & 0 \\ 0 & 0 & 1 \end{bmatrix} \begin{bmatrix} 0 & 0 & 1 \\ 0 & 1 & 0 \\ 1 & 0 & 0 \end{bmatrix} \begin{bmatrix} 4 & 0 & 0 \\ 0 & 1 & 0 \\ 0 & 0 & 1 \end{bmatrix}$

Find a matrix P such that the given equality holds.

11. $P \begin{bmatrix} a_1 & b_1 \\ a_2 & b_2 \\ a_3 & b_3 \end{bmatrix} = \begin{bmatrix} 5a_1 & 5b_1 \\ 4a_2 & 4b_2 \\ 2a_3 & 2b_3 \end{bmatrix}$

12. $P \begin{bmatrix} a_1 \\ a_2 \\ a_3 \\ a_4 \end{bmatrix} = \begin{bmatrix} 3a_1 \\ 5a_2 \\ 4a_3 \\ 2a_4 \end{bmatrix}$

13. $P \begin{bmatrix} a_1 & b_1 \\ a_2 & b_2 \\ a_3 & b_3 \\ a_4 & b_4 \end{bmatrix} = \begin{bmatrix} a_3 & b_3 \\ a_2 & b_2 \\ a_1 & b_1 \\ a_4 & b_4 \end{bmatrix}$

17. $P \begin{bmatrix} a_1 & b_1 \\ a_2 & b_2 \\ a_3 & b_3 \\ a_4 & b_4 \end{bmatrix} = \begin{bmatrix} a_1 + ja_2 & b_1 + jb_2 \\ a_2 & b_2 \\ a_3 + ka_2 & b_3 + kb_2 \\ a_4 + ma_2 & b_4 + mb_2 \end{bmatrix}$

14. $P \begin{bmatrix} a_1 \\ a_2 \\ a_3 \\ a_4 \end{bmatrix} = \begin{bmatrix} a_1 \\ a_4 \\ a_3 \\ a_2 \end{bmatrix}$

18. $P \begin{bmatrix} a_1 \\ a_2 \\ a_3 \\ a_4 \\ a_5 \end{bmatrix} = \begin{bmatrix} a_1 + 3a_4 \\ a_2 + a_4 \\ a_3 + 5a_4 \\ a_4 \\ a_5 + 7a_4 \end{bmatrix}$

15. $P \begin{bmatrix} a_1 & b_1 \\ a_2 & b_2 \\ a_3 & b_3 \end{bmatrix} = \begin{bmatrix} a_2 & b_2 \\ a_3 & b_3 \\ a_1 & b_1 \end{bmatrix}$

19. $P \begin{bmatrix} a_1 \\ a_2 \\ a_3 \\ a_4 \\ a_5 \end{bmatrix} = \begin{bmatrix} a_1 + 2a_2 \\ a_2 \\ a_3 + 7a_2 \\ a_4 + 8a_2 \\ a_5 + 3a_2 \end{bmatrix}$

16. $P \begin{bmatrix} a_1 \\ a_2 \\ a_3 \\ a_4 \end{bmatrix} = \begin{bmatrix} a_3 \\ a_1 \\ a_4 \\ a_2 \end{bmatrix}$

20. $P \begin{bmatrix} a_1 & b_1 & c_1 \\ a_2 & b_2 & c_2 \\ a_3 & b_3 & c_3 \\ a_4 & b_4 & c_4 \end{bmatrix} = \begin{bmatrix} a_1 + 5a_2 & b_1 + 5b_2 & c_1 + 5a_2 \\ a_2 & b_2 & c_2 \\ a_3 & b_3 & c_3 \\ a_4 & b_4 & c_4 \end{bmatrix}$

Find a matrix P such that matrix PA is the reduced row echelon form of A.

21. $A = \begin{bmatrix} -4 & 4 \\ 11 & -3 \\ 1 & -1 \end{bmatrix}$

23. $A = \begin{bmatrix} 1 & 1 & 3 \\ -1 & 1 & 1 \\ 1 & -1 & -1 \end{bmatrix}$

22. $A = \begin{bmatrix} 1 & 1 \\ -1 & 1 \\ 0 & -2 \end{bmatrix}$

24. $\begin{bmatrix} 0 & 1 & 1 \\ -2 & 2 & -2 \\ 1 & -1 & 1 \end{bmatrix}$

25. Write the inverse of the matrix

$$A = \begin{bmatrix} 1 & 0 & 0 \\ 0 & 2 & 0 \\ 0 & 0 & 1 \end{bmatrix} \begin{bmatrix} 0 & 0 & 1 \\ 0 & 1 & 0 \\ 1 & 0 & 0 \end{bmatrix}$$

as a product of 2 elementary matrices. Calculate A and A^{-1} and verify the result.

26. Write the inverse of the matrix

$$A = \begin{bmatrix} 1 & 0 & 0 \\ 0 & 0 & 1 \\ 0 & 1 & 0 \end{bmatrix} \begin{bmatrix} 1 & 0 & 0 \\ 0 & 1 & 0 \\ 0 & 0 & 3 \end{bmatrix}$$

as a product of 2 elementary matrices. Calculate A and A^{-1} and verify the result.

27. Write the inverse of the matrix

$$A = \begin{bmatrix} 1 & 0 & 4 \\ 0 & 1 & 7 \\ 0 & 0 & 1 \end{bmatrix} \begin{bmatrix} 1 & 0 & 0 \\ 0 & 2 & 0 \\ 0 & 0 & 1 \end{bmatrix} \begin{bmatrix} 1 & 0 & 0 \\ 0 & 0 & 1 \\ 0 & 1 & 0 \end{bmatrix}$$

as a product of 3 simple matrices. Calculate A and A^{-1} and verify the result.

28. Write the inverse of the matrix

$$A = \begin{bmatrix} 2 & 0 & 0 \\ 0 & 1 & 0 \\ 0 & 0 & 1 \end{bmatrix} \begin{bmatrix} 1 & 0 & 0 \\ 0 & 2 & 0 \\ 0 & 0 & 1 \end{bmatrix} \begin{bmatrix} 1 & 0 & 0 \\ 1 & 1 & 0 \\ 3 & 0 & 1 \end{bmatrix}$$

as a product of 3 simple matrices. Calculate A and A^{-1} and verify the result.

29. Write the inverse of the matrix

$$A = \begin{bmatrix} 1 & 0 & 0 & 0 \\ 0 & 3 & 0 & 0 \\ 0 & 0 & 1 & 0 \\ 0 & 0 & 0 & 1 \end{bmatrix} \begin{bmatrix} 0 & 0 & 1 & 0 \\ 0 & 1 & 0 & 0 \\ 1 & 0 & 0 & 0 \\ 0 & 0 & 0 & 1 \end{bmatrix}$$

as a product of 2 elementary matrices. Calculate A and A^{-1} and verify the result.

30. Write the inverse of the matrix

$$A = \begin{bmatrix} 1 & 0 & 0 & 0 \\ 0 & 0 & 0 & 1 \\ 0 & 0 & 1 & 0 \\ 0 & 1 & 0 & 0 \end{bmatrix} \begin{bmatrix} 1 & 4 & 0 & 0 \\ 0 & 1 & 0 & 0 \\ 0 & 0 & 1 & 0 \\ 0 & 0 & 0 & 1 \end{bmatrix}$$

as a product of 2 elementary matrices. Calculate A and A^{-1} and verify the result.

31. Write the inverse of the matrix

$$A = \begin{bmatrix} 1 & 0 & 0 & 0 \\ 0 & 1 & 0 & 0 \\ 0 & 0 & 0 & 1 \\ 0 & 0 & 1 & 0 \end{bmatrix} \begin{bmatrix} 1 & 0 & 0 & 3 \\ 0 & 1 & 0 & 5 \\ 0 & 0 & 1 & 2 \\ 0 & 0 & 0 & 1 \end{bmatrix}$$

as a product of 2 simple matrices. Calculate A and A^{-1} and verify the result.

32. Write the inverse of the matrix

$$A = \begin{bmatrix} 1 & 3 & 0 & 0 \\ 0 & 1 & 0 & 0 \\ 0 & 2 & 0 & 1 \\ 0 & 4 & 1 & 0 \end{bmatrix} \begin{bmatrix} 1 & 0 & 2 & 0 \\ 0 & 1 & 1 & 0 \\ 0 & 0 & 1 & 0 \\ 0 & 0 & 1 & 1 \end{bmatrix}$$

as a product of 2 simple matrices. Calculate A and A^{-1} and verify the result.

33. Find a matrix $\begin{bmatrix} x & p & s \\ y & q & t \end{bmatrix}$ such that

$$\begin{bmatrix} 2 & 1 \\ 3 & 4 \end{bmatrix}\begin{bmatrix} x & p & s \\ y & q & t \end{bmatrix} = \begin{bmatrix} 2 & 1 & 1 \\ 3 & 1 & 2 \end{bmatrix}.$$

34. Find a matrix $\begin{bmatrix} x & p & s \\ y & q & t \end{bmatrix}$ such that

$$\begin{bmatrix} 3 & 1 \\ 1 & 1 \end{bmatrix}\begin{bmatrix} x & p & s \\ y & q & t \end{bmatrix} = \begin{bmatrix} 2 & 5 & 1 \\ 0 & 4 & 2 \end{bmatrix}.$$

35. Find a matrix $\begin{bmatrix} x & p \\ y & q \\ z & r \end{bmatrix}$ such that

$$\begin{bmatrix} 2 & 3 & 1 \\ 3 & 4 & 1 \\ 1 & 0 & 2 \end{bmatrix}\begin{bmatrix} x & p \\ y & q \\ z & r \end{bmatrix} = \begin{bmatrix} 1 & 1 \\ 1 & 1 \\ 1 & 0 \end{bmatrix}.$$

36. Find a matrix $\begin{bmatrix} x & p \\ y & q \\ z & r \end{bmatrix}$ such that

$$\begin{bmatrix} 3 & 1 & 1 \\ 1 & 2 & 1 \\ 1 & 2 & 2 \end{bmatrix}\begin{bmatrix} x & p \\ y & q \\ z & r \end{bmatrix} = \begin{bmatrix} 0 & 2 \\ 2 & 0 \\ 0 & 0 \end{bmatrix}.$$

Show that the given matrix A is invertible and find its inverse.

37. $A = \begin{bmatrix} 2 & 1 & 1 \\ 1 & 3 & 1 \\ -1 & 1 & 1 \end{bmatrix}$

40. $A = \begin{bmatrix} 1 & 1 & 1 \\ 1 & 1 & 2 \\ 1 & 2 & 3 \end{bmatrix}$

38. $A = \begin{bmatrix} 4 & 3 & 3 \\ 3 & 4 & 3 \\ 3 & 3 & 4 \end{bmatrix}$

41. $A = \begin{bmatrix} 1 & 1 & 3 & 1 \\ 1 & 1 & 1 & 2 \\ 3 & 1 & 1 & 1 \\ 1 & 0 & 1 & 1 \end{bmatrix}$

39. $A = \begin{bmatrix} 2 & 1 & 1 \\ 2 & 1 & 2 \\ 2 & 2 & 1 \end{bmatrix}$

42. $A = \begin{bmatrix} 1 & 1 & 1 & 1 \\ 1 & 2 & 1 & 1 \\ 1 & 1 & 2 & 1 \\ 1 & 1 & 1 & 2 \end{bmatrix}$

Show that the given matrix A is not invertible.

43. $A = \begin{bmatrix} 1 & 1 & 2 & 0 \\ 1 & 2 & 1 & 2 \\ 2 & 1 & 1 & 2 \\ 1 & 1 & 1 & 1 \end{bmatrix}$

44. $A = \begin{bmatrix} 1 & 2 & 2 & 1 \\ 1 & 1 & 1 & 1 \\ 3 & 4 & 1 & 2 \\ 3 & 5 & 2 & 2 \end{bmatrix}$

Express the given matrix A as a product of simple matrices.

45. $A = \begin{bmatrix} 1 & 1 & 1 \\ 1 & 1 & 2 \\ 2 & 1 & 1 \end{bmatrix}$

46. $A = \begin{bmatrix} 1 & 0 & 1 \\ 1 & 2 & 0 \\ 0 & 1 & 1 \end{bmatrix}$

Express the inverse of the given matrix A as a product of simple matrices and then find A^{-1}.

47. $A = \begin{bmatrix} 2 & 4 & 1 \\ 1 & 2 & 2 \\ 3 & 5 & 2 \end{bmatrix}$

48. $A = \begin{bmatrix} 2 & 1 & 1 \\ 2 & 1 & 2 \\ 2 & 2 & 1 \end{bmatrix}$

Determine the LU-decomposition of the given matrix.

49. $\begin{bmatrix} 1 & 0 & 1 \\ 1 & 1 & 1 \\ 1 & 1 & 0 \end{bmatrix}$

51. $\begin{bmatrix} 2 & 1 & 1 \\ 1 & 2 & 1 \\ 1 & 1 & 2 \end{bmatrix}$

50. $\begin{bmatrix} 2 & 3 & 0 \\ 0 & 1 & 1 \\ 1 & 0 & 2 \end{bmatrix}$

52. $\begin{bmatrix} 1 & 2 & -1 \\ 2 & 1 & 2 \\ 1 & 1 & 1 \end{bmatrix}$

Chapter 3

The vector space $\mathbb{R}^2$

The main object of interest in this chapter are matrices of the form $\begin{bmatrix} u_1 \\ u_2 \end{bmatrix}$. We are interested in their algebraic properties and their geometric interpretations. Many ideas introduced in this chapter will be generalized in the rest of this book and in the second book of linear algebra. A solid understanding of these ideas in the familiar context of $\mathbb{R}^2$ will make it easier to understand their generalizations.

Matrices of the form $\begin{bmatrix} u_1 \\ u_2 \end{bmatrix}$ are usually called vectors. The set of all such vectors is denoted by $\mathbb{R}^2$. Elements of $\mathbb{R}^2$ are often denoted by (u_1, u_2).

We note that

$$\begin{bmatrix} u_1 \\ u_2 \end{bmatrix} = (u_1, u_2) \neq \begin{bmatrix} u_1 & u_2 \end{bmatrix}.$$

In other words, (u_1, u_2) is not the matrix $\begin{bmatrix} u_1 & u_2 \end{bmatrix}$.

We will use "a vector in $\mathbb{R}^2$", "an element of $\mathbb{R}^2$", "a 2×1 matrix", and "a point in $\mathbb{R}^2$" as interchangeable phrases. Depending on the context, we choose the one that seems most intuitive and thus best facilitating understanding of the discussed idea.

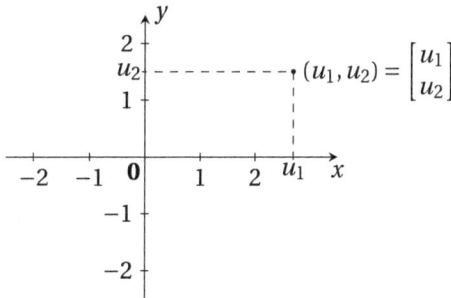

Figure 3.1: Cartesian coordinates.

In order to interpret $\mathbb{R}^2$ geometrically, we identify elements of $\mathbb{R}^2$ with points on the plane. This identification is done by the introduction on the plane of two perpendicular lines, usually called the **x-axis** and the **y-axis**. The point of intersection of the axes, called the **origin** and denoted by **0**, is identified with $\begin{bmatrix} 0 \\ 0 \end{bmatrix}$. On each axis we choose a unit of length (usually the same on both axes). Then every point on the plane can be described by a unique pair of numbers u_1 and u_2 as illustrated on Figure 3.1. The numbers u_1 and u_2 are called the **Cartesian coordinates** of the point.

3.1   Vectors in $\mathbb{R}^2$

Algebraic operations and vector lines in $\mathbb{R}^2$

Since elements of $\mathbb{R}^2$ are matrices, they can be added and multiplied by numbers:

$$\begin{bmatrix} u_1 \\ u_2 \end{bmatrix} + \begin{bmatrix} v_1 \\ v_2 \end{bmatrix} = \begin{bmatrix} u_1 + v_1 \\ u_2 + v_2 \end{bmatrix} \quad \text{and} \quad t \begin{bmatrix} u_1 \\ u_2 \end{bmatrix} = \begin{bmatrix} t u_1 \\ t u_2 \end{bmatrix}.$$

Example 3.1.1. We have

$$\begin{bmatrix} 3 \\ 2 \end{bmatrix} + \begin{bmatrix} 1 \\ 5 \end{bmatrix} = \begin{bmatrix} 4 \\ 7 \end{bmatrix} \quad \text{and} \quad 2 \begin{bmatrix} 3 \\ 5 \end{bmatrix} = \begin{bmatrix} 6 \\ 10 \end{bmatrix}.$$

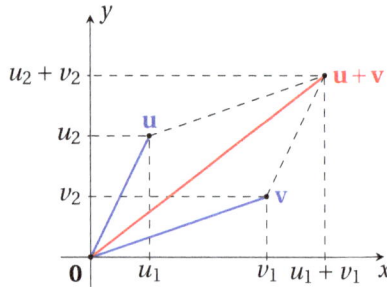

Figure 3.2: Addition in $\mathbb{R}^2$.

Addition in $\mathbb{R}^2$ has a clear geometric interpretation, as shown in Figure 3.2. Note that we label points as **u** and **v** instead of $\begin{bmatrix} u_1 \\ u_2 \end{bmatrix}$ and $\begin{bmatrix} v_1 \\ v_2 \end{bmatrix}$. We will do that often, when it is not essential to mention the coordinates. Observe that the geometric interpretation of addition allows us to add **u** and **v** without knowing the coordinates. It suffices to know their position relative to the origin (see Fig. 3.3).

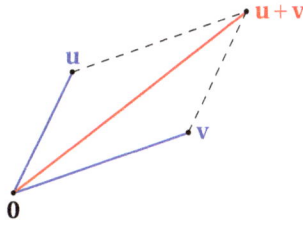

Figure 3.3: To add **u** and **v** it suffices to know the position of **u** and **v** relative to the origin.

The geometric interpretation of multiplication by numbers is illustrated in Figure 3.4. Again, it is not necessary to know the coordinates of **u**, but only the position of **u** relative to the origin.

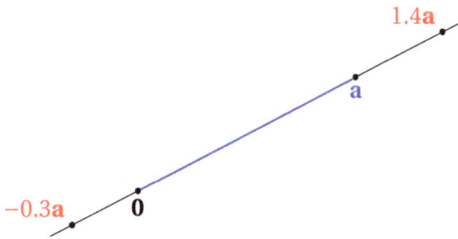

Figure 3.4: Multiplication by numbers.

Example 3.1.2. Choose arbitrary **a** and **b** in $\mathbb{R}^2$ and draw **b** − **a**.

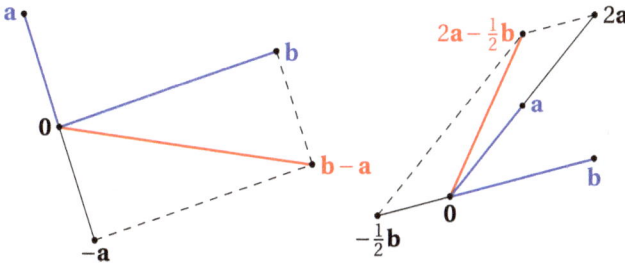

Figure 3.5: Solutions for Examples 3.1.2 and 3.1.3.

Example 3.1.3. Choose arbitrary **a** and **b** i $\mathbb{R}^2$ and draw $2\mathbf{a} - \frac{1}{2}\mathbf{b}$.

Example 3.1.4. Choose arbitrary **a**, **b**, and **c** i $\mathbb{R}^2$ and draw $\frac{1}{2}\mathbf{a} - \mathbf{b} + \frac{3}{2}\mathbf{c}$.

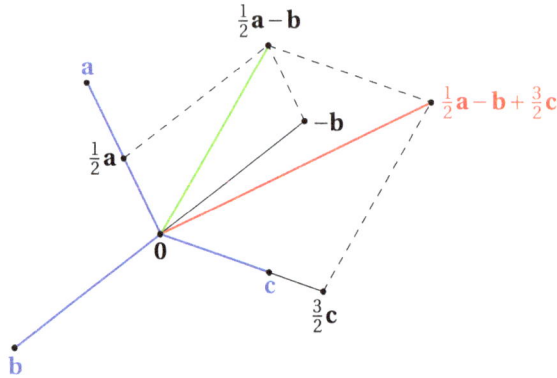

Figure 3.6: A solution for Example 3.1.4.

Using addition and multiplication by numbers we can describe lines in a convenient way. First consider a point **u** different from the origin, that is $\mathbf{u} \neq \mathbf{0}$. Note that when the real number t varies, then the point $t\mathbf{u}$ moves along the line through **u** and the origin. When we let t take all real values, then we obtain the entire line through **u** and the origin.

Definition 3.1.5. Let **u** be a vector in $\mathbb{R}^2$. The set of all vectors of the form $t\mathbf{u}$, where t is an arbitrary real number is called the ***vector subspace*** spanned by **u** and is denoted by Span{**u**}. That is,

$$\text{Span}\{\mathbf{u}\} = \{t\mathbf{u} : t \text{ in } \mathbb{R}\}.$$

If **u** is different from the origin, then Span{**u**} will be called a ***vector line***.

Figure 3.7: A segment of the vector line Span{**u**}.

In the definition of vector lines we have to assume that **u** is different from the origin, because otherwise Span{**u**} would not be a line, but a point, namely the origin. We adopt the convention that, when we say "a vector line Span{**u**}," we always implicitly assume that $\mathbf{u} \neq \mathbf{0}$.

Theorem 3.1.6. *For any two vectors* $\mathbf{u} \neq \mathbf{0}$ *and* $\mathbf{v} \neq \mathbf{0}$ *in* $\mathbb{R}^2$ *the following conditions are equivalent:*

(a) $\text{Span}\{\mathbf{u}\} = \text{Span}\{\mathbf{v}\}$;

(b) $\mathbf{v} = x\mathbf{u}$ *for some real number* $x \neq 0$.

Proof. If $\text{Span}\{\mathbf{u}\} = \text{Span}\{\mathbf{v}\}$, then $\mathbf{v}$ is in $\text{Span}\{\mathbf{u}\}$. This means that there is a real number x such that $\mathbf{v} = x\mathbf{u}$. Note that x cannot be 0, because then we would have $\mathbf{v} = x\mathbf{u} = 0\mathbf{u} = \mathbf{0}$.

If $\mathbf{v} = x\mathbf{u}$ for some real number $x \neq 0$, then we also have $\mathbf{u} = \frac{1}{x}\mathbf{v}$. Now, if $\mathbf{w}$ is in $\text{Span}\{\mathbf{v}\}$, then there is a number s such that $\mathbf{w} = s\mathbf{v}$ and thus $\mathbf{w} = (sx)\mathbf{u}$, which means that $\mathbf{w}$ is in $\text{Span}\{\mathbf{u}\}$.

On the other hand, if $\mathbf{w}$ is in $\text{Span}\{\mathbf{u}\}$, then there is a number t such that $\mathbf{w} = t\mathbf{u}$ and thus $\mathbf{w} = \frac{t}{x}\mathbf{v}$, which means that $\mathbf{w}$ is in $\text{Span}\{\mathbf{v}\}$. Consequently, $\text{Span}\{\mathbf{u}\} = \text{Span}\{\mathbf{v}\}$. □

Definition 3.1.7. Let $\mathbf{u}$ be a vector in $\mathbb{R}^2$. If $\mathbf{v}$ is a vector in $\text{Span}\{\mathbf{u}\}$ different from the origin, then $\{\mathbf{v}\}$ is called a **basis** of $\text{Span}\{\mathbf{u}\}$.

Linearly dependent and independent vectors in $\mathbb{R}^2$

If the vector $\mathbf{u}$ is in $\text{Span}\{\mathbf{v}\}$, then there is a real number s such that $\mathbf{u} = s\mathbf{v}$. In this case we say that the vector $\mathbf{u}$ is linearly dependent of the vector $\mathbf{v}$. Similarly, if the vector $\mathbf{v}$ is in $\text{Span}\{\mathbf{u}\}$, the vector $\mathbf{v}$ is linearly dependent of the vector $\mathbf{u}$.

Definition 3.1.8. Vectors $\mathbf{u}$ and $\mathbf{v}$ from $\mathbb{R}^2$ are *linearly dependent* if at least one of the following conditions is true:

(a) the vector $\mathbf{u}$ is in $\text{Span}\{\mathbf{v}\}$;

(b) the vector $\mathbf{v}$ is in $\text{Span}\{\mathbf{u}\}$.

Figure 3.8: Linearly dependent vectors.

Intuitively, vectors $\mathbf{u}$ and $\mathbf{v}$ in $\mathbb{R}^2$ are linearly dependent if $\mathbf{u}$, $\mathbf{v}$, and $\mathbf{0}$ are on the same line.

Example 3.1.9. Since

$$\begin{bmatrix} 6 \\ 3 \end{bmatrix} = 3 \begin{bmatrix} 2 \\ 1 \end{bmatrix},$$

the vector $\begin{bmatrix} 6 \\ 3 \end{bmatrix}$ is in $\text{Span}\left\{ \begin{bmatrix} 2 \\ 1 \end{bmatrix} \right\}$ and thus the vectors $\begin{bmatrix} 6 \\ 3 \end{bmatrix}$ and $\begin{bmatrix} 2 \\ 1 \end{bmatrix}$ are linearly dependent.

Example 3.1.10. The vectors $\mathbf{u} = \begin{bmatrix} 4 \\ -8 \end{bmatrix}$ and $\mathbf{v} = \begin{bmatrix} -7 \\ 14 \end{bmatrix}$ are linearly dependent, because we have $\mathbf{u} = -\frac{4}{7}\mathbf{v}$ and consequently $\mathbf{u}$ is in $\text{Span}\{\mathbf{v}\}$.

Example 3.1.11. The vectors $\mathbf{u} = \begin{bmatrix} 0 \\ 0 \end{bmatrix}$ and $\mathbf{v} = \begin{bmatrix} v_1 \\ v_2 \end{bmatrix}$ are linearly dependent, because we have $\mathbf{u} = 0\mathbf{v}$.

Similarly, the vectors $\mathbf{u} = \begin{bmatrix} u_1 \\ u_2 \end{bmatrix}$ and $\mathbf{v} = \begin{bmatrix} 0 \\ 0 \end{bmatrix}$ are linearly dependent, because we have $\mathbf{v} = 0\mathbf{u}$.

In other words, if at least one of the vectors $\mathbf{u}$ and $\mathbf{v}$ is the zero vector, then $\mathbf{u}$ and $\mathbf{v}$ are linearly dependent.

Example 3.1.12. The vectors $\mathbf{u} = \begin{bmatrix} 1 \\ 2 \end{bmatrix}$ and $\mathbf{v} = \begin{bmatrix} 3 \\ 4 \end{bmatrix}$ are not linearly dependent.

Indeed, if $\mathbf{u}$ was in $\text{Span}\{\mathbf{v}\}$, then we would have $\mathbf{u} = a\mathbf{v}$ for some real number a. But then that number would have to satisfy both equations $1 = 3a$ and $2 = 4a$. This is not possible, since it would mean that $a = \frac{1}{3}$ and at the same time $a = \frac{1}{2}$. This shows that $\mathbf{u}$ is not in $\text{Span}\{\mathbf{v}\}$.

Arguing in a similar way we can show that $\mathbf{v}$ is not in $\text{Span}\{\mathbf{u}\}$.

The next theorem gives us a useful method for verifying linear dependence without using the spans.

Theorem 3.1.13. *Vectors* **u** *and* **v** *in* $\mathbb{R}^2$ *are linearly dependent if and only if one of the following conditions holds:*

(a) **u** $= \mathbf{0}$ *or*

(b) **u** $\neq \mathbf{0}$ *and* **v** $= x\mathbf{u}$ *for a real number* x.

Proof. If either (a) or (b) holds, then it is clear that the vectors **u** and **v** are linearly dependent.

Now suppose that the vectors **u** and **v** are linearly dependent. To show that either (a) or (b) holds, it suffices to show that, if **u** $\neq \mathbf{0}$, then the equation **v** $= x\mathbf{u}$ has a solution.

If **v** is in Span{**u**}, then there is a number x such that **v** $= x\mathbf{u}$ and we are done.

If **u** is in Span{**v**}, there is a number y such that **u** $= y\mathbf{v}$. Since **u** $\neq \mathbf{0}$, we must have $y \neq 0$ and thus **v** $= \frac{1}{y}\mathbf{u}$.

$\square$

The following theorem shows that linear dependence of nonzero vectors is equivalent to the conditions in Theorem 3.1.6.

Theorem 3.1.14. *For any two vectors* **u** $\neq \mathbf{0}$ *and* **v** $\neq \mathbf{0}$ *in* $\mathbb{R}^2$ *the following conditions are equivalent:*

(a) *The vectors* **u** *and* **v** *are linearly dependent;*

(b) *Span{**u**} = Span{**v**};*

(c) **v** $= x\mathbf{u}$ *for a real number* $x \neq 0$.

Proof. This theorem is a consequence of Theorems 3.1.6 and 3.1.13. $\square$

The following theorem expresses linear dependence in a more algebraic language, namely, in terms of solutions of vector equations and determinants.

Theorem 3.1.15. *Let* $\mathbf{u} = \begin{bmatrix} u_1 \\ u_2 \end{bmatrix}$ *and* $\mathbf{v} = \begin{bmatrix} v_1 \\ v_2 \end{bmatrix}$. *The following conditions are equivalent:*

(a) *The vectors* $\mathbf{u}$ *and* $\mathbf{v}$ *are linearly dependent;*

(b) *The equation*

$$x \begin{bmatrix} u_1 \\ u_2 \end{bmatrix} + y \begin{bmatrix} v_1 \\ v_2 \end{bmatrix} = \begin{bmatrix} 0 \\ 0 \end{bmatrix} \tag{3.1}$$

has a nontrivial solution, that is, a solution different from the trivial solution $x = y = 0$;

(c)

$$\det \begin{bmatrix} u_1 & v_1 \\ u_2 & v_2 \end{bmatrix} = 0. \tag{3.2}$$

Proof. Assume that $\mathbf{u}$ and $\mathbf{v}$ are linearly dependent. If $\mathbf{u}$ is in Span$\{\mathbf{v}\}$, then $\mathbf{u} = a\mathbf{v}$ for some real number a. Then $\mathbf{u} - a\mathbf{v} = \mathbf{0}$ and thus $x = 1$ and $y = -a$ is a nontrivial solution of the equation (3.1). The case when $\mathbf{v}$ is in Span$\{\mathbf{u}\}$ can be treated in a similar way. Therefore (a) implies (b).

Now assume that the equation (3.1) has a nontrivial solution. If $x \neq 0$, then we have

$$\begin{bmatrix} u_1 \\ u_2 \end{bmatrix} = -\frac{y}{x} \begin{bmatrix} v_1 \\ v_2 \end{bmatrix}$$

or, equivalently,

$$u_1 = -\frac{y}{x} v_1 \quad \text{and} \quad u_2 = -\frac{y}{x} v_2.$$

Hence

$$\det \begin{bmatrix} u_1 & v_1 \\ u_2 & v_2 \end{bmatrix} = \det \begin{bmatrix} -\frac{y}{x} v_1 & v_1 \\ -\frac{y}{x} v_2 & v_2 \end{bmatrix} = -\frac{y}{x} v_1 v_2 + \frac{y}{x} v_2 v_1 = 0.$$

If $y \neq 0$ we use y instead of x and modify the proof accordingly. Therefore (b) implies (c).

Finally we assume that (3.2) holds. If $\mathbf{u} = \mathbf{0}$, then $\mathbf{u}$ is in Span$\{\mathbf{v}\}$ and $\mathbf{u}$ and $\mathbf{v}$ are linearly dependent. If $\mathbf{u} \neq \mathbf{0}$, then at least one of the numbers u_1 and u_2 must be different from 0. If $u_1 \neq 0$, then from (3.2) we get $v_2 = \frac{v_1}{u_1} u_2$ and, consequently,

$$\begin{bmatrix} v_1 \\ v_2 \end{bmatrix} = \frac{v_1}{u_1} \begin{bmatrix} u_1 \\ u_2 \end{bmatrix}$$

because the equality $v_1 = \frac{v_1}{u_1} u_1$ is obvious.

This means that $\mathbf{v}$ is in Span$\{\mathbf{u}\}$ and consequently $\mathbf{u}$ and $\mathbf{v}$ are linearly dependent. If $u_2 \neq 0$ we use u_2 instead of u_1 and modify the proof accordingly. Therefore (c) implies (a). $\square$

Example 3.1.16. The vectors $\begin{bmatrix} 3 \\ -9 \end{bmatrix}$ and $\begin{bmatrix} -4 \\ 12 \end{bmatrix}$ are linearly dependent, because

$$\det \begin{bmatrix} 3 & -4 \\ -9 & 12 \end{bmatrix} = 3 \cdot 12 - (-4) \cdot (-9) = 0.$$

Since

$$\begin{bmatrix} 3 \\ -9 \end{bmatrix} = 3 \begin{bmatrix} 1 \\ -3 \end{bmatrix} \quad \text{and} \quad \begin{bmatrix} -4 \\ 12 \end{bmatrix} = -4 \begin{bmatrix} 1 \\ -3 \end{bmatrix},$$

we have

$$\begin{bmatrix} -4 \\ 12 \end{bmatrix} = -\frac{4}{3} \begin{bmatrix} 3 \\ -9 \end{bmatrix}.$$

Example 3.1.17. In Example 3.1.12 we argue that the vectors $\mathbf{u} = \begin{bmatrix} 1 \\ 2 \end{bmatrix}$ and $\mathbf{v} = \begin{bmatrix} 3 \\ 4 \end{bmatrix}$ are not linearly dependent. We can accomplish that easier if we use the determinant:

$$\det \begin{bmatrix} 1 & 2 \\ 3 & 4 \end{bmatrix} = 1 \cdot 4 - 2 \cdot 3 = -2 \neq 0.$$

Definition 3.1.18. If the vectors $\mathbf{u}$ and $\mathbf{v}$ are not linearly dependent, we say that they are *linearly independent*. In other words, the vectors $\mathbf{u}$ and $\mathbf{v}$ are linearly independent if $\mathbf{u}$ is **not** in Span$\{\mathbf{v}\}$ and $\mathbf{v}$ is **not** in Span$\{\mathbf{u}\}$.

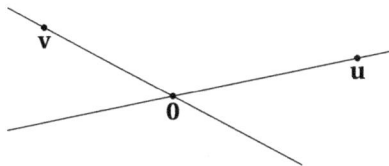

Figure 3.9: Linearly independent vectors.

Vectors $\mathbf{u}$ and $\mathbf{v}$ in Example 3.1.17 are linearly independent.

The following theorem is a reformulation of Theorem 3.1.15 in terms of linear independence.

Theorem 3.1.19. *Let* $\mathbf{u} = \begin{bmatrix} u_1 \\ u_2 \end{bmatrix}$ *and* $\mathbf{v} = \begin{bmatrix} v_1 \\ v_2 \end{bmatrix}$. *The following conditions are equivalent:*

(a) *The vectors* $\mathbf{u}$ *and* $\mathbf{v}$ *are linearly independent;*

(b) *The equation*

$$x \begin{bmatrix} u_1 \\ u_2 \end{bmatrix} + y \begin{bmatrix} v_1 \\ v_2 \end{bmatrix} = \begin{bmatrix} 0 \\ 0 \end{bmatrix}$$

has only the trivial solution, that is, $x = y = 0$;

(c)

$$\det \begin{bmatrix} u_1 & v_1 \\ u_2 & v_2 \end{bmatrix} \neq 0.$$

Theorem 1.3.9 connects invertibility of a matrix with a condition on its determinant. The above theorem relates the determinant of a matrix to linear independence of its columns. When we put these two facts together we obtain the following interesting and useful result.

Theorem 3.1.20. *Let* $\mathbf{a} = \begin{bmatrix} a_1 \\ a_2 \end{bmatrix}$ *and* $\mathbf{b} = \begin{bmatrix} b_1 \\ b_2 \end{bmatrix}$. *The matrix* $\begin{bmatrix} a_1 & b_1 \\ a_2 & b_2 \end{bmatrix} = \begin{bmatrix} \mathbf{a} & \mathbf{b} \end{bmatrix}$ *is invertible if and only if the column vectors* $\mathbf{a}$ *and* $\mathbf{b}$ *are linearly independent.*

Bases in $\mathbb{R}^2$

If $\mathbf{v}$ is any nonzero vector on the vector line Span$\{\mathbf{u}\}$, then any other vector from Span$\{\mathbf{u}\}$ can be written as $a\mathbf{v}$ for some real number a. In other words, the whole vector line can be reconstructed from any nonzero vector on that vector line. Now we are going to consider a similar question for the whole plane $\mathbb{R}^2$ instead of a vector line in $\mathbb{R}^2$. It leads to the idea of a basis. As we will see later, this is an idea of fundamental importance in linear algebra.

Definition 3.1.21. A pair of vectors $\{\mathbf{a}, \mathbf{b}\}$ in $\mathbb{R}^2$ is called a ***basis*** of $\mathbb{R}^2$ if the vectors satisfy the following two conditions:

(a) $\mathbf{a}$ and $\mathbf{b}$ are linearly independent;

(b) For every $\mathbf{c}$ in $\mathbb{R}^2$ there are real numbers x and y such that $\mathbf{c} = x\mathbf{a} + y\mathbf{b}$.

In other words, by a basis of $\mathbb{R}^2$ we mean a pair of linearly independent vectors $\mathbf{a}$ and $\mathbf{b}$ such that every vector in $\mathbb{R}^2$ can be written in the form $x\mathbf{a} + y\mathbf{b}$, so that the whole plane can be constructed this way from **a** and **b**.

The expression of the form $x\mathbf{a} + y\mathbf{b}$ is referred to as a ***linear combination*** of vectors **a** and **b**. Condition (2) in the above definition can be stated as "every vector in $\mathbb{R}^2$ is a linear combination **a** and **b**."

It turns out that it is not necessary to assume both conditions in the definition of a basis in $\mathbb{R}^2$. The next theorem shows that each one of the conditions implies the other one.

Theorem 3.1.22. *Let* **a** *and* **b** *be vectors in* $\mathbb{R}^2$. *The following two conditions are equivalent:*

(a) **a** *and* **b** *are linearly independent;*

(b) *For every* **c** *in* $\mathbb{R}^2$ *there are real numbers* x *and* y *such that* $\mathbf{c} = x\mathbf{a} + y\mathbf{b}$.

Proof. Let $\mathbf{a} = \begin{bmatrix} a_1 \\ a_2 \end{bmatrix}$ and $\mathbf{b} = \begin{bmatrix} b_1 \\ b_2 \end{bmatrix}$.

First we observe that, by Theorem 3.1.19, the vectors **a** and **b** are linearly independent if and only if $\det \begin{bmatrix} a_1 & b_1 \\ a_2 & b_2 \end{bmatrix} \neq 0$. Then we note that, if $\det \begin{bmatrix} a_1 & b_1 \\ a_2 & b_2 \end{bmatrix} \neq 0$, then the matrix equation

$$\begin{bmatrix} a_1 & b_1 \\ a_2 & b_2 \end{bmatrix} \begin{bmatrix} x \\ y \end{bmatrix} = \begin{bmatrix} c_1 \\ c_2 \end{bmatrix}$$

has a unique solution for every $\mathbf{c} = \begin{bmatrix} c_1 \\ c_2 \end{bmatrix}$, by Theorem 1.3.10. But this means that for every **c** in $\mathbb{R}^2$ there are real numbers x and y such that $\mathbf{c} = x\mathbf{a} + y\mathbf{b}$.

Now suppose that for every **c** in $\mathbb{R}^2$ there are real numbers x and y such that $\mathbf{c} = x\mathbf{a} + y\mathbf{b}$. Then there are real numbers x_1, y_1, x_2, y_2 such that

$$\begin{bmatrix} 1 \\ 0 \end{bmatrix} = x_1\mathbf{a} + y_1\mathbf{b} \quad \text{and} \quad \begin{bmatrix} 0 \\ 1 \end{bmatrix} = x_2\mathbf{a} + y_2\mathbf{b},$$

which can be written as

$$\begin{bmatrix} 1 & 0 \\ 0 & 1 \end{bmatrix} = \begin{bmatrix} \mathbf{a} & \mathbf{b} \end{bmatrix} \begin{bmatrix} x_1 & x_2 \\ y_1 & y_2 \end{bmatrix}.$$

But this means, by Theorem 1.2.18, that the matrix $\begin{bmatrix} \mathbf{a} & \mathbf{b} \end{bmatrix}$ is invertible, which is equivalent to the fact that the vectors **a** and **b** are linearly independent, by Theorem 3.1.20. □

Note that from the above proof it follows that, if the vectors **a** and **b** are linearly independent, then the numbers x and y such that $\mathbf{c} = x\mathbf{a} + y\mathbf{b}$ are unique for every **c** in $\mathbb{R}^2$. Those unique numbers x and y are called the ***coordinates*** of the vector **c** with respect to the basis $\{\mathbf{a}, \mathbf{b}\}$.

3.1.1 Exercises

1. Determine $\mathbf{d}$ such that $\mathbf{b}-\mathbf{a}=\mathbf{d}-\mathbf{c}$ where $\mathbf{a}=\begin{bmatrix} 4 \\ 1 \end{bmatrix}$, $\mathbf{b}=\begin{bmatrix} 2 \\ 3 \end{bmatrix}$, and $\mathbf{c}=\begin{bmatrix} 3 \\ 0 \end{bmatrix}$.
   Draw $\mathbf{a}$, $\mathbf{b}$, $\mathbf{c}$ and $\mathbf{d}$.

2. Determine $\mathbf{d}$ such that $\mathbf{b}-\mathbf{a}=\mathbf{d}-\mathbf{c}$ where $\mathbf{a}=\begin{bmatrix} 0 \\ 2 \end{bmatrix}$, $\mathbf{b}=\begin{bmatrix} 3 \\ 7 \end{bmatrix}$, and $\mathbf{c}=\begin{bmatrix} 2 \\ -1 \end{bmatrix}$.
   Draw $\mathbf{a}$, $\mathbf{b}$, $\mathbf{c}$ and $\mathbf{d}$.

Draw vectors $\mathbf{a}$ and $\mathbf{b}$ and then draw the given vector.

3. $2\mathbf{a}+\frac{1}{3}\mathbf{b}$

6. $-\frac{3}{2}\mathbf{a}+\frac{1}{2}\mathbf{b}$

4. $\frac{1}{2}\mathbf{a}-\mathbf{b}$

7. $-0.5\mathbf{a}-0.75\mathbf{b}$

5. $-\frac{4}{3}\mathbf{a}+2\mathbf{b}$

8. $2(\mathbf{a}+\mathbf{b})-\frac{1}{2}(\mathbf{a}-\mathbf{b})$

Draw vectors $\mathbf{a}$, $\mathbf{b}$, and $\mathbf{c}$ and then draw the given vector.

9. $\frac{1}{2}\mathbf{a}+\mathbf{b}-\frac{5}{3}\mathbf{c}$

11. $-\frac{3}{4}\mathbf{a}+\frac{1}{3}\mathbf{b}+3\mathbf{c}$

10. $\mathbf{a}-2\mathbf{b}-\frac{1}{2}\mathbf{c}$

12. $0.2\mathbf{a}-0.5\mathbf{b}+0.7\mathbf{c}$

Show that the given vectors $\mathbf{u}$ and $\mathbf{v}$ are linearly dependent.

13. $\mathbf{u}=\begin{bmatrix} 5 \\ -5 \end{bmatrix}$, $\mathbf{v}=\begin{bmatrix} -7 \\ 7 \end{bmatrix}$

14. $\mathbf{u}=\begin{bmatrix} 8 \\ -4 \end{bmatrix}$, $\mathbf{v}=\begin{bmatrix} -10 \\ 5 \end{bmatrix}$

Show that the given vectors $\mathbf{u}$ and $\mathbf{v}$ are linearly independent.

15. $\mathbf{u}=\begin{bmatrix} 5 \\ -5 \end{bmatrix}$, $\mathbf{v}=\begin{bmatrix} 7 \\ 7 \end{bmatrix}$

16. $\mathbf{u}=\begin{bmatrix} 8 \\ -2 \end{bmatrix}$, $\mathbf{v}=\begin{bmatrix} -10 \\ 5 \end{bmatrix}$

Find a real number a such that the given vectors $\mathbf{u}$ and $\mathbf{v}$ are linearly dependent.

17. $\mathbf{u}=\begin{bmatrix} a \\ 3 \end{bmatrix}$, $\mathbf{v}=\begin{bmatrix} 7 \\ -2 \end{bmatrix}$

21. $\mathbf{u}=\begin{bmatrix} a \\ 3 \end{bmatrix}$, $\mathbf{v}=\begin{bmatrix} 3 \\ 8+a \end{bmatrix}$

18. $\mathbf{u}=\begin{bmatrix} a \\ 2 \end{bmatrix}$, $\mathbf{v}=\begin{bmatrix} 8 \\ a \end{bmatrix}$

22. $\mathbf{u}=\begin{bmatrix} 2-a \\ 2 \end{bmatrix}$, $\mathbf{v}=\begin{bmatrix} 2 \\ 5-a \end{bmatrix}$

19. $\mathbf{u}=\begin{bmatrix} a-7 \\ 5 \end{bmatrix}$, $\mathbf{v}=\begin{bmatrix} -2 \\ a \end{bmatrix}$

23. $\mathbf{u}=\begin{bmatrix} 5-a \\ 2 \end{bmatrix}$, $\mathbf{v}=\begin{bmatrix} 3 \\ 4-a \end{bmatrix}$.

20. $\mathbf{u}=\begin{bmatrix} a \\ 3a \end{bmatrix}$, $\mathbf{v}=\begin{bmatrix} 7 \\ a \end{bmatrix}$

24. $\mathbf{u}=\begin{bmatrix} a \\ a^2-4 \end{bmatrix}$, $\mathbf{v}=\begin{bmatrix} 3a \\ a+2 \end{bmatrix}$

25. Show that $\left\{ \begin{bmatrix} 5 \\ 2 \end{bmatrix}, \begin{bmatrix} -1 \\ 4 \end{bmatrix} \right\}$ is a basis in $\mathbb{R}^2$.

26. Show that $\left\{ \begin{bmatrix} a \\ 1 \end{bmatrix}, \begin{bmatrix} -1 \\ a \end{bmatrix} \right\}$ is a basis in $\mathbb{R}^2$ for any real number a.

27. Show that $\left\{ \begin{bmatrix} a \\ b \end{bmatrix}, \begin{bmatrix} c \\ a \end{bmatrix} \right\}$ is a basis in $\mathbb{R}^2$ for any $b < 0$ and $c > 0$.

28. Show that $\left\{ \begin{bmatrix} a \\ b \end{bmatrix}, \begin{bmatrix} -b \\ a \end{bmatrix} \right\}$ is a basis in $\mathbb{R}^2$ for any $b \in \mathbb{R}$ such that $b \neq 0$.

Find the coordinates of the given vector $\mathbf{u}$ with respect to the given basis $\mathcal{B}$.

29. $\mathbf{u} = \begin{bmatrix} 1 \\ 2 \end{bmatrix}$, $\mathcal{B} = \left\{ \begin{bmatrix} 1 \\ 1 \end{bmatrix}, \begin{bmatrix} 1 \\ 0 \end{bmatrix} \right\}$

31. $\mathbf{u} = \begin{bmatrix} 1 \\ -2 \end{bmatrix}$, $\mathcal{B} = \left\{ \begin{bmatrix} 3 \\ 1 \end{bmatrix}, \begin{bmatrix} 0 \\ 1 \end{bmatrix} \right\}$

30. $\mathbf{u} = \begin{bmatrix} 3 \\ 1 \end{bmatrix}$, $\mathcal{B} = \left\{ \begin{bmatrix} 1 \\ -1 \end{bmatrix}, \begin{bmatrix} 1 \\ 2 \end{bmatrix} \right\}$

32. $\mathbf{u} = \begin{bmatrix} 1 \\ 2 \end{bmatrix}$, $\mathcal{B} = \left\{ \begin{bmatrix} 3 \\ 1 \end{bmatrix}, \begin{bmatrix} 2 \\ 4 \end{bmatrix} \right\}$

33. Find the coordinates of the vector $\begin{bmatrix} 1 \\ 0 \end{bmatrix}$ with respect to the basis $\left\{ \begin{bmatrix} 1 \\ 1 \end{bmatrix}, \begin{bmatrix} 0 \\ 1 \end{bmatrix} \right\}$ and to the basis $\left\{ \begin{bmatrix} 1 \\ 1 \end{bmatrix}, \begin{bmatrix} 1 \\ 3 \end{bmatrix} \right\}$.

34. Find the coordinates of the vector $\begin{bmatrix} 2 \\ -1 \end{bmatrix}$ with respect to the basis $\left\{ \begin{bmatrix} 2 \\ 1 \end{bmatrix}, \begin{bmatrix} 1 \\ 0 \end{bmatrix} \right\}$ and in the basis $\left\{ \begin{bmatrix} 2 \\ 1 \end{bmatrix}, \begin{bmatrix} 0 \\ 1 \end{bmatrix} \right\}$.

3.2   The dot product and the projection on a vector line in $\mathbb{R}^2$

The dot product in $\mathbb{R}^2$

Definition 3.2.1. For a vector $\mathbf{u} = \begin{bmatrix} u_1 \\ u_2 \end{bmatrix}$ we define

$$\|\mathbf{u}\| = \left\| \begin{bmatrix} u_1 \\ u_2 \end{bmatrix} \right\| = \sqrt{u_1^2 + u_2^2}.$$

The number $\|\mathbf{u}\|$ is called the ***norm*** of $\mathbf{u}$.

Geometrically $\|\mathbf{u}\|$ is the distance from an arbitrary point $\mathbf{u} = \begin{bmatrix} u_1 \\ u_2 \end{bmatrix}$ to the origin. The number $\|\mathbf{u} - \mathbf{v}\|$ is the distance from the point $\mathbf{u} = \begin{bmatrix} u_1 \\ u_2 \end{bmatrix}$ to the point $\mathbf{v} = \begin{bmatrix} v_1 \\ v_2 \end{bmatrix}$. Indeed,

$$\|\mathbf{u} - \mathbf{v}\| = \left\| \begin{bmatrix} u_1 \\ u_2 \end{bmatrix} - \begin{bmatrix} v_1 \\ v_2 \end{bmatrix} \right\| = \left\| \begin{bmatrix} u_1 - v_1 \\ v_2 - v_2 \end{bmatrix} \right\| = \sqrt{(u_1 - v_1)^2 + (u_2 - v_2)^2}.$$

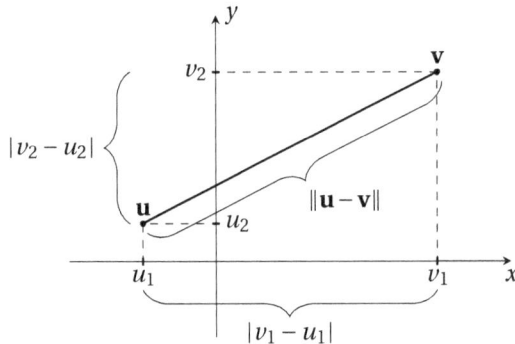

Figure 3.10: Norm as the distance between points.

If $\|\mathbf{u}\| = 1$, then we say that $\mathbf{u}$ is a ***unit vector***. If $\mathbf{u}$ is a nonzero vector, then the vector $\frac{1}{\|\mathbf{u}\|}\mathbf{u}$ is a unit vector. If we multiply a nonzero vector $\mathbf{u}$ by $\frac{1}{\|\mathbf{u}\|}$, we say that we ***normalize*** the vector $\mathbf{u}$.

Example 3.2.2.
$$\left\| \begin{bmatrix} 3 \\ 4 \end{bmatrix} \right\| = \sqrt{3^2 + 4^2} = \sqrt{9 + 16} = \sqrt{25} = 5$$

Example 3.2.3. The distance between $\begin{bmatrix} 5 \\ 1 \end{bmatrix}$ and $\begin{bmatrix} 3 \\ 4 \end{bmatrix}$ is

$$\left\| \begin{bmatrix} 5 \\ 1 \end{bmatrix} - \begin{bmatrix} 3 \\ 4 \end{bmatrix} \right\| = \left\| \begin{bmatrix} 2 \\ -3 \end{bmatrix} \right\| = \sqrt{2^2 + (-3)^2} = \sqrt{4 + 9} = \sqrt{13}.$$

Addition of vectors in $\mathbb{R}^2$, multiplication of a vector in $\mathbb{R}^2$ by a number, and the norm of a vector in $\mathbb{R}^2$ are similar to the familiar concepts from the algebra of real numbers. Now we are going to introduce an operation of "multiplication" of two vectors in $\mathbb{R}^2$. While it generalizes multiplication of real numbers, it is different from that operation, because the product of two vectors is not a vector, but a number.

Definition 3.2.4. For arbitrary **u** and **v** in $\mathbb{R}^2$, by the **dot product** of **u** and **v** we mean the real number **u**·**v** defined as follows

$$\mathbf{u} \cdot \mathbf{v} = \begin{bmatrix} u_1 \\ u_2 \end{bmatrix} \cdot \begin{bmatrix} v_1 \\ v_2 \end{bmatrix} = u_1 v_1 + u_2 v_2.$$

Example 3.2.5.

$$\begin{bmatrix} 3 \\ 5 \end{bmatrix} \cdot \begin{bmatrix} 2 \\ -6 \end{bmatrix} = 3 \cdot 2 + 5 \cdot (-6) = 6 - 30 = -24.$$

At this point the dot product has no obvious geometric meaning, but later we will learn that it has an important geometric interpretation.

Since

$$\mathbf{u} \cdot \mathbf{0} = 0,$$

for every **u** in $\mathbb{R}^2$, the vector **0** seems to play a role similar to the role of zero in multiplication of numbers. The dot product has other properties that are similar to the properties of multiplication of numbers. For example, it is easy to verify that

---

$$\mathbf{u} \cdot \mathbf{v} = \mathbf{v} \cdot \mathbf{u}$$

$$\mathbf{u} \cdot (\mathbf{v} + \mathbf{w}) = \mathbf{u} \cdot \mathbf{v} + \mathbf{u} \cdot \mathbf{w}$$

$$t(\mathbf{u} \cdot \mathbf{v}) = (t\mathbf{u}) \cdot \mathbf{v} = \mathbf{u} \cdot (t\mathbf{v})$$

---

for arbitrary **u**, **v**, and **w** in $\mathbb{R}^2$ and real number t.

On the other hand,

$$\begin{bmatrix} 1 \\ 2 \end{bmatrix} \cdot \begin{bmatrix} -6 \\ 3 \end{bmatrix} = 0,$$

which is very different from our experience with real numbers, where the product of two nonzero numbers is always different from zero.

Here is a nice and useful connection between the dot product and the norm:

$$\mathbf{u} \cdot \mathbf{u} = \|\mathbf{u}\|^2 \quad \text{or} \quad \|\mathbf{u}\| = \sqrt{\mathbf{u} \cdot \mathbf{u}}.$$

Definition 3.2.6. Two vectors **u** and **v** in $\mathbb{R}^2$ are called ***orthogonal vectors*** if

$$||\mathbf{u} - \mathbf{v}|| = ||\mathbf{u} + \mathbf{v}||.$$

Orthogonality is closely related to perpendicularity of lines. In elementary geometry, lines are called perpendicular if they intersect at a right angle. If **u** and **v** are different from the origin, then **u** and **v** are orthogonal if and only if the vector lines Span{**u**} and Span{**v**} are perpendicular, see Figure 3.11.

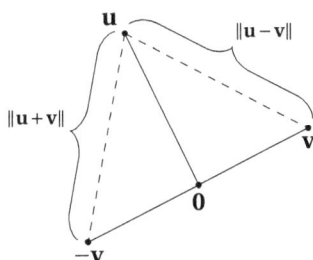

Figure 3.11: The lines Span{**u**} and Span{**v**} are perpendicular.

Example 3.2.7. $\begin{bmatrix} 3 \\ 2 \end{bmatrix}$ and $\begin{bmatrix} -6 \\ 9 \end{bmatrix}$ are orthogonal because

$$\left|\left|\begin{bmatrix} 3 \\ 2 \end{bmatrix} - \begin{bmatrix} -6 \\ 9 \end{bmatrix}\right|\right| = \left|\left|\begin{bmatrix} 9 \\ -7 \end{bmatrix}\right|\right| = \sqrt{9^2 + (-7)^2} = \sqrt{81 + 49} = \sqrt{130}$$

and

$$\left|\left|\begin{bmatrix} 3 \\ 2 \end{bmatrix} + \begin{bmatrix} -6 \\ 9 \end{bmatrix}\right|\right| = \left|\left|\begin{bmatrix} -3 \\ 11 \end{bmatrix}\right|\right| = \sqrt{(-3)^2 + 11^2} = \sqrt{9 + 121} = \sqrt{130}.$$

Here is another interesting connection between the dot product and the norm.

Theorem 3.2.8 (Parallelogram law). *For arbitrary* **u** *and* **v** *in* $\mathbb{R}^2$ *we have*

$$\mathbf{u} \cdot \mathbf{v} = \frac{1}{4}\left(||\mathbf{u} + \mathbf{v}||^2 - ||\mathbf{u} - \mathbf{v}||^2\right).$$

Proof. Since
$$||\mathbf{u} + \mathbf{v}||^2 = (\mathbf{u} + \mathbf{v}) \cdot (\mathbf{u} + \mathbf{v}) = \mathbf{u} \cdot \mathbf{u} + 2(\mathbf{u} \cdot \mathbf{v}) + \mathbf{v} \cdot \mathbf{v}$$

and

$$||\mathbf{u} - \mathbf{v}||^2 = (\mathbf{u} - \mathbf{v}) \cdot (\mathbf{u} - \mathbf{v}) = \mathbf{u} \cdot \mathbf{u} - 2(\mathbf{u} \cdot \mathbf{v}) + \mathbf{v} \cdot \mathbf{v},$$

we have

$$||\mathbf{u} + \mathbf{v}||^2 - ||\mathbf{u} - \mathbf{v}||^2 = 4\mathbf{u} \cdot \mathbf{v}.$$

$\square$

From the above lemma we obtain a theorem which gives us the first hint of the geometric meaning of the dot product.

Theorem 3.2.9. *Vectors* $\mathbf{u}$ *and* $\mathbf{v}$ *are orthogonal if and only if* $\mathbf{u} \cdot \mathbf{v} = 0$.

Proof. It is obvious from Theorem 3.2.8 that $\mathbf{u} \cdot \mathbf{v} = 0$ if and only if $||\mathbf{u} - \mathbf{v}|| = ||\mathbf{u} + \mathbf{v}||$.

$\square$

Theorem 3.2.10 (The Pythagorean Theorem). *If the vectors* $\mathbf{u}$ *and* $\mathbf{v}$ *are orthogonal, then*

$$\|\mathbf{u} + \mathbf{v}\|^2 = \|\mathbf{u}\|^2 + \|\mathbf{v}\|^2.$$

Proof. If $\mathbf{u}$ and $\mathbf{v}$ are orthogonal, then $\mathbf{u} \cdot \mathbf{v} = 0$ and thus

$$\|\mathbf{u} + \mathbf{v}\|^2 = (\mathbf{u} + \mathbf{v}) \cdot (\mathbf{u} + \mathbf{v}) = \mathbf{u} \cdot \mathbf{u} + 2(\mathbf{u} \cdot \mathbf{v}) + \mathbf{v} \cdot \mathbf{v} = \|\mathbf{u}\|^2 + \|\mathbf{v}\|^2.$$

$\square$

It is clear from the above proof and Theorem 3.2.9 that

$\mathbf{u}$ *and* $\mathbf{v}$ *are orthogonal if and only if* $\|\mathbf{u} + \mathbf{v}\|^2 = \|\mathbf{u}\|^2 + \|\mathbf{v}\|^2$.

Example 3.2.11. In Example 3.2.7 we have seen that $\begin{bmatrix} 3 \\ 2 \end{bmatrix}$ and $\begin{bmatrix} -6 \\ 9 \end{bmatrix}$ are orthogonal. As expected from the Pythagorean Theorem, we have

$$\left\| \begin{bmatrix} 3 \\ 2 \end{bmatrix} + \begin{bmatrix} -6 \\ 9 \end{bmatrix} \right\|^2 = \left\| \begin{bmatrix} -3 \\ 11 \end{bmatrix} \right\|^2 = (-3)^2 + 11^2 = 130$$

and

$$\left\| \begin{bmatrix} 3 \\ 2 \end{bmatrix} \right\|^2 + \left\| \begin{bmatrix} -6 \\ 9 \end{bmatrix} \right\|^2 = (3^2 + 2^2) + ((-6)^2 + 9^2) = 130.$$

Now we consider $\begin{bmatrix} 1 \\ 2 \end{bmatrix}$ and $\begin{bmatrix} 3 \\ 5 \end{bmatrix}$. Since

$$\begin{bmatrix} 1 \\ 2 \end{bmatrix} \cdot \begin{bmatrix} 3 \\ 5 \end{bmatrix} = 1 \cdot 3 + 2 \cdot 5 = 13,$$

the vectors $\begin{bmatrix} 1 \\ 2 \end{bmatrix}$ and $\begin{bmatrix} 3 \\ 5 \end{bmatrix}$ are not orthogonal. As expected, in this case the numbers

$$\left\| \begin{bmatrix} 1 \\ 2 \end{bmatrix} + \begin{bmatrix} 3 \\ 5 \end{bmatrix} \right\|^2 = \left\| \begin{bmatrix} 4 \\ 7 \end{bmatrix} \right\|^2 = 65$$

and

$$\left\| \begin{bmatrix} 1 \\ 2 \end{bmatrix} \right\|^2 + \left\| \begin{bmatrix} 3 \\ 5 \end{bmatrix} \right\|^2 = 5 + 34 = 39$$

are different.

Note that the dot product can be interpreted as a product of matrices:

$$\begin{bmatrix} u_1 \\ u_2 \end{bmatrix} \cdot \begin{bmatrix} v_1 \\ v_2 \end{bmatrix} = \begin{bmatrix} u_1 & u_2 \end{bmatrix} \begin{bmatrix} v_1 \\ v_2 \end{bmatrix} = \begin{bmatrix} v_1 & v_2 \end{bmatrix} \begin{bmatrix} u_1 \\ u_2 \end{bmatrix}. \qquad (3.3)$$

The above property can also be written as $\mathbf{u} \cdot \mathbf{v} = \mathbf{u}^T \mathbf{v} = \mathbf{v}^T \mathbf{u}$. This simple observation is often useful in calculations, see for example the proof of Theorem 3.2.17.

The projection on a vector line in $\mathbb{R}^2$

Consider a vector line Span$\{\mathbf{u}\}$ and a point $\mathbf{b}$ not on the line, see Figure 3.12. For every point $\mathbf{a}$ on the vector line we can measure the distance between $\mathbf{b}$ and $\mathbf{a}$. There must be a point $\mathbf{p}$ on the vector line for which the distance is the smallest. If $\mathbf{p}$ is such a point, we say that $\mathbf{p}$ minimizes the distance from $\mathbf{b}$ and we call $\mathbf{p}$ the best approximation to the point $\mathbf{b}$ by elements of the vector line Span$\{\mathbf{u}\}$. It turns out that linear algebra provides useful tools for dealing with such problems.

Theorem 3.2.12. *Consider a nonzero vector $\mathbf{u}$ in $\mathbb{R}^2$. For any point $\mathbf{b}$ in $\mathbb{R}^2$ there is a unique $\mathbf{p}$ on the vector line* Span$\{\mathbf{u}\}$ *such that*

$$\|\mathbf{b} - \mathbf{p}\| \leq \|\mathbf{b} - t\mathbf{u}\| \qquad (3.4)$$

for all t in $\mathbb{R}$. That unique point is

$$\mathbf{p} = \frac{\mathbf{b} \cdot \mathbf{u}}{\|\mathbf{u}\|^2} \mathbf{u}.$$

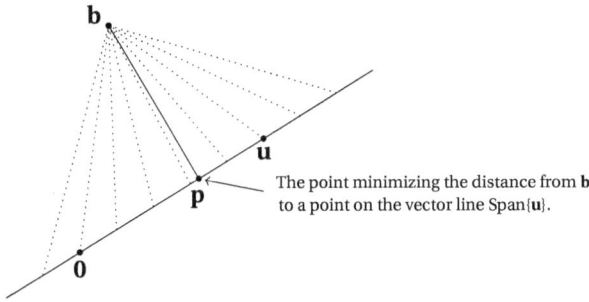

Figure 3.12: Best approximation to a point by elements of a vector line.

Proof. Since

$$\|\mathbf{b} - t\mathbf{u}\|^2 = (\mathbf{b} - t\mathbf{u}) \cdot (\mathbf{b} - t\mathbf{u}) = \|\mathbf{b}\|^2 - 2t\mathbf{b} \cdot \mathbf{u} + t^2 \|\mathbf{u}\|^2$$

$$= \|\mathbf{b}\|^2 - \frac{(\mathbf{b} \cdot \mathbf{u})^2}{\|\mathbf{u}\|^2} + \frac{(\mathbf{b} \cdot \mathbf{u})^2}{\|\mathbf{u}\|^2} - 2t\mathbf{b} \cdot \mathbf{u} + t^2 \|\mathbf{u}\|^2$$

$$= \|\mathbf{b}\|^2 - \frac{(\mathbf{b} \cdot \mathbf{u})^2}{\|\mathbf{u}\|^2} + \left(\frac{\mathbf{b} \cdot \mathbf{u}}{\|\mathbf{u}\|} - t\|\mathbf{u}\| \right)^2,$$

the norm $\|\mathbf{b} - t\mathbf{u}\|$ is minimized when $\frac{\mathbf{b} \cdot \mathbf{u}}{\|\mathbf{u}\|} - t\|\mathbf{u}\| = 0$. Solving for t we get $t = \frac{\mathbf{b} \cdot \mathbf{u}}{\|\mathbf{u}\|^2}$.
Moreover, since

$$\left\| \mathbf{b} - \frac{\mathbf{b} \cdot \mathbf{u}}{\|\mathbf{u}\|^2} \mathbf{u} \right\| < \|\mathbf{b} - t\mathbf{u}\|$$

for all $t \neq \frac{\mathbf{b} \cdot \mathbf{u}}{\|\mathbf{u}\|^2}$, the point minimizing the distance is unique.   □

The point

$$\mathbf{p} = \frac{\mathbf{b} \cdot \mathbf{u}}{\|\mathbf{u}\|^2} \mathbf{u}$$

is called the **best approximation** to the point $\mathbf{b}$ by elements of the vector line Span$\{\mathbf{u}\}$.
In calculating the best approximation we often use the identity

$$\frac{\mathbf{b} \cdot \mathbf{u}}{\|\mathbf{u}\|^2} \mathbf{u} = \frac{\mathbf{b} \cdot \mathbf{u}}{\mathbf{u} \cdot \mathbf{u}} \mathbf{u}.$$

We note that if the point $\mathbf{b}$ is on the vector line Span$\{\mathbf{u}\}$ then the best approximation
to the point $\mathbf{b}$ by elements of the vector line Span$\{\mathbf{u}\}$ is $\mathbf{b}$.

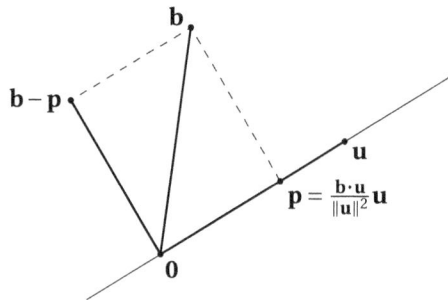

Figure 3.13: $(\mathbf{b} - \mathbf{p}) \cdot \mathbf{u} = 0$.

Theorem 3.2.13. *Consider a point* $\mathbf{b}$ *and a vector line* $\mathrm{Span}\{\mathbf{u}\}$ *in* $\mathbb{R}^2$. *The best approximation* $\mathbf{p}$ *to the point* $\mathbf{b}$ *by elements of the vector line* $\mathrm{Span}\{\mathbf{u}\}$ *can be characterized as the point* $\mathbf{p}$ *in* $\mathrm{Span}\{\mathbf{u}\}$ *satisfying the equation*

$$(\mathbf{b} - \mathbf{p}) \cdot \mathbf{u} = 0.$$

In other words, $\mathbf{p}$ *is the best approximation to* $\mathbf{b}$ *by elements of the line* $\mathrm{Span}\{\mathbf{u}\}$ *if and only if* $(\mathbf{b} - \mathbf{p}) \cdot \mathbf{u} = 0$.

Proof. The point $\mathbf{p}$, being on the vector line $\mathrm{Span}\{\mathbf{u}\}$, is of the form $t\mathbf{u}$. Note that

$$(\mathbf{b} - t\mathbf{u}) \cdot \mathbf{u} = 0$$

if and only if

$$\mathbf{b} \cdot \mathbf{u} - t\|\mathbf{u}\|^2 = 0$$

if and only if

$$t = \frac{\mathbf{b} \cdot \mathbf{u}}{\|\mathbf{u}\|^2}.$$

Hence $(\mathbf{b} - \mathbf{p}) \cdot \mathbf{u} = 0$ if and only if

$$\mathbf{p} = \frac{\mathbf{b} \cdot \mathbf{u}}{\|\mathbf{u}\|^2}\mathbf{u},$$

which is the point obtained in Theorem 3.2.12. $\qquad\square$

In view of Theorem 3.2.13, the best approximation to the point $\mathbf{b}$ by elements of the vector line $\mathrm{Span}\{\mathbf{u}\}$ is also called the **_projection_** of $\mathbf{b}$ on the vector line $\mathrm{Span}\{\mathbf{u}\}$ and is denoted by $\mathrm{proj}_{\mathrm{Span}\{\mathbf{u}\}}\mathbf{b}$ or simply $\mathrm{proj}_{\mathbf{u}}\mathbf{b}$:

$$\mathrm{proj}_{\mathbf{u}}\mathbf{b} = \frac{\mathbf{b} \cdot \mathbf{u}}{\|\mathbf{u}\|^2}\mathbf{u} = \frac{\mathbf{b} \cdot \mathbf{u}}{\mathbf{u} \cdot \mathbf{u}}\mathbf{u}$$

Example 3.2.14. Determine the projection of $\begin{bmatrix} 1 \\ 4 \end{bmatrix}$ on the vector line Span$\left\{ \begin{bmatrix} 3 \\ 1 \end{bmatrix} \right\}$.

Solution. The projection is

$$
\frac{\begin{bmatrix} 1 \\ 4 \end{bmatrix} \cdot \begin{bmatrix} 3 \\ 1 \end{bmatrix}}{\left\| \begin{bmatrix} 3 \\ 1 \end{bmatrix} \right\|^2} \begin{bmatrix} 3 \\ 1 \end{bmatrix} = \frac{7}{10} \begin{bmatrix} 3 \\ 1 \end{bmatrix} = \begin{bmatrix} \frac{21}{10} \\ \frac{7}{10} \end{bmatrix}.
$$

□

Example 3.2.15. Let $\mathbf{q}$ and $\mathbf{u}$ be points in $\mathbb{R}^2$ with $\mathbf{u} \neq \mathbf{0}$. Find a point $\mathbf{s}$ such that $\mathbf{p} = \frac{1}{2}(\mathbf{q} + \mathbf{s})$ where $\mathbf{p}$ is the projection of the point $\mathbf{q}$ on Span$\{\mathbf{u}\}$. Choose $\mathbf{q}$ and $\mathbf{u}$ and draw $\mathbf{s}$.

Solution. Since the projection $\mathbf{p}$ of the point $\mathbf{q}$ on the vector line Span$\{\mathbf{u}\}$ is $\frac{\mathbf{q} \cdot \mathbf{u}}{\mathbf{u} \cdot \mathbf{u}} \mathbf{u}$, the point $\mathbf{s}$ must satisfy the equation

$$
\frac{1}{2}(\mathbf{q} + \mathbf{s}) = \frac{\mathbf{q} \cdot \mathbf{u}}{\mathbf{u} \cdot \mathbf{u}} \mathbf{u}.
$$

Solving for $\mathbf{s}$ we get

$$
\mathbf{s} = 2 \frac{\mathbf{q} \cdot \mathbf{u}}{\mathbf{u} \cdot \mathbf{u}} \mathbf{u} - \mathbf{q}.
$$

□

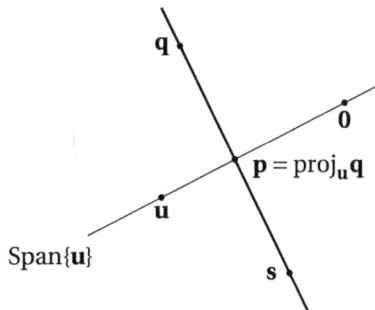

Figure 3.14: A solution for Example 3.2.15.

Definition 3.2.16. Let $\mathbf{q}$ and $\mathbf{u}$ be points in $\mathbb{R}^2$ with $\mathbf{u} \neq \mathbf{0}$. The point $\mathbf{s}$ such that

$$\text{proj}_{\mathbf{u}} \mathbf{q} = \frac{1}{2}(\mathbf{q} + \mathbf{s})$$

is called the ***reflection*** of the point $\mathbf{q}$ across the vector line Span$\{\mathbf{u}\}$.

The projection matrix on a vector line in $\mathbb{R}^2$

In the following theorem we give another way to calculate the projection of a point on a vector line. It uses the equality (3.3).

Theorem 3.2.17. *Let $\mathbf{u}$ be a nonzero vector in $\mathbb{R}^2$ and let*

$$A = \frac{1}{\|\mathbf{u}\|^2} \mathbf{u}\mathbf{u}^T.$$

Then for any $\mathbf{b}$ in $\mathbb{R}^2$ we have

$$\text{proj}_{\mathbf{u}} \mathbf{b} = A\mathbf{b}.$$

Moreover, the matrix A is the unique matrix with this property.

Proof. Let $\mathbf{u} = \begin{bmatrix} u_1 \\ u_2 \end{bmatrix} \neq \begin{bmatrix} 0 \\ 0 \end{bmatrix}$ and $\mathbf{b} = \begin{bmatrix} b_1 \\ b_2 \end{bmatrix}$. Then

$$\text{proj}_{\mathbf{u}} \mathbf{b} = = \frac{\mathbf{b} \cdot \mathbf{u}}{\|\mathbf{u}\|^2} \mathbf{u}$$

$$= \frac{\mathbf{u} \cdot \mathbf{b}}{\|\mathbf{u}\|^2} \mathbf{u}$$

$$= \frac{\begin{bmatrix} u_1 & u_2 \end{bmatrix} \begin{bmatrix} b_1 \\ b_2 \end{bmatrix}}{\|\mathbf{u}\|^2} \begin{bmatrix} u_1 \\ u_2 \end{bmatrix}$$

$$= \frac{1}{\|\mathbf{u}\|^2} \begin{bmatrix} u_1 \\ u_2 \end{bmatrix} \left(\begin{bmatrix} u_1 & u_2 \end{bmatrix} \begin{bmatrix} b_1 \\ b_2 \end{bmatrix} \right)$$

$$= \frac{1}{\|\mathbf{u}\|^2} \left(\begin{bmatrix} u_1 \\ u_2 \end{bmatrix} \begin{bmatrix} u_1 & u_2 \end{bmatrix} \right) \begin{bmatrix} b_1 \\ b_2 \end{bmatrix}$$

$$= \frac{1}{\|\mathbf{u}\|^2} (\mathbf{u}\mathbf{u}^T)\mathbf{b}.$$

The uniqueness part of the theorem follows from the fact that, if $A\mathbf{x} = B\mathbf{x}$ for every vector $\mathbf{x}$ from $\mathbb{R}^2$, then $A = B$, by Theorem 2.1.18. $\qquad \square$

Definition 3.2.18. Let $\mathbf{u}$ be a nonzero vector in $\mathbb{R}^2$. The matrix

$$\frac{1}{\|\mathbf{u}\|^2}(\mathbf{u}\mathbf{u}^T) \tag{3.5}$$

is called the **projection matrix** on the vector line Span{$\mathbf{u}$}.

Example 3.2.19. Let $\mathbf{u} = \begin{bmatrix} 1 \\ 3 \end{bmatrix}$. Determine the projection matrix on Span{$\mathbf{u}$} and use

it to calculate the projection of the vector $\mathbf{b} = \begin{bmatrix} 2 \\ 1 \end{bmatrix}$ on Span{$\mathbf{u}$}.

Solution. The projection matrix on Span{$\mathbf{u}$} is

$$\frac{1}{\|\mathbf{u}\|^2}(\mathbf{u}\mathbf{u}^T) = \frac{1}{10}\begin{bmatrix} 1 \\ 3 \end{bmatrix}\begin{bmatrix} 1 & 3 \end{bmatrix} = \frac{1}{10}\begin{bmatrix} 1 & 3 \\ 3 & 9 \end{bmatrix}$$

and the projection of $\mathbf{b} = \begin{bmatrix} 2 \\ 1 \end{bmatrix}$ on Span{$\mathbf{u}$} is

$$\text{proj}_{\mathbf{u}}\mathbf{b} = \frac{1}{\|\mathbf{u}\|^2}(\mathbf{u}\mathbf{u}^T)\mathbf{b} = \frac{1}{10}\begin{bmatrix} 1 & 3 \\ 3 & 9 \end{bmatrix}\begin{bmatrix} 2 \\ 1 \end{bmatrix} = \frac{1}{10}\begin{bmatrix} 5 \\ 15 \end{bmatrix} = \frac{5}{10}\begin{bmatrix} 1 \\ 3 \end{bmatrix} = \begin{bmatrix} \frac{1}{2} \\ \frac{3}{2} \end{bmatrix}.$$

$\square$

The perp operation

Now we introduce another simple but useful operation in $\mathbb{R}^2$.

Definition 3.2.20. For an arbitrary vector $\begin{bmatrix} a_1 \\ a_2 \end{bmatrix}$ in $\mathbb{R}^2$ we define

$$\begin{bmatrix} a_1 \\ a_2 \end{bmatrix}^{\llcorner} = \begin{bmatrix} -a_2 \\ a_1 \end{bmatrix}.$$

This operation is called the **perp operation** and $\mathbf{a}^{\llcorner}$ is read "a perp."

From the definition of the perp operation we easily obtain the following identities:

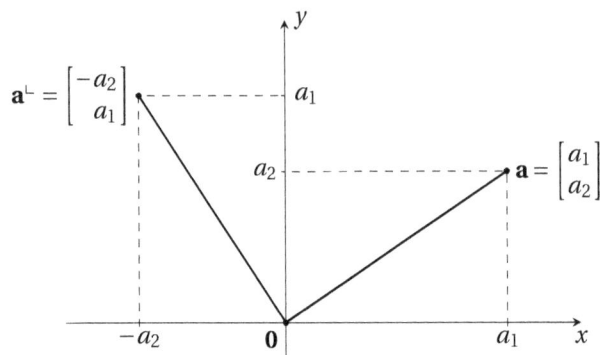

Figure 3.15: The perp operation.

$$\|\mathbf{a}^{\llcorner}\| = \|\mathbf{a}\|, \quad \mathbf{a} \cdot \mathbf{a}^{\llcorner} = 0, \quad \text{and} \quad (\mathbf{a}^{\llcorner})^{\llcorner} = -\mathbf{a}.$$

Example 3.2.21.

$$\begin{bmatrix} 1 \\ 2 \end{bmatrix}^{\llcorner} = \begin{bmatrix} -2 \\ 1 \end{bmatrix},$$

$$\left\| \begin{bmatrix} 1 \\ 2 \end{bmatrix} \right\| = \sqrt{5} = \left\| \begin{bmatrix} -2 \\ 1 \end{bmatrix} \right\|,$$

$$\begin{bmatrix} 1 \\ 2 \end{bmatrix} \cdot \begin{bmatrix} 1 \\ 2 \end{bmatrix}^{\llcorner} = \begin{bmatrix} 1 \\ 2 \end{bmatrix} \cdot \begin{bmatrix} -2 \\ 1 \end{bmatrix} = 1 \cdot (-2) + 2 \cdot 1 = 0,$$

$$\left(\begin{bmatrix} 1 \\ 2 \end{bmatrix}^{\llcorner} \right)^{\llcorner} = \begin{bmatrix} -2 \\ 1 \end{bmatrix}^{\llcorner} = \begin{bmatrix} -1 \\ -2 \end{bmatrix}.$$

The identity $\mathbf{a} \cdot \mathbf{a}^{\llcorner} = 0$ suggests that the perp operation is related to orthogonality. The following theorem exploits this property further.

Theorem 3.2.22. *Let* $\mathbf{n}$ *be a vector in* $\mathbb{R}^2$ *different from the origin. For any vector* $\mathbf{x}$ *in* $\mathbb{R}^2$ *the following conditions are equivalent:*

(a) $\mathbf{x} \cdot \mathbf{n} = 0$;

(b) $\mathbf{x}$ *is in* $\text{Span}\{\mathbf{n}^{\llcorner}\}$.

Proof. Let $\mathbf{n} = \begin{bmatrix} n_1 \\ n_2 \end{bmatrix}$ and $\mathbf{x} = \begin{bmatrix} x_1 \\ x_2 \end{bmatrix}$ and assume that $\mathbf{x} \cdot \mathbf{n} = 0$. Since $\mathbf{n}$ is different from the origin, at least one of the numbers n_1 and n_2 must be different from 0. Suppose $n_1 \neq 0$. Since

$$\mathbf{x} \cdot \mathbf{n} = \begin{bmatrix} x_1 \\ x_2 \end{bmatrix} \cdot \begin{bmatrix} n_1 \\ n_2 \end{bmatrix} = x_1 n_1 + x_2 n_2 = 0,$$

we have

$$x_1 = \frac{x_2}{n_1}(-n_2).$$

This, combined with the obvious equality

$$x_2 = \frac{x_2}{n_1} n_1,$$

gives us

$$\mathbf{x} = \begin{bmatrix} x_1 \\ x_2 \end{bmatrix} = \begin{bmatrix} \frac{x_2}{n_1}(-n_2) \\ \frac{x_2}{n_1} n_1 \end{bmatrix} = \frac{x_2}{n_1} \begin{bmatrix} -n_2 \\ n_1 \end{bmatrix} = \frac{x_2}{n_1} \mathbf{n}^{\perp},$$

which means that $\mathbf{x}$ is in $\mathrm{Span}\{\mathbf{n}^{\perp}\}$. For the case when $n_2 \neq 0$ the above argument requires only minor modifications. Therefore (a) implies (b).

Now we assume that $\mathbf{x}$ is in $\mathrm{Span}\{\mathbf{n}^{\perp}\}$. Then $\mathbf{x} = t\mathbf{n}^{\perp}$ for some real number t and consequently

$$\mathbf{x} \cdot \mathbf{n} = (t\mathbf{n}^{\perp}) \cdot \mathbf{n} = t(\mathbf{n}^{\perp} \cdot \mathbf{n}) = 0.$$

Therefore (b) implies (a).   □

Corollary 3.2.23. *Let* $\mathbf{u}$ *be a vector in* $\mathbb{R}^2$ *different from the origin. Then* $\mathbf{x}$ *is in* $\mathrm{Span}\{\mathbf{u}\}$ *if and only if* $\mathbf{x} \cdot \mathbf{u}^{\perp} = 0$.

Proof. By Theorem 3.2.22, $\mathbf{x} \cdot \mathbf{u}^{\perp} = 0$ if and only if $\mathbf{x}$ is in $\mathrm{Span}\{(\mathbf{u}^{\perp})^{\perp}\}$. This proves the result since

$$\mathrm{Span}\{(\mathbf{u}^{\perp})^{\perp}\} = \mathrm{Span}\{(-\mathbf{u})\} = \mathrm{Span}\{\mathbf{u}\}.$$

□

Example 3.2.24. Describe all solutions of the equation

$$2x + 3y = 0$$

as a the vector line $\mathrm{Span}\{\mathbf{u}\}$.

Solution. The equation $2x + 3y = 0$ can be written as

$$\begin{bmatrix} x \\ y \end{bmatrix} \cdot \begin{bmatrix} 2 \\ 3 \end{bmatrix} = 0.$$

According to Theorem 3.2.22, $\begin{bmatrix} x \\ y \end{bmatrix}$ satisfies the above equation if and only if $\begin{bmatrix} x \\ y \end{bmatrix}$ is in the vector line

$$\text{Span}\left\{\begin{bmatrix} 2 \\ 3 \end{bmatrix}^{\llcorner}\right\} = \text{Span}\left\{\begin{bmatrix} -3 \\ 2 \end{bmatrix}\right\}.$$

$\square$

Theorem 3.2.25. *Two nonzero orthogonal vectors* **a** *and* **b**, *that is two nonzero vectors* **a** *and* **b** *such that* $\mathbf{a} \cdot \mathbf{b} = 0$, *form a basis of* $\mathbb{R}^2$. *Such a basis is called an* **orthogonal basis**.

Proof. According to Theorem 3.2.22, if $\mathbf{a} \cdot \mathbf{b} = 0$, then $\mathbf{b} = t\mathbf{a}^{\llcorner}$, where t is a nonzero real number. It is easy to verify that

$$\det\begin{bmatrix} \mathbf{a} & \mathbf{b} \end{bmatrix} = \det\begin{bmatrix} \mathbf{a} & t\mathbf{a}^{\llcorner} \end{bmatrix} = t\det\begin{bmatrix} \mathbf{a} & \mathbf{a}^{\llcorner} \end{bmatrix} = t\|\mathbf{a}\|^2 \neq 0,$$

because $\mathbf{a} \neq \mathbf{0}$. The result is now a consequence of Theorems 3.1.19 and 3.1.22. $\square$

Area of a triangle in $\mathbb{R}^2$

Now we discuss the geometric meaning of the determinant of a 2×2 matrix. First we need to define the distance from a point to a line.

Definition 3.2.26. Let $\mathbf{a} = \begin{bmatrix} a_1 \\ a_2 \end{bmatrix}$ and $\mathbf{b} = \begin{bmatrix} b_1 \\ b_2 \end{bmatrix}$ be points in $\mathbb{R}^2$ such that $\mathbf{b} \neq \mathbf{0}$.
By the *distance from* **a** *to the vector line* Span{**b**} we mean the number

$$\|\mathbf{a} - \text{proj}_{\mathbf{b}}\mathbf{a}\|.$$

From Theorem 3.2.12 we obtain a formula for the distance of a point from a vector line in terms of a determinant.

Theorem 3.2.27. *Let* $\mathbf{a} = \begin{bmatrix} a_1 \\ a_2 \end{bmatrix}$ *and* $\mathbf{b} = \begin{bmatrix} b_1 \\ b_2 \end{bmatrix}$ *be points in* $\mathbb{R}^2$ *such that* $\mathbf{b} \neq \mathbf{0}$.
The distance from **a** *to the vector line* Span{**b**} *is*

$$\frac{1}{\|\mathbf{b}\|}\left|\det\begin{bmatrix} a_1 & b_1 \\ a_2 & b_2 \end{bmatrix}\right|.$$

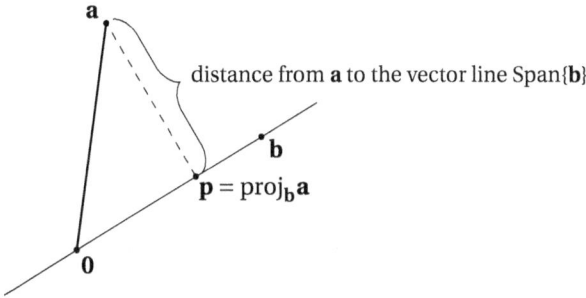

Figure 3.16: The distance from a point to a line.

Proof. Let $\mathbf{p}$ be the projection of $\mathbf{a}$ on the vector line Span$\{\mathbf{b}\}$, see Fig. 3.16. Then

$$\mathbf{p} = \mathrm{proj}_{\mathbf{b}}\mathbf{a} = \frac{\mathbf{a}\cdot\mathbf{b}}{\|\mathbf{b}\|^2}\mathbf{b},$$

by Theorem 3.2.12. If $\mathbf{a} = \begin{bmatrix} a_1 \\ a_2 \end{bmatrix}$ and $\mathbf{b} = \begin{bmatrix} b_1 \\ b_2 \end{bmatrix}$, then

$$
\begin{aligned}
\|\mathbf{a}-\mathbf{p}\|^2 &= \left\|\mathbf{a} - \frac{\mathbf{a}\cdot\mathbf{b}}{\|\mathbf{b}\|^2}\mathbf{b}\right\|^2 \\
&= \left(\mathbf{a} - \frac{\mathbf{a}\cdot\mathbf{b}}{\|\mathbf{b}\|^2}\mathbf{b}\right)\cdot\left(\mathbf{a} - \frac{\mathbf{a}\cdot\mathbf{b}}{\|\mathbf{b}\|^2}\mathbf{b}\right) \\
&= \|\mathbf{a}\|^2 - 2\frac{(\mathbf{a}\cdot\mathbf{b})^2}{\|\mathbf{b}\|^2} + \frac{(\mathbf{a}\cdot\mathbf{b})^2}{\|\mathbf{b}\|^4}\|\mathbf{b}\|^2 \\
&= \|\mathbf{a}\|^2 - \frac{(\mathbf{a}\cdot\mathbf{b})^2}{\|\mathbf{b}\|^2} \\
&= \frac{1}{\|\mathbf{b}\|^2}\left(\|\mathbf{a}\|^2\|\mathbf{b}\|^2 - (\mathbf{a}\cdot\mathbf{b})^2\right) \\
&= \frac{1}{\|\mathbf{b}\|^2}\left((a_1^2 + a_2^2)(b_1^2 + b_2^2) - (a_1 b_1 + a_2 b_2)^2\right) \\
&= \frac{1}{\|\mathbf{b}\|^2}(a_1 b_2 - a_2 b_1)^2 \\
&= \frac{1}{\|\mathbf{b}\|^2}\left(\det\begin{bmatrix} a_1 & b_1 \\ a_2 & b_2 \end{bmatrix}\right)^2.
\end{aligned}
$$

Consequently, $\|\mathbf{a}-\mathbf{p}\| = \frac{1}{\|\mathbf{b}\|}\left|\det\begin{bmatrix} a_1 & b_1 \\ a_2 & b_2 \end{bmatrix}\right|.$ □

The above theorem gives us a convenient formula for calculating the area of a triangle defined by two vectors in $\mathbb{R}^2$.

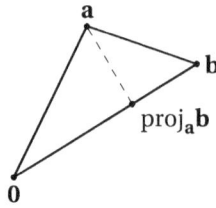

Figure 3.17: The triangle defined by vectors **a** and **b**.

Corollary 3.2.28. *Let* $\mathbf{a} = \begin{bmatrix} a_1 \\ a_2 \end{bmatrix}$ *and* $\mathbf{b} = \begin{bmatrix} b_1 \\ b_2 \end{bmatrix}$ *be nonzero vectors in* $\mathbb{R}^2$. *The area of the triangle* **0ab** *is*

$$\frac{1}{2}\left|\det\begin{bmatrix} a_1 & b_1 \\ a_2 & b_2 \end{bmatrix}\right|$$

Proof. Since the area of the triangle **0ab** is

$$\frac{1}{2} \cdot \text{(the length of the base \textbf{0b})} \cdot \text{(the height from \textbf{a})},$$

the result is an immediate consequence of Theorem 3.2.27.   □

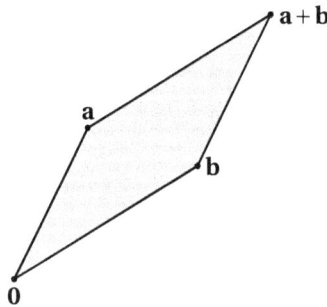

Figure 3.18: The area of the parallelogram is $|\det[\mathbf{a}\ \mathbf{b}]|$.

From the derived formula for the area of a triangle we obtain an explicit geometric interpretation of $|\det[\mathbf{a}\ \mathbf{b}]|$ as the area of the parallelogram with vertices **0**, **a**, **b**, and $\mathbf{a}+\mathbf{b}$, see Fig. 3.18.

Example 3.2.29. The area of the triangle $\mathbf{0ab}$ where $\mathbf{a} = \begin{bmatrix} 1 \\ 2 \end{bmatrix}$ and $\mathbf{b} = \begin{bmatrix} 3 \\ -5 \end{bmatrix}$ is

$$\frac{1}{2}\left| \det \begin{bmatrix} 1 & 3 \\ 2 & -5 \end{bmatrix} \right| = \frac{1}{2} \cdot |-11| = \frac{11}{2}.$$

3.2.1 Exercises

Calculate the norms of the given vectors.

1. $\begin{bmatrix} 3 \\ 4 \end{bmatrix}$

2. $\begin{bmatrix} 2 \\ 7 \end{bmatrix}$

3. $\begin{bmatrix} -5 \\ 3 \end{bmatrix}$

4. $\begin{bmatrix} \frac{1}{\sqrt{2}} \\ \frac{1}{\sqrt{2}} \end{bmatrix}$

5. $\begin{bmatrix} \frac{1}{\sqrt{a}} \\ \frac{1}{\sqrt{a}} \end{bmatrix}$, $a > 0$

6. $\begin{bmatrix} \sin\alpha \\ \cos\alpha \end{bmatrix}$

Calculate the distance between the given vectors.

7. $\begin{bmatrix} 5 \\ 4 \end{bmatrix}$ and $\begin{bmatrix} 7 \\ 1 \end{bmatrix}$

8. $\begin{bmatrix} 1 \\ 2 \end{bmatrix}$ and $\begin{bmatrix} 5 \\ -1 \end{bmatrix}$

9. $\begin{bmatrix} 0 \\ a \end{bmatrix}$ and $\begin{bmatrix} a \\ 0 \end{bmatrix}$

10. $\begin{bmatrix} a \\ b \end{bmatrix}$ and $\begin{bmatrix} -b \\ a \end{bmatrix}$

Calculate the following dot products.

11. $\begin{bmatrix} 1 \\ 2 \end{bmatrix} \cdot \begin{bmatrix} 7 \\ 2 \end{bmatrix}$

12. $\begin{bmatrix} 1 \\ 2 \end{bmatrix} \cdot \begin{bmatrix} 3 \\ 0 \end{bmatrix}$

13. $\begin{bmatrix} -4 \\ 1 \end{bmatrix} \cdot \begin{bmatrix} 1 \\ 4 \end{bmatrix}$

14. $\begin{bmatrix} 4 \\ 1 \end{bmatrix} \cdot \begin{bmatrix} -4 \\ 1 \end{bmatrix}$

For the given vectors $\mathbf{u}$ and $\mathbf{v}$ find a real number a such that $\mathbf{u} \cdot \mathbf{v} = 0$ and then draw $\mathbf{u}$ and $\mathbf{v}$.

15. $\mathbf{u} = \begin{bmatrix} 4 \\ a \end{bmatrix}$ and $\mathbf{v} = \begin{bmatrix} 1 \\ 2 \end{bmatrix}$

16. $\mathbf{u} = \begin{bmatrix} 1 \\ 1 \end{bmatrix}$ and $\mathbf{v} = \begin{bmatrix} a \\ -2 \end{bmatrix}$

17. $\mathbf{u} = \begin{bmatrix} a \\ -1 \end{bmatrix}$ and $\mathbf{v} = \begin{bmatrix} a \\ 1 \end{bmatrix}$

18. $\mathbf{u} = \begin{bmatrix} 4 \\ a \end{bmatrix}$ and $\mathbf{v} = \begin{bmatrix} -a \\ 2 \end{bmatrix}$

19. Let $\mathbf{a} = \begin{bmatrix} 1 \\ 1 \end{bmatrix}$, $\mathbf{b} = \begin{bmatrix} 2 \\ 1+a \end{bmatrix}$, $\mathbf{c} = \begin{bmatrix} 3 \\ 5 \end{bmatrix}$, and $\mathbf{d} = \begin{bmatrix} 7 \\ a \end{bmatrix}$. Find a real number a such that $(\mathbf{b} - \mathbf{a}) \cdot (\mathbf{d} - \mathbf{c}) = 0$ and then draw the vectors $\mathbf{a}$, $\mathbf{b}$, $\mathbf{c}$, and $\mathbf{d}$.

20. Let $\mathbf{a} = \begin{bmatrix} -1 \\ 2 \end{bmatrix}$, $\mathbf{b} = \begin{bmatrix} 1 \\ 2+a \end{bmatrix}$, $\mathbf{c} = \begin{bmatrix} 1 \\ 1 \end{bmatrix}$ and $\mathbf{d} = \begin{bmatrix} -1 \\ a-1 \end{bmatrix}$. Find a real number a such that $(\mathbf{b} - \mathbf{a}) \cdot (\mathbf{d} - \mathbf{c}) = 0$ and then draw the vectors $\mathbf{a}$, $\mathbf{b}$, $\mathbf{c}$, and $\mathbf{d}$.

21. For $\mathbf{a} = \begin{bmatrix} -1 \\ 2 \end{bmatrix}$ and $\mathbf{n} = \begin{bmatrix} 1 \\ 3 \end{bmatrix}$ draw all vectors $\mathbf{x}$ such that $(\mathbf{x} - \mathbf{a}) \cdot \mathbf{n} = 0$.

22. For $\mathbf{a} = \mathbf{n} = \begin{bmatrix} 4 \\ 1 \end{bmatrix}$ draw all vectors $\mathbf{x}$ such that $(\mathbf{x} - \mathbf{a}) \cdot \mathbf{n} = 0$.

For the given vectors $\mathbf{b}$ and $\mathbf{u}$ find $\text{proj}_{\mathbf{u}}\mathbf{b}$ using the formula in Theorem 3.2.12.

23. $\mathbf{b} = \begin{bmatrix} 2 \\ 1 \end{bmatrix}$ and $\mathbf{u} = \begin{bmatrix} 1 \\ 3 \end{bmatrix}$ 25. $\mathbf{b} = \begin{bmatrix} x \\ y \end{bmatrix}$ and $\mathbf{u} = \begin{bmatrix} 1 \\ 1 \end{bmatrix}$

24. $\mathbf{b} = \begin{bmatrix} 1 \\ -3 \end{bmatrix}$ and $\mathbf{u} = \begin{bmatrix} 2 \\ 5 \end{bmatrix}$ 26. $\mathbf{b} = \begin{bmatrix} x \\ y \end{bmatrix}$ and $\mathbf{u} = \begin{bmatrix} -3 \\ 2 \end{bmatrix}$

For the given vectors $\mathbf{b}$ and $\mathbf{u}$, find the projection matrix for the projection on the vector line Span$\{\mathbf{u}\}$ and then use it to calculate $\text{proj}_{\mathbf{u}}\mathbf{b}$.

27. $\mathbf{b} = \begin{bmatrix} 0 \\ 1 \end{bmatrix}$ and $\mathbf{u} = \begin{bmatrix} 1 \\ 1 \end{bmatrix}$ 29. $\mathbf{b} = \begin{bmatrix} x \\ y \end{bmatrix}$ and $\mathbf{u} = \begin{bmatrix} 2 \\ -1 \end{bmatrix}$

28. $\mathbf{b} = \begin{bmatrix} 1 \\ 1 \end{bmatrix}$ and $\mathbf{u} = \begin{bmatrix} -3 \\ 1 \end{bmatrix}$ 30. $\mathbf{b} = \begin{bmatrix} x \\ y \end{bmatrix}$ and $\mathbf{u} = \begin{bmatrix} 4 \\ -3 \end{bmatrix}$

Find the reflection of the point $\mathbf{b}$ across the vector line Span$\{\mathbf{u}\}$.

31. $\mathbf{b} = \begin{bmatrix} 2 \\ 1 \end{bmatrix}$ and $\mathbf{u} = \begin{bmatrix} 1 \\ 3 \end{bmatrix}$ 33. $\mathbf{b} = \begin{bmatrix} x \\ y \end{bmatrix}$ and $\mathbf{u} = \begin{bmatrix} 1 \\ 1 \end{bmatrix}$

32. $\mathbf{b} = \begin{bmatrix} 1 \\ -3 \end{bmatrix}$ and $\mathbf{u} = \begin{bmatrix} 2 \\ 5 \end{bmatrix}$ 34. $\mathbf{b} = \begin{bmatrix} x \\ y \end{bmatrix}$ and $\mathbf{u} = \begin{bmatrix} -3 \\ 2 \end{bmatrix}$

35. Find a matrix A such that the reflection of the point $\mathbf{b}$ across the vector line Span$\left\{\begin{bmatrix} -3 \\ 1 \end{bmatrix}\right\}$ is $A\mathbf{b}$.

36. Find a matrix A such that the reflection of the point $\mathbf{b}$ across the vector line Span$\left\{\begin{bmatrix} 2 \\ 5 \end{bmatrix}\right\}$ is $A\mathbf{b}$.

37. Describe all solutions of the equation $3x - 4y = 0$ as a vector line Span$\{\mathbf{u}\}$.

38. Describe all solutions of the equation $5x + 2y = 0$ as a vector line Span$\{\mathbf{u}\}$.

Find the equation of the vector line Span$\{\mathbf{u}\}$ for the given vector $\mathbf{u}$.

39. $\mathbf{u} = \begin{bmatrix} 1 \\ 1 \end{bmatrix}$

41. $\mathbf{u} = \begin{bmatrix} -3 \\ 1 \end{bmatrix}$

40. $\mathbf{u} = \begin{bmatrix} 2 \\ -1 \end{bmatrix}$

42. $\mathbf{u} = \begin{bmatrix} -1 \\ -1 \end{bmatrix}$

Find the distance from the given point $\mathbf{a}$ to the vector line Span$\{\mathbf{b}\}$.

43. $\mathbf{a} = \begin{bmatrix} 0 \\ 1 \end{bmatrix}$ and $\mathbf{b} = \begin{bmatrix} 1 \\ 1 \end{bmatrix}$

44. $\mathbf{a} = \begin{bmatrix} 2 \\ -1 \end{bmatrix}$ and $\mathbf{b} = \begin{bmatrix} 1 \\ 3 \end{bmatrix}$

45. Find the area of the triangle defined by vectors $\mathbf{a} = \begin{bmatrix} 3 \\ 1 \end{bmatrix}$ and $\mathbf{b} = \begin{bmatrix} 2 \\ 5 \end{bmatrix}$.

46. Find the area of the triangle defined by vectors $\mathbf{a} = \begin{bmatrix} -2 \\ -1 \end{bmatrix}$ and $\mathbf{b} = \begin{bmatrix} -3 \\ 7 \end{bmatrix}$.

47. Show that for arbitrary vectors $\mathbf{a}$ and $\mathbf{b}$ in $\mathbb{R}^2$ we have $\mathbf{b} \cdot \mathbf{a}^{\perp} = \det \begin{bmatrix} \mathbf{a} & \mathbf{b} \end{bmatrix}$.

48. Show that the following conditions are equivalent

   (a) $\mathbf{a}$ and $\mathbf{b}$ are linearly independent;

   (b) $\mathbf{b} \cdot \mathbf{a}^{\perp} \neq 0$;

   (c) $\mathbf{a} \cdot \mathbf{b}^{\perp} \neq 0$.

49. Show that $\det \begin{bmatrix} \mathbf{a}^{\perp} & \mathbf{b}^{\perp} \end{bmatrix} = \det \begin{bmatrix} \mathbf{a} & \mathbf{b} \end{bmatrix}$.

50. Show that the vectors $\mathbf{a}$ and $\mathbf{b}$ are linearly independent if and only if the vectors $\mathbf{a}^{\perp}$ and $\mathbf{b}^{\perp}$ are linearly independent.

51. Let $\mathbf{S}$ be the solution set of the equation

$$\begin{bmatrix} a_1 & b_1 \\ a_2 & b_2 \end{bmatrix} \begin{bmatrix} x \\ y \end{bmatrix} = \begin{bmatrix} 0 \\ 0 \end{bmatrix},$$

   that is,

$$\mathbf{S} = \left\{ \begin{bmatrix} x \\ y \end{bmatrix} \text{ in } \mathbb{R}^2 : \begin{bmatrix} a_1 & b_1 \\ a_2 & b_2 \end{bmatrix} \begin{bmatrix} x \\ y \end{bmatrix} = \begin{bmatrix} 0 \\ 0 \end{bmatrix} \right\}.$$

   Show that one of the following is true:

   (a) $\mathbf{S} = \left\{ \begin{bmatrix} 0 \\ 0 \end{bmatrix} \right\}$;

   (b) $\mathbf{S}$ is a vector line;

   (c) $\mathbf{S} = \mathbb{R}^2$.

52. Let $\mathbf{a}$ and $\mathbf{b}$ be linearly independent vectors in $\mathbb{R}^2$. Show that if $\mathbf{x} \cdot \mathbf{a} = 0$ and $\mathbf{x} \cdot \mathbf{b} = 0$, then $\mathbf{x} = \mathbf{0}$.

53. If $\mathbf{p} = \text{proj}_b \mathbf{a}$, show that

$$\frac{\mathbf{a} \cdot \mathbf{b}}{\|\mathbf{a}\| \|\mathbf{b}\|} = \frac{\|\mathbf{p}\|}{\|\mathbf{a}\|} \quad \text{if } \mathbf{a} \cdot \mathbf{b} \geq 0$$

and

$$\frac{\mathbf{a} \cdot \mathbf{b}}{\|\mathbf{a}\| \|\mathbf{b}\|} = -\frac{\|\mathbf{p}\|}{\|\mathbf{a}\|} \quad \text{if } \mathbf{a} \cdot \mathbf{b} \leq 0.$$

54. If $\mathbf{p} = \text{proj}_b \mathbf{a}$, show that

$$\frac{\det\begin{bmatrix} \mathbf{a} & \mathbf{b} \end{bmatrix}}{\|\mathbf{a}\| \|\mathbf{b}\|} = \frac{\|\mathbf{a} - \mathbf{p}\|}{\|\mathbf{a}\|} \quad \text{if } \det\begin{bmatrix} \mathbf{a} & \mathbf{b} \end{bmatrix} \geq 0$$

and

$$\frac{\det\begin{bmatrix} \mathbf{a} & \mathbf{b} \end{bmatrix}}{\|\mathbf{a}\| \|\mathbf{b}\|} = -\frac{\|\mathbf{a} - \mathbf{p}\|}{\|\mathbf{a}\|} \quad \text{if } \det\begin{bmatrix} \mathbf{a} & \mathbf{b} \end{bmatrix} \leq 0.$$

55. (Cramer's Rule revisited) Let $\mathbf{a}$ and $\mathbf{b}$ be linearly independent vectors in $\mathbb{R}^2$. For every $\mathbf{c}$ in $\mathbb{R}^2$ the equation

$$x\mathbf{a} + y\mathbf{b} = \mathbf{c}$$

has a unique solution

$$x = \frac{\mathbf{c} \cdot \mathbf{b}^{\perp}}{\mathbf{a} \cdot \mathbf{b}^{\perp}} \quad \text{and} \quad y = \frac{\mathbf{c} \cdot \mathbf{a}^{\perp}}{\mathbf{b} \cdot \mathbf{a}^{\perp}}.$$

56. Using exercise 55 solve the system

$$\begin{cases} 2x + 5y = 4 \\ x + 3y = 3 \end{cases}.$$

57. Find a matrix A such that the reflection of the point $\mathbf{b}$ across the vector line Span$\{\mathbf{u}\}$ is $A\mathbf{b}$. Show that A is a symmetric matrix.

3.3   Symmetric 2×2 matrices

Recall that a matrix A is called symmetric if $A^T = A$. In the case of 2×2 matrices, every symmetric matrix has the form

$$\begin{bmatrix} a & c \\ c & b \end{bmatrix}.$$

In this section we discuss eigenvectors of symmetric matrices. We start with the following interesting theorem.

Theorem 3.3.1. *Let A be a 2×2 matrix. If A has two orthogonal eigenvectors, then A is symmetric.*

Proof. Let $\begin{bmatrix} u_1 \\ u_2 \end{bmatrix}$ and $\begin{bmatrix} v_1 \\ v_2 \end{bmatrix}$ be orthogonal eigenvectors of the matrix A, that is,

$$A \begin{bmatrix} u_1 \\ u_2 \end{bmatrix} = \alpha \begin{bmatrix} u_1 \\ u_2 \end{bmatrix} \quad \text{and} \quad A \begin{bmatrix} v_1 \\ v_2 \end{bmatrix} = \beta \begin{bmatrix} v_1 \\ v_2 \end{bmatrix},$$

for some numbers α and β, and

$$\begin{bmatrix} u_1 \\ u_2 \end{bmatrix} \cdot \begin{bmatrix} v_1 \\ v_2 \end{bmatrix} = 0.$$

Let

$$\mathbf{q} = \frac{1}{\sqrt{u_1^2 + u_2^2}} \begin{bmatrix} u_1 \\ u_2 \end{bmatrix} = \begin{bmatrix} \dfrac{u_1}{\sqrt{u_1^2 + u_2^2}} \\ \dfrac{u_2}{\sqrt{u_1^2 + u_2^2}} \end{bmatrix} = \begin{bmatrix} q_1 \\ q_2 \end{bmatrix}$$

and

$$\mathbf{r} = \frac{1}{\sqrt{v_1^2 + v_2^2}} \begin{bmatrix} v_1 \\ v_2 \end{bmatrix} = \begin{bmatrix} \dfrac{v_1}{\sqrt{v_1^2 + v_2^2}} \\ \dfrac{v_2}{\sqrt{v_1^2 + v_2^2}} \end{bmatrix} = \begin{bmatrix} r_1 \\ r_2 \end{bmatrix}.$$

Note that

$$\|\mathbf{q}\| = \|\mathbf{r}\| = 1 \quad \text{and} \quad \mathbf{q} \cdot \mathbf{r} = 0$$

and

$$A\mathbf{q} = \alpha\mathbf{q} \quad \text{and} \quad A\mathbf{r} = \beta\mathbf{r}. \tag{3.6}$$

Equations (3.6) can be written as a single equation

$$A \begin{bmatrix} \mathbf{q} & \mathbf{r} \end{bmatrix} = \begin{bmatrix} \mathbf{q} & \mathbf{r} \end{bmatrix} \begin{bmatrix} \alpha & 0 \\ 0 & \beta \end{bmatrix}. \tag{3.7}$$

Now, if we let

$$P = \begin{bmatrix} \mathbf{q} & \mathbf{r} \end{bmatrix},$$

then we have

$$P^T P = \begin{bmatrix} \mathbf{q}^T \\ \mathbf{r}^T \end{bmatrix} \begin{bmatrix} \mathbf{q} & \mathbf{r} \end{bmatrix} = \begin{bmatrix} \mathbf{q} \cdot \mathbf{q} & \mathbf{q} \cdot \mathbf{r} \\ \mathbf{r} \cdot \mathbf{q} & \mathbf{r} \cdot \mathbf{r} \end{bmatrix} = \begin{bmatrix} 1 & 0 \\ 0 & 1 \end{bmatrix},$$

which means, by Theorem 1.2.18, that the matrix P is invertible and we have

$$P^{-1} = P^T.$$

Since equation (3.7) can be written as

$$AP = P \begin{bmatrix} \alpha & 0 \\ 0 & \beta \end{bmatrix},$$

we get

$$A = A \begin{bmatrix} 1 & 0 \\ 0 & 1 \end{bmatrix} = APP^{-1} = P \begin{bmatrix} \alpha & 0 \\ 0 & \beta \end{bmatrix} P^{-1} = P \begin{bmatrix} \alpha & 0 \\ 0 & \beta \end{bmatrix} P^T.$$

Now we can show that the matrix A is symmetric. Indeed, we have

$$A^T = \left(P\begin{bmatrix} \alpha & 0 \\ 0 & \beta \end{bmatrix} P^T\right)^T = (P^T)^T \begin{bmatrix} \alpha & 0 \\ 0 & \beta \end{bmatrix}^T P^T = P\begin{bmatrix} \alpha & 0 \\ 0 & \beta \end{bmatrix} P^T = A.$$

$\square$

Example 3.3.2. Find a matrix which has $\begin{bmatrix} 1 \\ 2 \end{bmatrix}$ as an eigenvector corresponding the eigenvalue $\lambda = 1$ and $\begin{bmatrix} -2 \\ 1 \end{bmatrix}$ as an eigenvector corresponding the eigenvalue $\lambda = 3$. Explain why the result is a symmetric matrix.

Solution. The matrix

$$\begin{bmatrix} 1 & -2 \\ 2 & 1 \end{bmatrix} \begin{bmatrix} 1 & 0 \\ 0 & 3 \end{bmatrix} \begin{bmatrix} 1 & -2 \\ 2 & 1 \end{bmatrix}^{-1}$$

will have the desired properties. Since

$$\begin{bmatrix} 1 & -2 \\ 2 & 1 \end{bmatrix}^{-1} = \frac{1}{5}\begin{bmatrix} 1 & 2 \\ -2 & 1 \end{bmatrix} = \begin{bmatrix} \frac{1}{5} & \frac{2}{5} \\ -\frac{2}{5} & \frac{1}{5} \end{bmatrix},$$

we have

$$\begin{bmatrix} 1 & -2 \\ 2 & 1 \end{bmatrix} \begin{bmatrix} 1 & 0 \\ 0 & 3 \end{bmatrix} \begin{bmatrix} 1 & -2 \\ 2 & 1 \end{bmatrix}^{-1} = \begin{bmatrix} 1 & -2 \\ 2 & 1 \end{bmatrix} \begin{bmatrix} 1 & 0 \\ 0 & 3 \end{bmatrix} \begin{bmatrix} \frac{1}{5} & \frac{2}{5} \\ -\frac{2}{5} & \frac{1}{5} \end{bmatrix} = \begin{bmatrix} \frac{13}{5} & -\frac{4}{5} \\ -\frac{4}{5} & \frac{7}{5} \end{bmatrix}.$$

This matrix is symmetric because the eigenvectors $\begin{bmatrix} 1 \\ 2 \end{bmatrix}$ and $\begin{bmatrix} -2 \\ 1 \end{bmatrix}$ are orthogonal.

$\square$

From the proof of Theorem 3.3.1 we can obtain the following useful result.

Theorem 3.3.3. *If $\mathbf{p}$ and $\mathbf{q}$ are vectors in $\mathbb{R}^2$ such that*

$$\|\mathbf{q}\| = \|\mathbf{r}\| = 1 \quad and \quad \mathbf{q} \cdot \mathbf{r} = 0,$$

then the matrix $P = \begin{bmatrix} \mathbf{q} & \mathbf{r} \end{bmatrix}$ is invertible and $P^{-1} = P^T$.

Matrices of the type described in the above theorem are important in theoretical considerations and practical applications.

Definition 3.3.4. A 2 × 2 matrix P is called an ***orthogonal matrix*** if it is invertible and
$$P^T = P^{-1}.$$

It turns out that every orthogonal matrix satisfies the conditions in Theorem 3.3.3, which explains the name.

Theorem 3.3.5. *If A is an orthogonal 2×2 matrix, then it has the form $A = \begin{bmatrix} \mathbf{q} & \mathbf{r} \end{bmatrix}$ with*
$$\|\mathbf{q}\| = \|\mathbf{r}\| = 1 \quad and \quad \mathbf{q} \cdot \mathbf{r} = 0.$$

Proof. If $A = \begin{bmatrix} \mathbf{q} & \mathbf{r} \end{bmatrix}$ is an orthogonal 2 × 2 matrix, then

$$\begin{bmatrix} 1 & 0 \\ 0 & 1 \end{bmatrix} = \begin{bmatrix} \mathbf{q} & \mathbf{r} \end{bmatrix}^T \begin{bmatrix} \mathbf{q} & \mathbf{r} \end{bmatrix} = \begin{bmatrix} \mathbf{q} \cdot \mathbf{q} & \mathbf{q} \cdot \mathbf{r} \\ \mathbf{r} \cdot \mathbf{q} & \mathbf{r} \cdot \mathbf{r} \end{bmatrix} = \begin{bmatrix} \|\mathbf{q}\|^2 & 0 \\ 0 & \|\mathbf{r}\|^2 \end{bmatrix}.$$

Consequently, $\|\mathbf{q}\| = \|\mathbf{r}\| = 1$ and $\mathbf{q} \cdot \mathbf{r} = 0$.

$\square$

Example 3.3.6. The columns of the matrix

$$A = \begin{bmatrix} 1 & -2 \\ 2 & 1 \end{bmatrix}$$

are orthogonal, but A is not an orthogonal matrix because

$$\left\| \begin{bmatrix} 1 \\ 2 \end{bmatrix} \right\| = \left\| \begin{bmatrix} -2 \\ 1 \end{bmatrix} \right\| = \sqrt{5}.$$

Since dividing every entry of the matrix by $\sqrt{5}$ does not affect the orthogonality of the columns of the matrix, the obtained matrix

$$\begin{bmatrix} \frac{1}{\sqrt{5}} & \frac{-2}{\sqrt{5}} \\ \frac{2}{\sqrt{5}} & \frac{1}{\sqrt{5}} \end{bmatrix}$$

is an orthogonal matrix.

Now we return to the main subject of this section, namely, symmetric matrices. We proved that every 2 × 2 matrix with two orthogonal eigenvectors is symmetric. It turns out that the converse is also true, that is, every symmetric 2 × 2 matrix has two orthogonal eigenvectors. To prove that we will use the following simple lemma.

> **Lemma 3.3.7.** *If A is a symmetric 2×2 matrix, then*
>
> $$(A\mathbf{u}) \cdot \mathbf{v} = \mathbf{u} \cdot (A\mathbf{v})$$
>
> *for every $\mathbf{u}$ and $\mathbf{v}$ in $\mathbb{R}^2$.*

Proof. The equality can be verified by direct calculations and is left as an exercise (Exercise 27). □

> **Theorem 3.3.8.** *Every symmetric 2×2 matrix has real eigenvalues and two orthogonal eigenvectors.*

Proof. Let

$$A = \begin{bmatrix} a & c \\ c & b \end{bmatrix}.$$

The roots of the equation

$$\det \begin{bmatrix} a - \lambda & c \\ c & b - \lambda \end{bmatrix} = (a - \lambda)(b - \lambda) - c^2 = \lambda^2 - \lambda(a + b) + ab - c^2 = 0$$

are

$$\alpha = \frac{1}{2}\left(a + b + \sqrt{(a - b)^2 + 4c^2}\right) \quad \text{and} \quad \beta = \frac{1}{2}\left(a + b - \sqrt{(a - b)^2 + 4c^2}\right).$$

Since $(a - b)^2 + 4c^2 \geq 0$, α and β are real numbers, so A has real eigenvalues.

Now we consider two cases: $\alpha \neq \beta$ and $\alpha = \beta$.

First we assume that $\alpha \neq \beta$. Let $\mathbf{u}$ be an eigenvector corresponding to the eigenvalue α and let $\mathbf{v}$ be an eigenvector corresponding to the eigenvalue β. Then

$$(\alpha - \beta)(\mathbf{u} \cdot \mathbf{v}) = (\alpha\mathbf{u}) \cdot \mathbf{v} - \mathbf{u} \cdot (\beta\mathbf{v}) = (A\mathbf{u}) \cdot \mathbf{v} - \mathbf{u} \cdot (A\mathbf{v}) = 0,$$

where the last equality follows from Lemma 3.3.7. Since $(\alpha - \beta)(\mathbf{u} \cdot \mathbf{v}) = 0$ and $\alpha \neq \beta$, we must have $\mathbf{u} \cdot \mathbf{v} = 0$.

Now we assume that $\alpha = \beta$. This means that $(a - b)^2 + 4c^2 = 0$ and consequently $a = b$ and $c = 0$. But then

$$A = \begin{bmatrix} a & 0 \\ 0 & a \end{bmatrix}.$$

This yields $\alpha = \beta = a$ and every nonzero vector is an eigenvector of A, so we can take, for example, $\mathbf{u} = \begin{bmatrix} 1 \\ 0 \end{bmatrix}$ and $\mathbf{v} = \begin{bmatrix} 0 \\ 1 \end{bmatrix}$. □

From the proof of the above theorem we obtain the following useful result.

Corollary 3.3.9. *Eigenvectors corresponding to different eigenvalues of a symmetric 2 × 2 matrix are orthogonal.*

Definition 3.3.10. A representation of a 2 × 2 matrix A in the form

$$A = P \begin{bmatrix} \alpha & 0 \\ 0 & \beta \end{bmatrix} P^T$$

where P is a 2 × 2 orthogonal matrix and α and β are real numbers is called an **orthogonal diagonalization** of A. If a matrix A has an orthogonal diagonalization, we say that A can be **orthogonally diagonalized.**

Example 3.3.11. Since, by Theorem 3.3.3, the inverse of an orthogonal matrix is equal to its transpose, we have

$$\begin{bmatrix} \frac{1}{\sqrt{5}} & \frac{-2}{\sqrt{5}} \\ \frac{2}{\sqrt{5}} & \frac{1}{\sqrt{5}} \end{bmatrix}^{-1} = \begin{bmatrix} \frac{1}{\sqrt{5}} & \frac{2}{\sqrt{5}} \\ \frac{-2}{\sqrt{5}} & \frac{1}{\sqrt{5}} \end{bmatrix}$$

and an orthogonal diagonalization of the matrix obtained in Example 3.3.2 is

$$\begin{bmatrix} \frac{1}{\sqrt{5}} & \frac{-2}{\sqrt{5}} \\ \frac{2}{\sqrt{5}} & \frac{1}{\sqrt{5}} \end{bmatrix} \begin{bmatrix} 1 & 0 \\ 0 & 3 \end{bmatrix} \begin{bmatrix} \frac{1}{\sqrt{5}} & \frac{2}{\sqrt{5}} \\ \frac{-2}{\sqrt{5}} & \frac{1}{\sqrt{5}} \end{bmatrix} = \begin{bmatrix} \frac{13}{5} & -\frac{4}{5} \\ -\frac{4}{5} & \frac{7}{5} \end{bmatrix}.$$

From the proof of Theorem 3.3.1 and Theorem 3.3.8 we obtain the following fundamental property of symmetric matrices.

Theorem 3.3.12. *A 2 × 2 matrix can be orthogonally diagonalized if and only if it is symmetric.*

Example 3.3.13. Orthogonally diagonalize the symmetric matrix

$$A = \begin{bmatrix} 7 & 3 \\ 3 & -1 \end{bmatrix}.$$

Solution. The eigenvalues of A are the roots of the equation

$$\det \begin{bmatrix} 7-\lambda & 3 \\ 3 & -1-\lambda \end{bmatrix} = \lambda^2 - 6\lambda - 16 = 0,$$

which are $\lambda = -2$ and $\lambda = 8$.

The eigenvectors corresponding to the eigenvalue $\lambda = -2$ are the solutions of the equation

$$\begin{bmatrix} 7 & 3 \\ 3 & -1 \end{bmatrix} \begin{bmatrix} x \\ y \end{bmatrix} = -2 \begin{bmatrix} x \\ y \end{bmatrix}$$

or, equivalently, of the equation

$$\begin{bmatrix} 9 & 3 \\ 3 & 1 \end{bmatrix} \begin{bmatrix} x \\ y \end{bmatrix} = \begin{bmatrix} 0 \\ 0 \end{bmatrix}.$$

This gives us $3x + y = 0$, which can be written as $y = -3x$. The solutions are of the form

$$\begin{bmatrix} x \\ y \end{bmatrix} = \begin{bmatrix} x \\ -3x \end{bmatrix} = x \begin{bmatrix} 1 \\ -3 \end{bmatrix},$$

so for an eigenvector corresponding to the eigenvalue $\lambda = -2$ we can take $\begin{bmatrix} 1 \\ -3 \end{bmatrix}$.

An eigenvector corresponding to the eigenvalue $\lambda = 8$ must be orthogonal to the vector $\begin{bmatrix} 1 \\ -3 \end{bmatrix}$, so we can take $\begin{bmatrix} 3 \\ 1 \end{bmatrix}$ because according to the Theorem 3.2.22 a vector orthogonal to the vector $\begin{bmatrix} 1 \\ -3 \end{bmatrix}$ is in $\mathrm{Span}\left\{ \begin{bmatrix} 3 \\ 1 \end{bmatrix} \right\}$.

Since

$$\left\| \begin{bmatrix} 1 \\ -3 \end{bmatrix} \right\| = \left\| \begin{bmatrix} 3 \\ 1 \end{bmatrix} \right\| = \sqrt{10},$$

we have

$$\begin{bmatrix} \frac{1}{\sqrt{10}} & \frac{3}{\sqrt{10}} \\ -\frac{3}{\sqrt{10}} & \frac{1}{\sqrt{10}} \end{bmatrix}^{-1} = \begin{bmatrix} \frac{1}{\sqrt{10}} & \frac{3}{\sqrt{10}} \\ -\frac{3}{\sqrt{10}} & \frac{1}{\sqrt{10}} \end{bmatrix}^{T} = \begin{bmatrix} \frac{1}{\sqrt{10}} & -\frac{3}{\sqrt{10}} \\ \frac{3}{\sqrt{10}} & \frac{1}{\sqrt{10}} \end{bmatrix}$$

and consequently

$$A = \begin{bmatrix} \frac{1}{\sqrt{10}} & \frac{3}{\sqrt{10}} \\ -\frac{3}{\sqrt{10}} & \frac{1}{\sqrt{10}} \end{bmatrix} \begin{bmatrix} -2 & 0 \\ 0 & 8 \end{bmatrix} \begin{bmatrix} \frac{1}{\sqrt{10}} & -\frac{3}{\sqrt{10}} \\ \frac{3}{\sqrt{10}} & \frac{1}{\sqrt{10}} \end{bmatrix}.$$

$\square$

The spectral decomposition of a 2×2 symmetric matrix

Now we turn our attention to the so-called spectral decomposition of a 2×2 symmetric matrix.

Definition 3.3.14. Let A be a 2×2 matrix. By a ***spectral decomposition*** of A we mean a representation of A in the form

$$A = \frac{\alpha}{\|\mathbf{u}\|^2}\mathbf{u}\mathbf{u}^T + \frac{\beta}{\|\mathbf{v}\|^2}\mathbf{v}\mathbf{v}^T,$$

where $\mathbf{u}$ and $\mathbf{v}$ are nonzero orthogonal vectors and α and β are real numbers.

Theorem 3.3.15. *If $\{\mathbf{u}, \mathbf{v}\}$ is a basis of orthogonal eigenvectors of the 2×2 symmetric matrix A, then*

$$A = \frac{\alpha}{\|\mathbf{u}\|^2}\mathbf{u}\mathbf{u}^T + \frac{\beta}{\|\mathbf{v}\|^2}\mathbf{v}\mathbf{v}^T,$$

where α is the eigenvalue corresponding to the eigenvector $\mathbf{u}$ and β is the eigenvalue corresponding to the eigenvector $\mathbf{v}$, that is,

$$A\mathbf{u} = \alpha\mathbf{u} \quad \text{and} \quad A\mathbf{v} = \beta\mathbf{v}.$$

Proof. Let α and β be the eigenvalues of A corresponding to $\mathbf{u}$ and $\mathbf{v}$, respectively. We define

$$\mathbf{q} = \frac{1}{\|\mathbf{u}\|}\mathbf{u} \quad \text{and} \quad \mathbf{r} = \frac{1}{\|\mathbf{v}\|}\mathbf{v}.$$

Then we get, as in the proof of Theorem 3.3.1,

$$A = \begin{bmatrix} \mathbf{q} & \mathbf{r} \end{bmatrix} \begin{bmatrix} \alpha & 0 \\ 0 & \beta \end{bmatrix} \begin{bmatrix} \mathbf{q}^T \\ \mathbf{r}^T \end{bmatrix},$$

and, for an arbitrary $\mathbf{x}$ in $\mathbb{R}^2$, we have

$$\begin{aligned}
A\mathbf{x} &= \begin{bmatrix} \mathbf{q} & \mathbf{r} \end{bmatrix} \begin{bmatrix} \alpha & 0 \\ 0 & \beta \end{bmatrix} \begin{bmatrix} \mathbf{q}^T \\ \mathbf{r}^T \end{bmatrix} \mathbf{x} \\
&= \begin{bmatrix} \mathbf{q} & \mathbf{r} \end{bmatrix} \begin{bmatrix} \alpha & 0 \\ 0 & \beta \end{bmatrix} \begin{bmatrix} \mathbf{q}^T\mathbf{x} \\ \mathbf{r}^T\mathbf{x} \end{bmatrix} \\
&= \begin{bmatrix} \mathbf{q} & \mathbf{r} \end{bmatrix} \begin{bmatrix} \alpha\mathbf{q}^T\mathbf{x} \\ \beta\mathbf{r}^T\mathbf{x} \end{bmatrix} \\
&= \alpha\mathbf{q}(\mathbf{q}^T\mathbf{x}) + \beta\mathbf{r}(\mathbf{r}^T\mathbf{x}) \\
&= \alpha(\mathbf{q}\mathbf{q}^T)\mathbf{x} + \beta(\mathbf{r}\mathbf{r}^T)\mathbf{x} \\
&= \left(\alpha(\mathbf{q}\mathbf{q}^T) + \beta(\mathbf{r}\mathbf{r}^T)\right)\mathbf{x}.
\end{aligned}$$

Consequently, by Theorem 2.1.18, we get

$$A = \alpha(\mathbf{q}\mathbf{q}^T) + \beta(\mathbf{r}\mathbf{r}^T) = \frac{\alpha}{\|\mathbf{u}\|^2}\mathbf{u}\mathbf{u}^T + \frac{\beta}{\|\mathbf{v}\|^2}\mathbf{v}\mathbf{v}^T.$$

□

The following theorem is a converse of Theorem 3.3.15.

Theorem 3.3.16. *If* $\{\mathbf{u}, \mathbf{v}\}$ *is a basis of orthogonal vectors of* $\mathbb{R}^2$ *and* α *and* β *are arbitrary real numbers, then the matrix*

$$A = \frac{\alpha}{\|\mathbf{u}\|^2}\mathbf{u}\mathbf{u}^T + \frac{\beta}{\|\mathbf{v}\|^2}\mathbf{v}\mathbf{v}^T$$

is symmetric and $\mathbf{u}$ *is an eigenvector of A corresponding to the eigenvalue* α *and* $\mathbf{v}$ *is an eigenvector of A corresponding to the eigenvalue* β.

Proof. Assume that

$$A = \frac{\alpha}{\|\mathbf{u}\|^2}\mathbf{u}\mathbf{u}^T + \frac{\beta}{\|\mathbf{v}\|^2}\mathbf{v}\mathbf{v}^T,$$

where $\mathbf{u}$ and $\mathbf{v}$ are nonzero orthogonal vectors and α and β are real numbers. Then

$$\begin{aligned}
A^T &= \left(\frac{\alpha}{\|\mathbf{u}\|^2}\mathbf{u}\mathbf{u}^T + \frac{\beta}{\|\mathbf{v}\|^2}\mathbf{v}\mathbf{v}^T\right)^T \\
&= \frac{\alpha}{\|\mathbf{u}\|^2}\left(\mathbf{u}\mathbf{u}^T\right)^T + \frac{\beta}{\|\mathbf{v}\|^2}\left(\mathbf{v}\mathbf{v}^T\right)^T \\
&= \frac{\alpha}{\|\mathbf{u}\|^2}\mathbf{u}\mathbf{u}^T + \frac{\beta}{\|\mathbf{v}\|^2}\mathbf{v}\mathbf{v}^T = A,
\end{aligned}$$

So A is a symmetric matrix. It is easy to verify that $A\mathbf{u} = \alpha\mathbf{u}$ and $A\mathbf{v} = \beta\mathbf{v}$.   □

Example 3.3.17. A symmetric 2×2 matrix A has the eigenvalues $\lambda = 3$ and $\lambda = 7$. Find the matrix A if $\begin{bmatrix} 4 \\ 1 \end{bmatrix}$ is an eigenvector of A corresponding to $\lambda = 3$.

Solution. Since the matrix A is symmetric, eigenvectors of A corresponding to the eigenvalue $\lambda = 7$ must be orthogonal to $\begin{bmatrix} 4 \\ 1 \end{bmatrix}$. For example, we can use $\begin{bmatrix} 1 \\ -4 \end{bmatrix}$ because according to the Theorem 3.2.22 a vector orthogonal to the vector $\begin{bmatrix} 4 \\ 1 \end{bmatrix}$ is in

Span $\left\{\begin{bmatrix} 1 \\ -4 \end{bmatrix}\right\}$. Consequently, the matrix is

$$\frac{3}{17}\begin{bmatrix} 4 \\ 1 \end{bmatrix}[4\ \ 1] + \frac{7}{17}\begin{bmatrix} 1 \\ -4 \end{bmatrix}[1\ \ -4] = \frac{3}{17}\begin{bmatrix} 16 & 4 \\ 4 & 1 \end{bmatrix} + \frac{7}{17}\begin{bmatrix} 1 & -4 \\ -4 & 16 \end{bmatrix}$$

$$= \frac{1}{17}\begin{bmatrix} 55 & -16 \\ -16 & 115 \end{bmatrix}$$

$$= \begin{bmatrix} \frac{55}{17} & -\frac{16}{17} \\ -\frac{16}{17} & \frac{115}{17} \end{bmatrix}.$$

□

Recall that for any nonzero vector $\mathbf{a}$ in $\mathbb{R}^2$ we have

$$\text{proj}_{\mathbf{a}}\mathbf{x} = \frac{1}{\|\mathbf{a}\|^2}\mathbf{a}\mathbf{a}^T\mathbf{x}$$

for all $\mathbf{x}$ in $\mathbb{R}^2$. From Theorems 3.3.15 and 3.3.16 we obtain the following version of the spectral decomposition.

Corollary 3.3.18.

(a) *Let A be a 2 × 2 matrix with two orthogonal eigenvectors* $\mathbf{u}$ *and* $\mathbf{v}$*. Then, for every vector* $\mathbf{x}$ *in* $\mathbb{R}^2$*, we have*

$$A\mathbf{x} = \alpha\,\text{proj}_{\mathbf{u}}\mathbf{x} + \beta\,\text{proj}_{\mathbf{v}}\mathbf{x},$$

where α *is the eigenvalue of A corresponding to* $\mathbf{u}$ *and* β *is the eigenvalue of A corresponding to* $\mathbf{v}$*.*

(b) *If* $\{\mathbf{u}, \mathbf{v}\}$ *is a basis of orthogonal vectors in* $\mathbb{R}^2$*,* α *and* β *are real numbers, and A is a 2 × 2 matrix such that for every vector* $\mathbf{x}$ *in* $\mathbb{R}^2$ *we have*

$$A\mathbf{x} = \alpha\,\text{proj}_{\mathbf{u}}\mathbf{x} + \beta\,\text{proj}_{\mathbf{v}}\mathbf{x},$$

then A is a symmetric matrix such that $\mathbf{u}$ *is an eigenvector of A corresponding to the eigenvalue* α *and* $\mathbf{v}$ *is an eigenvector of A corresponding to the eigenvalue* β*.*

The above version of spectral decomposition gives it a clear geometric meaning, see Fig. 3.19.

Example 3.3.19. Let α and β be two real numbers. Find a matrix A such that $\mathbf{u} =$

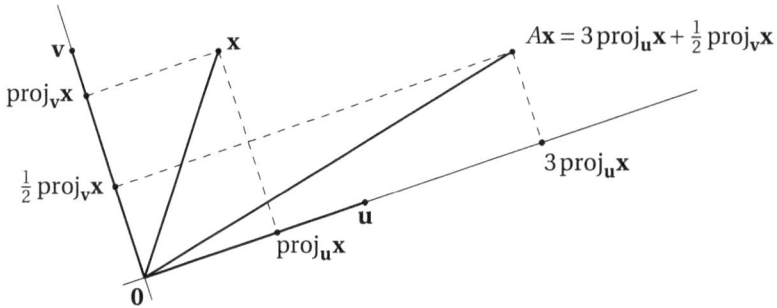

Figure 3.19: $A\mathbf{x} = 3\,\text{proj}_{\mathbf{u}}\mathbf{x} + \frac{1}{2}\,\text{proj}_{\mathbf{v}}\mathbf{x}$.

$\begin{bmatrix} 1 \\ 1 \end{bmatrix}$ is an eigenvector of A corresponding to an eigenvalue α and $\mathbf{v} = \begin{bmatrix} 1 \\ -1 \end{bmatrix}$ is an eigenvector of A corresponding to an eigenvalue β.

Solution 1. First we calculate $\text{proj}_{\mathbf{u}}\mathbf{x}$ and $\text{proj}_{\mathbf{v}}\mathbf{x}$ using the formula from Theorem 3.2.12, that is

$$\text{proj}_{\mathbf{w}}\mathbf{b} = \frac{\mathbf{b}\cdot\mathbf{w}}{\|\mathbf{w}\|^2}\mathbf{w} = \frac{\mathbf{b}\cdot\mathbf{w}}{\mathbf{w}\cdot\mathbf{w}}\mathbf{w}.$$

In our case we have

$$\text{proj}_{\mathbf{u}}\mathbf{x} = \frac{\begin{bmatrix} x \\ y \end{bmatrix}\cdot\begin{bmatrix} 1 \\ 1 \end{bmatrix}}{\begin{bmatrix} 1 \\ 1 \end{bmatrix}\cdot\begin{bmatrix} 1 \\ 1 \end{bmatrix}}\begin{bmatrix} 1 \\ 1 \end{bmatrix} = \frac{x+y}{2}\begin{bmatrix} 1 \\ 1 \end{bmatrix}$$

and

$$\text{proj}_{\mathbf{v}}\mathbf{x} = \frac{\begin{bmatrix} x \\ y \end{bmatrix}\cdot\begin{bmatrix} 1 \\ -1 \end{bmatrix}}{\begin{bmatrix} 1 \\ -1 \end{bmatrix}\cdot\begin{bmatrix} 1 \\ -1 \end{bmatrix}}\begin{bmatrix} 1 \\ -1 \end{bmatrix} = \frac{x+y}{2}\begin{bmatrix} 1 \\ -1 \end{bmatrix}.$$

Consequently,

$$A\begin{bmatrix} x \\ y \end{bmatrix} = \alpha\frac{x+y}{2}\begin{bmatrix} 1 \\ 1 \end{bmatrix} + \beta\frac{x-y}{2}\begin{bmatrix} 1 \\ -1 \end{bmatrix}.$$

The first column of the matrix A is

$$A\begin{bmatrix} 1 \\ 0 \end{bmatrix} = \alpha\frac{1}{2}\begin{bmatrix} 1 \\ 1 \end{bmatrix} + \beta\frac{1}{2}\begin{bmatrix} 1 \\ -1 \end{bmatrix} = \begin{bmatrix} \frac{\alpha}{2} + \frac{\beta}{2} \\ \frac{\alpha}{2} - \frac{\beta}{2} \end{bmatrix}$$

and the second column of the matrix A is

$$A \begin{bmatrix} 0 \\ 1 \end{bmatrix} = \alpha \frac{1}{2} \begin{bmatrix} 1 \\ 1 \end{bmatrix} + \beta \frac{-1}{2} \begin{bmatrix} 1 \\ -1 \end{bmatrix} = \begin{bmatrix} \frac{\alpha}{2} - \frac{\beta}{2} \\ \frac{\alpha}{2} + \frac{\beta}{2} \end{bmatrix}.$$

Hence

$$A = \begin{bmatrix} \frac{\alpha}{2} + \frac{\beta}{2} & \frac{\alpha}{2} - \frac{\beta}{2} \\ \frac{\alpha}{2} - \frac{\beta}{2} & \frac{\alpha}{2} + \frac{\beta}{2} \end{bmatrix}.$$

□

Solution 2. Using the formula obtained in Theorem 3.3.15 we get the same result, but the calculations are simpler:

$$A = \frac{\alpha}{2} \begin{bmatrix} 1 \\ 1 \end{bmatrix} \begin{bmatrix} 1 & 1 \end{bmatrix} + \frac{\beta}{2} \begin{bmatrix} 1 \\ -1 \end{bmatrix} \begin{bmatrix} 1 & -1 \end{bmatrix}$$

$$= \frac{\alpha}{2} \begin{bmatrix} 1 & 1 \\ 1 & 1 \end{bmatrix} + \frac{\beta}{2} \begin{bmatrix} 1 & -1 \\ -1 & 1 \end{bmatrix}$$

$$= \begin{bmatrix} \frac{\alpha}{2} + \frac{\beta}{2} & \frac{\alpha}{2} - \frac{\beta}{2} \\ \frac{\alpha}{2} - \frac{\beta}{2} & \frac{\alpha}{2} + \frac{\beta}{2} \end{bmatrix}.$$

The QR factorization of a 2 × 2 matrix

Now we consider a special factorization of 2 × 2 matrices that turns out to be useful in various computations, including solving systems of equations. This idea can be generalized to 3 × 3 square matrices (see the end of Chapter 7) and to square matrices of any size (see Core Topics in Linear Algebra). The QR factorization can also be used for non-square matrices. At the end of Chapter 4 we present QR factorization of 3 × 2 matrices.

Theorem 3.3.20. *A* 2 × 2 *matrix with linearly independent columns can written as a product of an orthogonal matrix and an upper triangular* 2 × 2 *matrix with positive entries on the main diagonal.*
More precisely, if the columns of the 2 × 2 *matrix* $A = \begin{bmatrix} \mathbf{c}_1 & \mathbf{c}_2 \end{bmatrix}$ *are linearly independent, then A can be represented in the form*

$$A = QR$$

where Q is a 2 × 2 *orthogonal matrix and* $R = \begin{bmatrix} r_{11} & r_{12} \\ 0 & r_{22} \end{bmatrix}$ *with* $r_{1,1} > 0$ *and* $r_{2,2} > 0$.

Proof. Let $A = \begin{bmatrix} \mathbf{c}_1 & \mathbf{c}_2 \end{bmatrix}$ be a 2 × 2 matrix with linearly independent columns. First we define

$$\mathbf{v}_1 = \mathbf{c}_1 \quad \text{and} \quad \mathbf{v}_2 = \mathbf{c}_2 - \text{proj}_{\mathbf{v}_1} \mathbf{c}_2 = \mathbf{c}_2 - \frac{\mathbf{c}_2 \cdot \mathbf{v}_1}{\mathbf{v}_1 \cdot \mathbf{v}_1} \mathbf{v}_1.$$

We note that $\mathbf{v}_2$ is nonzero (because the vectors $\mathbf{c}_1, \mathbf{c}_2$ are linearly independent), the vectors $\mathbf{v}_1$ and $\mathbf{v}_2$ are orthogonal, and we have

$$\mathbf{c}_2 = \mathbf{v}_2 + \frac{\mathbf{c}_2 \cdot \mathbf{v}_1}{\mathbf{v}_1 \cdot \mathbf{v}_1} \mathbf{v}_1.$$

Next we normalize the vectors $\mathbf{v}_1$ and $\mathbf{v}_2$:

$$\mathbf{u}_1 = \frac{1}{\|\mathbf{v}_1\|} \mathbf{v}_1 \quad \text{and} \quad \mathbf{u}_2 = \frac{1}{\|\mathbf{v}_2\|} \mathbf{v}_2.$$

If we denote

$$r_{1,1} = \|\mathbf{v}_1\|, \quad r_{1,2} = \|\mathbf{v}_1\| \frac{\mathbf{c}_2 \cdot \mathbf{v}_1}{\mathbf{v}_1 \cdot \mathbf{v}_1}, \quad \text{and} \quad r_{2,2} = \|\mathbf{v}_2\|,$$

then we have

$$\mathbf{c}_1 = r_{1,1} \mathbf{u}_1 \quad \text{and} \quad \mathbf{c}_2 = r_{1,2} \mathbf{u}_1 + r_{2,2} \mathbf{u}_2,$$

and consequently

$$\begin{bmatrix} \mathbf{c}_1 & \mathbf{c}_2 \end{bmatrix} = \begin{bmatrix} \mathbf{u}_1 & \mathbf{u}_2 \end{bmatrix} \begin{bmatrix} r_{11} & r_{12} \\ 0 & r_{22} \end{bmatrix}.$$

Note that $r_{1,1} > 0$ and $r_{2,2} > 0$. $\square$

Example 3.3.21. Find the QR factorization of the matrix $A = \begin{bmatrix} 2 & 3 \\ 4 & 1 \end{bmatrix}$.

Solution. We follow the construction used in the proof of Theorem 3.3.20.
   Since

$$\begin{bmatrix} 3 \\ 1 \end{bmatrix} - \frac{\begin{bmatrix} 3 \\ 1 \end{bmatrix} \cdot \begin{bmatrix} 2 \\ 4 \end{bmatrix}}{\begin{bmatrix} 2 \\ 4 \end{bmatrix} \cdot \begin{bmatrix} 2 \\ 4 \end{bmatrix}} \begin{bmatrix} 2 \\ 4 \end{bmatrix} = \begin{bmatrix} 3 \\ 1 \end{bmatrix} - \frac{1}{2} \begin{bmatrix} 2 \\ 4 \end{bmatrix} = \begin{bmatrix} 2 \\ -1 \end{bmatrix},$$

we have

$$\begin{bmatrix} 3 \\ 1 \end{bmatrix} = \frac{1}{2} \begin{bmatrix} 2 \\ 4 \end{bmatrix} + \begin{bmatrix} 2 \\ -1 \end{bmatrix}.$$

Next we calculate the norms

$$\left\| \begin{bmatrix} 2 \\ 4 \end{bmatrix} \right\| = 2\sqrt{5} \quad \text{and} \quad \left\| \begin{bmatrix} 2 \\ -1 \end{bmatrix} \right\| = \sqrt{5}$$

and let

$$\mathbf{u}_1 = \frac{1}{2\sqrt{5}} \begin{bmatrix} 2 \\ 4 \end{bmatrix} = \frac{1}{\sqrt{5}} \begin{bmatrix} 1 \\ 2 \end{bmatrix} \quad \text{and} \quad \mathbf{u}_2 = \frac{1}{\sqrt{5}} \begin{bmatrix} 2 \\ -1 \end{bmatrix}.$$

Then we have

$$\begin{bmatrix} 2 \\ 4 \end{bmatrix} = 2\sqrt{5}\mathbf{u}_1 \quad \text{and} \quad \begin{bmatrix} 3 \\ 1 \end{bmatrix} = \sqrt{5}\mathbf{u}_1 + \sqrt{5}\mathbf{u}_2.$$

Consequently,

$$A = \begin{bmatrix} \mathbf{u}_1 & \mathbf{u}_2 \end{bmatrix} \begin{bmatrix} 2\sqrt{5} & \sqrt{5} \\ 0 & \sqrt{5} \end{bmatrix} = \begin{bmatrix} \frac{1}{\sqrt{5}} & \frac{2}{\sqrt{5}} \\ \frac{2}{\sqrt{5}} & -\frac{1}{\sqrt{5}} \end{bmatrix} \begin{bmatrix} 2\sqrt{5} & \sqrt{5} \\ 0 & \sqrt{5} \end{bmatrix}.$$

□

Example 3.3.22. Find the QR factorization of the matrix $A = \begin{bmatrix} 2 & 1 \\ 1 & 5 \end{bmatrix}$.

Solution. Since

$$\begin{bmatrix} 1 \\ 5 \end{bmatrix} - \frac{\begin{bmatrix} 1 \\ 5 \end{bmatrix} \cdot \begin{bmatrix} 2 \\ 1 \end{bmatrix}}{\begin{bmatrix} 2 \\ 1 \end{bmatrix} \cdot \begin{bmatrix} 2 \\ 1 \end{bmatrix}} \begin{bmatrix} 2 \\ 1 \end{bmatrix} = \begin{bmatrix} 1 \\ 5 \end{bmatrix} - \frac{7}{5} \begin{bmatrix} 2 \\ 1 \end{bmatrix} = -\frac{9}{5} \begin{bmatrix} 1 \\ -2 \end{bmatrix},$$

we have

$$\begin{bmatrix} 1 \\ 5 \end{bmatrix} = \frac{7}{5} \begin{bmatrix} 2 \\ 1 \end{bmatrix} - \frac{9}{5} \begin{bmatrix} 1 \\ -2 \end{bmatrix} = \frac{7}{2} \begin{bmatrix} 2 \\ 1 \end{bmatrix} + \frac{9}{5} \begin{bmatrix} -1 \\ 2 \end{bmatrix}.$$

In the above sum we changed $\begin{bmatrix} 1 \\ -2 \end{bmatrix}$ to $\begin{bmatrix} -1 \\ 2 \end{bmatrix}$ because the coefficient in front of the vector $\mathbf{u}_2$ must be positive.

Next we calculate the norms

$$\left\| \begin{bmatrix} 2 \\ 1 \end{bmatrix} \right\| = \sqrt{5} \quad \text{and} \quad \left\| \begin{bmatrix} -1 \\ 2 \end{bmatrix} \right\| = \sqrt{5}$$

and define

$$\mathbf{u}_1 = \frac{1}{\sqrt{5}} \begin{bmatrix} 2 \\ 1 \end{bmatrix} \quad \text{and} \quad \mathbf{u}_2 = \frac{1}{\sqrt{5}} \begin{bmatrix} -1 \\ 2 \end{bmatrix}.$$

Consequently,

$$\begin{bmatrix} 2 \\ 1 \end{bmatrix} = \sqrt{5}\mathbf{u}_1 \quad \text{and} \quad \begin{bmatrix} 1 \\ 5 \end{bmatrix} = \frac{7}{5}\sqrt{5}\mathbf{u}_1 + \frac{9}{5}\sqrt{5}\mathbf{u}_2.$$

Now it is easy to obtain the QR factorization of the matrix A:

$$A = \begin{bmatrix} \mathbf{u}_1 & \mathbf{u}_2 \end{bmatrix} \begin{bmatrix} \sqrt{5} & \frac{7}{5}\sqrt{5} \\ 0 & \frac{9}{5}\sqrt{5} \end{bmatrix} = \begin{bmatrix} \frac{2}{\sqrt{5}} & -\frac{1}{\sqrt{5}} \\ \frac{1}{\sqrt{5}} & \frac{2}{\sqrt{5}} \end{bmatrix} \begin{bmatrix} \sqrt{5} & \frac{7}{5}\sqrt{5} \\ 0 & \frac{9}{5}\sqrt{5} \end{bmatrix}.$$

□

3.3.1   Exercises

1. A matrix A has $\begin{bmatrix} 3 \\ 2 \end{bmatrix}$ as an eigenvector corresponding the eigenvalue $\lambda = 2$ and $\begin{bmatrix} -2 \\ 3 \end{bmatrix}$ as an eigenvector corresponding the eigenvalue $\lambda = 5$. Explain why A is a symmetric matrix.

2. A matrix A has $\begin{bmatrix} a \\ b \end{bmatrix}$ as an eigenvector corresponding the eigenvalue α and $\begin{bmatrix} -b \\ a \end{bmatrix}$ as an eigenvector corresponding the eigenvalue β. Explain why A is a symmetric matrix.

For the given symmetric matrix A find matrices D and P such that D is a diagonal matrix, P is invertible, $P^{-1} = P^T$, and $A = PDP^T$.

3. $A = \begin{bmatrix} 8 & 2 \\ 2 & 5 \end{bmatrix}$

4. $A = \begin{bmatrix} 5 & 2 \\ 2 & 5 \end{bmatrix}$

5. $A = \begin{bmatrix} 9 & 8 \\ 8 & -3 \end{bmatrix}$

6. $A = \begin{bmatrix} 9 & 4 \\ 4 & 3 \end{bmatrix}$

7. $A = \begin{bmatrix} 27 & 5 \\ 5 & 3 \end{bmatrix}$

8. $A = \begin{bmatrix} 7 & 20 \\ 20 & 82 \end{bmatrix}$

9. $A = \begin{bmatrix} a+2 & a \\ a & a+2 \end{bmatrix}$

10. $A = \begin{bmatrix} a & a-k \\ a-k & a \end{bmatrix}$

Determine the spectral decomposition of the matrix A.

11. $A = \begin{bmatrix} 5 & 2 \\ 2 & 5 \end{bmatrix}$

12. $A = \begin{bmatrix} 9 & 4 \\ 4 & 3 \end{bmatrix}$

13. $A = \begin{bmatrix} 8 & 2 \\ 2 & 5 \end{bmatrix}$

14. $A = \begin{bmatrix} 9 & 8 \\ 8 & -3 \end{bmatrix}$

15. Find a symmetric 2×2 matrix A with eigenvalues 2 and -1 such that $\mathbf{u} = \begin{bmatrix} 1 \\ 1 \end{bmatrix}$ is an eigenvector of A corresponding to the eigenvalue 2.

16. Find a 2×2 matrix A with eigenvalues 5 and 2 such that $\mathbf{u} = \begin{bmatrix} 3 \\ 4 \end{bmatrix}$ is an eigenvector of A corresponding to the eigenvalue 5.

17. Let α and β be two real numbers. Find a 2×2 matrix A with eigenvalues α and β such that $\mathbf{u} = \begin{bmatrix} 2 \\ 1 \end{bmatrix}$ is an eigenvector of A corresponding to the eigenvalue α

and $\mathbf{v} = \begin{bmatrix} -1 \\ 2 \end{bmatrix}$ is an eigenvector of A corresponding to the eigenvalue β.

18. Let α and β are two real numbers. Find a 2×2 matrix A with eigenvalues α and β such that $\mathbf{u} = \begin{bmatrix} 3 \\ 4 \end{bmatrix}$ is an eigenvector of A corresponding to the eigenvalue α and $\mathbf{v} = \begin{bmatrix} 4 \\ -3 \end{bmatrix}$ is an eigenvector of A corresponding to the eigenvalue β.

Orthogonally diagonalize the projection matrix on the given vector line.

19. $\text{Span}\left\{ \begin{bmatrix} 1 \\ 1 \end{bmatrix} \right\}$

20. $\text{Span}\left\{ \begin{bmatrix} 2 \\ -1 \end{bmatrix} \right\}$

21. $\text{Span}\left\{ \begin{bmatrix} 1 \\ 3 \end{bmatrix} \right\}$

22. $\text{Span}\left\{ \begin{bmatrix} -4 \\ 1 \end{bmatrix} \right\}$

Orthogonally diagonalize the reflection matrix across the given vector line.

23. $\text{Span}\left\{ \begin{bmatrix} 4 \\ 1 \end{bmatrix} \right\}$

24. $\text{Span}\left\{ \begin{bmatrix} -3 \\ 1 \end{bmatrix} \right\}$

25. Find the eigenvalues and eigenvectors of the projection matrix on $\text{Span}\left\{ \begin{bmatrix} a \\ b \end{bmatrix} \right\}$ where $\begin{bmatrix} a \\ b \end{bmatrix} \neq \begin{bmatrix} 0 \\ 0 \end{bmatrix}$.

26. Suppose A is a symmetric 2×2 matrix with two different eigenvalues α and β and such that $\mathbf{u}$ is an eigenvector corresponding to the eigenvalue α. Show that $\mathbf{w} \neq \mathbf{0}$ is an eigenvector corresponding to the eigenvalue β if and only if $\mathbf{w}$ is in $\text{Span}\{\mathbf{u}^{\perp}\}$.

27. If A is a symmetric 2×2 matrix, show that $(A\mathbf{u}) \cdot \mathbf{v} = \mathbf{u} \cdot (A\mathbf{v})$ for every $\mathbf{u}$ and $\mathbf{v}$ in $\mathbb{R}^2$.

28. Let α and β are two real numbers. Find a matrix A such that $\mathbf{u} = \begin{bmatrix} u_1 \\ u_2 \end{bmatrix} \neq \begin{bmatrix} 0 \\ 0 \end{bmatrix}$ is an eigenvector of A corresponding to the eigenvalue α and $\mathbf{v} = \begin{bmatrix} u_2 \\ -u_1 \end{bmatrix}$ is an eigenvector of A corresponding to the eigenvalue β.

Find the QR factorization of the given matrix.

29. $\begin{bmatrix} 1 & 1 \\ 1 & 2 \end{bmatrix}$

30. $\begin{bmatrix} 3 & -1 \\ 1 & 0 \end{bmatrix}$

31. Find the spectral decomposition of the matrix $A = \begin{bmatrix} 1 & 0 \\ 0 & 1 \end{bmatrix}$.

Chapter 4

The vector space $\mathbb{R}^3$

In Chapter 3 we interpreted elements of $\mathbb{R}^2$ as 2×1 matrices and used tools of linear algebra to solve problems in $\mathbb{R}^2$. In this chapter we consider the space $\mathbb{R}^3$. We will see that some operations introduced in $\mathbb{R}^2$ generalize easily to $\mathbb{R}^3$ and have similar properties. On the other hand, there are same essential differences between $\mathbb{R}^2$ and $\mathbb{R}^3$.

Following the convention adopted in Chapter 3, elements of $\mathbb{R}^3$ will be denoted as 3×1 matrices $\begin{bmatrix} u_1 \\ u_2 \\ u_3 \end{bmatrix}$. In order to interpret algebraic operations in $\mathbb{R}^3$ geometrically, we identify elements of $\mathbb{R}^3$ with points in the space. This identification is done by applying the same idea that was used in the construction of Cartesian coordinates $\mathbb{R}^2$. We introduce in the space three mutually perpendicular lines, called the **x-axis**, **y-axis**, and **z-axis**, that intersect at one point called the **origin**. The origin is identified with the vector $\begin{bmatrix} 0 \\ 0 \\ 0 \end{bmatrix}$ and denoted by **0**. On each axis we choose a unit of length (usually the same on all three axes). Then every point in the space can be described by a unique triple of numbers u_1 u_2, and u_3 as illustrated in Fig. 4.1. These numbers are referred to as the **Cartesian coordinates** of the point.

Note that we use the same name "origin" and the same symbol **0** to denote $\begin{bmatrix} 0 \\ 0 \\ 0 \end{bmatrix}$ and $\begin{bmatrix} 0 \\ 0 \end{bmatrix}$. This slight inconsistency will not cause any problems. In the case of possible ambiguity we will write $\begin{bmatrix} 0 \\ 0 \\ 0 \end{bmatrix}$ or $\begin{bmatrix} 0 \\ 0 \end{bmatrix}$.

Depending on the context, we will say "an element of $\mathbb{R}^3$", "a 3×1 matrix", "a point in $\mathbb{R}^3$", or "a vector in $\mathbb{R}^3$". We will consider these expressions equivalent. Sometimes

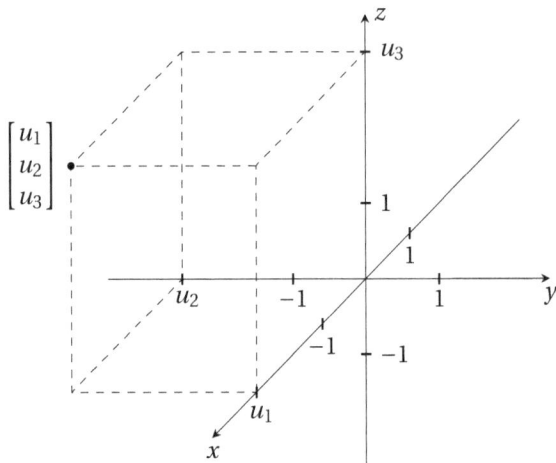

Figure 4.1: Cartesian coordinates in $\mathbb{R}^3$.

we will denote an element of $\mathbb{R}^3$ by (u_1, u_2, u_3). We note that

$$\begin{bmatrix} u_1 \\ u_2 \\ u_3 \end{bmatrix} = (u_1, u_2, u_3) \neq \begin{bmatrix} u_1 & u_2 & u_3 \end{bmatrix}.$$

In other words, (u_1, u_2, u_3) is not the matrix $\begin{bmatrix} u_1 & u_2 & u_3 \end{bmatrix}$.

4.1 Vectors in $\mathbb{R}^3$

Algebraic operations and the vector line in $\mathbb{R}^3$

Addition of vectors and multiplication of vectors by numbers are defined as for arbitrary matrices, that is:

$$\begin{bmatrix} u_1 \\ u_2 \\ u_3 \end{bmatrix} + \begin{bmatrix} v_1 \\ v_2 \\ v_3 \end{bmatrix} = \begin{bmatrix} u_1 + v_1 \\ u_2 + v_2 \\ u_3 + v_3 \end{bmatrix} \quad \text{and} \quad t \begin{bmatrix} u_1 \\ u_2 \\ u_3 \end{bmatrix} = \begin{bmatrix} tu_1 \\ tu_2 \\ tu_3 \end{bmatrix}.$$

These operations are natural extensions of the corresponding operations in $\mathbb{R}^2$ and have the same properties.

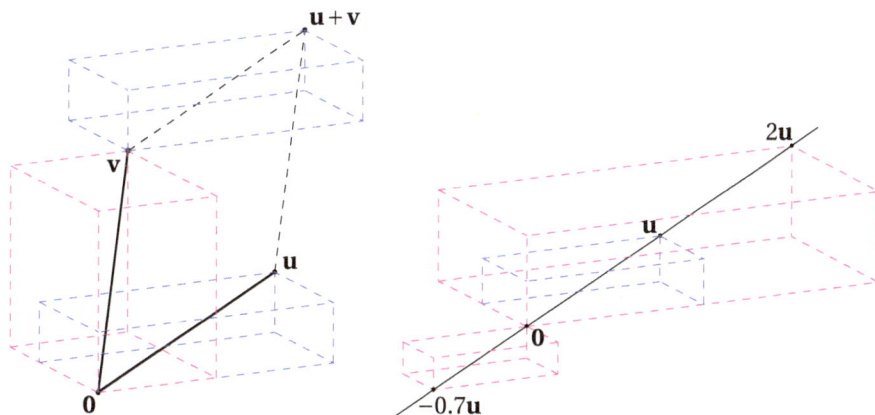

Figure 4.2: Addition and multiplication by a number in $\mathbb{R}^3$.

Definition 4.1.1. Let $\mathbf{u}$ be a vector in $\mathbb{R}^3$. The set of all vectors of the form $t\mathbf{u}$, where t is an arbitrary real number is called the ***vector subspace*** spanned by $\mathbf{u}$ and is denoted by Span$\{\mathbf{u}\}$. That is,

$$\mathrm{Span}\{\mathbf{u}\} = \{t\mathbf{u} : t \text{ in } \mathbb{R}\}.$$

If $\mathbf{u}$ is different from the origin, then Span$\{\mathbf{u}\}$ will be called a ***vector line***.

Note that the definitions of Span$\{\mathbf{u}\}$ and vector lines in $\mathbb{R}^3$ are identical with the definitions in $\mathbb{R}^2$. As before, we use the convention that when we say "vector line Span$\{\mathbf{u}\}$" we always implicitly assume that $\mathbf{u}$ is different from the origin.

Theorem 3.1.6 in Chapter 3 formulated for vectors in $\mathbb{R}^2$ is true for vectors in $\mathbb{R}^3$ and we can use the same proof.

Theorem 4.1.2. *For any two vectors* $\mathbf{u} \neq \mathbf{0}$ *and* $\mathbf{v} \neq \mathbf{0}$ *in* $\mathbb{R}^3$ *the following conditions are equivalent:*

(a) Span$\{\mathbf{u}\}$ = Span$\{\mathbf{v}\}$;

(b) $\mathbf{v} = x\mathbf{u}$ *for some real number* $x \neq 0$.

Definition 4.1.3. Let $\mathbf{u}$ be a vector in $\mathbb{R}^3$. If $\mathbf{v}$ is a vector in Span$\{\mathbf{u}\}$ different from the origin, then $\{\mathbf{v}\}$ is a ***basis*** of Span$\{\mathbf{u}\}$.

Linearly dependent and independent vectors in $\mathbb{R}^3$

Definition 3.1.8 of linear dependence of two vectors in $\mathbb{R}^2$ makes perfect sense in $\mathbb{R}^3$.

Definition 4.1.4. The vectors **u** and **v** from $\mathbb{R}^3$ are *linearly dependent* if at least one of the following conditions is true

(a) the vector **u** is in Span{**v**};

(b) the vector **v** is in Span{**u**}.

As in the case of vectors in $\mathbb{R}^2$, vectors **u** and **v** in $\mathbb{R}^3$ are linearly dependent if **u**, **v**, and **0** are on the same line.

Figure 4.3: Linearly dependent vectors.

The next two theorems are identical with the theorems formulated for vectors in $\mathbb{R}^2$ in Chapter 3 (Theorems 3.1.13 and 3.1.14). Moreover, the proofs presented there work in $\mathbb{R}^3$ without any modifications.

Theorem 4.1.5. *Vectors* **u** *and* **v** *in* $\mathbb{R}^3$ *are linearly dependent if and only if one of the following conditions holds:*

(a) **u** $= \mathbf{0}$ *or*

(b) **u** $\neq \mathbf{0}$ *and we have* **v** $= x\mathbf{u}$ *for a real number x.*

Theorem 4.1.6. *For any two vectors* **u** $\neq \mathbf{0}$ *and* **v** $\neq \mathbf{0}$ *in* $\mathbb{R}^3$ *the following conditions are equivalent:*

(a) *The vectors* **u** *and* **v** *are linearly dependent;*

(b) Span{**u**} $=$ Span{**v**}*;*

(c) **v** $= x\mathbf{u}$ *for some real number $x \neq 0$.*

Example 4.1.7. The vectors $\begin{bmatrix} 1 \\ 3 \\ -2 \end{bmatrix}$ and $\begin{bmatrix} -2 \\ -6 \\ 4 \end{bmatrix}$ are linearly dependent since

$$\begin{bmatrix} -2 \\ -6 \\ 4 \end{bmatrix} = -2 \begin{bmatrix} 1 \\ 3 \\ -2 \end{bmatrix}.$$

The following theorem, while very similar to Theorem 3.1.15, is not identical: condition (c) looks different. Like Theorem 3.1.15, this theorem describes practical methods for verifying linear dependence.

Theorem 4.1.8. *Let* $\mathbf{u} = \begin{bmatrix} u_1 \\ u_2 \\ u_3 \end{bmatrix}$ *and* $\mathbf{v} = \begin{bmatrix} v_1 \\ v_2 \\ v_3 \end{bmatrix}$. *The following conditions are equivalent*

(a) *The vectors* $\mathbf{u}$ *and* $\mathbf{v}$ *are linearly dependent;*

(b) *The equation*

$$x \begin{bmatrix} u_1 \\ u_2 \\ u_3 \end{bmatrix} + y \begin{bmatrix} v_1 \\ v_2 \\ v_3 \end{bmatrix} = \begin{bmatrix} 0 \\ 0 \\ 0 \end{bmatrix} \tag{4.1}$$

has a nontrivial solution, that is, a solution different from the trivial solution $x = y = 0$;

(c)

$$\det \begin{bmatrix} u_1 & v_1 \\ u_2 & v_2 \end{bmatrix} = \det \begin{bmatrix} u_2 & v_2 \\ u_3 & v_3 \end{bmatrix} = \det \begin{bmatrix} u_1 & v_1 \\ u_3 & v_3 \end{bmatrix} = 0. \tag{4.2}$$

Proof. The ideas used in the proof of Theorem 3.1.15 still work, but their implementation requires some modification.

First assume that $\mathbf{u}$ and $\mathbf{v}$ are linearly dependent. If $\mathbf{u}$ is in Span$\{\mathbf{v}\}$, then $\mathbf{u} = a\mathbf{v}$ for some real number a. Then $1 \cdot \mathbf{u} - a\mathbf{v} = \mathbf{0}$ and thus $x = 1$ and $y = -a$ is a nontrivial solution of the equation (4.1). The case when $\mathbf{v}$ is in Span$\{\mathbf{u}\}$ can be treated in a similar way. Therefore (a) implies (b).

Now assume that the equation (4.1) has a nontrivial solution. We can suppose that $x \neq 0$. Then we have

$$\begin{bmatrix} u_1 \\ u_2 \\ u_3 \end{bmatrix} = -\frac{y}{x} \begin{bmatrix} v_1 \\ v_2 \\ v_3 \end{bmatrix}$$

or, equivalently,

$$u_1 = -\frac{y}{x}v_1, \quad u_2 = -\frac{y}{x}v_2, \quad \text{and} \quad u_3 = -\frac{y}{x}v_3.$$

Hence

$$\det\begin{bmatrix} u_1 & v_1 \\ u_2 & v_2 \end{bmatrix} = \det\begin{bmatrix} -\frac{y}{x}v_1 & v_1 \\ -\frac{y}{x}v_2 & v_2 \end{bmatrix} = -\frac{y}{x}v_1v_2 + \frac{y}{x}v_2v_1 = 0,$$

$$\det\begin{bmatrix} u_2 & v_2 \\ u_3 & v_3 \end{bmatrix} = \det\begin{bmatrix} -\frac{y}{x}v_2 & v_2 \\ -\frac{y}{x}v_3 & v_3 \end{bmatrix} = -\frac{y}{x}v_2v_3 + \frac{y}{x}v_3v_2 = 0,$$

$$\det\begin{bmatrix} u_1 & v_1 \\ u_3 & v_3 \end{bmatrix} = \det\begin{bmatrix} -\frac{y}{x}v_1 & v_1 \\ -\frac{y}{x}v_3 & v_3 \end{bmatrix} = -\frac{y}{x}v_1v_3 + \frac{y}{x}v_3v_1 = 0.$$

The argument is similar if we have $y \neq 0$ instead of $x \neq 0$. Therefore (b) implies (c).

Finally assume that (4.2) holds. If $\mathbf{u} = \mathbf{0}$, then (a) is trivially true. If $\mathbf{u} \neq \mathbf{0}$, then at least one of the numbers u_1, u_2, u_3 must be different from 0. Without loss of generality we can assume that $u_1 \neq 0$. Then

$$v_1 = \frac{v_1}{u_1}u_1, \quad v_2 = \frac{v_1}{u_1}u_2, \quad \text{and} \quad v_3 = \frac{v_1}{u_1}u_3$$

and, consequently,

$$\begin{bmatrix} v_1 \\ v_2 \\ v_3 \end{bmatrix} = \frac{v_1}{u_1}\begin{bmatrix} u_1 \\ u_2 \\ u_3 \end{bmatrix}.$$

This means that $\mathbf{v}$ is in Span$\{\mathbf{u}\}$ and consequently $\mathbf{u}$ and $\mathbf{v}$ are linearly dependent. If we have $u_2 \neq 0$ or $u_3 \neq 0$ and we modify the proof accordingly. Therefore (c) implies (a). $\square$

Example 4.1.9. In Example 4.1.7 we found that the vectors $\begin{bmatrix} 1 \\ 3 \\ -2 \end{bmatrix}$ and $\begin{bmatrix} -2 \\ -6 \\ 4 \end{bmatrix}$ are linearly dependent. According to the above theorem the same conclusion can be derived from the following calculations:

$$\det\begin{bmatrix} 1 & -2 \\ 3 & -6 \end{bmatrix} = 1\cdot(-6) - (-2)\cdot3 = 0,$$

$$\det\begin{bmatrix} 3 & -6 \\ -2 & 4 \end{bmatrix} = 3\cdot4 - (-6)\cdot(-2) = 0,$$

$$\det\begin{bmatrix} 1 & -2 \\ -2 & 4 \end{bmatrix} = 1\cdot4 - (-2)\cdot(-2) = 0.$$

Example 4.1.10. The vectors $\begin{bmatrix} 1 \\ 3 \\ -2 \end{bmatrix}$ and $\begin{bmatrix} -2 \\ -5 \\ 4 \end{bmatrix}$ are not linearly dependent because

$$\det \begin{bmatrix} 1 & -2 \\ 3 & -5 \end{bmatrix} = 1 \cdot (-5) - (-2) \cdot 3 = 1 \neq 0.$$

There is no need to calculate $\det \begin{bmatrix} 3 & -6 \\ -2 & 4 \end{bmatrix}$ and $\det \begin{bmatrix} 1 & -2 \\ -2 & 4 \end{bmatrix}$.

Definition 4.1.11. If the vectors **u** and **v** are not linearly dependent, we say that they are *linearly independent.* In other words, the vectors **u** and **v** are linearly independent if **u** is **not** in Span{**v**} and **u** is **not** in Span{**v**}.

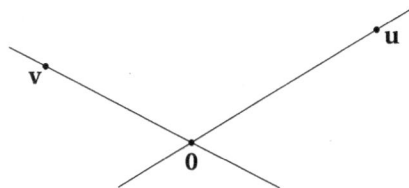

Figure 4.4: Linearly independent vectors.

Intuitively, vectors **u** and **v** are linearly independent if the only common point of the vector lines Span{**u**} and Span{**v**} is the origin **0**, see Fig. 4.4.

The following theorem characterizing linearly independent vectors is a direct consequence of Theorem 4.1.8.

Theorem 4.1.12. *Let* $\mathbf{u} = \begin{bmatrix} u_1 \\ u_2 \\ u_3 \end{bmatrix}$ *and* $\mathbf{v} = \begin{bmatrix} v_1 \\ v_2 \\ v_3 \end{bmatrix}$. *The following conditions are equivalent:*

(a) *The vectors* $\mathbf{u}$ *and* $\mathbf{v}$ *are linearly independent;*

(b) *The equation*

$$x \begin{bmatrix} u_1 \\ u_2 \\ u_3 \end{bmatrix} + y \begin{bmatrix} v_1 \\ v_2 \\ v_3 \end{bmatrix} = \begin{bmatrix} 0 \\ 0 \\ 0 \end{bmatrix}$$

has only the trivial solution, that is, $x = y = 0$;

(c) *At least one of the numbers*

$$\det \begin{bmatrix} u_1 & v_1 \\ u_2 & v_2 \end{bmatrix}, \quad \det \begin{bmatrix} u_2 & v_2 \\ u_3 & v_3 \end{bmatrix}, \quad \det \begin{bmatrix} u_1 & v_1 \\ u_3 & v_3 \end{bmatrix}$$

is different from 0.

The vector plane

Sets of all vectors in $\mathbb{R}^3$ that can be written as $s\mathbf{u} + t\mathbf{v}$ for some vectors $\mathbf{u}$ and $\mathbf{v}$ and arbitrary numbers s and t play an important role in linear algebra. They are a natural extension of Span$\{\mathbf{u}\}$ defined as the set of all vectors of the form $t\mathbf{u}$, where t is an arbitrary real number.

Definition 4.1.13. Let $\mathbf{u}$ and $\mathbf{v}$ be two vectors in $\mathbb{R}^3$. The set of all vectors in $\mathbb{R}^3$ of the form

$$s\mathbf{u} + t\mathbf{v},$$

where s and t are arbitrary real numbers, is called the **vector subspace** spanned by $\mathbf{u}$ and $\mathbf{v}$ and is denoted by Span$\{\mathbf{u}, \mathbf{v}\}$. That is,

$$\text{Span}\{\mathbf{u}, \mathbf{v}\} = \{s\mathbf{u} + t\mathbf{v} : s, t \text{ in } \mathbb{R}\}.$$

If the vectors $\mathbf{u}$ and $\mathbf{v}$ are linearly independent, then the vector subspace Span$\{\mathbf{u}, \mathbf{v}\}$ is called the **vector plane** spanned by the vectors $\mathbf{u}$ and $\mathbf{v}$.

Note that every vector plane is a vector subspace, but not every vector subspace is a vector plane.

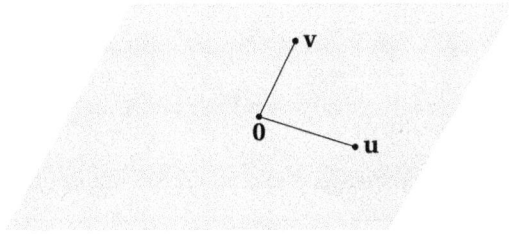

Figure 4.5: The vector plane spanned by vectors **u** and **v**.

Example 4.1.14. Since the vectors $\begin{bmatrix} 2 \\ 1 \\ 3 \end{bmatrix}$ and $\begin{bmatrix} 1 \\ -1 \\ 2 \end{bmatrix}$ are linearly independent,

$$\text{Span}\left\{ \begin{bmatrix} 2 \\ 1 \\ 3 \end{bmatrix}, \begin{bmatrix} 1 \\ -1 \\ 2 \end{bmatrix} \right\}$$

is a vector plane. On the other hand,

$$\text{Span}\left\{ \begin{bmatrix} 2 \\ -2 \\ 4 \end{bmatrix}, \begin{bmatrix} -1 \\ 1 \\ -2 \end{bmatrix} \right\}$$

is a vector subspace that is not a vector plane, because the vectors $\begin{bmatrix} 2 \\ -2 \\ 4 \end{bmatrix}$ and $\begin{bmatrix} -1 \\ 1 \\ -2 \end{bmatrix}$ are linearly dependent. Actually it is easy to verify that

$$\text{Span}\left\{ \begin{bmatrix} 2 \\ -2 \\ 4 \end{bmatrix}, \begin{bmatrix} -1 \\ 1 \\ -2 \end{bmatrix} \right\} = \text{Span}\left\{ \begin{bmatrix} 2 \\ -2 \\ 4 \end{bmatrix} \right\} = \text{Span}\left\{ \begin{bmatrix} -1 \\ 1 \\ -2 \end{bmatrix} \right\}.$$

As in the case of vector lines we adopt the convention that when we say "a vector plane Span{**u**, **v**}," we implicitly assume that **u** and **v** are linearly independent. When **u** and **v** are linearly dependent, then Span{**u**, **v**} is a vector line provided **u** ≠ **0** or **v** ≠ **0**. If **u** = **v** = **0**, then Span{**u**, **v**} = {**0**}.

The same vector subspace can be spanned by different pairs of vectors. How can we check that two pairs of vectors span the same vector subspace? We address this question in the remainder of this section.

Theorem 4.1.15. *Let* $\mathbf{a}, \mathbf{b}, \mathbf{u}, \mathbf{v}$ *be vectors in* $\mathbb{R}^3$. *The following two conditions are equivalent*

(a) Span$\{\mathbf{a}, \mathbf{b}\}$ = Span$\{\mathbf{u}, \mathbf{v}\}$;

(b) $\mathbf{a}, \mathbf{b}$ *are elements of* Span$\{\mathbf{u}, \mathbf{v}\}$ *and* $\mathbf{u}, \mathbf{v}$ *are elements of* Span$\{\mathbf{a}, \mathbf{b}\}$.

Proof. If Span$\{\mathbf{a}, \mathbf{b}\}$ = Span$\{\mathbf{u}, \mathbf{v}\}$, then clearly $\mathbf{a}, \mathbf{b}$ are elements of Span$\{\mathbf{u}, \mathbf{v}\}$ and $\mathbf{u}, \mathbf{v}$ are elements of Span$\{\mathbf{a}, \mathbf{b}\}$. This shows that (a) implies (b).

Now, if $\mathbf{a}, \mathbf{b}$ are elements of Span$\{\mathbf{u}, \mathbf{v}\}$, then there are real numbers q, r, s, t such that

$$\mathbf{a} = q\mathbf{u} + r\mathbf{v} \quad \text{and} \quad \mathbf{b} = s\mathbf{u} + t\mathbf{v}$$

and an arbitrary element from Span$\{\mathbf{a}, \mathbf{b}\}$ can be written in the form

$$x\mathbf{a} + y\mathbf{b} = x(q\mathbf{u} + r\mathbf{v}) + y(s\mathbf{u} + t\mathbf{v}) = (xq + ys)\mathbf{u} + (xy + yt)\mathbf{v}.$$

Consequently, an arbitrary element from Span$\{\mathbf{a}, \mathbf{b}\}$ is in Span$\{\mathbf{u}, \mathbf{v}\}$. Similarly, if $\mathbf{u}, \mathbf{v}$ are elements of Span$\{\mathbf{a}, \mathbf{b}\}$, then there are real numbers e, f, g, h such that

$$\mathbf{u} = e\mathbf{a} + f\mathbf{b} \quad \text{and} \quad \mathbf{v} = g\mathbf{a} + h\mathbf{b}$$

and an arbitrary element from Span$\{\mathbf{u}, \mathbf{v}\}$ can be written in the form

$$\alpha\mathbf{u} + \beta\mathbf{v} = \alpha(e\mathbf{a} + f\mathbf{b}) + \beta(g\mathbf{a} + h\mathbf{b}) = (\alpha e + \beta g)\mathbf{a} + (\alpha f + \beta h)\mathbf{b}.$$

Consequently, an arbitrary element from Span$\{\mathbf{u}, \mathbf{v}\}$ is in Span$\{\mathbf{a}, \mathbf{b}\}$. We can thus conclude that Span$\{\mathbf{a}, \mathbf{b}\}$ = Span$\{\mathbf{u}, \mathbf{v}\}$. This shows that (b) implies (a).   □

Example 4.1.16. Show that

$$\text{Span}\left\{\begin{bmatrix}1\\2\\3\end{bmatrix}, \begin{bmatrix}1\\-1\\2\end{bmatrix}\right\} = \text{Span}\left\{\begin{bmatrix}-1\\7\\0\end{bmatrix}, \begin{bmatrix}3\\0\\7\end{bmatrix}\right\}.$$

Solution. We use Theorem 4.1.15. Since

$$\begin{bmatrix}-1\\7\\0\end{bmatrix} = 2\begin{bmatrix}1\\2\\3\end{bmatrix} - 3\begin{bmatrix}1\\-1\\2\end{bmatrix} \quad \text{and} \quad \begin{bmatrix}3\\0\\7\end{bmatrix} = \begin{bmatrix}1\\2\\3\end{bmatrix} + 2\begin{bmatrix}1\\-1\\2\end{bmatrix},$$

the vectors $\begin{bmatrix}-1\\7\\0\end{bmatrix}$ and $\begin{bmatrix}3\\0\\7\end{bmatrix}$ are elements of the vector plane Span$\left\{\begin{bmatrix}1\\2\\3\end{bmatrix}, \begin{bmatrix}1\\-1\\2\end{bmatrix}\right\}$.

Moreover, since

$$\begin{bmatrix} 1 \\ 2 \\ 3 \end{bmatrix} = \frac{2}{7} \begin{bmatrix} -1 \\ 7 \\ 0 \end{bmatrix} + \frac{3}{7} \begin{bmatrix} 3 \\ 0 \\ 7 \end{bmatrix} \quad \text{and} \quad \begin{bmatrix} 1 \\ -1 \\ 2 \end{bmatrix} = -\frac{1}{7} \begin{bmatrix} -1 \\ 7 \\ 0 \end{bmatrix} + \frac{2}{7} \begin{bmatrix} 3 \\ 0 \\ 7 \end{bmatrix},$$

the vectors $\begin{bmatrix} 1 \\ 2 \\ 3 \end{bmatrix}$ and $\begin{bmatrix} 1 \\ -1 \\ 2 \end{bmatrix}$ are elements of the vector plane Span $\left\{ \begin{bmatrix} -1 \\ 7 \\ 0 \end{bmatrix}, \begin{bmatrix} 3 \\ 0 \\ 7 \end{bmatrix} \right\}$.

□

The following results are consequences of Theorem 4.1.15.

Theorem 4.1.17. *Let* **u** *and* **v** *be vectors in* $\mathbb{R}^3$.

(a) *For any real numbers s and t such that $s \neq 0$ and $t \neq 0$ we have*

$$\text{Span}\{\mathbf{u}, \mathbf{v}\} = \text{Span}\{s\mathbf{u}, t\mathbf{v}\};$$

(b) *For any real numbers s and t we have*

$$\text{Span}\{\mathbf{u}, \mathbf{v}\} = \text{Span}\{\mathbf{u} + s\mathbf{v}, \mathbf{v}\} = \text{Span}\{\mathbf{u}, \mathbf{v} + t\mathbf{u}\}.$$

Example 4.1.18. Let

$$\mathbf{u} = \begin{bmatrix} 1 \\ 2 \\ -1 \end{bmatrix} \quad \text{and} \quad \mathbf{v} = \begin{bmatrix} -1 \\ -1 \\ 2 \end{bmatrix}.$$

Since

$$2\mathbf{u} + 5\mathbf{v} = \begin{bmatrix} -3 \\ -1 \\ 8 \end{bmatrix},$$

we have

$$\text{Span} \left\{ \begin{bmatrix} 1 \\ 2 \\ -1 \end{bmatrix}, \begin{bmatrix} -1 \\ -1 \\ 2 \end{bmatrix} \right\} = \text{Span} \left\{ \begin{bmatrix} 2 \\ 4 \\ -2 \end{bmatrix}, \begin{bmatrix} -1 \\ -1 \\ 2 \end{bmatrix} \right\} = \text{Span} \left\{ \begin{bmatrix} -3 \\ -1 \\ 8 \end{bmatrix}, \begin{bmatrix} -1 \\ -1 \\ 2 \end{bmatrix} \right\}.$$

Bases of vector planes in $\mathbb{R}^3$

Vector planes in $\mathbb{R}^3$ have many properties that are similar to $\mathbb{R}^2$. For example, every vector plane in $\mathbb{R}^3$ has a basis consisting of two vectors.

Definition 4.1.19. Let **u** and **v** be linearly independent vectors. A pair of vectors {**a**, **b**} in Span{**u**, **v**} is called a ***basis*** of the vector plane Span{**u**, **v**} if the vectors satisfy the following two conditions:

1. **a** and **b** are linearly independent;

2. Span{**a**, **b**} = Span{**u**, **v**}.

An expression of the form $x\mathbf{a} + y\mathbf{b}$ is called a ***linear combination*** of vectors **a** and **b**. According to the above definition, linearly independent vectors **a** and **b** form a basis of the vector plane Span{**u**, **v**}, if every vector in Span{**u**, **v**} can be written as a linear combination of vectors **a** and **b**.

Clearly, {**u**, **v**} is a basis of the vector plane Span{**u**, **v**}.

Theorem 4.1.20. *Let **a** and **b** be linearly independent vectors and let **c** be an arbitrary vector in the vector plane* Span{**a**, **b**}. *Then the real numbers x and y such that*

$$\mathbf{c} = x\mathbf{a} + y\mathbf{b}$$

*are uniquely determined by the vector **c**.*

Proof. Suppose that

$$\mathbf{c} = x_1\mathbf{a} + y_1\mathbf{b} \quad \text{and} \quad \mathbf{c} = x_2\mathbf{a} + y_2\mathbf{b}.$$

Then

$$(x_1 - x_2)\mathbf{a} + (y_1 - y_2)\mathbf{b} = \mathbf{0}.$$

Since vectors **a** and **b** are linearly independent, we must have $x_1 - x_2 = 0$ and $y_1 - y_2 = 0$, by Theorem 4.1.12. Consequently, for every **c** in Span{**a**, **b**} there is only one pair of numbers x and y such that $\mathbf{c} = x\mathbf{a} + y\mathbf{b}$.   □

Definition 4.1.21. Let **c** be an arbitrary vector in Span{**a**, **b**}. The unique real numbers x and y such that $\mathbf{c} = x\mathbf{a} + y\mathbf{b}$ are called the ***coordinates*** of **c** in the basis {**a**, **b**}.

If {**a**, **b**} is a basis of the vector plane Span{**u**, **v**}, then Span{**u**, **v**} = Span{**a**, **b**}. In the next theorem we show that any two linearly independent vectors from a vector plane span that vector plane. It also shows that, if a vector subspace Span{**u**, **v**} contains two linearly independent vectors, then that vector subspace must be a vector plane.

Theorem 4.1.22. *If* **a** *and* **b** *are two linearly independent elements of a vector subspace* Span$\{$**u**,**v**$\}$, *then*

$$\text{Span}\{\mathbf{a}, \mathbf{b}\} = \text{Span}\{\mathbf{u}, \mathbf{v}\}$$

and the vectors **u** *and* **v** *are linearly independent.*

Proof. First we prove that Span$\{\mathbf{a}, \mathbf{b}\}$ = Span$\{\mathbf{u}, \mathbf{v}\}$. Let

$$\mathbf{a} = \begin{bmatrix} a_1 \\ a_2 \\ a_3 \end{bmatrix}, \mathbf{b} = \begin{bmatrix} b_1 \\ b_2 \\ b_3 \end{bmatrix}, \mathbf{u} = \begin{bmatrix} u_1 \\ u_2 \\ u_3 \end{bmatrix}, \mathbf{v} = \begin{bmatrix} v_1 \\ v_2 \\ v_3 \end{bmatrix}.$$

If **a**, **b** are in Span$\{\mathbf{u}, \mathbf{v}\}$, then

$$\begin{bmatrix} a_1 \\ a_2 \\ a_3 \end{bmatrix} = q \begin{bmatrix} u_1 \\ u_2 \\ u_3 \end{bmatrix} + r \begin{bmatrix} v_1 \\ v_2 \\ v_3 \end{bmatrix} \quad \text{and} \quad \begin{bmatrix} b_1 \\ b_2 \\ b_3 \end{bmatrix} = s \begin{bmatrix} u_1 \\ u_2 \\ u_3 \end{bmatrix} + t \begin{bmatrix} v_1 \\ v_2 \\ v_3 \end{bmatrix},$$

for some real numbers q, r, s, and t. These two equations can be written as a matrix equation:

$$\begin{bmatrix} a_1 & b_1 \\ a_2 & b_2 \\ a_3 & b_3 \end{bmatrix} = \begin{bmatrix} u_1 & v_1 \\ u_2 & v_2 \\ u_3 & v_3 \end{bmatrix} \begin{bmatrix} q & s \\ r & t \end{bmatrix}.$$

Consequently, if $\begin{bmatrix} x_1 \\ x_2 \end{bmatrix}$ is an arbitrary vector, then

$$\begin{bmatrix} a_1 & b_1 \\ a_2 & b_2 \\ a_3 & b_3 \end{bmatrix} \begin{bmatrix} x_1 \\ x_2 \end{bmatrix} = \begin{bmatrix} u_1 & v_1 \\ u_2 & v_2 \\ u_3 & v_3 \end{bmatrix} \begin{bmatrix} q & s \\ r & t \end{bmatrix} \begin{bmatrix} x_1 \\ x_2 \end{bmatrix}.$$

If $\begin{bmatrix} q & s \\ r & t \end{bmatrix} \begin{bmatrix} x_1 \\ x_2 \end{bmatrix} = \begin{bmatrix} 0 \\ 0 \end{bmatrix}$, then we have $\begin{bmatrix} a_1 & b_1 \\ a_2 & b_2 \\ a_3 & b_3 \end{bmatrix} \begin{bmatrix} x_1 \\ x_2 \end{bmatrix} = \begin{bmatrix} 0 \\ 0 \\ 0 \end{bmatrix}$ which can be written as

$$x_1 \begin{bmatrix} a_1 \\ a_2 \\ a_3 \end{bmatrix} + x_2 \begin{bmatrix} b_1 \\ b_2 \\ b_3 \end{bmatrix} = \begin{bmatrix} 0 \\ 0 \\ 0 \end{bmatrix}.$$

Consequently, $\begin{bmatrix} x_1 \\ x_2 \end{bmatrix} = \begin{bmatrix} 0 \\ 0 \end{bmatrix}$, because the vectors $\mathbf{a} = \begin{bmatrix} a_1 \\ a_2 \\ a_3 \end{bmatrix}$ and $\mathbf{b} = \begin{bmatrix} b_1 \\ b_2 \\ b_3 \end{bmatrix}$ are linearly independent. Thus the matrix $\begin{bmatrix} q & s \\ r & t \end{bmatrix}$ is invertible and we have

$$\begin{bmatrix} a_1 & b_1 \\ a_2 & b_2 \\ a_3 & b_3 \end{bmatrix} \begin{bmatrix} q & s \\ r & t \end{bmatrix}^{-1} = \begin{bmatrix} u_1 & v_1 \\ u_2 & v_2 \\ u_3 & v_3 \end{bmatrix} \begin{bmatrix} q & s \\ r & t \end{bmatrix} \begin{bmatrix} q & s \\ r & t \end{bmatrix}^{-1} = \begin{bmatrix} u_1 & v_1 \\ u_2 & v_2 \\ u_3 & v_3 \end{bmatrix}.$$

If we let

$$\begin{bmatrix} q & s \\ r & t \end{bmatrix}^{-1} = \begin{bmatrix} e & g \\ f & h \end{bmatrix},$$

then we have

$$\begin{bmatrix} a_1 & b_1 \\ a_2 & b_2 \\ a_3 & b_3 \end{bmatrix} \begin{bmatrix} e & g \\ f & h \end{bmatrix} = \begin{bmatrix} u_1 & v_1 \\ u_2 & v_2 \\ u_3 & v_3 \end{bmatrix},$$

which means that

$$\mathbf{u} = e\mathbf{a} + f\mathbf{b} \quad \text{and} \quad \mathbf{v} = g\mathbf{a} + h\mathbf{b}.$$

Now the equality Span$\{\mathbf{a}, \mathbf{b}\}$ = Span$\{\mathbf{u}, \mathbf{v}\}$ follows by Theorem 4.1.15.

Next we prove that the vectors $\mathbf{u}$ and $\mathbf{v}$ are linearly independent. The equation

$$x_1\mathbf{u} + x_2\mathbf{v} = \begin{bmatrix} \mathbf{u} & \mathbf{v} \end{bmatrix} \begin{bmatrix} x_1 \\ x_2 \end{bmatrix} = \begin{bmatrix} 0 \\ 0 \\ 0 \end{bmatrix}$$

is equivalent to the equation

$$\begin{bmatrix} \mathbf{a} & \mathbf{b} \end{bmatrix} \begin{bmatrix} q & s \\ r & t \end{bmatrix}^{-1} \begin{bmatrix} x_1 \\ x_2 \end{bmatrix} = \begin{bmatrix} 0 \\ 0 \\ 0 \end{bmatrix}.$$

Since the vectors $\mathbf{a}$ and $\mathbf{b}$ are linearly independent we get

$$\begin{bmatrix} q & s \\ r & t \end{bmatrix}^{-1} \begin{bmatrix} x_1 \\ x_2 \end{bmatrix} = \begin{bmatrix} 0 \\ 0 \end{bmatrix}.$$

Now we multiply both sides by $\begin{bmatrix} q & s \\ r & t \end{bmatrix}$ and obtain

$$\begin{bmatrix} x_1 \\ x_2 \end{bmatrix} = \begin{bmatrix} q & s \\ r & t \end{bmatrix} \begin{bmatrix} q & s \\ r & t \end{bmatrix}^{-1} \begin{bmatrix} x_1 \\ x_2 \end{bmatrix} = \begin{bmatrix} q & s \\ r & t \end{bmatrix} \begin{bmatrix} 0 \\ 0 \end{bmatrix} = \begin{bmatrix} 0 \\ 0 \end{bmatrix}.$$

This means that the vectors $\mathbf{u}$ and $\mathbf{v}$ are linearly independent.

$\square$

Note that Theorem 4.1.22 implies that the two conditions in the definition 4.1.19 are equivalent.

Theorem 4.1.23. *Let* $\mathbf{a}$ *and* $\mathbf{b}$ *be vectors in the vector plane* Span$\{\mathbf{u}, \mathbf{v}\}$. *The following two conditions are equivalent:*

(a) $\mathbf{a}$ *and* $\mathbf{b}$ *are linearly independent;*

(b) Span$\{\mathbf{a}, \mathbf{b}\}$ = Span$\{\mathbf{u}, \mathbf{v}\}$.

Proof. If the vectors **a** and **b** are linearly independent in the vector plane Span$\{\mathbf{u}, \mathbf{v}\}$ then we have

$$\text{Span}\{\mathbf{a}, \mathbf{b}\} = \text{Span}\{\mathbf{u}, \mathbf{v}\},$$

by Theorem 4.1.22.

If Span$\{\mathbf{u}, \mathbf{v}\}$ is a vector plane, then the vectors **u** and **v** are linearly independent. Consequently, if Span$\{\mathbf{a}, \mathbf{b}\}$ = Span$\{\mathbf{u}, \mathbf{v}\}$, then the vectors **a** and **b** are linearly independent, again by Theorem 4.1.22. □

Example 4.1.24. Let

$$\mathbf{u} = \begin{bmatrix} 1 \\ 2 \\ 3 \end{bmatrix} \quad \text{and} \quad \mathbf{v} = \begin{bmatrix} -1 \\ 0 \\ -2 \end{bmatrix}.$$

Consider

$$\mathbf{a} = 2\mathbf{u} - \mathbf{v} = 2 \begin{bmatrix} 1 \\ 2 \\ 3 \end{bmatrix} - \begin{bmatrix} -1 \\ 0 \\ -2 \end{bmatrix} = \begin{bmatrix} 3 \\ 4 \\ 8 \end{bmatrix}$$

and

$$\mathbf{b} = 3\mathbf{u} + 5\mathbf{v} = 3 \begin{bmatrix} 1 \\ 2 \\ 3 \end{bmatrix} + 5 \begin{bmatrix} -1 \\ 0 \\ -2 \end{bmatrix} = \begin{bmatrix} -2 \\ 6 \\ -1 \end{bmatrix}.$$

Show that $\{\mathbf{a}, \mathbf{b}\}$ is a basis of Span$\{\mathbf{u}, \mathbf{v}\}$.

Solution. Clearly, **a** and **b** are vectors in Span$\{\mathbf{u}, \mathbf{v}\}$. Moreover, since

$$\det \begin{bmatrix} 3 & -2 \\ 4 & 6 \end{bmatrix} = 26 \neq 0,$$

a and **b** are linearly independent, by Theorem 4.1.12. Consequently,

$$\text{Span}\{\mathbf{a}, \mathbf{b}\} = \text{Span}\{\mathbf{u}, \mathbf{v}\},$$

by Theorem 4.1.22, and $\{\mathbf{a}, \mathbf{b}\}$ is a basis of Span$\{\mathbf{u}, \mathbf{v}\}$. □

It is often necessary to switch from one basis of a vector plane to another basis of that vector plane. This transition between two bases is called a ***change of basis***. It can be conveniently described by a matrix. If $\{\mathbf{a}, \mathbf{b}\}$ and $\{\mathbf{u}, \mathbf{v}\}$ are two bases of a vector plane, since **a** and **b** are elements of Span$\{\mathbf{u}, \mathbf{v}\}$, there are real numbers q, r, s, and t such that

$$\mathbf{a} = q\mathbf{u} + r\mathbf{v} \quad \text{and} \quad \mathbf{b} = s\mathbf{u} + t\mathbf{v}.$$

But this means that

$$\begin{bmatrix} \mathbf{a} & \mathbf{b} \end{bmatrix} = \begin{bmatrix} \mathbf{u} & \mathbf{v} \end{bmatrix} \begin{bmatrix} q & s \\ r & t \end{bmatrix}.$$

Definition 4.1.25. Let $\{\mathbf{a}, \mathbf{b}\}$ and $\{\mathbf{u}, \mathbf{v}\}$ be two bases of a vector plane. Then the matrix $\begin{bmatrix} q & s \\ r & t \end{bmatrix}$ such that

$$[\mathbf{a} \ \ \mathbf{b}] = [\mathbf{u} \ \ \mathbf{v}] \begin{bmatrix} q & s \\ r & t \end{bmatrix}$$

is called the **transition matrix** from the basis $\{\mathbf{a}, \mathbf{b}\}$ to the basis $\{\mathbf{u}, \mathbf{v}\}$.

The next theorem shows that the transition matrix allows us to easily calculate the coordinates of any vector in the basis $\{\mathbf{u}, \mathbf{v}\}$ if we know its coordinates in the basis $\{\mathbf{a}, \mathbf{b}\}$.

Theorem 4.1.26. *Let* $\{\mathbf{a}, \mathbf{b}\}$ *and* $\{\mathbf{u}, \mathbf{v}\}$ *be two bases of a vector plane and let* $\begin{bmatrix} q & s \\ r & t \end{bmatrix}$ *be the transition matrix from the basis* $\{\mathbf{a}, \mathbf{b}\}$ *to the basis* $\{\mathbf{u}, \mathbf{v}\}$. *If*

$$\mathbf{c} = x\mathbf{a} + y\mathbf{b},$$

then

$$\mathbf{c} = x'\mathbf{u} + y'\mathbf{v}$$

where the real numbers x' *and* y' *are given by the equation*

$$\begin{bmatrix} x' \\ y' \end{bmatrix} = \begin{bmatrix} q & s \\ r & t \end{bmatrix} \begin{bmatrix} x \\ y \end{bmatrix}.$$

Proof. Note that the equation $\mathbf{c} = x\mathbf{a} + y\mathbf{b}$ can be written as

$$\mathbf{c} = [\mathbf{a} \ \ \mathbf{b}] \begin{bmatrix} x \\ y \end{bmatrix}$$

and the equation $\mathbf{c} = x'\mathbf{u} + y'\mathbf{v}$ can be written as

$$\mathbf{c} = [\mathbf{u} \ \ \mathbf{v}] \begin{bmatrix} x' \\ y' \end{bmatrix}.$$

Now to prove the desired property it suffices to note that

$$\mathbf{c} = [\mathbf{a} \ \ \mathbf{b}] \begin{bmatrix} x \\ y \end{bmatrix} = [\mathbf{u} \ \ \mathbf{v}] \begin{bmatrix} q & s \\ r & t \end{bmatrix} \begin{bmatrix} x \\ y \end{bmatrix} = [\mathbf{u} \ \ \mathbf{v}] \begin{bmatrix} x' \\ y' \end{bmatrix}.$$

$\square$

Example 4.1.27. Show that $\left\{ \begin{bmatrix} 3 \\ 1 \\ 2 \end{bmatrix}, \begin{bmatrix} 1 \\ 3 \\ 2 \end{bmatrix} \right\}$ and $\left\{ \begin{bmatrix} 1 \\ 1 \\ 1 \end{bmatrix}, \begin{bmatrix} 1 \\ -1 \\ 0 \end{bmatrix} \right\}$ are bases of the same

vector plane. Determine the transition matrix from the basis $\left\{ \begin{bmatrix} 3 \\ 1 \\ 2 \end{bmatrix}, \begin{bmatrix} 1 \\ 3 \\ 2 \end{bmatrix} \right\}$ to the

basis $\left\{ \begin{bmatrix} 1 \\ 1 \\ 1 \end{bmatrix}, \begin{bmatrix} 1 \\ -1 \\ 0 \end{bmatrix} \right\}$. Then use this matrix to determine the coordinates of the vec-

tor

$$\mathbf{w} = 3 \begin{bmatrix} 3 \\ 1 \\ 2 \end{bmatrix} - 2 \begin{bmatrix} 1 \\ 3 \\ 2 \end{bmatrix}$$

in the basis $\left\{ \begin{bmatrix} 1 \\ 1 \\ 1 \end{bmatrix}, \begin{bmatrix} 1 \\ -1 \\ 0 \end{bmatrix} \right\}$.

Solution. First we observe that

$$\begin{bmatrix} 3 \\ 1 \\ 2 \end{bmatrix} = 2 \begin{bmatrix} 1 \\ 1 \\ 1 \end{bmatrix} + \begin{bmatrix} 1 \\ -1 \\ 0 \end{bmatrix} \quad \text{and} \quad \begin{bmatrix} 1 \\ 3 \\ 2 \end{bmatrix} = 2 \begin{bmatrix} 1 \\ 1 \\ 1 \end{bmatrix} - \begin{bmatrix} 1 \\ -1 \\ 0 \end{bmatrix}. \tag{4.3}$$

Since the vectors $\begin{bmatrix} 3 \\ 1 \\ 2 \end{bmatrix}$ and $\begin{bmatrix} 1 \\ 3 \\ 2 \end{bmatrix}$ are linearly independent, we can conclude that

$$\text{Span}\left\{ \begin{bmatrix} 3 \\ 1 \\ 2 \end{bmatrix}, \begin{bmatrix} 1 \\ 3 \\ 2 \end{bmatrix} \right\} = \text{Span}\left\{ \begin{bmatrix} 1 \\ 1 \\ 1 \end{bmatrix}, \begin{bmatrix} 1 \\ -1 \\ 0 \end{bmatrix} \right\},$$

by Theorem 4.1.22.

Since the equalities in (4.3) can be written as the single matrix equality

$$\begin{bmatrix} 3 & 1 \\ 1 & 3 \\ 2 & 2 \end{bmatrix} = \begin{bmatrix} 1 & 1 \\ 1 & -1 \\ 1 & 0 \end{bmatrix} \begin{bmatrix} 2 & 2 \\ 1 & -1 \end{bmatrix},$$

the transition matrix from the basis $\left\{ \begin{bmatrix} 3 \\ 1 \\ 2 \end{bmatrix}, \begin{bmatrix} 1 \\ 3 \\ 2 \end{bmatrix} \right\}$ to the basis $\left\{ \begin{bmatrix} 1 \\ 1 \\ 1 \end{bmatrix}, \begin{bmatrix} 1 \\ -1 \\ 0 \end{bmatrix} \right\}$ is the

matrix $\begin{bmatrix} 2 & 2 \\ 1 & -1 \end{bmatrix}$.

Finally, since

$$\begin{bmatrix} 2 & 2 \\ 1 & -1 \end{bmatrix} \begin{bmatrix} 3 \\ -2 \end{bmatrix} = \begin{bmatrix} 2 \\ 5 \end{bmatrix},$$

we have

$$\mathbf{w} = 2 \begin{bmatrix} 1 \\ 1 \\ 1 \end{bmatrix} + 5 \begin{bmatrix} 1 \\ -1 \\ 0 \end{bmatrix}.$$

☐

4.1.1 Exercises

1. Show that $\begin{bmatrix} 1 \\ 2 \\ -1 \end{bmatrix}$ is on the vector line Span$\left\{ \begin{bmatrix} -3 \\ -6 \\ 3 \end{bmatrix} \right\}$.

2. Show that $\begin{bmatrix} 4 \\ 4 \\ 8 \end{bmatrix}$ is on the vector line Span$\left\{ \begin{bmatrix} 5 \\ 5 \\ 10 \end{bmatrix} \right\}$.

3. Show that $\begin{bmatrix} 1 \\ 2 \\ 3 \end{bmatrix}$ is not on the vector line Span$\left\{ \begin{bmatrix} 1 \\ 0 \\ -1 \end{bmatrix} \right\}$.

4. Show that $\begin{bmatrix} 3 \\ 7 \\ 1 \end{bmatrix}$ is not on the vector line Span$\left\{ \begin{bmatrix} 3 \\ 7 \\ 2 \end{bmatrix} \right\}$.

Show that the given vectors **u** and **v** are linearly dependent.

5. $\mathbf{u} = \begin{bmatrix} 3 \\ 1 \\ 2 \end{bmatrix}$ and $\mathbf{v} = \begin{bmatrix} 12 \\ 4 \\ 8 \end{bmatrix}$

7. $\mathbf{u} = \begin{bmatrix} 1 \\ -3 \\ 1 \end{bmatrix}$ and $\mathbf{v} = \begin{bmatrix} -1 \\ 3 \\ -1 \end{bmatrix}$

6. $\mathbf{u} = \begin{bmatrix} 1 \\ 2 \\ 4 \end{bmatrix}$ and $\mathbf{v} = \begin{bmatrix} \frac{1}{2} \\ 1 \\ 2 \end{bmatrix}$

8. $\mathbf{u} = \begin{bmatrix} 2a \\ 4a \\ 2a \end{bmatrix}$ and $\mathbf{v} = \begin{bmatrix} 1 \\ 2 \\ 1 \end{bmatrix}$

Show that the given vectors **u** and **v** are linearly independent.

9. $\mathbf{u} = \begin{bmatrix} 3 \\ 1 \\ 2 \end{bmatrix}$ and $\mathbf{v} = \begin{bmatrix} 3 \\ 1 \\ 3 \end{bmatrix}$

10. $\mathbf{u} = \begin{bmatrix} 1 \\ 2 \\ 4 \end{bmatrix}$ and $\mathbf{v} = \begin{bmatrix} 2 \\ 4 \\ 5 \end{bmatrix}$

11. $\mathbf{u} = \begin{bmatrix} 1 \\ -3 \\ 1 \end{bmatrix}$ and $\mathbf{v} = \begin{bmatrix} -1 \\ 3 \\ 2 \end{bmatrix}$

12. $\mathbf{u} = \begin{bmatrix} 2 \\ 3 \\ 2 \end{bmatrix}$ and $\mathbf{v} = \begin{bmatrix} 1 \\ 2 \\ 1 \end{bmatrix}$

Show the following equalities using Theorem 4.1.15.

13. $\text{Span}\left\{ \begin{bmatrix} 1 \\ -1 \\ 1 \end{bmatrix}, \begin{bmatrix} 3 \\ -1 \\ 3 \end{bmatrix} \right\} = \text{Span}\left\{ \begin{bmatrix} 1 \\ -1 \\ 1 \end{bmatrix}, \begin{bmatrix} 1 \\ 1 \\ 1 \end{bmatrix} \right\}$

14. $\text{Span}\left\{ \begin{bmatrix} 1 \\ -1 \\ 1 \end{bmatrix}, \begin{bmatrix} 4 \\ 3 \\ 4 \end{bmatrix} \right\} = \text{Span}\left\{ \begin{bmatrix} 1 \\ -1 \\ 1 \end{bmatrix}, \begin{bmatrix} 1 \\ 1 \\ 1 \end{bmatrix} \right\}$

15. $\text{Span}\left\{ \begin{bmatrix} 2 \\ 2 \\ 3 \end{bmatrix}, \begin{bmatrix} 1 \\ 0 \\ 2 \end{bmatrix} \right\} = \text{Span}\left\{ \begin{bmatrix} 2 \\ 2 \\ 3 \end{bmatrix}, \begin{bmatrix} 1 \\ 2 \\ 1 \end{bmatrix} \right\}$

16. $\text{Span}\left\{ \begin{bmatrix} 2 \\ 4 \\ 8 \end{bmatrix}, \begin{bmatrix} 1 \\ 1 \\ 2 \end{bmatrix} \right\} = \text{Span}\left\{ \begin{bmatrix} 1 \\ 2 \\ 4 \end{bmatrix}, \begin{bmatrix} 1 \\ 1 \\ 2 \end{bmatrix} \right\}$

Show that the vector $\mathbf{x}$ is in the vector plane Span $\{\mathbf{u}, \mathbf{v}\}$ and find the coordinates of $\mathbf{x}$ in the basis $\{\mathbf{u}, \mathbf{v}\}$.

17. $\mathbf{x} = \begin{bmatrix} 4 \\ 5 \\ 7 \end{bmatrix}, \mathbf{u} = \begin{bmatrix} 2 \\ 1 \\ 3 \end{bmatrix}, \mathbf{v} = \begin{bmatrix} 1 \\ 2 \\ 2 \end{bmatrix}$

20. $\mathbf{x} = \begin{bmatrix} 1 \\ 4 \\ 1 \end{bmatrix}, \mathbf{u} = \begin{bmatrix} 1 \\ -1 \\ 1 \end{bmatrix}, \mathbf{v} = \begin{bmatrix} 1 \\ 1 \\ 1 \end{bmatrix}$

18. $\mathbf{x} = \begin{bmatrix} 1 \\ 0 \\ 1 \end{bmatrix}, \mathbf{u} = \begin{bmatrix} 1 \\ -1 \\ 1 \end{bmatrix}, \mathbf{v} = \begin{bmatrix} 1 \\ 0 \\ 2 \end{bmatrix}$

21. $\mathbf{x} = \begin{bmatrix} 3 \\ -1 \\ 1 \end{bmatrix}, \mathbf{u} = \begin{bmatrix} 1 \\ -2 \\ 2 \end{bmatrix}, \mathbf{v} = \begin{bmatrix} 1 \\ 1 \\ -1 \end{bmatrix}$

19. $\mathbf{x} = \begin{bmatrix} 1 \\ 0 \\ 0 \end{bmatrix}, \mathbf{u} = \begin{bmatrix} 2 \\ 2 \\ 3 \end{bmatrix}, \mathbf{v} = \begin{bmatrix} 1 \\ 1 \\ 1 \end{bmatrix}$

22. $\mathbf{x} = \begin{bmatrix} -1 \\ 11 \\ 7 \end{bmatrix}, \mathbf{u} = \begin{bmatrix} 1 \\ 1 \\ 1 \end{bmatrix}, \mathbf{v} = \begin{bmatrix} -1 \\ 2 \\ 1 \end{bmatrix}$

23. Show that $\left\{ \begin{bmatrix} 3 \\ 5 \\ -1 \end{bmatrix}, \begin{bmatrix} 3 \\ 4 \\ 1 \end{bmatrix} \right\}$ is a basis in the vector plane Span $\left\{ \begin{bmatrix} 1 \\ 1 \\ 1 \end{bmatrix}, \begin{bmatrix} 1 \\ 2 \\ -1 \end{bmatrix} \right\}$.

24. Show that $\left\{ \begin{bmatrix} 1 \\ 7 \\ 5 \end{bmatrix}, \begin{bmatrix} 3 \\ 7 \\ -5 \end{bmatrix} \right\}$ is a basis in the vector plane Span $\left\{ \begin{bmatrix} 1 \\ 1 \\ 1 \end{bmatrix}, \begin{bmatrix} 1 \\ 2 \\ -1 \end{bmatrix} \right\}$.

25. Show that $\left\{ \begin{bmatrix} 5 \\ 7 \\ 1 \end{bmatrix}, \begin{bmatrix} 2 \\ 1 \\ 4 \end{bmatrix} \right\}$ is a basis in the vector plane Span $\left\{ \begin{bmatrix} 1 \\ 1 \\ 1 \end{bmatrix}, \begin{bmatrix} 1 \\ 2 \\ -1 \end{bmatrix} \right\}$.

26. Show that $\left\{ \begin{bmatrix} 3 \\ 1 \\ 7 \end{bmatrix}, \begin{bmatrix} 0 \\ -1 \\ 2 \end{bmatrix} \right\}$ is a basis in the vector plane Span $\left\{ \begin{bmatrix} 1 \\ 1 \\ 1 \end{bmatrix}, \begin{bmatrix} 1 \\ 2 \\ -1 \end{bmatrix} \right\}$.

27. Find the transition matrix from the basis $\left\{ \begin{bmatrix} 3 \\ 5 \\ -1 \end{bmatrix}, \begin{bmatrix} 3 \\ 4 \\ 1 \end{bmatrix} \right\}$ to the basis

$\left\{ \begin{bmatrix} 1 \\ 1 \\ 1 \end{bmatrix}, \begin{bmatrix} 1 \\ 2 \\ -1 \end{bmatrix} \right\}$ and use it to find the coordinates of the vector

$\mathbf{w} = 5 \begin{bmatrix} 3 \\ 5 \\ -1 \end{bmatrix} + 2 \begin{bmatrix} 3 \\ 4 \\ 1 \end{bmatrix}$ relative to the basis $\left\{ \begin{bmatrix} 1 \\ 1 \\ 1 \end{bmatrix}, \begin{bmatrix} 1 \\ 2 \\ -1 \end{bmatrix} \right\}$.

28. Find the transition matrix from the basis $\left\{ \begin{bmatrix} 1 \\ 7 \\ 5 \end{bmatrix}, \begin{bmatrix} 3 \\ 7 \\ -5 \end{bmatrix} \right\}$ to the basis Span

$\left\{ \begin{bmatrix} 1 \\ 1 \\ 1 \end{bmatrix}, \begin{bmatrix} 1 \\ 2 \\ -1 \end{bmatrix} \right\}$ and use it to find the coordinates of the vector

$\mathbf{w} = 3 \begin{bmatrix} 1 \\ 7 \\ 5 \end{bmatrix} - 2 \begin{bmatrix} 3 \\ 7 \\ -5 \end{bmatrix}$ relative to the basis $\left\{ \begin{bmatrix} 1 \\ 1 \\ 1 \end{bmatrix}, \begin{bmatrix} 1 \\ 2 \\ -1 \end{bmatrix} \right\}$.

29. Find the transition matrix from the basis $\left\{ \begin{bmatrix} 5 \\ 7 \\ 1 \end{bmatrix}, \begin{bmatrix} 2 \\ 1 \\ 4 \end{bmatrix} \right\}$ to the basis Span

$\left\{ \begin{bmatrix} 1 \\ 1 \\ 1 \end{bmatrix}, \begin{bmatrix} 1 \\ 2 \\ -1 \end{bmatrix} \right\}$ and use it to find the coordinates of the vector

$\mathbf{w} = a \begin{bmatrix} 5 \\ 7 \\ 1 \end{bmatrix} + b \begin{bmatrix} 2 \\ 1 \\ 4 \end{bmatrix}$ relative to the basis $\left\{ \begin{bmatrix} 1 \\ 1 \\ 1 \end{bmatrix}, \begin{bmatrix} 1 \\ 2 \\ -1 \end{bmatrix} \right\}$.

30. Find the transition matrix from the basis $\left\{ \begin{bmatrix} 3 \\ 1 \\ 7 \end{bmatrix}, \begin{bmatrix} 0 \\ -1 \\ 2 \end{bmatrix} \right\}$ to the basis

$\left\{ \begin{bmatrix} 1 \\ 1 \\ 1 \end{bmatrix}, \begin{bmatrix} 1 \\ 2 \\ -1 \end{bmatrix} \right\}$ and use it to find the coordinates of the vector

$\mathbf{w} = a \begin{bmatrix} 3 \\ 1 \\ 7 \end{bmatrix} + b \begin{bmatrix} 0 \\ -1 \\ 2 \end{bmatrix}$ relative to the basis $\left\{ \begin{bmatrix} 1 \\ 1 \\ 1 \end{bmatrix}, \begin{bmatrix} 1 \\ 2 \\ -1 \end{bmatrix} \right\}$.

31. Suppose that the vectors $\mathbf{a}$ and $\mathbf{b}$ in $\mathbb{R}^3$ are linearly independent and let $\mathbf{u} = 2\mathbf{a} + \mathbf{b}$ and $\mathbf{v} = 3\mathbf{a} + 5\mathbf{b}$. Find the transition matrix from the basis $\{\mathbf{a}, \mathbf{b}\}$ to the basis $\{\mathbf{u}, \mathbf{v}\}$ and from the basis $\{\mathbf{u}, \mathbf{v}\}$ to the basis $\{\mathbf{a}, \mathbf{b}\}$.

32. Suppose that the vectors **a** and **b** in $\mathbb{R}^3$ are linearly independent and let **u** = **a** + 3**b** and **v** = 3**a** + **b**. Find the transition matrix from the basis {**a**, **b**} to the basis {**u**, **v**} and from the basis {**u**, **v**} to the basis {**a**, **b**}.

33. Suppose that the vectors **a** and **b** in $\mathbb{R}^3$ are linearly independent and let **u** = 3**a** + 2**b** and **v** = **b**. Find the transition matrix from the basis {**a**, **b**} to the basis {**u**, **v**} and from the basis {**u**, **v**} to the basis {**a**, **b**}.

34. Suppose that the vectors **a** and **b** in $\mathbb{R}^3$ are linearly independent and let **u** = 5**a** + 2**b** and **v** = 7**a** + 3**b**. Find the transition matrix from the basis {**a**, **b**} to the basis {**u**, **v**} and from the basis {**u**, **v**} to the basis {**a**, **b**}.

35. Suppose that the vectors **a** and **b** in $\mathbb{R}^3$ are linearly independent and let **u** = 2**a** − **b** and **v** = **a** + 2**b**. Find the transition matrix from the basis {**a**, **b**} to the basis {**u**, **v**} and use it to find the coordinates of the vector **w** = **a** + **b** relative to the basis {**u**, **v**}.

36. Suppose that the vectors **a** and **b** in $\mathbb{R}^3$ are linearly independent and let **u** = 3**a** + **b** and **v** = 7**a** + 4**b**. Find the transition matrix from the basis {**a**, **b**} to the basis {**u**, **v**} and use it to find the coordinates of the vector **w** = 2**a** − **b** relative to the basis {**u**, **v**}.

37. Suppose that the vectors **a** and **b** in $\mathbb{R}^3$ are linearly independent and let **u** = 3**a** + 2**b** and **v** = **a** + 5**b**. Find the transition matrix from the basis {**a**, **b**} to the basis {**u**, **v**} and use it to find the coordinates of the vector **w** = 4**a** − 3**b** relative to the basis {**u**, **v**}.

38. Suppose that the vectors **a** and **b** in $\mathbb{R}^3$ are linearly independent and let **u** = **a** + **b** and **v** = 2**a**. Find the transition matrix from the basis {**a**, **b**} to the basis {**u**, **v**} and use it to find the coordinates of the vector **w** = x**a** + y**b** relative to the basis {**u**, **v**}.

39. Suppose that the vectors **a** and **b** in $\mathbb{R}^3$ are linearly independent and that the vector **u** ≠ **0** is in the vector plane Span{**a**, **b**}. Show that one of the following conditions holds:

    (a) {**a**, **u**} is a basis of the vector plane Span{**a**, **b**};

    (b) {**b**, **u**} is a basis of the vector plane Span{**a**, **b**}.

40. Let **u** and **v** be two linearly independent vectors in $\mathbb{R}^3$. Show that {p**u** + s**v**, **v**} and {**u**, q**v** + t**u**} are bases of the vector plane Span{**u**, **v**} for any real numbers p, q, s, and t such that p and q are different from 0.

4.2   Projections in $\mathbb{R}^3$

The dot product in $\mathbb{R}^3$

The definition of the dot product in $\mathbb{R}^2$ can be modified in an obvious way to work in $\mathbb{R}^3$:

$$\begin{bmatrix} u_1 \\ u_2 \\ u_3 \end{bmatrix} \cdot \begin{bmatrix} v_1 \\ v_2 \\ v_3 \end{bmatrix} = u_1 v_1 + u_2 v_2 + u_3 v_3.$$

Example 4.2.1.

$$\begin{bmatrix} 5 \\ 3 \\ -2 \end{bmatrix} \cdot \begin{bmatrix} 4 \\ 2 \\ 6 \end{bmatrix} = 5 \cdot 4 + 3 \cdot 2 + (-2) \cdot 6 = 14.$$

It is easy to verify that the dot product has the following properties

$$\mathbf{u} \cdot \mathbf{v} = \mathbf{v} \cdot \mathbf{u}$$

$$\mathbf{u} \cdot (\mathbf{v} + \mathbf{w}) = \mathbf{u} \cdot \mathbf{v} + \mathbf{u} \cdot \mathbf{w}$$

$$t(\mathbf{u} \cdot \mathbf{v}) = (t\mathbf{u}) \cdot \mathbf{v} = \mathbf{u} \cdot (t\mathbf{v})$$

for arbitrary $\mathbf{u}, \mathbf{v}, \mathbf{w}$ in $\mathbb{R}^3$ and t in $\mathbb{R}$.

As in the case of $\mathbb{R}^2$ we use the notation

$$\|\mathbf{u}\| = \left\| \begin{bmatrix} u_1 \\ u_2 \\ u_3 \end{bmatrix} \right\| = \sqrt{u_1^2 + u_2^2 + u_3^2}$$

and call $\|\mathbf{u}\|$ the ***norm*** of $\mathbf{u}$. Geometrically, $\|\mathbf{u}\|$ is the distance from the point $\mathbf{u}$ to the origin. If $\|\mathbf{u}\| = 1$, then we say that $\mathbf{u}$ is a ***unit vector***. If $\mathbf{u}$ is a non-zero vector, then the vector $\frac{1}{\|\mathbf{u}\|}\mathbf{u}$ is a unit vector. If we multiply a non-zero vector $\mathbf{u}$ by $\frac{1}{\|\mathbf{u}\|}$, we say that we ***normalize*** the vector.

Example 4.2.2.

$$\left\| \begin{bmatrix} 3 \\ -2 \\ 1 \end{bmatrix} \right\| = \sqrt{3^2 + (-2)^2 + 1^2} = \sqrt{14}.$$

The number $\|\mathbf{u} - \mathbf{v}\|$ is the distance from $\mathbf{u}$ to $\mathbf{v}$:

$$\|\mathbf{u} - \mathbf{v}\| = \left\| \begin{bmatrix} u_1 \\ u_2 \\ u_3 \end{bmatrix} - \begin{bmatrix} v_1 \\ v_2 \\ v_3 \end{bmatrix} \right\|$$

$$= \left\| \begin{bmatrix} u_1 - v_1 \\ v_2 - v_2 \\ u_3 - v_3 \end{bmatrix} \right\|$$

$$= \sqrt{(u_1 - v_1)^2 + (u_2 - v_2)^2 + (u_3 - v_3)^2}.$$

Example 4.2.3. The distance between $\begin{bmatrix} 5 \\ 1 \\ -2 \end{bmatrix}$ and $\begin{bmatrix} 3 \\ 4 \\ 0 \end{bmatrix}$ is

$$\left\| \begin{bmatrix} 5 \\ 1 \\ -2 \end{bmatrix} - \begin{bmatrix} 3 \\ 4 \\ 0 \end{bmatrix} \right\| = \left\| \begin{bmatrix} 2 \\ -3 \\ -2 \end{bmatrix} \right\| = \sqrt{2^2 + (-3)^2 + (-2)^2} = \sqrt{17}.$$

The useful connection between the dot product and the norm in $\mathbb{R}^2$ is also valid in $\mathbb{R}^3$:

$$\mathbf{u} \cdot \mathbf{u} = \|\mathbf{u}\|^2 \quad \text{or} \quad \|\mathbf{u}\| = \sqrt{\mathbf{u} \cdot \mathbf{u}}.$$

The same is true about the parallelogram law:

Theorem 4.2.4 (Parallelogram law). *For arbitrary* $\mathbf{u}$ *and* $\mathbf{v}$ *in* $\mathbb{R}^3$ *we have*

$$\mathbf{u} \cdot \mathbf{v} = \frac{1}{4} \left(\|\mathbf{u} + \mathbf{v}\|^2 - \|\mathbf{u} - \mathbf{v}\|^2 \right).$$

Proof. The proof is identical with the proof of Theorem 3.2.8. Note that the proof uses only the algebraic properties of the norm and the dot product that are the same in $\mathbb{R}^2$ and in $\mathbb{R}^3$. □

The definition of orthogonal vectors is the same as in $\mathbb{R}^2$:

Definition 4.2.5. Two vectors $\mathbf{u}$ and $\mathbf{v}$ in $\mathbb{R}^3$ are called ***orthogonal*** if

$$||\mathbf{u} - \mathbf{v}|| = ||\mathbf{u} + \mathbf{v}||.$$

From the Parallelogram law we obtain as in $\mathbb{R}^2$ the characterisation of orthogonal vectors:

Theorem 4.2.6. *Two vectors* $\mathbf{u}$ *and* $\mathbf{v}$ *in* $\mathbb{R}^3$ *are orthogonal if and only if* $\mathbf{u} \cdot \mathbf{v} = 0$.

Example 4.2.7. Since

$$\begin{bmatrix} 1 \\ 2 \\ 3 \end{bmatrix} \cdot \begin{bmatrix} 1 \\ 1 \\ -1 \end{bmatrix} = 1 + 2 - 3 = 0,$$

the vectors $\begin{bmatrix} 1 \\ 2 \\ 3 \end{bmatrix}$ and $\begin{bmatrix} 1 \\ 1 \\ -1 \end{bmatrix}$ are orthogonal.

The vector plane defined by an equation

Theorem 4.2.8. *If* $\mathbf{n}$ *is a vector in* $\mathbb{R}^3$ *different from the origin, then the equation*

$$\mathbf{n} \cdot \mathbf{x} = 0$$

defines a vector plane, that is, there are two linearly independent vectors $\mathbf{u}$ *and* $\mathbf{v}$ *in* $\mathbb{R}^3$ *such that* $\mathbf{n} \cdot \mathbf{x} = 0$ *if and only if* $\mathbf{x}$ *is in* Span$\{\mathbf{u}, \mathbf{v}\}$.

Proof. Let $\mathbf{n} = \begin{bmatrix} n_1 \\ n_2 \\ n_3 \end{bmatrix} \neq \mathbf{0}$ and $\mathbf{x} = \begin{bmatrix} x_1 \\ x_2 \\ x_3 \end{bmatrix}$. Since $\mathbf{n}$ is different from the origin, at least one of the numbers n_1, n_2, n_3 must be different from 0. Suppose that $n_3 \neq 0$. Let

$$\mathbf{u} = \begin{bmatrix} 1 \\ 0 \\ -\frac{n_1}{n_3} \end{bmatrix} \quad \text{and} \quad \mathbf{v} = \begin{bmatrix} 0 \\ 1 \\ -\frac{n_2}{n_3} \end{bmatrix}.$$

Clearly $\mathbf{u}$ and $\mathbf{v}$ are linearly independent.

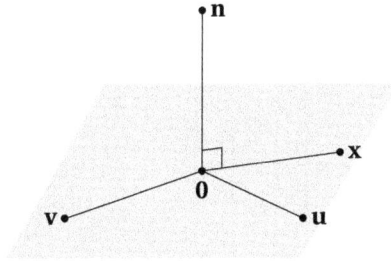

Figure 4.6: Illustration of Theorem 4.2.8.

If $\mathbf{n} \cdot \mathbf{x} = 0$, then

$$n_1 x_1 + n_2 x_2 + n_3 x_3 = 0,$$

which gives us

$$x_3 = -\frac{n_1}{n_3} x_1 + \left(-\frac{n_2}{n_3}\right) x_2.$$

Since

$$\begin{bmatrix} x_1 \\ x_2 \\ x_3 \end{bmatrix} = \begin{bmatrix} x_1 \\ x_2 \\ -\frac{n_1}{n_3} x_1 + \left(-\frac{n_2}{n_3}\right) x_2 \end{bmatrix} = x_1 \begin{bmatrix} 1 \\ 0 \\ -\frac{n_1}{n_3} \end{bmatrix} + x_2 \begin{bmatrix} 0 \\ 1 \\ -\frac{n_2}{n_3} \end{bmatrix} = x_1 \mathbf{u} + x_2 \mathbf{v},$$

$\mathbf{x}$ is in Span$\{\mathbf{u}, \mathbf{v}\}$.

Now assume that $\mathbf{x} = s\mathbf{u} + t\mathbf{v}$ for some numbers s and t. Then

$$\mathbf{n} \cdot \mathbf{x} = \begin{bmatrix} n_1 \\ n_2 \\ n_3 \end{bmatrix} \cdot \left(s \begin{bmatrix} 1 \\ 0 \\ -\frac{n_1}{n_3} \end{bmatrix} + t \begin{bmatrix} 0 \\ 1 \\ -\frac{n_2}{n_3} \end{bmatrix} \right)$$

$$= s \left(\begin{bmatrix} n_1 \\ n_2 \\ n_3 \end{bmatrix} \cdot \begin{bmatrix} 1 \\ 0 \\ -\frac{n_1}{n_3} \end{bmatrix} \right) + t \left(\begin{bmatrix} n_1 \\ n_2 \\ n_3 \end{bmatrix} \cdot \begin{bmatrix} 0 \\ 1 \\ -\frac{n_2}{n_3} \end{bmatrix} \right) = 0.$$

This completes the proof in the case when $n_3 \neq 0$. The cases when $n_1 \neq 0$ or $n_2 \neq 0$ are treated similarly with appropriate changes. $\qquad\square$

Example 4.2.9. Let

$$\mathbf{n} = \begin{bmatrix} 1 \\ -1 \\ 2 \end{bmatrix}.$$

Describe the vector plane defined by $\mathbf{n} \cdot \mathbf{x} = 0$ in the form Span$\{\mathbf{u}, \mathbf{v}\}$.

Proof. If we let $\mathbf{x} = \begin{bmatrix} x_1 \\ x_2 \\ x_3 \end{bmatrix}$, then

$$\mathbf{n} \cdot \mathbf{x} = \begin{bmatrix} 1 \\ -1 \\ 2 \end{bmatrix} \cdot \begin{bmatrix} x_1 \\ x_2 \\ x_3 \end{bmatrix} = x_1 - x_2 + 2x_3.$$

Now

$$\mathbf{n} \cdot \mathbf{x} = 0$$

is equivalent to

$$x_1 - x_2 + 2x_3 = 0$$

or

$$x_1 = x_2 - 2x_3$$

This yields

$$\begin{bmatrix} x_1 \\ x_2 \\ x_3 \end{bmatrix} = \begin{bmatrix} x_2 - 2x_3 \\ x_2 \\ x_3 \end{bmatrix} = x_2 \begin{bmatrix} 1 \\ 1 \\ 0 \end{bmatrix} + x_3 \begin{bmatrix} -2 \\ 0 \\ 1 \end{bmatrix}.$$

Note that the vectors $\begin{bmatrix} 1 \\ 1 \\ 0 \end{bmatrix}$ and $\begin{bmatrix} -2 \\ 0 \\ 1 \end{bmatrix}$ are linearly independent. If we define

$$\mathbf{u} = \begin{bmatrix} 1 \\ 1 \\ 0 \end{bmatrix} \quad \text{and} \quad \mathbf{v} = \begin{bmatrix} -2 \\ 0 \\ 1 \end{bmatrix},$$

then the above argument shows that $\mathbf{n} \cdot \mathbf{x} = 0$ if and only if $\mathbf{x} = s\mathbf{u} + t\mathbf{v}$ for some real numbers s and t. In other words, the equation $\mathbf{n} \cdot \mathbf{x} = 0$ defines the vector plane Span $\{\mathbf{u}, \mathbf{v}\}$. $\qquad\square$

The projection on a vector line in $\mathbb{R}^3$

We continue investigating how the ideas we considered in $\mathbb{R}^2$ extend to $\mathbb{R}^3$. Now we turn to the question of finding the best approximation to a point by elements of a vector line. As we see in the next theorem, the answer in $\mathbb{R}^3$ is identical with the case of $\mathbb{R}^2$.

Theorem 4.2.10. *Consider a nonzero vector* **u** *in* $\mathbb{R}^3$. *For any point* **b** *in* $\mathbb{R}^3$ *there is a unique point* **p** *on the vector line* Span{**u**} *such that*

$$\|\mathbf{b} - \mathbf{p}\| \le \|\mathbf{b} - t\mathbf{u}\| \tag{4.4}$$

for all t in $\mathbb{R}$. *That unique point is*

$$\mathbf{p} = \frac{\mathbf{b} \cdot \mathbf{u}}{\|\mathbf{u}\|^2} \mathbf{u} = \frac{\mathbf{b} \cdot \mathbf{u}}{\mathbf{u} \cdot \mathbf{u}} \mathbf{u}.$$

Proof. Since the proof of Theorem 3.2.12 uses only properties of the norm and the dot product, without referring to $\mathbb{R}^2$, it can simply be copied here without any changes. $\square$

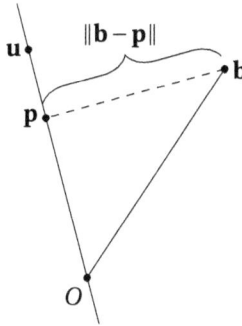

Figure 4.7: The point **p** minimizes the distance from **b** to the vector line Span{**u**}.

As in the case of $\mathbb{R}^2$, the point $\mathbf{p} = \dfrac{\mathbf{b} \cdot \mathbf{u}}{\|\mathbf{u}\|^2} \mathbf{u}$ is called the ***best approximation*** to the point **b** by elements of the vector line Span{**u**}.

Example 4.2.11. The point on the vector line Span$\left\{ \begin{bmatrix} 3 \\ 1 \\ 2 \end{bmatrix} \right\}$ closest to the point

$\begin{bmatrix} 2 \\ -1 \\ 1 \end{bmatrix}$ is

$$\mathbf{p} = \frac{\begin{bmatrix} 2 \\ -1 \\ 1 \end{bmatrix} \cdot \begin{bmatrix} 3 \\ 1 \\ 2 \end{bmatrix}}{\begin{bmatrix} 3 \\ 1 \\ 2 \end{bmatrix} \cdot \begin{bmatrix} 3 \\ 1 \\ 2 \end{bmatrix}} \begin{bmatrix} 3 \\ 1 \\ 2 \end{bmatrix} = \frac{7}{14} \begin{bmatrix} 3 \\ 1 \\ 2 \end{bmatrix} = \begin{bmatrix} \frac{3}{2} \\ \frac{1}{2} \\ 1 \end{bmatrix}.$$

Theorem 4.2.12. *Consider a point* $\mathbf{b}$ *and a vector line* $\mathrm{Span}\{\mathbf{u}\}$ *in* $\mathbb{R}^3$. *The best approximation* $\mathbf{p}$ *to the point* $\mathbf{b}$ *by elements of the vector line* $\mathrm{Span}\{\mathbf{u}\}$ *can be characterized as the point* $\mathbf{p}$ *in* $\mathrm{Span}\{\mathbf{u}\}$ *satisfying the equation*

$$(\mathbf{b} - \mathbf{p}) \cdot \mathbf{u} = 0.$$

In other words, the point $\mathbf{p}$ *in* $\mathrm{Span}\{\mathbf{u}\}$ *is the best approximation to* $\mathbf{b}$ *by elements of the vector line* $\mathrm{Span}\{\mathbf{u}\}$ *if and only if* $(\mathbf{b} - \mathbf{p}) \cdot \mathbf{u} = 0$.

Proof. Again, the proof is identical with the one given for Theorem 3.2.13. □

In view of Theorem 4.2.12, the best approximation to the point $\mathbf{b}$ by elements of the vector line $\mathrm{Span}\{\mathbf{u}\}$ is also called the ***projection*** of $\mathbf{b}$ on the vector line $\mathrm{Span}\{\mathbf{u}\}$ and is denoted by $\mathrm{proj}_{\mathrm{Span}\{\mathbf{u}\}}\mathbf{b}$ or simply by $\mathrm{proj}_{\mathbf{u}}\mathbf{b}$:

$$\mathrm{proj}_{\mathbf{u}}\mathbf{b} = \frac{\mathbf{b} \cdot \mathbf{u}}{\|\mathbf{u}\|^2}\mathbf{u}.$$

Finally, we show that Theorem 3.2.17 remains true in $\mathbb{R}^3$.

Theorem 4.2.13. *Let* $\mathbf{u}$ *be a nonzero vector in* $\mathbb{R}^3$ *and let*

$$A = \frac{1}{\|\mathbf{u}\|^2}\mathbf{u}\mathbf{u}^T.$$

Then for any $\mathbf{b}$ *in* $\mathbb{R}^3$ *we have*

$$\mathrm{proj}_{\mathbf{u}}\mathbf{b} = A\mathbf{b}.$$

Moreover, the matrix A *is the unique matrix with this property.*

Proof. Let $\mathbf{u} = \begin{bmatrix} u_1 \\ u_2 \\ u_3 \end{bmatrix}$. For any $\mathbf{b} = \begin{bmatrix} b_1 \\ b_2 \\ b_3 \end{bmatrix}$ we have

$$\text{proj}_{\mathbf{u}} \mathbf{b} = \frac{\mathbf{b} \cdot \mathbf{u}}{\|\mathbf{u}\|^2} \mathbf{u}$$

$$= \frac{1}{\|\mathbf{u}\|^2} \begin{bmatrix} u_1 \\ u_2 \\ u_3 \end{bmatrix} \left(\begin{bmatrix} u_1 & u_2 & u_3 \end{bmatrix} \begin{bmatrix} b_1 \\ b_2 \\ b_3 \end{bmatrix} \right)$$

$$= \frac{1}{\|\mathbf{u}\|^2} \left(\begin{bmatrix} u_1 \\ u_2 \\ u_3 \end{bmatrix} \begin{bmatrix} u_1 & u_2 & u_3 \end{bmatrix} \right) \begin{bmatrix} b_1 \\ b_2 \\ b_3 \end{bmatrix}$$

$$= A\mathbf{b}.$$

The uniqueness part of the theorem is a consequence of the fact that the equality $A\mathbf{x} = B\mathbf{x}$ for every $\mathbf{x}$ in $\mathbb{R}^3$ implies $A = B$, by Theorem 2.1.18. □

Definition 4.2.14. Let $\mathbf{u}$ be a nonzero vector in $\mathbb{R}^3$. The matrix

$$\frac{1}{\|\mathbf{u}\|^2} \mathbf{u}\mathbf{u}^T \tag{4.5}$$

is called the ***projection matrix*** on the vector line Span$\{\mathbf{u}\}$.

Example 4.2.15. Find the projection matrix on the vector line Span $\left\{ \begin{bmatrix} 1 \\ 3 \\ 1 \end{bmatrix} \right\}$.

Solution. Since

$$\mathbf{u}\mathbf{u}^T = \begin{bmatrix} 1 \\ 3 \\ 1 \end{bmatrix} \begin{bmatrix} 1 & 3 & 1 \end{bmatrix} = \begin{bmatrix} 1 & 3 & 1 \\ 3 & 9 & 3 \\ 1 & 3 & 1 \end{bmatrix},$$

and $\|\mathbf{u}\|^2 = 11$ the projection matrix on the vector line Span$\{\mathbf{u}\}$ is

$$\frac{1}{\|\mathbf{u}\|^2} \mathbf{u}\mathbf{u}^T = \frac{1}{11} \begin{bmatrix} 1 & 3 & 1 \\ 3 & 9 & 3 \\ 1 & 3 & 1 \end{bmatrix}.$$

□

We have a lot of flexibility in choosing a basis for a vector plane. As we will see, bases $\{\mathbf{a}, \mathbf{b}\}$ such that $\mathbf{a} \cdot \mathbf{b} = 0$ are often preferable.

Theorem 4.2.16. *Let* **a** *and* **b** *be two nonzero orthogonal vectors in* $\mathbb{R}^3$*, that is,* **a** $\neq$ **0**, **b** $\neq$ **0**, *and* **a**·**b** = 0. *Then the vectors* **a** *and* **b** *are linearly independent.*

Proof. If

$$x\mathbf{a} + y\mathbf{b} = \begin{bmatrix} 0 \\ 0 \\ 0 \end{bmatrix},$$

then

$$x\mathbf{a}\cdot\mathbf{a} + y\mathbf{b}\cdot\mathbf{a} = \begin{bmatrix} 0 \\ 0 \\ 0 \end{bmatrix}\cdot\mathbf{a}.$$

Since **a**·**a** = $\|\mathbf{a}\|^2$ and **b**·**a** = 0, the above can be written as $x\|\mathbf{a}\|^2 = 0$. This gives us $x = 0$, because $\|\mathbf{a}\| \neq 0$. In the same way we can show that $y = 0$. The result is now a consequence of Theorems 4.2.12 and 4.1.23.   □

Definition 4.2.17. A basis {**a**, **b**} of a vector plane is called an **orthogonal basis** if **a**·**b** = 0. A basis {**a**, **b**} of a vector plane is called an **orthonormal basis** if it is an orthogonal basis and $\|\mathbf{a}\| = \|\mathbf{b}\| = 1$.

If {**a**, **b**} is an orthogonal basis of a vector plane, it is easy to change it to an orthonormal basis. Clearly, we have

$$\text{Span}\{\mathbf{a}, \mathbf{b}\} = \text{Span}\left\{\frac{\mathbf{a}}{\|\mathbf{a}\|}, \frac{\mathbf{b}}{\|\mathbf{b}\|}\right\} \quad \text{and} \quad \left\|\frac{\mathbf{a}}{\|\mathbf{a}\|}\right\| = \left\|\frac{\mathbf{b}}{\|\mathbf{b}\|}\right\| = 1.$$

The next theorem shows that any basis of a vector plane can be transformed to an orthogonal basis of that vector plane.

Theorem 4.2.18. *Let* **u** *and* **v** *be two linearly independent vectors. There is a vector* **u**′ *such that* **u**′·**v** = 0 *and* Span{**u**′, **v**} = Span{**u**, **v**}*, that is* {**u**′, **v**} *is an orthogonal basis of the vector plane* Span{**u**, **v**}*.*
Similarly, there is a vector **v**′ *such that* **u**·**v**′ = 0 *and* Span{**u**, **v**′} = Span{**u**, **v**}*, that is* {**u**, **v**′} *is an orthogonal basis of* Span{**u**, **v**}*.*

Proof. We will show that the vector

$$\mathbf{u}' = \mathbf{u} - \frac{\mathbf{u}\cdot\mathbf{v}}{\mathbf{v}\cdot\mathbf{v}}\mathbf{v}$$

has the desired properties. First note that {**u**′, **v**} is a basis of the vector plane Span{**u**, **v**}, by Theorem 4.1.17. Moreover,

$$\mathbf{u}'\cdot\mathbf{v} = \mathbf{u}\cdot\mathbf{v} - \frac{\mathbf{u}\cdot\mathbf{v}}{\mathbf{v}\cdot\mathbf{v}}\mathbf{v}\cdot\mathbf{v} = 0,$$

completing the proof of the first part.

For the second part we use

$$\mathbf{v}' = \mathbf{v} - \frac{\mathbf{v} \cdot \mathbf{u}}{\mathbf{u} \cdot \mathbf{u}} \mathbf{u}$$

and proceed as in the first part of the proof.   □

Example 4.2.19. Determine two orthogonal bases in the vector plane

$$\text{Span}\left\{ \begin{bmatrix} 1 \\ 0 \\ 2 \end{bmatrix} \begin{bmatrix} 1 \\ -2 \\ 2 \end{bmatrix} \right\}.$$

Solution. Since

$$\begin{bmatrix} 1 \\ 0 \\ 2 \end{bmatrix} \cdot \begin{bmatrix} 1 \\ -2 \\ 2 \end{bmatrix} = 5,$$

$\left\{ \begin{bmatrix} 1 \\ 0 \\ 2 \end{bmatrix}, \begin{bmatrix} 1 \\ -2 \\ 2 \end{bmatrix} \right\}$ is not an orthogonal basis. We find that

$$\begin{bmatrix} 1 \\ -2 \\ 2 \end{bmatrix} \cdot \begin{bmatrix} 1 \\ -2 \\ 2 \end{bmatrix} = 9 \quad \text{and} \quad \begin{bmatrix} 1 \\ 0 \\ 2 \end{bmatrix} - \frac{5}{9} \begin{bmatrix} 1 \\ -2 \\ 2 \end{bmatrix} = \begin{bmatrix} \frac{4}{9} \\ \frac{10}{9} \\ \frac{8}{9} \end{bmatrix},$$

so $\left\{ \begin{bmatrix} \frac{4}{9} \\ \frac{10}{9} \\ \frac{8}{9} \end{bmatrix}, \begin{bmatrix} 1 \\ -2 \\ 2 \end{bmatrix} \right\}$ is an orthogonal basis in the vector plane Span $\left\{ \begin{bmatrix} 1 \\ 0 \\ 2 \end{bmatrix} \begin{bmatrix} 1 \\ -2 \\ 2 \end{bmatrix} \right\}.$

Note that we can use $\begin{bmatrix} 2 \\ 5 \\ 4 \end{bmatrix}$ instead of $\begin{bmatrix} \frac{4}{9} \\ \frac{10}{9} \\ \frac{8}{9} \end{bmatrix}$, because $\begin{bmatrix} \frac{4}{9} \\ \frac{10}{9} \\ \frac{8}{9} \end{bmatrix} = \frac{2}{9} \begin{bmatrix} 2 \\ 5 \\ 4 \end{bmatrix}$ and

$$\text{Span}\left\{ \begin{bmatrix} 2 \\ 5 \\ 4 \end{bmatrix}, \begin{bmatrix} 1 \\ -2 \\ 2 \end{bmatrix} \right\} = \text{Span}\left\{ \begin{bmatrix} \frac{4}{9} \\ \frac{10}{9} \\ \frac{8}{9} \end{bmatrix}, \begin{bmatrix} 1 \\ -2 \\ 2 \end{bmatrix} \right\} = \text{Span}\left\{ \begin{bmatrix} 1 \\ 0 \\ 2 \end{bmatrix} \begin{bmatrix} 1 \\ -2 \\ 2 \end{bmatrix} \right\},$$

by Theorem 4.1.17.

Thus $\left\{ \begin{bmatrix} 2 \\ 5 \\ 4 \end{bmatrix}, \begin{bmatrix} 1 \\ -2 \\ 2 \end{bmatrix} \right\}$ is an orthogonal basis in the vector plane

$$\text{Span}\left\{\begin{bmatrix}1\\0\\2\end{bmatrix}, \begin{bmatrix}1\\-2\\2\end{bmatrix}\right\}.$$

Another possibility is to use

$$\begin{bmatrix}1\\-2\\2\end{bmatrix} - \frac{\begin{bmatrix}1\\-2\\2\end{bmatrix}\cdot\begin{bmatrix}1\\0\\2\end{bmatrix}}{\begin{bmatrix}1\\0\\2\end{bmatrix}\cdot\begin{bmatrix}1\\0\\2\end{bmatrix}}\begin{bmatrix}1\\0\\2\end{bmatrix} = \begin{bmatrix}0\\-2\\0\end{bmatrix}$$

or just $\begin{bmatrix}0\\1\\0\end{bmatrix}$ because $\begin{bmatrix}0\\-2\\0\end{bmatrix} = -2\begin{bmatrix}0\\1\\0\end{bmatrix}$. Thus $\left\{\begin{bmatrix}1\\0\\2\end{bmatrix}, \begin{bmatrix}0\\1\\0\end{bmatrix}\right\}$ is another orthogonal ba-

sis in the vector plane $\text{Span}\left\{\begin{bmatrix}1\\0\\2\end{bmatrix}, \begin{bmatrix}1\\-2\\2\end{bmatrix}\right\}.$ □

The projection on a vector plane in $\mathbb{R}^3$

We have considered the problem of finding a point on a line minimizing the distance from a given point and observed that methods of linear algebra give us a simple and elegant solution to the problem. Now we would like to find a point on a vector plane minimizing the distance from a given point. Again we find that linear algebra can handle this problem well.

Theorem 4.2.20. *Let* **u** *and* **v** *be linearly independent vectors in* $\mathbb{R}^3$. *For every vector* **b** *in* $\mathbb{R}^3$ *there is a unique vector* **p** *in the vector plane* $\text{Span}\{\mathbf{u}, \mathbf{v}\}$ *such that the inequality*

$$\|\mathbf{b} - \mathbf{p}\| < \|\mathbf{b} - \mathbf{x}\|$$

holds for every **x** *in* $\text{Span}\{\mathbf{u}, \mathbf{v}\}$ *different from* **p**. *If the vectors* **u** *and* **v** *are orthogonal, then that unique point is*

$$\mathbf{p} = \frac{\mathbf{b}\cdot\mathbf{u}}{\|\mathbf{u}\|^2}\mathbf{u} + \frac{\mathbf{b}\cdot\mathbf{v}}{\|\mathbf{v}\|^2}\mathbf{v} = \text{proj}_{\mathbf{u}}\mathbf{b} + \text{proj}_{\mathbf{v}}\mathbf{b}. \qquad (4.6)$$

Proof. In view of Theorem 4.2.18, we can assume that the vectors **u** and **v** are orthogonal, that is $\mathbf{u}\cdot\mathbf{v} = 0$, because if they are not we can modify the basis of the vector plane so that it becomes an orthogonal basis. Now, using the fact that $\mathbf{u}\cdot\mathbf{v} = 0$, we rewrite the square of the distance from **b** to an arbitrary point on the vector plane $\text{Span}\{\mathbf{u}, \mathbf{v}\}$ in a way that may seem at first quite artificial:

$$\|b - su - tv\|^2 = (b - su - tv) \cdot (b - su - tv)$$

$$= \|b\|^2 + s^2\|u\|^2 + t^2\|v\|^2 - 2sb \cdot u - 2tb \cdot v + 2stu \cdot v$$

$$= \|b\|^2 - \frac{(b \cdot u)^2}{\|u\|^2} - \frac{(b \cdot v)^2}{\|v\|^2} + s^2\|u\|^2$$

$$- 2sb \cdot u + \frac{(b \cdot u)^2}{\|u\|^2} + t^2\|v\|^2 - 2tb \cdot v + \frac{(b \cdot v)^2}{\|v\|^2}$$

$$= \|b\|^2 - \frac{(b \cdot u)^2}{\|u\|^2} - \frac{(b \cdot v)^2}{\|v\|^2} + \left(s\|u\| - \frac{b \cdot u}{\|u\|}\right)^2 + \left(t\|v\| - \frac{b \cdot v}{\|v\|}\right)^2$$

Note that the part

$$\|b\|^2 - \frac{(b \cdot u)^2}{\|u\|^2} - \frac{(b \cdot v)^2}{\|v\|^2}$$

does not depend on t or s, so it is fixed for the point b and the vector plane. Consequently, to minimize $\|b - su - tv\|^2$, we need to minimize the remaining part, that is

$$\left(s\|u\| - \frac{b \cdot u}{\|u\|}\right)^2 + \left(t\|v\| - \frac{b \cdot v}{\|v\|}\right)^2.$$

Clearly, the minimum is attained when

$$s\|u\| - \frac{b \cdot u}{\|u\|} = 0 \quad \text{and} \quad t\|v\| - \frac{b \cdot v}{\|v\|} = 0.$$

Solving for t and s we get

$$s = \frac{b \cdot u}{\|u\|^2} \quad \text{and} \quad t = \frac{b \cdot v}{\|v\|^2}$$

and consequently

$$p = su + tv = \frac{b \cdot u}{\|u\|^2}u + \frac{b \cdot v}{\|v\|^2}v.$$

$\square$

The point p in the above theorem minimizes the distance from the point b to an arbitrary point on the vector plane Span$\{u, v\}$. For this reason p is called the **best approximation** to b by elements of the vector plane Span$\{u, v\}$.

Example 4.2.21. Determine the best approximation to the point $b = \begin{bmatrix} 3 \\ 1 \\ 1 \end{bmatrix}$ by elements of Span$\{u, v\}$ where $u = \begin{bmatrix} 1 \\ -2 \\ 1 \end{bmatrix}$ and $y = \begin{bmatrix} 1 \\ 1 \\ 1 \end{bmatrix}$.

Solution. Since the vectors $\mathbf{u} = \begin{bmatrix} 1 \\ -2 \\ 1 \end{bmatrix}$ and $\mathbf{u} = \begin{bmatrix} 1 \\ 1 \\ 1 \end{bmatrix}$ are orthogonal, we can use formula (4.6), which gives us

$$\mathbf{p} = \frac{\begin{bmatrix} 3 \\ 1 \\ 1 \end{bmatrix} \cdot \begin{bmatrix} 1 \\ -2 \\ 1 \end{bmatrix}}{\begin{bmatrix} 1 \\ -2 \\ 1 \end{bmatrix} \cdot \begin{bmatrix} 1 \\ -2 \\ 1 \end{bmatrix}} \begin{bmatrix} 1 \\ -2 \\ 1 \end{bmatrix} + \frac{\begin{bmatrix} 3 \\ 1 \\ 1 \end{bmatrix} \cdot \begin{bmatrix} 1 \\ 1 \\ 1 \end{bmatrix}}{\begin{bmatrix} 1 \\ 1 \\ 1 \end{bmatrix} \cdot \begin{bmatrix} 1 \\ 1 \\ 1 \end{bmatrix}} \begin{bmatrix} 1 \\ 1 \\ 1 \end{bmatrix} = \frac{1}{3} \begin{bmatrix} 1 \\ -2 \\ 1 \end{bmatrix} + \frac{5}{3} \begin{bmatrix} 1 \\ 1 \\ 1 \end{bmatrix} = \begin{bmatrix} 2 \\ 1 \\ 2 \end{bmatrix}.$$

$\square$

Example 4.2.22. Determine the best approximation to the point $\mathbf{b} = \begin{bmatrix} 1 \\ 1 \\ 1 \end{bmatrix}$ by elements of Span$\{\mathbf{x}, \mathbf{y}\}$ where $\mathbf{x} = \begin{bmatrix} 1 \\ 0 \\ 1 \end{bmatrix}$ and $\mathbf{y} = \begin{bmatrix} 1 \\ 1 \\ 0 \end{bmatrix}$.

Solution. Since $\mathbf{x} \cdot \mathbf{y} = 1$, the vectors $\mathbf{x}$ and $\mathbf{y}$ are not orthogonal. To find an orthogonal basis for Span$\{\mathbf{x}, \mathbf{y}\}$ we use the method presented in the proof of Theorem 4.2.18:

$$\text{Span}\{\mathbf{x}, \mathbf{y}\} = \text{Span}\left\{\mathbf{x} - \frac{\mathbf{x} \cdot \mathbf{y}}{\mathbf{y} \cdot \mathbf{y}} \mathbf{y}, \mathbf{y}\right\}$$

$$= \text{Span}\left\{ \begin{bmatrix} \frac{1}{2} \\ -\frac{1}{2} \\ 1 \end{bmatrix}, \begin{bmatrix} 1 \\ 1 \\ 0 \end{bmatrix} \right\}$$

$$= \text{Span}\left\{ \begin{bmatrix} 1 \\ -1 \\ 2 \end{bmatrix}, \begin{bmatrix} 1 \\ 1 \\ 0 \end{bmatrix} \right\}.$$

Since the vectors $\begin{bmatrix} 1 \\ -1 \\ 2 \end{bmatrix}$ and $\begin{bmatrix} 1 \\ 1 \\ 0 \end{bmatrix}$ are orthogonal, we can use formula (4.6) with

$\mathbf{u} = \begin{bmatrix} 1 \\ -1 \\ 2 \end{bmatrix}$ and $\mathbf{v} = \begin{bmatrix} 1 \\ 1 \\ 0 \end{bmatrix}$, which gives us

$$\mathbf{p} = \frac{\begin{bmatrix} 1 \\ 1 \\ 1 \end{bmatrix} \cdot \begin{bmatrix} 1 \\ -1 \\ 2 \end{bmatrix}}{\begin{bmatrix} 1 \\ -1 \\ 2 \end{bmatrix} \cdot \begin{bmatrix} 1 \\ -1 \\ 2 \end{bmatrix}} \begin{bmatrix} 1 \\ -1 \\ 2 \end{bmatrix} + \frac{\begin{bmatrix} 1 \\ 1 \\ 1 \end{bmatrix} \cdot \begin{bmatrix} 1 \\ 1 \\ 0 \end{bmatrix}}{\begin{bmatrix} 1 \\ 1 \\ 0 \end{bmatrix} \cdot \begin{bmatrix} 1 \\ 1 \\ 0 \end{bmatrix}} \begin{bmatrix} 1 \\ 1 \\ 0 \end{bmatrix} = \frac{1}{3} \begin{bmatrix} 1 \\ -1 \\ 2 \end{bmatrix} + \begin{bmatrix} 1 \\ 1 \\ 0 \end{bmatrix} = \frac{2}{3} \begin{bmatrix} 2 \\ 1 \\ 1 \end{bmatrix}.$$

$\square$

The following theorem is very similar to Theorem 4.2.12 that characterizes the best approximation to the point by elements of a vector line.

Theorem 4.2.23. *Consider a point* $\mathbf{b}$ *and a vector plane* Span$\{\mathbf{u}, \mathbf{v}\}$ *in* $\mathbb{R}^3$*. The best approximation to* $\mathbf{b}$ *by elements of* Span$\{\mathbf{u}, \mathbf{v}\}$ *can be characterized as the point* $\mathbf{p}$ *in* Span$\{\mathbf{u}, \mathbf{v}\}$ *such that*

$$(\mathbf{b} - \mathbf{p}) \cdot \mathbf{w} = 0 \quad \text{for every } \mathbf{w} \text{ in Span}\{\mathbf{u}, \mathbf{v}\}. \tag{4.7}$$

In other words, the point $\mathbf{p}$ *in* Span$\{\mathbf{u}, \mathbf{v}\}$ *is the best approximation to* $\mathbf{b}$ *by elements of the vector plane* Span$\{\mathbf{u}, \mathbf{v}\}$ *if and only if the vector* $\mathbf{b} - \mathbf{p}$ *is orthogonal to every vector in* Span$\{\mathbf{u}, \mathbf{v}\}$*.*

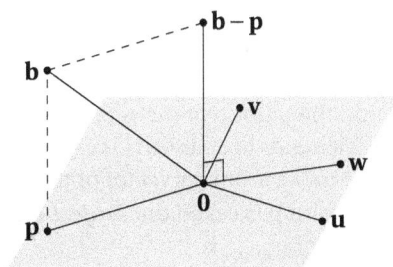

Figure 4.8: Illustration of Theorem 4.2.23.

Proof. Let $\mathbf{p}$ is the best approximation to $\mathbf{b}$ by elements of the vector plane Span$\{\mathbf{u}, \mathbf{v}\}$.

Without loss of generality we can assume that the vectors $\mathbf{u}$ and $\mathbf{v}$ are orthogonal. Then, by (4.6), we have

$$\mathbf{p} = \frac{\mathbf{b} \cdot \mathbf{u}}{\|\mathbf{u}\|^2} \mathbf{u} + \frac{\mathbf{b} \cdot \mathbf{v}}{\|\mathbf{v}\|^2} \mathbf{v}.$$

If $\mathbf{w}$ is in Span$\{\mathbf{u}, \mathbf{v}\}$, then $\mathbf{w} = s\mathbf{u} + t\mathbf{v}$ for some real numbers s and t and we have

$$(\mathbf{b} - \mathbf{p}) \cdot \mathbf{w} = \left(\mathbf{b} - \frac{\mathbf{b} \cdot \mathbf{u}}{\|\mathbf{u}\|^2} \mathbf{u} - \frac{\mathbf{b} \cdot \mathbf{v}}{\|\mathbf{v}\|^2} \mathbf{v} \right) \cdot (s\mathbf{u} + t\mathbf{v})$$

$$= s\mathbf{b} \cdot \mathbf{u} + t\mathbf{b} \cdot \mathbf{v} - s \frac{\mathbf{b} \cdot \mathbf{u}}{\|\mathbf{u}\|^2} \mathbf{u} \cdot \mathbf{u} - t \frac{\mathbf{b} \cdot \mathbf{v}}{\|\mathbf{v}\|^2} \mathbf{v} \cdot \mathbf{v} - t \frac{\mathbf{b} \cdot \mathbf{u}}{\|\mathbf{u}\|^2} \mathbf{u} \cdot \mathbf{v} - s \frac{\mathbf{b} \cdot \mathbf{v}}{\|\mathbf{v}\|^2} \mathbf{v} \cdot \mathbf{u}$$

$$= 0,$$

since $\mathbf{u} \cdot \mathbf{u} = \|\mathbf{u}\|^2$, $\mathbf{v} \cdot \mathbf{v} = \|\mathbf{v}\|^2$, and $\mathbf{u} \cdot \mathbf{v} = \mathbf{v} \cdot \mathbf{u} = 0$.

Now assume that the point $\mathbf{p}$ in Span$\{\mathbf{u}, \mathbf{v}\}$ satisfies the equality

$$(\mathbf{b} - \mathbf{p}) \cdot \mathbf{w} = 0$$

for every $\mathbf{w}$ in Span$\{\mathbf{u}, \mathbf{v}\}$. For any vector $\mathbf{x}$ in Span$\{\mathbf{u}, \mathbf{v}\}$ we have

$$\|\mathbf{b} - \mathbf{x}\|^2 = (\mathbf{b} - \mathbf{x}) \cdot (\mathbf{b} - \mathbf{x})$$

$$= (\mathbf{b} - \mathbf{p} + \mathbf{p} - \mathbf{x}) \cdot (\mathbf{b} - \mathbf{p} + \mathbf{p} - \mathbf{x})$$

$$= \|\mathbf{b} - \mathbf{p}\|^2 + 2(\mathbf{b} - \mathbf{p}) \cdot (\mathbf{p} - \mathbf{x}) + \|\mathbf{p} - \mathbf{x}\|^2$$

$$= \|\mathbf{b} - \mathbf{p}\|^2 + \|\mathbf{p} - \mathbf{x}\|^2,$$

because $\mathbf{w} = \mathbf{p} - \mathbf{x}$ is in Span$\{\mathbf{u}, \mathbf{v}\}$ and consequently $(\mathbf{b} - \mathbf{p}) \cdot (\mathbf{p} - \mathbf{x}) = 0$. Since

$$\|\mathbf{b} - \mathbf{x}\|^2 = \|\mathbf{b} - \mathbf{p}\|^2 + \|\mathbf{p} - \mathbf{x}\|^2,$$

we can conclude that

$$\|\mathbf{b} - \mathbf{x}\| > \|\mathbf{b} - \mathbf{p}\|$$

for every $\mathbf{x}$ in Span$\{\mathbf{u}, \mathbf{v}\}$ different from $\mathbf{p}$ which means that $\mathbf{p}$ is the best approximation to $\mathbf{b}$ by elements of the vector plane Span$\{\mathbf{u}, \mathbf{v}\}$.   □

Theorem 4.2.23 says that the point $\mathbf{p}$ of the vector plane Span$\{\mathbf{u}, \mathbf{v}\}$ which is the best approximation to $\mathbf{b}$ by elements of Span$\{\mathbf{u}, \mathbf{v}\}$ is exactly the point $\mathbf{p}$ for which the vector $\mathbf{b} - \mathbf{p}$ is perpendicular on an arbitrary vector of the vector plane Span$\{\mathbf{u}, \mathbf{v}\}$.

For this reason such the point $\mathbf{p}$ is called the ***projection*** of $\mathbf{b}$ on the vector plane Span$\{\mathbf{u}, \mathbf{v}\}$ and is denoted by proj$_{\text{Span}\{\mathbf{u},\mathbf{v}\}}\mathbf{b}$:

Let $\mathbf{u}$ and $\mathbf{v}$ be nonzero vectors in $\mathbb{R}^3$ such that $\mathbf{u} \cdot \mathbf{v} = 0$.

$$\text{proj}_{\text{Span}\{\mathbf{u},\mathbf{v}\}}\mathbf{b} = \frac{\mathbf{b} \cdot \mathbf{u}}{\|\mathbf{u}\|^2} \mathbf{u} + \frac{\mathbf{b} \cdot \mathbf{v}}{\|\mathbf{v}\|^2} \mathbf{v} = \text{proj}_{\mathbf{u}}\mathbf{b} + \text{proj}_{\mathbf{v}}\mathbf{b}.$$

Now we prove a result similar to the Theorem 4.2.13.

Theorem 4.2.24. *Let* **u** *and* **v** *be nonzero orthogonal vectors in* $\mathbb{R}^3$. *If*

$$A = \frac{1}{\|\mathbf{u}\|^2}(\mathbf{u}\mathbf{u}^T) + \frac{1}{\|\mathbf{v}\|^2}(\mathbf{v}\mathbf{v}^T),$$

then the projection of any vector **b** *in* $\mathbb{R}^3$ *on the vector plane* Span$\{\mathbf{u}, \mathbf{v}\}$ *is* $A\mathbf{b}$. *Moreover, the matrix* A *is the unique matrix with this property.*

Proof. Let $\mathbf{u} = \begin{bmatrix} u_1 \\ u_2 \\ u_3 \end{bmatrix}$ and $\mathbf{v} = \begin{bmatrix} v_1 \\ v_2 \\ v_3 \end{bmatrix}$. For any vector $\mathbf{b} = \begin{bmatrix} b_1 \\ b_2 \\ b_3 \end{bmatrix}$ in $\mathbb{R}^3$ we have

$$\text{proj}_{\text{Span}\{\mathbf{u},\mathbf{v}\}}\mathbf{b} = \frac{\mathbf{b}\cdot\mathbf{u}}{\|\mathbf{u}\|^2}\mathbf{u} + \frac{\mathbf{b}\cdot\mathbf{v}}{\|\mathbf{v}\|^2}\mathbf{v}$$

$$= \frac{1}{\|\mathbf{u}\|^2}\begin{bmatrix} u_1 \\ u_2 \\ u_3 \end{bmatrix}\left([u_1 \ u_2 \ u_3]\begin{bmatrix} b_1 \\ b_2 \\ b_3 \end{bmatrix}\right) + \frac{1}{\|\mathbf{v}\|^2}\begin{bmatrix} v_1 \\ v_2 \\ v_3 \end{bmatrix}\left([v_1 \ v_2 \ v_3]\begin{bmatrix} b_1 \\ b_2 \\ b_3 \end{bmatrix}\right)$$

$$= \frac{1}{\|\mathbf{u}\|^2}\left(\begin{bmatrix} u_1 \\ u_2 \\ u_3 \end{bmatrix}[u_1 \ u_2 \ u_3]\right)\begin{bmatrix} b_1 \\ b_2 \\ b_3 \end{bmatrix} + \frac{1}{\|\mathbf{v}\|^2}\left(\begin{bmatrix} v_1 \\ v_2 \\ v_3 \end{bmatrix}[v_1 \ v_2 \ v_3]\right)\begin{bmatrix} b_1 \\ b_2 \\ b_3 \end{bmatrix}$$

$$= \left(\frac{1}{\|\mathbf{u}\|^2}\begin{bmatrix} u_1 \\ u_2 \\ u_3 \end{bmatrix}[u_1 \ u_2 \ u_3] + \frac{1}{\|\mathbf{v}\|^2}\begin{bmatrix} v_1 \\ v_2 \\ v_3 \end{bmatrix}[v_1 \ v_2 \ v_3]\right)\begin{bmatrix} b_1 \\ b_2 \\ b_3 \end{bmatrix}$$

$$= \left(\frac{1}{\|\mathbf{u}\|^2}(\mathbf{u}\mathbf{u}^T) + \frac{1}{\|\mathbf{v}\|^2}(\mathbf{v}\mathbf{v}^T)\right)\mathbf{b}$$

$$= A\mathbf{b}.$$

The uniqueness part of the theorem is a consequence of the fact that the equality $A\mathbf{x} = B\mathbf{x}$ for every $\mathbf{x}$ in $\mathbb{R}^3$ implies $A = B$, by Theorem 2.1.18. $\qquad\square$

Definition 4.2.25. Let **u** and **v** be nonzero vectors in $\mathbb{R}^3$ such that $\mathbf{u}\cdot\mathbf{v} = 0$. The matrix

$$\frac{1}{\|\mathbf{u}\|^2}(\mathbf{u}\mathbf{u}^T) + \frac{1}{\|\mathbf{v}\|^2}(\mathbf{v}\mathbf{v}^T) \qquad (4.8)$$

is called the **projection matrix** on Span$\{\mathbf{u}, \mathbf{v}\}$.

Example 4.2.26. Determine the projection matrix on Span$\{\mathbf{x}, \mathbf{y}\}$, where $\mathbf{x} = \begin{bmatrix} 1 \\ 0 \\ 1 \end{bmatrix}$

and $\mathbf{y} = \begin{bmatrix} 1 \\ 1 \\ 0 \end{bmatrix}$.

Solution. From Example 4.2.22 we have

$$\text{Span}\{\mathbf{x}, \mathbf{y}\} = \text{Span}\left\{ \begin{bmatrix} 1 \\ -1 \\ 2 \end{bmatrix}, \begin{bmatrix} 1 \\ 1 \\ 0 \end{bmatrix} \right\}.$$

Since the vectors $\begin{bmatrix} 1 \\ -1 \\ 2 \end{bmatrix}$ and $\begin{bmatrix} 1 \\ 1 \\ 0 \end{bmatrix}$ are orthogonal, we can use formula (4.8) with

$\mathbf{u} = \begin{bmatrix} 1 \\ -1 \\ 2 \end{bmatrix}$ and $\mathbf{v} = \begin{bmatrix} 1 \\ 1 \\ 0 \end{bmatrix}$ which gives us

$$\frac{1}{6} \begin{bmatrix} 1 \\ -1 \\ 2 \end{bmatrix} \begin{bmatrix} 1 & -1 & 2 \end{bmatrix} + \frac{1}{2} \begin{bmatrix} 1 \\ 1 \\ 0 \end{bmatrix} \begin{bmatrix} 1 & 1 & 0 \end{bmatrix} = \frac{1}{6} \begin{bmatrix} 1 & -1 & 2 \\ -1 & 1 & -2 \\ 2 & -2 & 4 \end{bmatrix} + \frac{1}{2} \begin{bmatrix} 1 & 1 & 0 \\ 1 & 1 & 0 \\ 0 & 0 & 0 \end{bmatrix}$$

$$= \frac{1}{3} \begin{bmatrix} 2 & 1 & 1 \\ 1 & 2 & -1 \\ 1 & -1 & 2 \end{bmatrix}.$$

□

The least squares problem in $\mathbb{R}^3$

In this section we prove a theorem that shows how geometrical ideas can be used to solve a problem which does not seem to have anything to do with geometry.

Theorem 4.2.27. *The numbers x and y which minimize the sum*

$$(b_1 - u_1 x - v_1 y)^2 + (b_2 - u_2 x - v_2 y)^2 + (b_3 - u_3 x - v_3 y)^2$$

are the solution of the equation

$$A^T A \begin{bmatrix} x \\ y \end{bmatrix} = A^T \mathbf{b} \qquad (4.9)$$

where

$$A = \begin{bmatrix} u_1 & v_1 \\ u_2 & v_2 \\ u_3 & v_3 \end{bmatrix} \quad and \quad \mathbf{b} = \begin{bmatrix} b_1 \\ b_2 \\ b_3 \end{bmatrix}.$$

Proof. With A and $\mathbf{b}$ defined above the equation $A^T A \begin{bmatrix} x \\ y \end{bmatrix} = A^T \mathbf{b}$ is

$$\begin{bmatrix} u_1 & u_2 & u_3 \\ v_1 & v_2 & v_3 \end{bmatrix} \begin{bmatrix} u_1 & v_1 \\ u_2 & v_2 \\ u_3 & v_3 \end{bmatrix} \begin{bmatrix} x \\ y \end{bmatrix} = \begin{bmatrix} u_1 & u_2 & u_3 \\ v_1 & v_2 & v_3 \end{bmatrix} \begin{bmatrix} b_1 \\ b_2 \\ b_3 \end{bmatrix} \qquad (4.10)$$

and is equivalent to the equation

$$\begin{bmatrix} u_1 & u_2 & u_3 \\ v_1 & v_2 & v_3 \end{bmatrix} \left(\begin{bmatrix} b_1 \\ b_2 \\ b_3 \end{bmatrix} - \begin{bmatrix} u_1 & v_1 \\ u_2 & v_2 \\ u_3 & v_3 \end{bmatrix} \begin{bmatrix} x \\ y \end{bmatrix} \right) = \begin{bmatrix} 0 \\ 0 \end{bmatrix}.$$

This equation can be written as a system of equations:

$$\begin{cases} \begin{bmatrix} u_1 & u_2 & u_3 \end{bmatrix} \left(\begin{bmatrix} b_1 \\ b_2 \\ b_3 \end{bmatrix} - \begin{bmatrix} u_1 & v_1 \\ u_2 & v_2 \\ u_3 & v_3 \end{bmatrix} \begin{bmatrix} x \\ y \end{bmatrix} \right) = 0 \\[2em] \begin{bmatrix} v_1 & v_2 & v_3 \end{bmatrix} \left(\begin{bmatrix} b_1 \\ b_2 \\ b_3 \end{bmatrix} - \begin{bmatrix} u_1 & v_1 \\ u_2 & v_2 \\ u_3 & v_3 \end{bmatrix} \begin{bmatrix} x \\ y \end{bmatrix} \right) = 0 \end{cases}$$

If we let $\mathbf{p} = \begin{bmatrix} u_1 & v_1 \\ u_2 & v_2 \\ u_3 & v_3 \end{bmatrix} \begin{bmatrix} x \\ y \end{bmatrix}$, the system becomes

$$\begin{cases} \left(\begin{bmatrix} b_1 \\ b_2 \\ b_3 \end{bmatrix} - \mathbf{p} \right) \cdot \begin{bmatrix} u_1 \\ u_2 \\ u_3 \end{bmatrix} = 0 \\[2em] \left(\begin{bmatrix} b_1 \\ b_2 \\ b_3 \end{bmatrix} - \mathbf{p} \right) \cdot \begin{bmatrix} v_1 \\ v_2 \\ v_3 \end{bmatrix} = 0 \end{cases} \qquad (4.11)$$

If we denote $\mathbf{u} = \begin{bmatrix} u_1 \\ u_2 \\ u_3 \end{bmatrix}$ and $\mathbf{v} = \begin{bmatrix} v_1 \\ v_2 \\ v_3 \end{bmatrix}$ and $\mathbf{w} = s\mathbf{u} + t\mathbf{v}$, the system (4.11) yields

$$(\mathbf{b} - \mathbf{p}) \cdot \mathbf{w} = (\mathbf{b} - \mathbf{p}) \cdot (s\mathbf{u} + t\mathbf{v}) = s(\mathbf{b} - \mathbf{p}) \cdot \mathbf{u} + t(\mathbf{b} - \mathbf{p}) \cdot \mathbf{v} = 0.$$

If the vectors $\mathbf{u}$ and $\mathbf{v}$ are linearly independent, then, using Theorem 4.2.23, we can give a geometrical interpretation of the above system: the solution of the system (4.11) is the point $\mathbf{p} = \begin{bmatrix} u_1 & v_1 \\ u_2 & v_2 \\ u_3 & v_3 \end{bmatrix} \begin{bmatrix} x \\ y \end{bmatrix} = x\mathbf{u} + y\mathbf{v}$ that is the projection of $\mathbf{b}$ on the vector plane Span$\{\mathbf{u}, \mathbf{v}\}$. Consequently, if the vectors $\mathbf{u}$ and $\mathbf{v}$ are linearly independent, then the numbers x and y that satisfy the equation $A^T A \begin{bmatrix} x \\ y \end{bmatrix} = A^T \mathbf{b}$ minimize

$$\|\mathbf{b} - x\mathbf{u} - y\mathbf{v}\|^2 = (b_1 - u_1 x - v_1 y)^2 + (b_2 - u_2 x - v_2 y)^2 + (b_3 - u_3 x - v_3 y)^2.$$

If the vectors $\mathbf{u}$ and $\mathbf{v}$ are linearly dependent and $\mathbf{u} \neq \mathbf{0}$ or $\mathbf{v} \neq \mathbf{0}$, then $\text{Span}\{\mathbf{u}, \mathbf{v}\}$ is actually a vector line and, according to Theorem 4.2.12, the solution of the system (4.11) is the point $\mathbf{p}$ that is the projection of $\mathbf{b}$ on the vector line $\text{Span}\{\mathbf{u}, \mathbf{v}\}$. Note that in this case numbers x and y such that $\mathbf{p} = x\mathbf{u} + y\mathbf{v}$ are not unique. Nevertheless, any numbers x and y that satisfy the equation $A^T A \begin{bmatrix} x \\ y \end{bmatrix} = A^T \mathbf{b}$ minimize

$$\|\mathbf{b} - x\mathbf{u} - y\mathbf{v}\|^2 = (b_1 - u_1 x - v_1 y)^2 + (b_2 - u_2 x - v_2 y)^2 + (b_3 - u_3 x - v_3 y)^2.$$

Finally, if $\mathbf{u} = \mathbf{v} = \mathbf{0}$, then the sum

$$(b_1 - u_1 x - v_1 y)^2 + (b_2 - u_2 x - v_2 y)^2 + (b_3 - u_3 x - v_3 y)^2 = b_1^2 + b_2^2 + b_3^2$$

is a constant and there is nothing to prove.   ☐

Example 4.2.28. Find numbers x and y which minimize the sum

$$(1 - 2x - y)^2 + (2 - x + y)^2 + (1 + x + y)^2.$$

Solution. If we let $A = \begin{bmatrix} 2 & 1 \\ 1 & -1 \\ -1 & -1 \end{bmatrix}$ and $\mathbf{b} = \begin{bmatrix} 1 \\ 2 \\ 1 \end{bmatrix}$, then the numbers x and y which minimize $(1 - 2x - y)^2 + (2 - x + y)^2 + (1 + x + y)^2$ are the solution of the equation

$$A^T A \begin{bmatrix} x \\ y \end{bmatrix} = A^T \mathbf{b},$$

that is,

$$\begin{bmatrix} 2 & 1 & -1 \\ 1 & -1 & -1 \end{bmatrix} \begin{bmatrix} 2 & 1 \\ 1 & -1 \\ -1 & -1 \end{bmatrix} \begin{bmatrix} x \\ y \end{bmatrix} = \begin{bmatrix} 2 & 1 & -1 \\ 1 & -1 & -1 \end{bmatrix} \begin{bmatrix} 1 \\ 2 \\ 1 \end{bmatrix},$$

or

$$\begin{bmatrix} 6 & 2 \\ 2 & 3 \end{bmatrix} \begin{bmatrix} x \\ y \end{bmatrix} = \begin{bmatrix} 3 \\ -2 \end{bmatrix}.$$

This equation is equivalent to the system

$$\begin{cases} 6x + 2y = 3 \\ 2x + 3y = -2 \end{cases}$$

which can be solved using Cramer's rule:

$$x = \frac{\det \begin{bmatrix} 3 & 2 \\ -2 & 3 \end{bmatrix}}{\det \begin{bmatrix} 6 & 2 \\ 2 & 3 \end{bmatrix}} = \frac{13}{14} \quad \text{and} \quad y = \frac{\det \begin{bmatrix} 6 & 3 \\ 2 & -2 \end{bmatrix}}{\det \begin{bmatrix} 6 & 2 \\ 2 & 3 \end{bmatrix}} = -\frac{18}{14}.$$

☐

Note that the formula (4.9) in Theorem 4.2.27 gives us another way to calculate the point on a vector plane that minimizes the distance from a given point to an arbitrary point on that vector plane.

The system of equations (4.11) in the proof of Theorem 4.2.27 suggests a simple way to calculate the numbers x and y which minimize the sum

$$(b_1 - u_1 x - v_1 y)^2 + (b_2 - u_2 x - v_2 y)^2 + (b_3 - u_3 x - v_3 y)^2.$$

We formulate that observation in the next theorem.

Theorem 4.2.29. *The numbers* x *and* y *which minimize the sum*

$$(b_1 - u_1 x - v_1 y)^2 + (b_2 - u_2 x - v_2 y)^2 + (b_3 - u_3 x - v_3 y)^2$$

are the solution of the system

$$\begin{cases} \left(\begin{bmatrix} b_1 \\ b_2 \\ b_3 \end{bmatrix} - x \begin{bmatrix} u_1 \\ u_2 \\ u_3 \end{bmatrix} - y \begin{bmatrix} v_1 \\ v_2 \\ v_3 \end{bmatrix} \right) \cdot \begin{bmatrix} u_1 \\ u_2 \\ u_3 \end{bmatrix} = 0 \\[3mm] \left(\begin{bmatrix} b_1 \\ b_2 \\ b_3 \end{bmatrix} - x \begin{bmatrix} u_1 \\ u_2 \\ u_3 \end{bmatrix} - y \begin{bmatrix} v_1 \\ v_2 \\ v_3 \end{bmatrix} \right) \cdot \begin{bmatrix} v_1 \\ v_2 \\ v_3 \end{bmatrix} = 0 \end{cases} \tag{4.12}$$

Example 4.2.30. Find numbers x and y which minimize the sum

$$(1 - 2x - y)^2 + (2 - x + y)^2 + (1 + x + y)^2.$$

Solution. In this case the system (4.12) becomes

$$\begin{cases} \left(\begin{bmatrix} 1 \\ 2 \\ 1 \end{bmatrix} - x \begin{bmatrix} 2 \\ 1 \\ -1 \end{bmatrix} - y \begin{bmatrix} 1 \\ -1 \\ -1 \end{bmatrix} \right) \cdot \begin{bmatrix} 2 \\ 1 \\ -1 \end{bmatrix} = 0 \\[3mm] \left(\begin{bmatrix} 1 \\ 2 \\ 1 \end{bmatrix} - x \begin{bmatrix} 2 \\ 1 \\ -1 \end{bmatrix} - y \begin{bmatrix} 1 \\ -1 \\ -1 \end{bmatrix} \right) \cdot \begin{bmatrix} 1 \\ -1 \\ -1 \end{bmatrix} = 0 \end{cases}$$

After calculating the dot products and simplifying we obtain the system

$$\begin{cases} 6x + 2y = \ \ 3 \\ 2x + 3y = -2 \end{cases}.$$

The solutions are as in Example 4.2.28

$$x = \frac{13}{14} \quad \text{and} \quad y = -\frac{18}{14}.$$

□

Corollary 4.2.31. *Let* $\begin{bmatrix} u_1 \\ u_2 \\ u_3 \end{bmatrix}$, $\begin{bmatrix} v_1 \\ v_2 \\ v_3 \end{bmatrix}$, *and* $\begin{bmatrix} b_1 \\ b_2 \\ b_3 \end{bmatrix}$ *be vectors in* $\mathbb{R}^3$. *If numbers x and y satisfy the system of equations* (4.12), *then the vector*

$$x \begin{bmatrix} u_1 \\ u_2 \\ u_3 \end{bmatrix} + y \begin{bmatrix} v_1 \\ v_2 \\ v_3 \end{bmatrix}$$

is the projection of the vector $\begin{bmatrix} b_1 \\ b_2 \\ b_3 \end{bmatrix}$ *on the vector subspace*

$$\text{Span} \left\{ \begin{bmatrix} u_1 \\ u_2 \\ u_3 \end{bmatrix}, \begin{bmatrix} v_1 \\ v_2 \\ v_3 \end{bmatrix} \right\}.$$

Example 4.2.32. Let $\mathbf{b} = \begin{bmatrix} 1 \\ 1 \\ -1 \end{bmatrix}$, $\mathbf{u} = \begin{bmatrix} 1 \\ 2 \\ 2 \end{bmatrix}$, and $\mathbf{v} = \begin{bmatrix} 1 \\ 1 \\ 1 \end{bmatrix}$. Find the projection of the vector $\mathbf{b}$ on the vector plane Span $\{\mathbf{u}, \mathbf{v}\}$.

Proof. In our case the system (4.12) becomes

$$\begin{cases} \left(\begin{bmatrix} 1 \\ 1 \\ -1 \end{bmatrix} - x \begin{bmatrix} 1 \\ 2 \\ 2 \end{bmatrix} - y \begin{bmatrix} 1 \\ 1 \\ 1 \end{bmatrix} \right) \cdot \begin{bmatrix} 1 \\ 2 \\ 2 \end{bmatrix} = 0 \\ \left(\begin{bmatrix} 1 \\ 1 \\ -1 \end{bmatrix} - x \begin{bmatrix} 1 \\ 2 \\ 2 \end{bmatrix} - y \begin{bmatrix} 1 \\ 1 \\ 1 \end{bmatrix} \right) \cdot \begin{bmatrix} 1 \\ 1 \\ 1 \end{bmatrix} = 0 \end{cases}.$$

After calculating the dot products and simplifying we obtain the system

$$\begin{cases} 9x + 5y = 1 \\ 5x + 3y = 1 \end{cases}.$$

The solutions can be found using Cramer's Rule:

$$x = \frac{\det \begin{bmatrix} 1 & 5 \\ 1 & 3 \end{bmatrix}}{\det \begin{bmatrix} 9 & 5 \\ 5 & 3 \end{bmatrix}} = -1 \quad \text{and} \quad y = \frac{\det \begin{bmatrix} 9 & 1 \\ 5 & 1 \end{bmatrix}}{\det \begin{bmatrix} 9 & 5 \\ 5 & 3 \end{bmatrix}} = 2.$$

This means that the projection of the vector **b** on the vector plane Span $\{\mathbf{u}, \mathbf{v}\}$ is

$$-\mathbf{u} + 2\mathbf{v} = -\begin{bmatrix} 1 \\ 2 \\ 2 \end{bmatrix} + 2\begin{bmatrix} 1 \\ 1 \\ 1 \end{bmatrix} = \begin{bmatrix} 1 \\ 0 \\ 0 \end{bmatrix}.$$

□

The next result gives a formula to calculate the numbers x and y which minimize the sum

$$(b_1 - u_1 x - v_1 y)^2 + (b_2 - u_2 x - v_2 y)^2 + (b_3 - u_3 x - v_3 y)^2$$

and the projection matrix on Span$\{\mathbf{u}, \mathbf{v}\}$ when the vectors **u** and **v** are linearly independent.

Theorem 4.2.33. *Let* $\mathbf{u} = \begin{bmatrix} u_1 \\ u_2 \\ u_3 \end{bmatrix}$ *and* $\mathbf{v} = \begin{bmatrix} v_1 \\ v_2 \\ v_3 \end{bmatrix}$ *be linearly independent vectors*

and let $A = \begin{bmatrix} u_1 & v_1 \\ u_2 & v_2 \\ u_3 & v_3 \end{bmatrix}$. *Then*

(a) *The matrix* $A^T A$ *is invertible;*

(b) *If* $\mathbf{b} = \begin{bmatrix} b_1 \\ b_2 \\ b_3 \end{bmatrix}$ *is a vector in* $\mathbb{R}^3$ *the numbers* x *and* y *that minimize the sum*

$$(b_1 - u_1 x - v_1 y)^2 + (b_2 - u_2 x - v_2 y)^2 + (b_3 - u_3 x - v_3 y)^2$$

are given by the equation

$$\begin{bmatrix} x \\ y \end{bmatrix} = (A^T A)^{-1} A^T \mathbf{b}; \tag{4.13}$$

(c) *The projection matrix on the vector plane* Span$\{\mathbf{u}, \mathbf{v}\}$ *is*

$$A(A^T A)^{-1} A^T. \tag{4.14}$$

Proof. First we observe that

$$A^T A = \begin{bmatrix} u_1 & u_2 & u_3 \\ v_1 & v_2 & v_3 \end{bmatrix} \begin{bmatrix} u_1 & v_1 \\ u_2 & v_2 \\ u_3 & v_3 \end{bmatrix}$$

$$= \begin{bmatrix} u_1^2 + u_2^2 + u_3^2 & u_1 v_1 + u_2 v_2 + u_3 v_3 \\ v_1 u_1 + v_2 u_2 + v_3 u_3 & v_1^2 + v_2^2 + v_3^2 \end{bmatrix}.$$

Consequently,

$$\det A^T A = \det \begin{bmatrix} u_1^2 + u_2^2 + u_3^2 & u_1 v_1 + u_2 v_2 + u_3 v_3 \\ u_1 v_1 + u_2 v_2 + u_3 v_3 & v_1^2 + v_2^2 + v_3^2 \end{bmatrix}$$

$$= (u_1^2 + u_2^2 + u_3^2)(v_1^2 + v_2^2 + v_3^2) - (u_1 v_1 + u_2 v_2 + u_3 v_3)^2$$

$$= (u_1 v_2 - u_2 v_1)^2 + (u_2 v_3 - u_3 v_2)^2 + (u_1 v_3 - u_3 v_1)^2$$

$$= \left(\det \begin{bmatrix} u_1 & v_1 \\ u_2 & b_2 \end{bmatrix} \right)^2 + \left(\det \begin{bmatrix} u_2 & v_2 \\ u_3 & v_3 \end{bmatrix} \right)^2 + \left(\det \begin{bmatrix} u_1 & v_1 \\ u_3 & v_3 \end{bmatrix} \right)^2.$$

Since the vectors **u** and **v** are linearly independent, we get $\det A^T A \neq 0$, by Theorem 4.1.12, and that the matrix $A^T A$ is invertible, by Theorem 1.3.9.

Now, by the proof of Theorem 4.2.27, the projection of any vector **b** on the vector plane Span{**u**, **v**} is the point $x\mathbf{u} + y\mathbf{v}$ where x and y are the solutions of the equation

$$A^T A \begin{bmatrix} x \\ y \end{bmatrix} = A^T \mathbf{b}.$$

By multiplying this equation by the inverse of the matrix $A^T A$ we obtain

$$\begin{bmatrix} x \\ y \end{bmatrix} = (A^T A)^{-1} A^T \mathbf{b}.$$

Hence the projection of the vector **b** on the vector plane Span{**u**, **v**} is

$$x\mathbf{u} + y\mathbf{v} = \begin{bmatrix} u_1 & v_1 \\ u_2 & v_2 \\ u_3 & v_3 \end{bmatrix} \begin{bmatrix} x \\ y \end{bmatrix} = A \begin{bmatrix} x \\ y \end{bmatrix} = A(A^T A)^{-1} A^T \mathbf{b}.$$

Because the uniqueness of the projection matrix this means that

$$A(A^T A)^{-1} A^T$$

is the projection matrix on the vector plane Span{**u**, **v**}.   □

Example 4.2.34. We consider the vectors $\mathbf{u} = \begin{bmatrix} 1 \\ 0 \\ 1 \end{bmatrix}$ and $\mathbf{v} = \begin{bmatrix} 1 \\ 1 \\ 1 \end{bmatrix}$.

Determine the projection matrix on Span{**u**, **v**}.

Solution. Since the vectors are linearly independent, the projection matrix on the vector plane Span$\{\mathbf{u}, \mathbf{v}\}$ is

$$
\begin{bmatrix} 1 & 1 \\ 0 & 1 \\ 1 & 1 \end{bmatrix} \left(\begin{bmatrix} 1 & 0 & 1 \\ 1 & 1 & 1 \end{bmatrix} \begin{bmatrix} 1 & 1 \\ 0 & 1 \\ 1 & 1 \end{bmatrix} \right)^{-1} \begin{bmatrix} 1 & 0 & 1 \\ 1 & 1 & 1 \end{bmatrix} = \begin{bmatrix} 1 & 1 \\ 0 & 1 \\ 1 & 1 \end{bmatrix} \begin{bmatrix} 2 & 2 \\ 2 & 3 \end{bmatrix}^{-1} \begin{bmatrix} 1 & 0 & 1 \\ 1 & 1 & 1 \end{bmatrix}
$$

$$
= \begin{bmatrix} 1 & 1 \\ 0 & 1 \\ 1 & 1 \end{bmatrix} \left(\tfrac{1}{2} \begin{bmatrix} 3 & -2 \\ -2 & 2 \end{bmatrix} \right) \begin{bmatrix} 1 & 0 & 1 \\ 1 & 1 & 1 \end{bmatrix}
$$

$$
= \frac{1}{2} \begin{bmatrix} 1 & 0 \\ -2 & 2 \\ 1 & 0 \end{bmatrix} \begin{bmatrix} 1 & 0 & 1 \\ 1 & 1 & 1 \end{bmatrix}
$$

$$
= \frac{1}{2} \begin{bmatrix} 1 & 0 & 1 \\ 0 & 2 & 0 \\ 1 & 0 & 1 \end{bmatrix}.
$$

$\square$

The least-squares line

In this short section we present an example of application of Theorem 4.2.27 to statistics.

Let $(x_1, y_1), (x_2, y_2)$ and (x_3, y_3) be three points in $\mathbb{R}^2$ such that $x_1 < x_2 < x_3$. These three pairs of numbers are interpreted as the result of an experiment. The numbers x_1, x_2, x_3 are called the **parameter values**, since they represent some parameter that we can control, and the numbers y_1, y_2, y_3 are called the **observed values**, since they represent the observed outcomes for different values of the parameter.

We want to determine a line $y = b_0 + b_1 x$ which minimizes the sum

$$
(y_1 - b_0 - b_1 x_1)^2 + (y_2 - b_0 - b_1 x_2)^2 + (y_3 - b_0 - b_1 x_3)^2.
$$

The line $y = b_0 + b_1 x$ is called the **least-squares line**. It is the line that best fits the points (x_1, y_1), (x_2, y_2), and (x_3, y_3). In statistics it is usually called the **regression line** of y_1, y_2, y_3 on x_1, x_2, x_3. The numbers $b_0 + b_1 x_1$, $b_0 + b_1 x_2$, $b_0 + b_1 x_3$ are called the **predicted values** and the numbers $y_1 - (b_0 + b_1 x_1)$, $y_2 - (b_0 + b_1 x_2)$, $y_3 - (b_0 + b_1 x_3)$ are called the **residuals**.

To find the least square line we use Theorem 4.2.27. If we let

$$
X = \begin{bmatrix} 1 & x_1 \\ 1 & x_2 \\ 1 & x_3 \end{bmatrix},
$$

then the numbers b_0 and b_1 can be found as the solution of the equation

$$
X^T X \begin{bmatrix} b_0 \\ b_1 \end{bmatrix} = X^T \begin{bmatrix} y_1 \\ y_2 \\ y_3 \end{bmatrix}.
$$

Example 4.2.35. Determine the least square line that best fits the points $(1,3)$, $(3,4)$, and $(4,7)$.

Solution. Let $X = \begin{bmatrix} 1 & 1 \\ 1 & 3 \\ 1 & 4 \end{bmatrix}$. We have to solve the equation

$$X^T X \begin{bmatrix} b_0 \\ b_1 \end{bmatrix} = X^T \begin{bmatrix} 3 \\ 4 \\ 7 \end{bmatrix},$$

that is, the equation

$$\begin{bmatrix} 3 & 8 \\ 8 & 26 \end{bmatrix} \begin{bmatrix} b_0 \\ b_1 \end{bmatrix} = \begin{bmatrix} 14 \\ 43 \end{bmatrix}.$$

The solution is $b_0 = \frac{10}{7}$ and $b_1 = \frac{17}{14}$. The least squares line that best fits the points $(1,3)$, $(3,4)$, and $(4,7)$ is the line $y = \frac{10}{7} + \frac{17}{14} x$. $\qquad\square$

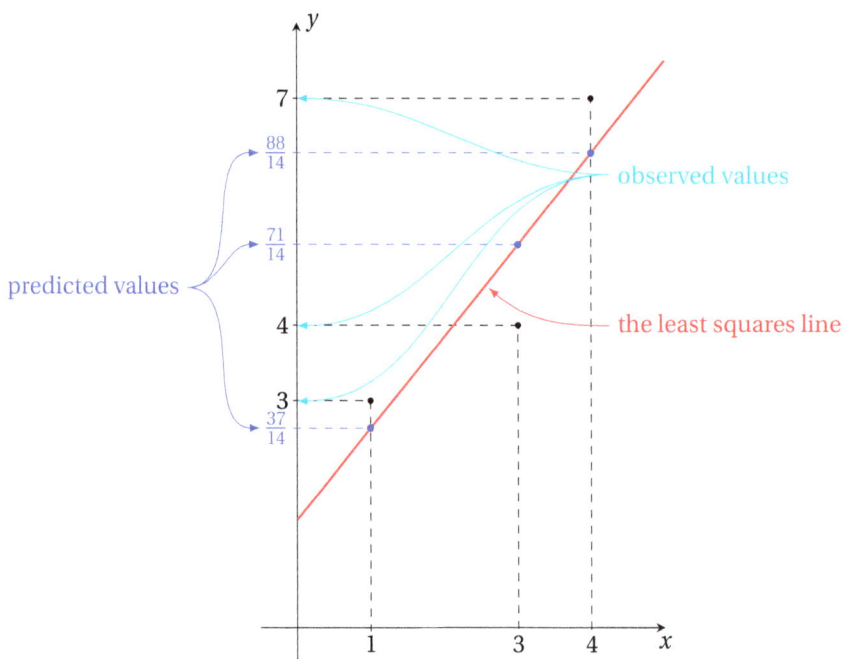

Figure 4.9: The least squares line.

The QR factorization of 3×2 matrices

The QR factorization of 2×2 matrices was introduced in Chapter 3. Here we present an extension of that idea to 3×2 matrices.

Theorem 4.2.36. *If the columns of a* 3×2 *matrix* $A = \begin{bmatrix} \mathbf{c}_1 & \mathbf{c}_2 \end{bmatrix}$ *are linearly independent, then* A *can be represented in the form*

$$A = QR,$$

where $Q = \begin{bmatrix} \mathbf{u}_1 & \mathbf{u}_2 \end{bmatrix}$ *is a* 3×2 *matrix such that the columns* $\mathbf{u}_1$ *and* $\mathbf{u}_2$ *are orthonormal vectors in* $\mathbb{R}^3$, *that is*

$$\|\mathbf{u}_1\| = \|\mathbf{u}_2\| = 1 \quad and \quad \mathbf{u}_1 \cdot \mathbf{u}_2 = 0,$$

and $R = \begin{bmatrix} r_{11} & r_{12} \\ 0 & r_{22} \end{bmatrix}$ *is an upper triangular* 2×2 *matrix such that* $r_{1,1} > 0$ *and* $r_{2,2} > 0$.
Moreover, we have

$$R = Q^T A.$$

Proof. Let $A = \begin{bmatrix} \mathbf{c}_1 & \mathbf{c}_2 \end{bmatrix}$ be a 3×2 matrix such that the vectors $\mathbf{c}_1$ and $\mathbf{c}_2$ are linearly independent. First we define

$$\mathbf{v}_1 = \mathbf{c}_1, \quad and \quad \mathbf{v}_2 = \mathbf{c}_2 - \text{proj}_{\mathbf{v}_1} \mathbf{c}_2 = \mathbf{c}_2 - \frac{\mathbf{c}_2 \cdot \mathbf{v}_1}{\mathbf{v}_1 \cdot \mathbf{v}_1} \mathbf{v}_1.$$

The vectors $\mathbf{v}_1$ and $\mathbf{v}_2$ are orthogonal and the vector $\mathbf{v}_2$ is nonzero, because the vectors $\mathbf{c}_1$ and $\mathbf{c}_2$ are linearly independent. Moreover, we have

$$\mathbf{c}_2 = \mathbf{v}_2 + \frac{\mathbf{c}_2 \cdot \mathbf{v}_1}{\mathbf{v}_1 \cdot \mathbf{v}_1} \mathbf{v}_1.$$

Next we define

$$\mathbf{u}_1 = \frac{1}{\|\mathbf{v}_1\|} \mathbf{v}_1 \quad and \quad \mathbf{u}_2 = \frac{1}{\|\mathbf{v}_2\|} \mathbf{v}_2,$$

and

$$r_{1,1} = \|\mathbf{v}_1\|, \quad r_{1,2} = \|\mathbf{v}_1\| \frac{\mathbf{c}_2 \cdot \mathbf{v}_1}{\mathbf{v}_1 \cdot \mathbf{v}_1}, \quad and \quad r_{2,2} = \|\mathbf{v}_2\|.$$

Note that the vectors $\mathbf{u}_1$ and $\mathbf{u}_2$ are orthonormal and we have $r_{1,1} > 0$ and $r_{2,2} > 0$. Since

$$\mathbf{c}_1 = r_{1,1} \mathbf{u}_1 \quad and \quad \mathbf{c}_2 = r_{1,2} \mathbf{u}_1 + r_{2,2} \mathbf{u}_2,$$

we have

$$A = \begin{bmatrix} \mathbf{c}_1 & \mathbf{c}_2 \end{bmatrix} = \begin{bmatrix} \mathbf{u}_1 & \mathbf{u}_2 \end{bmatrix} \begin{bmatrix} r_{11} & r_{12} \\ 0 & r_{22} \end{bmatrix},$$

which is the desired QR factorization of A.

Moreover, since

$$Q^T Q = \begin{bmatrix} \mathbf{u}_1^T \\ \mathbf{u}_2^T \end{bmatrix} \begin{bmatrix} \mathbf{u}_1 & \mathbf{u}_2 \end{bmatrix} = \begin{bmatrix} \mathbf{u}_1 \cdot \mathbf{u}_1 & \mathbf{u}_1 \cdot \mathbf{u}_2 \\ \mathbf{u}_1 \cdot \mathbf{u}_2 & \mathbf{u}_2 \cdot \mathbf{u}_2 \end{bmatrix} = \begin{bmatrix} 1 & 0 \\ 0 & 1 \end{bmatrix},$$

from the equality $A = QR$ we get

$$Q^T A = Q^T QR = \begin{bmatrix} 1 & 0 \\ 0 & 1 \end{bmatrix} R = R.$$

□

Example 4.2.37. Determine the QR factorization of the matrix $A = \begin{bmatrix} 1 & 1 \\ 1 & 1 \\ 1 & 2 \end{bmatrix}$.

Solution. Since

$$\begin{bmatrix} 1 \\ 1 \\ 2 \end{bmatrix} - \frac{\begin{bmatrix} 1 \\ 1 \\ 2 \end{bmatrix} \cdot \begin{bmatrix} 1 \\ 1 \\ 1 \end{bmatrix}}{\begin{bmatrix} 1 \\ 1 \\ 1 \end{bmatrix} \cdot \begin{bmatrix} 1 \\ 1 \\ 1 \end{bmatrix}} \begin{bmatrix} 1 \\ 1 \\ 1 \end{bmatrix} = \begin{bmatrix} 1 \\ 1 \\ 2 \end{bmatrix} - \frac{4}{3} \begin{bmatrix} 1 \\ 1 \\ 1 \end{bmatrix} = -\frac{1}{3} \begin{bmatrix} 1 \\ 1 \\ -2 \end{bmatrix},$$

we have

$$\begin{bmatrix} 1 \\ 1 \\ 2 \end{bmatrix} = \frac{4}{3} \begin{bmatrix} 1 \\ 1 \\ 1 \end{bmatrix} - \frac{1}{3} \begin{bmatrix} 1 \\ 1 \\ -2 \end{bmatrix} = \frac{4}{3} \begin{bmatrix} 1 \\ 1 \\ 1 \end{bmatrix} + \frac{1}{3} \begin{bmatrix} -1 \\ -1 \\ 2 \end{bmatrix}. \tag{4.15}$$

By a slight modification of the method from the proof of Theorem 4.2.36 we choose

$\mathbf{v}_1 = \begin{bmatrix} 1 \\ 1 \\ 1 \end{bmatrix}$ and $\mathbf{v}_2 = \begin{bmatrix} -1 \\ -1 \\ 2 \end{bmatrix}$. We have taken $\mathbf{v}_2 = \begin{bmatrix} -1 \\ -1 \\ 2 \end{bmatrix}$ and not $\mathbf{v}_2 = \begin{bmatrix} 1 \\ 1 \\ -2 \end{bmatrix}$, because the

last coefficient of the vector $\mathbf{v}_2$ in (4.15) must be positive. Now we calculate

$$\left\| \begin{bmatrix} 1 \\ 1 \\ 1 \end{bmatrix} \right\| = \sqrt{3} \quad \text{and} \quad \left\| \begin{bmatrix} -1 \\ -1 \\ 2 \end{bmatrix} \right\| = \sqrt{6}$$

and let

$$\mathbf{u}_1 = \frac{1}{\sqrt{3}} \begin{bmatrix} 1 \\ 1 \\ 1 \end{bmatrix} \quad \text{and} \quad \mathbf{u}_2 = \frac{1}{\sqrt{6}} \begin{bmatrix} -1 \\ -1 \\ 2 \end{bmatrix}.$$

Consequently

$$\begin{bmatrix} 1 \\ 1 \\ 1 \end{bmatrix} = \sqrt{3}\mathbf{u}_1 \quad \text{and} \quad \begin{bmatrix} 1 \\ 1 \\ 2 \end{bmatrix} = \frac{4\sqrt{3}}{3}\mathbf{u}_1 + \frac{\sqrt{6}}{3}\mathbf{u}_2.$$

Now we define

$$Q = \begin{bmatrix} \mathbf{u}_1 & \mathbf{u}_2 \end{bmatrix} \quad \text{and} \quad R = Q^T A = \begin{bmatrix} \mathbf{u}_1 & \mathbf{u}_2 \end{bmatrix}^T A = \begin{bmatrix} \sqrt{3} & \frac{4\sqrt{3}}{3} \\ 0 & \frac{\sqrt{6}}{3} \end{bmatrix}.$$

Thus the QR factorization of the matrix A is

$$A = \begin{bmatrix} \mathbf{u}_1 & \mathbf{u}_2 \end{bmatrix} \begin{bmatrix} \sqrt{3} & \frac{4\sqrt{3}}{3} \\ 0 & \frac{\sqrt{6}}{3} \end{bmatrix}.$$

$\square$

4.2.1 Exercises

Find the following dot products.

1. $\begin{bmatrix} 1 \\ 5 \\ 2 \end{bmatrix} \cdot \begin{bmatrix} 7 \\ 2 \\ 4 \end{bmatrix}$

2. $\begin{bmatrix} 3 \\ -4 \\ 1 \end{bmatrix} \cdot \begin{bmatrix} 2 \\ 1 \\ 4 \end{bmatrix}$

Find the norms of the given vectors.

3. $\begin{bmatrix} 2 \\ 1 \\ -5 \end{bmatrix}$

4. $\begin{bmatrix} -5 \\ 3 \\ 4 \end{bmatrix}$

Find the distance between the given points.

5. $\begin{bmatrix} 1 \\ 7 \\ 3 \end{bmatrix}$ and $\begin{bmatrix} -1 \\ 9 \\ 2 \end{bmatrix}$

6. $\begin{bmatrix} 2 \\ 1 \\ 5 \end{bmatrix}$ and $\begin{bmatrix} 7 \\ 2 \\ 4 \end{bmatrix}$

Describe the vector plane defined by $\mathbf{n} \cdot \mathbf{x} = 0$ in the form Span$\{\mathbf{a}, \mathbf{b}\}$.

7. $\mathbf{n} = \begin{bmatrix} 2 \\ 1 \\ -2 \end{bmatrix}$

8. $\mathbf{n} = \begin{bmatrix} 3 \\ 4 \\ -1 \end{bmatrix}$

Find the projection of the point $\mathbf{b}$ on the vector line Span$\{\mathbf{u}\}$ using the formula from Theorem 4.2.10.

9. $\mathbf{b} = \begin{bmatrix} 0 \\ 1 \\ 0 \end{bmatrix}$ and $\mathbf{u} = \begin{bmatrix} 5 \\ 1 \\ 1 \end{bmatrix}$ 11. $\mathbf{b} = \begin{bmatrix} 3 \\ 1 \\ 1 \end{bmatrix}$ and $\mathbf{u} = \begin{bmatrix} 1 \\ 2 \\ 2 \end{bmatrix}$

10. $\mathbf{b} = \begin{bmatrix} 2 \\ 1 \\ 1 \end{bmatrix}$ and $\mathbf{u} = \begin{bmatrix} 2 \\ -2 \\ 1 \end{bmatrix}$ 12. $\mathbf{b} = \begin{bmatrix} 1 \\ a \\ b \end{bmatrix}$ and $\mathbf{u} = \begin{bmatrix} 1 \\ 0 \\ -1 \end{bmatrix}$

13. Find x which minimizes the sum $(1-3x)^2 + (2-x)^2 + (2+x)^2$ using the projection on the vector line Span $\left\{ \begin{bmatrix} 3 \\ 1 \\ -1 \end{bmatrix} \right\}$.

14. Find x which minimizes the sum $(1+x)^2 + (1-x)^2 + (3-2x)^2$ using the projection on the vector line Span $\left\{ \begin{bmatrix} -1 \\ 1 \\ 2 \end{bmatrix} \right\}$.

Find the projection matrix on the given vector line.

15. Span $\left\{ \begin{bmatrix} 1 \\ 0 \\ 2 \end{bmatrix} \right\}$ 17. Span $\left\{ \begin{bmatrix} 3 \\ -2 \\ 5 \end{bmatrix} \right\}$

16. Span $\left\{ \begin{bmatrix} 1 \\ 1 \\ 1 \end{bmatrix} \right\}$ 18. Span $\left\{ \begin{bmatrix} 3 \\ 2 \\ -1 \end{bmatrix} \right\}$

Find the projection of $\mathbf{b}$ on the vector line Span$\{\mathbf{u}\}$ using Theorem 4.2.13.

19. $\mathbf{b} = \begin{bmatrix} 1 \\ 1 \\ 1 \end{bmatrix}$ and $\mathbf{u} = \begin{bmatrix} 1 \\ 1 \\ -2 \end{bmatrix}$ 21. $\mathbf{b} = \begin{bmatrix} x \\ y \\ z \end{bmatrix}$ and $\mathbf{u} = \begin{bmatrix} 1 \\ 2 \\ 2 \end{bmatrix}$

20. $\mathbf{b} = \begin{bmatrix} 1 \\ 3 \\ -2 \end{bmatrix}$ and $\mathbf{u} = \begin{bmatrix} 1 \\ 0 \\ -1 \end{bmatrix}$ 22. $\mathbf{b} = \begin{bmatrix} x \\ y \\ z \end{bmatrix}$ and $\mathbf{u} = \begin{bmatrix} 2 \\ 1 \\ -2 \end{bmatrix}$

Find two different orthogonal bases in the vector plane Span$\{\mathbf{u},\mathbf{v}\}$ (See Example 4.2.19).

23. $\mathbf{u} = \begin{bmatrix} 1 \\ 1 \\ -2 \end{bmatrix}$ and $\mathbf{v} = \begin{bmatrix} 1 \\ 0 \\ 2 \end{bmatrix}$ 25. $\mathbf{u} = \begin{bmatrix} 2 \\ 1 \\ 5 \end{bmatrix}$ and $\mathbf{v} = \begin{bmatrix} 3 \\ 2 \\ -4 \end{bmatrix}$

24. $\mathbf{u} = \begin{bmatrix} 2 \\ 2 \\ 1 \end{bmatrix}$ and $\mathbf{v} = \begin{bmatrix} -1 \\ -2 \\ 1 \end{bmatrix}$ 26. $\mathbf{u} = \begin{bmatrix} 3 \\ 1 \\ 1 \end{bmatrix}$ and $\mathbf{v} = \begin{bmatrix} 1 \\ 2 \\ 1 \end{bmatrix}$

Find an orthonormal basis in the given vector plane.

27. $2x + y - 2z = 0$ 29. $x - y + 2z = 0$

28. $3x - 2y + z = 0$ 30. $3x + y - z = 0$

Find the projection of the point **b** on Span$\{\mathbf{u}, \mathbf{v}\}$ where $\mathbf{u} \cdot \mathbf{v} = 0$.

31. $\mathbf{b} = \begin{bmatrix} 1 \\ 2 \\ 2 \end{bmatrix}, \mathbf{u} = \begin{bmatrix} 1 \\ -1 \\ -1 \end{bmatrix}, \mathbf{v} = \begin{bmatrix} 1 \\ -1 \\ 2 \end{bmatrix}$ 33. $\mathbf{b} = \begin{bmatrix} 0 \\ 0 \\ 1 \end{bmatrix}, \mathbf{u} = \begin{bmatrix} 1 \\ 1 \\ -4 \end{bmatrix}, \mathbf{v} = \begin{bmatrix} 5 \\ -1 \\ 1 \end{bmatrix}$

32. $\mathbf{b} = \begin{bmatrix} 1 \\ 0 \\ 0 \end{bmatrix}, \mathbf{u} = \begin{bmatrix} 1 \\ 0 \\ -1 \end{bmatrix}, \mathbf{v} = \begin{bmatrix} 1 \\ -1 \\ 1 \end{bmatrix}$ 34. $\mathbf{b} = \begin{bmatrix} 1 \\ 0 \\ 0 \end{bmatrix}, \mathbf{u} = \begin{bmatrix} 1 \\ 1 \\ 1 \end{bmatrix}, \mathbf{v} = \begin{bmatrix} 1 \\ -2 \\ 1 \end{bmatrix}$

Find the projection of the point **b** on Span$\{\mathbf{u}, \mathbf{v}\}$.

35. $\mathbf{b} = \begin{bmatrix} 1 \\ 0 \\ 1 \end{bmatrix}, \mathbf{u} = \begin{bmatrix} 1 \\ 2 \\ -1 \end{bmatrix}, \mathbf{v} = \begin{bmatrix} 1 \\ 1 \\ 1 \end{bmatrix}$ 37. $\mathbf{b} = \begin{bmatrix} 2 \\ 3 \\ 1 \end{bmatrix}, \mathbf{u} = \begin{bmatrix} 1 \\ 0 \\ 1 \end{bmatrix}, \mathbf{v} = \begin{bmatrix} 1 \\ 1 \\ 0 \end{bmatrix}$

36. $\mathbf{b} = \begin{bmatrix} 2 \\ 1 \\ -1 \end{bmatrix}, \mathbf{u} = \begin{bmatrix} 1 \\ 1 \\ 2 \end{bmatrix}, \mathbf{v} = \begin{bmatrix} 2 \\ -1 \\ 1 \end{bmatrix}$ 38. $\mathbf{b} = \begin{bmatrix} 3 \\ 2 \\ 2 \end{bmatrix}, \mathbf{u} = \begin{bmatrix} 1 \\ -1 \\ -1 \end{bmatrix}, \mathbf{v} = \begin{bmatrix} -1 \\ 1 \\ 1 \end{bmatrix}$

Find the distance of the point **b** to the vector plane Span$\{\mathbf{u}, \mathbf{v}\}$.

39. $\mathbf{b} = \begin{bmatrix} 2 \\ 3 \\ 1 \end{bmatrix}, \mathbf{u} = \begin{bmatrix} 1 \\ 2 \\ 1 \end{bmatrix}, \mathbf{v} = \begin{bmatrix} 2 \\ 1 \\ 2 \end{bmatrix}$ 41. $\mathbf{b} = \begin{bmatrix} 1 \\ 3 \\ 5 \end{bmatrix}, \mathbf{u} = \begin{bmatrix} 1 \\ 1 \\ 1 \end{bmatrix}, \mathbf{v} = \begin{bmatrix} 1 \\ -1 \\ 1 \end{bmatrix}$

40. $\mathbf{b} = \begin{bmatrix} 1 \\ 1 \\ 2 \end{bmatrix}, \mathbf{u} = \begin{bmatrix} 1 \\ 0 \\ 1 \end{bmatrix}, \mathbf{v} = \begin{bmatrix} 2 \\ 2 \\ -1 \end{bmatrix}$ 42. $\mathbf{b} = \begin{bmatrix} 1 \\ 1 \\ 1 \end{bmatrix}, \mathbf{u} = \begin{bmatrix} 2 \\ 1 \\ -1 \end{bmatrix}, \mathbf{v} = \begin{bmatrix} 1 \\ 1 \\ 3 \end{bmatrix}$

Find the projection matrix on Span$\{\mathbf{u}, \mathbf{v}\}$.

43. $\mathbf{u} = \begin{bmatrix} -1 \\ 1 \\ 1 \end{bmatrix}$ and $\mathbf{v} = \begin{bmatrix} 1 \\ 2 \\ -1 \end{bmatrix}$ 45. $\mathbf{u} = \begin{bmatrix} 1 \\ 2 \\ -1 \end{bmatrix}$ and $\mathbf{v} = \begin{bmatrix} 1 \\ 0 \\ 1 \end{bmatrix}$

44. $\mathbf{u} = \begin{bmatrix} 1 \\ 2 \\ 1 \end{bmatrix}$ and $\mathbf{v} = \begin{bmatrix} 2 \\ -1 \\ 0 \end{bmatrix}$ 46. $\mathbf{u} = \begin{bmatrix} 1 \\ 1 \\ 1 \end{bmatrix}$ and $\mathbf{v} = \begin{bmatrix} 2 \\ -1 \\ -1 \end{bmatrix}$

47. $\mathbf{u} = \begin{bmatrix} 1 \\ 1 \\ 2 \end{bmatrix}$ and $\mathbf{v} = \begin{bmatrix} 2 \\ 1 \\ 1 \end{bmatrix}$ 49. $\mathbf{u} = \begin{bmatrix} 1 \\ 2 \\ 3 \end{bmatrix}$ and $\mathbf{v} = \begin{bmatrix} 1 \\ 2 \\ 7 \end{bmatrix}$

48. $\mathbf{u} = \begin{bmatrix} 1 \\ 1 \\ 1 \end{bmatrix}$ and $\mathbf{v} = \begin{bmatrix} 1 \\ -1 \\ 1 \end{bmatrix}$ 50. $\mathbf{u} = \begin{bmatrix} 1 \\ 1 \\ 2 \end{bmatrix}$ and $\mathbf{v} = \begin{bmatrix} 1 \\ -2 \\ -1 \end{bmatrix}$

Using Theorem 4.2.27 find numbers x and y which minimize the sum.

51. $(2 - x + y)^2 + (1 - 2x + y)^2 + (1 + x - y)^2$

52. $(2 + y)^2 + (1 - x + y)^2 + (1 + x - y)^2$

53. $(1 + x - y)^2 + (1 + x + y)^2 + (1 - x - y)^2$

54. $(1 + x - y)^2 + (1 + y)^2 + (1 - x - y)^2$

55. $(2 + x - y)^2 + (1 + 2x - 2y)^2 + (1 - x + y)^2$ (Explain why the solution is not unique.)

56. $(1 - 2x + y)^2 + (1 + 2x - y)^2 + (1 - 4x + 2y)^2$ (Explain why the solution is not unique.)

Using (4.12) find numbers x and y which minimize the sum.

57. $(2 - y)^2 + (1 - x)^2 + (1 - 2x - y)^2$

58. $(2 - x - y)^2 + (1 - x + 3y)^2 + (1 - x - y)^2$

59. $(2 + x - y)^2 + (1 + x + y)^2 + (1 - x + y)^2$

60. $(1 - x - y)^2 + (1 - 2x + y)^2 + (1 + x - 3y)^2$

Find the projection of the point $\mathbf{b}$ on Span$\{\mathbf{u}, \mathbf{v}\}$. (See Example 4.2.32).

61. $\mathbf{b} = \begin{bmatrix} 1 \\ 2 \\ 1 \end{bmatrix}, \mathbf{u} = \begin{bmatrix} 1 \\ 1 \\ -1 \end{bmatrix}, \mathbf{v} = \begin{bmatrix} 3 \\ 2 \\ 2 \end{bmatrix}$ 63. $\mathbf{b} = \begin{bmatrix} 1 \\ 1 \\ 1 \end{bmatrix}, \mathbf{u} = \begin{bmatrix} 2 \\ 3 \\ 5 \end{bmatrix}, \mathbf{v} = \begin{bmatrix} 0 \\ 1 \\ 1 \end{bmatrix}$

62. $\mathbf{b} = \begin{bmatrix} 1 \\ 0 \\ 1 \end{bmatrix}, \mathbf{u} = \begin{bmatrix} 1 \\ 2 \\ 0 \end{bmatrix}, \mathbf{v} = \begin{bmatrix} 1 \\ 3 \\ 2 \end{bmatrix}$ 64. $\mathbf{b} = \begin{bmatrix} 1 \\ 1 \\ 1 \end{bmatrix}, \mathbf{u} = \begin{bmatrix} 2 \\ 3 \\ 5 \end{bmatrix}, \mathbf{v} = \begin{bmatrix} 0 \\ 1 \\ 1 \end{bmatrix}$

Using (4.13) find numbers x and y which minimize the sum.

65. $(2 - y)^2 + (1 - x)^2 + (1 - x - y)^2$ 66. $(1 - x - y)^2 + (1 - 2x + y)^2 + (1 - 2y)^2$

Find the projection matrix on Span$\{\mathbf{u}, \mathbf{v}\}$ using formula (4.14).

67. $\mathbf{u} = \begin{bmatrix} 1 \\ 1 \\ 0 \end{bmatrix}, \mathbf{v} = \begin{bmatrix} 1 \\ 0 \\ 2 \end{bmatrix}$
68. $\mathbf{u} = \begin{bmatrix} 0 \\ 1 \\ 0 \end{bmatrix}, \mathbf{v} = \begin{bmatrix} 1 \\ 5 \\ 2 \end{bmatrix}$

69. Assume that $\mathbf{u} \neq \mathbf{0}$, $\mathbf{v} \neq \mathbf{0}$, and $\mathbf{u} \cdot \mathbf{v} = 0$. Use Theorem 4.2.33 to show that the projection of the point $\mathbf{b} \in \mathbb{R}^3$ on the vector plane Span$\{\mathbf{u}, \mathbf{v}\}$ is

$$\frac{\mathbf{b} \cdot \mathbf{u}}{\mathbf{u} \cdot \mathbf{u}} \mathbf{u} + \frac{\mathbf{b} \cdot \mathbf{v}}{\mathbf{v} \cdot \mathbf{v}} \mathbf{v}.$$

70. Show that the intersection of two different vector planes is a vector line.

Find the least square line that best fits the given points.

71. $(1,7)$, $(2,4)$, and $(4,1)$

72. $(0,4)$, $(2,1)$, and $(3,-1)$

73. $(-1,0)$, $(2,1)$, and $(3,4)$

74. $(1,1)$, $(2,5)$, and $(5,3)$

Determine the QR factorization of the given matrix.

75. $\begin{bmatrix} 2 & 0 \\ 1 & -1 \\ 0 & 1 \end{bmatrix}$

76. $\begin{bmatrix} 1 & 2 \\ 1 & 0 \\ 1 & 1 \end{bmatrix}$

77. $\begin{bmatrix} 1 & 1 \\ 1 & 0 \\ 0 & 3 \end{bmatrix}$

Chapter 5

Determinants and bases in $\mathbb{R}^3$

5.1 The cross product

The definition of the cross product

The numbers

$$\det \begin{bmatrix} u_1 & v_1 \\ u_2 & v_2 \end{bmatrix}, \quad \det \begin{bmatrix} u_2 & v_2 \\ u_3 & v_3 \end{bmatrix}, \quad \det \begin{bmatrix} u_1 & v_1 \\ u_3 & v_3 \end{bmatrix}$$

that appear in Theorems 4.1.8 and 4.1.12 play a central role in the next result.

Theorem 5.1.1. *If the vectors* $\mathbf{u} = \begin{bmatrix} u_1 \\ u_2 \\ u_3 \end{bmatrix}$ *and* $\mathbf{v} = \begin{bmatrix} v_1 \\ v_2 \\ v_3 \end{bmatrix}$ *are linearly indepen-*

dent, then the solution of the system

$$\begin{cases} \mathbf{x} \cdot \mathbf{u} = 0 \\ \mathbf{x} \cdot \mathbf{v} = 0 \end{cases} \tag{5.1}$$

is

$$\mathbf{x} = t \begin{bmatrix} \det \begin{bmatrix} u_2 & v_2 \\ u_3 & v_3 \end{bmatrix} \\ -\det \begin{bmatrix} u_1 & v_1 \\ u_3 & v_3 \end{bmatrix} \\ \det \begin{bmatrix} u_1 & v_1 \\ u_2 & v_2 \end{bmatrix} \end{bmatrix}$$

where t is an arbitrary real number.

Proof. Since $\mathbf{u} = \begin{bmatrix} u_1 \\ u_2 \\ u_3 \end{bmatrix}$ and $\mathbf{v} = \begin{bmatrix} v_1 \\ v_2 \\ v_3 \end{bmatrix}$ are linearly independent at least one of the numbers

$$\det\begin{bmatrix} u_1 & v_1 \\ u_2 & v_2 \end{bmatrix}, \quad \det\begin{bmatrix} u_2 & v_2 \\ u_3 & v_3 \end{bmatrix}, \quad \det\begin{bmatrix} u_1 & v_1 \\ u_3 & v_3 \end{bmatrix}$$

must be different from 0, by (c) in Theorem 4.1.12. Suppose that

$$\det\begin{bmatrix} u_1 & v_1 \\ u_2 & v_2 \end{bmatrix} = \det\begin{bmatrix} u_1 & u_2 \\ v_1 & v_2 \end{bmatrix} \neq 0.$$

The system (5.1) can be written as

$$\begin{cases} u_1 x + u_2 y + u_3 z = 0 \\ v_1 x + v_2 y + v_3 z = 0 \end{cases}$$

or

$$\begin{cases} u_1 x + u_2 y = -u_3 z \\ v_1 x + v_2 y = -v_3 z \end{cases}.$$

From the Cramer's rule (Theorem 1.3.10), for every value of z we get a unique solution for x and y:

$$x = \frac{\det\begin{bmatrix} -u_3 z & u_2 \\ -v_3 z & v_2 \end{bmatrix}}{\det\begin{bmatrix} u_1 & u_2 \\ v_1 & v_2 \end{bmatrix}} = z\frac{\det\begin{bmatrix} u_2 & v_2 \\ u_3 & v_3 \end{bmatrix}}{\det\begin{bmatrix} u_1 & v_1 \\ u_2 & v_2 \end{bmatrix}}$$

and

$$y = \frac{\det\begin{bmatrix} u_1 & -z u_3 \\ v_1 & -z v_3 \end{bmatrix}}{\det\begin{bmatrix} u_1 & u_2 \\ v_1 & v_2 \end{bmatrix}} = -z\frac{\det\begin{bmatrix} u_1 & v_1 \\ u_3 & v_3 \end{bmatrix}}{\det\begin{bmatrix} u_1 & v_1 \\ u_2 & v_2 \end{bmatrix}}.$$

Note that we also have

$$z = z\frac{\det\begin{bmatrix} u_1 & v_1 \\ u_2 & v_2 \end{bmatrix}}{\det\begin{bmatrix} u_1 & v_1 \\ u_2 & v_2 \end{bmatrix}}.$$

If we denote

$$t = \frac{z}{\det\begin{bmatrix} u_1 & v_1 \\ u_2 & v_2 \end{bmatrix}},$$

then we have

$$x = t\det\begin{bmatrix} u_2 & v_2 \\ u_3 & v_3 \end{bmatrix}, \quad y = -t\det\begin{bmatrix} u_1 & v_1 \\ u_3 & v_3 \end{bmatrix}, \quad \text{and} \quad z = t\det\begin{bmatrix} u_1 & v_1 \\ u_2 & v_2 \end{bmatrix}.$$

It is easy to verify that if t is an arbitrary real number

$$x = t\det\begin{bmatrix} u_2 & v_2 \\ u_3 & v_3 \end{bmatrix}, \quad y = -t\det\begin{bmatrix} u_1 & v_1 \\ u_3 & v_3 \end{bmatrix}, \quad \text{and} \quad z = t\det\begin{bmatrix} u_1 & v_1 \\ u_2 & v_2 \end{bmatrix}$$

is a solution of the system (5.1).

Consequently, the solution of the system (5.1) is

$$\mathbf{x} = \begin{bmatrix} x \\ y \\ z \end{bmatrix} = t \begin{bmatrix} \det \begin{bmatrix} u_2 & v_2 \\ u_3 & v_3 \end{bmatrix} \\ -\det \begin{bmatrix} u_1 & v_1 \\ u_3 & v_3 \end{bmatrix} \\ \det \begin{bmatrix} u_1 & v_1 \\ u_2 & v_2 \end{bmatrix} \end{bmatrix}.$$

where t is an arbitrary real number.

If

$$\det \begin{bmatrix} u_2 & v_2 \\ u_3 & v_3 \end{bmatrix} \neq 0 \quad \text{or} \quad \det \begin{bmatrix} u_1 & v_1 \\ u_3 & v_3 \end{bmatrix} \neq 0,$$

the system is solved in a similar way, with obvious modifications. □

The above theorem can be rephrased as follows: x, y, and z solve the system

$$\begin{cases} \begin{bmatrix} x \\ y \\ z \end{bmatrix} \cdot \begin{bmatrix} u_1 \\ u_2 \\ u_3 \end{bmatrix} = 0 \\ \begin{bmatrix} x \\ y \\ z \end{bmatrix} \cdot \begin{bmatrix} v_1 \\ v_2 \\ v_3 \end{bmatrix} = 0 \end{cases} \tag{5.2}$$

if and only if $\begin{bmatrix} x \\ y \\ z \end{bmatrix}$ is a point on the vector line

$$\mathrm{Span}\left\{ \begin{bmatrix} \det \begin{bmatrix} u_2 & v_2 \\ u_3 & v_3 \end{bmatrix} \\ -\det \begin{bmatrix} u_1 & v_1 \\ u_3 & v_3 \end{bmatrix} \\ \det \begin{bmatrix} u_1 & v_1 \\ u_2 & v_2 \end{bmatrix} \end{bmatrix} \right\}. \tag{5.3}$$

Theorem 5.1.1 tells us that, if the vectors $\mathbf{u} = \begin{bmatrix} u_1 \\ u_2 \\ u_3 \end{bmatrix}$ and $\mathbf{v} = \begin{bmatrix} v_1 \\ v_2 \\ v_3 \end{bmatrix}$ are linearly independent, then the only vector line perpendicular to both Span$\{\mathbf{u}\}$ and Span$\{\mathbf{v}\}$ is the vector line (5.3). This geometric interpretation motivates the definition of the cross product, which is an important tool in $\mathbb{R}^3$.

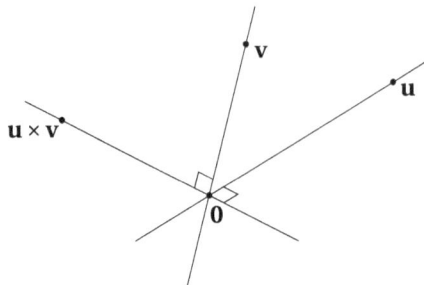

Figure 5.1: The vector line perpendicular to both Span{**u**} and Span{**v**}.

Definition 5.1.2. By the ***cross product*** of the vectors $\mathbf{u} = \begin{bmatrix} u_1 \\ u_2 \\ u_3 \end{bmatrix}$ and $\mathbf{v} = \begin{bmatrix} v_1 \\ v_2 \\ v_3 \end{bmatrix}$

we mean the vector

$$\mathbf{u} \times \mathbf{v} = \begin{bmatrix} \det \begin{bmatrix} u_2 & v_2 \\ u_3 & v_3 \end{bmatrix} \\ -\det \begin{bmatrix} u_1 & v_1 \\ v_3 & v_3 \end{bmatrix} \\ \det \begin{bmatrix} u_1 & v_1 \\ u_2 & v_2 \end{bmatrix} \end{bmatrix}.$$

Note that, unlike the dot product, the cross product of two elements from $\mathbb{R}^3$ is an element from $\mathbb{R}^3$.

Using the cross product, we can state Theorem 5.1.1 in a form that is easier to remember:

If the vectors **u** *and* **v** *in* $\mathbb{R}^3$ *are linearly independent, then the solution of the system*

$$\begin{cases} \mathbf{x} \cdot \mathbf{u} = 0 \\ \mathbf{x} \cdot \mathbf{v} = 0 \end{cases} \tag{5.4}$$

is $\mathbf{x} = t(\mathbf{u} \times \mathbf{v})$, *where t is an arbitrary real number.*

The theorem can be used to solve systems of two equations with three unknowns, as the next example illustrates.

Example 5.1.3. Solve the system

$$\begin{cases} 2x + y + z = 0 \\ x - y + 2z = 0 \end{cases}$$

Solution. First observe that the system can be written in the form

$$\begin{cases} \begin{bmatrix} x \\ y \\ z \end{bmatrix} \cdot \begin{bmatrix} 2 \\ 1 \\ 1 \end{bmatrix} = 0 \\ \begin{bmatrix} x \\ y \\ z \end{bmatrix} \cdot \begin{bmatrix} 1 \\ -1 \\ 2 \end{bmatrix} = 0 \end{cases}.$$

Consequently, the general solution is

$$\begin{bmatrix} x \\ y \\ z \end{bmatrix} = t \left(\begin{bmatrix} 2 \\ 1 \\ 1 \end{bmatrix} \times \begin{bmatrix} 1 \\ -1 \\ 2 \end{bmatrix} \right) = t \begin{bmatrix} 3 \\ -3 \\ -3 \end{bmatrix} = \begin{bmatrix} 3t \\ -3t \\ -3t \end{bmatrix},$$

where t is an arbitrary real number. Note that the same solution can be described in a simpler equivalent way as

$$\begin{bmatrix} x \\ y \\ z \end{bmatrix} = \begin{bmatrix} s \\ -s \\ -s \end{bmatrix},$$

where s is an arbitrary real number. □

The cross product gives us an elegant and useful characterization of linear independence of pairs of vectors in $\mathbb{R}^3$.

Theorem 5.1.4. *Vectors* **u** *and* **v** *in* $\mathbb{R}^3$ *are linearly independent if and only if* $\mathbf{u} \times \mathbf{v} \neq \mathbf{0}$.

Proof. The property is a direct consequence of Theorem 4.1.12. □

Example 5.1.5. Since

$$\begin{bmatrix} 1 \\ 0 \\ 1 \end{bmatrix} \times \begin{bmatrix} 2 \\ 1 \\ 1 \end{bmatrix} = \begin{bmatrix} -1 \\ 1 \\ 1 \end{bmatrix},$$

the vectors $\begin{bmatrix} 1 \\ 0 \\ 1 \end{bmatrix}$ and $\begin{bmatrix} 2 \\ 1 \\ 1 \end{bmatrix}$ are linearly independent.

On the other hand, since

$$\begin{bmatrix} -4 \\ -2 \\ -2 \end{bmatrix} \times \begin{bmatrix} 2 \\ 1 \\ 1 \end{bmatrix} = \begin{bmatrix} 0 \\ 0 \\ 0 \end{bmatrix},$$

the vectors $\begin{bmatrix} -4 \\ -2 \\ -2 \end{bmatrix}$ and $\begin{bmatrix} 2 \\ 1 \\ 1 \end{bmatrix}$ are linearly dependent.

In the next theorem we gather some algebraic properties of the cross product.

Theorem 5.1.6. *Let* **u**, **v**, *and* **w** *be arbitrary vectors in* $\mathbb{R}^3$ *and let x and y be arbitrary real numbers. Then*

(a) $\mathbf{u} \cdot (\mathbf{u} \times \mathbf{v}) = 0$ *and* $\mathbf{v} \cdot (\mathbf{u} \times \mathbf{v}) = 0$;

(b) $\mathbf{u} \times \mathbf{u} = \mathbf{0}$;

(c) $\mathbf{u} \times \mathbf{v} = -(\mathbf{v} \times \mathbf{u})$;

(d) $(x\mathbf{u} + y\mathbf{v}) \times \mathbf{w} = x(\mathbf{u} \times \mathbf{w}) + y(\mathbf{v} \times \mathbf{w})$;

(e) $\mathbf{u} \cdot (\mathbf{v} \times \mathbf{w}) = -\mathbf{v} \cdot (\mathbf{u} \times \mathbf{w})$,

(f) $\mathbf{u} \cdot (\mathbf{v} \times \mathbf{w}) = \mathbf{v} \cdot (\mathbf{w} \times \mathbf{u}) = \mathbf{w} \cdot (\mathbf{u} \times \mathbf{v})$.

Proof. Let $\mathbf{u} = \begin{bmatrix} u_1 \\ u_2 \\ u_3 \end{bmatrix}$, $\mathbf{v} = \begin{bmatrix} v_1 \\ v_2 \\ v_3 \end{bmatrix}$, and $\mathbf{w} = \begin{bmatrix} w_1 \\ w_2 \\ w_3 \end{bmatrix}$.

Part (a) is an immediate consequence of the Theorem 5.1.1.

For (b) it suffices to note that

$$\det \begin{bmatrix} u_2 & u_2 \\ u_3 & u_3 \end{bmatrix} = \det \begin{bmatrix} u_1 & u_1 \\ u_3 & u_3 \end{bmatrix} = \det \begin{bmatrix} u_1 & u_1 \\ u_2 & u_2 \end{bmatrix} = 0.$$

From

$$\mathbf{v} \times \mathbf{u} = \begin{bmatrix} \det \begin{bmatrix} v_2 & u_2 \\ v_3 & u_3 \end{bmatrix} \\ -\det \begin{bmatrix} v_1 & u_1 \\ v_3 & u_3 \end{bmatrix} \\ \det \begin{bmatrix} v_1 & u_1 \\ v_2 & u_2 \end{bmatrix} \end{bmatrix} = \begin{bmatrix} -\det \begin{bmatrix} u_2 & v_2 \\ u_3 & v_3 \end{bmatrix} \\ \det \begin{bmatrix} u_1 & v_1 \\ u_3 & v_3 \end{bmatrix} \\ -\det \begin{bmatrix} u_1 & v_1 \\ u_2 & v_2 \end{bmatrix} \end{bmatrix} = -\begin{bmatrix} \det \begin{bmatrix} u_2 & v_2 \\ u_3 & v_3 \end{bmatrix} \\ -\det \begin{bmatrix} u_1 & v_1 \\ u_3 & v_3 \end{bmatrix} \\ \det \begin{bmatrix} u_1 & v_1 \\ u_2 & v_2 \end{bmatrix} \end{bmatrix},$$

we get (c): $\mathbf{v} \times \mathbf{u} = -(\mathbf{u} \times \mathbf{v})$.

Since for any real numbers x and y we have

$$(x\mathbf{u} + y\mathbf{v}) \times \mathbf{w} = \left(x\begin{bmatrix} u_1 \\ u_2 \\ u_3 \end{bmatrix} + y\begin{bmatrix} v_1 \\ v_2 \\ v_3 \end{bmatrix} \right) \times \begin{bmatrix} w_1 \\ w_2 \\ w_3 \end{bmatrix}$$

$$= \begin{bmatrix} \det\begin{bmatrix} xu_2 + yv_2 & w_2 \\ xu_3 + yv_3 & w_3 \end{bmatrix} \\ -\det\begin{bmatrix} xu_1 + yv_1 & w_1 \\ xu_3 + yv_3 & w_3 \end{bmatrix} \\ \det\begin{bmatrix} xu_1 + yv_1 & w_1 \\ xu_2 + yv_2 & w_2 \end{bmatrix} \end{bmatrix}$$

$$= \begin{bmatrix} \det\begin{bmatrix} xu_2 & w_2 \\ xu_3 & w_3 \end{bmatrix} \\ -\det\begin{bmatrix} xu_1 & w_1 \\ xu_3 & w_3 \end{bmatrix} \\ \det\begin{bmatrix} xu_1 & w_1 \\ xu_2 & w_2 \end{bmatrix} \end{bmatrix} + \begin{bmatrix} \det\begin{bmatrix} yv_2 & w_2 \\ yv_3 & w_3 \end{bmatrix} \\ -\det\begin{bmatrix} yv_1 & w_1 \\ yv_3 & w_3 \end{bmatrix} \\ \det\begin{bmatrix} yv_1 & w_1 \\ yv_2 & w_2 \end{bmatrix} \end{bmatrix}$$

$$= x(\mathbf{u} \times \mathbf{w}) + y(\mathbf{v} \times \mathbf{w}),$$

we obtain (d).

Part (e) can be obtained easily from the properties of the dot product and the cross product already established. Indeed, since by (a),(b), and (c) we have

$$0 = (\mathbf{u} + \mathbf{v}) \cdot ((\mathbf{u} + \mathbf{v}) \times \mathbf{w})$$
$$= \mathbf{u} \cdot ((\mathbf{u} + \mathbf{v}) \times \mathbf{w}) + \mathbf{v} \cdot ((\mathbf{u} + \mathbf{v}) \times \mathbf{w})$$
$$= \mathbf{u} \cdot (\mathbf{u} \times \mathbf{w}) + \mathbf{u} \cdot (\mathbf{v} \times \mathbf{w}) + \mathbf{v} \cdot (\mathbf{u} \times \mathbf{w}) + \mathbf{v} \cdot (\mathbf{v} \times \mathbf{w})$$
$$= \mathbf{u} \cdot (\mathbf{v} \times \mathbf{w}) + \mathbf{v} \cdot (\mathbf{u} \times \mathbf{w}),$$

we have $\mathbf{u} \cdot (\mathbf{v} \times \mathbf{w}) = -\mathbf{v} \cdot (\mathbf{u} \times \mathbf{w})$.

To obtain (f) we use (c) and (e):

$$\mathbf{u} \cdot (\mathbf{v} \times \mathbf{w}) = -\mathbf{v} \cdot (\mathbf{u} \times \mathbf{w}) = \mathbf{v} \cdot (\mathbf{w} \times \mathbf{u}).$$

The equality $\mathbf{v} \cdot (\mathbf{w} \times \mathbf{u}) = \mathbf{w} \cdot (\mathbf{u} \times \mathbf{v})$ is obtained in the same way. □

The equation of a vector plane

The following theorem nicely complements Theorem 5.1.1. Both theorems provide some insight into the geometric interpretation of the cross product.

Theorem 5.1.7. *Let* **u** *and* **v** *be linearly independent vectors in* $\mathbb{R}^3$. *Then a vector* **x** *is in the vector plane* Span{**u**, **v**} *if and only if*

$$\mathbf{x} \cdot (\mathbf{u} \times \mathbf{v}) = 0 \tag{5.5}$$

Proof. The vector $\mathbf{x} = s\mathbf{u} + t\mathbf{v}$ satisfies (5.5) for any real numbers s and t, by Theorem 5.1.6.

Let

$$\mathbf{u} = \begin{bmatrix} u_1 \\ u_2 \\ u_3 \end{bmatrix} \quad \text{and} \quad \mathbf{v} = \begin{bmatrix} v_1 \\ v_2 \\ v_3 \end{bmatrix}$$

be linearly independent vectors and let $\mathbf{x} = \begin{bmatrix} x_1 \\ x_2 \\ x_3 \end{bmatrix}$ be such that (5.5) holds. Since **u** and **v** are linearly independent, at least one of the numbers

$$\det \begin{bmatrix} u_1 & v_1 \\ u_2 & v_2 \end{bmatrix}, \quad \det \begin{bmatrix} u_2 & v_2 \\ u_3 & v_3 \end{bmatrix}, \quad \det \begin{bmatrix} u_1 & v_1 \\ u_3 & v_3 \end{bmatrix}$$

must be different from 0, by (c) in Theorem 4.1.12. Suppose that

$$\det \begin{bmatrix} u_1 & v_1 \\ u_2 & v_2 \end{bmatrix} \neq 0.$$

Then the vectors $\begin{bmatrix} u_1 \\ u_2 \end{bmatrix}$ and $\begin{bmatrix} v_1 \\ v_2 \end{bmatrix}$ are linearly independent and there exist real numbers s and t such that

$$\begin{bmatrix} x_1 \\ x_2 \end{bmatrix} = s \begin{bmatrix} u_1 \\ u_2 \end{bmatrix} + t \begin{bmatrix} v_1 \\ v_2 \end{bmatrix},$$

by Theorem 3.1.22. Hence

$$x_1 = su_1 + tv_1 \quad \text{and} \quad x_2 = su_2 + tv_2. \tag{5.6}$$

Since $\mathbf{x} \cdot (\mathbf{u} \times \mathbf{v}) = 0$, $\mathbf{u} \cdot (\mathbf{u} \times \mathbf{v}) = 0$, and $\mathbf{v} \cdot (\mathbf{u} \times \mathbf{v}) = 0$, we have

$$\mathbf{x} \cdot (\mathbf{u} \times \mathbf{v}) - s\mathbf{u} \cdot (\mathbf{u} \times \mathbf{v}) - t\mathbf{v} \cdot (\mathbf{u} \times \mathbf{v}) = (\mathbf{x} - s\mathbf{u} - t\mathbf{v}) \cdot (\mathbf{u} \times \mathbf{v}) = 0.$$

Consequently, using (5.6), we obtain

$$0 = \begin{bmatrix} x_1 - su_1 + tv_1 \\ x_2 - su_2 + tv_2 \\ x_3 - su_3 + tv_3 \end{bmatrix} \cdot \begin{bmatrix} \det \begin{bmatrix} u_2 & v_2 \\ u_3 & v_3 \end{bmatrix} \\ -\det \begin{bmatrix} u_1 & v_1 \\ u_3 & v_3 \end{bmatrix} \\ \det \begin{bmatrix} u_1 & v_1 \\ u_2 & v_2 \end{bmatrix} \end{bmatrix}$$

$$= \begin{bmatrix} 0 \\ 0 \\ x_3 - su_3 - tv_3 \end{bmatrix} \cdot \begin{bmatrix} \det\begin{bmatrix} u_2 & v_2 \\ u_3 & v_3 \end{bmatrix} \\ -\det\begin{bmatrix} u_1 & v_1 \\ u_3 & v_3 \end{bmatrix} \\ \det\begin{bmatrix} u_1 & v_1 \\ u_2 & v_2 \end{bmatrix} \end{bmatrix}$$

$$= (x_3 - su_3 - tv_3)\det\begin{bmatrix} u_1 & v_1 \\ u_2 & v_2 \end{bmatrix}.$$

Since $\det\begin{bmatrix} u_1 & v_1 \\ u_2 & v_2 \end{bmatrix} \neq 0$, we must have

$$x_3 - su_3 - tv_3 = 0.$$

This, together with (5.6), gives us

$$\begin{cases} x_1 = su_1 + tv_1 \\ x_2 = su_2 + tv_2 \\ x_3 = su_3 + tv_3 \end{cases}$$

which is equivalent to

$$\begin{bmatrix} x_1 \\ x_2 \\ x_3 \end{bmatrix} = s\begin{bmatrix} u_1 \\ u_2 \\ u_3 \end{bmatrix} + t\begin{bmatrix} v_1 \\ v_2 \\ v_3 \end{bmatrix}.$$

This shows that, if $\det\begin{bmatrix} u_1 & v_1 \\ u_2 & v_2 \end{bmatrix} \neq 0$, then there are numbers s and t such that $\mathbf{x} = s\mathbf{u} + t\mathbf{v}$. The other cases, that is, when $\det\begin{bmatrix} u_2 & v_2 \\ u_3 & v_3 \end{bmatrix} \neq 0$ or $\det\begin{bmatrix} u_1 & v_1 \\ u_3 & v_3 \end{bmatrix} \neq 0$, are treated in a similar way with appropriate modifications. □

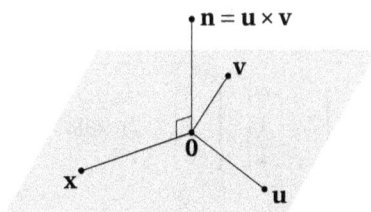

Figure 5.2: The vector plane Span{$\mathbf{u},\mathbf{v}$} is the set of all points $\mathbf{x}$ such that the angle $\angle\mathbf{n0x}$ is a right angle.

The above theorem gives the following geometric interpretation of a vector plane: If $\mathbf{u}$ and $\mathbf{v}$ are linearly independent, then the vector plane Span{$\mathbf{u},\mathbf{v}$} consists of all vectors perpendicular to the vector line Span{$\mathbf{u}\times\mathbf{v}$}. In other words, the vector plane Span{$\mathbf{u},\mathbf{v}$} is the set of all points $\mathbf{x}$ such that the angle $\angle\mathbf{n0x}$, where $\mathbf{n} = \mathbf{u}\times\mathbf{v}$, is a right angle, see Fig. 5.2.

Example 5.1.8. Find an equation of the vector plane which contains the vectors
$\mathbf{u} = \begin{bmatrix} 2 \\ 1 \\ 4 \end{bmatrix}$ and $\mathbf{v} = \begin{bmatrix} 1 \\ 5 \\ 3 \end{bmatrix}$.

Solution. Since $\mathbf{u} \times \mathbf{v} = \begin{bmatrix} -17 \\ -2 \\ 9 \end{bmatrix}$, the vector plane can be described by the equation

$$\begin{bmatrix} x \\ y \\ z \end{bmatrix} \cdot \begin{bmatrix} -17 \\ -2 \\ 9 \end{bmatrix} = 0$$

or $-17x - 2y + 9z = 0$. □

Example 5.1.9. We consider the vectors $\mathbf{u} = \begin{bmatrix} 1 \\ 2 \\ 2 \end{bmatrix}$ and $\mathbf{v} = \begin{bmatrix} 3 \\ 1 \\ 2 \end{bmatrix}$. Find a real number
a such that the vector and $\begin{bmatrix} a \\ 2a+1 \\ 1 \end{bmatrix}$ is in Span$\{\mathbf{u}, \mathbf{v}\}$.

Solution. First we find that $\mathbf{u} \times \mathbf{v} = \begin{bmatrix} 2 \\ 4 \\ -5 \end{bmatrix}$. According to the Theorem 5.1.7, the
vector $\begin{bmatrix} a \\ 2a+1 \\ 1 \end{bmatrix}$ is in Span$\{\mathbf{u}, \mathbf{v}\}$ if and only if

$$\begin{bmatrix} a \\ 2a+1 \\ 1 \end{bmatrix} \cdot (\mathbf{u} \times \mathbf{v}) = \begin{bmatrix} a \\ 2a+1 \\ 1 \end{bmatrix} \cdot \begin{bmatrix} 2 \\ 4 \\ -5 \end{bmatrix} = 2a + 8a + 4 - 5 = 10a - 1 = 0.$$

This gives us $a = 0.1$ and thus

$$\begin{bmatrix} a \\ 2a+1 \\ 1 \end{bmatrix} = \begin{bmatrix} 0.1 \\ 1.2 \\ 1 \end{bmatrix}.$$

□

Note that the vector $\begin{bmatrix} 1 \\ 12 \\ 10 \end{bmatrix}$ must also be an element of the vector plane Span$\{\mathbf{u}, \mathbf{v}\}$.

It is easy to check that

$$\begin{bmatrix} 1 \\ 12 \\ 10 \end{bmatrix} = 7 \begin{bmatrix} 1 \\ 2 \\ 2 \end{bmatrix} - 2 \begin{bmatrix} 3 \\ 1 \\ 2 \end{bmatrix}.$$

In the previous chapter we considered the question of when two pairs of vectors span the same vector subspace. The following theorem complements that discussion.

Theorem 5.1.10. *Let* **a**, **b**, **u**, *and* **v** *be vectors in* $\mathbb{R}^3$ *such that the vectors* **a** *and* **b** *are linearly independent and the vectors* **u** *and* **v** *are linearly independent. Then the following conditions are equivalent*

   (a) {**a**, **b**} *is a basis of the vector plane* Span{**u**, **v**};

   (b) *There is a real number* $\lambda \neq 0$ *such that* $\mathbf{a} \times \mathbf{b} = \lambda(\mathbf{u} \times \mathbf{v})$.

Proof. If **a**, **b** are elements of Span{**u**, **v**}, then there are real numbers q, r, s, t such that

$$\mathbf{a} = q\mathbf{u} + r\mathbf{v} \quad \text{and} \quad \mathbf{b} = s\mathbf{u} + t\mathbf{v}$$

and, using Theorem 5.1.6, we obtain

$$\begin{aligned} \mathbf{a} \times \mathbf{b} &= (q\mathbf{u} + r\mathbf{v}) \times (s\mathbf{u} + t\mathbf{v}) \\ &= qs(\mathbf{u} \times \mathbf{u}) + qt(\mathbf{u} \times \mathbf{v}) + rs(\mathbf{v} \times \mathbf{u}) + rt(\mathbf{v} \times \mathbf{v}) \\ &= qt(\mathbf{u} \times \mathbf{v}) + rs(\mathbf{v} \times \mathbf{u}) \\ &= (qt - rs)(\mathbf{u} \times \mathbf{v}). \end{aligned}$$

If we let $\lambda = qt - rs$, then we have $\mathbf{a} \times \mathbf{b} = \lambda(\mathbf{u} \times \mathbf{v})$. Moreover, if the vectors **a** and **b** are linearly independent, then we must have $\lambda \neq 0$, by Theorem 5.1.4. Thus (a) implies (b).

If there is a real number $\lambda \neq 0$ such that $\mathbf{a} \times \mathbf{b} = \lambda(\mathbf{u} \times \mathbf{v})$, then we have $\mathbf{x} \cdot (\mathbf{a} \times \mathbf{b}) = 0$ if and only if $\mathbf{x} \cdot (\mathbf{u} \times \mathbf{v}) = 0$. Consequently

$$\text{Span}\{\mathbf{a}, \mathbf{b}\} = \text{Span}\{\mathbf{u}, \mathbf{v}\},$$

by Theorem 5.1.7. Thus (b) implies (a). □

Example 5.1.11. Show that the set $\left\{ \begin{bmatrix} -1 \\ 7 \\ 0 \end{bmatrix}, \begin{bmatrix} 3 \\ 0 \\ 7 \end{bmatrix} \right\}$ is a basis of the vector plane

$\text{Span} \left\{ \begin{bmatrix} 1 \\ 2 \\ 3 \end{bmatrix}, \begin{bmatrix} 1 \\ -1 \\ 2 \end{bmatrix} \right\}.$

Solution. Since

$$\begin{bmatrix} 1 \\ 2 \\ 3 \end{bmatrix} \times \begin{bmatrix} 1 \\ -1 \\ 2 \end{bmatrix} = \begin{bmatrix} 7 \\ 1 \\ -3 \end{bmatrix}$$

and

$$\begin{bmatrix} -1 \\ 7 \\ 0 \end{bmatrix} \times \begin{bmatrix} 3 \\ 0 \\ 7 \end{bmatrix} = \begin{bmatrix} 49 \\ 7 \\ -21 \end{bmatrix} = 7 \begin{bmatrix} 7 \\ 1 \\ -3 \end{bmatrix},$$

$\left\{ \begin{bmatrix} -1 \\ 7 \\ 0 \end{bmatrix}, \begin{bmatrix} 3 \\ 0 \\ 7 \end{bmatrix} \right\}$ is a basis of the vector plane Span $\left\{ \begin{bmatrix} 1 \\ 2 \\ 3 \end{bmatrix}, \begin{bmatrix} 1 \\ -1 \\ 2 \end{bmatrix} \right\}$, by Theorem 5.1.10.

□

Determinants of 3×3 matrices

From Theorem 5.1.6 it follows that

$$\mathbf{a} \cdot (\mathbf{b} \times \mathbf{c}) = \mathbf{b} \cdot (\mathbf{c} \times \mathbf{a}) = \mathbf{c} \cdot (\mathbf{a} \times \mathbf{b})$$

for arbitrary vectors $\mathbf{a}, \mathbf{b}, \mathbf{c}$ in $\mathbb{R}^3$. At this point it is not obvious, but this mixed product extends the notion of the determinant of a 2×2 matrix to 3×3 matrices.

Definition 5.1.12. Let $\mathbf{a}, \mathbf{b}$, and $\mathbf{c}$ be arbitrary column vectors in $\mathbb{R}^3$. The common value

$$\mathbf{a} \cdot (\mathbf{b} \times \mathbf{c}) = \mathbf{b} \cdot (\mathbf{c} \times \mathbf{a}) = \mathbf{c} \cdot (\mathbf{a} \times \mathbf{b}) \tag{5.7}$$

is called the ***determinant*** of the 3×3 matrix $[\mathbf{a}\ \ \mathbf{b}\ \ \mathbf{c}]$ and is denoted by

$$\det[\mathbf{a}\ \ \mathbf{b}\ \ \mathbf{c}].$$

According to the definition of the determinant we have

$$\det \begin{bmatrix} a_1 & b_1 & c_1 \\ a_2 & b_2 & c_2 \\ a_3 & b_3 & c_3 \end{bmatrix} = \begin{bmatrix} a_1 \\ a_2 \\ a_3 \end{bmatrix} \cdot \left(\begin{bmatrix} b_1 \\ b_2 \\ b_3 \end{bmatrix} \times \begin{bmatrix} c_1 \\ c_2 \\ c_3 \end{bmatrix} \right)$$

$$= \begin{bmatrix} a_1 \\ a_2 \\ a_3 \end{bmatrix} \cdot \begin{bmatrix} \det \begin{bmatrix} b_2 & c_2 \\ b_3 & c_3 \end{bmatrix} \\ -\det \begin{bmatrix} b_1 & c_1 \\ b_3 & c_3 \end{bmatrix} \\ \det \begin{bmatrix} b_1 & c_1 \\ b_2 & c_2 \end{bmatrix} \end{bmatrix}$$

$$= a_1 \det \begin{bmatrix} b_2 & c_2 \\ b_3 & c_3 \end{bmatrix} - a_2 \det \begin{bmatrix} b_1 & c_1 \\ b_3 & c_3 \end{bmatrix} + a_3 \det \begin{bmatrix} b_1 & c_1 \\ b_2 & c_2 \end{bmatrix}.$$

The identity

$$
\det \begin{bmatrix} a_1 & b_1 & c_1 \\ a_2 & b_2 & c_2 \\ a_3 & b_3 & c_3 \end{bmatrix} = a_1 \det \begin{bmatrix} b_2 & c_2 \\ b_3 & c_3 \end{bmatrix} - a_2 \det \begin{bmatrix} b_1 & c_1 \\ b_3 & c_3 \end{bmatrix} + a_3 \det \begin{bmatrix} b_1 & c_1 \\ b_2 & c_2 \end{bmatrix} \qquad (5.8)
$$

connects determinants of 3×3 matrices with determinants of 2×2 matrices. It also gives us a practical way of calculating determinants of 3×3 matrices.

Example 5.1.13. For

$$
\mathbf{a} = \begin{bmatrix} 2 \\ 1 \\ 3 \end{bmatrix}, \quad \mathbf{b} = \begin{bmatrix} 1 \\ -2 \\ 2 \end{bmatrix}, \quad \text{and} \quad \mathbf{c} = \begin{bmatrix} -3 \\ 1 \\ -1 \end{bmatrix}.
$$

calculate the determinant $\det \begin{bmatrix} \mathbf{a} & \mathbf{b} & \mathbf{c} \end{bmatrix}$.

Solution.

$$
\det \begin{bmatrix} \mathbf{a} & \mathbf{b} & \mathbf{c} \end{bmatrix} = \det \begin{bmatrix} 2 & 1 & -3 \\ 1 & -2 & 1 \\ 3 & 2 & -1 \end{bmatrix}
$$

$$
= 2 \det \begin{bmatrix} -2 & 1 \\ 2 & -1 \end{bmatrix} - \det \begin{bmatrix} 1 & -3 \\ 2 & -1 \end{bmatrix} + 3 \begin{bmatrix} 1 & -3 \\ -2 & 1 \end{bmatrix}
$$

$$
= 2 \cdot 0 - 5 + 3 \cdot (-5)
$$

$$
= -20
$$

□

In the following theorem we list some useful properties of determinants.

Theorem 5.1.14. *For any* $\mathbf{a}, \mathbf{b}, \mathbf{c}, \mathbf{d}$ *in* $\mathbb{R}^3$ *and* s, t *in* $\mathbb{R}$ *we have*

(a) $\det\begin{bmatrix} \mathbf{a} & \mathbf{a} & \mathbf{b} \end{bmatrix} = \det\begin{bmatrix} \mathbf{a} & \mathbf{b} & \mathbf{a} \end{bmatrix} = \det\begin{bmatrix} \mathbf{a} & \mathbf{a} & \mathbf{b} \end{bmatrix} = 0;$

(b) $\det\begin{bmatrix} \mathbf{a} & \mathbf{b} & \mathbf{c} \end{bmatrix} = -\det\begin{bmatrix} \mathbf{a} & \mathbf{c} & \mathbf{b} \end{bmatrix} = -\det\begin{bmatrix} \mathbf{c} & \mathbf{b} & \mathbf{a} \end{bmatrix} = -\det\begin{bmatrix} \mathbf{b} & \mathbf{a} & \mathbf{c} \end{bmatrix};$

(c) $t\det\begin{bmatrix} \mathbf{a} & \mathbf{b} & \mathbf{c} \end{bmatrix} = \det\begin{bmatrix} t\mathbf{a} & \mathbf{b} & \mathbf{c} \end{bmatrix}) = \det\begin{bmatrix} \mathbf{a} & t\mathbf{b} & \mathbf{c} \end{bmatrix} = \det\begin{bmatrix} \mathbf{a} & \mathbf{b} & t\mathbf{c} \end{bmatrix};$

(d) $\det\begin{bmatrix} \mathbf{a}+\mathbf{d} & \mathbf{b} & \mathbf{c} \end{bmatrix} = \det\begin{bmatrix} \mathbf{a} & \mathbf{b} & \mathbf{c} \end{bmatrix} + \det\begin{bmatrix} \mathbf{d} & \mathbf{b} & \mathbf{c} \end{bmatrix};$

(e) $\det\begin{bmatrix} \mathbf{a} & \mathbf{b}+\mathbf{d} & \mathbf{c} \end{bmatrix} = \det\begin{bmatrix} \mathbf{a} & \mathbf{b} & \mathbf{c} \end{bmatrix} + \det\begin{bmatrix} \mathbf{a} & \mathbf{d} & \mathbf{c} \end{bmatrix};$

(f) $\det\begin{bmatrix} \mathbf{a} & \mathbf{b} & \mathbf{c}+\mathbf{d} \end{bmatrix} = \det\begin{bmatrix} \mathbf{a} & \mathbf{b} & \mathbf{c} \end{bmatrix} + \det\begin{bmatrix} \mathbf{a} & \mathbf{b} & \mathbf{d} \end{bmatrix};$

(g) $\det\begin{bmatrix} \mathbf{a}+s\mathbf{c} & \mathbf{b}+t\mathbf{c} & \mathbf{c} \end{bmatrix} = \det\begin{bmatrix} \mathbf{a} & \mathbf{b} & \mathbf{c} \end{bmatrix};$

(h) $\det\begin{bmatrix} \mathbf{a}+s\mathbf{b} & \mathbf{b} & \mathbf{c}+t\mathbf{b} \end{bmatrix} = \det\begin{bmatrix} \mathbf{a} & \mathbf{b} & \mathbf{c} \end{bmatrix};$

(i) $\det\begin{bmatrix} \mathbf{a} & \mathbf{b}+s\mathbf{a} & \mathbf{c}+t\mathbf{a} \end{bmatrix} = \det\begin{bmatrix} \mathbf{a} & \mathbf{b} & \mathbf{c} \end{bmatrix}.$

Proof. These identities follow easily from the definition of the determinant. We prove (g) and leave the other proofs as exercises. In the proof we are using the definition of the determinant and Theorem 5.1.6

$$\begin{aligned}
\det\begin{bmatrix} \mathbf{a}+s\mathbf{c} & \mathbf{b}+t\mathbf{c} & \mathbf{c} \end{bmatrix} &= \mathbf{c} \cdot ((\mathbf{a}+s\mathbf{c}) \times (\mathbf{b}+t\mathbf{c})) \\
&= \mathbf{c} \cdot (\mathbf{a} \times \mathbf{b} + t(\mathbf{a} \times \mathbf{c}) + s(\mathbf{c} \times \mathbf{b}) + st(\mathbf{c} \times \mathbf{c})) \\
&= \mathbf{c} \cdot (\mathbf{a} \times \mathbf{b} + t(\mathbf{a} \times \mathbf{c}) + s(\mathbf{c} \times \mathbf{b})) \\
&= \mathbf{c} \cdot (\mathbf{a} \times \mathbf{b}) + t(\mathbf{c} \cdot (\mathbf{a} \times \mathbf{c})) + s(\mathbf{c} \cdot (\mathbf{c} \times \mathbf{b})) \\
&= \mathbf{c} \cdot (\mathbf{a} \times \mathbf{b}) = \det\begin{bmatrix} \mathbf{a} & \mathbf{b} & \mathbf{c} \end{bmatrix}
\end{aligned}$$

$\square$

Here is another useful property of determinants.

Theorem 5.1.15. *For any* 3×3 *matrix A we have*

$$\det A^T = \det A,$$

that is,

$$\det \begin{bmatrix} a_1 & b_1 & c_1 \\ a_2 & b_2 & c_2 \\ a_3 & b_3 & c_3 \end{bmatrix} = \det \begin{bmatrix} a_1 & a_2 & a_3 \\ b_1 & b_2 & b_3 \\ c_1 & c_2 & c_3 \end{bmatrix}.$$

Proof. The identity can be obtained directly from the definition of the determinant by straightforward calculations. □

Now we show that the determinant of the product of two 3×3 matrices is equal to the product of determinants of those matrices. We proved the same property for determinants of 2×2 matrices in Theorem 1.3.4. In both cases the proofs are based on direct calculations. As expected, the proof for 3×3 matrices is much more tedious.

Theorem 5.1.16. *Let A and B be arbitrary 3×3 matrices. Then*

$$\det(AB) = \det(A)\det(B).$$

Proof. Let

$$A = \begin{bmatrix} a_1 & b_1 & c_1 \\ a_2 & b_2 & c_2 \\ a_3 & b_3 & c_3 \end{bmatrix} \quad \text{and} \quad B = \begin{bmatrix} s_1 & t_1 & u_1 \\ s_2 & t_2 & u_2 \\ s_3 & t_3 & u_3 \end{bmatrix}$$

and

$$\mathbf{a} = \begin{bmatrix} a_1 \\ a_2 \\ a_3 \end{bmatrix}, \quad \mathbf{b} = \begin{bmatrix} b_1 \\ b_2 \\ b_3 \end{bmatrix} \quad \text{and} \quad \mathbf{c} = \begin{bmatrix} c_1 \\ c_2 \\ c_3 \end{bmatrix}.$$

Then

$$\det(AB) = \det\left(\begin{bmatrix} a_1 & b_1 & c_1 \\ a_2 & b_2 & c_2 \\ a_3 & b_3 & c_3 \end{bmatrix}\begin{bmatrix} s_1 & t_1 & u_1 \\ s_2 & t_2 & u_2 \\ s_3 & t_3 & u_3 \end{bmatrix}\right)$$

$$= \det\begin{bmatrix} s_1\mathbf{a} + s_2\mathbf{b} + s_3\mathbf{c} & t_1\mathbf{a} + t_2\mathbf{b} + t_3\mathbf{c} & u_1\mathbf{a} + u_2\mathbf{b} + u_3\mathbf{c} \end{bmatrix}$$

$$= \det\begin{bmatrix} s_1\mathbf{a} & t_1\mathbf{a} + t_2\mathbf{b} + t_3\mathbf{c} & u_1\mathbf{a} + u_2\mathbf{b} + u_3\mathbf{c} \end{bmatrix}$$

$$\quad + \det\begin{bmatrix} s_2\mathbf{b} & t_1\mathbf{a} + t_2\mathbf{b} + t_3\mathbf{c} & u_1\mathbf{a} + u_2\mathbf{b} + u_3\mathbf{c} \end{bmatrix}$$

$$\quad + \det\begin{bmatrix} s_3\mathbf{c} & t_1\mathbf{a} + t_2\mathbf{b} + t_3\mathbf{c} & u_1\mathbf{a} + u_2\mathbf{b} + u_3\mathbf{c} \end{bmatrix}$$

$$= \det\begin{bmatrix} s_1\mathbf{a} & t_2\mathbf{b} + t_3\mathbf{c} & u_2\mathbf{b} + u_3\mathbf{c} \end{bmatrix}$$

$$\quad + \det\begin{bmatrix} s_2\mathbf{b} & t_1\mathbf{a} + t_3\mathbf{c} & u_1\mathbf{a} + u_3\mathbf{c} \end{bmatrix}$$

$$\quad + \det\begin{bmatrix} s_3\mathbf{c} & t_1\mathbf{a} + t_2\mathbf{b} & u_1\mathbf{a} + u_2\mathbf{b} \end{bmatrix}$$

$$= \det\begin{bmatrix} s_1\mathbf{a} & t_2\mathbf{b} & u_3\mathbf{c} \end{bmatrix} + \det\begin{bmatrix} s_1\mathbf{a} & t_3\mathbf{c} & u_2\mathbf{b} \end{bmatrix}$$

$$\quad + \det\begin{bmatrix} s_2\mathbf{b} & t_1\mathbf{a} & u_3\mathbf{c} \end{bmatrix} + \det\begin{bmatrix} s_2\mathbf{b} & t_3\mathbf{c} & u_1\mathbf{a} \end{bmatrix}$$

$$\quad + \det\begin{bmatrix} s_3\mathbf{c} & t_1\mathbf{a} & u_2\mathbf{b} \end{bmatrix} + \det\begin{bmatrix} s_3\mathbf{c} & t_2\mathbf{b} & u_1\mathbf{a} \end{bmatrix}$$

$$= \det\begin{bmatrix} s_1\mathbf{a} & t_2\mathbf{b} & u_3\mathbf{c} \end{bmatrix} - \det\begin{bmatrix} s_1\mathbf{a} & u_2\mathbf{b} & t_3\mathbf{c} \end{bmatrix}$$

$$\quad - \det\begin{bmatrix} t_1\mathbf{a} & s_2\mathbf{b} & u_3\mathbf{c} \end{bmatrix} + \det\begin{bmatrix} u_1\mathbf{a} & s_2\mathbf{b} & t_3\mathbf{c} \end{bmatrix}$$

$$\quad + \det\begin{bmatrix} t_1\mathbf{a} & u_2\mathbf{b} & s_3\mathbf{c} \end{bmatrix} - \det\begin{bmatrix} u_1\mathbf{a} & t_2\mathbf{b} & s_3\mathbf{c} \end{bmatrix}$$

$$= s_1 t_2 u_3 \det\begin{bmatrix} \mathbf{a} & \mathbf{b} & \mathbf{c} \end{bmatrix} - s_1 t_3 u_2 \det\begin{bmatrix} \mathbf{a} & \mathbf{b} & \mathbf{c} \end{bmatrix}$$

$$- s_2 t_1 u_3 \det \begin{bmatrix} \mathbf{a} & \mathbf{b} & \mathbf{c} \end{bmatrix} + s_2 t_3 u_1 \det \begin{bmatrix} \mathbf{a} & \mathbf{b} & \mathbf{c} \end{bmatrix}$$
$$+ s_3 t_1 u_2 \det \begin{bmatrix} \mathbf{a} & \mathbf{b} & \mathbf{c} \end{bmatrix} - s_3 t_2 u_1 \begin{bmatrix} \mathbf{a} & \mathbf{b} & \mathbf{c} \end{bmatrix}$$
$$= \det \begin{bmatrix} \mathbf{a} & \mathbf{b} & \mathbf{c} \end{bmatrix} (s_1(t_2 u_3 - t_3 u_2) - s_2(t_1 u_3 - t_3 u_1) + s_3(t_1 u_2 - t_2 u_1))$$
$$= \det \begin{bmatrix} \mathbf{a} & \mathbf{b} & \mathbf{c} \end{bmatrix} \det \begin{bmatrix} s_1 & t_1 & u_1 \\ s_2 & t_2 & u_2 \\ s_3 & t_3 & u_3 \end{bmatrix}$$
$$= \det(A)\det(B).$$

$\square$

In Section 3 we will see that the determinant $\det \begin{bmatrix} \mathbf{a} & \mathbf{b} & \mathbf{c} \end{bmatrix}$ tells us something important about the vectors $\mathbf{a}$, $\mathbf{b}$, $\mathbf{c}$ and has many useful properties. Later we will also show that the determinant of a 3×3 matrix can be interpreted as the volume of a tetrahedron.

5.1.1 Exercises

Calculate the given cross products.

1. $\begin{bmatrix} 5 \\ 2 \\ 7 \end{bmatrix} \times \begin{bmatrix} 1 \\ 4 \\ -2 \end{bmatrix}$

3. $\begin{bmatrix} 2 \\ 1 \\ 3 \end{bmatrix} \times \begin{bmatrix} 3 \\ -2 \\ 5 \end{bmatrix}$

2. $\begin{bmatrix} 3 \\ 2 \\ 1 \end{bmatrix} \times \begin{bmatrix} 1 \\ 2 \\ 3 \end{bmatrix}$

4. $\begin{bmatrix} 2 \\ 0 \\ 7 \end{bmatrix} \times \begin{bmatrix} 1 \\ 1 \\ 1 \end{bmatrix}$

Solve the given systems of equations.

5. $\begin{cases} x + 2y - 4z = 0 \\ x - 2y + 5z = 0 \end{cases}$

7. $\begin{cases} 5x + 2y + 7z = 0 \\ x + y + 2z = 0 \end{cases}$

6. $\begin{cases} 4x - y - z = 0 \\ x + y - 2z = 0 \end{cases}$

8. $\begin{cases} y + z = 0 \\ x + y + 5z = 0 \end{cases}$

Find an equation of the given vector planes.

9. $\text{Span} \left\{ \begin{bmatrix} 1 \\ 1 \\ 1 \end{bmatrix}, \begin{bmatrix} -1 \\ 2 \\ 1 \end{bmatrix} \right\}$

11. $\text{Span} \left\{ \begin{bmatrix} 2 \\ 1 \\ 5 \end{bmatrix}, \begin{bmatrix} 3 \\ 2 \\ -4 \end{bmatrix} \right\}$

10. $\text{Span} \left\{ \begin{bmatrix} 2 \\ 1 \\ 1 \end{bmatrix}, \begin{bmatrix} 5 \\ 2 \\ 1 \end{bmatrix} \right\}$

12. $\text{Span} \left\{ \begin{bmatrix} 1 \\ -2 \\ 2 \end{bmatrix}, \begin{bmatrix} 1 \\ 1 \\ -1 \end{bmatrix} \right\}$

Calculate the determinant $\det A$.

13. $A = \begin{bmatrix} 1 & 1 & 3 \\ 2 & 5 & -1 \\ 1 & 4 & 2 \end{bmatrix}$

15. $A = \begin{bmatrix} 4 & 2 & -3 \\ -2 & 1 & 2 \\ 3 & 7 & -1 \end{bmatrix}$

14. $A = \begin{bmatrix} 3 & 1 & 2 \\ 2 & 3 & 1 \\ 1 & 2 & 3 \end{bmatrix}$

16. $A = \begin{bmatrix} 2 & 1 & 1 \\ 1 & 2 & 1 \\ 1 & 1 & 2 \end{bmatrix}$

Show that the following identities hold for arbitrary vectors **a**, **b**, **c** and **d**, and for arbitrary numbers s and t.

17. $\det\begin{bmatrix} \mathbf{a} & \mathbf{a} & \mathbf{b} \end{bmatrix} = \det\begin{bmatrix} \mathbf{a} & \mathbf{b} & \mathbf{a} \end{bmatrix} = \det\begin{bmatrix} \mathbf{b} & \mathbf{a} & \mathbf{a} \end{bmatrix} = 0$

18. $\det\begin{bmatrix} \mathbf{a} & \mathbf{b} & \mathbf{c} \end{bmatrix} = -\det\begin{bmatrix} \mathbf{a} & \mathbf{c} & \mathbf{b} \end{bmatrix} = -\det\begin{bmatrix} \mathbf{c} & \mathbf{b} & \mathbf{a} \end{bmatrix} = -\det\begin{bmatrix} \mathbf{b} & \mathbf{a} & \mathbf{c} \end{bmatrix}$

19. $\det\begin{bmatrix} \mathbf{a}+\mathbf{d} & \mathbf{b} & \mathbf{c} \end{bmatrix} = \det\begin{bmatrix} \mathbf{a} & \mathbf{b} & \mathbf{c} \end{bmatrix} + \det\begin{bmatrix} \mathbf{d} & \mathbf{b} & \mathbf{c} \end{bmatrix}$

20. $\det\begin{bmatrix} \mathbf{a} & \mathbf{b} & \mathbf{c}+\mathbf{d} \end{bmatrix} = \det\begin{bmatrix} \mathbf{a} & \mathbf{b} & \mathbf{c} \end{bmatrix} + \det\begin{bmatrix} \mathbf{a} & \mathbf{b} & \mathbf{d} \end{bmatrix}$

21. $\det\begin{bmatrix} \mathbf{a}+s\mathbf{b} & \mathbf{b} & \mathbf{c}+t\mathbf{b} \end{bmatrix} = \det\begin{bmatrix} \mathbf{a} & \mathbf{b} & \mathbf{c} \end{bmatrix}$

22. $\det\begin{bmatrix} \mathbf{a} & \mathbf{b}+s\mathbf{a} & \mathbf{c}+t\mathbf{a} \end{bmatrix} = \det\begin{bmatrix} \mathbf{a} & \mathbf{b} & \mathbf{c} \end{bmatrix}$

23. $\det\begin{bmatrix} s\mathbf{a} & t\mathbf{b} & u\mathbf{c} \end{bmatrix} = stu\det\begin{bmatrix} \mathbf{a} & \mathbf{b} & \mathbf{c} \end{bmatrix}$

24. $\det\begin{bmatrix} \mathbf{a} & \mathbf{a}+\mathbf{b} & \mathbf{a}+\mathbf{b}+\mathbf{c} \end{bmatrix} = \det\begin{bmatrix} \mathbf{a} & \mathbf{b} & \mathbf{c} \end{bmatrix}$

25. Show that

$$\det\begin{bmatrix} a_1 & b_1 & c_1 \\ a_2 & b_2 & c_2 \\ a_3 & b_3 & c_3 \end{bmatrix} = \det\begin{bmatrix} a_1+sa_2 & b_1+sb_2 & c_1+sc_2 \\ a_2 & b_2 & c_2 \\ a_3+ta_2 & b_3+tb_2 & c_3+tc_2 \end{bmatrix}.$$

26. Show that

$$\det\begin{bmatrix} a_1 & b_1 & c_1 \\ a_2 & b_2 & c_2 \\ a_3 & b_3 & c_3 \end{bmatrix} = \det\begin{bmatrix} a_1 & b_1 & c_1 \\ a_2+sa_1 & b_2+sb_1 & c_2+sc_1 \\ a_3+ta_1 & b_3+tb_1 & c_3+tc_1 \end{bmatrix}.$$

27. Show that $\det\begin{bmatrix} 2 & 1 & 2 \\ 2 & 4 & 2 \\ 7 & 5 & 7 \end{bmatrix} = 0$ without calculating the determinant.

28. Show that $\det\begin{bmatrix} 1 & 2 & 1 \\ 2 & 4 & 1 \\ 5 & 10 & 2 \end{bmatrix} = 0$ without calculating the determinant.

If $\det\begin{bmatrix} a & b & c \\ p & q & r \\ x & y & z \end{bmatrix} = 33$, find the following determinants.

29. $\det \begin{bmatrix} a+9b & b & c \\ p+9q & q & r \\ x+9y & y & z \end{bmatrix}$

    31. $\det \begin{bmatrix} 3a+5b & b & c \\ 3p+5q & q & r \\ 3x+5y & y & z \end{bmatrix}$

30. $\det \begin{bmatrix} 5a & b & c \\ 5p & q & r \\ 5x & y & z \end{bmatrix}$

    32. $\det \begin{bmatrix} a+b+c & b & c \\ p+q+r & q & r \\ x+y+z & y & z \end{bmatrix}$

33. Show that

$$\det\left(\begin{bmatrix} 1 & s & 0 \\ 0 & 1 & 0 \\ 0 & t & 1 \end{bmatrix} \begin{bmatrix} a_1 & b_1 & c_1 \\ a_2 & b_2 & c_2 \\ a_3 & b_3 & c_3 \end{bmatrix} \right) = \det \begin{bmatrix} a_1 & b_1 & c_1 \\ a_2 & b_2 & c_2 \\ a_3 & b_3 & c_3 \end{bmatrix}$$

34. Show that

$$\det\left(\begin{bmatrix} a_1 & b_1 & c_1 \\ a_2 & b_2 & c_2 \\ a_3 & b_3 & c_3 \end{bmatrix} \begin{bmatrix} 1 & 0 & s \\ 0 & 1 & t \\ 0 & 0 & 1 \end{bmatrix} \right) = \det \begin{bmatrix} a_1 & b_1 & c_1 \\ a_2 & b_2 & c_2 \\ a_3 & b_3 & c_3 \end{bmatrix}.$$

5.2 Calculating inverses and determinants of 3×3 matrices

We have seen that solving problems often requires calculating determinants of matrices or inverse matrices. In this section we present some practical methods for calculating determinants and inverses of 3×3 matrices.

Recall that the product of two 3×3 matrices is defined as follows:

$$\begin{bmatrix} s_1 & t_1 & u_1 \\ s_2 & t_2 & u_2 \\ s_3 & t_3 & u_3 \end{bmatrix} \begin{bmatrix} a_1 & b_1 & c_1 \\ a_2 & b_2 & c_2 \\ a_3 & b_3 & c_3 \end{bmatrix} = \begin{bmatrix} \begin{bmatrix} s_1 & t_1 & u_1 \end{bmatrix} \begin{bmatrix} a_1 \\ a_2 \\ a_3 \end{bmatrix} & \begin{bmatrix} s_1 & t_1 & u_1 \end{bmatrix} \begin{bmatrix} b_1 \\ b_2 \\ b_3 \end{bmatrix} & \begin{bmatrix} s_1 & t_1 & u_1 \end{bmatrix} \begin{bmatrix} c_1 \\ c_2 \\ c_3 \end{bmatrix} \\ \begin{bmatrix} s_2 & t_2 & u_2 \end{bmatrix} \begin{bmatrix} a_1 \\ a_2 \\ a_3 \end{bmatrix} & \begin{bmatrix} s_2 & t_2 & u_2 \end{bmatrix} \begin{bmatrix} b_1 \\ b_2 \\ b_3 \end{bmatrix} & \begin{bmatrix} s_2 & t_2 & u_2 \end{bmatrix} \begin{bmatrix} c_1 \\ c_2 \\ c_3 \end{bmatrix} \\ \begin{bmatrix} s_3 & t_3 & u_3 \end{bmatrix} \begin{bmatrix} a_1 \\ a_2 \\ a_3 \end{bmatrix} & \begin{bmatrix} s_3 & t_3 & u_3 \end{bmatrix} \begin{bmatrix} b_1 \\ b_2 \\ b_3 \end{bmatrix} & \begin{bmatrix} s_3 & t_3 & u_3 \end{bmatrix} \begin{bmatrix} c_1 \\ c_2 \\ c_3 \end{bmatrix} \end{bmatrix}$$

or equivalently

$$
\begin{bmatrix} s_1 & t_1 & u_1 \\ s_2 & t_2 & u_2 \\ s_3 & t_3 & u_3 \end{bmatrix} \begin{bmatrix} a_1 & b_1 & c_1 \\ a_2 & b_2 & c_2 \\ a_3 & b_3 & c_3 \end{bmatrix} = \begin{bmatrix} \begin{bmatrix} s_1 \\ t_1 \\ u_1 \end{bmatrix} \cdot \begin{bmatrix} a_1 \\ a_2 \\ a_3 \end{bmatrix} & \begin{bmatrix} s_1 \\ t_1 \\ u_1 \end{bmatrix} \cdot \begin{bmatrix} b_1 \\ b_2 \\ b_3 \end{bmatrix} & \begin{bmatrix} s_1 \\ t_1 \\ u_1 \end{bmatrix} \cdot \begin{bmatrix} c_1 \\ c_2 \\ c_3 \end{bmatrix} \\ \begin{bmatrix} s_2 \\ t_2 \\ u_2 \end{bmatrix} \cdot \begin{bmatrix} a_1 \\ a_2 \\ a_3 \end{bmatrix} & \begin{bmatrix} s_2 \\ t_2 \\ u_2 \end{bmatrix} \cdot \begin{bmatrix} b_1 \\ b_2 \\ b_3 \end{bmatrix} & \begin{bmatrix} s_2 \\ t_2 \\ u_2 \end{bmatrix} \cdot \begin{bmatrix} c_1 \\ c_2 \\ c_3 \end{bmatrix} \\ \begin{bmatrix} s_3 \\ t_3 \\ u_3 \end{bmatrix} \cdot \begin{bmatrix} a_1 \\ a_2 \\ a_3 \end{bmatrix} & \begin{bmatrix} s_3 \\ t_3 \\ u_3 \end{bmatrix} \cdot \begin{bmatrix} b_1 \\ b_2 \\ b_3 \end{bmatrix} & \begin{bmatrix} s_3 \\ t_3 \\ u_3 \end{bmatrix} \cdot \begin{bmatrix} c_1 \\ c_2 \\ c_3 \end{bmatrix} \end{bmatrix}. \tag{5.9}
$$

Theorem 5.2.1. *Let*

$$
A = \begin{bmatrix} a_{11} & a_{12} & a_{13} \\ a_{21} & a_{22} & a_{23} \\ a_{31} & a_{32} & a_{33} \end{bmatrix}
$$

be an arbitrary 3 × 3 *matrix and let*

$$
B = \begin{bmatrix} \begin{bmatrix} a_{21} \\ a_{22} \\ a_{23} \end{bmatrix} \times \begin{bmatrix} a_{31} \\ a_{32} \\ a_{33} \end{bmatrix} & \begin{bmatrix} a_{31} \\ a_{32} \\ a_{33} \end{bmatrix} \times \begin{bmatrix} a_{11} \\ a_{12} \\ a_{13} \end{bmatrix} & \begin{bmatrix} a_{11} \\ a_{12} \\ a_{13} \end{bmatrix} \times \begin{bmatrix} a_{21} \\ a_{22} \\ a_{23} \end{bmatrix} \end{bmatrix}.
$$

Then

$$
AB = BA = \begin{bmatrix} \det A & 0 & 0 \\ 0 & \det A & 0 \\ 0 & 0 & \det A \end{bmatrix}.
$$

Proof. To simplify the calculations we denote

$$
A_1 = \begin{bmatrix} a_{11} \\ a_{12} \\ a_{13} \end{bmatrix}, A_2 = \begin{bmatrix} a_{21} \\ a_{22} \\ a_{23} \end{bmatrix}, \text{ and } A_3 = \begin{bmatrix} a_{31} \\ a_{32} \\ a_{33} \end{bmatrix}.
$$

Now we can write

$$
A = \begin{bmatrix} A_1^T \\ A_2^T \\ A_3^T \end{bmatrix}, \quad A^T = \begin{bmatrix} A_1 & A_2 & A_3 \end{bmatrix}, \quad \text{and} \quad B = \begin{bmatrix} A_2 \times A_3 & A_3 \times A_1 & A_1 \times A_2 \end{bmatrix}.
$$

From the definition of the determinant of a 3 × 3 matrix we have

$$
\det \begin{bmatrix} A_1 & A_2 & A_3 \end{bmatrix} = A_1 \cdot (A_2 \times A_3) = A_2 \cdot (A_3 \times A_1) = A_3 \cdot (A_1 \times A_2)
$$

and, by Theorem 5.1.15, we have

$$
\det \begin{bmatrix} A_1 & A_2 & A_3 \end{bmatrix} = \det A^T = \det A.
$$

Now, from (5.9) it follows that

$$AB = \begin{bmatrix} A_1 \cdot (A_2 \times A_3) & A_1 \cdot (A_3 \times A_1) & A_1 \cdot (A_1 \times A_2) \\ A_2 \cdot (A_2 \times A_3) & A_2 \cdot (A_3 \times A_1) & A_2 \cdot (A_1 \times A_2) \\ A_3 \cdot (A_2 \times A_3) & A_3 \cdot (A_3 \times A_1) & A_3 \cdot (A_1 \times A_2) \end{bmatrix}$$

$$= \begin{bmatrix} A_1 \cdot (A_2 \times A_3) & 0 & 0 \\ 0 & A_2 \cdot (A_3 \times A_1) & 0 \\ 0 & 0 & A_3 \cdot (A_1 \times A_2) \end{bmatrix}$$

$$= \begin{bmatrix} \det A & 0 & 0 \\ 0 & \det A & 0 \\ 0 & 0 & \det A \end{bmatrix}.$$

Next we show that

$$BA = \begin{bmatrix} \det A & 0 & 0 \\ 0 & \det A & 0 \\ 0 & 0 & \det A \end{bmatrix}.$$

For this part of the proof we define

$$C_1 = \begin{bmatrix} a_{11} \\ a_{21} \\ a_{31} \end{bmatrix}, \ C_2 = \begin{bmatrix} a_{12} \\ a_{22} \\ a_{32} \end{bmatrix}, \ C_3 = \begin{bmatrix} a_{13} \\ a_{23} \\ a_{33} \end{bmatrix}$$

and note that

$$B = \begin{bmatrix} \begin{bmatrix} a_{21} \\ a_{22} \\ a_{23} \end{bmatrix} \times \begin{bmatrix} a_{31} \\ a_{32} \\ a_{33} \end{bmatrix} & \begin{bmatrix} a_{31} \\ a_{32} \\ a_{33} \end{bmatrix} \times \begin{bmatrix} a_{11} \\ a_{12} \\ a_{13} \end{bmatrix} & \begin{bmatrix} a_{11} \\ a_{12} \\ a_{13} \end{bmatrix} \times \begin{bmatrix} a_{21} \\ a_{22} \\ a_{23} \end{bmatrix} \end{bmatrix}$$

$$= \begin{bmatrix} \det \begin{bmatrix} a_{22} & a_{23} \\ a_{32} & a_{33} \end{bmatrix} & -\det \begin{bmatrix} a_{12} & a_{13} \\ a_{32} & a_{33} \end{bmatrix} & \det \begin{bmatrix} a_{12} & a_{13} \\ a_{22} & a_{23} \end{bmatrix} \\ -\det \begin{bmatrix} a_{21} & a_{23} \\ a_{31} & a_{33} \end{bmatrix} & \det \begin{bmatrix} a_{11} & a_{13} \\ a_{31} & a_{33} \end{bmatrix} & -\det \begin{bmatrix} a_{11} & a_{13} \\ a_{21} & a_{23} \end{bmatrix} \\ \det \begin{bmatrix} a_{21} & a_{22} \\ a_{31} & a_{32} \end{bmatrix} & -\det \begin{bmatrix} a_{11} & a_{12} \\ a_{31} & a_{32} \end{bmatrix} & \det \begin{bmatrix} a_{11} & a_{12} \\ a_{21} & a_{22} \end{bmatrix} \end{bmatrix}$$

$$= \begin{bmatrix} C_2 \times C_3 & C_3 \times C_1 & C_1 \times C_2 \end{bmatrix}^T.$$

Since

$$A^T = \begin{bmatrix} C_1^T \\ C_2^T \\ C_3^T \end{bmatrix} \quad \text{and} \quad C_1 \cdot (C_2 \times C_3) = C_2 \cdot (C_3 \times C_1) = C_3 \cdot (C_1 \times C_2) = \det A,$$

we have

$$BA = \begin{bmatrix} C_2 \times C_3 & C_3 \times C_1 & C_1 \times C_2 \end{bmatrix}^T \begin{bmatrix} C_1^T \\ C_2^T \\ C_3^T \end{bmatrix}^T$$

$$= \left(\begin{bmatrix} C_1^T \\ C_2^T \\ C_3^T \end{bmatrix} \begin{bmatrix} C_2 \times C_3 & C_3 \times C_1 & C_1 \times C_2 \end{bmatrix} \right)^T$$

$$= \left(\begin{bmatrix} C_1 \cdot (C_2 \times C_3) & C_1 \cdot (C_3 \times C_1) & C_1 \cdot (C_1 \times C_2) \\ C_2 \cdot (C_2 \times C_3) & C_2 \cdot (C_3 \times C_1) & C_2 \cdot (C_1 \times C_2) \\ C_3 \cdot (C_2 \times C_3) & C_3 \cdot (C_3 \times C_1) & C_3 \cdot (C_1 \times C_2) \end{bmatrix} \right)^T$$

$$= \left(\begin{bmatrix} C_1 \cdot (C_2 \times C_3) & 0 & 0 \\ 0 & C_2 \cdot (C_3 \times C_1) & 0 \\ 0 & 0 & C_3 \cdot (C_1 \times C_2) \end{bmatrix} \right)^T$$

$$= \left(\begin{bmatrix} \det A & 0 & 0 \\ 0 & \det A & 0 \\ 0 & 0 & \det A \end{bmatrix} \right)^T$$

$$= \begin{bmatrix} \det A & 0 & 0 \\ 0 & \det A & 0 \\ 0 & 0 & \det A \end{bmatrix}.$$

□

Example 5.2.2. We consider the matrix $\begin{bmatrix} 3 & 1 & 2 \\ 1 & -1 & 4 \\ 1 & 0 & 3 \end{bmatrix}$. Since

$$\begin{bmatrix} 3 \\ 1 \\ 2 \end{bmatrix} \times \begin{bmatrix} 1 \\ -1 \\ 4 \end{bmatrix} = \begin{bmatrix} 6 \\ -10 \\ -4 \end{bmatrix},$$

$$\begin{bmatrix} 1 \\ -1 \\ 4 \end{bmatrix} \times \begin{bmatrix} 1 \\ 0 \\ 3 \end{bmatrix} = \begin{bmatrix} -3 \\ 1 \\ 1 \end{bmatrix},$$

$$\begin{bmatrix} 1 \\ 0 \\ 3 \end{bmatrix} \times \begin{bmatrix} 3 \\ 1 \\ 2 \end{bmatrix} = \begin{bmatrix} -3 \\ 7 \\ 1 \end{bmatrix},$$

and

$$\det \begin{bmatrix} 3 & 1 & 2 \\ 1 & -1 & 4 \\ 1 & 0 & 3 \end{bmatrix} = -6,$$

we have

$$\begin{bmatrix} 3 & 1 & 2 \\ 1 & -1 & 4 \\ 1 & 0 & 3 \end{bmatrix} \begin{bmatrix} -3 & -3 & 6 \\ 1 & 7 & -10 \\ 1 & 1 & -4 \end{bmatrix} = \begin{bmatrix} -3 & -3 & 6 \\ 1 & 7 & -10 \\ 1 & 1 & -4 \end{bmatrix} \begin{bmatrix} 3 & 1 & 2 \\ 1 & -1 & 4 \\ 1 & 0 & 3 \end{bmatrix} = \begin{bmatrix} -6 & 0 & 0 \\ 0 & -6 & 0 \\ 0 & 0 & -6 \end{bmatrix}.$$

Calculating inverses

From Theorem 5.3.10 we obtain that a 3×3 matrix A is invertible if and only if $\det A \neq 0$.

Now Theorem 5.2.1 gives us a practical method of calculating A^{-1}.

Theorem 5.2.3. *Let*

$$A = \begin{bmatrix} a_{11} & a_{12} & a_{13} \\ a_{21} & a_{22} & a_{23} \\ a_{31} & a_{32} & a_{33} \end{bmatrix}.$$

If the matrix A is invertible we have

$$A^{-1} = \frac{1}{\det A} \left(\left[\begin{bmatrix} a_{21} \\ a_{22} \\ a_{23} \end{bmatrix} \times \begin{bmatrix} a_{31} \\ a_{32} \\ a_{33} \end{bmatrix} \quad \begin{bmatrix} a_{31} \\ a_{32} \\ a_{33} \end{bmatrix} \times \begin{bmatrix} a_{11} \\ a_{12} \\ a_{13} \end{bmatrix} \quad \begin{bmatrix} a_{11} \\ a_{12} \\ a_{13} \end{bmatrix} \times \begin{bmatrix} a_{21} \\ a_{22} \\ a_{23} \end{bmatrix} \right] \right).$$

Proof. The theorem is a direct consequence of Theorem 5.2.1.   □

Definition 5.2.4. For an arbitrary 3×3 matrix

$$A = \begin{bmatrix} a_{11} & a_{12} & a_{13} \\ a_{21} & a_{22} & a_{23} \\ a_{31} & a_{32} & a_{33} \end{bmatrix}$$

the matrix

$$\operatorname{adj} A = \begin{bmatrix} \det \begin{bmatrix} a_{22} & a_{23} \\ a_{32} & a_{33} \end{bmatrix} & -\det \begin{bmatrix} a_{12} & a_{13} \\ a_{32} & a_{33} \end{bmatrix} & \det \begin{bmatrix} a_{12} & a_{13} \\ a_{22} & a_{23} \end{bmatrix} \\ -\det \begin{bmatrix} a_{21} & a_{23} \\ a_{31} & a_{33} \end{bmatrix} & \det \begin{bmatrix} a_{11} & a_{13} \\ a_{31} & a_{33} \end{bmatrix} & -\det \begin{bmatrix} a_{11} & a_{13} \\ a_{21} & a_{23} \end{bmatrix} \\ \det \begin{bmatrix} a_{21} & a_{22} \\ a_{31} & a_{32} \end{bmatrix} & -\det \begin{bmatrix} a_{11} & a_{12} \\ a_{31} & a_{32} \end{bmatrix} & \det \begin{bmatrix} a_{11} & a_{12} \\ a_{21} & a_{22} \end{bmatrix} \end{bmatrix}$$

is called the ***adjoint*** of the matrix A.

We note that this is the same matrix that was used in Theorem 5.2.1, where it was written as

$$\left[\begin{bmatrix} a_{21} \\ a_{22} \\ a_{23} \end{bmatrix} \times \begin{bmatrix} a_{31} \\ a_{32} \\ a_{33} \end{bmatrix} \quad \begin{bmatrix} a_{31} \\ a_{32} \\ a_{33} \end{bmatrix} \times \begin{bmatrix} a_{11} \\ a_{12} \\ a_{13} \end{bmatrix} \quad \begin{bmatrix} a_{11} \\ a_{12} \\ a_{13} \end{bmatrix} \times \begin{bmatrix} a_{21} \\ a_{22} \\ a_{23} \end{bmatrix} \right].$$

We are going to examine the construction of this matrix more carefully. To this end we first denote by A_{ij} the 2 × 2 matrix obtained from the matrix A by deleting the i-th row and the j-th column, that is,

$$A_{11} = \begin{bmatrix} a_{22} & a_{23} \\ a_{32} & a_{33} \end{bmatrix}, \qquad A_{21} = \begin{bmatrix} a_{12} & a_{13} \\ a_{32} & a_{33} \end{bmatrix}, \qquad A_{31} = \begin{bmatrix} a_{12} & a_{13} \\ a_{22} & a_{23} \end{bmatrix},$$

$$A_{12} = \begin{bmatrix} a_{21} & a_{23} \\ a_{31} & a_{33} \end{bmatrix}, \qquad A_{22} = \begin{bmatrix} a_{11} & a_{13} \\ a_{31} & a_{33} \end{bmatrix}, \qquad A_{32} = \begin{bmatrix} a_{11} & a_{13} \\ a_{21} & a_{23} \end{bmatrix},$$

$$A_{13} = \begin{bmatrix} a_{21} & a_{22} \\ a_{31} & a_{32} \end{bmatrix}, \qquad A_{23} = \begin{bmatrix} a_{11} & a_{12} \\ a_{31} & a_{32} \end{bmatrix}, \qquad A_{33} = \begin{bmatrix} a_{11} & a_{12} \\ a_{21} & a_{22} \end{bmatrix}.$$

Then we consider the matrix

$$\begin{bmatrix} \det A_{11} & \det A_{12} & \det A_{13} \\ \det A_{21} & \det A_{22} & \det A_{23} \\ \det A_{31} & \det A_{32} & \det A_{33} \end{bmatrix}$$

and change the sign of every other entry of this matrix according to the following pattern

$$\begin{bmatrix} + & - & + \\ - & + & - \\ + & - & + \end{bmatrix}$$

to obtain

$$\begin{bmatrix} \det A_{11} & -\det A_{12} & \det A_{13} \\ -\det A_{21} & \det A_{22} & -\det A_{23} \\ \det A_{31} & -\det A_{32} & \det A_{33} \end{bmatrix}.$$

Finally we transpose the above matrix and obtain the adjoint matrix A:

$$\operatorname{adj} A = \begin{bmatrix} \det A_{11} & -\det A_{21} & \det A_{31} \\ -\det A_{12} & \det A_{22} & -\det A_{32} \\ \det A_{13} & -\det A_{23} & \det A_{33} \end{bmatrix}.$$

In Theorem 5.2.1 we show that

$$AB = BA = \begin{bmatrix} \det A & 0 & 0 \\ 0 & \det A & 0 \\ 0 & 0 & \det A \end{bmatrix},$$

where $B = \operatorname{adj} A$. Consequently, for any matrix A such that $\det A \neq 0$, we have

$$A^{-1} = \frac{1}{\det A} \operatorname{adj} A. \tag{5.10}$$

Example 5.2.5. We consider the matrix

$$A = \begin{bmatrix} 2 & 4 & 3 \\ 5 & 1 & -1 \\ 7 & 3 & -2 \end{bmatrix}.$$

First we find

$$\det A_{11} = \det \begin{bmatrix} 1 & -1 \\ 3 & -2 \end{bmatrix} = 1, \quad \det A_{12} = \det \begin{bmatrix} 5 & -1 \\ 7 & -2 \end{bmatrix} = -3, \quad \det A_{13} = \det \begin{bmatrix} 5 & 1 \\ 7 & 3 \end{bmatrix} = 8,$$

$$\det A_{21} = \det \begin{bmatrix} 4 & 3 \\ 3 & -2 \end{bmatrix} = -17, \quad \det A_{22} = \det \begin{bmatrix} 2 & 3 \\ 7 & -2 \end{bmatrix} = -25, \quad \det A_{23} = \det \begin{bmatrix} 2 & 4 \\ 7 & 3 \end{bmatrix} = -22,$$

$$\det A_{31} = \det \begin{bmatrix} 4 & 3 \\ 1 & -1 \end{bmatrix} = -7, \quad \det A_{32} = \det \begin{bmatrix} 2 & 3 \\ 5 & -1 \end{bmatrix} = 17, \quad \det A_{33} = \det \begin{bmatrix} 2 & 4 \\ 5 & 1 \end{bmatrix} = -17.$$

This means that

$$\begin{bmatrix} \det A_{11} & \det A_{12} & \det A_{13} \\ \det A_{21} & \det A_{22} & \det A_{23} \\ \det A_{31} & \det A_{32} & \det A_{33} \end{bmatrix} = \begin{bmatrix} 1 & -3 & 8 \\ -17 & -25 & -22 \\ -7 & -17 & -18 \end{bmatrix}.$$

Now we change the sign of every other entry of this matrix and obtain

$$\begin{bmatrix} \det A_{11} & -\det A_{12} & \det A_{13} \\ -\det A_{21} & \det A_{22} & -\det A_{23} \\ \det A_{31} & -\det A_{32} & \det A_{33} \end{bmatrix} = \begin{bmatrix} 1 & 3 & 8 \\ 17 & -25 & 22 \\ -7 & 17 & -18 \end{bmatrix}.$$

The adjoint of the matrix A is the transpose of the above matrix:

$$\operatorname{adj} A = \begin{bmatrix} 1 & 3 & 8 \\ 17 & -25 & 22 \\ -7 & 17 & -18 \end{bmatrix}^T = \begin{bmatrix} 1 & 17 & -7 \\ 3 & -25 & 17 \\ 8 & 22 & -18 \end{bmatrix}.$$

Since

$$\begin{bmatrix} 2 & 4 & 3 \\ 5 & 1 & -1 \\ 7 & 3 & -2 \end{bmatrix} \begin{bmatrix} 1 & 17 & -7 \\ 3 & -25 & 17 \\ 8 & 22 & -18 \end{bmatrix} = \begin{bmatrix} 1 & 17 & -7 \\ 3 & -25 & 17 \\ 8 & 22 & -18 \end{bmatrix} \begin{bmatrix} 2 & 4 & 3 \\ 5 & 1 & -1 \\ 7 & 3 & -2 \end{bmatrix} = \begin{bmatrix} 38 & 0 & 0 \\ 0 & 38 & 0 \\ 0 & 0 & 38 \end{bmatrix},$$

we have

$$\begin{bmatrix} 2 & 4 & 3 \\ 5 & 1 & -1 \\ 7 & 3 & -2 \end{bmatrix}^{-1} = \frac{1}{38} \begin{bmatrix} 1 & 17 & -7 \\ 3 & -25 & 17 \\ 8 & 22 & -18 \end{bmatrix}.$$

Example 5.2.6. The adjoint of the matrix

$$\begin{bmatrix} 3 & -1 & 2 \\ 1 & -2 & 1 \\ 2 & 1 & 6 \end{bmatrix}$$

is

$$\begin{bmatrix} -13 & 8 & 3 \\ -4 & 14 & -1 \\ 5 & -5 & -5 \end{bmatrix}$$

and

$$\det \begin{bmatrix} 3 & -1 & 2 \\ 1 & -2 & 1 \\ 2 & 1 & 6 \end{bmatrix} = -25.$$

Consequently,

$$\begin{bmatrix} 3 & -1 & 2 \\ 1 & -2 & 1 \\ 2 & 1 & 6 \end{bmatrix}^{-1} = -\frac{1}{25} \begin{bmatrix} -13 & 8 & 3 \\ -4 & 14 & -1 \\ 5 & -5 & -5 \end{bmatrix}.$$

Calculating determinants of 3 × 3 matrices

Let

$$A = \begin{bmatrix} a_{11} & a_{12} & a_{13} \\ a_{21} & a_{22} & a_{23} \\ a_{31} & a_{32} & a_{33} \end{bmatrix}.$$

The identity

$$\begin{bmatrix} a_{11} & a_{12} & a_{13} \\ a_{21} & a_{22} & a_{23} \\ a_{31} & a_{32} & a_{33} \end{bmatrix} \begin{bmatrix} \det A_{11} & -\det A_{21} & \det A_{31} \\ -\det A_{12} & \det A_{22} & -\det A_{32} \\ \det A_{13} & -\det A_{23} & \det A_{33} \end{bmatrix} = \begin{bmatrix} \det A & 0 & 0 \\ 0 & \det A & 0 \\ 0 & 0 & \det A \end{bmatrix}$$

gives us a convenient method for evaluating the determinant. Since the entry in the upper left corner of the matrix $\begin{bmatrix} \det A & 0 & 0 \\ 0 & \det A & 0 \\ 0 & 0 & \det A \end{bmatrix}$ is the matrix product of the first row of A and the first column of A^{-1}, we obtain the following formula for the determinant of A:

$$\det A = a_{11} \det A_{11} - a_{12} \det A_{12} + a_{13} \det A_{13}$$

$$= a_{11} \det \begin{bmatrix} a_{22} & a_{23} \\ a_{32} & a_{33} \end{bmatrix} - a_{12} \det \begin{bmatrix} a_{21} & a_{23} \\ a_{31} & a_{33} \end{bmatrix} + a_{13} \det \begin{bmatrix} a_{21} & a_{22} \\ a_{31} & a_{32} \end{bmatrix}$$

$$\underbrace{\hspace{9cm}}_{\text{the expansion across the first row}}$$

If we use the second row of A and the second column of A^{-1} we obtain

$$\det A = -a_{21} \det A_{21} + a_{22} \det A_{22} - a_{23} \det A_{23}$$

$$= -a_{21} \det \begin{bmatrix} a_{12} & a_{13} \\ a_{22} & a_{23} \end{bmatrix} + a_{22} \det \begin{bmatrix} a_{11} & a_{13} \\ a_{31} & a_{33} \end{bmatrix} - a_{23} \det \begin{bmatrix} a_{11} & a_{12} \\ a_{31} & a_{32} \end{bmatrix}$$

$$\underbrace{\hspace{9cm}}_{\text{the expansion across the second row}}$$

and if we use the third row of A and the third column of A^{-1} we get

$$\det A = a_{31} \det A_{31} - a_{32} \det A_{32} + a_{33} \det A_{33}$$

$$= a_{31} \det \begin{bmatrix} a_{12} & a_{13} \\ a_{22} & a_{23} \end{bmatrix} - a_{32} \det \begin{bmatrix} a_{11} & a_{13} \\ a_{21} & a_{23} \end{bmatrix} + a_{33} \det \begin{bmatrix} a_{11} & a_{12} \\ a_{21} & a_{22} \end{bmatrix}.$$

$$\underbrace{\hspace{9cm}}_{\text{the expansion across the third row}}$$

Similarly, if we use

$$\begin{bmatrix} \det A_{11} & -\det A_{21} & \det A_{31} \\ -\det A_{12} & \det A_{22} & -\det A_{32} \\ \det A_{13} & -\det A_{23} & \det A_{33} \end{bmatrix} \begin{bmatrix} a_{11} & a_{12} & a_{13} \\ a_{21} & a_{22} & a_{23} \\ a_{31} & a_{32} & a_{33} \end{bmatrix} = \begin{bmatrix} \det A & 0 & 0 \\ 0 & \det A & 0 \\ 0 & 0 & \det A \end{bmatrix},$$

we get

$$\det A = a_{11} \det A_{11} - a_{21} \det A_{21} + a_{31} \det A_{31}$$

$$= a_{11} \det \begin{bmatrix} a_{22} & a_{23} \\ a_{32} & a_{33} \end{bmatrix} - a_{21} \det \begin{bmatrix} a_{12} & a_{13} \\ a_{22} & a_{23} \end{bmatrix} + a_{31} \det \begin{bmatrix} a_{12} & a_{13} \\ a_{22} & a_{23} \end{bmatrix},$$

$$\underbrace{\hspace{9cm}}_{\text{the expansion down the first column}}$$

$$\det A = -a_{12} \det A_{12} + a_{22} \det A_{22} - a_{32} \det A_{32}$$

$$= -a_{12} \det \begin{bmatrix} a_{21} & a_{23} \\ a_{31} & a_{33} \end{bmatrix} + a_{22} \det \begin{bmatrix} a_{11} & a_{13} \\ a_{31} & a_{33} \end{bmatrix} - a_{32} \det \begin{bmatrix} a_{11} & a_{13} \\ a_{21} & a_{23} \end{bmatrix},$$

$$\underbrace{\hspace{9cm}}_{\text{the expansion down the second column}}$$

and

$$\det A = a_{13} \det A_{13} - a_{23} \det A_{23} + a_{33} \det A_{33}$$

$$= a_{13} \det \begin{bmatrix} a_{21} & a_{22} \\ a_{31} & a_{32} \end{bmatrix} - a_{23} \det \begin{bmatrix} a_{11} & a_{12} \\ a_{31} & a_{32} \end{bmatrix} + a_{33} \det \begin{bmatrix} a_{11} & a_{12} \\ a_{21} & a_{22} \end{bmatrix}.$$

$$\underbrace{\hspace{9cm}}_{\text{the expansion down the third column}}$$

We observe that in all these expansions we multiply a_{ij} by $(-1)^{i+j} \det A_{ij}$. This means that when $i + j$ is an even number we have $(-1)^{i+j} a_{ij} \det A_{ij} = a_{ij} \det A_{ij}$ and that when $i + j$ is an odd number we have $(-1)^{i+j} a_{ij} \det A_{ij} = -a_{ij} \det A_{ij}$.

Example 5.2.7. For the matrix

$$A = \begin{bmatrix} 2 & 4 & 3 \\ 5 & 1 & -1 \\ 7 & 3 & -2 \end{bmatrix}$$

we have

$$\det A = \underbrace{a_{11} \det A_{11} - a_{12} \det A_{12} + a_{13} \det A_{13}}_{\text{the expansion across the first row}}$$
$$= 2 \cdot 1 - 4(-3) + 3 \cdot 8 = 38,$$

$$\det A = \underbrace{-a_{21} \det A_{21} + a_{22} \det A_{22} - a_{23} \det A_{23}}_{\text{the expansion across the second row}}$$
$$= (-5) \cdot (-17) + 1 \cdot (-25) - (-1)(-22) = 38,$$

$$\det A = \underbrace{a_{31} \det A_{31} - a_{32} \det A_{32} + a_{33} \det A_{33}}_{\text{the expansion across the third row}}$$
$$= 7 \cdot (-7) - 3 \cdot (-17) + (-2) \cdot (-18) = 38,$$

$$\det A = \underbrace{a_{11} \det A_{11} - a_{21} \det A_{21} + a_{31} \det A_{31}}_{\text{the expansion down the first column}}$$
$$= 2 \cdot 1 - 5 \cdot (-17) + 7 \cdot (-7) = 38$$

$$\det A = \underbrace{-a_{12} \det A_{12} + a_{22} \det A_{22} - a_{32} \det A_{32}}_{\text{the expansion down the second column}}$$
$$= -4 \cdot (-3) + 1 \cdot (-25) - 3 \cdot (-17) = 38,$$

$$\det A = \underbrace{a_{13} \det A_{13} - a_{23} \det A_{23} + a_{33} \det A_{33}}_{\text{the expansion down the third column}}$$
$$= 3 \cdot 8 - (-1) \cdot (-22) + (-2) \cdot (-18) = 38.$$

When it is necessary to find the determinant of a 3 × 3 matrix, we can use one of the six formulas. Since they all give the same result, it does not matter which one we choose. In practice, we look for the one where the calculations are simpler. In the above example it seems that

$$\det A = \underbrace{-a_{21} \det A_{21} + a_{22} \det A_{22} - a_{23} \det A_{23}}_{\text{the expansion across the second row}}$$

$$= (-5) \cdot (-17) + 1 \cdot (-25) - (-1)(-22) = 38$$

is probably the best choice, since we have 1 and −1 in the second row of A and multiplication by 1 or −1 is easy. The situation is even simpler, if the matrix A has some zero entries.

Example 5.2.8. For the matrix

$$A = \begin{bmatrix} 1 & 2 & 3 \\ 0 & 1 & 2 \\ 3 & 5 & -2 \end{bmatrix}$$

we find

$$\det A = \det A_{22} - 2 \det A_{23} = \det \begin{bmatrix} 1 & 3 \\ 3 & -2 \end{bmatrix} - 2 \det \begin{bmatrix} 1 & 2 \\ 3 & 5 \end{bmatrix} = -11 - 2 \cdot (-1) = -9.$$

For the matrix

$$A = \begin{bmatrix} 1 & 5 & 7 \\ 0 & 3 & -2 \\ 0 & 1 & 2 \end{bmatrix}$$

we have

$$\det A = \det A_{11} = \det \begin{bmatrix} 3 & -2 \\ 1 & 2 \end{bmatrix} = 8.$$

Cramer's Rule

We close this section with the statement of Cramer's Rule for systems of three equations with three variables. In the proof we use the method for calculating the inverse matrix introduced in this section.

Theorem 5.2.9 (Cramer's Rule). *Let*

$$A = \begin{bmatrix} a_{11} & a_{12} & a_{13} \\ a_{21} & a_{22} & a_{23} \\ a_{31} & a_{32} & a_{33} \end{bmatrix}$$

be an arbitrary 3×3 *matrix. If* $\det A \neq 0$, *then the system of equations*

$$\begin{cases} a_{11}x + a_{12}y + a_{13}z = b_1 \\ a_{21}x + a_{22}y + a_{23}z = b_2 \\ a_{31}x + a_{32}y + a_{33}z = b_3 \end{cases}$$

has a unique solution for any real numbers b_1, b_2, *and* b_3. *The solution is*

$$x = \frac{\det \begin{bmatrix} b_1 & a_{12} & a_{13} \\ b_2 & a_{22} & a_{23} \\ b_3 & a_{32} & a_{33} \end{bmatrix}}{\det A}, \quad y = \frac{\det \begin{bmatrix} a_{11} & b_1 & a_{13} \\ a_{21} & b_2 & a_{23} \\ a_{31} & b_3 & a_{33} \end{bmatrix}}{\det A}, \quad z = \frac{\det \begin{bmatrix} a_{11} & a_{12} & b_1 \\ a_{21} & a_{22} & b_2 \\ a_{31} & a_{32} & b_3 \end{bmatrix}}{\det A}.$$

Proof. If $\det A \neq 0$ then the matrix A is invertible and the system has a unique solution, by Theorem 2.3.22. The unique solution is

$$\begin{bmatrix} x \\ y \\ z \end{bmatrix} = \begin{bmatrix} a_{11} & a_{12} & a_{13} \\ a_{21} & a_{22} & a_{23} \\ a_{31} & a_{32} & a_{33} \end{bmatrix}^{-1} \begin{bmatrix} b_1 \\ b_2 \\ b_3 \end{bmatrix}$$

$$= \frac{1}{\det A} \begin{bmatrix} \det A_{11} & -\det A_{21} & \det A_{31} \\ -\det A_{12} & \det A_{22} & -\det A_{32} \\ \det A_{13} & -\det A_{23} & \det A_{33} \end{bmatrix} \begin{bmatrix} b_1 \\ b_2 \\ b_3 \end{bmatrix}$$

$$= \frac{1}{\det A} \begin{bmatrix} b_1 \det A_{11} - b_2 \det A_{21} + b_3 \det A_{31} \\ -b_1 \det A_{12} + b_2 \det A_{22} - b_3 \det A_{32} \\ b_1 \det A_{13} - b_2 \det A_{23} + b_3 \det A_{33} \end{bmatrix}.$$

Since

$$b_1 \det A_{11} - b_2 \det A_{21} + b_3 \det A_{31} = \det \begin{bmatrix} b_1 & a_{12} & a_{13} \\ b_2 & a_{22} & a_{23} \\ b_3 & a_{32} & a_{33} \end{bmatrix},$$

$$-b_1 \det A_{12} + b_2 \det A_{22} - b_3 \det A_{32} = \det \begin{bmatrix} a_{11} & b_1 & a_{13} \\ a_{21} & b_2 & a_{23} \\ a_{31} & b_3 & a_{33} \end{bmatrix},$$

and

$$b_1 \det A_{13} - b_2 \det A_{23} + b_3 \det A_{33} = \det \begin{bmatrix} a_{11} & a_{12} & b_1 \\ a_{21} & a_{22} & b_2 \\ a_{31} & a_{32} & b_3 \end{bmatrix},$$

we obtain

$$x = \frac{\det \begin{bmatrix} b_1 & a_{12} & a_{13} \\ b_2 & a_{22} & a_{23} \\ b_3 & a_{32} & a_{33} \end{bmatrix}}{\det A}, \quad y = \frac{\det \begin{bmatrix} a_{11} & b_1 & a_{13} \\ a_{21} & b_2 & a_{23} \\ a_{31} & b_3 & a_{33} \end{bmatrix}}{\det A}, \quad \text{and} \quad z = \frac{\det \begin{bmatrix} a_{11} & a_{12} & b_1 \\ a_{21} & a_{22} & b_2 \\ a_{31} & a_{32} & b_3 \end{bmatrix}}{\det A}.$$

$\square$

5.2.1   Exercises

For a given matrix A find a matrix B such that $AB = \begin{bmatrix} \det A & 0 & 0 \\ 0 & \det A & 0 \\ 0 & 0 & \det A \end{bmatrix}$, using Theorem 5.2.1.

1. $A = \begin{bmatrix} 2 & 1 & 1 \\ 1 & 3 & 2 \\ 1 & 2 & 1 \end{bmatrix}$

3. $A = \begin{bmatrix} 3 & 4 & 2 \\ 2 & 3 & 1 \\ 4 & 1 & 2 \end{bmatrix}$

2. $A = \begin{bmatrix} 1 & 2 & 1 \\ 2 & 1 & 2 \\ 1 & 1 & 2 \end{bmatrix}$

4. $A = \begin{bmatrix} 3 & 1 & 2 \\ 2 & 1 & 1 \\ 5 & 2 & 3 \end{bmatrix}$

Find the inverse of the given matrix A using Theorem 5.2.3.

5. $A = \begin{bmatrix} 1 & 1 & 1 \\ 2 & 1 & 1 \\ 1 & 1 & 5 \end{bmatrix}$

6. $A = \begin{bmatrix} 4 & 2 & 1 \\ 1 & 1 & 1 \\ 2 & 1 & 3 \end{bmatrix}$

Find the adjoint of the given matrix A using Definition 5.2.4.

7. $A = \begin{bmatrix} 4 & 1 & 1 \\ 2 & 2 & 1 \\ 3 & 1 & 5 \end{bmatrix}$

8. $A = \begin{bmatrix} 1 & 1 & 1 \\ 1 & 0 & 1 \\ 1 & 2 & 0 \end{bmatrix}$

Find the inverse of the given matrix A using (5.10).

9. $A = \begin{bmatrix} 1 & 1 & 1 \\ 1 & 0 & 1 \\ 1 & 1 & 0 \end{bmatrix}$

10. $A = \begin{bmatrix} 1 & 3 & 1 \\ 2 & 1 & 1 \\ 1 & 1 & 2 \end{bmatrix}$

11. Calculate the determinant of the matrix $A = \begin{bmatrix} 4 & 3 & 2 \\ 2 & 1 & 5 \\ 1 & 1 & 3 \end{bmatrix}$ using the expansion across the first row.

12. Calculate the determinant of the matrix $A = \begin{bmatrix} 4 & 3 & 2 \\ 2 & 1 & 5 \\ 1 & 1 & 3 \end{bmatrix}$ using the expansion across the third row.

13. Calculate the determinant of the matrix $A = \begin{bmatrix} 4 & 3 & 2 \\ 2 & 1 & 5 \\ 1 & 1 & 3 \end{bmatrix}$ using the expansion down the second column.

14. Calculate the determinant of the matrix $A = \begin{bmatrix} 4 & 1 & 1 \\ 1 & 0 & 1 \\ 2 & 3 & 1 \end{bmatrix}$ using the expansion down the first column.

15. Calculate the determinant of the matrix $A = \begin{bmatrix} 4 & 1 & 1 \\ 1 & 0 & 1 \\ 2 & 3 & 1 \end{bmatrix}$ using the expansion across the second row.

16. Calculate the determinant of the matrix $A = \begin{bmatrix} 4 & 1 & 1 \\ 1 & 0 & 1 \\ 2 & 3 & 1 \end{bmatrix}$ using the expansion across the first row.

17. Calculate the determinant of the matrix $A = \begin{bmatrix} 2 & 1 & 1 \\ 3 & 5 & 1 \\ 0 & 3 & 0 \end{bmatrix}$ using the expansion across the third row.

18. Calculate the determinant of the matrix $A = \begin{bmatrix} 2 & 1 & 1 \\ 3 & 5 & 1 \\ 0 & 3 & 0 \end{bmatrix}$ using the expansion across the second row.

19. Calculate the determinant of the matrix $A = \begin{bmatrix} 4 & 0 & 1 \\ 3 & 7 & 1 \\ 1 & 0 & 2 \end{bmatrix}$ using the expansion down the second column.

20. Calculate the determinant of the matrix $A = \begin{bmatrix} 4 & 0 & 1 \\ 3 & 7 & 1 \\ 1 & 0 & 2 \end{bmatrix}$ using the expansion down the first column.

Solve the given system of linear equations using Theorem 5.2.9.

21. $\begin{cases} 2x + y + 2z = 1 \\ x + 3y + z = 0 \\ 3x + 2y + 4z = 0 \end{cases}$

22. $\begin{cases} x + 4y + z = 0 \\ 2x + 3y + z = 1 \\ x + 2y + 3z = 1 \end{cases}$

23. $\begin{cases} x + y + 2z = 0 \\ x + 2y + z = 1 \\ 3x + 2y + 2z = 0 \end{cases}$

24. $\begin{cases} x + z = 0 \\ y + 2z = 0 \\ x + 2y + 3z = 5 \end{cases}$

5.3 Linear dependence of three vectors in $\mathbb{R}^3$

Linear dependence and independence of vectors was first considered in Section 3.1 in the context of vectors in $\mathbb{R}^2$. In Section 4.1 we observed that the definition used in $\mathbb{R}^2$ makes perfect sense in $\mathbb{R}^3$. However, in $\mathbb{R}^3$ there are possibilities that did not exist in $\mathbb{R}^2$. In particular, in $\mathbb{R}^3$ it makes sense to talk about linear dependence and independence of three vectors.

If the vector $\mathbf{u}$ is in Span$\{\mathbf{v}, \mathbf{w}\}$, then there are real numbers s and t such that $\mathbf{u} = s\mathbf{v} + t\mathbf{w}$ which makes the vector $\mathbf{u}$ dependent of the vectors $\mathbf{v}$ and $\mathbf{w}$. In the same way, if the vector $\mathbf{v}$ is in Span$\{\mathbf{u}, \mathbf{w}\}$, then the vector $\mathbf{v}$ is dependent of the vectors $\mathbf{u}$ and $\mathbf{w}$ and, if the vector $\mathbf{w}$ is in Span$\{\mathbf{u}, \mathbf{v}\}$, then the vector $\mathbf{w}$ is dependent of the vectors $\mathbf{u}$ and $\mathbf{w}$. This suggests the following extension of the definition of linear dependence of vectors to three vectors.

Definition 5.3.1. Vectors $\mathbf{u}$, $\mathbf{v}$, and $\mathbf{w}$ in $\mathbb{R}^3$ are ***linearly dependent*** if at least one of the following conditions holds:

(a) the vector $\mathbf{u}$ is in Span$\{\mathbf{v}, \mathbf{w}\}$;

(b) the vector $\mathbf{v}$ is in Span$\{\mathbf{u}, \mathbf{w}\}$;

(c) the vector $\mathbf{w}$ is in Span$\{\mathbf{u}, \mathbf{v}\}$.

The definition says that the vectors $\mathbf{u}$, $\mathbf{v}$, and $\mathbf{w}$ in $\mathbb{R}^3$ are linearly dependent if there are real numbers a and b such that $\mathbf{u} = a\mathbf{v} + b\mathbf{w}$ or there are real numbers c and d such that $\mathbf{v} = c\mathbf{u} + d\mathbf{w}$ or there are real numbers e and f such that $\mathbf{w} = e\mathbf{u} + f\mathbf{v}$.

Example 5.3.2. Since

$$\begin{bmatrix} -4 \\ 2 \\ 3 \end{bmatrix} = 2 \begin{bmatrix} 1 \\ 1 \\ 0 \end{bmatrix} - 3 \begin{bmatrix} 2 \\ 0 \\ -1 \end{bmatrix},$$

the vector $\begin{bmatrix} -4 \\ 2 \\ 3 \end{bmatrix}$ is in Span$\left\{ \begin{bmatrix} 1 \\ 1 \\ 0 \end{bmatrix}, \begin{bmatrix} 2 \\ 0 \\ -1 \end{bmatrix} \right\}$ and consequently the vectors $\begin{bmatrix} 1 \\ 1 \\ 0 \end{bmatrix}, \begin{bmatrix} -2 \\ 0 \\ 1 \end{bmatrix}$,

and $\begin{bmatrix} -4 \\ 2 \\ 3 \end{bmatrix}$ are linearly dependent.

The following theorem is a version of Theorem 4.1.8 for three vectors. Note the similarities and the differences between these two theorems.

> **Theorem 5.3.3.** *Let* $\mathbf{u}$, $\mathbf{v}$, *and* $\mathbf{w}$ *be vectors in* $\mathbb{R}^3$. *The following conditions are equivalent*
>
> (a) *The vectors* $\mathbf{u}, \mathbf{v}$ *and* $\mathbf{w}$ *are linearly dependent;*
>
> (b) *The equation*
> $$x\mathbf{u} + y\mathbf{v} + z\mathbf{w} = \mathbf{0}$$
> *has a nontrivial solution, that is, a solution different from the trivial solution* $x = y = z = 0$;
>
> (c)
> $$\det\begin{bmatrix}\mathbf{u} & \mathbf{v} & \mathbf{w}\end{bmatrix} = 0.$$

Proof. First we prove that conditions (a) and (b) are equivalent. If the vectors $\mathbf{u}$, $\mathbf{v}$, and $\mathbf{w}$ are linearly dependent, then one of the vectors is in the span of the remaining two. If $\mathbf{u}$ is in Span$\{\mathbf{v}, \mathbf{w}\}$, then $\mathbf{u} = s\mathbf{v} + t\mathbf{w}$ for some real numbers s and t. This means that

$$-\mathbf{u} + s\mathbf{v} + t\mathbf{w} = \mathbf{0},$$

so the equation $x\mathbf{u} + y\mathbf{v} + z\mathbf{w} = \mathbf{0}$ has a nontrivial solution, namely $x = -1$, $y = s$, and $z = t$. The cases when $\mathbf{v}$ is in Span$\{\mathbf{u}, \mathbf{w}\}$ or $\mathbf{w}$ is in Span$\{\mathbf{u}, \mathbf{v}\}$ are proved in the same way.

Now suppose that $x\mathbf{u} + y\mathbf{v} + z\mathbf{w} = \mathbf{0}$ for some x, y, and z, not all equal to 0. If $x \neq 0$, then

$$\mathbf{u} = -\frac{y}{x}\mathbf{v} - \frac{z}{x}\mathbf{w},$$

which means that $\mathbf{u}$ is in Span$\{\mathbf{v}, \mathbf{w}\}$ and thus the vectors $\mathbf{u}, \mathbf{v}$ and $\mathbf{w}$ are linearly dependent. If $y \neq 0$ or $z \neq 0$, then we modify the argument in the obvious way.

Now we prove that conditions (a) and (c) are equivalent.

If $\mathbf{u}$ is in Span$\{\mathbf{v}, \mathbf{w}\}$, then $\mathbf{u} = s\mathbf{v} + t\mathbf{w}$ for some real numbers s and t and consequently

$$\det\begin{bmatrix}\mathbf{u} & \mathbf{v} & \mathbf{w}\end{bmatrix} = \mathbf{u} \cdot (\mathbf{v} \times \mathbf{w}) = (s\mathbf{v} + t\mathbf{w}) \cdot (\mathbf{v} \times \mathbf{w}) = s(\mathbf{v} \cdot (\mathbf{v} \times \mathbf{w})) + t(\mathbf{w} \cdot (\mathbf{v} \times \mathbf{w})) = 0.$$

If $\mathbf{v}$ is in Span$\{\mathbf{u}, \mathbf{w}\}$, then $\mathbf{v} = s\mathbf{u} + t\mathbf{w}$ for some real numbers s and t and consequently

$$\det\begin{bmatrix}\mathbf{u} & \mathbf{v} & \mathbf{w}\end{bmatrix} = \mathbf{u} \cdot (\mathbf{v} \times \mathbf{w}) = \mathbf{u} \cdot ((s\mathbf{u} + t\mathbf{w}) \times \mathbf{w}) = s(\mathbf{u} \cdot (\mathbf{u} \times \mathbf{w})) + t(\mathbf{u} \cdot (\mathbf{w} \times \mathbf{w})) = 0.$$

If $\mathbf{w}$ is in Span$\{\mathbf{u}, \mathbf{v}\}$, then $\mathbf{w} = s\mathbf{u} + t\mathbf{v}$ for some real numbers s and t and consequently

$$\det\begin{bmatrix}\mathbf{u} & \mathbf{v} & \mathbf{w}\end{bmatrix} = \mathbf{u} \cdot (\mathbf{v} \times \mathbf{w}) = \mathbf{u} \cdot (\mathbf{v} \times (s\mathbf{u} + t\mathbf{v})) = s(\mathbf{u} \cdot (\mathbf{v} \times \mathbf{u})) + t(\mathbf{u} \cdot (\mathbf{v} \times \mathbf{v})) = 0.$$

Now suppose that $\det\begin{bmatrix}\mathbf{u} & \mathbf{v} & \mathbf{w}\end{bmatrix} = \mathbf{u} \cdot (\mathbf{v} \times \mathbf{w}) = 0$. If $\mathbf{v}$ and $\mathbf{w}$ are linearly independent, then $\mathbf{u} = s\mathbf{v} + t\mathbf{w}$, according to Theorem 5.1.7. But this means that $\mathbf{u}$ is in Span$\{\mathbf{v}, \mathbf{w}\}$. If $\mathbf{v}$ and $\mathbf{w}$ are linearly dependent, then the vector $\mathbf{v}$ is in Span$\{\mathbf{w}\}$ and consequently in Span$\{\mathbf{u}, \mathbf{w}\}$ or the vector $\mathbf{w}$ is in Span$\{\mathbf{v}\}$ and consequently in Span$\{\mathbf{u}, \mathbf{v}\}$. In any case, the vectors $\mathbf{u}, \mathbf{v}$ and $\mathbf{w}$ are linearly dependent. $\qquad\square$

Example 5.3.4. Show that the vectors $\begin{bmatrix} 1 \\ 0 \\ 1 \end{bmatrix}$, $\begin{bmatrix} 2 \\ -1 \\ 3 \end{bmatrix}$, and $\begin{bmatrix} 1 \\ 1 \\ 0 \end{bmatrix}$ are linearly dependent.

Solution. Since

$$\begin{bmatrix} 1 \\ 0 \\ 1 \end{bmatrix} \times \begin{bmatrix} 2 \\ -1 \\ 3 \end{bmatrix} = \begin{bmatrix} \det \begin{bmatrix} 0 & -1 \\ 1 & 3 \end{bmatrix} \\ -\det \begin{bmatrix} 1 & 2 \\ 1 & 3 \end{bmatrix} \\ \det \begin{bmatrix} 1 & 2 \\ 0 & -1 \end{bmatrix} \end{bmatrix} = \begin{bmatrix} 1 \\ -1 \\ -1 \end{bmatrix},$$

we have

$$\det \begin{bmatrix} 1 & 1 & 2 \\ 1 & 0 & -1 \\ 0 & 1 & 3 \end{bmatrix} = \begin{bmatrix} 1 \\ 1 \\ 0 \end{bmatrix} \cdot \left(\begin{bmatrix} 1 \\ 0 \\ 1 \end{bmatrix} \times \begin{bmatrix} 2 \\ -1 \\ 3 \end{bmatrix} \right) = \begin{bmatrix} 1 \\ 1 \\ 0 \end{bmatrix} \cdot \begin{bmatrix} 1 \\ -1 \\ -1 \end{bmatrix} = 0$$

and thus the vectors $\begin{bmatrix} 1 \\ 0 \\ 1 \end{bmatrix}$, $\begin{bmatrix} 2 \\ -1 \\ 3 \end{bmatrix}$, and $\begin{bmatrix} 1 \\ 1 \\ 0 \end{bmatrix}$ are linearly dependent.

We can also show dependence of these vectors by observing that

$$3 \begin{bmatrix} 1 \\ 0 \\ 1 \end{bmatrix} - \begin{bmatrix} 2 \\ -1 \\ 3 \end{bmatrix} - \begin{bmatrix} 1 \\ 1 \\ 0 \end{bmatrix} = \begin{bmatrix} 0 \\ 0 \\ 0 \end{bmatrix}.$$

□

The next theorem offers a somewhat different perspective on linear dependence of three vectors.

Theorem 5.3.5. *Vectors* **u**, **v**, *and* **w** *in* $\mathbb{R}^3$ *are linearly dependent if and only if one of the following conditions holds:*

(a) $\mathbf{u} = \mathbf{0}$;

(b) $\mathbf{u} \neq \mathbf{0}$ *and the equation*

$$x\mathbf{u} = \mathbf{v}$$

has a solution;

(c) *the vectors* **u** *and* **v** *are linearly independent and the equation*

$$x\mathbf{u} + y\mathbf{v} = \mathbf{w}$$

has a solution.

Proof. If one of the conditions (a), (b), or (c), holds, then it is clear that the vectors **u**, **v**, and **w** are linearly dependent.

Now assume that the vectors **u**, **v** and **w** are linearly dependent.

If the vectors **u** and **v** are linearly dependent, then we have (a) or (b), by Theorem 5.3.5. If the vectors **u** and **v** are linearly independent, then **w** is in Span{**u**, **v**} or **u** is in Span{**v**, **w**} or **v** is in Span{**u**, **w**}.

If **w** is in Span{**u**, **v**}, we have nothing to prove.

If **u** is in Span{**v**, **w**}, then there are real numbers b and c such that $\mathbf{u} = b\mathbf{v} + c\mathbf{w}$. We must have $c \neq 0$, because the vectors **u** and **v** are linearly independent. This means that we have $\mathbf{w} = \frac{1}{c}\mathbf{u} - \frac{b}{c}\mathbf{v}$.

The case when **v** is in Span{**u**, **w**} is similar. □

Sometimes it is beneficial to think of linear dependence of vectors **u**, **v**, and **w** in terms of properties of the matrix $\begin{bmatrix} \mathbf{u} & \mathbf{v} & \mathbf{w} \end{bmatrix}$. We saw the first indication of that in part (c) of Theorem 5.3.3. The next theorem connects linear dependence of vectors **u**, **v**, and **w** with the shape of the reduced row echelon form of the matrix $\begin{bmatrix} \mathbf{u} & \mathbf{v} & \mathbf{w} \end{bmatrix}$. The theorem is a direct consequence of Theorem 5.3.5.

Theorem 5.3.6. *Vectors* **u**, **v**, *and* **w** *in* $\mathbb{R}^3$ *are linearly dependent if and only if one of the following conditions holds:*

(a) *The first column of the reduced row echelon form of the matrix* $\begin{bmatrix} \mathbf{u} & \mathbf{v} & \mathbf{w} \end{bmatrix}$

is $\begin{bmatrix} 0 \\ 0 \\ 0 \end{bmatrix}$;

(b) *The first two columns of the reduced row echelon form of the matrix*

$\begin{bmatrix} \mathbf{u} & \mathbf{v} & \mathbf{w} \end{bmatrix}$ *are* $\begin{bmatrix} 1 & x \\ 0 & 0 \\ 0 & 0 \end{bmatrix}$;

(c) *The reduced row echelon form of the matrix* $\begin{bmatrix} \mathbf{u} & \mathbf{v} & \mathbf{w} \end{bmatrix}$ *is* $\begin{bmatrix} 1 & 0 & x \\ 0 & 1 & y \\ 0 & 0 & 0 \end{bmatrix}$.

Example 5.3.7. Show that the vectors $\begin{bmatrix} 3 \\ 1 \\ 1 \end{bmatrix}$, $\begin{bmatrix} 2 \\ 2 \\ 1 \end{bmatrix}$, and $\begin{bmatrix} 1 \\ 3 \\ 1 \end{bmatrix}$ are linearly dependent

and write $\begin{bmatrix} 1 \\ 3 \\ 1 \end{bmatrix}$ as a linear combination of the vectors $\begin{bmatrix} 3 \\ 1 \\ 1 \end{bmatrix}$ and $\begin{bmatrix} 2 \\ 2 \\ 1 \end{bmatrix}$.

Solution. We note that the vectors $\begin{bmatrix} 3 \\ 1 \\ 1 \end{bmatrix}$ and $\begin{bmatrix} 2 \\ 2 \\ 1 \end{bmatrix}$ are linearly independent. We need to show that the equation

$$x\begin{bmatrix} 3 \\ 1 \\ 1 \end{bmatrix} + y\begin{bmatrix} 2 \\ 2 \\ 1 \end{bmatrix} = \begin{bmatrix} 1 \\ 3 \\ 1 \end{bmatrix}$$

has a solution. Since

$$\begin{bmatrix} 3 & 2 & 1 \\ 1 & 2 & 3 \\ 1 & 1 & 1 \end{bmatrix} \sim \begin{bmatrix} 1 & 0 & -1 \\ 0 & 1 & 2 \\ 0 & 0 & 0 \end{bmatrix},$$

the solution is $x = -1$ and $y = 2$. $\qquad\square$

Example 5.3.8. Consider the vectors $\begin{bmatrix} 2 \\ -1 \\ 1 \end{bmatrix}$, $\begin{bmatrix} 1 \\ 1 \\ 1 \end{bmatrix}$, and $\begin{bmatrix} 1 \\ 2a \\ a+2 \end{bmatrix}$. Find a number a such that these vectors are linearly dependent and then write the vector $\begin{bmatrix} 1 \\ 2a \\ a+2 \end{bmatrix}$ as a linear combination of the vectors $\begin{bmatrix} 2 \\ -1 \\ 1 \end{bmatrix}$ and $\begin{bmatrix} 1 \\ 1 \\ 1 \end{bmatrix}$.

Solution. Since the vectors $\begin{bmatrix} 2 \\ -1 \\ 1 \end{bmatrix}$ and $\begin{bmatrix} 1 \\ 1 \\ 1 \end{bmatrix}$ are linearly independent we need to show that the equation

$$x\begin{bmatrix} 2 \\ -1 \\ 1 \end{bmatrix} + y\begin{bmatrix} 1 \\ 1 \\ 1 \end{bmatrix} = \begin{bmatrix} 1 \\ 2a \\ a+2 \end{bmatrix}$$

has a solution. We find that

$$\begin{bmatrix} 2 & 1 & 1 \\ -1 & 1 & 2a \\ 1 & 1 & a+2 \end{bmatrix} \sim \begin{bmatrix} 1 & 0 & -a-1 \\ 0 & 1 & 2a+3 \\ 0 & 0 & -a-4 \end{bmatrix}.$$

Consequently, the equation has a solution if and only if $-a - 4 = 0$ or $a = -4$. If $a = -4$, then

$$\begin{bmatrix} 2 & 1 & 1 \\ -1 & 1 & 2a \\ 1 & 1 & a+2 \end{bmatrix} \sim \begin{bmatrix} 1 & 0 & 3 \\ 0 & 1 & -5 \\ 0 & 0 & 0 \end{bmatrix},$$

which means that the solution is $x = 3$ and $y = -5$. $\qquad\square$

Definition 5.3.9. If vectors **u**, **v**, and **w** are not linearly dependent, then we say that they are ***linearly independent.***

In other words, vectors **u**, **v**, and **w** are linearly independent, if the vector **u** is not in Span{**v**, **w**}, the vector **v** is not in Span{**u**, **w**}, and the vector **w** is not in Span{**u**, **v**}. The following theorem is a direct consequence of Theorems 5.3.3 and 2.3.17.

Theorem 5.3.10. *Let* **u**, **v**, *and* **w** *be vectors from* $\mathbb{R}^3$. *The following conditions are equivalent:*

(a) **u**, **v**, *and* **w** *are linearly independent;*

(b) *The only solution of the equation*

$$x\mathbf{u} + y\mathbf{v} + z\mathbf{w} = \mathbf{0}$$

is the trivial solution $x = y = z = 0$;

(c) $\det\begin{bmatrix} \mathbf{u} & \mathbf{v} & \mathbf{w} \end{bmatrix} \neq 0$;

(d) *The matrix* $\begin{bmatrix} \mathbf{u} & \mathbf{v} & \mathbf{w} \end{bmatrix}$ *is invertible;*

(e) *The reduced row echelon form of the matrix* $\begin{bmatrix} \mathbf{u} & \mathbf{v} & \mathbf{w} \end{bmatrix}$ *is* $\begin{bmatrix} 1 & 0 & 0 \\ 0 & 1 & 0 \\ 0 & 0 & 1 \end{bmatrix}$.

Note that if three vectors in $\mathbb{R}^3$ are linearly independent, then any two of them are linearly independent. However, the converse is not true. In Example 5.3.4 we show that the vectors

$$\begin{bmatrix} 1 \\ 0 \\ 1 \end{bmatrix}, \begin{bmatrix} 2 \\ -1 \\ 3 \end{bmatrix}, \text{ and } \begin{bmatrix} 1 \\ 1 \\ 0 \end{bmatrix}$$

are linearly dependent, but any two of them are linearly independent.

Example 5.3.11. Show that the vectors

$$\begin{bmatrix} 1 \\ 0 \\ 1 \end{bmatrix}, \begin{bmatrix} 2 \\ -1 \\ 3 \end{bmatrix}, \text{ and } \begin{bmatrix} 1 \\ 1 \\ 1 \end{bmatrix}.$$

are linearly independent.

Solution. Since

$$\begin{bmatrix} 1 \\ 0 \\ 1 \end{bmatrix} \times \begin{bmatrix} 2 \\ -1 \\ 3 \end{bmatrix} = \begin{bmatrix} 1 \\ -1 \\ -1 \end{bmatrix} \quad \text{and} \quad \begin{bmatrix} 1 \\ 1 \\ 1 \end{bmatrix} \cdot \begin{bmatrix} 1 \\ -1 \\ -1 \end{bmatrix} = -1 \neq 0,$$

we have

$$\det \begin{bmatrix} 1 & 2 & 1 \\ 0 & -1 & 1 \\ 1 & 3 & 1 \end{bmatrix} = -1$$

and the vectors $\begin{bmatrix} 1 \\ 0 \\ 1 \end{bmatrix}$, $\begin{bmatrix} 2 \\ -1 \\ 3 \end{bmatrix}$, and $\begin{bmatrix} 1 \\ 1 \\ 1 \end{bmatrix}$ are linearly independent.

We can also show linear independence using

$$\begin{bmatrix} 1 & 2 & 1 \\ 0 & -1 & 1 \\ 1 & 3 & 1 \end{bmatrix} \sim \begin{bmatrix} 1 & 0 & 0 \\ 0 & 1 & 0 \\ 0 & 0 & 1 \end{bmatrix},$$

which means that the equation

$$x \begin{bmatrix} 1 \\ 0 \\ 1 \end{bmatrix} + y \begin{bmatrix} 2 \\ -1 \\ 3 \end{bmatrix} + z \begin{bmatrix} 1 \\ 1 \\ 1 \end{bmatrix} = \begin{bmatrix} 0 \\ 0 \\ 0 \end{bmatrix}$$

has only the trivial solution, that is, $x = y = z = 0$.   □

Bases in $\mathbb{R}^3$

In Chapter 3 we introduced the notion of a basis in $\mathbb{R}^2$ as a pair of linearly independent vectors $\{\mathbf{a}, \mathbf{b}\}$ in $\mathbb{R}^2$ such that for any $\mathbf{c}$ in $\mathbb{R}^2$ we have $\mathbf{c} = x\mathbf{a} + y\mathbf{b}$ for some numbers x and y. Now we are going to define a basis in $\mathbb{R}^3$. As expected, this time three linearly independent vectors are needed, but otherwise the definition is the same.

Definition 5.3.12. A set of three vectors $\{\mathbf{a}, \mathbf{b}, \mathbf{c}\}$ in $\mathbb{R}^3$ is called a **basis** in $\mathbb{R}^3$ if the vectors satisfy the following two conditions:

  (i) $\mathbf{a}, \mathbf{b}$ and $\mathbf{c}$ are linearly independent;

  (ii) For every vector $\mathbf{d}$ in $\mathbb{R}^3$ there exist real numbers x, y, and z such that $\mathbf{d} = x\mathbf{a} + y\mathbf{b} + z\mathbf{c}$.

An expression of the form $x\mathbf{a} + y\mathbf{b} + z\mathbf{c}$ is called a **linear combination** of vectors $\mathbf{a}, \mathbf{b}$, and $\mathbf{c}$. According to the above definition, linearly independent vectors $\mathbf{a}, \mathbf{b}$, and

c form a basis in $\mathbb{R}^3$, if every vector in $\mathbb{R}^3$ can be written as a linear combination of vectors **a**, **b**, and **c**.

Definition 5.3.13. Let **a**, **b**, and **c** be vectors in $\mathbb{R}^3$. The set of all possible linear combinations of the form $x\mathbf{a} + y\mathbf{b} + z\mathbf{c}$ is denoted by Span$\{\mathbf{a}, \mathbf{b}, \mathbf{c}\}$ and is called the *vector subspace* spanned by the vectors **a**, **b**, and **c**. In symbols,

$$\text{Span}\{\mathbf{a}, \mathbf{b}, \mathbf{c}\} = \{x\mathbf{a} + y\mathbf{b} + z\mathbf{c} : x, y, z \text{ in } \mathbb{R}\}.$$

Note that instead of the condition (2) in Definition 5.3.12 we could simply say that Span$\{\mathbf{a}, \mathbf{b}, \mathbf{c}\} = \mathbb{R}^3$.

Theorem 5.3.14. *Let* $\mathbf{a}, \mathbf{b}, \mathbf{c}, \mathbf{u}, \mathbf{v}, \mathbf{w}$ *be vectors in* $\mathbb{R}^3$. *The following two conditions are equivalent:*

(a) Span$\{\mathbf{a}, \mathbf{b}, \mathbf{c}\}$ = Span$\{\mathbf{u}, \mathbf{v}, \mathbf{w}\}$;

(b) $\mathbf{a}, \mathbf{b}, \mathbf{c}$ *are elements of* Span$\{\mathbf{u}, \mathbf{v}, \mathbf{w}\}$ *and* $\mathbf{u}, \mathbf{v}, \mathbf{w}$ *are elements of* Span$\{\mathbf{a}, \mathbf{b}, \mathbf{c}\}$.

Proof. The proof is similar to the proof of Theorem 4.1.15 and we leave it as an exercise. □

As in the case of bases in $\mathbb{R}^2$, the representation of any vector from $\mathbb{R}^3$ as a linear combination of vectors from a basis is unique.

Theorem 5.3.15. *Let* $\{\mathbf{a}, \mathbf{b}, \mathbf{c}\}$ *be a basis in* $\mathbb{R}^3$ *and let* **d** *be an arbitrary vector in* $\mathbb{R}^3$. *The real numbers* x, y, *and* z *such that*

$$\mathbf{d} = x\mathbf{a} + y\mathbf{b} + z\mathbf{c}$$

are uniquely determined by the vector **d**.

Proof. If
$$\mathbf{d} = x\mathbf{a} + y\mathbf{b} + z\mathbf{c}$$
and
$$\mathbf{d} = x'\mathbf{a} + y'\mathbf{b} + z'\mathbf{c},$$
then
$$\mathbf{0} = (x' - x)\mathbf{a} + (y' - y)\mathbf{b} + (z' - z)\mathbf{c},$$
which implies that $x' - x = y' - y = z' - z = 0$, because the vectors **a**, **b**, and **c** are linearly independent. □

Definition 5.3.16. Let $\{\mathbf{a}, \mathbf{b}, \mathbf{c}\}$ be a basis in $\mathbb{R}^3$ and let $\mathbf{d}$ be an arbitrary vector in $\mathbb{R}^3$. The unique real numbers x, y, and z such that

$$\mathbf{d} = x\mathbf{a} + y\mathbf{b} + z\mathbf{c}$$

are called the **coordinates** of $\mathbf{d}$ in the basis $\{\mathbf{a}, \mathbf{b}, \mathbf{c}\}$.

In the definition of a basis of $\mathbb{R}^3$ we are assuming that the vectors are linearly independent and that every vector in $\mathbb{R}^3$ is a linear combination of vectors from the basis. It turns out that it is sufficient to assume only one of the conditions. Actually, the conditions are equivalent, which is a consequence of the following important theorem.

Theorem 5.3.17. *Let $\mathbf{c}_1$, $\mathbf{c}_2$, $\mathbf{c}_3$ be linearly independent vectors in $\mathbb{R}^3$ and let $\mathbf{b}_1$, $\mathbf{b}_2$, $\mathbf{b}_3$ be arbitrary vectors in $\mathbb{R}^3$. If the vectors $\mathbf{c}_1$, $\mathbf{c}_2$, $\mathbf{c}_3$ are elements of the vector subspace* $\mathrm{Span}\{\mathbf{b}_1, \mathbf{b}_2, \mathbf{b}_3\}$, *then*

$$\mathrm{Span}\{\mathbf{c}_1, \mathbf{c}_2, \mathbf{c}_3\} = \mathrm{Span}\{\mathbf{b}_1, \mathbf{b}_2, \mathbf{b}_3\} = \mathbb{R}^3$$

and the vectors $\mathbf{b}_1$, $\mathbf{b}_2$, $\mathbf{b}_3$ are linearly independent.

Proof. Since the vectors $\mathbf{c}_1$, $\mathbf{c}_2$, $\mathbf{c}_3$ are elements of $\mathrm{Span}\{\mathbf{b}_1, \mathbf{b}_2, \mathbf{b}_3\}$, for $j = 1, 2, 3$ there are real numbers a_{1j}, a_{2j}, a_{3j} such that

$$\mathbf{c}_j = a_{1j}\mathbf{b}_1 + a_{2j}\mathbf{b}_2 + a_{3j}\mathbf{b}_3.$$

This can be written as a matrix product equation

$$\begin{bmatrix} \mathbf{c}_1 & \mathbf{c}_2 & \mathbf{c}_3 \end{bmatrix} = \begin{bmatrix} \mathbf{b}_1 & \mathbf{b}_2 & \mathbf{b}_3 \end{bmatrix} \begin{bmatrix} \mathbf{a}_1 & \mathbf{a}_2 & \mathbf{a}_3 \end{bmatrix}, \tag{5.11}$$

where $\mathbf{a}_j = \begin{bmatrix} a_{1j} \\ a_{2j} \\ a_{3j} \end{bmatrix}$. Consequently, if $\begin{bmatrix} x_1 \\ x_2 \\ x_3 \end{bmatrix}$ is an arbitrary vector, then

$$\begin{bmatrix} \mathbf{c}_1 & \mathbf{c}_2 & \mathbf{c}_3 \end{bmatrix} \begin{bmatrix} x_1 \\ x_2 \\ x_3 \end{bmatrix} = \begin{bmatrix} \mathbf{b}_1 & \mathbf{b}_2 & \mathbf{b}_3 \end{bmatrix} \begin{bmatrix} \mathbf{a}_1 & \mathbf{a}_2 & \mathbf{a}_3 \end{bmatrix} \begin{bmatrix} x_1 \\ x_2 \\ x_3 \end{bmatrix}.$$

If $\begin{bmatrix} \mathbf{a}_1 & \mathbf{a}_2 & \mathbf{a}_3 \end{bmatrix} \begin{bmatrix} x_1 \\ x_2 \\ x_3 \end{bmatrix} = \begin{bmatrix} 0 \\ 0 \\ 0 \end{bmatrix}$, then we have $\begin{bmatrix} \mathbf{c}_1 & \mathbf{c}_2 & \mathbf{c}_3 \end{bmatrix} \begin{bmatrix} x_1 \\ x_2 \\ x_3 \end{bmatrix} = \begin{bmatrix} 0 \\ 0 \\ 0 \end{bmatrix}$ and consequently $\begin{bmatrix} x_1 \\ x_2 \\ x_3 \end{bmatrix} = \begin{bmatrix} 0 \\ 0 \\ 0 \end{bmatrix}$, because the vectors $\mathbf{c}_1$, $\mathbf{c}_2$, $\mathbf{c}_3$ are linearly independent. This implies that the 3×3 matrix $[\mathbf{a}_1 \ \mathbf{a}_2 \ \mathbf{a}_3]$ is invertible, by Theorem 2.3.17. Moreover, from (5.11) we obtain

$$[\mathbf{c}_1 \ \mathbf{c}_2 \ \mathbf{c}_3][\mathbf{a}_1 \ \mathbf{a}_2 \ \mathbf{a}_3]^{-1} = [\mathbf{b}_1 \ \mathbf{b}_2 \ \mathbf{b}_3],$$

which means that the vectors $\mathbf{b}_1, \mathbf{b}_2, \mathbf{b}_3$ are elements of $\text{Span}\{\mathbf{c}_1, \mathbf{c}_2, \mathbf{c}_3\}$ and, consequently, we have

$$\text{Span}\{\mathbf{c}_1, \mathbf{c}_2, \mathbf{c}_3\} = \text{Span}\{\mathbf{b}_1, \mathbf{b}_2, \mathbf{b}_3\}$$

by Theorem 5.3.14.

The equation

$$[\mathbf{b}_1 \ \mathbf{b}_2 \ \mathbf{b}_3]\begin{bmatrix} x_1 \\ x_2 \\ x_3 \end{bmatrix} = \begin{bmatrix} 0 \\ 0 \\ 0 \end{bmatrix}$$

is equivalent to the equation

$$[\mathbf{c}_1 \ \mathbf{c}_2 \ \mathbf{c}_3][\mathbf{a}_1 \ \mathbf{a}_2 \ \mathbf{a}_3]^{-1}\begin{bmatrix} x_1 \\ x_2 \\ x_3 \end{bmatrix} = \begin{bmatrix} 0 \\ 0 \\ 0 \end{bmatrix}.$$

Because the vectors $\mathbf{c}_1, \mathbf{c}_2, \mathbf{c}_3$ are linearly independent we get

$$[\mathbf{a}_1 \ \mathbf{a}_2 \ \mathbf{a}_3]^{-1}\begin{bmatrix} x_1 \\ x_2 \\ x_3 \end{bmatrix} = \begin{bmatrix} 0 \\ 0 \\ 0 \end{bmatrix}.$$

Now we multiply both sides by $[\mathbf{a}_1 \ \mathbf{a}_2 \ \mathbf{a}_3]$ and obtain

$$\begin{bmatrix} x_1 \\ x_2 \\ x_3 \end{bmatrix} = [\mathbf{a}_1 \ \mathbf{a}_2 \ \mathbf{a}_3][\mathbf{a}_1 \ \mathbf{a}_2 \ \mathbf{a}_3]^{-1}\begin{bmatrix} x_1 \\ x_2 \\ x_3 \end{bmatrix} = [\mathbf{a}_1 \ \mathbf{a}_2 \ \mathbf{a}_3]\begin{bmatrix} 0 \\ 0 \\ 0 \end{bmatrix} = \begin{bmatrix} 0 \\ 0 \\ 0 \end{bmatrix}.$$

This means that the vectors $\mathbf{b}_1, \mathbf{b}_2, \mathbf{b}_3$ are linearly independent.
Since all linearly independent vectors $\mathbf{c}_1, \mathbf{c}_2, \mathbf{c}_3$ are in

$$\text{Span}\left\{ \begin{bmatrix} 1 \\ 0 \\ 0 \end{bmatrix}, \begin{bmatrix} 0 \\ 1 \\ 0 \end{bmatrix}, \begin{bmatrix} 0 \\ 0 \\ 1 \end{bmatrix} \right\} = \mathbb{R}^3$$

we conclude that

$$\text{Span}\{\mathbf{c}_1, \mathbf{c}_2, \mathbf{c}_3\} = \text{Span}\{\mathbf{b}_1, \mathbf{b}_2, \mathbf{b}_3\} = \mathbb{R}^3.$$

□

The property described in Theorem 5.3.17 is not a special property of $\mathbb{R}^3$. A general version of the theorem will be presented in the book Core Topics in Linear Algebra.

Theorem 5.3.18. *For arbitrary vectors* **a**, **b**, **c** *in* $\mathbb{R}^3$, *the following two conditions are equivalent:*

   (a) **a**, **b**, **c** *are linearly independent;*

   (b) Span$\{$**a**, **b**, **c**$\} = \mathbb{R}^3$.

Proof. If **a**, **b**, **c** are linearly independent vectors in $\mathbb{R}^3$, then we must have

$$\text{Span}\{\mathbf{a}, \mathbf{b}, \mathbf{c}\} = \text{Span}\left\{ \begin{bmatrix} 1 \\ 0 \\ 0 \end{bmatrix}, \begin{bmatrix} 0 \\ 1 \\ 0 \end{bmatrix}, \begin{bmatrix} 0 \\ 0 \\ 1 \end{bmatrix} \right\} = \mathbb{R}^3,$$

by Theorem 5.3.17.

If Span$\{$**a**, **b**, **c**$\} = \mathbb{R}^3$, then

$$\text{Span}\left\{ \begin{bmatrix} 1 \\ 0 \\ 0 \end{bmatrix}, \begin{bmatrix} 0 \\ 1 \\ 0 \end{bmatrix}, \begin{bmatrix} 0 \\ 0 \\ 1 \end{bmatrix} \right\} = \text{Span}\{\mathbf{a}, \mathbf{b}, \mathbf{c}\}$$

and the vectors **a**, **b**, and **c** must be linearly independent vectors by Theorem 5.3.17.

$\square$

Corollary 5.3.19. *Any set of three linearly independent vectors in* $\mathbb{R}^3$ *is a basis of* $\mathbb{R}^3$.

Corollary 5.3.20. *Let* $\mathbf{v}_1, \mathbf{v}_2,$ *and* $\mathbf{v}_3$ *be three nonzero orthogonal vectors in* $\mathbb{R}^3$, *that is,*
$$\mathbf{v}_1 \cdot \mathbf{v}_2 = \mathbf{v}_1 \cdot \mathbf{v}_3 = \mathbf{v}_1 \cdot \mathbf{v}_3 = 0.$$
Then $\{\mathbf{v}_1, \mathbf{v}_2, \mathbf{v}_3\}$ *is a basis of* $\mathbb{R}^3$.

Proof. It is enough to show that the vectors $\mathbf{v}_1$, $\mathbf{v}_2$, and $\mathbf{v}_3$ are linearly independent. If

$$x_1\mathbf{v}_1 + x_2\mathbf{v}_2 + x_3\mathbf{v}_3 = \begin{bmatrix} 0 \\ 0 \\ 0 \end{bmatrix},$$

then

$$x_1\mathbf{v}_1 \cdot \mathbf{v}_1 + x_2\mathbf{v}_2 \cdot \mathbf{v}_1 + x_3\mathbf{v}_3 \cdot \mathbf{v}_1 = \begin{bmatrix} 0 \\ 0 \\ 0 \end{bmatrix} \cdot \mathbf{v}_1.$$

Since $\mathbf{v}_1 \cdot \mathbf{v}_1 = \|\mathbf{v}_1\|^2$, $\mathbf{v}_2 \cdot \mathbf{v}_1 = 0$, and $\mathbf{v}_3 \cdot \mathbf{v}_1 = 0$, the above can be written as $x\|\mathbf{v}_1\|^2 = 0$. This gives us $x = 0$, because $\|\mathbf{v}_1\| \neq 0$. In the same way we can show that $y = 0$ and $z = 0$. $\qquad\square$

A basis $\{\mathbf{v}_1, \mathbf{v}_2, \mathbf{v}_3\}$ of $\mathbb{R}^3$ such that the vectors $\mathbf{v}_1, \mathbf{v}_2$ and $\mathbf{v}_3$ are orthogonal is called an ***orthogonal basis***.

Corollary 5.3.21. *For arbitrary vectors* $\mathbf{a}, \mathbf{b}, \mathbf{c}$, *and* $\mathbf{d}$ *in* $\mathbb{R}^3$ *the equation*

$$x_1\mathbf{a} + x_2\mathbf{b} + x_3\mathbf{c} + x_4\mathbf{d} = \mathbf{0}$$

has a nontrivial solution, that is, a solution such that at least one of the numbers x_1, x_2, x_3, *or* x_4 *is different from* 0.

Proof. If the vectors $\mathbf{a}$, $\mathbf{b}$, and $\mathbf{c}$ are linearly dependent, then the result is a direct consequence of the definition of linear dependence. If the vectors $\mathbf{a}, \mathbf{b}, \mathbf{c}$ are linearly independent this is a consequence of Corollary 5.3.19. $\qquad\square$

The definition of linear dependence of three vectors in $\mathbb{R}^3$ (Definition 5.3.1) can be extended to more than three vectors. For example, we can say that vectors $\mathbf{a}$, $\mathbf{b}$, $\mathbf{c}$, and $\mathbf{d}$ are linearly dependent, if one of these vectors is in the vector subspace spanned by the remaining three vectors, that is, we have $\mathbf{a}$ is in $\mathrm{Span}\{\mathbf{b}, \mathbf{c}, \mathbf{d}\}$ or $\mathbf{b}$ is in $\mathrm{Span}\{\mathbf{a}, \mathbf{c}, \mathbf{d}\}$ or $\mathbf{c}$ is in $\mathrm{Span}\{\mathbf{a}, \mathbf{b}, \mathbf{d}\}$ or $\mathbf{d}$ is in $\mathrm{Span}\{\mathbf{a}, \mathbf{b}, \mathbf{c}\}$. Note that, using linear dependence of four vectors, Corollary 5.3.21 can be equivalently stated as follows.

Corollary 5.3.22. *Any four vectors in* $\mathbb{R}^3$ *are linearly dependent.*

The following simple theorem is often useful in proofs.

Theorem 5.3.23. *Let A and B be 3×3 matrices and let $\{\mathbf{v}_1, \mathbf{v}_2, \mathbf{v}_3\}$ be a basis in* $\mathbb{R}^3$. *If*

$$A\mathbf{v}_1 = B\mathbf{v}_1, \quad A\mathbf{v}_2 = B\mathbf{v}_2, \quad \text{and} \quad A\mathbf{v}_3 = B\mathbf{v}_3,$$

then

$$A = B.$$

Proof. An arbitrary vector in $\mathbb{R}^3$ is of the form $x\mathbf{v}_1 + y\mathbf{v}_2 + z\mathbf{v}_3$. Since

$$A(x\mathbf{v}_1 + y\mathbf{v}_2 + z\mathbf{v}_3) = xA\mathbf{v}_1 + yA\mathbf{v}_2 + zA\mathbf{v}_3 = xB\mathbf{v}_1 + yB\mathbf{v}_2 + zB\mathbf{v}_3 = B(x\mathbf{v}_1 + y\mathbf{v}_2 + z\mathbf{v}_3),$$

the result is a consequence of Theorem 2.1.18. $\qquad\square$

Example 5.3.24. Let $\{v_1, v_2, v_3\}$ be an orthogonal basis of $\mathbb{R}^3$, that is, such that $v_1 \cdot v_2 = v_1 \cdot v_3 = v_2 \cdot v_3 = 0$. Show that

$$\frac{1}{\|v_1\|^2} v_1 v_1^T + \frac{1}{\|v_2\|^2} v_2 v_2^T + \frac{1}{\|v_3\|^2} v_3 v_3^T = \begin{bmatrix} 1 & 0 & 0 \\ 0 & 1 & 0 \\ 0 & 0 & 1 \end{bmatrix}.$$

Solution. Since $\{v_1, v_2, v_3\}$ is an orthogonal basis of $\mathbb{R}^3$, it is easy to verify that

$$\left(\frac{1}{\|v_1\|^2} v_1 v_1^T + \frac{1}{\|v_2\|^2} v_2 v_2^T + \frac{1}{\|v_3\|^2} v_3 v_3^T \right) v_1 = v_1,$$

$$\left(\frac{1}{\|v_1\|^2} v_1 v_1^T + \frac{1}{\|v_2\|^2} v_2 v_2^T + \frac{1}{\|v_3\|^2} v_3 v_3^T \right) v_2 = v_2,$$

and

$$\left(\frac{1}{\|v_1\|^2} v_1 v_1^T + \frac{1}{\|v_2\|^2} v_2 v_2^T + \frac{1}{\|v_3\|^2} v_3 v_3^T \right) v_3 = v_3.$$

Now, because we also have

$$\begin{bmatrix} 1 & 0 & 0 \\ 0 & 1 & 0 \\ 0 & 0 & 1 \end{bmatrix} v_1 = v_1, \quad \begin{bmatrix} 1 & 0 & 0 \\ 0 & 1 & 0 \\ 0 & 0 & 1 \end{bmatrix} v_2 = v_2, \quad \text{and} \quad \begin{bmatrix} 1 & 0 & 0 \\ 0 & 1 & 0 \\ 0 & 0 & 1 \end{bmatrix} v_3 = v_3,$$

the desired equality is a consequence of Theorem 5.3.23.   □

Characterization of the vector subspaces of $\mathbb{R}^3$

Theorem 5.3.25. *Let* **u**, **v**, *and* **w** *be arbitrary vectors in* $\mathbb{R}^3$ *and let* V *be* $\mathrm{Span}\{u\}$ *or* $\mathrm{Span}\{u, v\}$ *or* $\mathrm{Span}\{u, v, w\}$. *Then*

(i) *If* **a** *is a vector in* V *and c a real number, then the vector c**a** is in* V, *and*

(ii) *If* **a** *and* **b** *are vectors in* V, *then the vector* **a** + **b** *is in* V.

Proof. We give the proof in the case when V is the vector subspace $\mathrm{Span}\{u, v\}$. The other cases are similar.

If **a** is a vector in V, then

$$a = xu + yv,$$

for some real numbers x and y. For any real number c we have

$$ca = c(xu + yv) = (cx)u + (cy)v,$$

which shows that the vector $c\mathbf{a}$ is in V.

If $\mathbf{a}$ and $\mathbf{b}$ are vectors in V, then

$$\mathbf{a} = x_1\mathbf{u} + y_1\mathbf{v} \quad \text{and} \quad \mathbf{b} = x_2\mathbf{v}_1 + y_2\mathbf{v},$$

for some real numbers x_1, x_2, y_1, and y_2. Since

$$\mathbf{a} + \mathbf{b} = x_1\mathbf{u} + y_1\mathbf{v} + x_2\mathbf{v}_1 + y_2\mathbf{v} = (x_1 + y_1)\mathbf{u} + (x_2 + y_2)\mathbf{v},$$

the vector $\mathbf{a} + \mathbf{b}$ is in V.

$\square$

It turns out that any subset of $\mathbb{R}^3$ that satisfies conditions (i) and (ii) in Theorem 5.3.25 must be one of the special sets listed in Theorem 5.3.25.

Theorem 5.3.26. *Let V be a subset of $\mathbf{R}^3$ such that*

   (i) *If $\mathbf{a}$ is a vector in V and c a real number, then the vector $c\mathbf{a}$ is in V, and*

   (ii) *If $\mathbf{a}$ and $\mathbf{b}$ are vectors in V, then the vector $\mathbf{a} + \mathbf{b}$ is in V.*

Then V is the set $\left\{ \begin{bmatrix} 0 \\ 0 \\ 0 \end{bmatrix} \right\}$ *or a vector line in $\mathbb{R}^3$ or a vector plane in $\mathbb{R}^3$ or all of* $\mathbb{R}^3$.

Proof. If V contains a vector $\mathbf{u} \neq \mathbf{0}$, then V must contain Span$\{\mathbf{u}\}$. If $V = \text{Span}\{\mathbf{u}\}$, then we are done.

If $V \neq \text{Span}\{\mathbf{u}\}$, then V contains a vector $\mathbf{v}$ which is not in Span$\{\mathbf{u}\}$ and the vectors $\mathbf{u}$ and $\mathbf{v}$ are linearly independent, by part (b) in Theorem 4.1.5. The subset V must contain Span$\{\mathbf{u}, \mathbf{v}\}$. Now there are two possibilities: either $V = \text{Span}\{\mathbf{u}, \mathbf{v}\}$ or $V \neq \text{Span}\{\mathbf{u}, \mathbf{v}\}$.

If $V = \text{Span}\{\mathbf{u}, \mathbf{v}\}$, then we are done.

If $V \neq \text{Span}\{\mathbf{u}, \mathbf{v}\}$, then V contains a vector $\mathbf{w}$ which is not in Span$\{\mathbf{u}, \mathbf{v}\}$ and the vectors $\mathbf{u}$, $\mathbf{v}$, and $\mathbf{w}$ are linearly independent, by part (c) in Theorem 5.3.5. The subset V must contain Span$\{\mathbf{u}, \mathbf{v}, \mathbf{w}\}$. Consequently, by Theorem 5.3.18, $V = \text{Span}\{\mathbf{u}, \mathbf{v}, \mathbf{w}\} = \mathbb{R}^3$.

$\square$

From Theorems 5.3.25 and 5.3.26 we get the following corollary that characterizes all vector subspaces of $\mathbb{R}^3$.

Corollary 5.3.27. *A subset V is a vector subspace spanned by one, two, or three vectors if and only if V satisfies the following two conditions*

   (i) *If $\mathbf{a}$ is a vector in V and c a real number then the vector $c\mathbf{a}$ is in V, and;*

   (ii) *If $\mathbf{a}$ and $\mathbf{b}$ are vectors in V, then the vector $\mathbf{a} + \mathbf{b}$ is in V.*

5.3.1 Exercises

Determine if the vectors **a**, **b**, **c** are linearly dependent or linearly independent.

1. $\mathbf{a} = \begin{bmatrix} 3 \\ 3 \\ 1 \end{bmatrix}, \mathbf{b} = \begin{bmatrix} 1 \\ 3 \\ 3 \end{bmatrix}, \mathbf{c} = \begin{bmatrix} 3 \\ 1 \\ 3 \end{bmatrix}$

5. $\mathbf{a} = \begin{bmatrix} 1 \\ 1 \\ 1 \end{bmatrix}, \mathbf{b} = \begin{bmatrix} 1 \\ 1 \\ 0 \end{bmatrix}, \mathbf{c} = \begin{bmatrix} 1 \\ 0 \\ 0 \end{bmatrix}$

2. $\mathbf{a} = \begin{bmatrix} 2 \\ 2 \\ 1 \end{bmatrix}, \mathbf{b} = \begin{bmatrix} 1 \\ 2 \\ 2 \end{bmatrix}, \mathbf{c} = \begin{bmatrix} 1 \\ 1 \\ 3 \end{bmatrix}$

6. $\mathbf{a} = \begin{bmatrix} 4 \\ 1 \\ 3 \end{bmatrix}, \mathbf{b} = \begin{bmatrix} 1 \\ 1 \\ 1 \end{bmatrix}, \mathbf{c} = \begin{bmatrix} 8 \\ 5 \\ 7 \end{bmatrix}$

3. $\mathbf{a} = \begin{bmatrix} 1 \\ 3 \\ 2 \end{bmatrix}, \mathbf{b} = \begin{bmatrix} 3 \\ 2 \\ -1 \end{bmatrix}, \mathbf{c} = \begin{bmatrix} 1 \\ 2 \\ 1 \end{bmatrix}$

7. $\mathbf{a} = \begin{bmatrix} 2 \\ 1 \\ 9 \end{bmatrix}, \mathbf{b} = \begin{bmatrix} 1 \\ 3 \\ 7 \end{bmatrix}, \mathbf{c} = \begin{bmatrix} 1 \\ 4 \\ 8 \end{bmatrix}$

4. $\mathbf{a} = \begin{bmatrix} 1 \\ 2 \\ 3 \end{bmatrix}, \mathbf{b} = \begin{bmatrix} 4 \\ 3 \\ 2 \end{bmatrix}, \mathbf{c} = \begin{bmatrix} 1 \\ 1 \\ 1 \end{bmatrix}$

8. $\mathbf{a} = \begin{bmatrix} 2 \\ 3 \\ 1 \end{bmatrix}, \mathbf{b} = \begin{bmatrix} 3 \\ 1 \\ 1 \end{bmatrix}, \mathbf{c} = \begin{bmatrix} 1 \\ 4 \\ 5 \end{bmatrix}$

9. Suppose that the vectors **b** and **c** are linearly independent and the vectors **a**, **b**, **c** are linearly dependent. Find the reduced row echelon forms of the matrix $\begin{bmatrix} \mathbf{a} & \mathbf{b} & \mathbf{c} \end{bmatrix}$.

10. Suppose that the vectors **a** and **c** are linearly independent and the vectors **a**, **b**, **c** are linearly dependent. Find the reduced row echelon forms of the matrix $\begin{bmatrix} \mathbf{a} & \mathbf{b} & \mathbf{c} \end{bmatrix}$.

Show that the given vectors **a**, **b**, **c** are in the same vector plane and determine the equation of this plane. Are the vectors linearly dependent?

11. $\mathbf{a} = \begin{bmatrix} 1 \\ 3 \\ 2 \end{bmatrix}, \mathbf{b} = \begin{bmatrix} 3 \\ 2 \\ -1 \end{bmatrix}, \mathbf{c} = \begin{bmatrix} 1 \\ 2 \\ 1 \end{bmatrix}$

13. $\mathbf{a} = \begin{bmatrix} 2 \\ 1 \\ 9 \end{bmatrix}, \mathbf{b} = \begin{bmatrix} 1 \\ 3 \\ 7 \end{bmatrix}, \mathbf{c} = \begin{bmatrix} 1 \\ 4 \\ 8 \end{bmatrix}$

12. $\mathbf{a} = \begin{bmatrix} 1 \\ 1 \\ 2 \end{bmatrix}, \mathbf{b} = \begin{bmatrix} 3 \\ 1 \\ 1 \end{bmatrix}, \mathbf{c} = \begin{bmatrix} 7 \\ 5 \\ 9 \end{bmatrix}$

14. $\mathbf{a} = \begin{bmatrix} 1 \\ 2 \\ 2 \end{bmatrix}, \mathbf{b} = \begin{bmatrix} 2 \\ 1 \\ 1 \end{bmatrix}, \mathbf{c} = \begin{bmatrix} 1 \\ 1 \\ 1 \end{bmatrix}$

15. Let $\mathbf{a} = \begin{bmatrix} 1 \\ 2 \\ 3 \end{bmatrix}$ and $\mathbf{b} = \begin{bmatrix} 1 \\ 1 \\ 1 \end{bmatrix}$. Determine all vectors **c** such that the vectors **a**, **b**, **c** are linearly dependent.

16. Let $\mathbf{a} = \begin{bmatrix} 3 \\ 2 \\ 3 \end{bmatrix}$ and $\mathbf{b} = \begin{bmatrix} 5 \\ 1 \\ 4 \end{bmatrix}$. Determine all vectors **c** such that the vectors **a**, **b**, **c** are linearly dependent.

17. Find a number a such that the vectors $\begin{bmatrix} 1 \\ 1 \\ 0 \end{bmatrix}$, $\begin{bmatrix} 1 \\ 0 \\ 1 \end{bmatrix}$, and $\begin{bmatrix} a \\ 1 \\ 1 \end{bmatrix}$ are linearly depen-

    dent and then with this value of a write the vector $\begin{bmatrix} a \\ 1 \\ 1 \end{bmatrix}$ as a linear combination

    of vectors $\begin{bmatrix} 1 \\ 1 \\ 0 \end{bmatrix}$ and $\begin{bmatrix} 1 \\ 0 \\ 1 \end{bmatrix}$.

18. Find a number a such that the vectors $\begin{bmatrix} 5 \\ 4 \\ 3 \end{bmatrix}$, $\begin{bmatrix} 2 \\ 3 \\ 4 \end{bmatrix}$, and $\begin{bmatrix} a \\ 1 \\ a \end{bmatrix}$ are linearly depen-

    dent and then with this value of a write the vector $\begin{bmatrix} a \\ 1 \\ a \end{bmatrix}$ as a linear combination

    of vectors $\begin{bmatrix} 5 \\ 4 \\ 3 \end{bmatrix}$ and $\begin{bmatrix} 2 \\ 3 \\ 4 \end{bmatrix}$.

19. Find a number a such that the vectors $\begin{bmatrix} 1 \\ -1 \\ 1 \end{bmatrix}$, $\begin{bmatrix} 2 \\ 1 \\ 1 \end{bmatrix}$, and $\begin{bmatrix} a \\ a \\ 1 \end{bmatrix}$ are linearly depen-

    dent and then with this value of a write the vector $\begin{bmatrix} a \\ a \\ 1 \end{bmatrix}$ as a linear combination

    of vectors $\begin{bmatrix} 1 \\ -1 \\ 1 \end{bmatrix}$ and $\begin{bmatrix} 2 \\ 1 \\ 1 \end{bmatrix}$.

20. Find a number a such that the vectors $\begin{bmatrix} 1 \\ 2 \\ 1 \end{bmatrix}$, $\begin{bmatrix} 3 \\ 5 \\ 2 \end{bmatrix}$, and $\begin{bmatrix} 5 \\ 9 \\ a \end{bmatrix}$ are linearly depen-

    dent and then with this value of a write the vector $\begin{bmatrix} 5 \\ 9 \\ a \end{bmatrix}$ as a linear combination

    of vectors $\begin{bmatrix} 1 \\ 2 \\ 1 \end{bmatrix}$ and $\begin{bmatrix} 3 \\ 5 \\ 2 \end{bmatrix}$.

Find a number a such that the given vectors are linearly independent.

21. $\begin{bmatrix} 2 \\ 2 \\ 1 \end{bmatrix}$, $\begin{bmatrix} 1 \\ 0 \\ 1 \end{bmatrix}$, $\begin{bmatrix} a \\ 2 \\ 1 \end{bmatrix}$ 22. $\begin{bmatrix} 3 \\ 1 \\ 1 \end{bmatrix}$, $\begin{bmatrix} 2 \\ 1 \\ 2 \end{bmatrix}$, $\begin{bmatrix} 1 \\ a \\ 1 \end{bmatrix}$

23. $\begin{bmatrix} 1 \\ 2 \\ 1 \end{bmatrix}, \begin{bmatrix} 2 \\ 1 \\ 1 \end{bmatrix}, \begin{bmatrix} a \\ 2a \\ 1 \end{bmatrix}$ 24. $\begin{bmatrix} 2 \\ 2 \\ 1 \end{bmatrix}, \begin{bmatrix} 1 \\ 3 \\ 1 \end{bmatrix}, \begin{bmatrix} 5 \\ 7 \\ a \end{bmatrix}$

For the given vectors **a**, **b**, **c**, and **x** show that {**a**,**b**,**c**} is a basis of $\mathbb{R}^3$ and find the coordinates of **x** in the basis {**a**,**b**,**c**}.

25. $\mathbf{a} = \begin{bmatrix} 1 \\ 1 \\ 1 \end{bmatrix}, \mathbf{b} = \begin{bmatrix} 3 \\ 1 \\ 0 \end{bmatrix}, \mathbf{c} = \begin{bmatrix} 2 \\ 0 \\ 0 \end{bmatrix}$, and $\mathbf{x} = \begin{bmatrix} 1 \\ 2 \\ 3 \end{bmatrix}$

26. $\mathbf{a} = \begin{bmatrix} 1 \\ 0 \\ 1 \end{bmatrix}, \mathbf{b} = \begin{bmatrix} 1 \\ 1 \\ 0 \end{bmatrix}, \mathbf{c} = \begin{bmatrix} 0 \\ 1 \\ 1 \end{bmatrix}$, and $\mathbf{x} = \begin{bmatrix} 1 \\ 1 \\ 1 \end{bmatrix}$

27. $\mathbf{a} = \begin{bmatrix} 2 \\ 1 \\ 1 \end{bmatrix}, \mathbf{b} = \begin{bmatrix} 1 \\ 2 \\ 1 \end{bmatrix}, \mathbf{c} = \begin{bmatrix} 1 \\ 1 \\ 2 \end{bmatrix}$, and $\mathbf{x} = \begin{bmatrix} 1 \\ 0 \\ 0 \end{bmatrix}$

28. $\mathbf{a} = \begin{bmatrix} 2 \\ 1 \\ 0 \end{bmatrix}, \mathbf{b} = \begin{bmatrix} 1 \\ 2 \\ 0 \end{bmatrix}, \mathbf{c} = \begin{bmatrix} 1 \\ 3 \\ 2 \end{bmatrix}$, and $\mathbf{x} = \begin{bmatrix} 1 \\ 1 \\ 1 \end{bmatrix}$

29. Find a number a such that the columns of the matrix $\begin{bmatrix} 1 & 1 & 2 \\ 1 & 5-a & 4 \\ 2 & 3 & 7-a \end{bmatrix}$ are linearly dependent.

30. Find a number a such that the columns of the matrix $\begin{bmatrix} 4-a & 1 & 2 \\ 3 & 2-a & 2 \\ 2 & 3 & 1 \end{bmatrix}$ are linearly dependent.

31. Find a number a such that the system

$$\begin{cases} 2x + 3y + z = 0 \\ 3x + y + 2z = 0 \\ 2x + ay + 3z = 0 \end{cases}$$

has a nontrivial solution and then solve the system.

32. Find a number a such that the system

$$\begin{cases} 3x + y + z = 0 \\ 2x + 3y + 2z = 0 \\ x + 2y + az = 0 \end{cases}$$

has a nontrivial solution and then solve the system.

33. Let $\mathbf{a}, \mathbf{b}, \mathbf{c}, \mathbf{u}, \mathbf{v}$ be vectors in $\mathbb{R}^3$. Show that the following two conditions are equivalent:

(a) Span$\{\mathbf{a}, \mathbf{b}, \mathbf{c}\}$ = Span$\{\mathbf{u}, \mathbf{v}\}$;

(b) $\mathbf{a}, \mathbf{b}, \mathbf{c}$ are elements of Span$\{\mathbf{u}, \mathbf{v}\}$ and $\mathbf{u}, \mathbf{v}$ are elements of Span$\{\mathbf{a}, \mathbf{b}, \mathbf{c}\}$.

34. Let $\mathbf{a}, \mathbf{b}, \mathbf{c}, \mathbf{u}, \mathbf{v}, \mathbf{w}$ be vectors in $\mathbb{R}^3$. Show that the following two conditions are equivalent:

(a) Span$\{\mathbf{a}, \mathbf{b}, \mathbf{v}\}$ = Span$\{\mathbf{u}, \mathbf{v}, \mathbf{w}\}$;

(b) $\mathbf{a}, \mathbf{b}, \mathbf{c}$ are elements of Span$\{\mathbf{u}, \mathbf{v}, \mathbf{w}\}$ and $\mathbf{u}, \mathbf{v}, \mathbf{w}$ are elements of Span$\{\mathbf{a}, \mathbf{b}, \mathbf{c}\}$.

35. Show that $\det\begin{bmatrix}\mathbf{a} & \mathbf{b} & \mathbf{a} \times \mathbf{b}\end{bmatrix} = \|\mathbf{a} \times \mathbf{b}\|^2$ for all vectors $\mathbf{a}, \mathbf{b}, \mathbf{c}$ in $\mathbb{R}^3$.

36. Show that $\|\mathbf{a} \times \mathbf{b}\|^2 = \|\mathbf{a}\|^2 \|\mathbf{b}\|^2 - (\mathbf{a} \cdot \mathbf{b})^2$ for all vectors $\mathbf{a}, \mathbf{b}, \mathbf{c}$ in $\mathbb{R}^3$.

37. Show that the set $\{\mathbf{a}, \mathbf{b}, \mathbf{a} \times \mathbf{b}\}$ is a basis in $\mathbb{R}^3$ for any two linearly independent vectors $\mathbf{a}$ and $\mathbf{b}$ in $\mathbb{R}^3$.

38. Find the coordinates of the vector $\begin{bmatrix}1\\2\\1\end{bmatrix}$ in the basis $\{\mathbf{a}, \mathbf{b}, \mathbf{a} \times \mathbf{b}\}$ if $\mathbf{a} = \begin{bmatrix}1\\3\\0\end{bmatrix}$ and $\mathbf{b} = \begin{bmatrix}1\\0\\2\end{bmatrix}$.

39. Let $\mathbf{a}$ and $\mathbf{b}$ be two linearly independent vectors in $\mathbb{R}^3$ and let $\mathbf{c}$ be an arbitrary vector in $\mathbb{R}^3$. If
$$\mathbf{c} = r\mathbf{a} + s\mathbf{b} + t(\mathbf{a} \times \mathbf{b}),$$
for some numbers r, s, and t, show that $r\mathbf{a} + s\mathbf{b}$ is the projection of $\mathbf{c}$ on the vector plane Span$\{\mathbf{a}, \mathbf{b}\}$ and $t(\mathbf{a} \times \mathbf{b})$ is the projection of $\mathbf{c}$ on the vector line Span$\{\mathbf{a} \times \mathbf{b}\}$.

40. Let $\mathbf{a}$ and $\mathbf{b}$ be two linearly independent vectors in $\mathbb{R}^3$ and let $\mathbf{c}$ be an arbitrary vector in $\mathbb{R}^3$. Show that the projection $\mathbf{p}$ of the point $\mathbf{c}$ on the vector plane Span$\{\mathbf{a}, \mathbf{b}\}$ is
$$\mathbf{p} = \mathbf{c} - \frac{\mathbf{c} \cdot (\mathbf{a} \times \mathbf{b})}{\|\mathbf{a} \times \mathbf{b}\|^2}(\mathbf{a} \times \mathbf{b}). \tag{5.12}$$

41. Let $\mathbf{a}$ and $\mathbf{b}$ be two linearly independent vectors in $\mathbb{R}^3$ and let $\mathbf{c}$ be an arbitrary vector in $\mathbb{R}^3$. Show that the distance from $\mathbf{c}$ to the vector plane Span$\{\mathbf{a}, \mathbf{b}\}$ is
$$\frac{|\det\begin{bmatrix}\mathbf{a} & \mathbf{b} & \mathbf{c}\end{bmatrix}|}{\|\mathbf{a} \times \mathbf{b}\|}.$$

42. Show that the distance from a point $\mathbf{a}$ in $\mathbb{R}^3$ to a vector line Span$\{\mathbf{b}\}$ in $\mathbb{R}^3$ is
$$\frac{\|\mathbf{a} \times \mathbf{b}\|}{\|\mathbf{b}\|}.$$

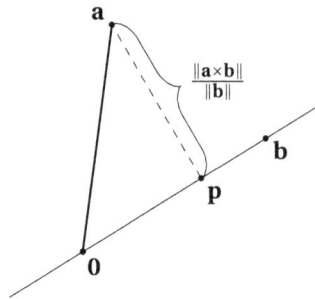

Figure 5.3: The projection of **a** on the vector line Span{**b**} in $\mathbb{R}^3$.

43. Let **a** and **b** be nonzero vectors in $\mathbb{R}^3$. Using the formula for the area of a triangle

$$A = \frac{1}{2} \cdot (\text{the length of the base}) \cdot (\text{the height}),$$

show that the area of the triangle **0ab** is $\frac{1}{2} \|\mathbf{a} \times \mathbf{b}\|$.

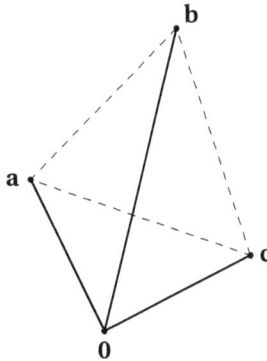

Figure 5.4: The tetrahedron defined by vectors **a**, **b**, and **c**.

44. Let **a**, **b**, and **c** be linearly independent vectors in $\mathbb{R}^3$. Using the formula for the volume of a tetrahedron

$$V = \frac{1}{3} \cdot (\text{the area of the base}) \cdot (\text{the height}),$$

show the the volume of the tetrahedron **0abc** is $\frac{1}{6} \left| \det \begin{bmatrix} \mathbf{a} & \mathbf{b} & \mathbf{c} \end{bmatrix} \right|$.

45. Suppose that the set {**a**, **b**, **c**} is a basis of $\mathbb{R}^3$ and that the **u** is nonzero vector in $\mathbb{R}^3$. Show that one of the following is true:

    (a) {**a**, **b**, **u**} is a basis in $\mathbb{R}^3$;

    (b) $\{\mathbf{b}, \mathbf{c}, \mathbf{u}\}$ is a basis in $\mathbb{R}^3$;

    (c) $\{\mathbf{a}, \mathbf{c}, \mathbf{u}\}$ is a basis in $\mathbb{R}^3$.

46. Let $\{\mathbf{a}, \mathbf{b}, \mathbf{c}\}$ be a basis of $\mathbb{R}^3$. If $\mathbf{u}$ and $\mathbf{v}$ are linearly independent vectors in $\mathbb{R}^3$, show that one of the following is true:

    (a) $\{\mathbf{a}, \mathbf{u}, \mathbf{v}\}$ is a basis in $\mathbb{R}^3$;

    (b) $\{\mathbf{b}, \mathbf{u}, \mathbf{v}\}$ is a basis in $\mathbb{R}^3$;

    (c) $\{\mathbf{c}, \mathbf{u}, \mathbf{v}\}$ is a basis in $\mathbb{R}^3$.

5.4   The dimension of a vector subspace of $\mathbb{R}^3$

The dimension of a vector subspace is one of the fundamental ideas of linear algebra. We motivate the definition with the following three simple theorems.

Theorem 5.4.1. *Let V be a vector line.*

(a) *If a vector $\mathbf{u} \neq \mathbf{0}$ is in V, then $\{\mathbf{u}\}$ is a basis of V.*

(b) *If vectors $\mathbf{u}$ and $\mathbf{v}$ are in V, then they are linearly dependent.*

Proof. Part (a) is a consequence of Theorem 4.1.6.

    If vectors $\mathbf{u}$ and $\mathbf{v}$ are in the vector line $V = \text{Span}\{\mathbf{w}\}$, then $\mathbf{u} = s\mathbf{w}$ and $\mathbf{v} = t\mathbf{w}$. If $t = 0$, then $\mathbf{v} = \mathbf{0}$ and, if $t \neq 0$, then $t\mathbf{u} - s\mathbf{v} = \mathbf{0}$. In both cases the vectors $\mathbf{u}$ and $\mathbf{v}$ are linearly dependent. This proves part (b).   □

Theorem 5.4.2. *Let V be a vector plane.*

(a) *If vectors $\mathbf{u}$ and $\mathbf{v}$ are in V and are linearly independent, then $\{\mathbf{u}, \mathbf{v}\}$ is a basis of V.*

(b) *If vectors $\mathbf{u}$, $\mathbf{v}$, and $\mathbf{w}$ are in V, then they are linearly dependent.*

(c) *If $\mathbf{u}$ is an arbitrary vector in $\mathbb{R}^3$, then $V \neq \text{Span}\{\mathbf{u}\}$.*

Proof. Part (a) is a consequence of Theorem 4.1.22.

    Let $\mathbf{u}$, $\mathbf{v}$, and $\mathbf{w}$ be vectors in V. If the vectors $\mathbf{u}$ and $\mathbf{v}$ are linearly dependent, we are done. If the vectors $\mathbf{u}, \mathbf{v}$ are linearly independent then $V = \text{Span}\{\mathbf{u}, \mathbf{v}\}$, by Theorem 4.1.22, and $\mathbf{w}$ is in $\text{Span}\{\mathbf{u}, \mathbf{v}\}$. But then the vectors $\mathbf{u}, \mathbf{v}$ and $\mathbf{w}$ are linearly dependent, completing the proof of part (b).

    Now suppose that $V = \text{Span}\{\mathbf{v}, \mathbf{w}\}$ and $V = \text{Span}\{\mathbf{u}\}$. Then the vectors $\mathbf{v}$ and $\mathbf{w}$ are linearly dependent, by Theorem 5.4.1, which is not true. This proves part (c).   □

Theorem 5.4.3.

   (a) *If vectors* **u**, **v**, *and* **w** *are three linearly independent vectors in* $\mathbb{R}^3$, *then* $\{\mathbf{u},\mathbf{v},\mathbf{w}\}$ *is a basis of* $\mathbb{R}^3$;

   (b) *Any four vectors in* $\mathbb{R}^3$ *are linearly dependent;*

   (c) *If* **u** *and* **v** *are arbitrary vectors in* $\mathbb{R}^3$, *then* $\operatorname{Span}\{\mathbf{u},\mathbf{v}\} \neq \mathbb{R}^3$.

Proof. Part (a) is a consequence of Theorem 5.3.18.

   Part (b) is a consequence of the Corollary 5.3.21.

   To prove part (c) we observe that $\mathbb{R}^3 = \operatorname{Span}\{\mathbf{u},\mathbf{v}\}$ would imply that $\begin{bmatrix} 1 \\ 0 \\ 0 \end{bmatrix}, \begin{bmatrix} 0 \\ 1 \\ 0 \end{bmatrix}, \begin{bmatrix} 0 \\ 0 \\ 1 \end{bmatrix}$

are linearly dependent, by Theorem 5.4.2 (b).

$\square$

Note that the above three theorems show that the minimum number of nonzero vectors that span a vector subspace V of $\mathbb{R}^3$ is the same as the maximum number of linearly independent vectors in V and is the same as the number of vectors in an arbitrary basis of V.

Definition 5.4.4. Let V be a nontrivial vector subspace of $\mathbb{R}^3$, that is, a vector subspace $V \neq \left\{ \begin{bmatrix} 0 \\ 0 \\ 0 \end{bmatrix} \right\}$. By the ***dimension*** of V, denoted by $\dim V$, we mean the minimum number of nonzero vectors that span V. Consequently,

   (a) A vector line V has dimension 1 and we write $\dim V = 1$;

   (b) A vector plane V has dimension 2 and we write $\dim V = 2$;

   (c) The vector space $\mathbb{R}^3$ has dimension 3 and we write $\dim \mathbb{R}^3 = 3$.

The trivial vector subspace $\operatorname{Span}\left\{ \begin{bmatrix} 0 \\ 0 \\ 0 \end{bmatrix} \right\} = \left\{ \begin{bmatrix} 0 \\ 0 \\ 0 \end{bmatrix} \right\}$ has dimension 0, that is,

$\dim \left\{ \begin{bmatrix} 0 \\ 0 \\ 0 \end{bmatrix} \right\} = 0$.

Example 5.4.5. Determine the dimension of Span$\left\{ \begin{bmatrix} 1 \\ 1 \\ 3 \end{bmatrix}, \begin{bmatrix} 2 \\ 1 \\ 5 \end{bmatrix}, \begin{bmatrix} 1 \\ 2 \\ 4 \end{bmatrix} \right\}$.

Solution. Since the reduced row echelon form of the matrix $\begin{bmatrix} 1 & 2 & 1 \\ 1 & 1 & 2 \\ 3 & 5 & 4 \end{bmatrix}$ is $\begin{bmatrix} 1 & 0 & 3 \\ 0 & 1 & -1 \\ 0 & 0 & 0 \end{bmatrix}$,

the vector $\begin{bmatrix} 1 \\ 2 \\ 4 \end{bmatrix}$ is in Span$\left\{ \begin{bmatrix} 1 \\ 1 \\ 3 \end{bmatrix}, \begin{bmatrix} 2 \\ 1 \\ 5 \end{bmatrix} \right\}$ and we have

$$\text{Span}\left\{ \begin{bmatrix} 1 \\ 1 \\ 3 \end{bmatrix}, \begin{bmatrix} 2 \\ 1 \\ 5 \end{bmatrix} \right\} = \text{Span}\left\{ \begin{bmatrix} 1 \\ 1 \\ 3 \end{bmatrix}, \begin{bmatrix} 2 \\ 1 \\ 5 \end{bmatrix}, \begin{bmatrix} 1 \\ 2 \\ 4 \end{bmatrix} \right\}.$$

Consequently,

$$\dim\left(\text{Span}\left\{ \begin{bmatrix} 1 \\ 1 \\ 3 \end{bmatrix}, \begin{bmatrix} 2 \\ 1 \\ 5 \end{bmatrix}, \begin{bmatrix} 1 \\ 2 \\ 4 \end{bmatrix} \right\} \right) = 2.$$

$\square$

The Rank Theorem for 3×3 matrices

Theorem 5.4.6. *Let* $\begin{bmatrix} a_1 & b_1 & c_1 \\ a_2 & b_2 & c_2 \\ a_3 & b_3 & c_3 \end{bmatrix}$ *be an arbitrary* 3×3 *matrix. Then*

$$\dim\left(\text{Span}\left\{ \begin{bmatrix} a_1 \\ a_2 \\ a_3 \end{bmatrix}, \begin{bmatrix} b_1 \\ b_2 \\ b_3 \end{bmatrix}, \begin{bmatrix} c_1 \\ c_2 \\ c_3 \end{bmatrix} \right\} \right) = \dim\left(\text{Span}\left\{ \begin{bmatrix} a_1 \\ b_1 \\ c_1 \end{bmatrix}, \begin{bmatrix} a_2 \\ b_2 \\ c_2 \end{bmatrix}, \begin{bmatrix} a_3 \\ b_3 \\ c_3 \end{bmatrix} \right\} \right)$$

Proof. Since

$$\det \begin{bmatrix} a_1 & b_1 & c_1 \\ a_2 & b_2 & c_2 \\ a_3 & b_3 & c_3 \end{bmatrix} = \det \begin{bmatrix} a_1 & a_2 & a_3 \\ b_1 & b_2 & b_3 \\ c_1 & c_2 & c_3 \end{bmatrix},$$

by Theorem 5.1.15, we have

$$\dim\left(\text{Span}\left\{ \begin{bmatrix} a_1 \\ a_2 \\ a_3 \end{bmatrix}, \begin{bmatrix} b_1 \\ b_2 \\ b_3 \end{bmatrix}, \begin{bmatrix} c_1 \\ c_2 \\ c_3 \end{bmatrix} \right\} \right) = 3$$

if and only if

$$\dim\left(\operatorname{Span}\left\{\begin{bmatrix} a_1 \\ b_1 \\ c_1 \end{bmatrix}, \begin{bmatrix} a_2 \\ b_2 \\ c_2 \end{bmatrix}, \begin{bmatrix} a_3 \\ b_3 \\ c_3 \end{bmatrix}\right\}\right) = 3.$$

Now suppose that

$$\dim\left(\operatorname{Span}\left\{\begin{bmatrix} a_1 \\ a_2 \\ a_3 \end{bmatrix}, \begin{bmatrix} b_1 \\ b_2 \\ b_3 \end{bmatrix}, \begin{bmatrix} c_1 \\ c_2 \\ c_3 \end{bmatrix}\right\}\right) = 2.$$

This means that the vectors $\begin{bmatrix} a_1 \\ a_2 \\ a_3 \end{bmatrix}$, $\begin{bmatrix} b_1 \\ b_2 \\ b_3 \end{bmatrix}$, and $\begin{bmatrix} c_1 \\ c_2 \\ c_3 \end{bmatrix}$ are linearly dependent and two

vectors from $\left\{\begin{bmatrix} a_1 \\ a_2 \\ a_3 \end{bmatrix}, \begin{bmatrix} b_1 \\ b_2 \\ b_3 \end{bmatrix}, \begin{bmatrix} c_1 \\ c_2 \\ c_3 \end{bmatrix}\right\}$ are linearly independent. Suppose that the vec-

tors $\begin{bmatrix} a_1 \\ a_2 \\ a_3 \end{bmatrix}$ and $\begin{bmatrix} b_1 \\ b_2 \\ b_3 \end{bmatrix}$ are linearly independent. Then $\det\begin{bmatrix} a_1 & b_1 \\ a_2 & b_2 \end{bmatrix} \neq 0$ or $\det\begin{bmatrix} a_1 & b_1 \\ a_3 & b_3 \end{bmatrix} \neq$

0 or $\det\begin{bmatrix} a_2 & b_2 \\ a_3 & b_3 \end{bmatrix} \neq 0$, by Theorem 4.1.12. The proof is similar in all three cases. If we

suppose that $\det\begin{bmatrix} a_1 & b_1 \\ a_2 & b_2 \end{bmatrix} \neq 0$, then the vectors $\begin{bmatrix} a_1 \\ b_1 \\ c_1 \end{bmatrix}$ and $\begin{bmatrix} a_2 \\ b_2 \\ c_2 \end{bmatrix}$ are linearly indepen-

dent, because $\det\begin{bmatrix} a_1 & b_1 \\ a_2 & b_2 \end{bmatrix} = \det\begin{bmatrix} a_1 & a_2 \\ b_1 & b_2 \end{bmatrix}$.

By Theorem 5.1.15 we have

$$\det\begin{bmatrix} a_1 & a_2 & a_3 \\ b_1 & b_2 & b_3 \\ c_1 & c_2 & c_3 \end{bmatrix} = \det\begin{bmatrix} a_1 & b_1 & c_1 \\ a_2 & b_2 & c_2 \\ a_3 & b_3 & c_3 \end{bmatrix} = 0$$

and thus the vectors $\begin{bmatrix} a_1 \\ b_1 \\ c_1 \end{bmatrix}$, $\begin{bmatrix} a_2 \\ b_2 \\ c_2 \end{bmatrix}$, $\begin{bmatrix} a_3 \\ b_3 \\ c_3 \end{bmatrix}$ are linearly dependent, by Theorem 5.3.3,

and we have

$$\begin{bmatrix} a_3 \\ b_3 \\ c_3 \end{bmatrix} = \alpha \begin{bmatrix} a_1 \\ b_1 \\ c_1 \end{bmatrix} + \beta \begin{bmatrix} a_2 \\ b_2 \\ c_2 \end{bmatrix},$$

for some real numbers α and β. Consequently

$$\operatorname{Span}\left\{\begin{bmatrix} a_1 \\ b_1 \\ c_1 \end{bmatrix}, \begin{bmatrix} a_2 \\ b_2 \\ c_2 \end{bmatrix}, \begin{bmatrix} a_3 \\ b_3 \\ c_3 \end{bmatrix}\right\} = \operatorname{Span}\left\{\begin{bmatrix} a_1 \\ b_1 \\ c_1 \end{bmatrix}, \begin{bmatrix} a_2 \\ b_2 \\ c_2 \end{bmatrix}\right\},$$

which means that $\left\{\begin{bmatrix} a_1 \\ b_1 \\ c_1 \end{bmatrix}, \begin{bmatrix} a_2 \\ b_2 \\ c_2 \end{bmatrix}\right\}$ is a basis of $\operatorname{Span}\left\{\begin{bmatrix} a_1 \\ b_1 \\ c_1 \end{bmatrix}, \begin{bmatrix} a_2 \\ b_2 \\ c_2 \end{bmatrix}, \begin{bmatrix} a_3 \\ b_3 \\ c_3 \end{bmatrix}\right\}$, because the vectors $\begin{bmatrix} a_1 \\ b_1 \\ c_1 \end{bmatrix}$ and $\begin{bmatrix} a_2 \\ b_2 \\ c_2 \end{bmatrix}$ are linearly independent. This gives us

$$\dim \mathrm{Span}\left\{ \begin{bmatrix} a_1 \\ b_1 \\ c_1 \end{bmatrix}, \begin{bmatrix} a_2 \\ b_2 \\ c_2 \end{bmatrix}, \begin{bmatrix} a_3 \\ b_3 \\ c_3 \end{bmatrix} \right\} = 2.$$

The cases when the vectors $\begin{bmatrix} a_1 \\ a_2 \\ a_3 \end{bmatrix}$ and $\begin{bmatrix} c_1 \\ c_2 \\ c_3 \end{bmatrix}$ are linearly independent or when the

vectors $\begin{bmatrix} b_1 \\ b_2 \\ b_3 \end{bmatrix}$ and $\begin{bmatrix} c_1 \\ c_2 \\ c_3 \end{bmatrix}$ are linearly independent are treated in a similar way.

Finally we consider the case when

$$\dim\left(\mathrm{Span}\left\{ \begin{bmatrix} a_1 \\ a_2 \\ a_3 \end{bmatrix}, \begin{bmatrix} b_1 \\ b_2 \\ b_3 \end{bmatrix}, \begin{bmatrix} c_1 \\ c_2 \\ c_3 \end{bmatrix} \right\} \right) = 1.$$

Then at least one vector from $\left\{ \begin{bmatrix} a_1 \\ a_2 \\ a_3 \end{bmatrix}, \begin{bmatrix} b_1 \\ b_2 \\ b_3 \end{bmatrix}, \begin{bmatrix} c_1 \\ c_2 \\ c_3 \end{bmatrix} \right\}$ is different from $\begin{bmatrix} 0 \\ 0 \\ 0 \end{bmatrix}$. If $\begin{bmatrix} a_1 \\ a_2 \\ a_3 \end{bmatrix} \neq$

$\begin{bmatrix} 0 \\ 0 \\ 0 \end{bmatrix}$, then we must have $a_1 \neq 0$ or $a_2 \neq 0$ or $a_3 \neq 0$. If $a_1 \neq 0$, then we also have

$\begin{bmatrix} a_1 \\ b_1 \\ c_1 \end{bmatrix} \neq \begin{bmatrix} 0 \\ 0 \\ 0 \end{bmatrix}$ and, because $\left\{ \begin{bmatrix} a_1 \\ a_2 \\ a_3 \end{bmatrix} \right\}$ is a basis of $\mathrm{Span}\left\{ \begin{bmatrix} a_1 \\ a_2 \\ a_3 \end{bmatrix}, \begin{bmatrix} b_1 \\ b_2 \\ b_3 \end{bmatrix}, \begin{bmatrix} c_1 \\ c_2 \\ c_3 \end{bmatrix} \right\}$, there

are real numbers s and t such that $s\begin{bmatrix} a_1 \\ a_2 \\ a_3 \end{bmatrix} = \begin{bmatrix} b_1 \\ b_2 \\ b_3 \end{bmatrix}$ and $t\begin{bmatrix} a_1 \\ a_2 \\ a_3 \end{bmatrix} = \begin{bmatrix} c_1 \\ c_2 \\ c_3 \end{bmatrix}$. This gives us

$\frac{a_2}{a_1}\begin{bmatrix} a_1 \\ b_1 \\ c_1 \end{bmatrix} = \begin{bmatrix} a_2 \\ b_2 \\ c_2 \end{bmatrix}$ and $\frac{a_3}{a_1}\begin{bmatrix} a_1 \\ b_1 \\ c_1 \end{bmatrix} = \begin{bmatrix} a_3 \\ b_3 \\ c_3 \end{bmatrix}$ and, consequently,

$$\mathrm{Span}\left\{ \begin{bmatrix} a_1 \\ b_1 \\ c_1 \end{bmatrix}, \begin{bmatrix} a_2 \\ b_2 \\ c_2 \end{bmatrix}, \begin{bmatrix} a_3 \\ b_3 \\ c_3 \end{bmatrix} \right\} = \mathrm{Span}\left\{ \begin{bmatrix} a_1 \\ b_1 \\ c_1 \end{bmatrix} \right\},$$

which means that

$$\dim\left(\mathrm{Span}\left\{ \begin{bmatrix} a_1 \\ b_1 \\ c_1 \end{bmatrix}, \begin{bmatrix} a_2 \\ b_2 \\ c_2 \end{bmatrix}, \begin{bmatrix} a_3 \\ b_3 \\ c_3 \end{bmatrix} \right\} \right) = 1,$$

because $\begin{bmatrix} a_1 \\ b_1 \\ c_1 \end{bmatrix} \neq \begin{bmatrix} 0 \\ 0 \\ 0 \end{bmatrix}$. The proof when $a_2 \neq 0$ or $a_3 \neq 0$ is similar.

The cases when $\begin{bmatrix} b_1 \\ b_2 \\ b_3 \end{bmatrix} \neq \begin{bmatrix} 0 \\ 0 \\ 0 \end{bmatrix}$ and $\begin{bmatrix} c_1 \\ c_2 \\ c_3 \end{bmatrix} \neq \begin{bmatrix} 0 \\ 0 \\ 0 \end{bmatrix}$ are treated in a similar way.

Note that from the above proof it follows that if $\dim\left(\mathrm{Span}\left\{ \begin{bmatrix} a_1 \\ b_1 \\ c_1 \end{bmatrix}, \begin{bmatrix} a_2 \\ b_2 \\ c_2 \end{bmatrix}, \begin{bmatrix} a_3 \\ b_3 \\ c_3 \end{bmatrix} \right\}\right) =$

2, then $\dim\left(\mathrm{Span}\left\{ \begin{bmatrix} a_1 \\ a_2 \\ a_3 \end{bmatrix}, \begin{bmatrix} b_1 \\ b_2 \\ b_3 \end{bmatrix}, \begin{bmatrix} c_1 \\ c_2 \\ c_3 \end{bmatrix} \right\}\right) = 2$, and if $\dim\left(\mathrm{Span}\left\{ \begin{bmatrix} a_1 \\ b_1 \\ c_1 \end{bmatrix}, \begin{bmatrix} a_2 \\ b_2 \\ c_2 \end{bmatrix}, \begin{bmatrix} a_3 \\ b_3 \\ c_3 \end{bmatrix} \right\}\right) =$

1, then $\dim\left(\mathrm{Span}\left\{ \begin{bmatrix} a_1 \\ a_2 \\ a_3 \end{bmatrix}, \begin{bmatrix} b_1 \\ b_2 \\ b_3 \end{bmatrix}, \begin{bmatrix} c_1 \\ c_2 \\ c_3 \end{bmatrix} \right\}\right) = 1$.

Since in the case when $\begin{vmatrix} a_1 & a_2 & a_3 \\ b_1 & b_2 & b_3 \\ c_1 & c_2 & c_3 \end{vmatrix} = \begin{bmatrix} 0 & 0 & 0 \\ 0 & 0 & 0 \\ 0 & 0 & 0 \end{bmatrix}$ there is nothing to prove, the

proof is now complete.   □

Definition 5.4.7. Let A be the 3×3 matrix $\begin{bmatrix} a_1 & b_1 & c_1 \\ a_2 & b_2 & c_2 \\ a_3 & b_3 & c_3 \end{bmatrix}$. The vector subspace

$\mathrm{Span}\left\{ \begin{bmatrix} a_1 \\ a_2 \\ a_3 \end{bmatrix}, \begin{bmatrix} b_1 \\ b_2 \\ b_3 \end{bmatrix}, \begin{bmatrix} c_1 \\ c_2 \\ c_3 \end{bmatrix} \right\}$ is called the **column space** of A and the vector

subspace $\mathrm{Span}\left\{ \begin{bmatrix} a_1 \\ b_1 \\ c_1 \end{bmatrix}, \begin{bmatrix} a_2 \\ b_2 \\ c_2 \end{bmatrix}, \begin{bmatrix} a_3 \\ b_3 \\ c_3 \end{bmatrix} \right\}$ is called the **row space** of A. The di-

mension of these vector subspaces is called the **rank** of the matrix A.

Example 5.4.8. Verify the Rank Theorem for the matrix $\begin{bmatrix} 5 & 2 & 3 \\ 1 & 1 & 2 \\ 3 & 0 & -1 \end{bmatrix}$.

Solution. The result is a consequence of the fact that the reduced row echelon

form of the matrix $\begin{bmatrix} 5 & 2 & 3 \\ 1 & 1 & 2 \\ 3 & 0 & -1 \end{bmatrix}$ is $\begin{bmatrix} 1 & 0 & -\frac{1}{3} \\ 0 & 1 & \frac{7}{3} \\ 0 & 0 & 0 \end{bmatrix}$ and the reduced row echelon form

of the matrix $\begin{bmatrix} 5 & 1 & 3 \\ 2 & 1 & 0 \\ 3 & 2 & -1 \end{bmatrix}$ is $\begin{bmatrix} 1 & 0 & 1 \\ 0 & 1 & -2 \\ 0 & 0 & 0 \end{bmatrix}$.   □

> ### 5.4.1 Exercises

Determine the dimension of the given subspace of $\mathbb{R}^3$.

1. Span $\left\{ \begin{bmatrix} 2 \\ 5 \\ 1 \end{bmatrix}, \begin{bmatrix} 1 \\ 2 \\ 3 \end{bmatrix} \right\}$

2. Span $\left\{ \begin{bmatrix} 1 \\ 1 \\ 0 \end{bmatrix}, \begin{bmatrix} 1 \\ 1 \\ 1 \end{bmatrix} \right\}$

3. Span $\left\{ \begin{bmatrix} 2 \\ -1 \\ 4 \end{bmatrix}, \begin{bmatrix} -4 \\ 2 \\ -8 \end{bmatrix} \right\}$

4. Span $\left\{ \begin{bmatrix} 4 \\ -4 \\ 8 \end{bmatrix}, \begin{bmatrix} 5 \\ -5 \\ 10 \end{bmatrix} \right\}$

5. Span $\left\{ \begin{bmatrix} 3 \\ 1 \\ 7 \end{bmatrix}, \begin{bmatrix} 2 \\ 3 \\ 7 \end{bmatrix}, \begin{bmatrix} 1 \\ 2 \\ 4 \end{bmatrix} \right\}$

6. Span $\left\{ \begin{bmatrix} 3 \\ 1 \\ 1 \end{bmatrix}, \begin{bmatrix} 1 \\ 3 \\ 1 \end{bmatrix}, \begin{bmatrix} 2 \\ 2 \\ 1 \end{bmatrix} \right\}$

7. Span $\left\{ \begin{bmatrix} 10 \\ 5 \\ 5 \end{bmatrix}, \begin{bmatrix} 4 \\ 2 \\ 2 \end{bmatrix}, \begin{bmatrix} 2 \\ 1 \\ 1 \end{bmatrix} \right\}$

8. Span $\left\{ \begin{bmatrix} 3 \\ 9 \\ 3 \end{bmatrix}, \begin{bmatrix} 1 \\ 3 \\ 1 \end{bmatrix}, \begin{bmatrix} -4 \\ -12 \\ -4 \end{bmatrix} \right\}$

9. Span $\left\{ \begin{bmatrix} 2 \\ 1 \\ 2 \end{bmatrix}, \begin{bmatrix} 1 \\ 2 \\ 2 \end{bmatrix}, \begin{bmatrix} 2 \\ 2 \\ 1 \end{bmatrix} \right\}$

10. Span $\left\{ \begin{bmatrix} 1 \\ 1 \\ 3 \end{bmatrix}, \begin{bmatrix} 4 \\ 1 \\ 2 \end{bmatrix}, \begin{bmatrix} 1 \\ 0 \\ 5 \end{bmatrix} \right\}$

Verify the Rank Theorem for the given matrix.

11. $\begin{bmatrix} 1 & 7 & 4 \\ 2 & 2 & 3 \\ -1 & 5 & 1 \end{bmatrix}$

12. $\begin{bmatrix} 1 & 2 & 3 \\ 1 & 1 & 4 \\ 2 & 5 & 5 \end{bmatrix}$

13. $\begin{bmatrix} 1 & 1 & 0 \\ 1 & 1 & 1 \\ 1 & 3 & 1 \end{bmatrix}$

14. $\begin{bmatrix} 3 & 2 & 1 \\ 4 & -3 & 5 \\ 2 & 2 & 0 \end{bmatrix}$

Chapter 6

Singular value decomposition of 3×2 matrices

In Chapter 3 we discussed orthogonal diagonalization and spectral decomposition of 2×2 matrices. We proved that a 2×2 matrix can be orthogonally diagonalized and has a spectral decomposition if and only if it is symmetric. In this chapter we obtain similar results for 3×2 matrices. The method described here generalizes to matrices of any dimension and has numerous practical applications, including, data compression, noise reduction, or data analysis.

We start by observing that for an arbitrary 3×2 matrix A, the 2×2 matrix $A^T A$ is symmetric, so it is easier to work with and it gives us useful information about the original matrix A.

Theorem 6.1. *Let*
$$A = \begin{bmatrix} a_1 & b_1 \\ a_2 & b_2 \\ a_3 & b_3 \end{bmatrix}$$

and let λ_1 and λ_2 be the eigenvalues of the symmetric matrix $A^T A$, not necessarily different. Let $\mathbf{v}_1$ and $\mathbf{v}_2$ be orthonormal eigenvectors of the matrix $A^T A$ corresponding to the eigenvalues λ_1 and λ_2, respectively. Then

(a) $\lambda_1 \geq 0$ *and* $\lambda_2 \geq 0$,

(b) $\| A\mathbf{v}_1 \| = \sqrt{\lambda_1}$ *and* $\| A\mathbf{v}_2 \| = \sqrt{\lambda_2}$,

(c) $A\mathbf{v}_1 \cdot A\mathbf{v}_2 = 0$,

(d) $\mathrm{Span}\{A\mathbf{v}_1, A\mathbf{v}_2\} = \mathrm{Span}\left\{ \begin{bmatrix} a_1 \\ a_2 \\ a_3 \end{bmatrix}, \begin{bmatrix} b_1 \\ b_2 \\ b_3 \end{bmatrix} \right\}.$

Proof. Since

$$\|A\mathbf{v}_1\|^2 = (A\mathbf{v}_1) \cdot (A\mathbf{v}_1) = (A\mathbf{v}_1)^T (A\mathbf{v}_1)$$
$$= \mathbf{v}_1^T A^T (A\mathbf{v}_1) = (\mathbf{v}_1^T)(A^T A\mathbf{v}_1)$$
$$= (\mathbf{v}_1^T)(\lambda_1 \mathbf{v}_1) = \lambda_1 \mathbf{v}_1^T \mathbf{v}_1 = \lambda_1,$$

we obtain $\lambda_1 \geq 0$ and $\|A\mathbf{v}_1\| = \sqrt{\lambda_1}$. In the same way we get $\lambda_2 \geq 0$ and $\|A\mathbf{v}_2\| = \sqrt{\lambda_2}$.
Since

$$A\mathbf{v}_1 \cdot A\mathbf{v}_2 = \mathbf{v}_1^T (A^T A\mathbf{v}_2) = \lambda_2 \mathbf{v}_1 \cdot \mathbf{v}_2 = 0,$$

we obtain (c).

Now let x_1 and x_2 be arbitrary real numbers. Since the set $\{\mathbf{v}_1, \mathbf{v}_2\}$ is a basis in $\mathbb{R}^2$, we have

$$\begin{bmatrix} x_1 \\ x_2 \end{bmatrix} = s\mathbf{v}_1 + t\mathbf{v}_2,$$

for some real numbers s and t. Then

$$x_1 \begin{bmatrix} a_1 \\ a_2 \\ a_3 \end{bmatrix} + x_2 \begin{bmatrix} b_1 \\ b_2 \\ b_3 \end{bmatrix} = A \begin{bmatrix} x_1 \\ x_2 \end{bmatrix} = A(s\mathbf{v}_1 + t\mathbf{v}_2) = sA\mathbf{v}_1 + tA\mathbf{v}_2,$$

which implies

$$\text{Span}\{A\mathbf{v}_1, A\mathbf{v}_2\} = \text{Span}\left\{ \begin{bmatrix} a_1 \\ a_2 \\ a_3 \end{bmatrix}, \begin{bmatrix} b_1 \\ b_2 \\ b_3 \end{bmatrix} \right\},$$

because it is obvious that $A(\mathbf{v}_1)$ and $A(\mathbf{v}_2)$ are in $\text{Span}\left\{ \begin{bmatrix} a_1 \\ a_2 \\ a_3 \end{bmatrix}, \begin{bmatrix} b_1 \\ b_2 \\ b_3 \end{bmatrix} \right\}$.   □

Definition 6.2. Let

$$A = \begin{bmatrix} a_1 & b_1 \\ a_2 & b_2 \\ a_3 & b_3 \end{bmatrix}$$

and let λ_1 and λ_2 be the eigenvalues of the matrix $A^T A$. The numbers $\sigma_1 = \sqrt{\lambda_1}$ and $\sigma_2 = \sqrt{\lambda_2}$ are called the ***singular values*** of the matrix A.

It is customary to label the eigenvalues λ_1 and λ_2 of the matrix $A^T A$ so that we have $\lambda_1 \geq \lambda_2$ and, consequently, $\sigma_1 \geq \sigma_2$.

Example 6.3. Find the singular values of the matrix $A = \begin{bmatrix} 4 & 2 \\ 1 & -7 \\ 4 & 2 \end{bmatrix}$.

Solution. Since

$$A^T A = \begin{bmatrix} 33 & 9 \\ 9 & 57 \end{bmatrix},$$

the eigenvalues of the matrix $A^T A$ are the roots of the equation

$$\det \begin{bmatrix} 33 - \lambda & 9 \\ 9 & 57 - \lambda \end{bmatrix} = \lambda^2 - 90\lambda + 1800 = 0.$$

Consequently, the eigenvalues of the matrix $A^T A$ are 60 and 30 and the singular values of the matrix A are $\sqrt{60}$ and $\sqrt{30}$. $\qquad\square$

Theorem 6.4. *Let*

$$A = \begin{bmatrix} a_1 & b_1 \\ a_2 & b_2 \\ a_3 & b_3 \end{bmatrix}.$$

Let $\mathbf{v}_1$ *and* $\mathbf{v}_2$ *be orthonormal eigenvectors of the matrix* $A^T A$ *and let* σ_1 *and* σ_2 *are the singular values of the matrix A. The following conditions are equivalent:*

(a) *The vectors* $\begin{bmatrix} a_1 \\ a_2 \\ a_3 \end{bmatrix}$ *and* $\begin{bmatrix} b_1 \\ b_2 \\ b_3 \end{bmatrix}$ *are linearly independent;*

(b) *The vectors* $A\mathbf{v}_1$ *and* $A\mathbf{v}_2$ *are linearly independent;*

(c) $\{A\mathbf{v}_1, A\mathbf{v}_2\}$ *is a basis for* $\mathrm{Span}\left\{ \begin{bmatrix} a_1 \\ a_2 \\ a_3 \end{bmatrix}, \begin{bmatrix} b_1 \\ b_2 \\ b_3 \end{bmatrix} \right\};$

(d) $\sigma_1 > 0$ *and* $\sigma_2 > 0.$

Proof. Equivalence of (a), (b), and (c) follows immediately from part (d) of Theorem 6.1 and Theorem 4.1.22.

If the vectors $A\mathbf{v}_1$ and $A\mathbf{v}_2$ are linearly independent, then $A\mathbf{v}_1 \neq \mathbf{0}$ and $A\mathbf{v}_2 \neq \mathbf{0}$ and consequently $\sigma_1 = \| A\mathbf{v}_1 \| > 0$ and $\sigma_2 = \| A\mathbf{v}_2 \| > 0$. On the other hand, if $\| A\mathbf{v}_1 \| = \sigma_1 > 0$ and $\| A\mathbf{v}_2 \| = \sigma_2 > 0$, then the vectors $A\mathbf{v}_1$ and $A\mathbf{v}_2$ are linearly independent, because they are orthogonal. Indeed, if

$$x A(\mathbf{v}_1) + y A(\mathbf{v}_2) = \begin{bmatrix} 0 \\ 0 \\ 0 \end{bmatrix},$$

then

$$x A(\mathbf{v}_1) \cdot A(\mathbf{v}_1) + y A(\mathbf{v}_2) \cdot A(\mathbf{v}_1) = \begin{bmatrix} 0 \\ 0 \\ 0 \end{bmatrix} \cdot A(\mathbf{v}_1).$$

Since $A(\mathbf{v}_2) \cdot A(\mathbf{v}_1) = 0$, the above can be written as

$$x \| A(\mathbf{v}_1) \|^2 = 0.$$

This gives us $x = 0$, because $\| A(\mathbf{v}_1) \| \neq 0$. In the same way we can show that $y = 0$. □

Example 6.5. Find an orthogonal basis for $\mathrm{Span}\left\{ \begin{bmatrix} 4 \\ 1 \\ 4 \end{bmatrix}, \begin{bmatrix} 2 \\ -7 \\ 2 \end{bmatrix} \right\}$.

Solution. Let $A = \begin{bmatrix} 4 & 2 \\ 1 & -7 \\ 4 & 2 \end{bmatrix}$. First we find that the vectors $\begin{bmatrix} 1 \\ 3 \end{bmatrix}$ and $\begin{bmatrix} -3 \\ 1 \end{bmatrix}$ are a basis

of eigenvectors of the matrix $A^T A$. Since $A \begin{bmatrix} 1 \\ 3 \end{bmatrix} = \begin{bmatrix} 10 \\ -20 \\ 10 \end{bmatrix}$ and $A \begin{bmatrix} -3 \\ 1 \end{bmatrix} = \begin{bmatrix} 10 \\ 10 \\ 10 \end{bmatrix}$, the

set $\left\{ \begin{bmatrix} 1 \\ -2 \\ 1 \end{bmatrix}, \begin{bmatrix} 1 \\ 1 \\ 1 \end{bmatrix} \right\}$ is an orthogonal basis for $\mathrm{Span}\left\{ \begin{bmatrix} 4 \\ 1 \\ 4 \end{bmatrix}, \begin{bmatrix} 2 \\ -7 \\ 2 \end{bmatrix} \right\}$. □

Now we consider the case when the columns of A are linearly dependent.

Theorem 6.6. *Let*

$$A = \begin{bmatrix} a_1 & b_1 \\ a_2 & b_2 \\ a_3 & b_3 \end{bmatrix}$$

be a nonzero matrix. Let $\mathbf{v}_1$ and $\mathbf{v}_2$ be orthonormal eigenvectors of the matrix $A^T A$ and let σ_1 and σ_2 be the singular values of the matrix A such that $\sigma_1 \geq \sigma_2$. The following conditions are equivalent:

(a) *The vectors $\begin{bmatrix} a_1 \\ a_2 \\ a_3 \end{bmatrix}$ and $\begin{bmatrix} b_1 \\ b_2 \\ b_3 \end{bmatrix}$ are linearly dependent;*

(b) *The vectors $A\mathbf{v}_1$ and $A\mathbf{v}_2$ are linearly dependent;*

(c) *$\{A\mathbf{v}_1\}$ is a basis for $\mathrm{Span}\left\{ \begin{bmatrix} a_1 \\ a_2 \\ a_3 \end{bmatrix}, \begin{bmatrix} b_1 \\ b_2 \\ b_3 \end{bmatrix} \right\}$;*

(d) *$\sigma_1 > 0$ and $\sigma_2 = 0$.*

Proof. Equivalence of (a) and (b) follows immediately from the equivalence of (a) and (b) in Theorem 6.1.

Since the vectors $A\mathbf{v}_1$ and $A\mathbf{v}_2$ are orthogonal, they are linearly dependent if and only if one of them is the zero vector. If one of the vectors is the zero vector, then it must be $A\mathbf{v}_2$ because we have

$$\|A\mathbf{v}_1\| = \sigma_1 \geq \sigma_2 = \|A\mathbf{v}_2\|.$$

This shows that the vectors $A\mathbf{v}_1$ and $A\mathbf{v}_2$ are linearly dependent if and only if $A\mathbf{v}_2 = \mathbf{0}$. If $A\mathbf{v}_2 = \mathbf{0}$, then

$$\text{Span}\{A\mathbf{v}_1\} = \text{Span}\{A\mathbf{v}_1, A\mathbf{v}_2\} = \text{Span}\left\{ \begin{bmatrix} a_1 \\ a_2 \\ a_3 \end{bmatrix}, \begin{bmatrix} b_1 \\ b_2 \\ b_3 \end{bmatrix} \right\},$$

so $\{A\mathbf{v}_1\}$ is a basis for $\text{Span}\left\{ \begin{bmatrix} a_1 \\ a_2 \\ a_3 \end{bmatrix}, \begin{bmatrix} b_1 \\ b_2 \\ b_3 \end{bmatrix} \right\}.$

Now, if $\{A\mathbf{v}_1\}$ is a basis for $\text{Span}\left\{ \begin{bmatrix} a_1 \\ a_2 \\ a_3 \end{bmatrix}, \begin{bmatrix} b_1 \\ b_2 \\ b_3 \end{bmatrix} \right\}$, then we must have

$$\sigma_1 = \|A\mathbf{v}_1\| > 0,$$

because the matrix A is a nonzero matrix, and

$$\sigma_2 = \|A\mathbf{v}_2\| = 0,$$

because the vectors $A\mathbf{v}_1$ and $A\mathbf{v}_2$ are orthogonal and

$$\text{Span}\{A\mathbf{v}_1\} = \text{Span}\left\{ \begin{bmatrix} a_1 \\ a_2 \\ a_3 \end{bmatrix}, \begin{bmatrix} b_1 \\ b_2 \\ b_3 \end{bmatrix} \right\} = \text{Span}\{A\mathbf{v}_1, A\mathbf{v}_2\}.$$

Finally, if $\sigma_2 = 0$, then $\|A\mathbf{v}_2\| = \sigma_2 = 0$ and thus $A\mathbf{v}_2 = \mathbf{0}$, so the vectors $A\mathbf{v}_1$ and $A\mathbf{v}_2$ are linearly dependent. □

In the next lemma we prove a simple identity for orthonormal vectors in $\mathbb{R}^2$ that will be used in the proof of Theorem 6.8.

Lemma 6.7. *If the vectors $\mathbf{v}_1$ and $\mathbf{v}_2$ in $\mathbb{R}^2$ are orthonormal, then*

$$\begin{bmatrix} 1 & 0 \\ 0 & 1 \end{bmatrix} = \mathbf{v}_1\mathbf{v}_1^T + \mathbf{v}_2\mathbf{v}_2^T.$$

Proof. If $\mathbf{v}_1 = \begin{bmatrix} a \\ b \end{bmatrix}$, then $\mathbf{v}_2 = \begin{bmatrix} -b \\ a \end{bmatrix}$ or $\mathbf{v}_2 = \begin{bmatrix} b \\ -a \end{bmatrix}.$

If $\mathbf{v}_2 = \begin{bmatrix} -b \\ a \end{bmatrix}$, then we have

$$\mathbf{v}_1\mathbf{v}_1^T + \mathbf{v}_2\mathbf{v}_2^T = \begin{bmatrix} a \\ b \end{bmatrix}\begin{bmatrix} a & b \end{bmatrix} + \begin{bmatrix} -b \\ a \end{bmatrix}\begin{bmatrix} -b & a \end{bmatrix}$$

$$= \begin{bmatrix} a^2 & ab \\ ab & b^2 \end{bmatrix} + \begin{bmatrix} b^2 & -ab \\ -ab & a^2 \end{bmatrix}$$

$$= \begin{bmatrix} a^2 + b^2 & 0 \\ 0 & a^2 + b^2 \end{bmatrix}$$

$$= \begin{bmatrix} 1 & 0 \\ 0 & 1 \end{bmatrix},$$

because $a^2 + b^2 = \|\mathbf{v}_1\| = 1$.

In the case $\mathbf{v}_2 = \begin{bmatrix} b \\ -a \end{bmatrix}$ the above argument requires only some obvious modifications.   □

Now we prove a result similar to the spectral decomposition for symmetric matrices.

Theorem 6.8. *Let* $A = \begin{bmatrix} a_1 & b_1 \\ a_2 & b_2 \\ a_3 & b_3 \end{bmatrix}$ *be a nonzero matrix.*

(a) *If the vectors* $\begin{bmatrix} a_1 \\ a_2 \\ a_3 \end{bmatrix}$ *and* $\begin{bmatrix} b_1 \\ b_2 \\ b_3 \end{bmatrix}$ *are linearly independent, then there are two orthonormal vectors* $\mathbf{u}_1$ *and* $\mathbf{u}_2$ *in* $\mathbb{R}^3$ *and two orthonormal vectors* $\mathbf{v}_1$ *and* $\mathbf{v}_2$ *in* $\mathbb{R}^2$ *such that*

$$A = \sigma_1\mathbf{u}_1\mathbf{v}_1^T + \sigma_2\mathbf{u}_2\mathbf{v}_2^T,$$

where σ_1 *and* σ_2 *are the singular values of the matrix A.*

(b) *If the vectors* $\begin{bmatrix} a_1 \\ a_2 \\ a_3 \end{bmatrix}$ *and* $\begin{bmatrix} b_1 \\ b_2 \\ b_3 \end{bmatrix}$ *are linearly dependent, then there are a unit vector* $\mathbf{u}_1$ *in* $\mathbb{R}^3$ *and a unit vector* $\mathbf{v}_1$ *in* $\mathbb{R}^2$ *such that*

$$A = \sigma_1\mathbf{u}_1\mathbf{v}_1^T,$$

where σ_1 *is the singular value of the matrix A such that* $\sigma_1 > 0$.

Proof. Let $\mathbf{v}_1$ and $\mathbf{v}_2$ be orthonormal eigenvectors of the matrix $A^T A$. Since

$$\begin{bmatrix} 1 & 0 \\ 0 & 1 \end{bmatrix} = \mathbf{v}_1\mathbf{v}_1^T + \mathbf{v}_2\mathbf{v}_2^T,$$

we have

$$A = A \begin{bmatrix} 1 & 0 \\ 0 & 1 \end{bmatrix} = A\mathbf{v}_1\mathbf{v}_1^T + A\mathbf{v}_2\mathbf{v}_2^T. \tag{6.1}$$

If the vectors $\begin{bmatrix} a_1 \\ a_2 \\ a_3 \end{bmatrix}$ and $\begin{bmatrix} b_1 \\ b_2 \\ b_3 \end{bmatrix}$ are linearly independent, then $A\mathbf{v}_1 \neq \mathbf{0}$ and $A\mathbf{v}_2 \neq \mathbf{0}$, by Theorem 6.4, and (6.1) can be written as

$$A = \|A\mathbf{v}_1\| \left(\tfrac{1}{\|A\mathbf{v}_1\|} A\mathbf{v}_1 \right) \mathbf{v}_1^T + \|A\mathbf{v}_2\| \left(\tfrac{1}{\|A\mathbf{v}_2\|} A\mathbf{v}_2 \right) \mathbf{v}_2^T = \sigma_1\mathbf{u}_1\mathbf{v}_1^T + \sigma_2\mathbf{u}_2\mathbf{v}_2^T,$$

where $\sigma_1 = \|A\mathbf{v}_1\|$, $\mathbf{u}_1 = \tfrac{1}{\|A\mathbf{v}_1\|} A\mathbf{v}_1$, $\sigma_2 = \|A\mathbf{v}_2\|$, and $\mathbf{u}_2 = \tfrac{1}{\|A\mathbf{v}_2\|} A\mathbf{v}_2$.

If the vectors $\begin{bmatrix} a_1 \\ a_2 \\ a_3 \end{bmatrix}$ and $\begin{bmatrix} b_1 \\ b_2 \\ b_3 \end{bmatrix}$ are linearly dependent, then $A\mathbf{v}_1 \neq \mathbf{0}$ and $A\mathbf{v}_2 = \mathbf{0}$, by Theorem 6.4, and (6.1) becomes

$$A = A\mathbf{v}_1\mathbf{v}_1^T = \|A\mathbf{v}_1\| \left(\tfrac{1}{\|A\mathbf{v}_1\|} A\mathbf{v}_1 \right) \mathbf{v}_1^T = \sigma_1\mathbf{u}_1\mathbf{v}_1^T$$

where $\sigma_1 = \|A\mathbf{v}_1\|$ and $\mathbf{u}_1 = \tfrac{1}{\|A\mathbf{v}_1\|} A\mathbf{v}_1$.   □

Note that $A = \sigma_1\mathbf{u}_1\mathbf{v}_1^T$ can be viewed as a special case of $A = \sigma_1\mathbf{u}_1\mathbf{v}_1^T + \sigma_2\mathbf{u}_2\mathbf{v}_2^T$ with $\sigma_2 = 0$.

The representation

$$A = \sigma_1\mathbf{u}_1\mathbf{v}_1^T + \sigma_2\mathbf{u}_2\mathbf{v}_2^T$$

is called the **outer product expansion** for the matrix A.

In the proof of Theorem 6.8 we assume from the beginning that the eigenvectors $\mathbf{v}_1$ and $\mathbf{v}_2$ are unit vectors. When we find eigenvectors of a specific matrix, they are usually not unit vectors. The "recipe" for finding the outer product expansion of a matrix 3 × 2 matrix given below takes this fact into account.

Step 1 Calculate the matrix $A^T A$.

Step 2 Find the eigenvalues $\lambda_1 \geq \lambda_2$ and eigenvectors $\mathbf{V}_1, \mathbf{V}_2$ of $A^T A$.

Step 3 Write

$$\begin{bmatrix} 1 & 0 \\ 0 & 1 \end{bmatrix} = \frac{1}{\|\mathbf{V}_1\|^2} \mathbf{V}_1\mathbf{V}_1^T + \frac{1}{\|\mathbf{V}_2\|^2} \mathbf{V}_2\mathbf{V}_2^T.$$

Step 4 Multiply both sides of the equality by the matrix A to get

$$A = A \begin{bmatrix} 1 & 0 \\ 0 & 1 \end{bmatrix} = \frac{1}{\|\mathbf{V}_1\|^2} (A\mathbf{V}_1)\mathbf{V}_1^T + \frac{1}{\|\mathbf{V}_2\|^2} (A\mathbf{V}_2)\mathbf{V}_2^T.$$

Step 5 If the columns of A are linearly independent, then the outer product expansion of the matrix A is

$$A = \sigma_1 \left(\frac{1}{\sigma_1 \|\mathbf{V}_1\|} A\mathbf{V}_1 \right) \left(\frac{1}{\|\mathbf{V}_1\|} \mathbf{V}_1^T \right) + \sigma_2 \left(\frac{1}{\sigma_2 \|\mathbf{V}_2\|} A\mathbf{V}_2 \right) \left(\frac{1}{\|\mathbf{V}_2\|} \mathbf{V}_2^T \right)$$

or

$$A = \sigma_1 \mathbf{u}_1 \mathbf{v}_1^T + \sigma_2 \mathbf{u}_2 \mathbf{v}_2^T,$$

where

$$\mathbf{v}_1 = \frac{1}{\|\mathbf{V}_1\|} \mathbf{V}_1^T, \qquad\qquad \mathbf{v}_2 = \frac{1}{\|\mathbf{V}_2\|} \mathbf{V}_2^T,$$

$$\mathbf{u}_1 = \frac{1}{\sigma_1 \|\mathbf{V}_1\|} A\mathbf{V}_1, \qquad\qquad \mathbf{u}_2 = \frac{1}{\sigma_2 \|\mathbf{V}_2\|} A\mathbf{V}_2.$$

To get this expansion we used the fact that the singular values are

$$\sigma_1 = \sqrt{\lambda_1} = \frac{\|A\mathbf{V}_1\|}{\|\mathbf{V}_1\|} = \left\| A\left(\frac{\mathbf{V}_1}{\|\mathbf{V}_1\|} \right) \right\| \quad \text{and} \quad \sigma_2 = \sqrt{\lambda_2} = \frac{\|A\mathbf{V}_2\|}{\|\mathbf{V}_2\|} = \left\| A\left(\frac{\mathbf{V}_2}{\|\mathbf{V}_2\|} \right) \right\|.$$

If the columns of A are linearly dependent, then the outer product expansion of the matrix A is

$$A = \sigma_1 \left(\frac{1}{\sigma_1 \|\mathbf{V}_1\|} A\mathbf{V}_1 \right) \left(\frac{1}{\|\mathbf{V}_1\|} \mathbf{V}_1^T \right) = \sigma_1 \mathbf{u}_1 \mathbf{v}_1^T,$$

where

$$\mathbf{v}_1 = \frac{1}{\|\mathbf{V}_1\|} \mathbf{V}_1^T, \quad \mathbf{u}_1 = \frac{1}{\sigma_1 \|\mathbf{V}_1\|} A\mathbf{V}_1.$$

To get this expansion we used the fact that

$$\sigma_1 = \sqrt{\lambda_1} = \frac{\|A\mathbf{V}_1\|}{\|\mathbf{V}_1\|} = \left\| A\left(\frac{\mathbf{V}_1}{\|\mathbf{V}_1\|} \right) \right\|.$$

Example 6.9. Find the outer product expansion for the matrix $A = \begin{bmatrix} 1 & 4 \\ 1 & 1 \\ 1 & -1 \end{bmatrix}$.

Solution.

   Step 1 We calculate the matrix $A^T A$:

$$A^T A = \begin{bmatrix} 1 & 1 & 1 \\ 4 & 1 & -1 \end{bmatrix} \begin{bmatrix} 1 & 4 \\ 1 & 1 \\ 1 & -1 \end{bmatrix} = \begin{bmatrix} 3 & 4 \\ 4 & 18 \end{bmatrix}.$$

   Step 2 The eigenvalues of the matrix $A^T A$ are $\lambda_1 = 19$ and $\lambda_2 = 2$, $\begin{bmatrix} 1 \\ 4 \end{bmatrix}$ is an eigenvector corresponding to the eigenvalue 19, and $\begin{bmatrix} 4 \\ -1 \end{bmatrix}$ an eigenvector corresponding to the eigenvalue 2.

Step 3

$$\begin{bmatrix} 1 & 0 \\ 0 & 1 \end{bmatrix} = \frac{1}{17}\begin{bmatrix} 1 \\ 4 \end{bmatrix}\begin{bmatrix} 1 & 4 \end{bmatrix} + \frac{1}{17}\begin{bmatrix} 4 \\ -1 \end{bmatrix}\begin{bmatrix} 4 & -1 \end{bmatrix}.$$

Step 4

$$A = \begin{bmatrix} 1 & 4 \\ 1 & 1 \\ 1 & -1 \end{bmatrix}\begin{bmatrix} 1 & 0 \\ 0 & 1 \end{bmatrix}$$

$$= \frac{1}{17}\begin{bmatrix} 1 & 4 \\ 1 & 1 \\ 1 & -1 \end{bmatrix}\begin{bmatrix} 1 \\ 4 \end{bmatrix}\begin{bmatrix} 1 & 4 \end{bmatrix} + \frac{1}{17}\begin{bmatrix} 1 & 4 \\ 1 & 1 \\ 1 & -1 \end{bmatrix}\begin{bmatrix} 4 \\ -1 \end{bmatrix}\begin{bmatrix} 4 & -1 \end{bmatrix}$$

$$= \frac{1}{17}\begin{bmatrix} 17 \\ 5 \\ -3 \end{bmatrix}\begin{bmatrix} 1 & 4 \end{bmatrix} + \frac{1}{17}\begin{bmatrix} 0 \\ 3 \\ 5 \end{bmatrix}\begin{bmatrix} 4 & -1 \end{bmatrix}$$

Step 5

$$A = \sqrt{19}\left(\frac{1}{\sqrt{19\cdot17}}\begin{bmatrix} 17 \\ 5 \\ -3 \end{bmatrix}\right)\left(\frac{1}{\sqrt{17}}\begin{bmatrix} 1 & 4 \end{bmatrix}\right) + \sqrt{2}\left(\frac{1}{\sqrt{2\cdot17}}\begin{bmatrix} 0 \\ 3 \\ 5 \end{bmatrix}\right)\left(\frac{1}{\sqrt{17}}\begin{bmatrix} 4 & -1 \end{bmatrix}\right)$$

or

$$A = \sigma_1\mathbf{u}_1\mathbf{v}_1^T + \sigma_2\mathbf{u}_2\mathbf{v}_2^T,$$

where

$$\mathbf{u}_1 = \frac{1}{\sqrt{19\cdot17}}\begin{bmatrix} 17 \\ 5 \\ -3 \end{bmatrix}, \mathbf{u}_2 = \frac{1}{\sqrt{2\cdot17}}\begin{bmatrix} 0 \\ 3 \\ 5 \end{bmatrix}, \mathbf{v}_1 = \frac{1}{\sqrt{17}}\begin{bmatrix} 1 \\ 4 \end{bmatrix}, \mathbf{v}_2 = \frac{1}{\sqrt{17}}\begin{bmatrix} 4 \\ -1 \end{bmatrix},$$

and the singular values are $\sigma_1 = \sqrt{19}$ and $\sigma_2 = \sqrt{2}$. □

Example 6.10. Find the outer product expansion of the matrix $A = \begin{bmatrix} 1 & 2 \\ -1 & -2 \\ 1 & 2 \end{bmatrix}$.

Solution.

Step 1

$$A^TA = \begin{bmatrix} 3 & 6 \\ 6 & 12 \end{bmatrix}.$$

Step 2 The eigenvalues of the matrix A^TA are $\lambda_1 = 15$ and $\lambda_2 = 0$, $\begin{bmatrix} 1 \\ 2 \end{bmatrix}$ is an eigenvector corresponding to the eigenvalue 15, and $\begin{bmatrix} 2 \\ -1 \end{bmatrix}$ an eigenvector corresponding to the eigenvalue 0.

Step 3

$$\begin{bmatrix} 1 & 0 \\ 0 & 1 \end{bmatrix} = \frac{1}{5}\begin{bmatrix} 1 \\ 2 \end{bmatrix}\begin{bmatrix} 1 & 2 \end{bmatrix} + \frac{1}{5}\begin{bmatrix} 2 \\ -1 \end{bmatrix}\begin{bmatrix} 2 & -1 \end{bmatrix}.$$

Step 4

$$A = \begin{bmatrix} 1 & 2 \\ -1 & -2 \\ 1 & 2 \end{bmatrix}\begin{bmatrix} 1 & 0 \\ 0 & 1 \end{bmatrix}$$

$$= \frac{1}{5}\begin{bmatrix} 1 & 2 \\ -1 & -2 \\ 1 & 2 \end{bmatrix}\begin{bmatrix} 1 \\ 2 \end{bmatrix}\begin{bmatrix} 1 & 2 \end{bmatrix} + \frac{1}{5}\begin{bmatrix} 1 & 2 \\ -1 & -2 \\ 1 & 2 \end{bmatrix}\begin{bmatrix} 2 \\ -1 \end{bmatrix}\begin{bmatrix} 2 & -1 \end{bmatrix}$$

$$= \frac{1}{5}\begin{bmatrix} 5 \\ -5 \\ 5 \end{bmatrix}\begin{bmatrix} 1 & 2 \end{bmatrix} + \frac{1}{5}\begin{bmatrix} 0 \\ 0 \\ 0 \end{bmatrix}\begin{bmatrix} 2 & -1 \end{bmatrix}$$

$$= \frac{1}{5}\begin{bmatrix} 5 \\ -5 \\ 5 \end{bmatrix}\begin{bmatrix} 1 & 2 \end{bmatrix}$$

$$= \begin{bmatrix} 1 \\ -1 \\ 1 \end{bmatrix}\begin{bmatrix} 1 & 2 \end{bmatrix}$$

Step 5

$$A = \sqrt{15}\left(\frac{1}{\sqrt{3}}\begin{bmatrix} 1 \\ -1 \\ 1 \end{bmatrix}\right)\left(\frac{1}{\sqrt{5}}\begin{bmatrix} 1 & 2 \end{bmatrix}\right) = \sigma_1 \mathbf{u}_1 \mathbf{v}_1^T$$

where

$$\mathbf{u}_1 = \frac{1}{\sqrt{3}}\begin{bmatrix} 1 \\ -1 \\ 1 \end{bmatrix}, \quad \mathbf{v}_1 = \frac{1}{\sqrt{5}}[, 1 \ 2]$$

and $\sigma_1 = \sqrt{15}$.

□

Definition 6.11. Let A be a 3×2 matrix. By a ***singular value decomposition*** of A we mean the representation of A in the form

$$A = \begin{bmatrix} \mathbf{u}_1 & \mathbf{u}_2 & \mathbf{u}_3 \end{bmatrix} \begin{bmatrix} \sigma_1 & 0 \\ 0 & \sigma_2 \\ 0 & 0 \end{bmatrix} \begin{bmatrix} \mathbf{v}_1^T \\ \mathbf{v}_2^T \end{bmatrix},$$

where $\{\mathbf{u}_1, \mathbf{u}_2, \mathbf{u}_3\}$ is an orthonormal basis in $\mathbb{R}^3$, $\{\mathbf{v}_1, \mathbf{v}_2\}$ is an orthogonal basis in $\mathbb{R}^2$, and σ_1 and σ_2 are the singular values of the matrix A.

The singular value decomposition should be thought of as a version of diagonalization of 2×2 matrices for 3×2 matrices. The following theorem implies that every 3×2 matrix has a singular value decomposition.

Theorem 6.12. *Let*

$$A = \begin{bmatrix} a_1 & b_1 \\ a_2 & b_2 \\ a_3 & b_3 \end{bmatrix}$$

be a nonzero matrix.

(a) *If the vectors* $\begin{bmatrix} a_1 \\ a_2 \\ a_3 \end{bmatrix}$ *and* $\begin{bmatrix} b_1 \\ b_2 \\ b_3 \end{bmatrix}$ *are linearly independent, then there is an orthonormal basis* $\{\mathbf{u}_1, \mathbf{u}_2, \mathbf{u}_3\}$ *in* $\mathbb{R}^3$ *and an orthogonal basis* $\{\mathbf{v}_1, \mathbf{v}_2\}$ *in* $\mathbb{R}^2$ *such that*

$$A = \begin{bmatrix} \mathbf{u}_1 & \mathbf{u}_2 & \mathbf{u}_3 \end{bmatrix} \begin{bmatrix} \sigma_1 & 0 \\ 0 & \sigma_2 \\ 0 & 0 \end{bmatrix} \begin{bmatrix} \mathbf{v}_1^T \\ \mathbf{v}_2^T \end{bmatrix},$$

where σ_1 *and* σ_2 *are the singular values of the matrix* A.

(b) *If the vectors* $\begin{bmatrix} a_1 \\ a_2 \\ a_3 \end{bmatrix}$ *and* $\begin{bmatrix} b_1 \\ b_2 \\ b_3 \end{bmatrix}$ *are linearly dependent, then there is an orthonormal basis* $\{\mathbf{u}_1, \mathbf{u}_2, \mathbf{u}_3\}$ *in* $\mathbb{R}^3$ *and an orthogonal basis* $\{\mathbf{v}_1, \mathbf{v}_2\}$ *in* $\mathbb{R}^2$ *such that*

$$A = \begin{bmatrix} \mathbf{u}_1 & \mathbf{u}_2 & \mathbf{u}_3 \end{bmatrix} \begin{bmatrix} \sigma_1 & 0 \\ 0 & 0 \\ 0 & 0 \end{bmatrix} \begin{bmatrix} \mathbf{v}_1^T \\ \mathbf{v}_2^T \end{bmatrix},$$

where σ_1 *is the singular value of the matrix* A *such that* $\sigma_1 > 0$.

Proof. If the vectors $\begin{bmatrix} a_1 \\ a_2 \\ a_3 \end{bmatrix}$ and $\begin{bmatrix} b_1 \\ b_2 \\ b_3 \end{bmatrix}$ are linearly independent, then there are two orthonormal vectors $\mathbf{u}_1, \mathbf{u}_2$ in $\mathbb{R}^3$ and two orthonormal vectors $\mathbf{v}_1$ and $\mathbf{v}_2$ in $\mathbb{R}^2$ such that

$$A = \sigma_1 \mathbf{u}_1 \mathbf{v}_1^T + \sigma_2 \mathbf{u}_2 \mathbf{v}_2^T, \tag{6.2}$$

where σ_1 and σ_2 are the singular values of the matrix A, by Theorem 6.8. The equality (6.2) can be written as

$$A = \begin{bmatrix} \mathbf{u}_1 & \mathbf{u}_2 \end{bmatrix} \begin{bmatrix} \sigma_1 & 0 \\ 0 & \sigma_2 \end{bmatrix} \begin{bmatrix} \mathbf{v}_1^T \\ \mathbf{v}_2^T \end{bmatrix},$$

because for every vector $\mathbf{x}$ of $\mathbf{R}^2$ we have

$$\begin{aligned} A\mathbf{x} &= (\sigma_1 \mathbf{u}_1 \mathbf{v}_1^T + \sigma_2 \mathbf{u}_2 \mathbf{v}_2^T)\mathbf{x} \\ &= \begin{bmatrix} \mathbf{u}_1 & \mathbf{u}_2 \end{bmatrix} \begin{bmatrix} \sigma_1 \mathbf{v}_1^T \mathbf{x} \\ \sigma_2 \mathbf{v}_2^T \mathbf{x} \end{bmatrix} \\ &= \begin{bmatrix} \mathbf{u}_1 & \mathbf{u}_2 \end{bmatrix} \begin{bmatrix} \sigma_1 & 0 \\ 0 & \sigma_2 \end{bmatrix} \begin{bmatrix} \mathbf{v}_1^T \mathbf{x} \\ \mathbf{v}_2^T \mathbf{x} \end{bmatrix} \\ &= \begin{bmatrix} \mathbf{u}_1 & \mathbf{u}_2 \end{bmatrix} \begin{bmatrix} \sigma_1 & 0 \\ 0 & \sigma_2 \end{bmatrix} \begin{bmatrix} \mathbf{v}_1^T \\ \mathbf{v}_2^T \end{bmatrix} \mathbf{x}. \end{aligned}$$

Note that, if $\mathbf{u}_3$ is an arbitrary vector in $\mathbb{R}^3$, then we have

$$\begin{bmatrix} \mathbf{u}_1 & \mathbf{u}_2 \end{bmatrix} \begin{bmatrix} \sigma_1 & 0 \\ 0 & \sigma_2 \end{bmatrix} = \begin{bmatrix} \mathbf{u}_1 & \mathbf{u}_2 & \mathbf{u}_3 \end{bmatrix} \begin{bmatrix} \sigma_1 & 0 \\ 0 & \sigma_2 \\ 0 & 0 \end{bmatrix}.$$

Since we want that $\{\mathbf{u}_1, \mathbf{u}_2, \mathbf{u}_3\}$ to be an orthonormal basis in $\mathbb{R}^3$, we choose

$$\mathbf{u}_3 = \frac{1}{\|\mathbf{u}_1 \times \mathbf{u}_2\|} (\mathbf{u}_1 \times \mathbf{u}_2).$$

We note that $\mathbf{u}_1 \times \mathbf{u}_2 \neq \mathbf{0}$, because the vectors $\mathbf{u}_1$ and $\mathbf{u}_2$ are linearly independent. The desired representation of the matrix A is

$$A = \begin{bmatrix} \mathbf{u}_1 & \mathbf{u}_2 & \mathbf{u}_3 \end{bmatrix} \begin{bmatrix} \sigma_1 & 0 \\ 0 & \sigma_2 \\ 0 & 0 \end{bmatrix} \begin{bmatrix} \mathbf{v}_1^T \\ \mathbf{v}_2^T \end{bmatrix}.$$

Now we assume that the vectors $\begin{bmatrix} a_1 \\ a_2 \\ a_3 \end{bmatrix}$ and $\begin{bmatrix} b_1 \\ b_2 \\ b_3 \end{bmatrix}$ are linearly dependent. By Theorem 6.8, there are a unit vector $\mathbf{u}_1$ in $\mathbb{R}^3$ and a unit vector $\mathbf{v}_1$ in $\mathbb{R}^2$ such that

$$A = \sigma_1 \mathbf{u}_1 \mathbf{v}_1^T, \tag{6.3}$$

where σ_1 is the singular value of the matrix A such that $\sigma_1 > 0$. Note that, if $\mathbf{u}_2$ and $\mathbf{u}_3$ are arbitrary vectors in $\mathbb{R}^3$, then we have

$$\sigma_1 \mathbf{u}_1 = \begin{bmatrix} \mathbf{u}_1 & \mathbf{u}_2 & \mathbf{u}_3 \end{bmatrix} \begin{bmatrix} \sigma_1 \\ 0 \\ 0 \end{bmatrix}.$$

We want $\{\mathbf{u}_1, \mathbf{u}_2, \mathbf{u}_3\}$ to be an orthonormal basis in $\mathbb{R}^3$, so we choose $\mathbf{u}_2$ and $\mathbf{u}_3$ such that $\{\mathbf{u}_2, \mathbf{u}_3\}$ is an orthonormal basis of the vector plane $\mathbf{u}_1 \cdot \mathbf{x} = 0$. Since

$$\begin{bmatrix} \sigma_1 \\ 0 \\ 0 \end{bmatrix} \mathbf{v}_1^T = \begin{bmatrix} \sigma_1 & 0 \\ 0 & 0 \\ 0 & 0 \end{bmatrix} \begin{bmatrix} \mathbf{v}_1^T \\ \mathbf{v}_2^T \end{bmatrix},$$

we obtain the desired representation of the matrix A:

$$A = \begin{bmatrix} \mathbf{u}_1 & \mathbf{u}_2 & \mathbf{u}_3 \end{bmatrix} \begin{bmatrix} \sigma_1 & 0 \\ 0 & 0 \\ 0 & 0 \end{bmatrix} \begin{bmatrix} \mathbf{v}_1^T \\ \mathbf{v}_2^T \end{bmatrix}.$$

$\square$

Example 6.13. Find the singular value decomposition of the matrix

$$A = \begin{bmatrix} 1 & 3 \\ \sqrt{2} & 0 \\ 0 & \sqrt{2} \end{bmatrix}.$$

Solution. The outer product expansion of A is

$$A = \sigma_1 \mathbf{u}_1 \mathbf{v}_1^T + \sigma_2 \mathbf{u}_2 \mathbf{v}_2^T,$$

where

$$\mathbf{v}_1 = \frac{1}{\sqrt{10}} \begin{bmatrix} 1 \\ 3 \end{bmatrix}, \ \mathbf{v}_2 = \frac{1}{\sqrt{10}} \begin{bmatrix} 3 \\ -1 \end{bmatrix}, \ \mathbf{u}_1 = \frac{1}{2\sqrt{30}} \begin{bmatrix} 10 \\ \sqrt{2} \\ 3\sqrt{2} \end{bmatrix}, \ \mathbf{u}_2 = \frac{1}{2\sqrt{5}} \begin{bmatrix} 0 \\ 3\sqrt{2} \\ -\sqrt{2} \end{bmatrix},$$

and the singular values are $\sigma_1 = 2\sqrt{3}$ and $\sigma_2 = \sqrt{2}$.

To complete $\{\mathbf{u}_1, \mathbf{u}_2\}$ to an orthonormal basis in $\mathbb{R}^3$ we calculate the cross product

$$\begin{bmatrix} 0 \\ 3\sqrt{2} \\ -\sqrt{2} \end{bmatrix} \times \begin{bmatrix} 10 \\ \sqrt{2} \\ 3\sqrt{2} \end{bmatrix} = \begin{bmatrix} 20 \\ -10\sqrt{2} \\ -30\sqrt{2} \end{bmatrix} = 10\sqrt{2} \begin{bmatrix} \sqrt{2} \\ -1 \\ -3 \end{bmatrix}$$

and the norm

$$\left\| \begin{bmatrix} \sqrt{2} \\ -1 \\ -3 \end{bmatrix} \right\| = \sqrt{12} = 2\sqrt{3}$$

and then define

$$\mathbf{u}_3 = \frac{1}{2\sqrt{3}} \begin{bmatrix} \sqrt{2} \\ -1 \\ -3 \end{bmatrix} = \begin{bmatrix} \frac{1}{\sqrt{6}} \\ -\frac{1}{2\sqrt{3}} \\ -\frac{3}{2\sqrt{3}} \end{bmatrix}.$$

Now $\{\mathbf{u}_1, \mathbf{u}_2, \mathbf{u}_3\}$ is an orthonormal basis in $\mathbb{R}^3$ and the singular value decomposition of the matrix A is

$$A = \begin{bmatrix} \mathbf{u}_1 & \mathbf{u}_2 & \mathbf{u}_3 \end{bmatrix} \begin{bmatrix} 2\sqrt{3} & 0 \\ 0 & \sqrt{2} \\ 0 & 0 \end{bmatrix} \begin{bmatrix} \mathbf{v}_1^T \\ \mathbf{v}_2^T \end{bmatrix},$$

that is,

$$\begin{bmatrix} 1 & 3 \\ \sqrt{2} & 0 \\ 0 & \sqrt{2} \end{bmatrix} = \begin{bmatrix} \frac{\sqrt{10}}{2\sqrt{3}} & 0 & \frac{1}{\sqrt{6}} \\ \frac{1}{2\sqrt{15}} & \frac{3}{\sqrt{10}} & -\frac{1}{2\sqrt{3}} \\ \frac{3}{2\sqrt{15}} & -\frac{1}{\sqrt{10}} & -\frac{3}{2\sqrt{3}} \end{bmatrix} \begin{bmatrix} 2\sqrt{3} & 0 \\ 0 & \sqrt{2} \\ 0 & 0 \end{bmatrix} \begin{bmatrix} \frac{1}{\sqrt{10}} & \frac{3}{\sqrt{10}} \\ \frac{3}{\sqrt{10}} & 1\frac{1}{\sqrt{10}} \end{bmatrix}.$$

$\square$

Example 6.14. Find the singular value decomposition of the matrix

$$A = \begin{bmatrix} 1 & -1 \\ -3 & 3 \\ -1 & 1 \end{bmatrix}.$$

Solution. The outer product expansion of A is

$$A = \sigma_1 \mathbf{u}_1 \mathbf{v}_1^T,$$

where

$$\mathbf{v}_1 = \begin{bmatrix} \frac{1}{\sqrt{2}} \\ -\frac{1}{\sqrt{2}} \end{bmatrix}, \quad \mathbf{u}_1 = \frac{1}{\sqrt{11}} \begin{bmatrix} 1 \\ -3 \\ -1 \end{bmatrix} = \begin{bmatrix} \frac{1}{\sqrt{11}} \\ -\frac{3}{\sqrt{11}} \\ -\frac{1}{\sqrt{11}} \end{bmatrix}$$

and $\sigma_1 = \sqrt{22}$.

Consequently,

$$A = \begin{bmatrix} \frac{1}{\sqrt{11}} \\ -\frac{3}{\sqrt{11}} & \mathbf{u}_2 & \mathbf{u}_3 \\ -\frac{1}{\sqrt{11}} \end{bmatrix} \begin{bmatrix} \sqrt{22} & 0 \\ 0 & 0 \\ 0 & 0 \end{bmatrix} \begin{bmatrix} \frac{1}{\sqrt{2}} & -\frac{1}{\sqrt{2}} \\ \frac{1}{\sqrt{2}} & \frac{1}{\sqrt{2}} \end{bmatrix},$$

where the vectors $\mathbf{u}_2$ and $\mathbf{u}_3$ form an orthonormal basis of the vector plane $\mathbf{u}_1 \cdot \mathbf{x} = 0$, that is, the vector plane

$$x - 3y - z = 0.$$

Since the vector $\begin{bmatrix} 3 \\ 1 \\ 0 \end{bmatrix}$ and the vector

$$\begin{bmatrix} 1 \\ -3 \\ -1 \end{bmatrix} \times \begin{bmatrix} 3 \\ 1 \\ 0 \end{bmatrix} = \begin{bmatrix} 1 \\ -3 \\ 10 \end{bmatrix}$$

are orthogonal vectors in that plane, we can take

$$\mathbf{u}_2 = \begin{bmatrix} \frac{3}{\sqrt{10}} \\ \frac{1}{\sqrt{10}} \\ 0 \end{bmatrix} \quad \text{and} \quad \mathbf{u}_3 = \begin{bmatrix} \frac{1}{\sqrt{110}} \\ -\frac{3}{\sqrt{110}} \\ -\frac{10}{\sqrt{110}} \end{bmatrix}.$$

Consequently, the singular value decomposition of the matrix A is

$$\begin{bmatrix} 1 & -1 \\ -3 & 3 \\ -1 & 1 \end{bmatrix} = \begin{bmatrix} \frac{1}{\sqrt{11}} & \frac{3}{\sqrt{10}} & \frac{1}{\sqrt{110}} \\ -\frac{3}{\sqrt{11}} & \frac{1}{\sqrt{10}} & -\frac{3}{\sqrt{110}} \\ -\frac{1}{\sqrt{11}} & 0 & -\frac{10}{\sqrt{110}} \end{bmatrix} \begin{bmatrix} \sqrt{22} & 0 \\ 0 & 0 \\ 0 & 0 \end{bmatrix} \begin{bmatrix} \frac{1}{\sqrt{2}} & -\frac{1}{\sqrt{2}} \\ \frac{1}{\sqrt{2}} & \frac{1}{\sqrt{2}} \end{bmatrix}.$$

□

6.1 Exercises

Find the singular values of the given matrix.

1. $\begin{bmatrix} 2 & 1 \\ -1 & 2 \\ 1 & 1 \end{bmatrix}$

2. $\begin{bmatrix} 1 & 3 \\ 3 & -1 \\ -1 & 1 \end{bmatrix}$

3. $\begin{bmatrix} 3 & 1 \\ -1 & 3 \\ -2 & 6 \end{bmatrix}$

4. $\begin{bmatrix} 2 & 0 \\ 3 & -1 \\ -2 & -4 \end{bmatrix}$

Use Theorem 6.1 to find an orthogonal basis in the given vector plane.

5. Span $\left\{ \begin{bmatrix} 2 \\ 3 \\ -2 \end{bmatrix}, \begin{bmatrix} 0 \\ -1 \\ -4 \end{bmatrix} \right\}$

7. Span $\left\{ \begin{bmatrix} 2 \\ 1 \\ 1 \end{bmatrix}, \begin{bmatrix} 1 \\ 2 \\ -2 \end{bmatrix} \right\}$

6. Span $\left\{ \begin{bmatrix} -3 \\ 1 \\ -5 \end{bmatrix}, \begin{bmatrix} 4 \\ 2 \\ 0 \end{bmatrix} \right\}$

8. Span $\left\{ \begin{bmatrix} 1 \\ 3 \\ -1 \end{bmatrix}, \begin{bmatrix} 3 \\ -1 \\ 1 \end{bmatrix} \right\}$

Find the outer product expansion of the given matrix.

9. $\begin{bmatrix} 3 & 1 \\ 1 & 2 \\ -1 & 3 \end{bmatrix}$

11. $\begin{bmatrix} 1 & -2 \\ -2 & -1 \\ 3 & -1 \end{bmatrix}$

10. $\begin{bmatrix} 3 & 1 \\ -3 & 4 \\ 3 & 1 \end{bmatrix}$

12. $\begin{bmatrix} -3 & 4 \\ 1 & 2 \\ -5 & 0 \end{bmatrix}$

Find the singular value decomposition of the given matrix.

13. $\begin{bmatrix} 3 & 1 \\ 1 & 2 \\ -1 & 3 \end{bmatrix}$

15. $\begin{bmatrix} 1 & -2 \\ -2 & -1 \\ 3 & -1 \end{bmatrix}$

14. $\begin{bmatrix} 3 & 1 \\ -3 & 4 \\ 3 & 1 \end{bmatrix}$

16. $\begin{bmatrix} 1 & 3 \\ 3 & -1 \\ -1 & 1 \end{bmatrix}$

17. Let A be a 3×2 matrix. Suppose that $\mathbf{u}_1$ and $\mathbf{u}_2$ are two orthonormal vectors in $\mathbb{R}^3$ and $\mathbf{v}_1$ and $\mathbf{v}_2$ are two orthonormal vectors in $\mathbb{R}^2$. If $A = \sigma_1 \mathbf{u}_1 \mathbf{v}_1^T + \sigma_2 \mathbf{u}_2 \mathbf{v}_2^T$ is the outer product expansion of A, show that $A^T = \sigma_1 \mathbf{v}_1 \mathbf{u}_1^T + \sigma_2 \mathbf{v}_2 \mathbf{u}_2^T$.

18. Let A be a 3×2 matrix. Suppose that $\mathbf{u}_1$ and $\mathbf{u}_2$ are two orthonormal vectors in $\mathbb{R}^3$ and $\mathbf{v}_1$ and $\mathbf{v}_2$ are two orthonormal vectors in $\mathbb{R}^2$. If $A = \sigma_1 \mathbf{u}_1 \mathbf{v}_1^T + \sigma_2 \mathbf{u}_2 \mathbf{v}_2^T$ is the outer product expansion of A, show that $A^T A = \sigma_1^2 \mathbf{v}_1 \mathbf{v}_1^T + \sigma_2^2 \mathbf{v}_2 \mathbf{v}_2^T$.

19. Let A be a 3×2 matrix. Suppose that $\mathbf{u}_1$ and $\mathbf{u}_2$ are two orthonormal vectors in $\mathbb{R}^3$ and $\mathbf{v}_1$ and $\mathbf{v}_2$ are two orthonormal vectors in $\mathbb{R}^2$. If $A = \sigma_1 \mathbf{u}_1 \mathbf{v}_1^T + \sigma_2 \mathbf{u}_2 \mathbf{v}_2^T$ for some real numbers $\sigma_1 \geq \sigma_2 \geq 0$, show that the numbers σ_1 and σ_2 are the singular values of A.

20. Let A be a 3×2 matrix. Suppose that $\mathbf{u}_1$ and $\mathbf{u}_2$ are two orthonormal vectors in $\mathbb{R}^3$ and $\mathbf{v}_1$ and $\mathbf{v}_2$ are two orthonormal vectors in $\mathbb{R}^2$. If $A = \sigma_1 \mathbf{u}_1 \mathbf{v}_1^T + \sigma_2 \mathbf{u}_2 \mathbf{v}_2^T$ is the outer product expansion of A, show that the vectors $\mathbf{u}_1$ and $\mathbf{u}_2$ are eigenvectors of the matrix $A A^T$ and determine the corresponding eigenvalues.

Chapter 7

Diagonalization of 3×3 matrices

In Chapters 1 and 3 we discussed diagonalization of 2×2 matrices, which is one of the most important ideas in linear algebra with numerous applications. In this chapter we consider diagonalization of 3×3 matrices. While there are many similarities, things are more complicated in the case of 3×3 matrices.

7.1 Eigenvalues and eigenvectors of 3×3 matrices

We begin this section with some preliminary results.

The solution of the equation $Ax = 0$

Solving the equation $\begin{bmatrix} a_1 & b_1 & c_1 \\ a_2 & b_2 & c_2 \\ a_3 & b_3 & c_3 \end{bmatrix} \begin{bmatrix} x \\ y \\ z \end{bmatrix} = \begin{bmatrix} 0 \\ 0 \\ 0 \end{bmatrix}$, especially in the case when the matrix $\begin{bmatrix} a_1 & b_1 & c_1 \\ a_2 & b_2 & c_2 \\ a_3 & b_3 & c_3 \end{bmatrix}$ is not invertible, plays a crucial role in this chapter. In the first theorem we give a complete description of possible solutions of such equations. Moreover, the presented proof gives us a method of solving such equations. The method is used in many examples in this chapter.

Theorem 7.1.1. *Let* $\begin{bmatrix} a_1 & b_1 & c_1 \\ a_2 & b_2 & c_2 \\ a_3 & b_3 & c_3 \end{bmatrix}$ *be an arbitrary 3×3 matrix. The general solution of the equation*

$$\begin{bmatrix} a_1 & b_1 & c_1 \\ a_2 & b_2 & c_2 \\ a_3 & b_3 & c_3 \end{bmatrix} \begin{bmatrix} x \\ y \\ z \end{bmatrix} = \begin{bmatrix} 0 \\ 0 \\ 0 \end{bmatrix}, \tag{7.1}$$

is one of the following:

(a) *The vector* $\begin{bmatrix} 0 \\ 0 \\ 0 \end{bmatrix}$;

(b) *A vector line in* $\mathbb{R}^3$;

(c) *A vector plane in* $\mathbb{R}^3$;

(d) *All of* $\mathbb{R}^3$.

Proof. Let $A = \begin{bmatrix} a_1 & b_1 & c_1 \\ a_2 & b_2 & c_2 \\ a_3 & b_3 & c_3 \end{bmatrix}$.

If $\det \begin{bmatrix} a_1 & b_1 & c_1 \\ a_2 & b_2 & c_2 \\ a_3 & b_3 & c_3 \end{bmatrix} \neq 0$, then the unique solution of (7.1) is $\begin{bmatrix} x \\ y \\ z \end{bmatrix} = \begin{bmatrix} 0 \\ 0 \\ 0 \end{bmatrix}$, by Theorem 5.3.10.

Now suppose that $\det \begin{bmatrix} a_1 & b_1 & c_1 \\ a_2 & b_2 & c_2 \\ a_3 & b_3 & c_3 \end{bmatrix} = 0$. Since

$$\det \begin{bmatrix} a_1 & a_2 & a_3 \\ b_1 & b_2 & b_3 \\ c_1 & c_2 & c_3 \end{bmatrix} = \det \begin{bmatrix} a_1 & b_1 & c_1 \\ a_2 & b_2 & c_2 \\ a_3 & b_3 & c_3 \end{bmatrix} = 0,$$

by Theorem 5.1.15, the vectors $\begin{bmatrix} a_1 \\ b_1 \\ c_1 \end{bmatrix}$, $\begin{bmatrix} a_2 \\ b_2 \\ c_2 \end{bmatrix}$, and $\begin{bmatrix} a_3 \\ b_3 \\ c_3 \end{bmatrix}$ are linearly dependent.

Suppose that the vectors $\begin{bmatrix} a_1 \\ b_1 \\ c_1 \end{bmatrix}$ and $\begin{bmatrix} a_2 \\ b_2 \\ c_2 \end{bmatrix}$ are linearly independent. The equation (7.1) is equivalent to the following three equations

$$\begin{bmatrix} a_1 \\ b_1 \\ c_1 \end{bmatrix} \cdot \begin{bmatrix} x \\ y \\ z \end{bmatrix} = 0, \quad \begin{bmatrix} a_2 \\ b_2 \\ c_2 \end{bmatrix} \cdot \begin{bmatrix} x \\ y \\ z \end{bmatrix} = 0, \quad \text{and} \quad \begin{bmatrix} a_3 \\ b_3 \\ c_3 \end{bmatrix} \cdot \begin{bmatrix} x \\ y \\ z \end{bmatrix} = 0.$$

Since the vectors $\begin{bmatrix} a_1 \\ b_1 \\ c_1 \end{bmatrix}$, $\begin{bmatrix} a_2 \\ b_2 \\ c_2 \end{bmatrix}$, and $\begin{bmatrix} a_3 \\ b_3 \\ c_3 \end{bmatrix}$ are linearly dependent, we have

$$\begin{bmatrix} a_3 \\ b_3 \\ c_3 \end{bmatrix} = \alpha \begin{bmatrix} a_1 \\ b_1 \\ c_1 \end{bmatrix} + \beta \begin{bmatrix} a_2 \\ b_2 \\ c_2 \end{bmatrix},$$

for some real numbers α and β. Consequently, the equation $\begin{bmatrix} a_3 \\ b_3 \\ c_3 \end{bmatrix} \cdot \begin{bmatrix} x \\ y \\ z \end{bmatrix} = 0$ is a consequance of the equations

$$\begin{bmatrix} a_1 \\ b_1 \\ c_1 \end{bmatrix} \cdot \begin{bmatrix} x \\ y \\ z \end{bmatrix} = 0 \quad \text{and} \quad \begin{bmatrix} a_2 \\ b_2 \\ c_2 \end{bmatrix} \cdot \begin{bmatrix} x \\ y \\ z \end{bmatrix} = 0.$$

This means that the equation (7.1) is equivalent to the following two equations

$$\begin{bmatrix} a_1 \\ b_1 \\ c_1 \end{bmatrix} \cdot \begin{bmatrix} x \\ y \\ z \end{bmatrix} = 0 \quad \text{and} \quad \begin{bmatrix} a_2 \\ b_2 \\ c_2 \end{bmatrix} \cdot \begin{bmatrix} x \\ y \\ z \end{bmatrix} = 0$$

and the general solution of (7.1) is

$$\begin{bmatrix} x \\ y \\ z \end{bmatrix} = t \left(\begin{bmatrix} a_1 \\ b_1 \\ c_1 \end{bmatrix} \times \begin{bmatrix} a_2 \\ b_2 \\ c_2 \end{bmatrix} \right)$$

where t is an arbitrary number, by Theorem 5.1.1. This means that $\begin{bmatrix} x \\ y \\ z \end{bmatrix}$ is a solution of (7.1) if it is in the vector line Span $\left\{ \begin{bmatrix} a_1 \\ b_1 \\ c_1 \end{bmatrix} \times \begin{bmatrix} a_2 \\ b_2 \\ c_2 \end{bmatrix} \right\}$.

The cases when the vectors $\begin{bmatrix} a_1 \\ b_1 \\ c_1 \end{bmatrix}$ and $\begin{bmatrix} a_3 \\ b_3 \\ c_3 \end{bmatrix}$ are linearly independent or when the vectors $\begin{bmatrix} a_2 \\ b_2 \\ c_2 \end{bmatrix}$ and $\begin{bmatrix} a_3 \\ b_3 \\ c_3 \end{bmatrix}$ are linearly independent are treated in a similar way.

Suppose now that any two vectors from $\begin{bmatrix} a_1 \\ b_1 \\ c_1 \end{bmatrix}$, $\begin{bmatrix} a_2 \\ b_2 \\ c_2 \end{bmatrix}$, and $\begin{bmatrix} a_3 \\ b_3 \\ c_3 \end{bmatrix}$ are linearly dependent. If $\begin{bmatrix} a_1 \\ b_1 \\ c_1 \end{bmatrix} \neq \begin{bmatrix} 0 \\ 0 \\ 0 \end{bmatrix}$, then $\begin{bmatrix} a_2 \\ b_2 \\ c_2 \end{bmatrix} = \alpha \begin{bmatrix} a_1 \\ b_1 \\ c_1 \end{bmatrix}$ and $\begin{bmatrix} a_3 \\ b_3 \\ c_3 \end{bmatrix} = \beta \begin{bmatrix} a_1 \\ b_1 \\ c_1 \end{bmatrix}$ for some real numbers α and β. Consequently, the equations

$$\begin{bmatrix} a_2 \\ b_2 \\ c_2 \end{bmatrix} \cdot \begin{bmatrix} x \\ y \\ z \end{bmatrix} = 0 \quad \text{and} \quad \begin{bmatrix} a_3 \\ b_3 \\ c_3 \end{bmatrix} \cdot \begin{bmatrix} x \\ y \\ z \end{bmatrix} = 0$$

follow from the equation

$$\begin{bmatrix} a_1 \\ b_1 \\ c_1 \end{bmatrix} \cdot \begin{bmatrix} x \\ y \\ z \end{bmatrix} = 0$$

In other words, the equation (7.1) is equivalent to the above equation, which is a equation of a vector plane, by Theorem 4.2.8.

The cases when $\begin{bmatrix} a_2 \\ b_2 \\ c_2 \end{bmatrix} \neq \begin{bmatrix} 0 \\ 0 \\ 0 \end{bmatrix}$ or $\begin{bmatrix} a_3 \\ b_3 \\ c_3 \end{bmatrix} \neq \begin{bmatrix} 0 \\ 0 \\ 0 \end{bmatrix}$ are similar.

Finally we note that every vector $\begin{bmatrix} x \\ y \\ z \end{bmatrix}$ in $\mathbb{R}^3$ satisfies the equation

$$\begin{bmatrix} 0 & 0 & 0 \\ 0 & 0 & 0 \\ 0 & 0 & 0 \end{bmatrix} \begin{bmatrix} x \\ y \\ z \end{bmatrix} = \begin{bmatrix} 0 \\ 0 \\ 0 \end{bmatrix}.$$

□

Example 7.1.2. We solve the equation

$$\begin{bmatrix} 2 & 1 & 3 \\ 1 & 3 & 2 \\ 3 & -1 & 4 \end{bmatrix} \begin{bmatrix} x \\ y \\ z \end{bmatrix} = \begin{bmatrix} 0 \\ 0 \\ 0 \end{bmatrix},$$

which is equivalent to the system

$$\begin{cases} 2x + y + 3z = 0 \\ x + 3y + 2z = 0 \\ 3x - y + 4z = 0 \end{cases} \tag{7.2}$$

Since

$$\det \begin{bmatrix} 2 & 1 & 3 \\ 1 & 3 & 2 \\ 3 & -1 & 4 \end{bmatrix} = 0$$

and the vectors $\begin{bmatrix} 2 \\ 1 \\ 3 \end{bmatrix}$ and $\begin{bmatrix} 1 \\ 3 \\ 2 \end{bmatrix}$ are linearly independent, the system (7.2) is equivalent to the system

$$\begin{cases} 2x + y + 3z = 0 \\ x + 3y + 2z = 0 \end{cases}.$$

The general solution of this system is

$$\begin{bmatrix} x \\ y \\ z \end{bmatrix} = t \left(\begin{bmatrix} 2 \\ 1 \\ 3 \end{bmatrix} \times \begin{bmatrix} 1 \\ 3 \\ 2 \end{bmatrix} \right) = t \begin{bmatrix} -7 \\ -1 \\ 5 \end{bmatrix},$$

where t is an arbitrary real number. This means that all solutions of the system are on the vector line

$$\mathrm{Span}\left\{ \begin{bmatrix} -7 \\ -1 \\ 5 \end{bmatrix} \right\}.$$

Note that the system (7.2) is also equivalent to the system

$$\begin{cases} 2x + y + 3z = 0 \\ 3x - y + 4z = 0 \end{cases}.$$

The general solution of this system is

$$\begin{bmatrix} x \\ y \\ z \end{bmatrix} = t \left(\begin{bmatrix} 2 \\ 1 \\ 3 \end{bmatrix} \times \begin{bmatrix} 3 \\ -1 \\ 4 \end{bmatrix} \right) = t \begin{bmatrix} 7 \\ 1 \\ -5 \end{bmatrix},$$

where t is an arbitrary real number. This means that all solutions of the system are on the vector line

$$\mathrm{Span}\left\{ \begin{bmatrix} 7 \\ 1 \\ -5 \end{bmatrix} \right\} = \mathrm{Span}\left\{ \begin{bmatrix} -7 \\ -1 \\ 5 \end{bmatrix} \right\}.$$

Finally we note that the system (7.2) is equivalent to the system

$$\begin{cases} x + 3y + 2z = 0 \\ 3x - y + 4z = 0 \end{cases}.$$

The general solution of this system is

$$\begin{bmatrix} x \\ y \\ z \end{bmatrix} = t \left(\begin{bmatrix} 1 \\ 3 \\ 2 \end{bmatrix} \times \begin{bmatrix} 3 \\ -1 \\ 4 \end{bmatrix} \right) = t \begin{bmatrix} 14 \\ 2 \\ -10 \end{bmatrix},$$

where t is an arbitrary real number. This means that all solutions of the system are on the vector line

$$\mathrm{Span}\left\{ \begin{bmatrix} 14 \\ 2 \\ -10 \end{bmatrix} \right\} = \mathrm{Span}\left\{ \begin{bmatrix} -7 \\ -1 \\ 5 \end{bmatrix} \right\}.$$

From Theorem 7.1.1 and its proof we can easily obtain an important result called the rank-nullity theorem. First we need a new definition.

Definition 7.1.3. Let $A = \begin{bmatrix} a_1 & b_1 & c_1 \\ a_2 & b_2 & c_2 \\ a_3 & b_3 & c_3 \end{bmatrix}$ be an arbitrary 3 × 3 matrix. By the

nullspace of A, denoted by $\mathbf{N}(A)$, we mean the set of all vectors $\begin{bmatrix} x \\ y \\ z \end{bmatrix}$ such

that

$$\begin{bmatrix} a_1 & b_1 & c_1 \\ a_2 & b_2 & c_2 \\ a_3 & b_3 & c_3 \end{bmatrix} \begin{bmatrix} x \\ y \\ z \end{bmatrix} = \begin{bmatrix} 0 \\ 0 \\ 0 \end{bmatrix}.$$

From Theorem 7.1.1, or by verifying the conditions in Corollary 5.3.27, it follows that $\mathbf{N}(A)$, the nullspace of A, is a vector subspace of $\mathbb{R}^3$. The dimension of the subspace $\mathbf{N}(A)$ is called the ***nullity*** of A and is denoted by nullity(A).

Theorem 7.1.4 (The rank-nullity theorem for 3 × 3 matrices). *If A is an arbitrary* 3 × 3 *matrix, then*

$$\text{rank}(A) + \text{nullity}(A) = 3.$$

Determinants of 3 × 3 matrices revisited

In this chapter we often have to calculate determinants in order to find eigenvalues. The next result list properties of determinants which facilitate such calculations. Most of these results are in the exercises in Chapter 5.

Theorem 7.1.5. *For an arbitrary* 3×3 *matrix* $A = \begin{bmatrix} a_1 & b_1 & c_1 \\ a_2 & b_2 & c_2 \\ a_3 & b_3 & c_3 \end{bmatrix}$ *we have:*

(a)

$$\det \begin{bmatrix} ta_1 & b_1 & c_1 \\ ta_2 & b_2 & c_2 \\ ta_3 & b_3 & c_3 \end{bmatrix} = \det \begin{bmatrix} a_1 & tb_1 & c_1 \\ a_2 & tb_2 & c_2 \\ a_3 & tb_3 & c_3 \end{bmatrix} = \det \begin{bmatrix} a_1 & b_1 & tc_1 \\ a_2 & b_2 & tc_2 \\ a_3 & b_3 & tc_3 \end{bmatrix} = t \det A$$

(b)

$$\det \begin{bmatrix} ta_1 & tb_1 & tc_1 \\ a_2 & b_2 & c_2 \\ a_3 & b_3 & c_3 \end{bmatrix} = \det \begin{bmatrix} a_1 & b_1 & c_1 \\ ta_2 & tb_2 & tc_2 \\ a_3 & b_3 & c_3 \end{bmatrix} = \det \begin{bmatrix} a_1 & b_1 & c_1 \\ a_2 & b_2 & c_2 \\ ta_3 & tb_3 & tc_3 \end{bmatrix} = t \det A$$

(c)

$$\det \begin{bmatrix} a_1 & b_1 + sa_1 & c_1 + ta_1 \\ a_2 & b_2 + sa_2 & c_2 + ta_2 \\ a_3 & b_3 + sa_3 & c_3 + ta_3 \end{bmatrix} = \det \begin{bmatrix} a_1 + sb_1 & b_1 & c_1 + tb_1 \\ a_2 + sb_2 & b_2 & c_2 + tb_2 \\ a_3 + sb_3 & b_3 & c_3 + tb_3 \end{bmatrix}$$

$$= \det \begin{bmatrix} a_1 + sc_1 & b_1 + tc_1 & c_1 \\ a_2 + sc_2 & b_2 + tc_2 & c_2 \\ a_3 + sc_3 & b_3 + tc_3 & c_3 \end{bmatrix} = \det A$$

(d)

$$\det \begin{bmatrix} a_1 & b_1 & c_1 \\ a_2 + sa_1 & b_2 + sb_1 & c_2 + sc_1 \\ a_3 + ta_1 & b_3 + tb_1 & c_3 + tc_1 \end{bmatrix} = \det \begin{bmatrix} a_1 + sa_2 & b_1 + sb_2 & c_1 + sc_2 \\ a_2 & b_2 & c_2 \\ a_3 + ta_2 & b_3 + tb_2 & c_3 + tc_2 \end{bmatrix}$$

$$= \det \begin{bmatrix} a_1 + sa_3 & b_1 + sb_3 & c_1 + sc_3 \\ a_2 + ta_3 & b_2 + tb_3 & c_2 + tc_3 \\ a_3 & b_3 & c_3 \end{bmatrix} = \det A$$

Proof. We can verify these equalities by direct calculations or using the cross product. To illustrate this method we show that

$$\det \begin{bmatrix} a_1 & b_1 & c_1 \\ a_2 + sa_1 & b_2 + sb_1 & c_2 + sc_1 \\ a_3 + ta_1 & b_3 + tb_1 & c_3 + tc_1 \end{bmatrix} = \det \begin{bmatrix} a_1 & b_1 & c_1 \\ a_2 & b_2 & c_2 \\ a_3 & b_3 & c_3 \end{bmatrix}.$$

First we note that

$$\det \begin{bmatrix} a_1 & b_1 & c_1 \\ a_2 + sa_1 & b_2 + sb_1 & c_2 + sc_1 \\ a_3 + ta_1 & b_3 + tb_1 & c_3 + tc_1 \end{bmatrix} = \det \begin{bmatrix} a_1 & a_2 + sa_1 & a_3 + ta_1 \\ b_1 & b_2 + sb_1 & b_3 + tb_1 \\ c_1 & c_2 + sc_1 & c_3 + tc_1 \end{bmatrix}$$

because the determinant of a matrix is equal to the determinant of its transpose, by

Theorem 5.1.15. If we let $\mathbf{e} = \begin{bmatrix} a_1 \\ b_1 \\ c_1 \end{bmatrix}$, $\mathbf{f} = \begin{bmatrix} a_2 \\ b_2 \\ c_2 \end{bmatrix}$, and $\mathbf{g} = \begin{bmatrix} a_3 \\ b_3 \\ c_3 \end{bmatrix}$, then we have

$$\det \begin{bmatrix} a_1 & a_2 + sa_1 & a_3 + ta_1 \\ b_1 & b_2 + sb_1 & b_3 + tb_1 \\ c_1 & c_2 + sc_1 & c_3 + tc_1 \end{bmatrix} = \det \begin{bmatrix} \mathbf{e} & \mathbf{f} + s\mathbf{e} & \mathbf{g} + t\mathbf{e} \end{bmatrix}$$

$$= \mathbf{e} \cdot \big((\mathbf{f} + s\mathbf{e}) \times (\mathbf{g} + t\mathbf{e}) \big)$$
$$= \mathbf{e} \cdot (\mathbf{f} \times \mathbf{g}) + s(\mathbf{e} \cdot (\mathbf{e} \times \mathbf{g})) + t(\mathbf{e} \cdot (\mathbf{f} \times \mathbf{e})) + st(\mathbf{e} \cdot (\mathbf{e} \times \mathbf{e}))$$
$$= \mathbf{e} \cdot (\mathbf{f} \times \mathbf{g})$$
$$= \det \begin{bmatrix} \mathbf{e} & \mathbf{f} & \mathbf{g} \end{bmatrix}$$

To complete the proof we use again the fact that the determinant of a matrix is equal to the determinant of its transpose and obtain

$$\det \begin{bmatrix} \mathbf{e} & \mathbf{f} & \mathbf{g} \end{bmatrix} = \det \begin{bmatrix} a_1 & a_2 & a_3 \\ b_1 & b_2 & b_3 \\ c_1 & c_2 & c_3 \end{bmatrix} = \det \begin{bmatrix} a_1 & b_1 & c_1 \\ a_2 & b_2 & c_2 \\ a_3 & b_3 & c_3 \end{bmatrix} = \det A.$$

$\square$

Eigenvalues of 3×3 matrices

In this chapter we generalize the ideas introduced in the previous chapters to 3×3 matrices. In particular, we are interested in representing a 3×3 matrix A in the form $A = PDP^{-1}$ where P is an invertible matrix and D is a diagonal matrix. As in the case of 2×2 matrices, eigenvalues play a fundamental role.

The definition of eigenvalues for 3×3 matrices is similar to the definition of eigenvalues for 2×2 matrices (Definition 1.4.2).

Definition 7.1.6. A real number λ is an *eigenvalue* of the matrix $\begin{bmatrix} a & b & c \\ b & d & e \\ c & e & f \end{bmatrix}$

if the equation

$$\begin{bmatrix} a & b & c \\ b & d & e \\ c & e & f \end{bmatrix} \begin{bmatrix} x \\ y \\ z \end{bmatrix} = \lambda \begin{bmatrix} x \\ y \\ z \end{bmatrix}$$

has a solution $\begin{bmatrix} x \\ y \\ z \end{bmatrix} \neq \begin{bmatrix} 0 \\ 0 \\ 0 \end{bmatrix}$.

Example 7.1.7. The number 4 is an eigenvalue of the matrix $\begin{bmatrix} 7 & 3 & 3 \\ 1 & 5 & 1 \\ 5 & 5 & 9 \end{bmatrix}$ because we have

$$\begin{bmatrix} 7 & 3 & 3 \\ 1 & 5 & 1 \\ 5 & 5 & 9 \end{bmatrix} \begin{bmatrix} 2 \\ -5 \\ 3 \end{bmatrix} = 4 \begin{bmatrix} 2 \\ -5 \\ 3 \end{bmatrix}.$$

When finding eigenvalues of a 3 × 3 matrix we use the following theorem which is an immediate consequence of Theorem 7.1.1.

Theorem 7.1.8. *A real number λ is an eigenvalue of the matrix* $\begin{bmatrix} a & b & c \\ b & d & e \\ c & e & f \end{bmatrix}$ *if and only if*

$$\det \begin{bmatrix} a-\lambda & b & c \\ b & d-\lambda & e \\ c & e & f-\lambda \end{bmatrix} = 0.$$

Definition 7.1.9. The polynomial

$$P(\lambda) = \det \begin{bmatrix} a-\lambda & b & c \\ b & d-\lambda & e \\ c & e & f-\lambda \end{bmatrix}$$

is called the ***characteristic polynomial*** of the matrix $\begin{bmatrix} a & b & c \\ b & d & e \\ c & e & f \end{bmatrix}$.

Example 7.1.10. Determine the eigenvalues of the matrix $\begin{bmatrix} 7 & 2 & 2 \\ 4 & 5 & 2 \\ 2 & 1 & 4 \end{bmatrix}$.

Solution. The eigenvalues are the roots of the equation

$$\det \begin{bmatrix} 7-\lambda & 2 & 2 \\ 4 & 5-\lambda & 2 \\ 2 & 1 & 4-\lambda \end{bmatrix} = 0.$$

To make solving this equation easier we first multiply the third row by −2 and add to the second row and multiply the third row by $\frac{\lambda-7}{2}$ and add to the first row and obtain

$$\det\begin{bmatrix} 7-\lambda+2\cdot\frac{\lambda-7}{2} & 2+\frac{\lambda-7}{2} & 2+(\frac{\lambda-7}{2})(4-\lambda) \\ 4-4 & 5-\lambda-2 & 2-2(4-\lambda) \\ 2 & 1 & 4-\lambda \end{bmatrix} = 0.$$

Since

$$\det\begin{bmatrix} 0 & \frac{\lambda-3}{2} & \frac{-\lambda^2+11\lambda-24}{2} \\ 0 & 3-\lambda & 2\lambda-6 \\ 2 & 1 & 4-\lambda \end{bmatrix} = 2\det\begin{bmatrix} \frac{\lambda-3}{2} & \frac{-\lambda^2+11\lambda-24}{2} \\ 3-\lambda & 2\lambda-6 \end{bmatrix} = 0,$$

it suffices to solve the equation

$$\det\begin{bmatrix} \lambda-3 & -\lambda^2+11\lambda-24 \\ 3-\lambda & 2\lambda-6 \end{bmatrix} = 0.$$

The above equation is

$$(\lambda-3)(2\lambda-6)-(-\lambda^2+11\lambda-24)(3-\lambda)=0,$$

which can be simplified to

$$(3-\lambda)(-2\lambda+6+\lambda^2-11\lambda+24)=0$$

or

$$(3-\lambda)(\lambda^2-13\lambda+30)=0.$$

Because the roots of the quadratic equation $\lambda^2-13\lambda+30=0$ are 3 and 10, the eigenvalues are 3 and 10, with 3 being a double eigenvalue.

We note that we can calculate the determinant

$$\det\begin{bmatrix} 7-\lambda & 2 & 2 \\ 4 & 5-\lambda & 2 \\ 2 & 1 & 4-\lambda \end{bmatrix}$$

in many different ways. For example, we can subtract the third column from the second one and get

$$\det\begin{bmatrix} 7-\lambda & 2 & 0 \\ 4 & 5-\lambda & -3+\lambda \\ 2 & 1 & 3-\lambda \end{bmatrix} = (3-\lambda)\det\begin{bmatrix} 7-\lambda & 2 & 0 \\ 4 & 5-\lambda & -1 \\ 2 & 1 & 1 \end{bmatrix}.$$

Now we add in the new matrix the third row to the second one and get

$$(3-\lambda)\det\begin{bmatrix} 7-\lambda & 2 & 0 \\ 4 & 5-\lambda & -1 \\ 2 & 1 & 1 \end{bmatrix} = (3-\lambda)\det\begin{bmatrix} 7-\lambda & 2 & 0 \\ 6 & 6-\lambda & 0 \\ 2 & 1 & 1 \end{bmatrix} = (3-\lambda)(\lambda^2-13\lambda+30).$$

$\square$

The first method for calculating the above determinant is not the shortest but does not demand tricks. Generally, we have to use properties of determinants in order to calculate the characteristic polynomial. Then we need to find the roots of this polynomial. Since the characteristic polynomial of a 3 × 3 matrix is a polynomial of degree 3, it is often not obvious how to find the roots.

Eigenvectors of 3 × 3 matrices

Definition 7.1.11. Let λ be an eigenvalue of the matrix $\begin{bmatrix} a & b & c \\ b & d & e \\ c & e & f \end{bmatrix}$. A vector $\begin{bmatrix} x \\ y \\ z \end{bmatrix} \neq \begin{bmatrix} 0 \\ 0 \\ 0 \end{bmatrix}$ such that

$$\begin{bmatrix} a & b & c \\ b & d & e \\ c & e & f \end{bmatrix} \begin{bmatrix} x \\ y \\ z \end{bmatrix} = \lambda \begin{bmatrix} x \\ y \\ z \end{bmatrix}$$

is called an **eigenvector** corresponding to the eigenvalue λ.

Example 7.1.12. The vector $\begin{bmatrix} 2 \\ -5 \\ 3 \end{bmatrix}$ is an eigenvector of the matrix $\begin{bmatrix} 7 & 3 & 3 \\ 1 & 5 & 1 \\ 5 & 5 & 9 \end{bmatrix}$ corresponding to the eigenvalue 4 because $\begin{bmatrix} 7 & 3 & 3 \\ 1 & 5 & 1 \\ 5 & 5 & 9 \end{bmatrix} \begin{bmatrix} 2 \\ -5 \\ 3 \end{bmatrix} = 4 \begin{bmatrix} 2 \\ -5 \\ 3 \end{bmatrix}$.

Note that eigenvectors corresponding two different eigenvalues of a matrix have to be different. Indeed, if $\begin{bmatrix} x \\ y \\ z \end{bmatrix}$ was an eigenvector corresponding to the eigenvalue α and, at the same time, to the eigenvalue β, then we would have

$$\alpha \begin{bmatrix} x \\ y \\ z \end{bmatrix} = \begin{bmatrix} a & b & c \\ b & d & e \\ c & e & f \end{bmatrix} \begin{bmatrix} x \\ y \\ z \end{bmatrix} = \beta \begin{bmatrix} x \\ y \\ z \end{bmatrix}$$

and consequently

$$(\alpha - \beta) \begin{bmatrix} x \\ y \\ z \end{bmatrix} = \begin{bmatrix} 0 \\ 0 \\ 0 \end{bmatrix}.$$

But this implies $\alpha = \beta$, since $\begin{bmatrix} x \\ y \\ z \end{bmatrix} \neq \begin{bmatrix} 0 \\ 0 \\ 0 \end{bmatrix}$.

Definition 7.1.13. If λ is an eigenvalue of the matrix $\begin{bmatrix} a & b & c \\ b & d & e \\ c & e & f \end{bmatrix}$, then the set

of all vectors $\begin{bmatrix} x \\ y \\ z \end{bmatrix}$ which satisfy the equation

$$\begin{bmatrix} a & b & c \\ b & d & e \\ c & e & f \end{bmatrix} \begin{bmatrix} x \\ y \\ z \end{bmatrix} = \lambda \begin{bmatrix} x \\ y \\ z \end{bmatrix}$$

is called the **eigenspace** corresponding to the eigenvalue λ and is denoted by $\mathcal{E}_\lambda$.

The eigenspace $\mathcal{E}_\lambda$ consists of all eigenvectors corresponding to λ and the vector $\begin{bmatrix} 0 \\ 0 \\ 0 \end{bmatrix}$, which is not an eigenvector. Note that any eigenspace of a 3 × 3 matrix is a solution of an equation considered in Theorem 7.1.1 and, consequently, is a vector subspace of $\mathbb{R}^3$.

Example 7.1.14. Show that 8 is an eigenvalue of the matrix $A = \begin{bmatrix} 3 & 2 & 2 \\ 3 & 4 & 3 \\ 2 & 2 & 3 \end{bmatrix}$ and then find the eigenspace corresponding to the eigenvalue 8.

Solution. It is easy to verify that 8 is an eigenvalue of the matrix because

$$\det \begin{bmatrix} 3-8 & 2 & 2 \\ 3 & 4-8 & 3 \\ 2 & 2 & 3-8 \end{bmatrix} = \det \begin{bmatrix} -5 & 2 & 2 \\ 3 & -4 & 3 \\ 2 & 2 & -5 \end{bmatrix} = 0.$$

To find the eigenvectors corresponding to the eigenvalue 8 we solve the equation

$$\begin{bmatrix} 3 & 2 & 2 \\ 3 & 4 & 3 \\ 2 & 2 & 3 \end{bmatrix} \begin{bmatrix} x \\ y \\ z \end{bmatrix} = 8 \begin{bmatrix} x \\ y \\ z \end{bmatrix}$$

which is equivalent to the system of equations

$$\begin{cases} -5x + 2y + 2z = 0 \\ 3x - 4y + 3z = 0 \\ 2x + 2y - 5z = 0 \end{cases}.$$

This system can be solved with Gauss elimination or using the cross product as follows. By Theorem 7.1.1, the system reduces to

$$\begin{cases} 3x - 4y + 3z = 0 \\ 2x + 2y - 5z = 0 \end{cases}.$$

The general solution of this system is

$$\begin{bmatrix} x \\ y \\ z \end{bmatrix} = t \left(\begin{bmatrix} 3 \\ -4 \\ 3 \end{bmatrix} \times \begin{bmatrix} 2 \\ 2 \\ -5 \end{bmatrix} \right) = t \begin{bmatrix} 14 \\ 21 \\ 14 \end{bmatrix} = 7t \begin{bmatrix} 2 \\ 3 \\ 2 \end{bmatrix},$$

where t is an arbitrary real number.

This means that the eigenspace corresponding to the eigenvalue $\lambda = 8$ is the

vector line Span $\left\{ \begin{bmatrix} 2 \\ 3 \\ 2 \end{bmatrix} \right\}$. $\qquad\square$

Definition 7.1.15. A 3×3 matrix D is called ***diagonal*** if there are real numbers α, β, and γ such that

$$D = \begin{bmatrix} \alpha & 0 & 0 \\ 0 & \beta & 0 \\ 0 & 0 & \gamma \end{bmatrix}.$$

Definition 7.1.16. A 3×3 matrix A is called ***diagonalizable*** if there is a diagonal matrix D and an invertible matrix P such that

$$A = PDP^{-1}.$$

This means that the matrix

$$\begin{bmatrix} a_1 & b_1 & c_1 \\ a_2 & b_2 & c_2 \\ a_3 & b_3 & c_3 \end{bmatrix}$$

is diagonalizable if there are real numbers α, β, and γ and an invertible matrix

$$\begin{bmatrix} q_1 & r_1 & s_1 \\ q_2 & r_2 & s_2 \\ q_3 & r_3 & s_3 \end{bmatrix}$$

such that

$$\begin{bmatrix} a_1 & b_1 & c_1 \\ a_2 & b_2 & c_2 \\ a_3 & b_3 & c_3 \end{bmatrix} = \begin{bmatrix} q_1 & r_1 & s_1 \\ q_2 & r_2 & s_2 \\ q_3 & r_3 & s_3 \end{bmatrix} \begin{bmatrix} \alpha & 0 & 0 \\ 0 & \beta & 0 \\ 0 & 0 & \gamma \end{bmatrix} \begin{bmatrix} q_1 & r_1 & s_1 \\ q_2 & r_2 & s_2 \\ q_3 & r_3 & s_3 \end{bmatrix}^{-1}.$$

Theorem 7.1.17. *A 3×3 matrix is diagonalizable if and only if the matrix has 3 linearly independent eigenvectors.*

Proof. Let $A = \begin{bmatrix} a_1 & b_1 & c_1 \\ a_2 & b_2 & c_2 \\ a_3 & b_3 & c_3 \end{bmatrix}$. We need to show that A is diagonalizable if and only if there exist real numbers α, β, and γ, not necessarily different, and linearly indepen-

dent vectors $\begin{bmatrix} u_1 \\ u_2 \\ u_3 \end{bmatrix}$, $\begin{bmatrix} v_1 \\ v_2 \\ v_3 \end{bmatrix}$ and $\begin{bmatrix} w_1 \\ w_2 \\ w_3 \end{bmatrix}$ such that

$$\begin{bmatrix} a_1 & b_1 & c_1 \\ a_2 & b_2 & c_2 \\ a_3 & b_3 & c_3 \end{bmatrix} \begin{bmatrix} u_1 \\ u_2 \\ u_3 \end{bmatrix} = \alpha \begin{bmatrix} u_1 \\ u_2 \\ u_3 \end{bmatrix},$$

$$\begin{bmatrix} a_1 & b_1 & c_1 \\ a_2 & b_2 & c_2 \\ a_3 & b_3 & c_3 \end{bmatrix} \begin{bmatrix} v_1 \\ v_2 \\ v_3 \end{bmatrix} = \beta \begin{bmatrix} v_1 \\ v_2 \\ v_3 \end{bmatrix},$$

and

$$\begin{bmatrix} a_1 & b_1 & c_1 \\ a_2 & b_2 & c_2 \\ a_3 & b_3 & c_3 \end{bmatrix} \begin{bmatrix} w_1 \\ w_2 \\ w_3 \end{bmatrix} = \gamma \begin{bmatrix} w_1 \\ w_2 \\ w_3 \end{bmatrix}.$$

We first note that the above three equations can be written as a single equation

$$\begin{bmatrix} a_1 & b_1 & c_1 \\ a_2 & b_2 & c_2 \\ a_3 & b_3 & c_3 \end{bmatrix} \begin{bmatrix} u_1 & v_1 & w_1 \\ u_2 & v_2 & w_2 \\ u_1 & v_1 & w_3 \end{bmatrix} = \begin{bmatrix} u_1 & v_1 & w_1 \\ u_2 & v_2 & w_2 \\ u_1 & v_1 & w_3 \end{bmatrix} \begin{bmatrix} \alpha & 0 & 0 \\ 0 & \beta & 0 \\ 0 & 0 & \gamma \end{bmatrix}.$$

By Theorem 5.3.10, the vectors $\begin{bmatrix} u_1 \\ u_2 \\ u_3 \end{bmatrix}$, $\begin{bmatrix} v_1 \\ v_2 \\ v_3 \end{bmatrix}$ and $\begin{bmatrix} w_1 \\ w_2 \\ w_3 \end{bmatrix}$ are linearly independent if

and only if the matrix $P = \begin{bmatrix} u_1 & v_1 & w_1 \\ u_2 & v_2 & w_2 \\ u_1 & v_1 & w_3 \end{bmatrix}$ is invertible. Consequently, to prove the

theorem it suffices to show that a matrix A is diagonalizable if and only if there exists an invertible matrix P and a diagonal matrix D such that $AP = PD$.

If A is diagonalizable, then there exists an invertible matrix P and a diagonal matrix D such that $A = PDP^{-1}$. Consequently,

$$AP = PDP^{-1}P = PD.$$

Now, if there exists an invertible matrix P and a diagonal matrix D such that $AP = PD$, then

$$PDP^{-1} = APP^{-1} = A,$$

so A is diagonalizable. □

Note that the above proof explains the relationship between the linearly independent eigenvectors of a matrix $A = \begin{bmatrix} a_1 & b_1 & c_1 \\ a_2 & b_2 & c_2 \\ a_3 & b_3 & c_3 \end{bmatrix}$ and the invertible matrix P in the equation $A = PDP^{-1}$: the eigenvectors are the column vectors of P. More precisely, if

$$\begin{bmatrix} a_1 & b_1 & c_1 \\ a_2 & b_2 & c_2 \\ a_3 & b_3 & c_3 \end{bmatrix} \begin{bmatrix} u_1 \\ u_2 \\ u_3 \end{bmatrix} = \alpha \begin{bmatrix} u_1 \\ u_2 \\ u_3 \end{bmatrix},$$

$$\begin{bmatrix} a_1 & b_1 & c_1 \\ a_2 & b_2 & c_2 \\ a_3 & b_3 & c_3 \end{bmatrix} \begin{bmatrix} v_1 \\ v_2 \\ v_3 \end{bmatrix} = \beta \begin{bmatrix} v_1 \\ v_2 \\ v_3 \end{bmatrix},$$

and

$$\begin{bmatrix} a_1 & b_1 & c_1 \\ a_2 & b_2 & c_2 \\ a_3 & b_3 & c_3 \end{bmatrix} \begin{bmatrix} w_1 \\ w_2 \\ w_3 \end{bmatrix} = \gamma \begin{bmatrix} w_1 \\ w_2 \\ w_3 \end{bmatrix},$$

then for

$$P = \begin{bmatrix} u_1 & v_1 & w_1 \\ u_2 & v_2 & w_2 \\ u_1 & v_1 & w_3 \end{bmatrix} \quad \text{and} \quad D = \begin{bmatrix} \alpha & 0 & 0 \\ 0 & \beta & 0 \\ 0 & 0 & \gamma \end{bmatrix}$$

we have $A = PDP^{-1}$.

Theorem 7.1.18. *If a 3×3 matrix A has 3 different eigenvalues, then A is diagonalizable.*

Proof. Let

$$A = \begin{bmatrix} a_1 & b_1 & c_1 \\ a_2 & b_2 & c_2 \\ a_3 & b_3 & c_3 \end{bmatrix}$$

and let

$$\det \begin{bmatrix} a_1 - \lambda & b_1 & c_1 \\ a_2 & b_2 - \lambda & c_2 \\ a_3 & b_3 & c_3 - \lambda \end{bmatrix} = -(\lambda - \alpha)(\lambda - \beta)(\lambda - \gamma),$$

where α, β, and γ are three different real numbers.

The equation

$$\begin{bmatrix} a_1 & b_1 & c_1 \\ a_2 & b_2 & c_2 \\ a_3 & b_3 & c_3 \end{bmatrix} \begin{bmatrix} x \\ y \\ z \end{bmatrix} = \alpha \begin{bmatrix} x \\ y \\ z \end{bmatrix}$$

has a nontrivial solution $\mathbf{u} = \begin{bmatrix} u_1 \\ u_2 \\ u_3 \end{bmatrix}$ because

$$\det \begin{bmatrix} a_1 - \alpha & b_1 & c_1 \\ a_2 & b_2 - \alpha & c_2 \\ a_3 & b_3 & c_3 - \alpha \end{bmatrix} = 0.$$

In the same way the equation

$$\begin{bmatrix} a_1 & b_1 & c_1 \\ a_2 & b_2 & c_2 \\ a_3 & b_3 & c_3 \end{bmatrix} \begin{bmatrix} x \\ y \\ z \end{bmatrix} = \beta \begin{bmatrix} x \\ y \\ z \end{bmatrix}$$

has a nontrivial solution $\mathbf{v} = \begin{bmatrix} v_1 \\ v_2 \\ v_3 \end{bmatrix}$ and the equation

$$\begin{bmatrix} a_1 & b_1 & c_1 \\ a_2 & b_2 & c_2 \\ a_3 & b_3 & c_3 \end{bmatrix} \begin{bmatrix} x \\ y \\ z \end{bmatrix} = \gamma \begin{bmatrix} x \\ y \\ z \end{bmatrix}$$

has a nontrivial solution $\mathbf{w} = \begin{bmatrix} w_1 \\ w_2 \\ w_3 \end{bmatrix}$. We have to prove that the matrix $\begin{bmatrix} u_1 & v_1 & w_1 \\ u_2 & v_2 & w_2 \\ u_1 & v_1 & w_3 \end{bmatrix}$ is invertible which is equivalent to the fact that the vectors $\mathbf{u}$, $\mathbf{v}$, and $\mathbf{w}$ are linearly independent.

We show first that the vectors $\mathbf{u}$ and $\mathbf{v}$ are linearly independent.

We have to show that

$$p\mathbf{u} + q\mathbf{v} = \mathbf{0}$$

implies $p = q = 0$.

If $p\mathbf{u} + q\mathbf{v} = \mathbf{0}$, then

$$p\alpha\mathbf{u} + q\beta\mathbf{v} = pA\mathbf{u} + qA\mathbf{v} = A(p\mathbf{u} + q\mathbf{v}) = A\mathbf{0} = \mathbf{0}.$$

Since

$$(p\alpha\mathbf{u} + q\beta\mathbf{v}) - \beta(p\mathbf{u} + q\mathbf{v}) = p(\alpha - \beta)\mathbf{u} + q(\beta - \beta)\mathbf{v} = p(\alpha - \beta)\mathbf{u} = \mathbf{0},$$

$\alpha - \beta \neq 0$, and $\mathbf{u} \neq \mathbf{0}$, we conclude that $p = 0$. Similarly, since

$$(p\alpha\mathbf{u} + q\beta\mathbf{v}) - \alpha(p\mathbf{u} + q\mathbf{v}) = p(\alpha - \alpha)\mathbf{u} + q(\beta - \alpha)\mathbf{v} = q(\beta - \alpha)\mathbf{v} = \mathbf{0},$$

$\beta - \alpha \neq 0$, and $\mathbf{v} \neq \mathbf{0}$, we conclude that $q = 0$.

Now we prove that the vectors $\mathbf{u}$, $\mathbf{v}$, and $\mathbf{w}$ are linearly independent.

To this end we will prove that $r\mathbf{u} + s\mathbf{v} + t\mathbf{w} = \mathbf{0}$ implies that $r = s = t = 0$.

If $r\mathbf{u} + s\mathbf{v} + t\mathbf{w} = \mathbf{0}$, then

$$r\alpha\mathbf{u} + s\beta\mathbf{v} + t\gamma\mathbf{w} = rA\mathbf{u} + sA\mathbf{v} + tA\mathbf{w} = A(r\mathbf{u} + s\mathbf{v} + t\mathbf{w}) = \mathbf{0}$$

and hence

$$(r\alpha\mathbf{u} + s\beta\mathbf{v} + t\gamma\mathbf{w}) - \gamma(r\mathbf{u} + s\mathbf{v} + t\mathbf{w}) = r(\alpha - \gamma)\mathbf{u} + s(\beta - \gamma)\mathbf{v} = \mathbf{0}.$$

Since $\alpha - \gamma \neq 0$ and $\beta - \gamma \neq 0$ and because the vectors $\mathbf{u}$ and $\mathbf{v}$ are linearly independent, we obtain that $r = s = 0$. Now, since $\mathbf{w} \neq \mathbf{0}$, we also obtain that $t = 0$, which completes the proof. $\qquad\square$

Example 7.1.19. If possible, diagonalize the matrix $\begin{bmatrix} 2 & 1 & -1 \\ 2 & 3 & 1 \\ -2 & -1 & 1 \end{bmatrix}$.

Solution. The eigenvalues of the matrix are the roots of the equation

$$\det \begin{bmatrix} 2-\lambda & 1 & -1 \\ 2 & 3-\lambda & 1 \\ -2 & -1 & 1-\lambda \end{bmatrix} = 0.$$

We add the second row to the third one and then we add the second row multiplied by $-\frac{2-\lambda}{2} = \frac{\lambda-2}{2}$ to the first row and get

$$\det \begin{bmatrix} 0 & 1+\frac{\lambda-2}{2}(3-\lambda) & -1+\frac{\lambda-2}{2} \\ 2 & 3-\lambda & 1 \\ 0 & 2-\lambda & 2-\lambda \end{bmatrix} = \lambda(2-\lambda)(\lambda-4) = 0.$$

Consequently, the egenvalues are 0, 2, and 4.

We can also calculate the determinant in the following simpler way. If we add the first row to the last one, the determinant becomes

$$\det \begin{bmatrix} 2-\lambda & 1 & -1 \\ 2 & 3-\lambda & 1 \\ -\lambda & 0 & -\lambda \end{bmatrix} = -\lambda\det \begin{bmatrix} 2-\lambda & 1 & -1 \\ 2 & 3-\lambda & 1 \\ 1 & 0 & 1 \end{bmatrix}.$$

Next we subtract the first column from the third column and the determinant becomes

$$-\lambda\det \begin{bmatrix} 2-\lambda & 1 & -3+\lambda \\ 2 & 3-\lambda & -1 \\ 1 & 0 & 0 \end{bmatrix} = -\lambda(\lambda^2 - 6\lambda + 8) = 0.$$

Now we find a basis of eigenvectors for the matrix. First we determine the eigenvectors which correspond to the eigenvalue 0. We have to solve the equation

$$\begin{bmatrix} 2 & 1 & -1 \\ 2 & 3 & 1 \\ -2 & -1 & 1 \end{bmatrix} \begin{bmatrix} x \\ y \\ z \end{bmatrix} = 0 \begin{bmatrix} x \\ y \\ z \end{bmatrix}.$$

This equation is equivalent to the system

$$\begin{cases} 2x + \ y - z = 0 \\ 2x + 3y + z = 0 \\ -2x - \ y + z = 0 \end{cases}.$$

which is equivalent to

$$\begin{cases} 2x + \ y - z = 0 \\ 2x + 3y + z = 0 \end{cases}.$$

The general solution of this system is

$$\begin{bmatrix} x \\ y \\ z \end{bmatrix} = t \begin{bmatrix} 4 \\ -4 \\ 4 \end{bmatrix} = 4t \begin{bmatrix} 1 \\ -1 \\ 1 \end{bmatrix},$$

where t is an arbitrary real number. This means that the eigenspace corresponding to the eigenvalue 0 is the vector line Span $\left\{ \begin{bmatrix} 1 \\ -1 \\ 1 \end{bmatrix} \right\}$.

Now we determine the eigenvectors which correspond to the eigenvalue 2. We have to solve the equation

$$\begin{bmatrix} 2 & 1 & -1 \\ 2 & 3 & 1 \\ -2 & -1 & 1 \end{bmatrix} \begin{bmatrix} x \\ y \\ z \end{bmatrix} = 2 \begin{bmatrix} x \\ y \\ z \end{bmatrix}.$$

This equation is equivalent to the system

$$\begin{cases} 2x + \ y - z = 2x \\ 2x + 3y + z = 2y \\ -2x - \ y + z = 2z \end{cases}$$

or

$$\begin{cases} \ y - z = 0 \\ 2x + y + z = 0 \end{cases}.$$

The general solution of this system is

$$\begin{bmatrix} x \\ y \\ z \end{bmatrix} = t \begin{bmatrix} 2 \\ -2 \\ -2 \end{bmatrix} = -2t \begin{bmatrix} -1 \\ 1 \\ 1 \end{bmatrix},$$

where t is an arbitrary real number. This means that the eigenspace corresponding to the eigenvalue 2 is the vector line Span $\left\{ \begin{bmatrix} -1 \\ 1 \\ 1 \end{bmatrix} \right\}$.

Finally we determine the eigenvectors which correspond to the eigenvalue 4. We have to solve the equation

$$\begin{bmatrix} 2 & 1 & -1 \\ 2 & 3 & 1 \\ -2 & -1 & 1 \end{bmatrix} \begin{bmatrix} x \\ y \\ z \end{bmatrix} = 4 \begin{bmatrix} x \\ y \\ z \end{bmatrix}$$

or the system

$$\begin{cases} 2x + y - z = 4x \\ 2x + 3y + z = 4y \\ -2x - y + z = 4z \end{cases}$$

which is equivalent to

$$\begin{cases} -2x + y - z = 0 \\ 2x - y + z = 0 \\ -2x - y - 3z = 0 \end{cases}$$

or

$$\begin{cases} 2x - y + z = 0 \\ -2x - y - 3z = 0 \end{cases}$$

The general solution of this system is

$$\begin{bmatrix} x \\ y \\ z \end{bmatrix} = t \begin{bmatrix} 4 \\ 4 \\ -4 \end{bmatrix} = 4t \begin{bmatrix} 1 \\ 1 \\ -1 \end{bmatrix},$$

where t is an arbitrary real number. This means that the eigenspace corresponding to the eigenvalue 4 is the vector line $\text{Span} \left\{ \begin{bmatrix} 1 \\ 1 \\ -1 \end{bmatrix} \right\}$.

From our calculations we can conclude that

$$\begin{bmatrix} 2 & 1 & -1 \\ 2 & 3 & 1 \\ -2 & -1 & 1 \end{bmatrix} = \begin{bmatrix} -1 & 1 & 1 \\ 1 & -1 & 1 \\ 1 & 1 & -1 \end{bmatrix} \begin{bmatrix} 2 & 0 & 0 \\ 0 & 0 & 0 \\ 0 & 0 & 4 \end{bmatrix} \begin{bmatrix} -1 & 1 & 1 \\ 1 & -1 & 1 \\ 1 & 1 & -1 \end{bmatrix}^{-1}.$$

Note that the order of the eigenvalues in the matrix

$$\begin{bmatrix} 2 & 0 & 0 \\ 0 & 0 & 0 \\ 0 & 0 & 4 \end{bmatrix}$$

corresponds to the order of the corresponding eigenvectors in the matrix

$$\begin{bmatrix} -1 & 1 & 1 \\ 1 & -1 & 1 \\ 1 & 1 & -1 \end{bmatrix}.$$

□

The matrix in the example above had three different eigenvalues. Now we are going to investigate what happens when a matrix has only two different eigenvalues. From Theorem 7.1.17 we know that a 3×3 matrix is diagonalizable if and only if the matrix has 3 linearly independent eigenvectors. Therefore, in our case we expect to find two linearly independent eigenvectors for one of the two eigenvalues.

Example 7.1.20. If possible, diagonalize the matrix $A = \begin{bmatrix} 4 & 1 & 2 \\ 2 & 3 & 2 \\ 2 & 1 & 4 \end{bmatrix}$.

Solution. The eigenvalues are given by the equation

$$\det \begin{bmatrix} 4-\lambda & 1 & 2 \\ 2 & 3-\lambda & 2 \\ 2 & 1 & 4-\lambda \end{bmatrix} = 0.$$

In order to calculate this determinant we subtract the second row from the third one and then we add the second row multiplied by $\frac{\lambda-4}{2}$ to the first one. We get

$$\det \begin{bmatrix} 0 & \frac{-\lambda^2+7\lambda-10}{2} & \lambda-2 \\ 2 & 3-\lambda & 2 \\ 0 & \lambda-2 & 2-\lambda \end{bmatrix} = (\lambda-2)^2(7-\lambda) = 0.$$

The same result can be obtained if we subtract the second row from the first one and get

$$\det \begin{bmatrix} 2-\lambda & \lambda-2 & 0 \\ 2 & 3-\lambda & 2 \\ 2 & 1 & 4-\lambda \end{bmatrix} = (\lambda-2)\det \begin{bmatrix} -1 & 1 & 0 \\ 2 & 3-\lambda & 2 \\ 2 & 1 & 4-\lambda \end{bmatrix}.$$

and then we add the first column to the second one and get

$$(\lambda-2)\det \begin{bmatrix} -1 & 0 & 0 \\ 2 & 5-\lambda & 2 \\ 2 & 3 & 4-\lambda \end{bmatrix} = (\lambda-2)^2(7-\lambda).$$

Consequently, the matrix A has two eigenvalues: 2 and 7.

First we consider the eigenvalue 2. The equation

$$\begin{bmatrix} 4 & 1 & 2 \\ 2 & 3 & 2 \\ 2 & 1 & 4 \end{bmatrix} \begin{bmatrix} x \\ y \\ z \end{bmatrix} = 2 \begin{bmatrix} x \\ y \\ z \end{bmatrix}$$

can be written as

$$\begin{bmatrix} 2 & 1 & 2 \\ 2 & 1 & 2 \\ 2 & 1 & 2 \end{bmatrix} \begin{bmatrix} x \\ y \\ z \end{bmatrix} = \begin{bmatrix} 0 \\ 0 \\ 0 \end{bmatrix}$$

and it thus reduces to a single equation

$$2x + y + 2z = 0$$

or

$$y = -2x - 2z.$$

The general solution of this equation is

$$\begin{bmatrix} x \\ y \\ z \end{bmatrix} = \begin{bmatrix} x \\ -2x - 2z \\ z \end{bmatrix} = x \begin{bmatrix} 1 \\ -2 \\ 0 \end{bmatrix} + z \begin{bmatrix} 0 \\ -2 \\ 1 \end{bmatrix}$$

for arbitrary real numbers x and z. This means that both vectors $\begin{bmatrix} 1 \\ -2 \\ 0 \end{bmatrix}$ and $\begin{bmatrix} 0 \\ -2 \\ 1 \end{bmatrix}$

are eigenvectors of A corresponding to the eigenvalue 2. In this case the eigenspace

is the vector plane Span $\left\{ \begin{bmatrix} 1 \\ -2 \\ 0 \end{bmatrix} , \begin{bmatrix} 0 \\ -2 \\ 1 \end{bmatrix} \right\}$.

Now we consider the eigenvalue 7. The equation

$$\begin{bmatrix} 4 & 1 & 2 \\ 2 & 3 & 2 \\ 2 & 1 & 4 \end{bmatrix} \begin{bmatrix} x \\ y \\ z \end{bmatrix} = 7 \begin{bmatrix} x \\ y \\ z \end{bmatrix}$$

can be written as

$$\begin{bmatrix} -3 & 1 & 2 \\ 2 & -4 & 2 \\ 2 & 1 & -3 \end{bmatrix} \begin{bmatrix} x \\ y \\ z \end{bmatrix} = \begin{bmatrix} 0 \\ 0 \\ 0 \end{bmatrix}.$$

The equation is equivalent to the system

$$\begin{cases} -3x + y + 2z = 0 \\ 2x - 4y + 2z = 0 \,, \\ 2x + y - 3z = 0 \end{cases}$$

which can be solve using Gauss elimination or Theorem 7.1.1. The system reduces to a system of two equations

$$\begin{cases} -3x + y + 2z = 0 \\ x - 2y + z = 0 \end{cases}.$$

The general solution of this system is

$$\begin{bmatrix} x \\ y \\ z \end{bmatrix} = t \left(\begin{bmatrix} -3 \\ 1 \\ 2 \end{bmatrix} \times \begin{bmatrix} 1 \\ -2 \\ 1 \end{bmatrix} \right) = t \begin{bmatrix} 5 \\ 5 \\ 5 \end{bmatrix} = 5t \begin{bmatrix} 1 \\ 1 \\ 1 \end{bmatrix}$$

where t is an arbitrary real number. This means that the eigenspace corresponding to the eigenvalue 7 is the vector line $\text{Span}\left\{ \begin{bmatrix} 1 \\ 1 \\ 1 \end{bmatrix} \right\}$. Since

$$\det \begin{bmatrix} 1 & 0 & 1 \\ -2 & -2 & 1 \\ 0 & 1 & 1 \end{bmatrix} = -5,$$

the vectors $\begin{bmatrix} 1 \\ -2 \\ 0 \end{bmatrix}$, $\begin{bmatrix} 0 \\ -2 \\ 1 \end{bmatrix}$, and $\begin{bmatrix} 1 \\ 1 \\ 1 \end{bmatrix}$ are linearly independent and thus the matrix A is diagonalizable, by Theorem 7.1.17:

$$\begin{bmatrix} 4 & 1 & 2 \\ 2 & 3 & 2 \\ 2 & 1 & 4 \end{bmatrix} = \begin{bmatrix} 1 & 0 & 1 \\ -2 & -2 & 1 \\ 0 & 1 & 1 \end{bmatrix} \begin{bmatrix} 2 & 0 & 0 \\ 0 & 2 & 0 \\ 0 & 0 & 7 \end{bmatrix} \begin{bmatrix} 1 & 0 & 1 \\ -2 & -2 & 1 \\ 0 & 1 & 1 \end{bmatrix}^{-1}.$$

Note that we have other choices for diagonalization of A, for example,

$$\begin{bmatrix} 4 & 1 & 2 \\ 2 & 3 & 2 \\ 2 & 1 & 4 \end{bmatrix} = \begin{bmatrix} 1 & 1 & 0 \\ -2 & 1 & -2 \\ 0 & 1 & 1 \end{bmatrix} \begin{bmatrix} 2 & 0 & 0 \\ 0 & 7 & 0 \\ 0 & 0 & 2 \end{bmatrix} \begin{bmatrix} 1 & 1 & 0 \\ -2 & 1 & -2 \\ 0 & 1 & 1 \end{bmatrix}^{-1}.$$

or

$$\begin{bmatrix} 4 & 1 & 2 \\ 2 & 3 & 2 \\ 2 & 1 & 4 \end{bmatrix} = \begin{bmatrix} 1 & 1 & 0 \\ 1 & -2 & -2 \\ 1 & 0 & 1 \end{bmatrix} \begin{bmatrix} 7 & 0 & 0 \\ 0 & 2 & 0 \\ 0 & 0 & 2 \end{bmatrix} \begin{bmatrix} 1 & 1 & 0 \\ 1 & -2 & -2 \\ 1 & 0 & 1 \end{bmatrix}^{-1}.$$

□

The matrix in the above example has two eigenvalues: 2 and 7. We found two linearly independent eigenvectors corresponding to 2, namely $\begin{bmatrix} 1 \\ -2 \\ 0 \end{bmatrix}$ and $\begin{bmatrix} 0 \\ -2 \\ 1 \end{bmatrix}$, and one eigenvector corresponding to 7, namely $\begin{bmatrix} 1 \\ 1 \\ 1 \end{bmatrix}$. We used the determinant to check that these three vectors are linearly independent. It turns out that it was not necessary to check that.

Theorem 7.1.21. *Let A be a 3 × 3 matrix with two different eigenvalues α and β. If **u** and **v** are linearly independent eigenvectors corresponding to the eigenvalue α and **w** is an eigenvector corresponding to the eigenvalue β, then the vectors **u**, **v**, and **w** are linearly independent.*

Proof. To show that the vectors **u**, **v**, and **w** are linearly independent assume that

$$r\mathbf{u} + s\mathbf{v} + t\mathbf{w} = 0.$$

We need to show that $r = s = t = 0$.

If $t \neq 0$, then

$$\mathbf{w} = \frac{r}{t}\mathbf{u} + \frac{s}{t}\mathbf{v}$$

which means, because an eigenspace is a vector subspace, that **w** is in the eigenspace corresponding to the eigenvalue α. But this is not possible, since **w** is in the eigenspace corresponding to the eigenvalue $\lambda = \beta$ and $\alpha \neq \beta$. Thus $t = 0$ and the equation $r\mathbf{u} + s\mathbf{v} + t\mathbf{w} = 0$ reduces to the equation

$$r\mathbf{u} + s\mathbf{v} = 0.$$

Since the vectors **u** and **v** are linearly independent, we must have $r = s = 0$. $\square$

Example 7.1.22. If possible, diagonalize the matrix $A = \begin{bmatrix} 4 & 1 & -2 \\ 3 & 2 & -2 \\ 4 & 2 & -2 \end{bmatrix}$.

Solution. The eigenvalues are 1 and 2. The eigenspace corresponding to the eigenvalue 1 is $\text{Span}\left\{ \begin{bmatrix} 1 \\ 1 \\ 2 \end{bmatrix} \right\}$ and the eigenspace corresponding to the eigenvalue 2 is $\text{Span}\left\{ \begin{bmatrix} 2 \\ 2 \\ 3 \end{bmatrix} \right\}$. Since the matrix does not have three linearly independent eigenvectors, it is not diagonalizable, by Theorem 7.1.17. $\square$

7.1.1 Exercises

Find an eigenvalue of the matrix A without calculating the characteristic polynomial of A.

1. $A = \begin{bmatrix} 7 & 4 & 9 \\ 1 & 5 & 1 \\ 1 & 2 & 4 \end{bmatrix}$

5. $A = \begin{bmatrix} 3 & 4 & 1 \\ 5 & 7 & 9 \\ 2 & 4 & 2 \end{bmatrix}$

2. $A = \begin{bmatrix} 5 & 4 & 2 \\ 1 & 2 & 0 \\ 1 & 1 & 1 \end{bmatrix}$

6. $A = \begin{bmatrix} 5 & 4 & 1 \\ 1 & 7 & 4 \\ 1 & 2 & 9 \end{bmatrix}$

3. $A = \begin{bmatrix} 2 & 1 & 1 \\ 1 & 2 & 1 \\ 8 & 3 & 3 \end{bmatrix}$

7. $A = \begin{bmatrix} 7 & 8 & 4 \\ 2 & 9 & 3 \\ 2 & 4 & 5 \end{bmatrix}$

4. $A = \begin{bmatrix} 4 & 2 & 1 \\ 1 & 5 & 1 \\ 1 & 2 & 4 \end{bmatrix}$

8. $A = \begin{bmatrix} 11 & 5 & 15 \\ 2 & 2 & 3 \\ 8 & 9 & 7 \end{bmatrix}$

Find two eigenvalues of the given matrix without calculating its characteristic polynomial.

9. $\begin{bmatrix} 5 & 4 & 4 \\ 4 & 5 & 4 \\ 4 & 1 & 8 \end{bmatrix}$

10. $\begin{bmatrix} 3 & 3 & 1 \\ 1 & 5 & 1 \\ 1 & 2 & 4 \end{bmatrix}$

Verify that the given λ is an eigenvalue of the given matrix A and calculate the eigenspace corresponding to the eigenvalue λ.

11. $\lambda = 7$ and $A = \begin{bmatrix} 3 & 3 & 1 \\ 1 & 5 & 1 \\ 2 & 2 & 3 \end{bmatrix}$

15. $\lambda = 3$ and $A = \begin{bmatrix} 4 & 2 & 5 \\ 1 & 5 & 5 \\ 1 & 2 & 8 \end{bmatrix}$

12. $\lambda = 1$ and $A = \begin{bmatrix} 3 & 3 & 1 \\ 2 & 4 & 1 \\ 2 & 3 & 2 \end{bmatrix}$

16. $\lambda = 1$ and $A = \begin{bmatrix} 2 & 4 & 1 \\ 1 & 5 & 1 \\ 1 & 4 & 2 \end{bmatrix}$

13. $\lambda = 11$ and $A = \begin{bmatrix} 4 & 2 & 5 \\ 1 & 5 & 5 \\ 1 & 2 & 8 \end{bmatrix}$

17. $\lambda = 2$ and $A = \begin{bmatrix} 3 & 3 & 1 \\ 1 & 5 & 1 \\ 2 & 2 & 3 \end{bmatrix}$

14. $\lambda = 4$ and $A = \begin{bmatrix} 5 & 1 & 1 \\ 3 & 7 & 3 \\ 1 & 5 & 2 \end{bmatrix}$

18. $\lambda = 7$ and $A = \begin{bmatrix} 8 & 1 & 1 \\ 3 & 10 & 3 \\ 1 & 1 & 8 \end{bmatrix}$

Find the eigenvalues of the given matrix.

19. $\begin{bmatrix} 1 & 2 & 3 \\ -1 & 2 & 1 \\ 1 & 1 & 2 \end{bmatrix}$

21. $\begin{bmatrix} 2 & 1 & 1 \\ 1 & 2 & 1 \\ 1 & 0 & 2 \end{bmatrix}$

20. $\begin{bmatrix} 4 & 1 & 1 \\ 1 & 4 & 1 \\ 1 & 3 & 2 \end{bmatrix}$

22. $\begin{bmatrix} 3 & 1 & 2 \\ 2 & 4 & 2 \\ 2 & 1 & 3 \end{bmatrix}$

Write, if possible, the given matrix A in the form $A = PDP^{-1}$ where P is an invertible matrix and D is a diagonal matrix.

23. $A = \begin{bmatrix} 1 & 0 & 0 \\ 1 & 2 & 0 \\ 2 & 1 & 0 \end{bmatrix}$

27. $A = \begin{bmatrix} 4 & 2 & 2 \\ 2 & 7 & 4 \\ 2 & 4 & 7 \end{bmatrix}$

24. $A = \begin{bmatrix} 3 & 0 & 2 \\ 1 & 2 & 2 \\ 4 & 4 & 9 \end{bmatrix}$

28. $A = \begin{bmatrix} 3 & 1 & 5 \\ 1 & 3 & 5 \\ 1 & 1 & 7 \end{bmatrix}$

25. $A = \begin{bmatrix} 3 & 4 & 2 \\ 1 & 3 & 1 \\ 1 & 2 & 2 \end{bmatrix}$

29. $A = \begin{bmatrix} 1 & 2 & 0 \\ -1 & 4 & 0 \\ -1 & 3 & 2 \end{bmatrix}$

26. $A = \begin{bmatrix} 2 & 1 & 3 \\ 1 & 2 & 3 \\ 3 & 3 & 10 \end{bmatrix}$

30. $A = \begin{bmatrix} 5 & 2 & -3 \\ 2 & 5 & -3 \\ 5 & 3 & -3 \end{bmatrix}$

7.2   Symmetric 3×3 matrices

Recall that a matrix is called symmetric if $A^T = A$.

Definition 7.2.1. A 3×3 matrix P is called **orthogonal matrix** if it is invertible and we have
$$P^T = P^{-1}.$$

Note that we can also say that a 3×3 matrix A is orthogonal if

$$P^T P = \begin{bmatrix} 1 & 0 & 0 \\ 0 & 1 & 0 \\ 0 & 0 & 1 \end{bmatrix}.$$

Example 7.2.2. Since

$$\begin{bmatrix} \frac{1}{3} & -\frac{2}{3} & \frac{2}{3} \\ \frac{2}{3} & -\frac{1}{3} & -\frac{2}{3} \\ \frac{2}{3} & \frac{2}{3} & \frac{1}{3} \end{bmatrix}^T \begin{bmatrix} \frac{1}{3} & -\frac{2}{3} & \frac{2}{3} \\ \frac{2}{3} & -\frac{1}{3} & -\frac{2}{3} \\ \frac{2}{3} & \frac{2}{3} & \frac{1}{3} \end{bmatrix} = \begin{bmatrix} \frac{1}{3} & \frac{2}{3} & \frac{2}{3} \\ -\frac{2}{3} & -\frac{1}{3} & \frac{2}{3} \\ \frac{2}{3} & -\frac{2}{3} & \frac{1}{3} \end{bmatrix} \begin{bmatrix} \frac{1}{3} & -\frac{2}{3} & \frac{2}{3} \\ \frac{2}{3} & -\frac{1}{3} & -\frac{2}{3} \\ \frac{2}{3} & \frac{2}{3} & \frac{1}{3} \end{bmatrix} = \begin{bmatrix} 1 & 0 & 0 \\ 0 & 1 & 0 \\ 0 & 0 & 1 \end{bmatrix},$$

the matrix

$$\begin{bmatrix} \frac{1}{3} & -\frac{2}{3} & \frac{2}{3} \\ \frac{2}{3} & -\frac{1}{3} & -\frac{2}{3} \\ \frac{2}{3} & \frac{2}{3} & \frac{1}{3} \end{bmatrix}$$

is orthogonal.

The following theorem gives an indication why orthogonal matrices are called orthogonal.

> **Theorem 7.2.3.** *If the columns of a 3 × 3 matrix P are pairwise orthogonal unit vectors, then P is an orthogonal matrix.*

Proof. Let $P = \begin{bmatrix} \mathbf{p}_1 & \mathbf{p}_2 & \mathbf{p}_3 \end{bmatrix}$ where $\mathbf{p}_1$, $\mathbf{p}_2$, and $\mathbf{p}_3$ are vectors such that

$$\mathbf{p}_1 \cdot \mathbf{p}_3 = \mathbf{p}_2 \cdot \mathbf{p}_3 = \mathbf{p}_3 \cdot \mathbf{p}_1 = 0 \quad \text{and} \quad \|\mathbf{p}_1\| = \|\mathbf{p}_2\| = \|\mathbf{p}_3\| = 1.$$

Then

$$P^T P = \begin{bmatrix} \mathbf{p}_1^T \\ \mathbf{p}_2^T \\ \mathbf{p}_3^T \end{bmatrix} \begin{bmatrix} \mathbf{p}_1 & \mathbf{p}_2 & \mathbf{p}_2 \end{bmatrix} = \begin{bmatrix} \mathbf{p}_1^T \mathbf{p}_1 & \mathbf{p}_1^T \mathbf{p}_2 & \mathbf{p}_1^T \mathbf{p}_3 \\ \mathbf{p}_2^T \mathbf{p}_1 & \mathbf{p}_2^T \mathbf{p}_2 & \mathbf{p}_2^T \mathbf{p}_3 \\ \mathbf{p}_3^T \mathbf{p}_1 & \mathbf{p}_3^T \mathbf{p}_2 & \mathbf{p}_3^T \mathbf{p}_3 \end{bmatrix} = \begin{bmatrix} \mathbf{p}_1 \cdot \mathbf{p}_1 & \mathbf{p}_1 \cdot \mathbf{p}_2 & \mathbf{p}_1 \cdot \mathbf{p}_3 \\ \mathbf{p}_2 \cdot \mathbf{p}_1 & \mathbf{p}_2 \cdot \mathbf{p}_2 & \mathbf{p}_2 \cdot \mathbf{p}_3 \\ \mathbf{p}_3 \cdot \mathbf{p}_1 & \mathbf{p}_3 \cdot \mathbf{p}_2 & \mathbf{p}_3 \cdot \mathbf{p}_3 \end{bmatrix} = \begin{bmatrix} 1 & 0 & 0 \\ 0 & 1 & 0 \\ 0 & 0 & 1 \end{bmatrix},$$

which means that $P^T = P^{-1}$, that is, P is an orthogonal matrix. □

Note that the columns of the matrix in Example 7.2.2, that is,

$$\begin{bmatrix} \frac{1}{3} & -\frac{2}{3} & \frac{2}{3} \\ \frac{2}{3} & -\frac{1}{3} & -\frac{2}{3} \\ \frac{2}{3} & \frac{2}{3} & \frac{1}{3} \end{bmatrix},$$

are pairwise orthogonal unit vectors.

> **Corollary 7.2.4.** *Let $\mathbf{p}_1$, $\mathbf{p}_2$, and $\mathbf{p}_3$ be three vectors in $\mathbb{R}^3$ such that the matrix $\begin{bmatrix} \mathbf{p}_1 & \mathbf{p}_2 & \mathbf{p}_3 \end{bmatrix}$ is an orthogonal matrix. Then $\{\mathbf{p}_1, \mathbf{p}_2, \mathbf{p}_3\}$ is a basis of $\mathbb{R}^3$.*

Proof. The set $\{\mathbf{p}_1, \mathbf{p}_2, \mathbf{p}_3\}$ is a basis of $\mathbb{R}^3$ because the matrix $[\mathbf{p}_1 \quad \mathbf{p}_2 \quad \mathbf{p}_3]$ is invertible.

$\square$

A basis $\{\mathbf{p}_1, \mathbf{p}_2, \mathbf{p}_3\}$ of $\mathbb{R}^3$ such that the vectors $\mathbf{p}_1$, $\mathbf{p}_2$, and $\mathbf{p}_3$ are orthonormal is called an **orthonormal basis**.

Definition 7.2.5. We say that a matrix A is **orthogonally diagonalizable** if there are an orthogonal matrix P and a diagonal matrix D such that

$$A = PDP^{-1} = PDP^T.$$

Example 7.2.6. In the previous example we show that the matrix

$$\begin{bmatrix} \frac{1}{3} & -\frac{2}{3} & \frac{2}{3} \\ \frac{2}{3} & -\frac{1}{3} & -\frac{2}{3} \\ \frac{2}{3} & \frac{2}{3} & \frac{1}{3} \end{bmatrix}$$

is orthogonal. Consequently, the matrix

$$\begin{bmatrix} \frac{1}{3} & -\frac{2}{3} & \frac{2}{3} \\ \frac{2}{3} & -\frac{1}{3} & -\frac{2}{3} \\ \frac{2}{3} & \frac{2}{3} & \frac{1}{3} \end{bmatrix} \begin{bmatrix} 2 & 0 & 0 \\ 0 & -1 & 0 \\ 0 & 0 & 3 \end{bmatrix} \begin{bmatrix} \frac{1}{3} & \frac{2}{3} & \frac{2}{3} \\ -\frac{2}{3} & -\frac{1}{3} & \frac{2}{3} \\ \frac{2}{3} & -\frac{2}{3} & \frac{1}{3} \end{bmatrix} = \begin{bmatrix} \frac{2}{9} & -\frac{2}{9} & \frac{10}{9} \\ -\frac{2}{9} & \frac{11}{9} & \frac{8}{9} \\ \frac{10}{9} & \frac{8}{9} & \frac{5}{9} \end{bmatrix}$$

is orthogonally diagonalizable.

We know that the matrix

$$\begin{bmatrix} \frac{2}{9} & -\frac{2}{9} & \frac{10}{9} \\ -\frac{2}{9} & \frac{11}{9} & \frac{8}{9} \\ \frac{10}{9} & \frac{8}{9} & \frac{5}{9} \end{bmatrix}$$

in the example above is orthogonally diagonalizable because it is constructed as a product PDP^T where P is an orthogonal matrix P and D is a diagonal matrix, but how could we check if a matrix is orthogonally diagonalizable? The following theorem can help with this question.

Theorem 7.2.7. *If a* 3 × 3 *matrix A has* 3 *orthogonal eigenvectors, then A is symmetric and orthogonally diagonalizable.*

Proof. Let $\mathbf{v}_1, \mathbf{v}_2$, and $\mathbf{v}_3$ be the orthogonal eigenvectors of the matrix A corresponding to the eigenvalues α_1, α_2, and α_3, respectively. This means that

$$A\mathbf{v}_1 = \alpha_1 \mathbf{v}_1, \quad A\mathbf{v}_2 = \alpha_2 \mathbf{v}_2, \quad A\mathbf{v}_3 = \alpha_3 \mathbf{v}_3, \quad \text{and} \quad \mathbf{v}_1 \cdot \mathbf{v}_2 = \mathbf{v}_2 \cdot \mathbf{v}_3 = \mathbf{v}_1 \cdot \mathbf{v}_3 = 0.$$

If we let

$$\mathbf{p}_1 = \frac{1}{\|\mathbf{v}_1\|}\mathbf{v}_1, \quad \mathbf{p}_2 = \frac{1}{\|\mathbf{v}_2\|}\mathbf{v}_2, \quad \text{and} \quad \mathbf{p}_3 = \frac{1}{\|\mathbf{v}_3\|}\mathbf{v}_3,$$

then we have

$$\mathbf{p}_1 \cdot \mathbf{p}_2 = \mathbf{p}_2 \cdot \mathbf{p}_3 = \mathbf{p}_1 \cdot \mathbf{p}_3 = 0 \quad \text{and} \quad \|\mathbf{p}_1\| = \|\mathbf{p}_2\| = \|\mathbf{p}_3\| = 1.$$

Let P be the matrix whose columns are the vectors $\mathbf{p}_1$, $\mathbf{p}_2$, and $\mathbf{p}_3$, that is,

$$P = \begin{bmatrix} \mathbf{p}_1 & \mathbf{p}_2 & \mathbf{p}_3 \end{bmatrix}.$$

Then P is an orthogonal matrix, by Theorem 7.2.3.

The vectors $\mathbf{p}_1$, $\mathbf{p}_2$, and $\mathbf{p}_3$ satisfy the equations

$$A\mathbf{p}_1 = \alpha_1 \mathbf{p}_1, \quad A\mathbf{p}_2 = \alpha_2 \mathbf{p}_2, \quad A\mathbf{p}_3 = \alpha_3 \mathbf{p}_3,$$

that can be written as a single equation

$$A\begin{bmatrix} \mathbf{p}_1 & \mathbf{p}_2 & \mathbf{p}_3 \end{bmatrix} = \begin{bmatrix} \mathbf{p}_1 & \mathbf{p}_2 & \mathbf{p}_3 \end{bmatrix} \begin{bmatrix} \alpha_1 & 0 & 0 \\ 0 & \alpha_2 & 0 \\ 0 & 0 & \alpha_3 \end{bmatrix}$$

or

$$AP = P\begin{bmatrix} \alpha_1 & 0 & 0 \\ 0 & \alpha_2 & 0 \\ 0 & 0 & \alpha_3 \end{bmatrix}.$$

Hence

$$A = A\begin{bmatrix} 1 & 0 & 0 \\ 0 & 1 & 0 \\ 0 & 0 & 1 \end{bmatrix} = APP^{-1} = P\begin{bmatrix} \alpha_1 & 0 & 0 \\ 0 & \alpha_2 & 0 \\ 0 & 0 & \alpha_3 \end{bmatrix}P^{-1} = P\begin{bmatrix} \alpha_1 & 0 & 0 \\ 0 & \alpha_2 & 0 \\ 0 & 0 & \alpha_3 \end{bmatrix}P^{T}.$$

Since P is an orthogonal matrix, the matrix A is orthogonally diagonalizable. Moreover, since

$$A^{T} = \left(P\begin{bmatrix} \alpha & 0 & 0 \\ 0 & \beta & 0 \\ 0 & 0 & \gamma \end{bmatrix}P^{T} \right)^{T} = (P^{T})^{T}\begin{bmatrix} \alpha_1 & 0 & 0 \\ 0 & \alpha_2 & 0 \\ 0 & 0 & \alpha_3 \end{bmatrix}^{T}P^{T} = P\begin{bmatrix} \alpha_1 & 0 & 0 \\ 0 & \alpha_2 & 0 \\ 0 & 0 & \alpha_3 \end{bmatrix}P^{T} = A,$$

the matrix A is symmetric.   □

From Theorem 7.2.7 and its proof we obtain the following useful corollary.

Corollary 7.2.8. *If A is a* 3 × 3 *matrix with* 3 *orthogonal eigenvectors* $\mathbf{v}_1$, $\mathbf{v}_2$, $\mathbf{v}_3$ *corresponding to the real eigenvalues* α_1, α_2, α_3, *that is*

$$A\mathbf{v}_1 = \alpha_1\mathbf{v}_1, \; A\mathbf{v}_2 = \alpha_2\mathbf{v}_2, \; A\mathbf{v}_3 = \alpha_3\mathbf{v}_3,$$

then

$$A = PDP^T,$$

where $P = \begin{bmatrix} \mathbf{p}_1 & \mathbf{p}_2 & \mathbf{p}_3 \end{bmatrix}$ *is the orthogonal matrix with columns*

$$\mathbf{p}_1 = \frac{1}{\|\mathbf{v}_1\|}\mathbf{v}_1, \; \mathbf{p}_2 = \frac{1}{\|\mathbf{v}_2\|}\mathbf{v}_2, \; \mathbf{p}_3 = \frac{1}{\|\mathbf{v}_3\|}\mathbf{v}_3$$

and

$$D = \begin{bmatrix} \alpha_1 & 0 & 0 \\ 0 & \alpha_2 & 0 \\ 0 & 0 & \alpha_3 \end{bmatrix}.$$

Calculating the diagonal form of a 3 × 3 symmetric matrix

We first state and prove the following useful property of the dot product.

Lemma 7.2.9. *If A is an arbitrary* 3 × 3 *matrix, then*

$$A\mathbf{u} \cdot \mathbf{v} = \mathbf{u} \cdot A^T\mathbf{v}$$

for any vectors $\mathbf{u}$ *and* $\mathbf{v}$ *in* $\mathbb{R}^3$.

Proof. The proof uses the connection between the dot product and the product of matrices, namely

$$\mathbf{x} \cdot \mathbf{y} = \mathbf{x}^T\mathbf{y},$$

associativity of matrix multiplication, and a property of the transpose operation:

$$A\mathbf{u} \cdot \mathbf{v} = (A\mathbf{u})^T\mathbf{v} = (\mathbf{u}^T A^T)\mathbf{v} = \mathbf{u}^T(A^T\mathbf{v}) = \mathbf{u} \cdot A^T\mathbf{v}.$$

$\square$

Note that for symmetric matrices we have

$$A\mathbf{u} \cdot \mathbf{v} = \mathbf{u} \cdot A\mathbf{v}.$$

In the next two theorems we formulate some properties of symmetric matrices that will help us diagonalize such matrices. These facts are important in their own right and belong to the fundamental properties of symmetric matrices.

> **Theorem 7.2.10.** *Eigenvectors corresponding to different eigenvalues of a symmetric matrix are orthogonal.*

Proof. If $A\mathbf{u} = \alpha\mathbf{u}$ and $A\mathbf{v} = \beta\mathbf{v}$, then

$$(\alpha - \beta)(\mathbf{u}\cdot\mathbf{v}) = \alpha(\mathbf{u}\cdot\mathbf{v}) - \beta(\mathbf{u}\cdot\mathbf{v}) = (\alpha\mathbf{u})\cdot\mathbf{v} - \mathbf{u}\cdot\beta(\mathbf{v}) = A\mathbf{u}\cdot\mathbf{v} - \mathbf{u}\cdot A\mathbf{v}.$$

For a symmetric matrix A, we have

$$A\mathbf{u}\cdot\mathbf{v} - \mathbf{u}\cdot A\mathbf{v} = 0$$

and consequently

$$(\alpha - \beta)(\mathbf{u}\cdot\mathbf{v}) = 0.$$

If $\alpha \neq \beta$, we must have $\mathbf{u}\cdot\mathbf{v} = 0$. $\qquad\square$

> **Theorem 7.2.11.** *If A is a symmetric 3×3 matrix with three different eigenvalues, then there is an orthonormal basis of $\mathbb{R}^3$ consisting of eigenvectors of A.*

Proof. Let

$$A = \begin{bmatrix} a & b & c \\ b & d & e \\ c & e & f \end{bmatrix}$$

and let

$$\det \begin{bmatrix} a-\lambda & b & c \\ b & d-\lambda & e \\ c & e & f-\lambda \end{bmatrix} = -(\lambda - \alpha)(\lambda - \beta)(\lambda - \gamma),$$

where α, β, and γ are three different real numbers. Since A has three different eigenvalues, it is diagonalizable, by Theorem 7.1.18, and thus it has three linearly independent eigenvectors, by Theorem 7.1.17. Let $\mathbf{u}$ be an eigenvector corresponding to the eigenvalue α, $\mathbf{v}$ an eigenvector corresponding to the eigenvalue β, and $\mathbf{w}$ an eigenvector corresponding to the eigenvalue γ. By Corollary 5.3.19, $\{\mathbf{u}, \mathbf{v}, \mathbf{w}\}$ is a basis in $\mathbb{R}^3$. Since the vectors $\mathbf{u}$, $\mathbf{v}$, $\mathbf{w}$ are orthogonal, by Theorem 7.2.10, $\{\mathbf{u}, \mathbf{v}, \mathbf{w}\}$ is an orthogonal basis in $\mathbb{R}^3$. Consequently,

$$\left\{ \frac{\mathbf{u}}{\|\mathbf{u}\|}, \frac{\mathbf{v}}{\|\mathbf{v}\|}, \frac{\mathbf{w}}{\|\mathbf{w}\|} \right\}$$

is an orthonormal basis in $\mathbb{R}^3$ consisting of eigenvectors of the matrix A.

$\qquad\square$

Example 7.2.12. Orthogonally diagonalize the matrix $A = \begin{bmatrix} 4 & 1 & 2 \\ 1 & 5 & 1 \\ 2 & 1 & 4 \end{bmatrix}$.

Solution. The eigenvalues are the roots of the equation

$$\det \begin{bmatrix} 4-\lambda & 1 & 2 \\ 1 & 5-\lambda & 1 \\ 2 & 1 & 4-\lambda \end{bmatrix} = 0.$$

We multiply the second row by $\lambda - 4$ and add to the first and then multiply the second row by -2 and add to the third row and get

$$\det \begin{bmatrix} 0 & (5-\lambda)(\lambda-4)+1 & \lambda-2 \\ 1 & 5-\lambda & 1 \\ 0 & -9+2\lambda & 2-\lambda \end{bmatrix} = (2-\lambda)(\lambda^2 - 11\lambda + 28) = 0.$$

Instead, we can subtract the third row from the first one and get

$$\det \begin{bmatrix} 4-\lambda & 1 & 2 \\ 1 & 5-\lambda & 1 \\ 2 & 1 & 4-\lambda \end{bmatrix} = \det \begin{bmatrix} 2-\lambda & 0 & \lambda-2 \\ 1 & 5-\lambda & 1 \\ 2 & 1 & 4-\lambda \end{bmatrix}$$

$$= (2-\lambda)\det \begin{bmatrix} 1 & 0 & -1 \\ 1 & 5-\lambda & 1 \\ 2 & 1 & 4-\lambda \end{bmatrix}$$

In the last determinant we add the first column to the third one and get

$$= (2-\lambda)\det \begin{bmatrix} 1 & 0 & 0 \\ 1 & 5-\lambda & 2 \\ 2 & 1 & 6-\lambda \end{bmatrix}$$

$$= (2-\lambda)(\lambda^2 - 11\lambda + 28).$$

Consequently, the eigenvalues are 2, 4, and 7. Now we calculate the eigenspaces. The eigenvectors corresponding to the eigenvalue 2 are given by the equation

$$\begin{bmatrix} 4 & 1 & 2 \\ 1 & 5 & 1 \\ 2 & 1 & 4 \end{bmatrix} \begin{bmatrix} x \\ y \\ z \end{bmatrix} = 2 \begin{bmatrix} x \\ y \\ z \end{bmatrix}$$

which can be written as

$$\begin{cases} 4x + y + 2z = 2x \\ x + 5y + z = 2y \\ 2x + y + 4z = 2z \end{cases}$$

or

$$\begin{cases} 2x + \ y + 2z = 0 \\ \ x + 3y + \ z = 0 \ . \\ 2x + \ y + 2z = 0 \end{cases}$$

This system is equivalent to the system

$$\begin{cases} 2x + \ y + 2z = 0 \\ \ x + 3y + \ z = 0 \end{cases}$$

and its general solution is

$$\begin{bmatrix} x \\ y \\ z \end{bmatrix} = t \begin{bmatrix} -5 \\ 0 \\ 5 \end{bmatrix} = -5t \begin{bmatrix} 1 \\ 0 \\ -1 \end{bmatrix} ,$$

where t is an arbitrary real number. This means that the eigenspace corresponding to the eigenvalue 2 is Span $\left\{ \begin{bmatrix} 1 \\ 0 \\ -1 \end{bmatrix} \right\}$.

The eigenvectors corresponding to the eigenvalue 4 are given by the equation

$$\begin{bmatrix} 4 & 1 & 2 \\ 1 & 5 & 1 \\ 2 & 1 & 4 \end{bmatrix} \begin{bmatrix} x \\ y \\ z \end{bmatrix} = 4 \begin{bmatrix} x \\ y \\ z \end{bmatrix} .$$

This equation is equivalent to the system

$$\begin{cases} 4x + \ y + 2z = 4x \\ \ x + 5y + \ z = 4y \\ 2x + \ y + 4z = 4z \end{cases}$$

or

$$\begin{cases} \quad\ y + 2z = 0 \\ \ x + y + \ z = 0 \ . \\ 2x + y \quad\quad = 0 \end{cases}$$

The above system reduces to the system

$$\begin{cases} \quad\ y + 2z = 0 \\ 2x + y \quad\quad = 0 \end{cases}$$

and its general solution is

$$\begin{bmatrix} x \\ y \\ z \end{bmatrix} = t \begin{bmatrix} -2 \\ 4 \\ -2 \end{bmatrix} = -2t \begin{bmatrix} 1 \\ -2 \\ 1 \end{bmatrix} ,$$

where t is an arbitrary real number. Consequently, the eigenspace corresponding to the eigenvalue 4 is $\text{Span}\left\{\begin{bmatrix} 1 \\ -2 \\ 1 \end{bmatrix}\right\}$.

The eigenvectors corresponding to the eigenvalue 7 are given by the equation

$$\begin{bmatrix} 4 & 1 & 2 \\ 1 & 5 & 1 \\ 2 & 1 & 4 \end{bmatrix} \begin{bmatrix} x \\ y \\ z \end{bmatrix} = 7 \begin{bmatrix} x \\ y \\ z \end{bmatrix}$$

This equation is equivalent to the system

$$\begin{cases} 4x + y + 2z = 7x \\ x + 5y + z = 7y \\ 2x + y + 4z = 7z \end{cases}$$

or

$$\begin{cases} -3x + y + 2z = 0 \\ x - 2y + z = 0 \\ 2x + y - 3z = 0 \end{cases}.$$

The above system reduces to the system

$$\begin{cases} x - 2y + z = 0 \\ 2x + y - 3z = 0 \end{cases}$$

and its general solution is

$$\begin{bmatrix} x \\ y \\ z \end{bmatrix} = t \begin{bmatrix} 5 \\ 5 \\ 5 \end{bmatrix} = 5t \begin{bmatrix} 1 \\ 1 \\ 1 \end{bmatrix},$$

where t is an arbitrary real number. Thus the eigenspace corresponding to the eigenvalue 7 is $\text{Span}\left\{\begin{bmatrix} 1 \\ 1 \\ 1 \end{bmatrix}\right\}$.

Note that we can obtain that the eigenspace corresponding to the eigenvalue 7 as a consequence of the fact that the eigenvectors corresponding to the eigenvalue 7 have to be orthogonal on the eigenvectors corresponding to the eigenvalues 2 and 4. Since

$$\begin{bmatrix} 1 \\ 0 \\ -1 \end{bmatrix} \times \begin{bmatrix} 1 \\ -2 \\ 1 \end{bmatrix} = \begin{bmatrix} -2 \\ -2 \\ -2 \end{bmatrix} = -2 \begin{bmatrix} 1 \\ 1 \\ 1 \end{bmatrix},$$

$\text{Span}\left\{\begin{bmatrix} 1 \\ 1 \\ 1 \end{bmatrix}\right\}$ must be the eigenspace corresponding to the eigenvalue 7.

Since the vectors $\begin{bmatrix} 1 \\ 0 \\ -1 \end{bmatrix}$, $\begin{bmatrix} 1 \\ -2 \\ 1 \end{bmatrix}$, and $\begin{bmatrix} 1 \\ 1 \\ 1 \end{bmatrix}$ are orthogonal eigenvectors corre-

sponding to eigenvalues 2, 4 and 7, respectively, they can be used to orthogonally diagonalize the matrix A. The last necessary step is normalization of these eigenvectors. Since

$$\left\| \begin{bmatrix} 1 \\ 0 \\ -1 \end{bmatrix} \right\| = \sqrt{2}, \quad \left\| \begin{bmatrix} 1 \\ -2 \\ 1 \end{bmatrix} \right\| = \sqrt{6}, \quad \text{and} \quad \left\| \begin{bmatrix} 1 \\ 1 \\ 1 \end{bmatrix} \right\| = \sqrt{3},$$

we conclude that

$$\begin{bmatrix} 4 & 1 & 2 \\ 1 & 5 & 1 \\ 2 & 1 & 4 \end{bmatrix} = \begin{bmatrix} \frac{1}{\sqrt{2}} & \frac{1}{\sqrt{6}} & \frac{1}{\sqrt{3}} \\ 0 & -\frac{2}{\sqrt{6}} & \frac{1}{\sqrt{3}} \\ -\frac{1}{\sqrt{2}} & \frac{1}{\sqrt{6}} & \frac{1}{\sqrt{3}} \end{bmatrix} \begin{bmatrix} 2 & 0 & 0 \\ 0 & 4 & 0 \\ 0 & 0 & 7 \end{bmatrix} \begin{bmatrix} \frac{1}{\sqrt{2}} & 0 & -\frac{1}{\sqrt{2}} \\ \frac{1}{\sqrt{6}} & -\frac{2}{\sqrt{6}} & \frac{1}{\sqrt{6}} \\ \frac{1}{\sqrt{3}} & \frac{1}{\sqrt{3}} & \frac{1}{\sqrt{3}} \end{bmatrix}$$

is an orthogonal diagonalization of A.   □

In the next theorem we discuss the general case of 3×3 symmetric matrices with two different eigenvalues.

Theorem 7.2.13. *Let A be a symmetric 3×3 matrix. If the characteristic polynomial of A has the form*

$$P(\lambda) = -(\lambda - \alpha)^2 (\lambda - \beta),$$

where α and β are two different real numbers, then the matrix A has an orthogonal basis of eigenvectors consisting of two eigenvectors corresponding to the eigenvalue α and one eigenvector corresponding to the eigenvalue β.

Proof. Let $A = \begin{bmatrix} a & b & c \\ b & d & e \\ c & e & f \end{bmatrix}$. The equation

$$\begin{bmatrix} a - \alpha & b & c \\ b & d - \alpha & e \\ c & e & f - \alpha \end{bmatrix} \begin{bmatrix} x \\ y \\ z \end{bmatrix} = \begin{bmatrix} 0 \\ 0 \\ 0 \end{bmatrix}$$

has a nontrivial solution $\begin{bmatrix} u_1 \\ u_2 \\ u_3 \end{bmatrix}$ because

$$\det \begin{bmatrix} a - \alpha & b & c \\ b & d - \alpha & e \\ c & e & f - \alpha \end{bmatrix} = 0.$$

Similarly, the equation

$$\begin{bmatrix} a-\beta & b & c \\ b & d-\beta & e \\ c & e & f-\beta \end{bmatrix} \begin{bmatrix} x \\ y \\ z \end{bmatrix} = \begin{bmatrix} 0 \\ 0 \\ 0 \end{bmatrix}$$

has a nontrivial solution $\begin{bmatrix} v_1 \\ v_2 \\ v_3 \end{bmatrix}$. By Theorem 7.2.10, we have $\begin{bmatrix} u_1 \\ u_2 \\ u_3 \end{bmatrix} \cdot \begin{bmatrix} v_1 \\ v_2 \\ v_3 \end{bmatrix} = 0$. Let

$$\begin{bmatrix} w_1 \\ w_2 \\ w_3 \end{bmatrix} = \begin{bmatrix} u_1 \\ u_2 \\ u_3 \end{bmatrix} \times \begin{bmatrix} v_1 \\ v_2 \\ v_3 \end{bmatrix}.$$

We have $\begin{bmatrix} w_1 \\ w_2 \\ w_3 \end{bmatrix} \neq \begin{bmatrix} 0 \\ 0 \\ 0 \end{bmatrix}$ because the vectors $\begin{bmatrix} u_1 \\ u_2 \\ u_3 \end{bmatrix}$ and $\begin{bmatrix} v_1 \\ v_2 \\ v_3 \end{bmatrix}$ are linearly independent being nonzero and orthogonal. By Lemma 7.2.9, we have

$$\left(\begin{bmatrix} a & b & c \\ b & d & e \\ c & e & f \end{bmatrix} \begin{bmatrix} w_1 \\ w_2 \\ w_3 \end{bmatrix} \right) \cdot \begin{bmatrix} u_1 \\ u_2 \\ u_3 \end{bmatrix} = \begin{bmatrix} w_1 \\ w_2 \\ w_3 \end{bmatrix} \cdot \left(\begin{bmatrix} a & b & c \\ b & d & e \\ c & e & f \end{bmatrix} \begin{bmatrix} u_1 \\ u_2 \\ u_3 \end{bmatrix} \right)$$

$$= \begin{bmatrix} w_1 \\ w_2 \\ w_3 \end{bmatrix} \cdot \alpha \begin{bmatrix} u_1 \\ u_2 \\ u_3 \end{bmatrix} = \alpha \left(\begin{bmatrix} w_1 \\ w_2 \\ w_3 \end{bmatrix} \cdot \begin{bmatrix} u_1 \\ u_2 \\ u_3 \end{bmatrix} \right) = 0$$

and

$$\left(\begin{bmatrix} a & b & c \\ b & d & e \\ c & e & f \end{bmatrix} \begin{bmatrix} w_1 \\ w_2 \\ w_3 \end{bmatrix} \right) \cdot \begin{bmatrix} v_1 \\ v_2 \\ v_3 \end{bmatrix} = \begin{bmatrix} w_1 \\ w_2 \\ w_3 \end{bmatrix} \cdot \left(\begin{bmatrix} a & b & c \\ b & d & e \\ c & e & f \end{bmatrix} \begin{bmatrix} v_1 \\ v_2 \\ v_3 \end{bmatrix} \right)$$

$$= \begin{bmatrix} w_1 \\ w_2 \\ w_3 \end{bmatrix} \cdot \beta \begin{bmatrix} v_1 \\ v_2 \\ v_3 \end{bmatrix} = \beta \left(\begin{bmatrix} w_1 \\ w_2 \\ w_3 \end{bmatrix} \cdot \begin{bmatrix} v_1 \\ v_2 \\ v_3 \end{bmatrix} \right) = 0.$$

Hence

$$\begin{bmatrix} a & b & c \\ b & d & e \\ c & e & f \end{bmatrix} \begin{bmatrix} w_1 \\ w_2 \\ w_3 \end{bmatrix} = \gamma \begin{bmatrix} w_1 \\ w_2 \\ w_3 \end{bmatrix}$$

for some real number γ by Theorem 5.1.1. So γ is an eigenvalue of A and we must have

$$-(\lambda - \alpha)^2 (\lambda - \beta) = -(\lambda - \alpha)(\lambda - \beta)(\lambda - \gamma),$$

so $\gamma = \alpha$ and thus $\begin{bmatrix} w_1 \\ w_2 \\ w_3 \end{bmatrix}$ is another eigenvector corresponding to α. □

Example 7.2.14. Orthogonally diagonalize the matrix $A = \begin{bmatrix} 2 & 1 & 2 \\ 1 & 2 & 2 \\ 2 & 2 & 5 \end{bmatrix}$.

Solution. First we need to find the eigenvalues of A, that is, the values of λ satisfying the equation

$$\det \begin{bmatrix} 2-\lambda & 1 & 2 \\ 1 & 2-\lambda & 2 \\ 2 & 2 & 5-\lambda \end{bmatrix} = 0.$$

We add the second row multiplied by -2 to the third one and then we multiply the second row by $\lambda - 2$ and add to the first row. We get

$$\det \begin{bmatrix} 0 & 1+(\lambda-2)(2-\lambda) & -2+2\lambda \\ 1 & 2-\lambda & 2 \\ 0 & -2+2\lambda & 1-\lambda \end{bmatrix} = 0$$

or

$$\det \begin{bmatrix} 0 & -\lambda^2+4\lambda-3 & -2+2\lambda \\ 1 & 2-\lambda & 2 \\ 0 & -2+2\lambda & 1-\lambda \end{bmatrix} = 0.$$

Since

$$\det \begin{bmatrix} 0 & -\lambda^2+4\lambda-3 & -2+2\lambda \\ 1 & 2-\lambda & 2 \\ 0 & -2+2\lambda & 1-\lambda \end{bmatrix} = (1-\lambda)(\lambda^2-8\lambda+7),$$

the eigenvalues are 1, 1, and 7.

Alternatively, we can subtract the second row from the first one and get

$$\det \begin{bmatrix} 1-\lambda & \lambda-1 & 0 \\ 1 & 2-\lambda & 2 \\ 2 & 2 & 5-\lambda \end{bmatrix} = (1-\lambda)\begin{bmatrix} 1 & -1 & 0 \\ 1 & 2-\lambda & 2 \\ 2 & 2 & 5-\lambda \end{bmatrix}$$

$$= (1-\lambda)(\lambda^2-8\lambda+7)$$
$$= (1-\lambda)^2(7-\lambda).$$

The eigenvectors corresponding to the eigenvalue 1 are given by the equation

$$\begin{bmatrix} 2 & 1 & 2 \\ 1 & 2 & 2 \\ 2 & 2 & 5 \end{bmatrix} \begin{bmatrix} x \\ y \\ z \end{bmatrix} = \begin{bmatrix} x \\ y \\ z \end{bmatrix}.$$

This equation is equivalent to the system

$$\begin{cases} 2x + y + 2z = x \\ x + 2y + 2z = y \\ 2x + 2y + 5z = z \end{cases}$$

which reduces to the equation

$$x + y + 2z = 0$$

or

$$x = -y - 2z.$$

Consequently, the general solution of the system is

$$\begin{bmatrix} x \\ y \\ z \end{bmatrix} = \begin{bmatrix} -y - 2z \\ y \\ z \end{bmatrix} = y \begin{bmatrix} -1 \\ 1 \\ 0 \end{bmatrix} + z \begin{bmatrix} -2 \\ 0 \\ 1 \end{bmatrix}$$

and the eigenspace corresponding to the eigenvalue 1 is $\text{Span}\left\{ \begin{bmatrix} -1 \\ 1 \\ 0 \end{bmatrix}, \begin{bmatrix} -2 \\ 0 \\ 1 \end{bmatrix} \right\}$.

Since the vector

$$\begin{bmatrix} -2 \\ 0 \\ 1 \end{bmatrix} - \frac{\begin{bmatrix} -2 \\ 0 \\ 1 \end{bmatrix} \cdot \begin{bmatrix} -1 \\ 1 \\ 0 \end{bmatrix}}{\begin{bmatrix} -1 \\ 1 \\ 0 \end{bmatrix} \cdot \begin{bmatrix} -1 \\ 1 \\ 0 \end{bmatrix}} \begin{bmatrix} -1 \\ 1 \\ 0 \end{bmatrix} = \begin{bmatrix} -1 \\ -1 \\ 1 \end{bmatrix}$$

is an eigenvector corresponding to the eigenvalue 1 which is orthogonal to $\begin{bmatrix} -1 \\ 1 \\ 0 \end{bmatrix}$,

the vectors $\begin{bmatrix} -1 \\ 1 \\ 0 \end{bmatrix}$ and $\begin{bmatrix} -1 \\ -1 \\ 1 \end{bmatrix}$ are orthogonal and

$$\text{Span}\left\{ \begin{bmatrix} -1 \\ 1 \\ 0 \end{bmatrix}, \begin{bmatrix} -2 \\ 0 \\ 1 \end{bmatrix} \right\} = \text{Span}\left\{ \begin{bmatrix} -1 \\ 1 \\ 0 \end{bmatrix}, \begin{bmatrix} -1 \\ -1 \\ 1 \end{bmatrix} \right\}.$$

Now we find the eigenspace corresponding to the eigenvalue 7 by noting that an eigenvector corresponding to the eigenvalue 7 is orthogonal to all the eigenvectors corresponding to the eigenvalue 1. We have

$$\begin{bmatrix} -1 \\ 1 \\ 0 \end{bmatrix} \times \begin{bmatrix} -2 \\ 0 \\ 1 \end{bmatrix} = \begin{bmatrix} 1 \\ 1 \\ 2 \end{bmatrix},$$

so Span $\left\{ \begin{bmatrix} 1 \\ 1 \\ 2 \end{bmatrix} \right\}$ must be the eigenspace corresponding to the eigenvalue 7. Since

$$\left\| \begin{bmatrix} 1 \\ 1 \\ 2 \end{bmatrix} \right\|^2 = 6, \quad \left\| \begin{bmatrix} -1 \\ 1 \\ 0 \end{bmatrix} \right\|^2 = 2, \quad \text{and} \quad \left\| \begin{bmatrix} -1 \\ -1 \\ 1 \end{bmatrix} \right\|^2 = 3,$$

normalizing the eigenvectors we obtain a basis of orthonormal eigenvectors

$$\begin{bmatrix} \frac{1}{\sqrt{6}} \\ \frac{1}{\sqrt{6}} \\ \frac{2}{\sqrt{6}} \end{bmatrix}, \quad \begin{bmatrix} -\frac{1}{\sqrt{2}} \\ \frac{1}{\sqrt{2}} \\ 0 \end{bmatrix}, \quad \begin{bmatrix} -\frac{1}{\sqrt{3}} \\ -\frac{1}{\sqrt{3}} \\ \frac{1}{\sqrt{3}} \end{bmatrix},$$

corresponding to the eigenvalues $\lambda = 7$, $\lambda = 1$, $\lambda = 1$. Now we are ready to present an orthogonal diagonalization of A:

$$A = \begin{bmatrix} \frac{1}{\sqrt{6}} & -\frac{1}{\sqrt{2}} & -\frac{1}{\sqrt{3}} \\ \frac{1}{\sqrt{6}} & \frac{1}{\sqrt{2}} & -\frac{1}{\sqrt{3}} \\ \frac{2}{\sqrt{6}} & 0 & \frac{1}{\sqrt{3}} \end{bmatrix} \begin{bmatrix} 7 & 0 & 0 \\ 0 & 1 & 0 \\ 0 & 0 & 1 \end{bmatrix} \begin{bmatrix} \frac{1}{\sqrt{6}} & \frac{1}{\sqrt{6}} & \frac{2}{\sqrt{6}} \\ -\frac{1}{\sqrt{2}} & \frac{1}{\sqrt{2}} & 0 \\ -\frac{1}{\sqrt{3}} & -\frac{1}{\sqrt{3}} & \frac{1}{\sqrt{3}} \end{bmatrix}.$$

$\square$

The spectral decomposition of a symmetric matrix

The representation of a matrix presented in the next theorem is a consequence of the representation $A = PDP^{-1}$, but is expressed in a very different form. This form explains the geometric meaning of orthogonal diagonalization and is useful in applications.

Theorem 7.2.15. *If $\{\mathbf{v}_1, \mathbf{v}_2, \mathbf{v}_3\}$ be a basis of orthogonal eigenvectors of a 3 × 3 symmetric matrix A, then*

$$A = \alpha_1 \frac{1}{\|\mathbf{v}_1\|^2} \mathbf{v}_1 \mathbf{v}_1^T + \alpha_2 \frac{1}{\|\mathbf{v}_2\|^2} \mathbf{v}_2 \mathbf{v}_2^T + \alpha_3 \frac{1}{\|\mathbf{v}_3\|^2} \mathbf{v}_3 \mathbf{v}_3^T,$$

where α_1, α_2, and α_3 are the eigenvalues of A corresponding to the eigenvectors $\mathbf{v}_1$, $\mathbf{v}_2$, and $\mathbf{v}_3$, respectively, that is,

$$A\mathbf{v}_1 = \alpha_1 \mathbf{v}_1, \quad A\mathbf{v}_2 = \alpha_2 \mathbf{v}_2, \quad \text{and} \quad A\mathbf{v}_3 = \alpha_3 \mathbf{v}_3.$$

Proof. We give two proofs.

Proof 1. Let

$$\mathbf{p}_1 = \frac{1}{\|\mathbf{v}_1\|}\mathbf{v}_1, \quad \mathbf{p}_2 = \frac{1}{\|\mathbf{v}_2\|}\mathbf{v}_2, \quad \text{and} \quad \mathbf{p}_3 = \frac{1}{\|\mathbf{v}_3\|}\mathbf{v}_3.$$

First we note that

$$A = \begin{bmatrix} \mathbf{p}_1 & \mathbf{p}_2 & \mathbf{p}_3 \end{bmatrix} \begin{bmatrix} \alpha_1 & 0 & 0 \\ 0 & \alpha_2 & 0 \\ 0 & 0 & \alpha_3 \end{bmatrix} \begin{bmatrix} \mathbf{p}_1^T \\ \mathbf{p}_2^T \\ \mathbf{p}_3^T \end{bmatrix}.$$

Now, if $\mathbf{x}$ is an arbitrary vector from $\mathbb{R}^3$, then we have

$$A\mathbf{x} = \begin{bmatrix} \mathbf{p}_1 & \mathbf{p}_2 & \mathbf{p}_3 \end{bmatrix} \begin{bmatrix} \alpha_1 & 0 & 0 \\ 0 & \alpha_2 & 0 \\ 0 & 0 & \alpha_3 \end{bmatrix} \begin{bmatrix} \mathbf{p}_1^T \\ \mathbf{p}_2^T \\ \mathbf{p}_3^T \end{bmatrix} \mathbf{x}$$

$$= \begin{bmatrix} \mathbf{p}_1 & \mathbf{p}_2 & \mathbf{p}_3 \end{bmatrix} \begin{bmatrix} \alpha_1 & 0 & 0 \\ 0 & \alpha_2 & 0 \\ 0 & 0 & \alpha_3 \end{bmatrix} \begin{bmatrix} \mathbf{p}_1^T\mathbf{x} \\ \mathbf{p}_2^T\mathbf{x} \\ \mathbf{p}_3^T\mathbf{x} \end{bmatrix}$$

$$= \begin{bmatrix} \mathbf{p}_1 & \mathbf{p}_2 & \mathbf{p}_3 \end{bmatrix} \begin{bmatrix} \alpha_1\mathbf{p}_1^T\mathbf{x} \\ \alpha_2\mathbf{p}_2^T\mathbf{x} \\ \alpha_3\mathbf{p}_3^T\mathbf{x} \end{bmatrix}$$

$$= \alpha_1\mathbf{p}_1\left(\mathbf{p}_1^T\mathbf{x}\right) + \alpha_2\mathbf{p}_2\left(\mathbf{p}_2^T\mathbf{x}\right) + \alpha_3\mathbf{p}_3\left(\mathbf{p}_3^T\mathbf{x}\right)$$

$$= \alpha_1\left(\mathbf{p}_1\mathbf{p}_1^T\right)\mathbf{x} + \alpha_2\left(\mathbf{p}_2\mathbf{p}_2^T\right)\mathbf{x} + \alpha_3\left(\mathbf{p}_3\mathbf{p}_3^T\right)\mathbf{x}$$

$$= \left(\alpha_1\mathbf{p}_1\mathbf{p}_1^T + \alpha_2\mathbf{p}_2\mathbf{p}_2^T + \alpha_3\mathbf{p}_3\mathbf{p}_3^T\right)\mathbf{x}.$$

Consequently, by Theorem 2.1.18, we have

$$A = \alpha_1\mathbf{p}_1\mathbf{p}_1^T + \alpha_2\mathbf{p}_2\mathbf{p}_2^T + \alpha_3\mathbf{p}_3\mathbf{p}_3^T,$$

which can be written as

$$A = \alpha_1\frac{1}{\|\mathbf{v}_1\|^2}\mathbf{v}_1\mathbf{v}_1^T + \alpha_2\frac{1}{\|\mathbf{v}_2\|^2}\mathbf{v}_2\mathbf{v}_2^T + \alpha_3\frac{1}{\|\mathbf{v}_3\|^2}\mathbf{v}_3\mathbf{v}_3^T.$$

Proof 2. We have shown in Example 5.3.24 that

$$\frac{1}{\|\mathbf{v}_1\|^2}\mathbf{v}_1\mathbf{v}_1^T + \frac{1}{\|\mathbf{v}_2\|^2}\mathbf{v}_2\mathbf{v}_2^T + \frac{1}{\|\mathbf{v}_3\|^2}\mathbf{v}_3\mathbf{v}_3^T = \begin{bmatrix} 1 & 0 & 0 \\ 0 & 1 & 0 \\ 0 & 0 & 1 \end{bmatrix},$$

for any orthogonal basis $\{\mathbf{v}_1, \mathbf{v}_2, \mathbf{v}_3\}$ in $\mathbb{R}^3$. Hence

$$A\left(\frac{1}{\|\mathbf{v}_1\|^2}\mathbf{v}_1\mathbf{v}_1^T + \frac{1}{\|\mathbf{v}_2\|^2}\mathbf{v}_2\mathbf{v}_2^T + \frac{1}{\|\mathbf{v}_3\|^2}\mathbf{v}_3\mathbf{v}_3^T\right) = A\begin{bmatrix} 1 & 0 & 0 \\ 0 & 1 & 0 \\ 0 & 0 & 1 \end{bmatrix} = A,$$

which can be written as

$$\frac{1}{\|\mathbf{v}_1\|^2}A\mathbf{v}_1\mathbf{v}_1^T + \frac{1}{\|\mathbf{v}_2\|^2}A\mathbf{v}_2\mathbf{v}_2^T + \frac{1}{\|\mathbf{v}_3\|^2}A\mathbf{v}_3\mathbf{v}_3^T = A.$$

Now, because

$$A\mathbf{v}_1 = \alpha_1\mathbf{v}_1, \quad A\mathbf{v}_2 = \alpha_2\mathbf{v}_2, \quad A\mathbf{v}_3 = \alpha_3\mathbf{v}_3,$$

we obtain the desired spectral decomposition

$$A = \alpha_1 \frac{1}{\|\mathbf{v}_1\|^2}\mathbf{v}_1\mathbf{v}_1^T + \alpha_2 \frac{1}{\|\mathbf{v}_2\|^2}\mathbf{v}_2\mathbf{v}_2^T + \alpha_3 \frac{1}{\|\mathbf{v}_3\|^2}\mathbf{v}_3\mathbf{v}_3^T.$$

$\square$

Definition 7.2.16. Let A be a 3 × 3 matrix. By a *spectral decomposition* of A we mean a representation of A in the form

$$A = \frac{\alpha_1}{\|\mathbf{v}_1\|^2}\mathbf{v}_1\mathbf{v}_1^T + \frac{\alpha_2}{\|\mathbf{v}_2\|^2}\mathbf{v}_2\mathbf{v}_2^T + \frac{\alpha_3}{\|\mathbf{v}_3\|^2}\mathbf{v}_3\mathbf{v}_3^T,$$

where $\mathbf{v}_1$, $\mathbf{v}_2$, and $\mathbf{v}_3$ are nonzero orthogonal vectors and α_1, α_2, and α_3 are real numbers.

Recall that, for any nonzero vector $\mathbf{u}$ in $\mathbb{R}^3$, the matrix $\frac{1}{\|\mathbf{u}\|^2}\mathbf{u}\mathbf{u}^T$ is the projection matrix on the vector line Span$\{\mathbf{u}\}$, (see Theorem 4.2.13). The spectral decomposition of a symmetric matrix is thus a representation of the matrix in terms of projection matrices on vector lines spanned by the eigenvectors of that matrix.

The following theorem is a converse of Theorem 7.2.15.

Theorem 7.2.17. *If* $\{\mathbf{v}_1, \mathbf{v}_2, \mathbf{v}_3\}$ *is a basis of orthogonal vectors in* $\mathbb{R}^3$, *then the matrix*

$$A = \alpha_1 \frac{1}{\|\mathbf{v}_1\|^2}\mathbf{v}_1\mathbf{v}_1^T + \alpha_2 \frac{1}{\|\mathbf{v}_2\|^2}\mathbf{v}_2\mathbf{v}_2^T + \alpha_3 \frac{1}{\|\mathbf{v}_3\|^2}\mathbf{v}_3\mathbf{v}_3^T,$$

is a symmetric matrix such that

$$A\mathbf{v}_1 = \alpha_1\mathbf{v}_1, \quad A\mathbf{v}_2 = \alpha_2\mathbf{v}_2, \quad and \quad A\mathbf{v}_3 = \alpha_3\mathbf{v}_3.$$

Proof. The proof is an easy consequence of the fact that the matrix $\mathbf{u}\mathbf{u}^T$ is symmetric for any $\mathbf{u}$ in $\mathbb{R}^3$. $\square$

Example 7.2.18. Find the spectral decomposition of the matrix $A = \begin{bmatrix} 2 & 1 & 2 \\ 1 & 2 & 2 \\ 2 & 2 & 5 \end{bmatrix}$.

Solution. The matrix A is the matrix considered in Example 7.2.14 where we found that its eigenvalues are 1, 1, and 7, and corresponding orthogonal eigenvectors are

$\begin{bmatrix} -1 \\ 1 \\ 0 \end{bmatrix}, \begin{bmatrix} -1 \\ -1 \\ 1 \end{bmatrix}$, and $\begin{bmatrix} 1 \\ 1 \\ 2 \end{bmatrix}$. Since

$$\frac{1}{\left\| \begin{bmatrix} -1 \\ 1 \\ 0 \end{bmatrix} \right\|^2} = \frac{1}{2}, \quad \frac{1}{\left\| \begin{bmatrix} -1 \\ -1 \\ 1 \end{bmatrix} \right\|^2} = \frac{1}{3}, \quad \text{and} \quad \frac{7}{\left\| \begin{bmatrix} 1 \\ 1 \\ 2 \end{bmatrix} \right\|^2} = \frac{7}{6},$$

the spectral decomposition of A is

$$\begin{bmatrix} 2 & 1 & 2 \\ 1 & 2 & 2 \\ 2 & 2 & 5 \end{bmatrix} = \frac{1}{2} \begin{bmatrix} -1 \\ 1 \\ 0 \end{bmatrix} [-1 \ 1 \ 0] + \frac{1}{3} \begin{bmatrix} -1 \\ -1 \\ 1 \end{bmatrix} [-1 \ -1 \ 1] + \frac{7}{6} \begin{bmatrix} 1 \\ 1 \\ 2 \end{bmatrix} [1 \ 1 \ 2].$$

□

Example 7.2.19. Find a 3 × 3 symmetric matrix A with the eigenvalues α, β and γ such that

$\begin{bmatrix} 1 \\ 1 \\ 1 \end{bmatrix}$ is an eigenvector of A corresponding to the eigenvalue α,

$\begin{bmatrix} 1 \\ -2 \\ 1 \end{bmatrix}$ is an eigenvector of A corresponding to the eigenvalue β.

Solution. Since

$$\begin{bmatrix} 1 \\ 1 \\ 1 \end{bmatrix} \times \begin{bmatrix} 1 \\ -2 \\ 1 \end{bmatrix} = \begin{bmatrix} 3 \\ 0 \\ -3 \end{bmatrix} = 3 \begin{bmatrix} 1 \\ 0 \\ -1 \end{bmatrix}$$

and

$$\left\| \begin{bmatrix} 1 \\ 1 \\ 1 \end{bmatrix} \right\|^2 = 3, \quad \left\| \begin{bmatrix} 1 \\ -2 \\ 1 \end{bmatrix} \right\|^2 = 6, \quad \text{and} \quad \left\| \begin{bmatrix} 1 \\ 0 \\ -1 \end{bmatrix} \right\|^2 = 2,$$

we have

$$A = \frac{\alpha}{3} \begin{bmatrix} 1 \\ 1 \\ 1 \end{bmatrix} [1 \ 1 \ 1] + \frac{\beta}{6} \begin{bmatrix} 1 \\ -2 \\ 1 \end{bmatrix} [1 \ -2 \ 1] + \frac{\gamma}{2} \begin{bmatrix} 1 \\ 0 \\ -1 \end{bmatrix} [1 \ 0 \ -1]$$

$$= \frac{\alpha}{3} \begin{bmatrix} 1 & 1 & 1 \\ 1 & 1 & 1 \\ 1 & 1 & 1 \end{bmatrix} + \frac{\beta}{6} \begin{bmatrix} 1 & -2 & 1 \\ -2 & 4 & -2 \\ 1 & -2 & 1 \end{bmatrix} + \frac{\gamma}{2} \begin{bmatrix} 1 & 0 & -1 \\ 0 & 0 & 0 \\ -1 & 0 & 1 \end{bmatrix}$$

$$= \begin{bmatrix} \frac{\alpha}{3} + \frac{\beta}{6} + \frac{\gamma}{2} & \frac{\alpha}{3} - \frac{\beta}{3} & \frac{\alpha}{3} + \frac{\beta}{6} - \frac{\gamma}{2} \\ \frac{\alpha}{3} - \frac{\beta}{3} & \frac{\alpha}{3} + \frac{2\beta}{3} & \frac{\alpha}{3} - \frac{\beta}{3} \\ \frac{\alpha}{3} + \frac{\beta}{6} - \frac{\gamma}{2} & \frac{\alpha}{3} - \frac{\beta}{3} & \frac{\alpha}{3} + \frac{\beta}{6} + \frac{\gamma}{2} \end{bmatrix}.$$

$\square$

The QR factorization of a 3×3 matrix

In Chapter 3 we introduced the QR factorization for 2 × 2 matrices and in Chapter 4 for 3 × 2 matrices. Here we present an extension of the idea to 3 × 3 matrices.

Theorem 7.2.20. *If a* 3×3 *matrix* $A = \begin{bmatrix} \mathbf{c}_1 & \mathbf{c}_2 & \mathbf{c}_3 \end{bmatrix}$ *has linearly independent columns, then A can be represented in the form*

$$A = QR,$$

where Q is a 3×3 *orthogonal matrix and R is an upper triangular* 3×3 *matrix, that is,*

$$R = \begin{bmatrix} r_{11} & r_{12} & r_{13} \\ 0 & r_{22} & r_{23} \\ 0 & 0 & r_{33} \end{bmatrix}$$

with $r_{1,1} > 0$, $r_{2,2} > 0$, *and* $r_{3,3} > 0$.

Proof. First we define

$$\mathbf{v}_1 = \mathbf{c}_1$$

$$\mathbf{v}_2 = \mathbf{c}_2 - \text{proj}_{\mathbf{v}_1} \mathbf{c}_2 = \mathbf{c}_2 - \frac{\mathbf{c}_2 \cdot \mathbf{v}_1}{\mathbf{v}_1 \cdot \mathbf{v}_1} \mathbf{v}_1$$

$$\mathbf{v}_3 = \mathbf{c}_3 - \text{proj}_{\text{Span}\{\mathbf{v}_1, \mathbf{v}_2\}} \mathbf{c}_3 = \mathbf{c}_3 - \text{proj}_{\mathbf{v}_1} \mathbf{c}_3 - \text{proj}_{\mathbf{v}_2} \mathbf{c}_3 = \mathbf{c}_3 - \frac{\mathbf{c}_3 \cdot \mathbf{v}_1}{\mathbf{v}_1 \cdot \mathbf{v}_1} \mathbf{v}_1 - \frac{\mathbf{c}_3 \cdot \mathbf{v}_2}{\mathbf{v}_2 \cdot \mathbf{v}_2} \mathbf{v}_2$$

Note that the vectors $\mathbf{v}_1$, $\mathbf{v}_2$, and $\mathbf{v}_3$ are nonzero vectors ($\mathbf{v}_2$ is nonzero because the vectors $\mathbf{c}_1, \mathbf{c}_2$ are linearly independent and $\mathbf{v}_3$ is nonzero because the vectors $\mathbf{c}_1, \mathbf{c}_2, \mathbf{c}_3$ are linearly independent),

$$\mathbf{v}_1 \cdot \mathbf{v}_2 = \mathbf{v}_1 \cdot \mathbf{v}_3 = \mathbf{v}_2 \cdot \mathbf{v}_3 = 0,$$

and

$$\mathbf{c}_2 = \frac{\mathbf{c}_2 \cdot \mathbf{v}_1}{\mathbf{v}_1 \cdot \mathbf{v}_1} \mathbf{v}_1 + \mathbf{v}_2 \quad \text{and} \quad \mathbf{c}_3 = \frac{\mathbf{c}_3 \cdot \mathbf{v}_1}{\mathbf{v}_1 \cdot \mathbf{v}_1} \mathbf{v}_1 + \frac{\mathbf{c}_3 \cdot \mathbf{v}_2}{\mathbf{v}_2 \cdot \mathbf{v}_2} \mathbf{v}_2 + \mathbf{v}_3.$$

Now we normalize the vectors $\mathbf{v}_1, \mathbf{v}_2, \mathbf{v}_3$:

$$\mathbf{u}_1 = \frac{1}{\|\mathbf{v}_1\|} \mathbf{v}_1, \quad \mathbf{u}_2 = \frac{1}{\|\mathbf{v}_2\|} \mathbf{v}_2, \quad \text{and} \quad \mathbf{u}_3 = \frac{1}{\|\mathbf{v}_3\|} \mathbf{v}_3$$

and denote

$$r_{1,1} = \|\mathbf{v}_1\|, \quad r_{1,2} = \|\mathbf{v}_1\|\frac{\mathbf{c}_2 \cdot \mathbf{v}_1}{\mathbf{v}_1 \cdot \mathbf{v}_1}, \quad r_{1,3} = \|\mathbf{v}_1\|\frac{\mathbf{c}_3 \cdot \mathbf{v}_1}{\mathbf{v}_1 \cdot \mathbf{v}_1},$$

$$r_{2,3} = \|\mathbf{v}_2\|\frac{\mathbf{c}_3 \cdot \mathbf{v}_2}{\mathbf{v}_2 \cdot \mathbf{v}_2}, \quad r_{2,2} = \|\mathbf{v}_2\|,$$

$$r_{3,3} = \|\mathbf{v}_3\|.$$

Then we have

$$\mathbf{c}_1 = r_{1,1}\mathbf{u}_1,$$

$$\mathbf{c}_2 = r_{1,2}\mathbf{u}_1 + r_{2,2}\mathbf{u}_2,$$

$$\mathbf{c}_3 = r_{1,3}\mathbf{u}_1 + r_{2,3}\mathbf{u}_2 + + r_{3,3}\mathbf{u}_3,$$

and consequently

$$\begin{bmatrix} \mathbf{c}_1 & \mathbf{c}_2 & \mathbf{c}_3 \end{bmatrix} = \begin{bmatrix} \mathbf{u}_1 & \mathbf{u}_2 & \mathbf{u}_3 \end{bmatrix} \begin{bmatrix} r_{11} & r_{12} & r_{13} \\ 0 & r_{22} & r_{23} \\ 0 & 0 & r_{33} \end{bmatrix}.$$

Note that $r_{1,1} > 0$, $r_{2,2} > 0$, and $r_{3,3} > 0$. □

The method used in the proof of the above result to construct orthogonal vectors is called ***Gram-Schimdt process.***

Example 7.2.21. Determine the QR factorization of the matrix $A = \begin{bmatrix} 1 & 0 & 1 \\ 1 & 1 & 0 \\ 1 & 1 & 1 \end{bmatrix}$.

Solution. From the equality

$$\begin{bmatrix} 0 \\ 1 \\ 1 \end{bmatrix} - \frac{\begin{bmatrix} 0 \\ 1 \\ 1 \end{bmatrix} \cdot \begin{bmatrix} 1 \\ 1 \\ 1 \end{bmatrix}}{\begin{bmatrix} 1 \\ 1 \\ 1 \end{bmatrix} \cdot \begin{bmatrix} 1 \\ 1 \\ 1 \end{bmatrix}} \begin{bmatrix} 1 \\ 1 \\ 1 \end{bmatrix} = \begin{bmatrix} 0 \\ 1 \\ 1 \end{bmatrix} - \frac{2}{3}\begin{bmatrix} 1 \\ 1 \\ 1 \end{bmatrix} = \frac{1}{3}\begin{bmatrix} -2 \\ 1 \\ 1 \end{bmatrix}$$

we get

$$\begin{bmatrix} 0 \\ 1 \\ 1 \end{bmatrix} = \frac{2}{3}\begin{bmatrix} 1 \\ 1 \\ 1 \end{bmatrix} + \frac{1}{3}\begin{bmatrix} -2 \\ 1 \\ 1 \end{bmatrix} \tag{7.3}$$

and from the equality

$$\begin{bmatrix} 1 \\ 0 \\ 1 \end{bmatrix} - \frac{2}{3}\begin{bmatrix} 1 \\ 1 \\ 1 \end{bmatrix} + \frac{1}{6}\begin{bmatrix} -2 \\ 1 \\ 1 \end{bmatrix} = -\frac{1}{2}\begin{bmatrix} 0 \\ 1 \\ -1 \end{bmatrix}$$

we get

$$\begin{bmatrix} 1 \\ 0 \\ 1 \end{bmatrix} = \frac{2}{3}\begin{bmatrix} 1 \\ 1 \\ 1 \end{bmatrix} - \frac{1}{6}\begin{bmatrix} -2 \\ 1 \\ 1 \end{bmatrix} - \frac{1}{2}\begin{bmatrix} 0 \\ 1 \\ -1 \end{bmatrix} = \frac{2}{3}\begin{bmatrix} 1 \\ 1 \\ 1 \end{bmatrix} - \frac{1}{6}\begin{bmatrix} -2 \\ 1 \\ 1 \end{bmatrix} + \frac{1}{2}\begin{bmatrix} 0 \\ -1 \\ 1 \end{bmatrix}. \qquad (7.4)$$

With a slight modification of the method in the proof of Theorem 7.2.20 we let

$$\mathbf{v}_1 = \begin{bmatrix} 1 \\ 1 \\ 1 \end{bmatrix}, \quad \mathbf{v}_2 = \begin{bmatrix} -2 \\ 1 \\ 1 \end{bmatrix}, \quad \text{and} \quad \mathbf{v}_3 = \begin{bmatrix} 0 \\ -1 \\ 1 \end{bmatrix}.$$

We have taken $\mathbf{v}_3 = \begin{bmatrix} 0 \\ -1 \\ 1 \end{bmatrix}$ and not $\mathbf{v}_3 = \begin{bmatrix} 0 \\ 1 \\ -1 \end{bmatrix}$ because the third coefficient of the vector $\mathbf{v}_3$ in (7.4) must be positive. Similarly, if the second coefficient of the vector $\mathbf{v}_2$ in (7.3) was negative, we would have to replace $\mathbf{v}_2$ by $-\mathbf{v}_2$.

Since

$$\left\|\begin{bmatrix} 1 \\ 1 \\ 1 \end{bmatrix}\right\| = \sqrt{3}, \quad \left\|\begin{bmatrix} -2 \\ 1 \\ 1 \end{bmatrix}\right\| = \sqrt{6}, \quad \text{and} \quad \left\|\begin{bmatrix} 0 \\ -1 \\ 1 \end{bmatrix}\right\| = \sqrt{2},$$

we let

$$\mathbf{u}_1 = \frac{1}{\sqrt{3}}\begin{bmatrix} 1 \\ 1 \\ 1 \end{bmatrix}, \quad \mathbf{u}_2 = \frac{1}{\sqrt{6}}\begin{bmatrix} -2 \\ 1 \\ 1 \end{bmatrix}, \quad \text{and} \quad \mathbf{u}_3 = \frac{1}{\sqrt{2}}\begin{bmatrix} 0 \\ -1 \\ 1 \end{bmatrix}.$$

Consequently

$$\begin{bmatrix} 1 \\ 1 \\ 1 \end{bmatrix} = \sqrt{3}\,\mathbf{u}_1,$$

$$\begin{bmatrix} 0 \\ 1 \\ 1 \end{bmatrix} = \frac{2\sqrt{3}}{3}\mathbf{u}_1 + \frac{\sqrt{6}}{3}\mathbf{u}_2,$$

$$\begin{bmatrix} 1 \\ 0 \\ 1 \end{bmatrix} = \frac{2\sqrt{3}}{3}\mathbf{u}_1 - \frac{\sqrt{6}}{6}\mathbf{u}_2 + \frac{\sqrt{2}}{2}\mathbf{u}_3.$$

Now we define

$$Q = \begin{bmatrix} \mathbf{u}_1 & \mathbf{u}_2 & \mathbf{u}_3 \end{bmatrix}.$$

Since the matrix $Q = \begin{bmatrix} \mathbf{u}_1 & \mathbf{u}_2 & \mathbf{u}_3 \end{bmatrix}$ is an orthogonal matrix, from the equality $A = QR$ we get

$$Q^T A = Q^T Q R = R.$$

Hence

$$R = Q^T A = \begin{bmatrix} \mathbf{u}_1 & \mathbf{u}_2 & \mathbf{u}_3 \end{bmatrix}^T A = \begin{bmatrix} \sqrt{3} & \frac{2\sqrt{3}}{3} & \frac{2\sqrt{3}}{3} \\ 0 & \frac{\sqrt{6}}{3} & -\frac{\sqrt{6}}{6} \\ 0 & 0 & \frac{\sqrt{2}}{2} \end{bmatrix}.$$

This means that the QR factorization of the matrix A is

$$A = \begin{bmatrix} \mathbf{u}_1 & \mathbf{u}_2 & \mathbf{u}_3 \end{bmatrix} \begin{bmatrix} \sqrt{3} & \frac{2\sqrt{3}}{3} & \frac{2\sqrt{3}}{3} \\ 0 & \frac{\sqrt{6}}{3} & -\frac{\sqrt{6}}{6} \\ 0 & 0 & \frac{\sqrt{2}}{2} \end{bmatrix}.$$

□

7.2.1 Exercises

Orthogonally diagonalize the given matrix.

1. $\begin{bmatrix} 1 & 2 & 2 \\ 2 & 1 & 2 \\ 2 & 2 & 5 \end{bmatrix}$

2. $\begin{bmatrix} 1 & 4 & -2 \\ 4 & 1 & -2 \\ -2 & -2 & 7 \end{bmatrix}.$

3. $\begin{bmatrix} 13 & -4 & 1 \\ -4 & 10 & -4 \\ 1 & -4 & 13 \end{bmatrix}.$

4. $\begin{bmatrix} 1 & 3 & 4 \\ 3 & 1 & 4 \\ 4 & 4 & 8 \end{bmatrix}$

5. $\begin{bmatrix} 3 & 2 & 4 \\ 2 & 3 & 4 \\ 4 & 4 & 9 \end{bmatrix}$

6. $\begin{bmatrix} 2 & 4 & 2 \\ 4 & 17 & 8 \\ 2 & 8 & 5 \end{bmatrix}.$

7. $\begin{bmatrix} 27 & 9 & 9 \\ 9 & 3 & 3 \\ 9 & 3 & 3 \end{bmatrix}.$

8. $\begin{bmatrix} -2 & 6 & 3 \\ 6 & 3 & -2 \\ 3 & -2 & 6 \end{bmatrix}$

9. $\begin{bmatrix} 6 & -2 & 10 \\ -2 & 9 & 5 \\ 10 & 5 & -15 \end{bmatrix}$

10. $\begin{bmatrix} 9 & 0 & -3 \\ 0 & 6 & 0 \\ -3 & 0 & 9 \end{bmatrix}.$

Determine the spectral decomposition of the given matrix.

11. $\begin{bmatrix} 13 & -4 & 1 \\ -4 & 10 & -4 \\ 1 & -4 & 13 \end{bmatrix}.$

12. $\begin{bmatrix} 1 & 2 & 2 \\ 2 & 1 & 2 \\ 2 & 2 & 5 \end{bmatrix}$

13. $\begin{bmatrix} 6 & -2 & 10 \\ -2 & 9 & 5 \\ 10 & 5 & -15 \end{bmatrix}$
14. $\begin{bmatrix} 27 & 9 & 9 \\ 9 & 3 & 3 \\ 9 & 3 & 3 \end{bmatrix}$.

15. Find a 3×3 matrix A such that $\begin{bmatrix} 1 \\ 1 \\ 1 \end{bmatrix}$ is an eigenvector of the matrix A corre-

sponding to the eigenvalue 9, $\begin{bmatrix} 4 \\ -5 \\ 1 \end{bmatrix}$ is an eigenvector of the matrix A corre-

sponding to the eigenvalue 30, and $\begin{bmatrix} 2 \\ 1 \\ -3 \end{bmatrix}$ is an eigenvector of the matrix A

corresponding to the eigenvalue 28.

16. Find a 3×3 matrix A such that $\begin{bmatrix} 1 \\ 1 \\ 0 \end{bmatrix}$ is an eigenvector of the matrix A corre-

sponding to the eigenvalue 8, $\begin{bmatrix} 1 \\ -1 \\ 1 \end{bmatrix}$ is an eigenvector of the matrix A cor-

responding to the eigenvalue 3, and $\begin{bmatrix} 1 \\ -1 \\ -2 \end{bmatrix}$ is an eigenvector of the matrix A

corresponding to the eigenvalue 24.

17. Find a symmetric 3×3 matrix A with eigenvalues 4 and 33 such that $\begin{bmatrix} 1 \\ 1 \\ 1 \end{bmatrix}$ and

$\begin{bmatrix} 1 \\ 1 \\ 3 \end{bmatrix}$ are eigenvectors of the matrix A corresponding to the eigenvalue 33.

18. Find a symmetric 3×3 matrix A with eigenvalues 4 and 9 such that $\begin{bmatrix} 1 \\ 1 \\ 0 \end{bmatrix}$ and

$\begin{bmatrix} 1 \\ 0 \\ 1 \end{bmatrix}$ are eigenvectors of the matrix A corresponding to the eigenvalue 4.

Determine the QR-factorization of the given matrix.

19. $\begin{bmatrix} 1 & 2 & 0 \\ 1 & 0 & 1 \\ 0 & 1 & 1 \end{bmatrix}$
20. $\begin{bmatrix} 1 & 1 & 1 \\ -1 & 0 & 1 \\ 0 & 1 & -1 \end{bmatrix}$

21. Let A be a symmetric 3×3 matrix. Show that if $\det(A - \lambda I) = -(\lambda - \alpha)^3$, for some real number α, then $A = \begin{bmatrix} \alpha & 0 & 0 \\ 0 & \alpha & 0 \\ 0 & 0 & \alpha \end{bmatrix}$.

Chapter 8

Applications to geometry

In Chapters 3 and 4 we discussed vector lines in $\mathbb{R}^2$ and vector lines and vector planes in $\mathbb{R}^3$. Such lines and planes always contain the origin. In this chapter we generalize those considerations to general lines and planes. The proofs in this chapter have a more geometrical flavor and remind us of the classical presentations of analytical geometry, but are compatible with the proofs given so far in this book. This section gives us more opportunities to use the concepts of linear algebra and to understand the connections between geometry and linear algebra.

When discussing geometry we often call elements of $\mathbb{R}^2$ or $\mathbb{R}^3$ *points* rather than *vectors.* There is no mathematical difference between points and vectors, but in the context of geometry it is often more intuitive to talk about points.

8.1 Lines in $\mathbb{R}^2$

Definition 8.1.1. Let $\mathbf{u}, \mathbf{a}$ in $\mathbb{R}^2$. If $\mathbf{u}$ is different from the origin, then the set of all points of the form $\mathbf{a} + t\mathbf{u}$, where t is an arbitrary real number, will be called a *line* and denoted by $\mathbf{a} + \text{Span}\{\mathbf{u}\}$, that is,

$$\mathbf{a} + \text{Span}\{\mathbf{u}\} = \{\mathbf{a} + t\mathbf{u}, t \text{ in } \mathbb{R}\}.$$

We say that $\mathbf{a} + \text{Span}\{\mathbf{u}\}$ is a line that contains the point $\mathbf{a}$ and is parallel to the vector line $\text{Span}\{\mathbf{u}\}$. Note that a line $\mathbf{a} + \text{Span}\{\mathbf{u}\}$ is a vector line if and only if $\mathbf{a} = \mathbf{0}$. In the definition of lines we have to assume that $\mathbf{u}$ is different from the origin, because otherwise $\text{Span}\{\mathbf{u}\}$ would not be a line, but a point.

Now let $\mathbf{a}$ and $\mathbf{u}$ be two points in $\mathbb{R}^2$ such that $\mathbf{u}$ is different from the origin. Consider points of the form $\mathbf{a} + t\mathbf{u}$ for different values of t, see Figure 8.1. Observe that the points $\mathbf{a} + t\mathbf{u}$ lie on the line through $\mathbf{a}$ that is parallel to the line which contains $\mathbf{u}$ and the origin, that is, the vector line $\text{Span}\{\mathbf{u}\}$. If t takes all real values, then we obtain the entire line through $\mathbf{a}$ that is parallel to the vector line $\text{Span}\{\mathbf{u}\}$.

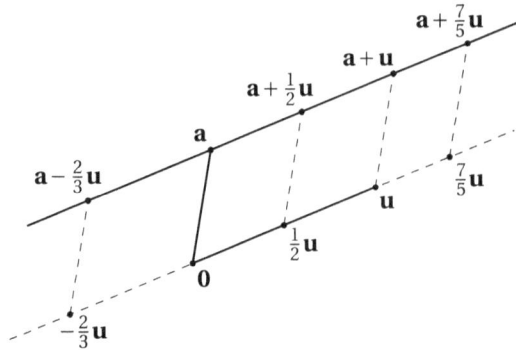

Figure 8.1: The line $\mathbf{a} + \mathrm{Span}\{\mathbf{u}\}$.

Example 8.1.2. The line through $\begin{bmatrix} -1 \\ 1 \end{bmatrix}$ and parallel to the vector line $\mathrm{Span}\left\{\begin{bmatrix} 8 \\ -2 \end{bmatrix}\right\}$ is

$$\begin{bmatrix} -1 \\ 1 \end{bmatrix} + \mathrm{Span}\left\{\begin{bmatrix} 8 \\ -2 \end{bmatrix}\right\} = \left\{\begin{bmatrix} -1 + 8t \\ 1 - 2t \end{bmatrix} : t \text{ in } \mathbb{R}\right\}.$$

As in the case of vector lines, we adopt the convention that when we say "a line $\mathbf{a} + \mathrm{Span}\{\mathbf{u}\}$," we always implicitly assume that $\mathbf{u} \neq \mathbf{0}$.

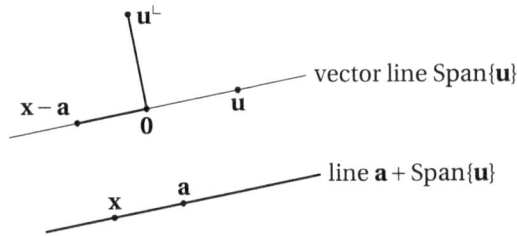

Figure 8.2: An illustration of Theorem 8.1.3.

Theorem 8.1.3. *Let* $\mathbf{u}$ *and* $\mathbf{a}$ *be vectors of* $\mathbb{R}^2$ *such that* $\mathbf{u}$ *is different from the origin. Then, for every* $\mathbf{x}$ *in* $\mathbb{R}^2$,

$$\mathbf{x} \text{ is on the line } \mathbf{a} + \mathrm{Span}\{\mathbf{u}\} \quad \textit{if and only if} \quad (\mathbf{x} - \mathbf{a}) \cdot \mathbf{u}^{\perp} = 0.$$

Proof. The proof is similar to the proof of Theorem 4.2.8.

If $\mathbf{x} = \mathbf{a} + t\mathbf{u}$ for some t in $\mathbb{R}$, then

$$(\mathbf{x} - \mathbf{a}) \cdot \mathbf{u}^\perp = ((\mathbf{a} + t\mathbf{u}) - \mathbf{a}) \cdot \mathbf{u}^\perp = t\mathbf{u} \cdot \mathbf{u}^\perp = 0.$$

Conversely, if $(\mathbf{x} - \mathbf{a}) \cdot \mathbf{u}^\perp = 0$, then $\mathbf{x} - \mathbf{a} = s(\mathbf{u}^\perp)^\perp = -s\mathbf{u}$ for some s in $\mathbb{R}$, by Theorem 3.2.22. Hence $\mathbf{x} = \mathbf{a} + (-s)\mathbf{u}$, which means that $\mathbf{x}$ is on the line $\mathbf{a} + \mathrm{Span}\{\mathbf{u}\}$. $\qquad\square$

Corollary 8.1.4. *Let* $\mathbf{u}$ *and* $\mathbf{a}$ *be vectors of* $\mathbb{R}^2$ *such that* $\mathbf{u}$ *is different from the origin. Then, for every* $\mathbf{x}$ *in* $\mathbb{R}^2$,

$$\mathbf{x} \text{ is on the line } \mathbf{a} + \mathrm{Span}\{\mathbf{u}\} \quad \textit{if and only if} \quad \det\begin{bmatrix} \mathbf{x} - \mathbf{a} & \mathbf{u} \end{bmatrix} = 0.$$

Proof. Since $(\mathbf{x} - \mathbf{a}) \cdot \mathbf{u}^\perp = \det\begin{bmatrix} \mathbf{u} & \mathbf{x} - \mathbf{a} \end{bmatrix} = -\det\begin{bmatrix} \mathbf{x} - \mathbf{a} & \mathbf{u} \end{bmatrix}$, Theorem 8.1.3 implies that $\mathbf{x}$ is on the line $\mathbf{a} + \mathrm{Span}\{\mathbf{u}\}$ if and only if $\det\begin{bmatrix} \mathbf{x} - \mathbf{a} & \mathbf{u} \end{bmatrix} = 0$. $\qquad\square$

Corollary 8.1.5. *Let* $\mathbf{n}$ *and* $\mathbf{a}$ *be vectors of* $\mathbb{R}^2$ *such that* $\mathbf{n}$ *is different from the origin. Then, for every* $\mathbf{x}$ *in* $\mathbb{R}^2$,

$$(\mathbf{x} - \mathbf{a}) \cdot \mathbf{n} = 0 \quad \textit{if and only if} \quad \mathbf{x} \text{ is on the line } \mathbf{a} + \mathrm{Span}\{\mathbf{n}^\perp\}.$$

Proof. By Theorem 8.1.3, the vector $\mathbf{x}$ is on the line $\mathbf{a} + \mathrm{Span}\{\mathbf{n}^\perp\}$ if and only if it satisfies the equation $(\mathbf{x} - \mathbf{a}) \cdot (\mathbf{n}^\perp)^\perp = 0$, wich is equivalent to the equation $(\mathbf{x} - \mathbf{a}) \cdot \mathbf{n} = 0$, since $(\mathbf{n}^\perp)^\perp = -\mathbf{n}$. $\qquad\square$

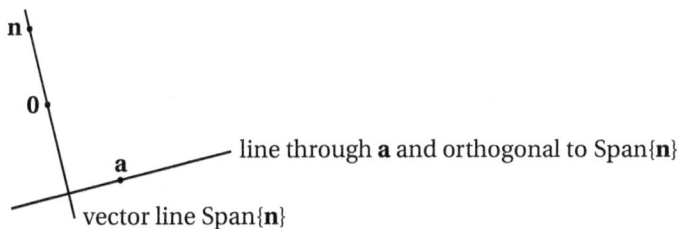

Figure 8.3: Line through the point $\mathbf{a}$ and orthogonal to the vector line $\mathrm{Span}\{\mathbf{n}\}$.

Definition 8.1.6. Let $\mathbf{n}$ be a nonzero element in $\mathbb{R}^2$ and $\mathbf{a}$ a point in $\mathbb{R}^2$. The set of all $\mathbf{x}$ such that

$$(\mathbf{x} - \mathbf{a}) \cdot \mathbf{n} = 0$$

is called the **line through the point** $\mathbf{a}$ **and orthogonal to the vector line** $\mathrm{Span}\{\mathbf{n}\}$.

Example 8.1.7. Write the equation $2x + 5y = 3$ in the form $(\mathbf{x} - \mathbf{a}) \cdot \mathbf{n} = 0$.

Solution. Since the equation $2x + 5y = 3$ can be written as

$$\begin{bmatrix} x \\ y \end{bmatrix} \cdot \begin{bmatrix} 2 \\ 5 \end{bmatrix} = 3,$$

we can take $\mathbf{n} = \begin{bmatrix} 2 \\ 5 \end{bmatrix}$. Next we notice that

$$3 = \begin{bmatrix} -1 \\ 1 \end{bmatrix} \cdot \begin{bmatrix} 2 \\ 5 \end{bmatrix},$$

so the equation $2x + 5y = 3$ can be written as

$$\begin{bmatrix} x \\ y \end{bmatrix} \cdot \begin{bmatrix} 2 \\ 5 \end{bmatrix} = \begin{bmatrix} -1 \\ 1 \end{bmatrix} \cdot \begin{bmatrix} 2 \\ 5 \end{bmatrix}$$

or

$$\left(\begin{bmatrix} x \\ y \end{bmatrix} - \begin{bmatrix} -1 \\ 1 \end{bmatrix} \right) \cdot \begin{bmatrix} 2 \\ 5 \end{bmatrix} = 0.$$

Consequently, the line defined by the equation $2x + 5y = 3$ can be described as the line through the point $\begin{bmatrix} -1 \\ 1 \end{bmatrix}$ and orthogonal to the vector line $\mathrm{Span}\left\{ \begin{bmatrix} 2 \\ 5 \end{bmatrix} \right\}$.

Note that the presented solution is not unique. For example, instead of $\begin{bmatrix} -1 \\ 1 \end{bmatrix}$ we could use $\begin{bmatrix} 1 \\ 0 \end{bmatrix}$. Since $2 = \begin{bmatrix} 1 \\ 0 \end{bmatrix} \cdot \begin{bmatrix} 2 \\ 5 \end{bmatrix}$, the equation $2x + 5y = 3$ can be written as

$$\left(\begin{bmatrix} x \\ y \end{bmatrix} - \tfrac{3}{2} \begin{bmatrix} 1 \\ 0 \end{bmatrix} \right) \cdot \begin{bmatrix} 2 \\ 5 \end{bmatrix} = 0.$$

A somewhat different solution is based on the observation that we have

$$3 = \frac{3}{\begin{bmatrix} 2 \\ 5 \end{bmatrix} \cdot \begin{bmatrix} 2 \\ 5 \end{bmatrix}} \begin{bmatrix} 2 \\ 5 \end{bmatrix} \cdot \begin{bmatrix} 2 \\ 5 \end{bmatrix} = \frac{3}{29} \begin{bmatrix} 2 \\ 5 \end{bmatrix} \cdot \begin{bmatrix} 2 \\ 5 \end{bmatrix}$$

and consequently the equation $2x + 5y = 3$ can be written as

$$\begin{bmatrix} x \\ y \end{bmatrix} \cdot \begin{bmatrix} 2 \\ 5 \end{bmatrix} = \frac{3}{29} \begin{bmatrix} 2 \\ 5 \end{bmatrix} \cdot \begin{bmatrix} 2 \\ 5 \end{bmatrix}$$

or

$$\left(\begin{bmatrix} x \\ y \end{bmatrix} - \frac{3}{29} \begin{bmatrix} 2 \\ 5 \end{bmatrix} \right) \cdot \begin{bmatrix} 2 \\ 5 \end{bmatrix} = 0.$$

This approach may seem artificial, but this solution is special because, unlike in other solutions, the point $\mathbf{a} = \frac{3}{29} \begin{bmatrix} 2 \\ 5 \end{bmatrix}$ is on the vector line $\mathrm{Span}\left\{ \begin{bmatrix} 2 \\ 5 \end{bmatrix} \right\}$, see Fig. 8.4. From this point of view, the last approach is quite natural. It is often used in arguments where $\mathbf{n}$ is not known, for example, in the proof of the next theorem. □

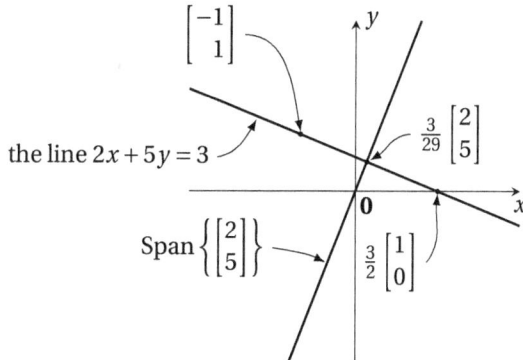

Figure 8.4: Different solutions in Example 8.1.7.

Theorem 8.1.8. *Let* $\mathbf{n}$ *be a nonzero vector in* $\mathbb{R}^2$. *The equation*

$$\mathbf{x} \cdot \mathbf{n} = d$$

defines a line in $\mathbb{R}^2$ *for any real number* d.

Proof. If $\mathbf{n}$ is a nonzero vector, then the equation

$$\mathbf{x} \cdot \mathbf{n} = d$$

can be written as

$$\left(\mathbf{x} - \tfrac{d}{\mathbf{n} \cdot \mathbf{n}} \mathbf{n} \right) \cdot \mathbf{n} = 0,$$

which is the equation of a line in $\mathbb{R}^2$ which contains the point $\frac{d}{\mathbf{n} \cdot \mathbf{n}} \mathbf{n}$ and is orthogonal to the vector line $\mathrm{Span}\{\mathbf{n}\}$. □

Theorem 8.1.9. *Let* $\mathbf{u}$ *and* $\mathbf{v}$ *be linearly independent vectors in* $\mathbb{R}^2$. *For arbitrary* $\mathbf{a}$ *and* $\mathbf{b}$ *in* $\mathbb{R}^2$ *the lines* $\mathbf{a} + \mathrm{Span}\{\mathbf{u}\}$ *and* $\mathbf{b} + \mathrm{Span}\{\mathbf{v}\}$ *have a unique common point.*

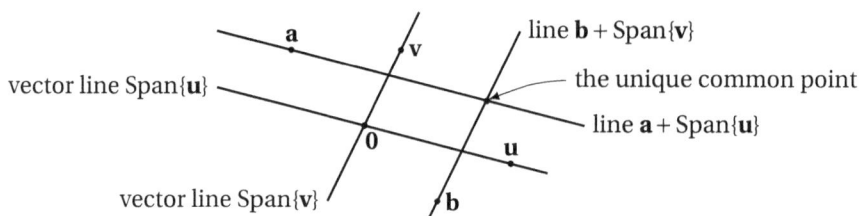

Figure 8.5: The unique common point the lines $\mathbf{a} + \mathrm{Span}\{\mathbf{u}\}$ and $\mathbf{b} + \mathrm{Span}\{\mathbf{v}\}$.

Proof. The common point of the lines $\mathbf{a} + \mathrm{Span}\{\mathbf{u}\}$ and $\mathbf{b} + \mathrm{Span}\{\mathbf{v}\}$ is given by the equation

$$\mathbf{a} + s\mathbf{u} = \mathbf{b} + t\mathbf{v},$$

which is equivalent to

$$s\mathbf{u} - t\mathbf{v} = \mathbf{b} - \mathbf{a}.$$

The uniqueness of the numbers s and t satisfying this equation is a consequence of the fact that the set $\{\mathbf{u}, \mathbf{v}\}$ is a basis of $\mathbb{R}^2$. $\square$

Example 8.1.10. We want to find the intersection point of the lines

$$\begin{bmatrix} 1 \\ 3 \end{bmatrix} + \mathrm{Span}\left\{ \begin{bmatrix} -1 \\ 2 \end{bmatrix} \right\} \quad \text{and} \quad \begin{bmatrix} 2 \\ 0 \end{bmatrix} + \mathrm{Span}\left\{ \begin{bmatrix} 1 \\ 1 \end{bmatrix} \right\}.$$

Since $\det \begin{bmatrix} -1 & 1 \\ 2 & 1 \end{bmatrix} = -3 \neq 0$, the lines have a unique common point. To find this point we have to solve the equation

$$\begin{bmatrix} 1 \\ 3 \end{bmatrix} + s \begin{bmatrix} -1 \\ 2 \end{bmatrix} = \begin{bmatrix} 2 \\ 0 \end{bmatrix} + t \begin{bmatrix} 1 \\ 1 \end{bmatrix}. \tag{8.1}$$

We could use Theorem 1.3.10 or proceed as follows. We apply the dot product with $\begin{bmatrix} 1 \\ 1 \end{bmatrix}^{\perp} = \begin{bmatrix} -1 \\ 1 \end{bmatrix}$ to both sides of the equation (8.1) and get

$$2 + 3s = -2.$$

Consequently, $s = -\frac{4}{3}$ and the point of intersection of the lines is

$$\begin{bmatrix} 1 \\ 3 \end{bmatrix} - \frac{4}{3} \begin{bmatrix} -1 \\ 2 \end{bmatrix} = \begin{bmatrix} \frac{7}{3} \\ \frac{1}{3} \end{bmatrix}.$$

Instead, we could apply the dot product with $-\begin{bmatrix} -1 \\ 2 \end{bmatrix}^{\llcorner} = \begin{bmatrix} 2 \\ 1 \end{bmatrix}$ to both sides of the equation (8.1) and get

$$5 = 4 + 3t.$$

Hence $t = \frac{1}{3}$ and the point of intersection of the lines is

$$\begin{bmatrix} 2 \\ 0 \end{bmatrix} + \frac{1}{3}\begin{bmatrix} 1 \\ 1 \end{bmatrix} = \begin{bmatrix} \frac{7}{3} \\ \frac{1}{3} \end{bmatrix}.$$

In both cases we got the same point of intersection, as expected.

The uniqueness property in the following theorem plays an important role in geometry.

Theorem 8.1.11. *If* $\mathbf{a}, \mathbf{b}$ *are in* $\mathbb{R}^2$ *and* $\mathbf{a} \neq \mathbf{b}$, *then there is a unique line which contains both points* $\mathbf{a}$ *and* $\mathbf{b}$. *That unique line can be described in any of the following ways:*

$$\mathbf{a} + \mathrm{Span}\{\mathbf{b} - \mathbf{a}\} = \mathbf{a} + \mathrm{Span}\{\mathbf{a} - \mathbf{b}\} = \mathbf{b} + \mathrm{Span}\{\mathbf{b} - \mathbf{a}\} = \mathbf{b} + \mathrm{Span}\{\mathbf{a} - \mathbf{b}\}.$$

Proof. If a line $\mathbf{c} + \mathrm{Span}\{\mathbf{u}\}$ contains the points $\mathbf{a}$ and $\mathbf{b}$, then

$$\mathbf{a} = \mathbf{c} + s\mathbf{u} \quad \text{and} \quad \mathbf{b} = \mathbf{c} + t\mathbf{u},$$

for some real numbers s and t. Hence

$$\mathbf{b} - \mathbf{a} = (t - s)\mathbf{u}$$

with $s \neq t$. Consequently,

$$\mathrm{Span}\{\mathbf{b} - \mathbf{a}\} = \mathrm{Span}\{\mathbf{u}\}.$$

and, since $\mathbf{a} = \mathbf{c} + s\mathbf{u}$ and $s\mathbf{u} + \mathrm{Span}\{\mathbf{u}\} = \mathrm{Span}\{\mathbf{u}\}$,

$$\mathbf{a} + \mathrm{Span}\{\mathbf{b} - \mathbf{a}\} = \mathbf{c} + s\mathbf{u} + \mathrm{Span}\{\mathbf{u}\} = \mathbf{c} + \mathrm{Span}\{\mathbf{u}\}.$$

The equalities

$$\mathbf{a} + \mathrm{Span}\{\mathbf{b} - \mathbf{a}\} = \mathbf{a} + \mathrm{Span}\{\mathbf{a} - \mathbf{b}\} = \mathbf{b} + \mathrm{Span}\{\mathbf{b} - \mathbf{a}\} = \mathbf{b} + \mathrm{Span}\{\mathbf{a} - \mathbf{b}\}$$

follow from the fact that all the above lines contain $\mathbf{a}$ and $\mathbf{b}$. □

Corollary 8.1.12. *If* $\mathbf{a}, \mathbf{b}$ *are in* $\mathbb{R}^2$ *and* $\mathbf{a} \neq \mathbf{b}$, *then the vector* $\mathbf{x}$ *is on the line which contains both points* $\mathbf{a}$ *and* $\mathbf{b}$ *if and only if*

$$\det\begin{bmatrix} \mathbf{x} - \mathbf{a} & \mathbf{b} - \mathbf{a} \end{bmatrix} = 0$$

Projections on lines in $\mathbb{R}^2$

Projections on vector lines in $\mathbb{R}^2$ were discussed in Chapter 3. It turns out that the situation does not change much when we consider projections on arbitrary lines in $\mathbb{R}^2$. For example, the next theorem is almost identical to Theorem 3.2.13 in $\mathbb{R}^2$.

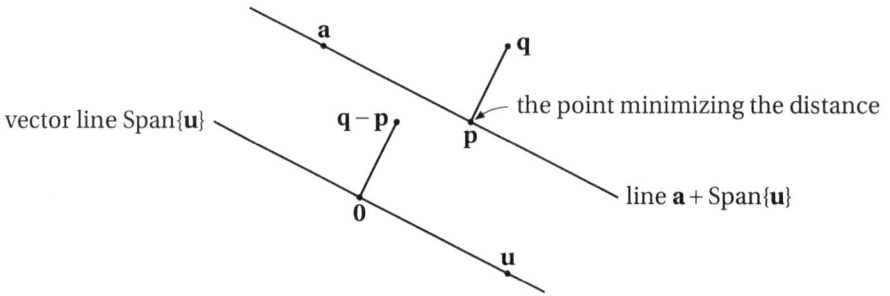

Figure 8.6: The point minimizing the distance from $\mathbf{q}$ to the line $\mathbf{a} + \text{Span}\{\mathbf{u}\}$.

> **Theorem 8.1.13.** *Let* $\mathbf{a}$, $\mathbf{q}$, *and* $\mathbf{u}$ *be elements of* $\mathbb{R}^2$, *where* $\mathbf{u}$ *is different from the origin. There is a unique point* $\mathbf{p}$ *on the line* $\mathbf{a} + \text{Span}\{\mathbf{u}\}$ *such that the distance from* $\mathbf{q}$ *to* $\mathbf{p}$ *is the shortest distance from* $\mathbf{q}$ *to any point on the line* $\mathbf{a} + \text{Span}\{\mathbf{u}\}$. *This point* $\mathbf{p}$ *is characterized as the point on the line* $\mathbf{a} + \text{Span}\{\mathbf{u}\}$ *satisfying the equation*
> $$(\mathbf{q} - \mathbf{p}) \cdot \mathbf{u} = 0.$$

Proof. Since

$$\|\mathbf{q} - \mathbf{a} - t\mathbf{u}\|^2 = \|\mathbf{q} - \mathbf{a}\|^2 - 2t(\mathbf{q} - \mathbf{a}) \cdot \mathbf{u} + t^2 \|\mathbf{u}\|^2$$

$$= \|\mathbf{q} - \mathbf{a}\|^2 - \left(\frac{(\mathbf{q} - \mathbf{a}) \cdot \mathbf{u}}{\|\mathbf{u}\|}\right)^2 + \left(\frac{(\mathbf{q} - \mathbf{a}) \cdot \mathbf{u}}{\|\mathbf{u}\|}\right)^2 - 2t\|\mathbf{u}\| \frac{(\mathbf{q} - \mathbf{a}) \cdot \mathbf{u}}{\|\mathbf{u}\|} + t^2 \|\mathbf{u}\|^2$$

$$= \|\mathbf{q} - \mathbf{a}\|^2 - \left(\frac{(\mathbf{q} - \mathbf{a}) \cdot \mathbf{u}}{\|\mathbf{u}\|}\right)^2 + \left(\frac{(\mathbf{q} - \mathbf{a}) \cdot \mathbf{u}}{\|\mathbf{u}\|} - t\|\mathbf{u}\|\right)^2,$$

the distance $\|\mathbf{q} - \mathbf{a} - t\mathbf{u}\|$ is minimized for the value of t for which $\left(\frac{(\mathbf{q}-\mathbf{a})\cdot\mathbf{u}}{\|\mathbf{u}\|} - t\|\mathbf{u}\|\right)^2 = 0$. Solving for t we get $t = \frac{(\mathbf{q}-\mathbf{a})\cdot\mathbf{u}}{\|\mathbf{u}\|^2}$ and consequently

$$\mathbf{p} = \mathbf{a} + \frac{(\mathbf{q} - \mathbf{a}) \cdot \mathbf{u}}{\|\mathbf{u}\|^2} \mathbf{u}.$$

It is easy to verify that for this $\mathbf{p}$ we have $(\mathbf{q} - \mathbf{p}) \cdot \mathbf{u} = 0$.

Now, if

$$(\mathbf{q} - \mathbf{a} - t\mathbf{u}) \cdot \mathbf{u} = 0,$$

then we must have $t = \frac{(\mathbf{q}-\mathbf{a})\cdot\mathbf{u}}{\|\mathbf{u}\|^2}$ which means that the point $\mathbf{p}$ is characterized by the equation $(\mathbf{q}-\mathbf{p})\cdot\mathbf{u} = 0$. $\qquad\qquad\qquad\square$

Definition 8.1.14. Let $\mathbf{a}, \mathbf{q}$ and $\mathbf{u}$ be elements of $\mathbb{R}^2$, where $\mathbf{u}$ is different from the origin.

(a) The unique point $\mathbf{p}$ on the line $\mathbf{a} + \mathrm{Span}\{\mathbf{u}\}$ that minimizes the distance from $\mathbf{q}$ to the line $\mathbf{a} + \mathrm{Span}\{\mathbf{u}\}$ is called the ***projection of*** $\mathbf{q}$ ***on the line*** $\mathbf{a} + \mathrm{Span}\{\mathbf{u}\}$.

(b) By the ***distance from the point*** $\mathbf{q}$ ***to the line*** $\mathbf{a} + \mathrm{Span}\{\mathbf{u}\}$ we mean the distance $\|\mathbf{q}-\mathbf{p}\|$ between the point $\mathbf{q}$ and its projection $\mathbf{p}$ on the line $\mathbf{a} + \mathrm{Span}\{\mathbf{u}\}$.

Example 8.1.15. Calculate the projection of the point $\mathbf{q} = \begin{bmatrix} 1 \\ 2 \end{bmatrix}$ on the line

$$\begin{bmatrix} 3 \\ -1 \end{bmatrix} + \mathrm{Span}\left\{\begin{bmatrix} 2 \\ 3 \end{bmatrix}\right\}.$$

Solution. The point

$$\begin{bmatrix} 3 \\ -1 \end{bmatrix} + t\begin{bmatrix} 2 \\ 3 \end{bmatrix} = \begin{bmatrix} 3+2t \\ -1+3t \end{bmatrix}$$

must satisfy the equation

$$\left(\begin{bmatrix} 3+2t \\ -1+3t \end{bmatrix} - \begin{bmatrix} 1 \\ 2 \end{bmatrix}\right)\cdot\begin{bmatrix} 2 \\ 3 \end{bmatrix} = 0$$

which is equivalent to the equation

$$2(2+2t) + 3(-3+3t) = 0.$$

Solving for t we get $t = \frac{5}{13}$. Thus the projection is

$$\begin{bmatrix} 3 \\ -1 \end{bmatrix} + \frac{5}{13}\begin{bmatrix} 2 \\ 3 \end{bmatrix} = \frac{1}{13}\begin{bmatrix} 49 \\ 2 \end{bmatrix}$$

$\qquad\qquad\qquad\square$

Recall that the equation $(\mathbf{x}-\mathbf{q})\cdot\mathbf{u} = 0$ describes a line through the point $\mathbf{q}$. Using this interpretation we can rephrase Theorem 8.1.13 in terms of intersecting lines.

Corollary 8.1.16. *Let* **u** *be a nonzero vector in* $\mathbb{R}^2$ *and let* **q** *be an arbitrary point in* $\mathbb{R}^3$. *The projection of* **q** *on the line* **a** + Span{**u**} *is the intersection of the line* **a** + Span{**u**} *and the line* (**x** − **q**) · **u** = 0.

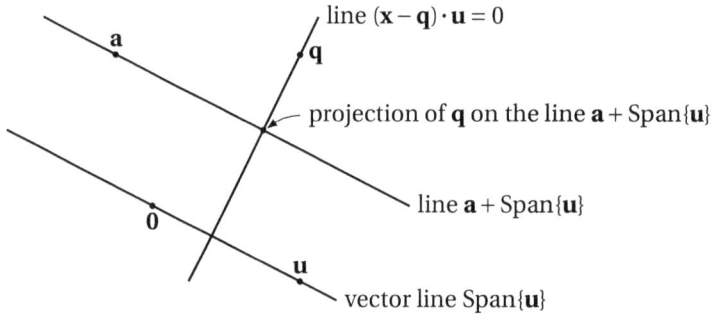

Figure 8.7: The projection of **q** on the line **a** + Span{**u**}.

The next theorem gives us a practical method for calculating the projection of a point on a line given by the equation **x** · **n** = *d*.

Theorem 8.1.17. *Let* **n** *be a nonzero vector in* $\mathbb{R}^2$ *and let* **q** *be an arbitrary point in* $\mathbb{R}^2$. *The lines* **x** · **n** = *d and* **q** + Span{**n**} *have a unique common point* **p**. *This point* **p** *is the projection of the point* **q** *on the line* **x** · **n** = *d*.

Proof. Every point on the line **q** + Span{**n**} is of the form **q** + *t***n** for some real number *t*. A point **q** + *t***n** is on the line **x** · **n** = *d* if

$$(\mathbf{q} + t\mathbf{n}) \cdot \mathbf{n} = d.$$

Solving for *t* gives us

$$t = \frac{d - \mathbf{q} \cdot \mathbf{n}}{\mathbf{n} \cdot \mathbf{n}}.$$

This means that the line **q** + Span{**n**} and the line **x** · **n** = *d* have a unique common point and that point is

$$\mathbf{p} = \mathbf{q} + \frac{d - \mathbf{q} \cdot \mathbf{n}}{\mathbf{n} \cdot \mathbf{n}} \mathbf{n}.$$

If **x** is an arbitrary point on the line **x** · **n** = *d*, then we have

$$(\mathbf{q} - \mathbf{p}) \cdot (\mathbf{x} - \mathbf{p}) = \frac{-d + \mathbf{q} \cdot \mathbf{n}}{\mathbf{n} \cdot \mathbf{n}} \mathbf{n} \cdot (\mathbf{x} - \mathbf{p}) = \frac{-d + \mathbf{q} \cdot \mathbf{n}}{\mathbf{n} \cdot \mathbf{n}} (\mathbf{n} \cdot \mathbf{x} - \mathbf{n} \cdot \mathbf{p}) = \frac{-d + \mathbf{q} \cdot \mathbf{n}}{\mathbf{n} \cdot \mathbf{n}} (d - d) = 0.$$

Hence

$$\|\mathbf{q}-\mathbf{x}\|^2 = \|\mathbf{q}-\mathbf{p}\|^2 - 2(\mathbf{q}-\mathbf{p})\cdot(\mathbf{x}-\mathbf{p}) + \|\mathbf{x}-\mathbf{p}\|^2 = \|\mathbf{q}-\mathbf{p}\|^2 + \|\mathbf{x}-\mathbf{p}\|^2.$$

This means that

$$\|\mathbf{q}-\mathbf{x}\|^2 \geq \|\mathbf{q}-\mathbf{p}\|^2$$

for every $\mathbf{x}$ on the line $\mathbf{x}\cdot\mathbf{n} = d$. Moreover,

$$\|\mathbf{q}-\mathbf{x}\|^2 > \|\mathbf{q}-\mathbf{p}\|^2$$

for every $\mathbf{x}$ on the line $\mathbf{x}\cdot\mathbf{n} = d$ that is different from $\mathbf{p}$. In other words, the distance from $\mathbf{q}$ to $\mathbf{p}$ is the shortest distance from $\mathbf{q}$ to any point in the line $\mathbf{x}\cdot\mathbf{n} = d$. $\qquad\square$

Example 8.1.18. Find the projection of the point $\mathbf{q} = \begin{bmatrix} 2 \\ 3 \end{bmatrix}$ on the line $x + 2y = 5$.

Solution. The line $x + 2y = 5$ can be written as $\mathbf{x}\cdot\begin{bmatrix} 1 \\ 2 \end{bmatrix} = 5$ where $\mathbf{x} = \begin{bmatrix} x \\ y \end{bmatrix}$.

The projection is

$$\mathbf{p} = \mathbf{q} + \frac{d - \mathbf{q}\cdot\mathbf{n}}{\mathbf{n}\cdot\mathbf{n}}\mathbf{n} = \begin{bmatrix} 2 \\ 3 \end{bmatrix} + \frac{5 - \begin{bmatrix} 2 \\ 3 \end{bmatrix}\cdot\begin{bmatrix} 1 \\ 2 \end{bmatrix}}{\begin{bmatrix} 1 \\ 2 \end{bmatrix}\cdot\begin{bmatrix} 1 \\ 2 \end{bmatrix}}\begin{bmatrix} 1 \\ 2 \end{bmatrix} = \begin{bmatrix} 2 \\ 3 \end{bmatrix} - \frac{3}{5}\begin{bmatrix} 1 \\ 2 \end{bmatrix} = \begin{bmatrix} \frac{7}{5} \\ \frac{9}{5} \end{bmatrix}$$

$\qquad\square$

The following theorem gives us a useful formula for calculating the distance from a point to a line.

Theorem 8.1.19. *Let $\mathbf{n}$ be a nonzero vector in $\mathbb{R}^2$ and let $\mathbf{q}$ be an arbitrary point in $\mathbb{R}^2$. The distance from the point $\mathbf{q}$ to the line $\mathbf{x}\cdot\mathbf{n} = d$ is*

$$\frac{|d - \mathbf{q}\cdot\mathbf{n}|}{\|\mathbf{n}\|}.$$

Proof. Since, as shown the proof of Theorem 8.1.17, the projection of the point $\mathbf{q}$ on the line $\mathbf{x}\cdot\mathbf{n} = d$ is

$$\mathbf{p} = \mathbf{q} + \frac{d - \mathbf{q}\cdot\mathbf{n}}{\mathbf{n}\cdot\mathbf{n}}\mathbf{n},$$

we have

$$\|\mathbf{q}-\mathbf{p}\| = \left\|\frac{d - \mathbf{q}\cdot\mathbf{n}}{\mathbf{n}\cdot\mathbf{n}}\mathbf{n}\right\| = \left|\frac{d - \mathbf{q}\cdot\mathbf{n}}{\|\mathbf{n}\|^2}\right|\|\mathbf{n}\| = \frac{|d - \mathbf{q}\cdot\mathbf{n}|}{\|\mathbf{n}\|}.$$

$\qquad\square$

Example 8.1.20. Find the distance from the point $\begin{bmatrix} 2 \\ 5 \end{bmatrix}$ to the line $\mathbf{x} \cdot \begin{bmatrix} 1 \\ -1 \end{bmatrix} = 3$.

Solution. In our case $\mathbf{q} = \begin{bmatrix} 2 \\ 5 \end{bmatrix}$, $\mathbf{n} = \begin{bmatrix} 1 \\ -1 \end{bmatrix}$, and $d = 3$. Consequently, the distance is

$$\frac{|d - \mathbf{q} \cdot \mathbf{n}|}{\|\mathbf{n}\|} = \frac{\left| 3 - \begin{bmatrix} 2 \\ 5 \end{bmatrix} \cdot \begin{bmatrix} 1 \\ -1 \end{bmatrix} \right|}{\left\| \begin{bmatrix} 1 \\ -1 \end{bmatrix} \right\|} = \frac{6}{\sqrt{2}}.$$

$\square$

Theorem 8.1.21. *The distance from a point $\mathbf{q}$ to the line $\mathbf{a} + \mathrm{Span}\{\mathbf{u}\}$ in $\mathbb{R}^2$ is*

$$\frac{\left| \det [\mathbf{q} - \mathbf{a} \quad \mathbf{u}] \right|}{\|\mathbf{u}\|}.$$

Proof. First we observe that for any vectors $\mathbf{v}$ and $\mathbf{w}$ in $\mathbb{R}^2$ we have

$$\left| \det [\mathbf{v} \quad \mathbf{w}] \right|^2 = \|\mathbf{v}\|^2 \|\mathbf{w}\|^2 - (\mathbf{v} \cdot \mathbf{w})^2,$$

which can be verified by direct calculations. Using the above identity and Theorem 8.1.17 we obtain

$$\begin{aligned}
\|\mathbf{q} - \mathbf{p}\| &= \sqrt{\left\| \mathbf{q} - \mathbf{a} - \frac{(\mathbf{q} - \mathbf{a}) \cdot \mathbf{u}}{\mathbf{u} \cdot \mathbf{u}} \mathbf{u} \right\|^2} \\
&= \sqrt{\frac{\|\mathbf{q} - \mathbf{a}\|^2 \|\mathbf{u}\|^2 - ((\mathbf{q} - \mathbf{a}) \cdot \mathbf{u})^2}{\|\mathbf{u}\|^2}} \\
&= \frac{\sqrt{\|\mathbf{q} - \mathbf{a}\|^2 \|\mathbf{u}\|^2 - ((\mathbf{q} - \mathbf{a}) \cdot \mathbf{u})^2}}{\|\mathbf{u}\|} \\
&= \frac{\left| \det [\mathbf{q} - \mathbf{a} \quad \mathbf{u}] \right|}{\|\mathbf{u}\|}.
\end{aligned}$$

$\square$

Example 8.1.22. Find the distance from then point $\begin{bmatrix} 1 \\ 2 \end{bmatrix}$ to the line $\begin{bmatrix} 2 \\ 3 \end{bmatrix} + \mathrm{Span}\left\{ \begin{bmatrix} 1 \\ -1 \end{bmatrix} \right\}$.

Solution. Since in this case $\mathbf{q} = \begin{bmatrix} 1 \\ 2 \end{bmatrix}$, $\mathbf{a} = \begin{bmatrix} 2 \\ 3 \end{bmatrix}$, and $\mathbf{u} = \begin{bmatrix} 1 \\ -1 \end{bmatrix}$, the distance is

$$\frac{\left| \det \begin{bmatrix} \mathbf{q} - \mathbf{a} & \mathbf{u} \end{bmatrix} \right|}{\|\mathbf{u}\|} = \frac{\left| \det \begin{bmatrix} -1 & 1 \\ -1 & -1 \end{bmatrix} \right|}{\left\| \begin{bmatrix} 1 \\ -1 \end{bmatrix} \right\|} = \frac{2}{\sqrt{2}} = \sqrt{2}.$$

□

Theorem 8.1.23. *The distance from a point* $\mathbf{q}$ *to the line through two distinct points* $\mathbf{a}$ *and* $\mathbf{b}$ *in* $\mathbb{R}^2$ *is*

$$\frac{\left| \det \begin{bmatrix} \mathbf{q} - \mathbf{a} & \mathbf{q} - \mathbf{b} \end{bmatrix} \right|}{\|\mathbf{b} - \mathbf{a}\|}.$$

Proof. The line through two distinct points $\mathbf{a}$ and $\mathbf{b}$ is $\mathbf{a} + \text{Span}\{\mathbf{b} - \mathbf{a}\}$. From Theorem 8.1.21, the distance from $\mathbf{q}$ to the line $\mathbf{a} + \text{Span}\{\mathbf{b} - \mathbf{a}\}$ is

$$\frac{\left| \det \begin{bmatrix} \mathbf{q} - \mathbf{a} & \mathbf{b} - \mathbf{a} \end{bmatrix} \right|}{\|\mathbf{b} - \mathbf{a}\|}.$$

Since

$$\begin{aligned} \det \begin{bmatrix} \mathbf{q} - \mathbf{a} & \mathbf{b} - \mathbf{a} \end{bmatrix} &= \det \begin{bmatrix} \mathbf{q} - \mathbf{a} & \mathbf{b} - \mathbf{q} + \mathbf{q} - \mathbf{a} \end{bmatrix} \\ &= \det \begin{bmatrix} \mathbf{q} - \mathbf{a} & \mathbf{b} - \mathbf{q} \end{bmatrix} + \det \begin{bmatrix} \mathbf{q} - \mathbf{a} & \mathbf{q} - \mathbf{a} \end{bmatrix} \\ &= \det \begin{bmatrix} \mathbf{q} - \mathbf{a} & \mathbf{b} - \mathbf{q} \end{bmatrix} \\ &= -\det \begin{bmatrix} \mathbf{q} - \mathbf{a} & \mathbf{q} - \mathbf{b} \end{bmatrix}, \end{aligned}$$

the distance can also be written as

$$\frac{\left| \det \begin{bmatrix} \mathbf{q} - \mathbf{a} & \mathbf{q} - \mathbf{b} \end{bmatrix} \right|}{\|\mathbf{b} - \mathbf{a}\|}.$$

□

Example 8.1.24. Find the distance from the point $\begin{bmatrix} 1 \\ 0 \end{bmatrix}$ to the line through the points $\begin{bmatrix} -1 \\ 2 \end{bmatrix}$ and $\begin{bmatrix} 2 \\ 3 \end{bmatrix}$.

Solution. Since in this case we have $\mathbf{q} = \begin{bmatrix} 1 \\ 0 \end{bmatrix}$, $\mathbf{a} = \begin{bmatrix} -1 \\ 2 \end{bmatrix}$, and $\mathbf{b} = \begin{bmatrix} 2 \\ 3 \end{bmatrix}$, the distance is

$$\frac{\left|\det\begin{bmatrix} \mathbf{q}-\mathbf{a} & \mathbf{q}-\mathbf{b} \end{bmatrix}\right|}{\|\mathbf{b}-\mathbf{a}\|} = \frac{\left|\det\begin{bmatrix} 2 & -1 \\ -2 & -5 \end{bmatrix}\right|}{\left\|\begin{bmatrix} 3 \\ 1 \end{bmatrix}\right\|} = \frac{12}{\sqrt{10}}.$$

□

Corollary 8.1.25. *Let* $\mathbf{a}, \mathbf{b}$ *and* $\mathbf{q}$ *be vectors in* $\mathbb{R}^2$ *such that the vectors* $\mathbf{q}-\mathbf{a}$ *and* $\mathbf{q}-\mathbf{b}$ *are linearly independent. The area of the triangle* $\mathbf{qab}$ *is*

$$\frac{1}{2}\left|\det\begin{bmatrix} \mathbf{q}-\mathbf{a} & \mathbf{q}-\mathbf{b} \end{bmatrix}\right|.$$

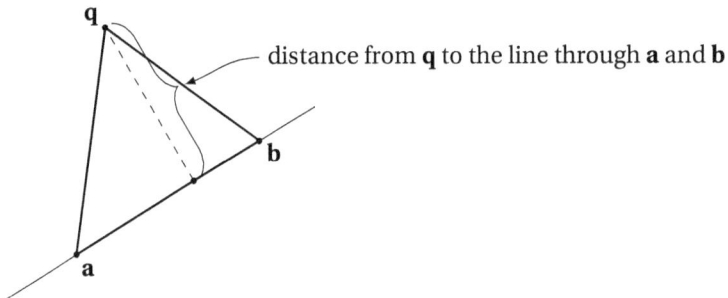

Figure 8.8: The triangle $\mathbf{qab}$.

Proof. Since the area of a triangle is

$$A = \frac{1}{2} \cdot (\text{the length of the base}) \cdot (\text{the height}),$$

we have

$$A = \frac{1}{2} \cdot \|\mathbf{b}-\mathbf{a}\| \cdot \frac{\left|\det\begin{bmatrix} \mathbf{q}-\mathbf{a} & \mathbf{q}-\mathbf{b} \end{bmatrix}\right|}{\|\mathbf{b}-\mathbf{a}\|} = \frac{1}{2}\left|\det\begin{bmatrix} \mathbf{q}-\mathbf{a} & \mathbf{q}-\mathbf{b} \end{bmatrix}\right|.$$

□

Example 8.1.26. Find the area of the triangle with vertices $\begin{bmatrix} 1 \\ -1 \end{bmatrix}$, $\begin{bmatrix} 2 \\ 2 \end{bmatrix}$, and $\begin{bmatrix} -2 \\ 3 \end{bmatrix}$.

Solution. We can set $\mathbf{q} = \begin{bmatrix} 1 \\ -1 \end{bmatrix}$, $\mathbf{a} = \begin{bmatrix} 2 \\ 2 \end{bmatrix}$, and $\mathbf{b} = \begin{bmatrix} -2 \\ 3 \end{bmatrix}$. Then the area of the triangle $\mathbf{qab}$ can be calculated as

$$\frac{1}{2} \left| \det \begin{bmatrix} \mathbf{q} - \mathbf{a} & \mathbf{q} - \mathbf{b} \end{bmatrix} \right| = \frac{1}{2} \left| \det \begin{bmatrix} -1 & 3 \\ -3 & -4 \end{bmatrix} \right| = \frac{13}{2}.$$

If we choose $\mathbf{q} = \begin{bmatrix} 2 \\ 2 \end{bmatrix}$, $\mathbf{a} = \begin{bmatrix} 1 \\ -1 \end{bmatrix}$, and $\mathbf{b} = \begin{bmatrix} -2 \\ 3 \end{bmatrix}$. Then for the area of the triangle $\mathbf{qab}$ we get

$$\frac{1}{2} \left| \det \begin{bmatrix} \mathbf{q} - \mathbf{a} & \mathbf{q} - \mathbf{b} \end{bmatrix} \right| = \frac{1}{2} \left| \det \begin{bmatrix} 1 & 4 \\ 3 & -1 \end{bmatrix} \right| = \frac{13}{2}.$$

As expected, the answers are the same. $\qquad\square$

8.1.1 Exercises

1. Find an equation of the line which contains the point $\begin{bmatrix} 0 \\ 1 \end{bmatrix}$ and is orthogonal to the vector line Span$\left\{ \begin{bmatrix} 2 \\ 1 \end{bmatrix} \right\}$.

2. Find an equation of the line which contains the point $\begin{bmatrix} 1 \\ 2 \end{bmatrix}$ and is orthogonal to the vector line Span$\left\{ \begin{bmatrix} 2 \\ 3 \end{bmatrix} \right\}$.

3. Write the equation of the line $\begin{bmatrix} 3 \\ 1 \end{bmatrix} +$ Span$\left\{ \begin{bmatrix} -2 \\ 3 \end{bmatrix} \right\}$ in the form $ax + by = c$.

4. Write the equation of the line $\begin{bmatrix} 2 \\ -1 \end{bmatrix} +$ Span$\left\{ \begin{bmatrix} 1 \\ 3 \end{bmatrix} \right\}$ in the form $ax + by = c$.

5. Write the equation $3x + y = 5$ in the form $(\mathbf{x} - \mathbf{a}) \cdot \mathbf{n} = 0$.

6. Write the equation $x - 2y = 1$ in the form $(\mathbf{x} - \mathbf{a}) \cdot \mathbf{n} = 0$.

7. Find the projection of $\begin{bmatrix} 1 \\ 1 \end{bmatrix}$ on the line $\mathbf{a} +$ Span$\{\mathbf{u}\}$ where $\mathbf{a} = \begin{bmatrix} 0 \\ 2 \end{bmatrix}$ and $\mathbf{u} = \begin{bmatrix} -3 \\ 1 \end{bmatrix}$.

8. Find the projection of $\begin{bmatrix} 2 \\ -1 \end{bmatrix}$ on the line $\mathbf{a} +$ Span$\{\mathbf{u}\}$ where $\mathbf{a} = \begin{bmatrix} 1 \\ 0 \end{bmatrix}$ and $\mathbf{u} = \begin{bmatrix} 3 \\ 2 \end{bmatrix}$.

9. Find the projection of $\begin{bmatrix} 1 \\ 3 \end{bmatrix}$ on the line $x + y = -1$.

10. Find the projection of $\begin{bmatrix} 1 \\ 4 \end{bmatrix}$ on the line $x - y = 2$.

11. Find the area of the triangle with the vertices $\begin{bmatrix} 1 \\ 1 \end{bmatrix}$, $\begin{bmatrix} 2 \\ -1 \end{bmatrix}$, and $\begin{bmatrix} 4 \\ 0 \end{bmatrix}$.

12. Find the area of the triangle with the vertices $\begin{bmatrix} 1 \\ 0 \end{bmatrix}$, $\begin{bmatrix} 0 \\ 2 \end{bmatrix}$, and $\begin{bmatrix} 3 \\ 5 \end{bmatrix}$.

13. Let $\mathbf{a} = \begin{bmatrix} a_1 \\ a_2 \end{bmatrix}$ and $\mathbf{b} = \begin{bmatrix} b_1 \\ b_2 \end{bmatrix}$ be points in $\mathbb{R}^2$ such that $\mathbf{a} \neq \mathbf{b}$. Show that the point $\mathbf{x} = \begin{bmatrix} x_1 \\ x_2 \end{bmatrix}$ is on the line containing the points $\mathbf{a}$ and $\mathbf{b}$ if and only if

$$\det \begin{bmatrix} 1 & 1 & 1 \\ x_1 & a_1 & b_1 \\ x_2 & a_2 & b_2 \end{bmatrix} = 0.$$

8.2 Lines and planes in $\mathbb{R}^3$

Lines in $\mathbb{R}^3$

Here we gather some definitions and theorems about lines in $\mathbb{R}^3$ that are the same for $\mathbb{R}^2$ and for $\mathbb{R}^2$. Even the proofs presented for lines in $\mathbb{R}^2$ are valid in $\mathbb{R}^3$.

Definition 8.2.1. Let $\mathbf{u}, \mathbf{a}$ be vectors in $\mathbb{R}^3$. If $\mathbf{u}$ is different from the origin, then the set of all points of the form $\mathbf{a} + t\mathbf{u}$, where t is an arbitrary real number, will be called a *line* and denoted by $\mathbf{a} + \text{Span}\{\mathbf{u}\}$, that is,

$$\mathbf{a} + \text{Span}\{\mathbf{u}\} = \{\mathbf{a} + t\mathbf{u}, t \text{ in } \mathbb{R}\}.$$

Theorem 8.2.2. *If $\mathbf{a}, \mathbf{b}$ are in $\mathbb{R}^3$ and $\mathbf{a} \neq \mathbf{b}$, then there is a unique line which contains both points $\mathbf{a}$ and $\mathbf{b}$. That unique line can be described in any of the following ways:*

$$\mathbf{a} + \text{Span}\{\mathbf{b} - \mathbf{a}\} = \mathbf{a} + \text{Span}\{\mathbf{a} - \mathbf{b}\} = \mathbf{b} + \text{Span}\{\mathbf{b} - \mathbf{a}\} = \mathbf{b} + \text{Span}\{\mathbf{a} - \mathbf{b}\}.$$

Theorem 8.2.3. *Let* $\mathbf{a}, \mathbf{q}$ *and* $\mathbf{u}$ *be elements of* $\mathbb{R}^3$, *where* $\mathbf{u}$ *is different from the origin. There is a unique point* $\mathbf{p}$ *on the line* $\mathbf{a} + \mathrm{Span}\{\mathbf{u}\}$ *such that the distance from* $\mathbf{q}$ *to* $\mathbf{p}$ *is the shortest distance from* $\mathbf{q}$ *to any point on the line* $\mathbf{a} + \mathrm{Span}\{\mathbf{u}\}$. *This point* $\mathbf{p}$ *is characterized as the point on the line* $\mathbf{a} + \mathrm{Span}\{\mathbf{u}\}$ *satisfying the equation*

$$(\mathbf{q} - \mathbf{p}) \cdot \mathbf{u} = 0.$$

Definition 8.2.4. Let $\mathbf{a}, \mathbf{q}$ and $\mathbf{u}$ be elements of $\mathbb{R}^3$, where $\mathbf{u}$ is different from the origin.

(a) The unique point $\mathbf{p}$ on the line $\mathbf{a} + \mathrm{Span}\{\mathbf{u}\}$ that minimizes the distance from $\mathbf{q}$ to the line $\mathbf{a} + \mathrm{Span}\{\mathbf{u}\}$ is called the ***projection of*** $\mathbf{q}$ ***on the line*** $\mathbf{a} + \mathrm{Span}\{\mathbf{u}\}$.

(b) By the ***distance from the point*** $\mathbf{q}$ ***to the line*** $\mathbf{a} + \mathrm{Span}\{\mathbf{u}\}$ we mean the distance $\|\mathbf{q} - \mathbf{p}\|$ between the point $\mathbf{q}$ and its projection $\mathbf{p}$ on the line $\mathbf{a} + \mathrm{Span}\{\mathbf{u}\}$.

Theorem 8.2.5. *Let* $\mathbf{a}, \mathbf{q}$ *and* $\mathbf{u}$ *be elements of* $\mathbb{R}^3$, *where* $\mathbf{u}$ *is different from the origin. The projection of the point* $\mathbf{q}$ *on the line* $\mathbf{a} + \mathrm{Span}\{\mathbf{u}\}$ *is*

$$\mathbf{p} = \mathbf{a} + \frac{(\mathbf{q} - \mathbf{a}) \cdot \mathbf{u}}{\mathbf{u} \cdot \mathbf{u}} \mathbf{u}.$$

Some theorems from Section 8.1 are not mentioned above, because of differences between $\mathbb{R}^2$ and $\mathbb{R}^3$. For example, the equation $(\mathbf{x} - \mathbf{a}) \cdot \mathbf{n} = 0$ does not define a line in $\mathbb{R}^3$ and the lines $\mathbf{a} + \mathrm{Span}\{\mathbf{u}\}$ and $\mathbf{b} + \mathrm{Span}\{\mathbf{v}\}$ need not intersect for arbitrary linearly independent $\mathbf{a}$ and $\mathbf{b}$ in $\mathbb{R}^3$.

Planes in $\mathbb{R}^3$

In this section we generalize some results on projections on vector planes in $\mathbb{R}^3$ presented in Chapter 4. These more general results can be easily obtained from similar results in Chapter 4 or can be proved by obvious modification of proofs presented there.

Definition 8.2.6. Let $\mathbf{u}, \mathbf{v}, \mathbf{a}$ be elements of $\mathbb{R}^3$. If $\mathbf{u}$ and $\mathbf{v}$ are linearly independent, then the set of all points of the form $\mathbf{a} + s\mathbf{u} + t\mathbf{v}$, where s and t are arbitrary real numbers, will be called a ***plane*** and denoted by $\mathbf{a} + \text{Span}\{\mathbf{u}, \mathbf{v}\}$, that is,

$$\mathbf{a} + \text{Span}\{\mathbf{u}, \mathbf{v}\} = \{\mathbf{a} + s\mathbf{u} + t\mathbf{v}, s, t \text{ in } \mathbb{R}\}.$$

Theorem 8.2.7. *Let $\mathbf{u}$ and $\mathbf{v}$ be linearly independent vectors in $\mathbb{R}^3$. Then a vector $\mathbf{x}$ is in the vector plane $\mathbf{a} + \text{Span}\{\mathbf{u}, \mathbf{v}\}$ if and only if*

$$(\mathbf{x} - \mathbf{a}) \cdot (\mathbf{u} \times \mathbf{v}) = 0. \tag{8.2}$$

Proof. The vector $\mathbf{x} = \mathbf{a} + s\mathbf{u} + t\mathbf{v}$ satisfies (8.2) for any real numbers s and t, by Theorem 5.1.6.

If $\mathbf{x}$ satisfies (8.2), then there are real number s and t such that

$$\mathbf{x} - \mathbf{a} = s\mathbf{u} + t\mathbf{v},$$

by Theorem 5.1.7. □

Theorem 8.2.8. *If $\mathbf{n}$ is a vector in $\mathbb{R}^3$ different from the origin, then the equation*

$$(\mathbf{x} - \mathbf{a}) \cdot \mathbf{n} = 0$$

defines a plane, that is, there are two linearly independent vectors $\mathbf{u}$ and $\mathbf{v}$ in $\mathbb{R}^3$ such that $(\mathbf{x} - \mathbf{a}) \cdot \mathbf{n} = 0$ if and only if $\mathbf{x}$ is in $\mathbf{a} + \text{Span}\{\mathbf{u}, \mathbf{v}\}$.

Proof. The proof is similar to the proof of Theorem 4.2.8. □

The equation $(\mathbf{x} - \mathbf{a}) \cdot \mathbf{n} = 0$ defines a line in $\mathbb{R}^2$. The same equation $(\mathbf{x} - \mathbf{a}) \cdot \mathbf{n} = 0$ defines a plane in $\mathbb{R}^3$. From the point of view of geometry these are very different objects. On the other hand, lines in $\mathbb{R}^2$ and planes in $\mathbb{R}^3$ share many algebraic properties.

Definition 8.2.9. Let $\mathbf{n}$ be a nonzero element in $\mathbb{R}^3$ and $\mathbf{a}$ a point in $\mathbb{R}^3$. The set of all $\mathbf{x}$ such that

$$(\mathbf{x} - \mathbf{a}) \cdot \mathbf{n} = 0$$

is called the ***plane through the point $\mathbf{a}$ and orthogonal to the vector line*** $\text{Span}\{\mathbf{n}\}$.

The equation of the plane that contains a point **a** and is orthogonal to the vector line Span{**n**} is the same as the plane which contains the projection of the point **a** on the vector line Span{**n**} and is orthogonal to the vector line Span{**n**}, see Fig. 8.9. In other words, if **p** is the projection of the point **a** on the vector line Span{**n**}, then the equations $(\mathbf{x} - \mathbf{p}) \cdot \mathbf{n} = 0$ and $(\mathbf{x} - \mathbf{a}) \cdot \mathbf{n} = 0$ are equivalent. This is a direct consequence of the fact that $(\mathbf{a} - \mathbf{p}) \cdot \mathbf{n} = 0$ and $\mathbf{x} - \mathbf{a} = \mathbf{p} - \mathbf{a} + \mathbf{x} - \mathbf{p}$. This observation makes the above definition more intuitive.

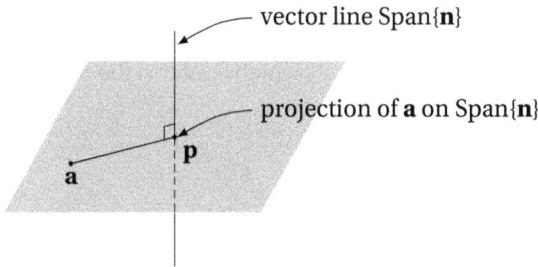

Figure 8.9: The plane that contains **a** and is orthogonal to the vector line Span{**n**}.

Example 8.2.10. The equation of the plane which contains the point $\begin{bmatrix} 2 \\ 1 \\ 4 \end{bmatrix}$ and is orthogonal to the vector line Span $\left\{ \begin{bmatrix} 5 \\ 8 \\ 3 \end{bmatrix} \right\}$ is

$$\left(\begin{bmatrix} x \\ y \\ z \end{bmatrix} - \begin{bmatrix} 2 \\ 1 \\ 4 \end{bmatrix} \right) \cdot \begin{bmatrix} 5 \\ 8 \\ 3 \end{bmatrix} = 0.$$

This vector equation can be written in the form

$$5(x - 2) + 8(y - 1) + 3(z - 4) = 0$$

or

$$5x + 8y + 3z = 30.$$

Theorem 8.2.11. *Let* **n** *be a nonzero vector in* $\mathbb{R}^3$*. The equation*

$$\mathbf{x} \cdot \mathbf{n} = d$$

defines a plane in $\mathbb{R}^3$*.*

Proof. If **n** is a nonzero vector, then the equation $\mathbf{x} \cdot \mathbf{n} = d$ can be written as

$$\left(\mathbf{x} - \tfrac{d}{\mathbf{n} \cdot \mathbf{n}} \mathbf{n}\right) \cdot \mathbf{n} = 0,$$

which is the equation of the plane in $\mathbb{R}^3$ which contains the point $\frac{d}{\mathbf{n} \cdot \mathbf{n}} \mathbf{n}$ and is orthogonal to the vector line Span{**n**}.   □

Example 8.2.12. Write the equation $2x + 5y + 3z = 4$ in the form $(\mathbf{x} - \mathbf{a}) \cdot \mathbf{n} = 0$.

Solution. This equation can be written as

$$\begin{bmatrix} x \\ y \\ z \end{bmatrix} \cdot \begin{bmatrix} 2 \\ 5 \\ 3 \end{bmatrix} = 4$$

or, equivalently, as

$$\begin{bmatrix} x \\ y \\ z \end{bmatrix} \cdot \begin{bmatrix} 2 \\ 5 \\ 3 \end{bmatrix} = \frac{4}{2^2 + 5^2 + 3^2} \begin{bmatrix} 2 \\ 5 \\ 3 \end{bmatrix} \cdot \begin{bmatrix} 2 \\ 5 \\ 3 \end{bmatrix}$$

and finally as

$$\left(\begin{bmatrix} x \\ y \\ z \end{bmatrix} - \tfrac{2}{19} \begin{bmatrix} 2 \\ 5 \\ 3 \end{bmatrix} \right) \cdot \begin{bmatrix} 2 \\ 5 \\ 3 \end{bmatrix} = 0.$$

The presented solution produces the only point **a** that is on the vector line Span{**n**}. If we are not interested in this particular solution, we can find a simpler solution. For example, we note that

$$\begin{bmatrix} 2 \\ 0 \\ 0 \end{bmatrix} \cdot \begin{bmatrix} 2 \\ 5 \\ 3 \end{bmatrix} = 4,$$

so the equation $2x + 5y + 3z = 4$ can be written as

$$\left(\begin{bmatrix} x \\ y \\ z \end{bmatrix} - \begin{bmatrix} 2 \\ 0 \\ 0 \end{bmatrix} \right) \cdot \begin{bmatrix} 2 \\ 5 \\ 3 \end{bmatrix} = 0.$$

□

Example 8.2.13. If $\begin{bmatrix} a \\ b \\ c \end{bmatrix} \neq \mathbf{0}$, write the equation $ax + by + cz = d$ in the form $(\mathbf{x} - \mathbf{a}) \cdot$ $\mathbf{n} = 0$.

Solution. The equation $ax + by + cz = d$ can be written as

$$\begin{bmatrix} x \\ y \\ z \end{bmatrix} \cdot \begin{bmatrix} a \\ b \\ c \end{bmatrix} = d$$

or, equivalently, as

$$\left(\begin{bmatrix} x \\ y \\ z \end{bmatrix} - \frac{d}{a^2+b^2+c^2} \begin{bmatrix} a \\ b \\ c \end{bmatrix} \right) \cdot \begin{bmatrix} a \\ b \\ c \end{bmatrix} = 0.$$

□

The above example shows that the equation $ax + by + cz = d$ describes all points $\begin{bmatrix} x \\ y \\ z \end{bmatrix}$ which are in the plane that is orthogonal to the vector line $\text{Span} \left\{ \begin{bmatrix} a \\ b \\ c \end{bmatrix} \right\}$ and intersects this line at the point $\frac{d}{a^2+b^2+c^2} \begin{bmatrix} a \\ b \\ c \end{bmatrix}$.

Projections on planes in $\mathbb{R}^3$

Now we turn our attention to the problem of finding the point on a plane that minimizes the distance from a given point.

Theorem 8.2.14. *Let* **n** *be a nonzero vector in* $\mathbb{R}^3$ *and let* **q** *be an arbitrary point in* $\mathbb{R}^3$.

The plane $\mathbf{x} \cdot \mathbf{n} = d$ *and the line* $\mathbf{q} + \mathrm{Span}\{\mathbf{n}\}$ *have a unique common point*
$$\mathbf{p} = \mathbf{q} + \frac{d - \mathbf{q} \cdot \mathbf{n}}{\mathbf{n} \cdot \mathbf{n}} \mathbf{n}.$$ *The distance from* **q** *to* **p** *is the shortest distance from* **q** *to any point in the plane* $\mathbf{x} \cdot \mathbf{n} = d$ *and* **p** *is the unique point with this property. The point* **p** *is characterized as the point in the plane* $\mathbf{x} \cdot \mathbf{n} = d$ *satisfying the equation*
$$(\mathbf{q} - \mathbf{p}) \cdot (\mathbf{x} - \mathbf{p}) = 0$$
for any point in the plane $\mathbf{x} \cdot \mathbf{n} = d$.

Proof. Every point on the line $\mathbf{q} + \mathrm{Span}\{\mathbf{n}\}$ is of the form $\mathbf{q} + t\mathbf{n}$ for some real number t. A point $\mathbf{q} + t\mathbf{n}$ is in the plane $\mathbf{x} \cdot \mathbf{n} = d$, if

$$(\mathbf{q} + t\mathbf{n}) \cdot \mathbf{n} = d.$$

Solving for t gives us
$$t = \frac{d - \mathbf{q} \cdot \mathbf{n}}{\mathbf{n} \cdot \mathbf{n}}.$$

This means that the line $\mathbf{q} + \mathrm{Span}\{\mathbf{n}\}$ and the plane $\mathbf{x} \cdot \mathbf{n} = d$ have a unique common point and that point is
$$\mathbf{p} = \mathbf{q} + \frac{d - \mathbf{q} \cdot \mathbf{n}}{\mathbf{n} \cdot \mathbf{n}} \mathbf{n}.$$

If $\mathbf{x}$ is an arbitrary point in the plane $\mathbf{x} \cdot \mathbf{n} = d$, then we have

$$(\mathbf{q} - \mathbf{p}) \cdot (\mathbf{x} - \mathbf{p}) = \frac{-d + \mathbf{q} \cdot \mathbf{n}}{\mathbf{n} \cdot \mathbf{n}} \mathbf{n} \cdot (\mathbf{x} - \mathbf{p}) = \frac{-d + \mathbf{q} \cdot \mathbf{n}}{\mathbf{n} \cdot \mathbf{n}} (\mathbf{n} \cdot \mathbf{x} - \mathbf{n} \cdot \mathbf{p}) = \frac{-d + \mathbf{q} \cdot \mathbf{n}}{\mathbf{n} \cdot \mathbf{n}} (d - d) = 0.$$

Hence

$$\|\mathbf{q} - \mathbf{x}\|^2 = \|\mathbf{q} - \mathbf{p}\|^2 - 2(\mathbf{q} - \mathbf{p}) \cdot (\mathbf{x} - \mathbf{p}) + \|\mathbf{x} - \mathbf{p}\|^2 = \|\mathbf{q} - \mathbf{p}\|^2 + \|\mathbf{x} - \mathbf{p}\|^2.$$

This means that
$$\|\mathbf{q} - \mathbf{x}\|^2 \geq \|\mathbf{q} - \mathbf{p}\|^2$$

for every $\mathbf{x}$ in the plane $\mathbf{x} \cdot \mathbf{n} = d$. Moreover,

$$\|\mathbf{q} - \mathbf{x}\|^2 = \|\mathbf{q} - \mathbf{p}\|^2$$

if and only if $\mathbf{x} = \mathbf{p}$. In other words, the distance from $\mathbf{q}$ to $\mathbf{p}$ is the shortest distance from $\mathbf{q}$ to any point in the plane $\mathbf{x} \cdot \mathbf{n} = d$.

It results from the above proof that the point $\mathbf{p}$ is characterized as the point in the plane $\mathbf{x} \cdot \mathbf{n} = d$ satisfying the equation

$$(\mathbf{q} - \mathbf{p}) \cdot (\mathbf{x} - \mathbf{p}) = 0.$$

$\square$

> **Definition 8.2.15.** Let $\mathbf{n}$ be a nonzero vector in $\mathbb{R}^3$ and let $\mathbf{q}$ be an arbitrary point in $\mathbb{R}^3$. The unique common point of the plane $\mathbf{x} \cdot \mathbf{n} = d$ and the line $\mathbf{q} + \mathrm{Span}\{\mathbf{n}\}$ is called the ***projection of*** $\mathbf{q}$ ***on the plane*** $\mathbf{x} \cdot \mathbf{n} = d$.

Example 8.2.16. Find the projection of the point $\mathbf{q} = \begin{bmatrix} 1 \\ 2 \\ 1 \end{bmatrix}$ on the plane

$x + y - z = 1$.

Solution. The plane $x + y - z = 1$ can be described by the equation

$$\begin{bmatrix} x \\ y \\ z \end{bmatrix} \cdot \begin{bmatrix} 1 \\ 1 \\ -1 \end{bmatrix} = 1,$$

so in this case we have $\mathbf{n} = \begin{bmatrix} 1 \\ 1 \\ -1 \end{bmatrix}$. We need to find the point common to the plane

and the line

$$\begin{bmatrix} 1 \\ 2 \\ 1 \end{bmatrix} + \mathrm{Span}\left\{ \begin{bmatrix} 1 \\ 1 \\ -1 \end{bmatrix} \right\}.$$

The point $\begin{bmatrix} 1 \\ 2 \\ 1 \end{bmatrix} + t \begin{bmatrix} 1 \\ 1 \\ -1 \end{bmatrix} = \begin{bmatrix} 1+t \\ 2+t \\ 1-t \end{bmatrix}$ is on the plane $x + y - z = 1$ if

$$1 + t + 2 + t - (1 - t) = 1.$$

Solving for t we obtain $t = -\frac{1}{3}$. Consequently, the projection of the point $\mathbf{q} = \begin{bmatrix} 1 \\ 2 \\ 1 \end{bmatrix}$

on the plane $x + y - z = 1$ is

$$\begin{bmatrix} 1 \\ 2 \\ 1 \end{bmatrix} - \frac{1}{3} \begin{bmatrix} 1 \\ 1 \\ -1 \end{bmatrix} = \begin{bmatrix} \frac{2}{3} \\ \frac{5}{3} \\ \frac{4}{3} \end{bmatrix}.$$

The problem can also be solved by an application of the formula given in

Theorem 8.2.14:

$$\mathbf{p} = \mathbf{q} + \frac{d - \mathbf{q} \cdot \mathbf{n}}{\mathbf{n} \cdot \mathbf{n}} \mathbf{n} = \begin{bmatrix} 1 \\ 2 \\ 1 \end{bmatrix} + \frac{1 - \begin{bmatrix} 1 \\ 2 \\ 1 \end{bmatrix} \cdot \begin{bmatrix} 1 \\ 1 \\ -1 \end{bmatrix}}{\begin{bmatrix} 1 \\ 1 \\ -1 \end{bmatrix} \cdot \begin{bmatrix} 1 \\ 1 \\ -1 \end{bmatrix}} \begin{bmatrix} 1 \\ 1 \\ -1 \end{bmatrix} = \begin{bmatrix} 1 \\ 2 \\ 1 \end{bmatrix} + \frac{-1}{3} \begin{bmatrix} 1 \\ 1 \\ -1 \end{bmatrix} = \begin{bmatrix} \frac{2}{3} \\ \frac{5}{3} \\ \frac{4}{3} \end{bmatrix}.$$

$\square$

Projections on lines in $\mathbb{R}^3$ revisited

In $\mathbb{R}^2$ the projection of a point $\mathbf{q}$ on the line $\mathbf{a} + \text{Span}\{\mathbf{u}\}$ is characterized by the equation $(\mathbf{q} - \mathbf{p}) \cdot \mathbf{u} = 0$ (see Theorem 8.1.13). It turns out that the same characterization works in $\mathbb{R}^3$, in spite of the fact that the geometric interpretation of the equation is different in $\mathbb{R}^2$ and in $\mathbb{R}^3$.

> **Theorem 8.2.17.** *Let* $\mathbf{a}$, $\mathbf{q}$, *and* $\mathbf{u}$ *be elements of* $\mathbb{R}^3$, *where* $\mathbf{u}$ *is different from the origin. The projection of* $\mathbf{q}$ *on the line* $\mathbf{a} + \text{Span}\{\mathbf{u}\}$ *is the intersection of the line and the plane* $(\mathbf{x} - \mathbf{q}) \cdot \mathbf{u} = 0$.

Proof. The intersection of the line $\mathbf{a} + \text{Span}\{\mathbf{u}\}$ and the plane $(\mathbf{x} - \mathbf{q}) \cdot \mathbf{u} = 0$ is a point $\mathbf{p} = \mathbf{a} + t\mathbf{u}$ such that $(\mathbf{a} + t\mathbf{u} - \mathbf{q}) \cdot \mathbf{u} = 0$. Solving for t we obtain

$$t = \frac{(\mathbf{q} - \mathbf{a}) \cdot \mathbf{u}}{\mathbf{u} \cdot \mathbf{u}}$$

and thus

$$\mathbf{p} = \mathbf{a} + \frac{(\mathbf{q} - \mathbf{a}) \cdot \mathbf{u}}{\mathbf{u} \cdot \mathbf{u}} \mathbf{u},$$

which is the projection of $\mathbf{q}$ on the line $\mathbf{a} + \text{Span}\{\mathbf{u}\}$. $\square$

Example 8.2.18. Find the projection of $\begin{bmatrix} 1 \\ 2 \\ 1 \end{bmatrix}$ on the line $\begin{bmatrix} 1 \\ 0 \\ 1 \end{bmatrix} + \text{Span}\left\{ \begin{bmatrix} 1 \\ 1 \\ 1 \end{bmatrix} \right\}$.

Solution. We solve the equation $(\mathbf{q} - \mathbf{p}) \cdot \mathbf{u} = 0$ for $\mathbf{p}$ with $\mathbf{q} = \begin{bmatrix} 1 \\ 2 \\ 1 \end{bmatrix}$, $\mathbf{u} = \begin{bmatrix} 1 \\ 1 \\ 1 \end{bmatrix}$, and

$$\mathbf{p} = \begin{bmatrix} 1 \\ 0 \\ 1 \end{bmatrix} + t \begin{bmatrix} 1 \\ 1 \\ 1 \end{bmatrix} = \begin{bmatrix} 1+t \\ t \\ 1+t \end{bmatrix}. \text{ From the equation}$$

$$\left(\begin{bmatrix} 1 \\ 2 \\ 1 \end{bmatrix} - \begin{bmatrix} 1+t \\ t \\ 1+t \end{bmatrix} \right) \cdot \begin{bmatrix} 1 \\ 1 \\ 1 \end{bmatrix} = \begin{bmatrix} -t \\ 2-t \\ -t \end{bmatrix} \cdot \begin{bmatrix} 1 \\ 1 \\ 1 \end{bmatrix} = 0$$

we find that $t = \frac{2}{3}$. Hence the projection is

$$\begin{bmatrix} 1+\frac{2}{3} \\ \frac{2}{3} \\ 1+\frac{2}{3} \end{bmatrix} = \begin{bmatrix} \frac{5}{3} \\ \frac{2}{3} \\ \frac{5}{3} \end{bmatrix}.$$

$\square$

The distance from a point to a plane

Definition 8.2.19. Let $\mathbf{n}$ be a nonzero vector in $\mathbb{R}^3$ and let $\mathbf{q}$ be an arbitrary point in $\mathbb{R}^3$. By the *distance from the point $\mathbf{q}$ to the plane* $\mathbf{x} \cdot \mathbf{n} = d$ we mean the distance $\|\mathbf{q} - \mathbf{p}\|$ between the point $\mathbf{q}$ and its projection $\mathbf{p}$ on the plane $\mathbf{x} \cdot \mathbf{n} = d$.

The following theorem gives us a useful formula for calculating the distance from a point to a plane.

Theorem 8.2.20. *Let $\mathbf{n}$ be a nonzero vector in $\mathbb{R}^3$ and let $\mathbf{q}$ be an arbitrary point in $\mathbb{R}^3$. The distance from the point $\mathbf{q}$ to the plane $(\mathbf{x} - \mathbf{a}) \cdot \mathbf{n} = 0$ is*

$$\frac{\left| (\mathbf{q} - \mathbf{a}) \cdot \mathbf{n} \right|}{\|\mathbf{n}\|}.$$

Proof. Since the projection of the point $\mathbf{q}$ to the plane $(\mathbf{x} - \mathbf{a}) \cdot \mathbf{n} = 0$ is

$$\mathbf{p} = \mathbf{q} + \frac{d - \mathbf{q} \cdot \mathbf{n}}{\mathbf{n} \cdot \mathbf{n}} \mathbf{n},$$

where $d = \mathbf{a} \cdot \mathbf{n}$, we have

$$\|\mathbf{q} - \mathbf{p}\| = \left\| \frac{d - \mathbf{q} \cdot \mathbf{n}}{\mathbf{n} \cdot \mathbf{n}} \mathbf{n} \right\| = \left\| \frac{\mathbf{a} \cdot \mathbf{n} - \mathbf{q} \cdot \mathbf{n}}{\mathbf{n} \cdot \mathbf{n}} \mathbf{n} \right\| = \left| \frac{\mathbf{a} \cdot \mathbf{n} - \mathbf{q} \cdot \mathbf{n}}{\|\mathbf{n}\|^2} \right| \|\mathbf{n}\| = \frac{\left| \mathbf{q} \cdot \mathbf{n} - \mathbf{a} \cdot \mathbf{n} \right|}{\|\mathbf{n}\|} = \frac{\left| (\mathbf{q} - \mathbf{a}) \cdot \mathbf{n} \right|}{\|\mathbf{n}\|}.$$

$\square$

Example 8.2.21. Find the distance from the point $\begin{bmatrix} 1 \\ 2 \\ 1 \end{bmatrix}$ to the plane $x + y - z = 1$.

Solution. The plane $x + y - z = 1$ can be described by the equation

$$\begin{bmatrix} x \\ y \\ z \end{bmatrix} \cdot \begin{bmatrix} 1 \\ 1 \\ -1 \end{bmatrix} = 1,$$

so in this case we have $\mathbf{n} = \begin{bmatrix} 1 \\ 1 \\ -1 \end{bmatrix}$. We need to find the point common to the plane

and the line

$$\begin{bmatrix} 1 \\ 2 \\ 1 \end{bmatrix} + \mathrm{Span}\left\{ \begin{bmatrix} 1 \\ 1 \\ -1 \end{bmatrix} \right\}.$$

The point $\begin{bmatrix} 1 \\ 2 \\ 1 \end{bmatrix} + t \begin{bmatrix} 1 \\ 1 \\ -1 \end{bmatrix} = \begin{bmatrix} 1+t \\ 2+t \\ 1-t \end{bmatrix}$ is on the plane $x + y - z = 1$ if

$$1 + t + 2 + t - (1 - t) = 1.$$

Solving for t we obtain $t = -\frac{1}{3}$. Consequently, the projection of the point $\mathbf{q} = \begin{bmatrix} 1 \\ 2 \\ 1 \end{bmatrix}$

on the plane $x + y - z = 1$ is

$$\begin{bmatrix} 1 \\ 2 \\ 1 \end{bmatrix} - \frac{1}{3} \begin{bmatrix} 1 \\ 1 \\ -1 \end{bmatrix} = \begin{bmatrix} \frac{2}{3} \\ \frac{5}{3} \\ \frac{4}{3} \end{bmatrix}.$$

Hence the distance from $\mathbf{q}$ to the plane $x + y - z = 1$ is

$$\left\| \begin{bmatrix} 1 \\ 2 \\ 1 \end{bmatrix} - \begin{bmatrix} \frac{2}{3} \\ \frac{5}{3} \\ \frac{4}{3} \end{bmatrix} \right\| = \left\| \begin{bmatrix} \frac{1}{3} \\ \frac{1}{3} \\ -\frac{1}{3} \end{bmatrix} \right\| = \frac{1}{\sqrt{3}}.$$

Since the hyperplane $x + y - z = 1$ is given by the equation

$$\left(\begin{bmatrix} x \\ y \\ z \end{bmatrix} - \frac{1}{3} \begin{bmatrix} 1 \\ 1 \\ -1 \end{bmatrix} \right) \cdot \begin{bmatrix} 1 \\ 1 \\ -1 \end{bmatrix} = 0,$$

using the formula in Theorem 8.2.20 we get

$$\frac{|(\mathbf{q}-\mathbf{a})\cdot\mathbf{n}|}{\|\mathbf{n}\|} = \frac{\left|\left(\begin{bmatrix}1\\2\\1\end{bmatrix} - \frac{1}{3}\begin{bmatrix}1\\1\\-1\end{bmatrix}\right)\cdot\begin{bmatrix}1\\1\\-1\end{bmatrix}\right|}{\left\|\begin{bmatrix}1\\1\\-1\end{bmatrix}\right\|} = \frac{\left|\begin{bmatrix}\frac{2}{3}\\\frac{5}{3}\\\frac{4}{3}\end{bmatrix}\cdot\begin{bmatrix}1\\1\\-1\end{bmatrix}\right|}{\left\|\begin{bmatrix}1\\1\\-1\end{bmatrix}\right\|} = \frac{1}{\sqrt{3}}.$$

$\square$

The equation of a plane through three points

We close this section with some formulas specific to $\mathbb{R}^3$. First we derive a formula for the plane through three points in $\mathbb{R}^3$.

Theorem 8.2.22. *Let* $\mathbf{a}$, $\mathbf{b}$, *and* $\mathbf{c}$ *be points in* $\mathbb{R}^3$ *such that* $\mathbf{b}-\mathbf{a}$ *and* $\mathbf{c}-\mathbf{a}$ *are linearly independent. There is a unique plane containing the points* $\mathbf{a}$, $\mathbf{b}$, *and* $\mathbf{c}$. *The plane is defined by the equation*

$$(\mathbf{x}-\mathbf{a})\cdot((\mathbf{b}-\mathbf{a})\times(\mathbf{c}-\mathbf{a})) = \det\begin{bmatrix}\mathbf{x}-\mathbf{a} & \mathbf{b}-\mathbf{a} & \mathbf{c}-\mathbf{a}\end{bmatrix} = 0.$$

Proof. First we note that the plane

$$(\mathbf{x}-\mathbf{a})\cdot((\mathbf{b}-\mathbf{a})\times(\mathbf{c}-\mathbf{a})) = 0$$

contains $\mathbf{a}$, $\mathbf{b}$, and $\mathbf{c}$ because we have

$$(\mathbf{a}-\mathbf{a})\cdot((\mathbf{b}-\mathbf{a})\times(\mathbf{c}-\mathbf{a})) = 0,$$
$$(\mathbf{b}-\mathbf{a})\cdot((\mathbf{b}-\mathbf{a})\times(\mathbf{c}-\mathbf{a})) = 0,$$
$$(\mathbf{c}-\mathbf{a})\cdot((\mathbf{b}-\mathbf{a})\times(\mathbf{c}-\mathbf{a})) = 0.$$

Now assume that the plane $\mathbf{x}\cdot\mathbf{n} = d$ contains $\mathbf{a}$, $\mathbf{b}$, and $\mathbf{c}$. Then

$$\mathbf{a}\cdot\mathbf{n} = \mathbf{b}\cdot\mathbf{n} = \mathbf{c}\cdot\mathbf{n} = d.$$

This implies

$$(\mathbf{b}-\mathbf{a})\cdot\mathbf{n} = 0 \quad\text{and}\quad (\mathbf{c}-\mathbf{a})\cdot\mathbf{n} = 0,$$

which means that $\mathbf{n}$ is of the form,

$$t((\mathbf{b}-\mathbf{a})\times(\mathbf{c}-\mathbf{a})), t \neq 0$$

by Theorem 5.1.1 and the remark after the proof of Theorem 5.1.6.

We can take $\mathbf{n} = (\mathbf{b} - \mathbf{a}) \times (\mathbf{c} - \mathbf{a})$.

The equation of the plane can be written as

$$\mathbf{x} \cdot ((\mathbf{b} - \mathbf{a}) \times (\mathbf{c} - \mathbf{a})) = d.$$

Since

$$d = \mathbf{a} \cdot \mathbf{n} = \mathbf{a} \cdot ((\mathbf{b} - \mathbf{a}) \times (\mathbf{c} - \mathbf{a})),$$

we obtain the equation

$$(\mathbf{x} - \mathbf{a}) \cdot ((\mathbf{b} - \mathbf{a}) \times (\mathbf{c} - \mathbf{a})) = 0.$$

Note that we have

$$\mathbf{n} = (\mathbf{b} - \mathbf{a}) \times (\mathbf{c} - \mathbf{a}) = -((\mathbf{a} - \mathbf{b}) \times (\mathbf{c} - \mathbf{b})) = (\mathbf{a} - \mathbf{c}) \times (\mathbf{b} - \mathbf{c}).$$

$\square$

Example 8.2.23. Find the equation of the plane which contains the points

$$\begin{bmatrix} 2 \\ 1 \\ 3 \end{bmatrix}, \begin{bmatrix} 3 \\ 1 \\ 4 \end{bmatrix}, \text{ and } \begin{bmatrix} 5 \\ 2 \\ 7 \end{bmatrix}.$$

Solution. Since

$$\left(\begin{bmatrix} x \\ y \\ z \end{bmatrix} - \begin{bmatrix} 2 \\ 1 \\ 3 \end{bmatrix} \right) \cdot \left(\left(\begin{bmatrix} 3 \\ 1 \\ 4 \end{bmatrix} - \begin{bmatrix} 2 \\ 1 \\ 3 \end{bmatrix} \right) \times \left(\begin{bmatrix} 5 \\ 2 \\ 7 \end{bmatrix} - \begin{bmatrix} 2 \\ 1 \\ 3 \end{bmatrix} \right) \right) = \begin{bmatrix} x - 2 \\ y - 1 \\ z - 3 \end{bmatrix} \cdot \left(\begin{bmatrix} 1 \\ 0 \\ 1 \end{bmatrix} \times \begin{bmatrix} 3 \\ 1 \\ 4 \end{bmatrix} \right)$$

$$= \begin{bmatrix} x - 2 \\ y - 1 \\ z - 3 \end{bmatrix} \cdot \begin{bmatrix} -1 \\ -1 \\ 1 \end{bmatrix}$$

$$= (-1) \cdot (x - 2) + (-1) \cdot (y - 1) + 1 \cdot (z - 3)$$

$$= -x - y + z,$$

the equation is

$$-x - y + z = 0.$$

$\square$

Example 8.2.24. We consider the points

$$\mathbf{a} = \begin{bmatrix} 1 \\ 1 \\ 2 \end{bmatrix}, \mathbf{b} = \begin{bmatrix} 3 \\ 2 \\ 3 \end{bmatrix}, \text{ and } \mathbf{c} = \begin{bmatrix} 7 \\ 4 \\ 5 \end{bmatrix}.$$

Since

$$\left(\begin{bmatrix} x \\ y \\ z \end{bmatrix} - \begin{bmatrix} 1 \\ 1 \\ 2 \end{bmatrix} \right) \cdot \left(\left(\begin{bmatrix} 3 \\ 2 \\ 3 \end{bmatrix} - \begin{bmatrix} 1 \\ 1 \\ 2 \end{bmatrix} \right) \times \left(\begin{bmatrix} 7 \\ 4 \\ 5 \end{bmatrix} - \begin{bmatrix} 1 \\ 1 \\ 2 \end{bmatrix} \right) \right) = \begin{bmatrix} x-2 \\ y-1 \\ z-3 \end{bmatrix} \cdot \left(\begin{bmatrix} 2 \\ 1 \\ 1 \end{bmatrix} \times \begin{bmatrix} 6 \\ 3 \\ 3 \end{bmatrix} \right)$$

$$= \begin{bmatrix} x-2 \\ y-1 \\ z-3 \end{bmatrix} \cdot \begin{bmatrix} 0 \\ 0 \\ 0 \end{bmatrix} = 0,$$

the points $\begin{bmatrix} 1 \\ 1 \\ 2 \end{bmatrix}$, $\begin{bmatrix} 3 \\ 2 \\ 3 \end{bmatrix}$, and $\begin{bmatrix} 7 \\ 4 \\ 5 \end{bmatrix}$ do not determine an unique plane because the equation is satisfied by all points $\begin{bmatrix} x \\ y \\ z \end{bmatrix}$ of $\mathbb{R}^3$. This happens because the vectors $\mathbf{b}-\mathbf{a}$ and $\mathbf{c}-\mathbf{a}$ are not linearly independent. Both vectors are on the line $\text{Span}\left\{ \begin{bmatrix} 2 \\ 1 \\ 1 \end{bmatrix} \right\}$.

Area of a triangle

The following two theorems give us formulas for calculating the distance from a point to a line in $\mathbb{R}^3$. These theorems are limited to $\mathbb{R}^3$ because they use the cross product that is not available outside of $\mathbb{R}^3$

Theorem 8.2.25. *The distance from a point* $\mathbf{q}$ *to a line* $\mathbf{a}+\text{Span}\{\mathbf{u}\}$ *in* $\mathbb{R}^3$ *is*

$$\frac{\|(\mathbf{q}-\mathbf{a}) \times \mathbf{u}\|}{\|\mathbf{u}\|}.$$

Proof. We have

$$\|\mathbf{q} - \mathbf{p}\| = \sqrt{\left\| \mathbf{q} - \mathbf{a} - \frac{(\mathbf{q}-\mathbf{a})\cdot\mathbf{u}}{\mathbf{u}\cdot\mathbf{u}}\mathbf{u} \right\|^2}$$

$$= \sqrt{\frac{\|\mathbf{q}-\mathbf{a}\|^2\|\mathbf{u}\|^2 - ((\mathbf{q}-\mathbf{a})\cdot\mathbf{u})^2}{\|\mathbf{u}\|^2}}$$

$$= \frac{\sqrt{\|\mathbf{q}-\mathbf{a}\|^2\|\mathbf{u}\|^2 - ((\mathbf{q}-\mathbf{a})\cdot\mathbf{u})^2}}{\|\mathbf{u}\|}$$

$$= \frac{\|(\mathbf{q}-\mathbf{a}) \times \mathbf{u}\|}{\|\mathbf{u}\|}.$$

The last equality is justified by the identity

$$\|\mathbf{v} \times \mathbf{w}\|^2 = \|\mathbf{v}\|^2 \|\mathbf{w}\|^2 - (\mathbf{v} \cdot \mathbf{w})^2.$$

□

> **Theorem 8.2.26.** *The distance from a point* $\mathbf{q}$ *to the line through two distinct points* $\mathbf{a}$ *and* $\mathbf{b}$ *in* $\mathbb{R}^3$ *is*
> $$\frac{\|(\mathbf{q} - \mathbf{a}) \times \mathbf{q} - \mathbf{a}\|}{\|\mathbf{b} - \mathbf{a}\|}.$$

Proof. The line through two distinct points $\mathbf{a}$ and $\mathbf{b}$ is $\mathbf{a} + \mathrm{Span}\{\mathbf{b} - \mathbf{a}\}$. From Theorem 8.2.25, the distance from $\mathbf{q}$ to the line $\mathbf{a} + \mathrm{Span}\{\mathbf{b} - \mathbf{a}\}$ is

$$\frac{\|(\mathbf{q} - \mathbf{a}) \times (\mathbf{b} - \mathbf{a})\|}{\|\mathbf{b} - \mathbf{a}\|}.$$

Since

$$\begin{aligned}
(\mathbf{q} - \mathbf{a}) \times (\mathbf{b} - \mathbf{a}) &= (\mathbf{q} - \mathbf{a}) \times (\mathbf{b} - \mathbf{q} + \mathbf{q} - \mathbf{a}) \\
&= (\mathbf{q} - \mathbf{a}) \times (\mathbf{b} - \mathbf{q}) + (\mathbf{q} - \mathbf{a}) \times (\mathbf{q} - \mathbf{a}) \\
&= (\mathbf{q} - \mathbf{a}) \times (\mathbf{b} - \mathbf{q}) \\
&= -(\mathbf{q} - \mathbf{a}) \times (\mathbf{q} - \mathbf{b}),
\end{aligned}$$

we have

$$\frac{\|(\mathbf{q} - \mathbf{a}) \times (\mathbf{b} - \mathbf{a})\|}{\|\mathbf{b} - \mathbf{a}\|} = \frac{\|(\mathbf{q} - \mathbf{a}) \times (\mathbf{q} - \mathbf{b})\|}{\|\mathbf{b} - \mathbf{a}\|}.$$

□

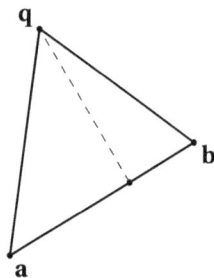

Figure 8.10: The triangle **qab**.

In Chapter 1 we obtained a formula for a triangle with one vertex at the origin. From Theorem 8.2.26 we obtain a simple formula for the area of an arbitrary triangle in $\mathbb{R}^3$:

> **Corollary 8.2.27.** *Let* **a**, **b** *and* **q** *be vectors in* $\mathbb{R}^3$ *such that the vectors* **q** − **a** *and* **q** − **b** *are linearly independent. The area of the triangle* **qab** *is*
> $$\frac{1}{2}\|(\mathbf{q}-\mathbf{a}) \times (\mathbf{q}-\mathbf{b})\|.$$

Proof.

The area of the triangle $\mathbf{qab} = \dfrac{1}{2}$ (the length of the base) $\cdot$ (the height)

$$= \frac{1}{2}\|\mathbf{b}-\mathbf{a}\|\frac{\|(\mathbf{q}-\mathbf{a}) \times (\mathbf{q}-\mathbf{b})\|}{\|\mathbf{b}-\mathbf{a}\|}$$

$$= \frac{1}{2}\|(\mathbf{q}-\mathbf{a}) \times (\mathbf{q}-\mathbf{b})\|.$$

$\square$

The volume of the tetrahedron

The following theorem gives us a direct formula for the distance of an arbitrary point **q** in $\mathbb{R}^3$ from a plane through points **a**, **b**, and **c**.

> **Theorem 8.2.28.** *Let* **a**, **b**, *and* **c** *be points in* $\mathbb{R}^3$ *such that* **b** − **a** *and* **c** − **a** *are linearly independent. The distance from a point* **q** *to the plane through points* **a**, **b**, *and* **c** *is*
> $$\frac{\left|\det\begin{bmatrix}\mathbf{q}-\mathbf{a} & \mathbf{q}-\mathbf{b} & \mathbf{q}-\mathbf{c}\end{bmatrix}\right|}{\|(\mathbf{b}-\mathbf{a}) \times (\mathbf{c}-\mathbf{a})\|}.$$

Proof. Since, by Theorem 8.2.20, the plane through the points **a**, **b**, and **c** is the plane given by the equation
$$(\mathbf{x}-\mathbf{a}) \cdot ((\mathbf{b}-\mathbf{a}) \times (\mathbf{c}-\mathbf{a})) = 0,$$

the distance is
$$\frac{|(\mathbf{q}-\mathbf{a}) \cdot \mathbf{n}|}{\|\mathbf{n}\|} = \frac{|(\mathbf{q}-\mathbf{a}) \cdot ((\mathbf{b}-\mathbf{a}) \times (\mathbf{c}-\mathbf{a}))|}{\|(\mathbf{b}-\mathbf{a}) \times (\mathbf{c}-\mathbf{a})\|}.$$

Now because

$$(\mathbf{q}-\mathbf{a}) \cdot ((\mathbf{b}-\mathbf{a}) \times (\mathbf{c}-\mathbf{a})) = (\mathbf{q}-\mathbf{a}) \cdot (\mathbf{b}-\mathbf{q}+\mathbf{q}-\mathbf{a}) \times (\mathbf{c}-\mathbf{q}+\mathbf{q}-\mathbf{a})$$

$$= (\mathbf{q}{-}\mathbf{a}){\cdot}((\mathbf{b}{-}\mathbf{q}){\times}(\mathbf{c}{-}\mathbf{q}))+(\mathbf{q}{-}\mathbf{a}){\cdot}((\mathbf{b}{-}\mathbf{q}){\times}(\mathbf{q}{-}\mathbf{a}))+(\mathbf{q}{-}\mathbf{a}){\cdot}((\mathbf{q}{-}\mathbf{a}){\times}(\mathbf{c}{-}\mathbf{q}))+(\mathbf{q}{-}\mathbf{a}){\cdot}((\mathbf{q}{-}\mathbf{a}){\times}(\mathbf{q}{-}\mathbf{a}))$$

$$= (\mathbf{q}-\mathbf{a}) \cdot ((\mathbf{b}-\mathbf{q}) \times (\mathbf{c}-\mathbf{q}))$$

$$= (\mathbf{q}-\mathbf{a}) \cdot ((\mathbf{q}-\mathbf{b}) \times (\mathbf{q}-\mathbf{c}))$$

$$= \det\begin{bmatrix} \mathbf{q}-\mathbf{a} & \mathbf{q}-\mathbf{b} & \mathbf{q}-\mathbf{c} \end{bmatrix}$$

the distance is

$$\frac{\left|\det\begin{bmatrix} \mathbf{q}-\mathbf{a} & \mathbf{q}-\mathbf{b} & \mathbf{q}-\mathbf{c} \end{bmatrix}\right|}{\|(\mathbf{b}-\mathbf{a})\times(\mathbf{c}-\mathbf{a})\|}.$$

□

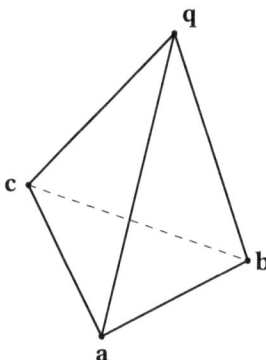

Figure 8.11: The tetrahedron **qabc**.

Theorem 8.2.29. *Let* **a**, **b**,**c** *and* **q** *be points in* $\mathbb{R}^3$ *such that* $\mathbf{q}-\mathbf{a}, \mathbf{q}-\mathbf{b}$ *and* $\mathbf{q}-\mathbf{c}$ *are linearly independent. Then the volume of the tetrahedron* **qabc** *is*

$$\frac{1}{6}\left|\det\begin{bmatrix} \mathbf{q}-\mathbf{a} & \mathbf{q}-\mathbf{b} & \mathbf{q}-\mathbf{c} \end{bmatrix}\right|.$$

Proof. We note that if the vectors $\mathbf{q}-\mathbf{a}, \mathbf{q}-\mathbf{b}$ and $\mathbf{q}-\mathbf{c}$ are linearly independent then the vectors $\mathbf{b}-\mathbf{a}$ and $\mathbf{c}-\mathbf{a}$ are linearly independent.

Now we have

$$\text{the volume of } \mathbf{qabc} = \frac{1}{3}(\text{the area of the base})\cdot(\text{the height})$$

$$= \frac{1}{3}\cdot\frac{1}{2}\|(\mathbf{b}-\mathbf{a})\times(\mathbf{c}-\mathbf{a})\|\cdot\frac{\left|\det\begin{bmatrix} \mathbf{q}-\mathbf{a} & \mathbf{q}-\mathbf{b} & \mathbf{q}-\mathbf{c} \end{bmatrix}\right|}{\|(\mathbf{b}-\mathbf{a})\times(\mathbf{c}-\mathbf{a})\|}$$

$$= \frac{1}{6}\left|\det\begin{bmatrix} \mathbf{q}-\mathbf{a} & \mathbf{q}-\mathbf{b} & \mathbf{q}-\mathbf{c} \end{bmatrix}\right|.$$

□

The results obtained in this section give us a simple formula for the volume of a tetrahedron defined by three linearly independent vectors in $\mathbb{R}^3$.

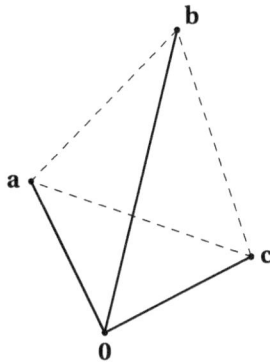

Figure 8.12: The tetrahedron defined by vectors **a**, **b**, and **c**.

Theorem 8.2.30. *Let* **a**, **b**, *and* **c** *be linearly independent vectors in* $\mathbb{R}^3$. *The volume of the tetrahedron* **0abc** *is*

$$\frac{1}{6}\left|\det\begin{bmatrix} \mathbf{a} & \mathbf{b} & \mathbf{c} \end{bmatrix}\right|.$$

8.2.1 Exercises

1. Find the projection of the point $\begin{bmatrix} 1 \\ 1 \\ 0 \end{bmatrix}$ on the line $\begin{bmatrix} 0 \\ 0 \\ 1 \end{bmatrix} + \text{Span}\left\{ \begin{bmatrix} 1 \\ 1 \\ 1 \end{bmatrix} \right\}$.

2. Find the projection of the point $\begin{bmatrix} 0 \\ 1 \\ 0 \end{bmatrix}$ on the line $\begin{bmatrix} 1 \\ 0 \\ 1 \end{bmatrix} + \text{Span}\left\{ \begin{bmatrix} 1 \\ 1 \\ 0 \end{bmatrix} \right\}$.

Write the given equation of the plane in the form $ax = by + cz = d$.

3. $\begin{bmatrix} 2 \\ 0 \\ -1 \end{bmatrix} + \text{Span}\left\{ \begin{bmatrix} 1 \\ 3 \\ 1 \end{bmatrix}, \begin{bmatrix} 2 \\ 2 \\ 1 \end{bmatrix} \right\}$

5. $\begin{bmatrix} 1 \\ 1 \\ -1 \end{bmatrix} + \text{Span}\left\{ \begin{bmatrix} 2 \\ 1 \\ 1 \end{bmatrix}, \begin{bmatrix} 2 \\ -1 \\ 3 \end{bmatrix} \right\}$

4. $\begin{bmatrix} 0 \\ 0 \\ 1 \end{bmatrix} + \text{Span}\left\{ \begin{bmatrix} 1 \\ 1 \\ 1 \end{bmatrix}, \begin{bmatrix} 2 \\ 1 \\ 1 \end{bmatrix} \right\}$

6. $\begin{bmatrix} 1 \\ 2 \\ 3 \end{bmatrix} + \text{Span}\left\{ \begin{bmatrix} 1 \\ 0 \\ 1 \end{bmatrix}, \begin{bmatrix} 0 \\ 1 \\ 1 \end{bmatrix} \right\}$

7. Find an equation of the plane which contains the point $\begin{bmatrix} 1 \\ 1 \\ 2 \end{bmatrix}$ and is orthogonal

to the vector line Span $\left\{ \begin{bmatrix} 2 \\ 3 \\ 2 \end{bmatrix} \right\}$.

8. Write the equation of the plane which contains the point $\begin{bmatrix} 2 \\ 0 \\ -1 \end{bmatrix}$ and is orthogonal to the vector line Span $\left\{ \begin{bmatrix} 1 \\ 1 \\ 1 \end{bmatrix} \right\}$.

Write the given equation of a plane in the form $(\mathbf{x} - \mathbf{a}) \cdot \mathbf{n} = 0$.

9. $3x - y + z = 2$ 11. $3x - y + z = 2$

10. $x - 2y + 5z = 1$ 12. $x - 2y + 5z = 1$

13. Find the projection of the point $\begin{bmatrix} 1 \\ 1 \\ 0 \end{bmatrix}$ on the plane $3x + y - 4z = 1$.

14. Find the projection of the point $\begin{bmatrix} 2 \\ 1 \\ 1 \end{bmatrix}$ on the plane $x - y + z = 1$.

15. Calculate the distance from the point $\begin{bmatrix} 1 \\ 1 \\ 0 \end{bmatrix}$ to the plane $3x + y - 4z = 1$.

16. Calculate the distance from the point $\begin{bmatrix} 2 \\ 1 \\ 1 \end{bmatrix}$ to the plane $x - y + z = 1$.

17. Find the projection of the point $\begin{bmatrix} 1 \\ 0 \\ 0 \end{bmatrix}$ on the line $\begin{bmatrix} 0 \\ 2 \\ 1 \end{bmatrix} + \text{Span} \left\{ \begin{bmatrix} 1 \\ -1 \\ 1 \end{bmatrix} \right\}$, using Theorem 8.2.17.

18. Find the projection of the point $\begin{bmatrix} 0 \\ 0 \\ 1 \end{bmatrix}$ on the line $\begin{bmatrix} 1 \\ 1 \\ 1 \end{bmatrix} + \text{Span} \left\{ \begin{bmatrix} 1 \\ 1 \\ 1 \end{bmatrix} \right\}$, using Theorem 8.2.17.

Find an equation of the plane which contains the given points.

19. $\begin{bmatrix} 1 \\ 1 \\ 1 \end{bmatrix}, \begin{bmatrix} 2 \\ 3 \\ 0 \end{bmatrix}, \begin{bmatrix} 1 \\ 2 \\ 2 \end{bmatrix}$ 20. $\begin{bmatrix} 2 \\ 2 \\ 1 \end{bmatrix}, \begin{bmatrix} 1 \\ 1 \\ 1 \end{bmatrix}, \begin{bmatrix} 1 \\ 2 \\ 1 \end{bmatrix}$

21. $\begin{bmatrix} 1 \\ 0 \\ 0 \end{bmatrix}, \begin{bmatrix} 0 \\ 1 \\ 0 \end{bmatrix}, \begin{bmatrix} 0 \\ 0 \\ 1 \end{bmatrix}$

22. $\begin{bmatrix} 1 \\ 0 \\ 1 \end{bmatrix}, \begin{bmatrix} 1 \\ 1 \\ 1 \end{bmatrix}, \begin{bmatrix} 1 \\ 0 \\ 0 \end{bmatrix}$

Find the area of the triangle with the given vertices.

23. $\begin{bmatrix} 1 \\ 1 \\ 0 \end{bmatrix}, \begin{bmatrix} 2 \\ 1 \\ 1 \end{bmatrix}, \begin{bmatrix} 3 \\ 2 \\ 2 \end{bmatrix}$

25. $\begin{bmatrix} 1 \\ 0 \\ 0 \end{bmatrix}, \begin{bmatrix} 0 \\ 1 \\ 0 \end{bmatrix}, \begin{bmatrix} 0 \\ 0 \\ 1 \end{bmatrix}$

24. $\begin{bmatrix} 2 \\ 2 \\ 1 \end{bmatrix}, \begin{bmatrix} 1 \\ 1 \\ 1 \end{bmatrix}, \begin{bmatrix} 1 \\ 2 \\ 1 \end{bmatrix}$

26. $\begin{bmatrix} 1 \\ 0 \\ 1 \end{bmatrix}, \begin{bmatrix} 0 \\ 1 \\ 1 \end{bmatrix}, \begin{bmatrix} 1 \\ 1 \\ 1 \end{bmatrix}$

Find the volume of the tetrahedron with the given vertices.

27. $\begin{bmatrix} 1 \\ 0 \\ 0 \end{bmatrix}, \begin{bmatrix} 2 \\ 1 \\ 1 \end{bmatrix}, \begin{bmatrix} 3 \\ -1 \\ 1 \end{bmatrix}, \begin{bmatrix} 4 \\ 1 \\ 1 \end{bmatrix}$

29. $\begin{bmatrix} 1 \\ 1 \\ 1 \end{bmatrix}, \begin{bmatrix} 1 \\ 2 \\ 1 \end{bmatrix}, \begin{bmatrix} 3 \\ 2 \\ 2 \end{bmatrix}, \begin{bmatrix} 3 \\ 2 \\ 8 \end{bmatrix}$

28. $\begin{bmatrix} 1 \\ 2 \\ 0 \end{bmatrix}, \begin{bmatrix} 2 \\ 1 \\ 1 \end{bmatrix}, \begin{bmatrix} 1 \\ 0 \\ 1 \end{bmatrix}, \begin{bmatrix} 1 \\ 1 \\ 1 \end{bmatrix}$

30. $\begin{bmatrix} 1 \\ 0 \\ 1 \end{bmatrix}, \begin{bmatrix} 1 \\ 1 \\ 1 \end{bmatrix}, \begin{bmatrix} 2 \\ 0 \\ 1 \end{bmatrix}, \begin{bmatrix} 1 \\ 0 \\ 2 \end{bmatrix}$

31. Show that **p** is the projection of the point **q** on the plane **a** + Span{**u**, **v**} if and only if
$$(\mathbf{p} - \mathbf{a}) \cdot (\mathbf{u} \times \mathbf{v}) = 0, (\mathbf{q} - \mathbf{p}) \cdot \mathbf{u} = 0 \quad \text{and} \quad (\mathbf{q} - \mathbf{p}) \cdot \mathbf{v} = 0.$$

Chapter 9

Rotations

9.1 Rotations in $\mathbb{R}^2$

Consider two vectors $\mathbf{a}$ and $\mathbf{b}$ in $\mathbb{R}^2$ such that $\|\mathbf{a}\| = \|\mathbf{b}\| = 1$. We can think of $\mathbf{b}$ as $\mathbf{a}$ rotated about the origin to a new position. This point of view turns out to be important in mathematics and many applications. In this chapter we describe the operation of rotating vectors about the origin in the language of linear algebra. As we will see, linear algebra provides an elegant description of rotations. Moreover, it will lead us in a natural way to trigonometric functions and allow us to give simple proofs of some basic formulas from trigonometry.

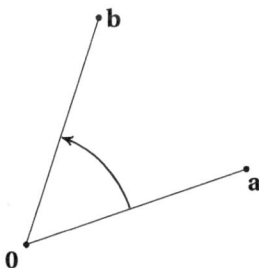

Figure 9.1: We can think of $\mathbf{b}$ as $\mathbf{a}$ rotated about the origin to a new position.

We start with a theorem that is the theoretical basis for this chapter.

> **Theorem 9.1.1.** *Let* **a** *and* **b** *be vectors in* $\mathbb{R}^2$. *If* $\|\mathbf{a}\| = \|\mathbf{b}\| = 1$, *then there are unique real numbers p and q such that*
>
> $$\mathbf{b} = p\mathbf{a} + q\mathbf{a}^{\perp}. \qquad (9.1)$$
>
> *Moreover,*
>
> (a) $p^2 + q^2 = 1$,
>
> (b) $p = \mathbf{a} \cdot \mathbf{b}$,
>
> (c) $q = \det[\mathbf{a} \ \ \mathbf{b}]$.

Proof. Existence and uniqueness of p and q follow from the fact that $\{\mathbf{a}, \mathbf{a}^{\perp}\}$ is a basis in $\mathbb{R}^2$, by Theorem 3.2.25.

From the Pythagorean theorem we get

$$\|\mathbf{b}\|^2 = p^2 \|\mathbf{a}\|^2 + q^2 \|\mathbf{a}^{\perp}\|^2 = (p^2 + q^2)\|\mathbf{a}\|^2.$$

Since $\|\mathbf{a}\| = \|\mathbf{b}\| = 1$, we have $p^2 + q^2 = 1$.

Finally, from (9.1) we obtain

$$\mathbf{a} \cdot \mathbf{b} = p\mathbf{a} \cdot \mathbf{a} + q\mathbf{a} \cdot \mathbf{a}^{\perp} = p\|\mathbf{a}\|^2 = p$$

and

$$\det[\mathbf{a} \ \ \mathbf{b}] = \mathbf{b} \cdot \mathbf{a}^{\perp} = p\mathbf{a} \cdot \mathbf{a}^{\perp} + q\mathbf{a}^{\perp} \cdot \mathbf{a}^{\perp} = q\|\mathbf{a}^{\perp}\|^2 = q\|\mathbf{a}\|^2 = q.$$

$\square$

Example 9.1.2. Let

$$\mathbf{a} = \begin{bmatrix} \frac{1}{\sqrt{2}} \\ \frac{1}{\sqrt{2}} \end{bmatrix} \quad \text{and} \quad \mathbf{b} = \begin{bmatrix} \frac{1}{\sqrt{5}} \\ \frac{2}{\sqrt{5}} \end{bmatrix}.$$

Note that $\|\mathbf{a}\| = \|\mathbf{b}\| = 1$. Since

$$\mathbf{a} \cdot \mathbf{b} = \begin{bmatrix} \frac{1}{\sqrt{2}} \\ \frac{1}{\sqrt{2}} \end{bmatrix} \cdot \begin{bmatrix} \frac{1}{\sqrt{5}} \\ \frac{2}{\sqrt{5}} \end{bmatrix} = \frac{1}{\sqrt{2}}\frac{1}{\sqrt{5}} + \frac{1}{\sqrt{2}}\frac{2}{\sqrt{5}} = \frac{3}{\sqrt{10}}$$

and

$$\det[\mathbf{a} \ \ \mathbf{b}] = \det \begin{bmatrix} \frac{1}{\sqrt{2}} & \frac{1}{\sqrt{5}} \\ \frac{1}{\sqrt{2}} & \frac{2}{\sqrt{5}} \end{bmatrix} = \frac{1}{\sqrt{2}}\frac{2}{\sqrt{5}} - \frac{1}{\sqrt{2}}\frac{1}{\sqrt{5}} = \frac{1}{\sqrt{10}},$$

we have

$$\mathbf{b} = p\mathbf{a} + q\mathbf{a}^{\perp} = \frac{3}{\sqrt{10}}\mathbf{a} + \frac{1}{\sqrt{10}}\mathbf{a}^{\perp} = \frac{3}{\sqrt{10}}\begin{bmatrix} \frac{1}{\sqrt{2}} \\ \frac{1}{\sqrt{2}} \end{bmatrix} + \frac{1}{\sqrt{10}}\begin{bmatrix} -\frac{1}{\sqrt{2}} \\ \frac{1}{\sqrt{2}} \end{bmatrix}.$$

We can easily verify that this is correct.

Every point $\mathbf{w}$ on the unit circle at the origin can be identified with a vector $\begin{bmatrix} p \\ q \end{bmatrix}$ such that $p^2 + q^2 = 1$.

Definition 9.1.3. For any point $\mathbf{w} = \begin{bmatrix} p \\ q \end{bmatrix}$ such that $p^2 + q^2 = 1$ we define a transformation $R_{\mathbf{w}}$ of $\mathbb{R}^2$, that is, a function from $\mathbb{R}^2$ to $\mathbb{R}^2$:

$$R_{\mathbf{w}}(\mathbf{a}) = p\mathbf{a} + q\mathbf{a}^{\perp}. \tag{9.2}$$

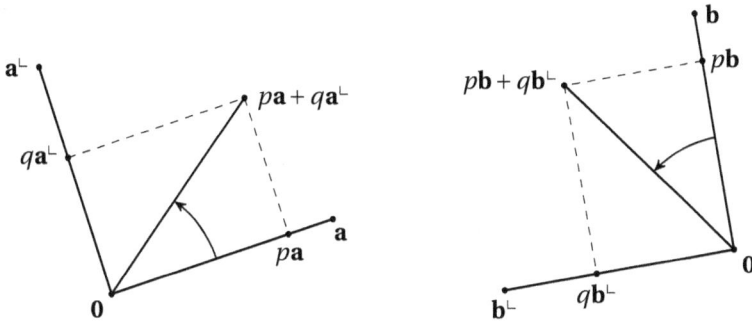

Figure 9.2: Rotations

The transformation defined by (9.2) is a rotation. This means that the result of its application to any vector $\mathbf{a}$ is counterclockwise rotation of $\mathbf{a}$ about the origin by an angle that is determined by p and q and does not depend on $\mathbf{a}$, as illustrated on Figure 9.2.

Since

$$R_{\mathbf{w}}\begin{bmatrix} 1 \\ 0 \end{bmatrix} = p\begin{bmatrix} 1 \\ 0 \end{bmatrix} + q\begin{bmatrix} 1 \\ 0 \end{bmatrix}^{\perp} = \begin{bmatrix} p \\ q \end{bmatrix},$$

we can see that $R_{\mathbf{w}}$ defined by (9.2) is determined by the image of $\begin{bmatrix} 1 \\ 0 \end{bmatrix}$. Actually, $R_{\mathbf{w}}$ is completely determined by the image of any nonzero $\mathbf{a}$. Indeed, from (9.2) we obtain

$$\mathbf{a}\cdot R_{\mathbf{w}}(\mathbf{a}) = p\mathbf{a}\cdot\mathbf{a} + q\mathbf{a}\cdot\mathbf{a}^{\perp} \quad \text{and} \quad \mathbf{a}^{\perp}\cdot R_{\mathbf{w}}(\mathbf{a}) = p\mathbf{a}^{\perp}\cdot\mathbf{a} + q\mathbf{a}^{\perp}\cdot\mathbf{a}^{\perp}.$$

Hence
$$\mathbf{a} \cdot R_{\mathbf{w}}(\mathbf{a}) = p\|\mathbf{a}\|^2 \quad \text{and} \quad \mathbf{a}^{\perp} \cdot R_{\mathbf{w}}(\mathbf{a}) = q\|\mathbf{a}^{\perp}\|^2 = q\|\mathbf{a}\|^2.$$

Solving for p and q we obtain:
$$p = \frac{\mathbf{a} \cdot R_{\mathbf{w}}(\mathbf{a})}{\|\mathbf{a}\|^2}$$

and
$$q = \frac{\mathbf{a}^{\perp} \cdot R_{\mathbf{w}}(\mathbf{a})}{\|\mathbf{a}\|^2} = \frac{1}{\|\mathbf{a}\|^2} \det\begin{bmatrix} \mathbf{a} & R_{\mathbf{w}}(\mathbf{a}) \end{bmatrix}$$

Since p and q do not depend on $\mathbf{a}$, for any other nonzero point $\mathbf{b}$ we must have
$$\frac{\mathbf{a} \cdot R_{\mathbf{w}}(\mathbf{a})}{\|\mathbf{a}\|^2} = \frac{\mathbf{b} \cdot R_{\mathbf{w}}(\mathbf{b})}{\|\mathbf{b}\|^2} \quad \text{and} \quad \frac{\mathbf{a}^{\perp} \cdot R_{\mathbf{w}}(\mathbf{a})}{\|\mathbf{a}\|^2} = \frac{\mathbf{b}^{\perp} \cdot R_{\mathbf{w}}(\mathbf{b})}{\|\mathbf{b}\|^2}.$$

For an arbitrary rotation $R_{\mathbf{w}}$ we define
$$C(R_{\mathbf{w}}) = \frac{\mathbf{a} \cdot R_{\mathbf{w}}(\mathbf{a})}{\|\mathbf{a}\|^2} \quad \text{and} \quad S(R_{\mathbf{w}}) = \frac{\mathbf{a}^{\perp} \cdot R_{\mathbf{w}}(\mathbf{a})}{\|\mathbf{a}\|^2},$$

where $\mathbf{a}$ is an arbitrary nonzero vector in $\mathbb{R}^2$. These definitions are consistent, since the defined values do not depend on $\mathbf{a}$, but only on $R_{\mathbf{w}}$. We can say that C and S are real-valued functions defined on the set of all rotations.

Theorem 9.1.4. *If*
$$\mathbf{b} = p\mathbf{a} + q\mathbf{a}^{\perp} \quad and \quad \mathbf{c} = s\mathbf{b} + t\mathbf{b}^{\perp},$$

then
$$\mathbf{c} = (pr - qs)\mathbf{a} + (ps + qr)\mathbf{a}^{\perp}.$$

Proof.
$$\begin{aligned}
\mathbf{c} &= s\mathbf{b} + t\mathbf{b}^{\perp} \\
&= s\left(p\mathbf{a} + q\mathbf{a}^{\perp}\right) + t\left(p\mathbf{a} + q\mathbf{a}^{\perp}\right)^{\perp} \\
&= ps\mathbf{a} + qs\mathbf{a}^{\perp} + pt\mathbf{a}^{\perp} - qt\mathbf{a} \\
&= (ps - qt)\mathbf{a} + (pt + qs)\mathbf{a}^{\perp}.
\end{aligned}$$

$\square$

The above theorem says that if R_1 is the rotation defined by $R_1(\mathbf{x}) = p\mathbf{x} + q\mathbf{x}^{\perp}$ and R_2 is the rotation defined by $R_2(\mathbf{x}) = s\mathbf{x} + t\mathbf{x}^{\perp}$ then we have
$$R_2 \circ R_1(\mathbf{a}) = R_2(R_1(\mathbf{a})) = (ps - qt)\mathbf{a} + (pt + qs)\mathbf{a}^{\perp}$$

Composition of functions is not a commutative operation. On the other hand, our intuition tells us that, when following one rotation about the origin by another one, the order should not matter. It is easy to verify that
$$R_1 \circ R_2(\mathbf{a}) = (pr - qs)\mathbf{a} + (ps + qr)\mathbf{a}^{\perp} = R_2 \circ R_1(\mathbf{a}) \tag{9.3}$$

which shows that our intuition is correct.

Since $p = C(R_1)$, $q = S(R_1)$, $s = C(R_2)$, and $t = S(R_2)$, from (9.3) we obtain the following identities:

$$C(R_1 \circ R_2) = C(R_1)C(R_2) - S(R_1)S(R_2) \tag{9.4}$$

and

$$S(R_1 \circ R_2) = C(R_1)S(R_2) + S(R_1)C(R_2). \tag{9.5}$$

You may recognize that these formulas are similar to the formulas for cosine and sine of the sum of two angles. As we will see in the next section this is not a coincidence. In fact, the functions C and S can be interpreted as the familiar cosine and sine functions.

Theorem 9.1.5. *If* $\|\mathbf{a}\| = \|\mathbf{b}\| = 1$ *and* $\mathbf{b} = p\mathbf{a} + q\mathbf{a}^\perp$, *then* $\mathbf{a} = p\mathbf{b} - q\mathbf{b}^\perp$.

Proof. If $\mathbf{b} = p\mathbf{a} + q\mathbf{a}^\perp$, then $\mathbf{b}^\perp = p\mathbf{a}^\perp - q\mathbf{a}$ and thus

$$p\mathbf{b} - q\mathbf{b}^\perp = p(p\mathbf{a} + q\mathbf{a}^\perp) - q(p\mathbf{a}^\perp - q\mathbf{a}) = (p^2 + q^2)\mathbf{a} = \mathbf{a}.$$

$\square$

The above theorem says that the rotation that reverses the effect of the rotation $R(\mathbf{a}) = p\mathbf{a} + q\mathbf{a}^\perp$ is the rotation defined by $p\mathbf{a} - q\mathbf{a}^\perp$. In other words, if $R(\mathbf{a}) = p\mathbf{a} + q\mathbf{a}^\perp$, then $R^{-1}(\mathbf{a}) = p\mathbf{a} - q\mathbf{a}^\perp$.

Trigonometry

Interpretation of the results in this chapter in the language of the familiar trigonometric functions requires assigning numerical values to angles so that every real number α corresponds to a rotation R_α in such a way that the following properties hold:

$$R_\alpha \circ R_\beta = R_{\alpha+\beta},$$
$$R_0 = I,$$
$$R_\alpha^{-1} = R_{-\alpha}.$$

The number α is interpreted as a measure of the angle of rotation R_α. While this association can be done in different ways, for example degrees or radians, it is important to note that trigonometric identities do not depend on how the measure is assigned to angles, as long as the above properties hold.

When a measure of angles is chosen, we can define

$$\cos \alpha = C(R_\alpha) \quad \text{and} \quad \sin \alpha = S(R_\alpha).$$

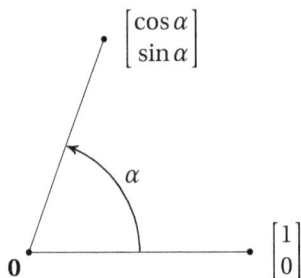

Figure 9.3: Vector rotated about the origin counterclockwise by an angle α.

If we rotate the vector $\begin{bmatrix} 1 \\ 0 \end{bmatrix}$ about the origin counterclockwise by an angle α, then the obtained vector will be

$$\cos\alpha \begin{bmatrix} 1 \\ 0 \end{bmatrix} + \sin\alpha \begin{bmatrix} 1 \\ 0 \end{bmatrix}^{\llcorner} = \cos\alpha \begin{bmatrix} 1 \\ 0 \end{bmatrix} + \sin\alpha \begin{bmatrix} 0 \\ 1 \end{bmatrix} = \begin{bmatrix} \cos\alpha \\ \sin\alpha \end{bmatrix}$$

More generally, if R_α is the counterclockwise rotation about the origin by an angle α, then

$$R_\alpha(\mathbf{a}) = \cos\alpha\,\mathbf{a} + \sin\alpha\,\mathbf{a}^{\llcorner}.$$

This interpretation of the numbers p and q in the expression $p\mathbf{a} + q\mathbf{a}^{\llcorner}$ allows us to obtain some properties of the sine and cosine functions from properties of rotations in $\mathbb{R}^2$.

Theorem 9.1.6. *For any α and β we have*

$$\sin(\alpha + \beta) = \sin\alpha \cos\beta + \cos\alpha \sin\beta$$

and

$$\cos(\alpha + \beta) = \cos\alpha \cos\beta - \sin\alpha \sin\beta.$$

Proof. Since

$$R_\alpha \circ R_\beta = R_{\alpha+\beta},$$

the identities follow immediately from (9.4) and (9.5). $\square$

Note that our algebraic proof of the above trigonometric identities is much simpler than the standard geometric proof. Moreover, a single proof gives us a pair of identities: one for sine and one for cosine. The same is true for the next theorem.

> **Theorem 9.1.7.** *For any* α *we have*
>
> $$\cos(-\alpha) = \cos\alpha \quad \text{and} \quad \sin(-\alpha) = -\sin\alpha.$$

Proof. This is a direct consequence of Theorem 9.1.5. □

In this book we choose to use radians as the measure of an angle, that is, the arc length in the unit circle. Consequently, the measure of the angle associated with the rotation

$$R(\mathbf{a}) = 0 \cdot \mathbf{a} + 1 \cdot \mathbf{a}^{\perp} = \mathbf{a}^{\perp}$$

is $\frac{\pi}{2}$. In other words,

$$R_{\frac{\pi}{2}}(\mathbf{a}) = \mathbf{a}^{\perp}.$$

Note that this gives us

$$\cos\frac{\pi}{2} = 0 \quad \text{and} \quad \sin\frac{\pi}{2} = 1.$$

If $0 < \alpha < \frac{\pi}{2}$ we have $\cos\alpha > 0$ and $\sin\alpha > 0$ and for every couple of numbers p and q such that $p^2 + q^2 = 1$ there is a unique number α such that $0 \le \alpha < 2\pi$ and

$$\cos\alpha = p \quad \text{and} \quad \sin\alpha = q.$$

We do not prove these results in this book.

9.1.1 Exercises

1. Let $\|\mathbf{a}\| = \|\mathbf{b}\| = 1$. Show that if $\mathbf{b} = p\mathbf{a} + q\mathbf{a}^{\perp}$ and $\mathbf{a}^{\perp} = s\mathbf{b} + t\mathbf{b}^{\perp}$, then $s = q$ and $t = p$.

2. Let $\mathbf{a}$ and $\mathbf{b}$ be vectors in $\mathbb{R}^2$ such that $\|\mathbf{a}\| = \|\mathbf{b}\|$. Show that there are unique real numbers p and q such that $\mathbf{b} = p\mathbf{a} + q\mathbf{a}^{\perp}$. Moreover,

    (a) $p^2 + q^2 = 1$,

    (b) $p = \dfrac{\mathbf{a} \cdot \mathbf{b}}{\|\mathbf{a}\| \|\mathbf{b}\|}$,

    (c) $q = \dfrac{\det[\mathbf{a}\ \mathbf{b}]}{\|\mathbf{a}\| \|\mathbf{b}\|}$.

3. Let $\mathbf{a}$ and $\mathbf{b}$ be vectors in $\mathbb{R}^2$ such that $\|\mathbf{a}\| = \|\mathbf{b}\| = 1$. Show that if $\mathbf{b} = p\mathbf{a} + q\mathbf{a}^{\perp}$ and $-\mathbf{a} = s\mathbf{b} + t\mathbf{b}^{\perp}$, then $p = -s$ and $q = t$.

4. Let $\mathbf{a} = \begin{bmatrix} 2 \\ 3 \end{bmatrix}$ and $\mathbf{b} = \begin{bmatrix} 3 \\ -2 \end{bmatrix}$. Find numbers p and q such that $\mathbf{b} = p\mathbf{a} + q\mathbf{a}^{\perp}$.

5. Let $\|\mathbf{a}\| = \|\mathbf{b}\| = \|\mathbf{c}\| = 1$. If $\mathbf{b} = p\mathbf{a} + q\mathbf{a}^{\perp}$ and $\mathbf{c} = p\mathbf{b} + q\mathbf{b}^{\perp}$, show that $\mathbf{c} = (2p^2 - 1)\mathbf{a} + 2pq\mathbf{a}^{\perp}$.

6. Let $R(\mathbf{a}) = p\mathbf{a} + q\mathbf{a}^{\llcorner}$ for some real numbers p and q such that $p^2 + q^2 = 1$. Show that

   (a) $R(\lambda\mathbf{a}) = \lambda R(\mathbf{a})$,

   (b) $R(\mathbf{a} + \mathbf{b}) = R(\mathbf{a}) + R(\mathbf{b})$,

   (c) $R(\mathbf{a}) \cdot R(\mathbf{b}) = \mathbf{a} \cdot \mathbf{b}$,

   (d) $\|R(\mathbf{a})\| = \|\mathbf{a}\|$.

7. Show that $\mathbf{b} = p\mathbf{a} + q\mathbf{a}^{\llcorner}$ if and only if $\mathbf{b} = \begin{bmatrix} p & -q \\ q & p \end{bmatrix}\mathbf{a}$.

8. Consider the functions $L : \mathbb{R}^2 \to \mathbb{R}^2$ and $M : \mathbb{R}^2 \to \mathbb{R}^2$ defined by

$$L(\mathbf{x}) = \frac{1}{2}\mathbf{x} + \frac{\sqrt{3}}{2}\mathbf{x}^{\llcorner} \quad \text{and} \quad M(\mathbf{x}) = \frac{\sqrt{3}}{2}\mathbf{x} + \frac{1}{2}\mathbf{x}^{\llcorner}.$$

   Find $L(M(\mathbf{x}))$.

9. Consider the functions $L : \mathbb{R}^2 \to \mathbb{R}^2$ and $M : \mathbb{R}^2 \to \mathbb{R}^2$ defined by $L(\mathbf{x}) = p\mathbf{x} + q\mathbf{x}^{\llcorner}$, for some real numbers p and q such that $p^2 + q^2 = 1$, and $M(\mathbf{x}) = \mathbf{x}^{\llcorner}$. Find $L(M(\mathbf{x}))$.

10. Let $\|\mathbf{a}\| = \|\mathbf{b}\| = \|\mathbf{c}\| = \|\mathbf{d}\| = 1$. Show that, if

$$\mathbf{b} = p\mathbf{a} + q\mathbf{a}^{\llcorner}, \quad \mathbf{c} = p\mathbf{b} + q\mathbf{b}^{\llcorner} \quad \text{and} \quad \mathbf{d} = p\mathbf{c} + q\mathbf{c}^{\llcorner},$$

   then $\mathbf{d} = (4p^3 - 3p)\mathbf{a} + (3q - 4q^3)\mathbf{a}^{\llcorner}$.

11. Let $\|\mathbf{a}\| = \|\mathbf{b}\| = \|\mathbf{c}\| = \|\mathbf{d}\| = 1$. If, for some real numbers p and q such that $p > 0, q > 0$ and $p^2 + q^2 = 1$,

$$\mathbf{b} = p\mathbf{a} + q\mathbf{a}^{\llcorner}, \quad \mathbf{c} = p\mathbf{b} + q\mathbf{b}^{\llcorner} \quad \text{and} \quad \mathbf{d} = p\mathbf{c} + q\mathbf{c}^{\llcorner} = -\mathbf{a},$$

   show that $\mathbf{b} = \frac{1}{2}\mathbf{a} + \frac{\sqrt{3}}{2}\mathbf{a}^{\llcorner}$.

12. Let $\|\mathbf{a}\| = \|\mathbf{b}\| = \|\mathbf{c}\| = \|\mathbf{d}\| = 1$. If, for some real numbers p and q such that $p > 0, q > 0$ and $p^2 + q^2 = 1$,

$$\mathbf{b} = p\mathbf{a} + q\mathbf{a}^{\llcorner}, \quad \mathbf{c} = p\mathbf{b} + q\mathbf{b}^{\llcorner} \quad \text{and} \quad \mathbf{d} = p\mathbf{c} + q\mathbf{c}^{\llcorner} = \mathbf{a}^{\llcorner},$$

   show that $\mathbf{b} = \frac{\sqrt{3}}{2}\mathbf{a} + \frac{1}{2}\mathbf{a}^{\llcorner}$.

13. Consider the function $L : \mathbb{R}^2 \to \mathbb{R}^2$ defined by $L(\mathbf{x}) = p\mathbf{x} + q\mathbf{x}^{\llcorner}$ for for some real numbers p and q such that $p > 0, q > 0$ and $p^2 + q^2 = 1$. Find p and q such that $L(L(\mathbf{a})) = \frac{1}{2}\mathbf{a} + \frac{\sqrt{3}}{2}\mathbf{a}^{\llcorner}$.

14. Consider the function $L : \mathbb{R}^2 \to \mathbb{R}^2$ defined by $L(\mathbf{x}) = \frac{\sqrt{2}}{2}\mathbf{x} + \frac{\sqrt{2}}{2}\mathbf{x}^{\llcorner}$. Show that L is a rotation about the origin and that $L(L(\mathbf{x})) = \mathbf{x}^{\llcorner}$.

15. Consider the function $L:\mathbb{R}^2 \to \mathbb{R}^2$ defined by $L(\mathbf{x}) = \frac{1}{2}\mathbf{x} + \frac{\sqrt{3}}{2}\mathbf{x}^{\llcorner}$. Show that L is a rotation about the origin and that $L(L(L(\mathbf{x}))) = -\mathbf{x}$.

16. Consider the function $L:\mathbb{R}^2 \to \mathbb{R}^2$ defined by $L(\mathbf{x}) = \frac{1}{2}\mathbf{x} + \frac{\sqrt{3}}{2}\mathbf{x}^{\llcorner}$. Prove that $L(L(L(L(L(\mathbf{x}))))) = \mathbf{x}$.

17. Consider the function $L:\mathbb{R}^2 \to \mathbb{R}^2$ defined by $L(\mathbf{x}) = p\mathbf{x} + q\mathbf{x}^{\llcorner}$ for some real numbers p and q such that $p > 0, q > 0$ and $p^2 + q^2 = 1$. Find p and q such that $L(L(\mathbf{x})) = -\mathbf{x}$.

18. Let $\|\mathbf{a}\| = \|\mathbf{b}\| = \|\mathbf{c}\| = 1$. If $\mathbf{b} = p\mathbf{a} + q\mathbf{a}^{\llcorner}$ for some real numbers p and q such that $p > 0, q > 0$ and $p^2 + q^2 = 1$ and $\mathbf{c} = p\mathbf{b} + q\mathbf{b}^{\llcorner} = \mathbf{a}^{\llcorner}$, show that $\mathbf{b} = \frac{1}{\sqrt{2}}\mathbf{a} + \frac{1}{\sqrt{2}}\mathbf{a}^{\llcorner}$.

19. Let $\mathbf{a}$ be an arbitrary vector in $\mathbb{R}^2$ and let $\mathbf{b} = \frac{1}{2}\mathbf{a} + \frac{\sqrt{3}}{2}\mathbf{a}^{\llcorner}$. Find $\|\mathbf{b} - \mathbf{a}\|$.

20. Consider the functions $L:\mathbb{R}^2 \to \mathbb{R}^2$ and $M:\mathbb{R}^2 \to \mathbb{R}^2$ defined by

$$L(\mathbf{x}) = \frac{1}{2}\mathbf{x} + \frac{\sqrt{3}}{2}\mathbf{x}^{\llcorner} \quad \text{and} \quad M(\mathbf{x}) = \frac{1}{2}\mathbf{x} - \frac{\sqrt{3}}{2}\mathbf{x}^{\llcorner}.$$

Find $L(M(\mathbf{x}))$.

Calculate the following values using the results from this section.

21. $\cos\frac{\pi}{3}$

22. $\sin\frac{\pi}{3}$

23. $\cos\frac{\pi}{6}$

24. $\sin\frac{\pi}{6}$

25. $\cos\frac{7\pi}{12}$

26. $\sin\frac{7\pi}{12}$

27. $\cos\frac{\pi}{12}$

28. $\sin\frac{\pi}{12}$

29. $\cos\frac{5\pi}{12}$

30. $\sin\frac{5\pi}{12}$

Prove the following trigonometric identities.

31. $\cos(\theta + \pi) = -\cos\theta$

32. $\sin(\theta + \pi) = -\sin\theta$

33. $\cos(3\alpha) = 4\cos^3(\alpha) - 3\cos(\alpha)$

34. $\sin(3\alpha) = 3\sin(\alpha) - 4\sin^3(\alpha)$

35. $\cos(\pi - \alpha) = -\cos\alpha$

36. $\sin(\pi - \alpha) = \sin\alpha$

37. $\cos\left(\frac{\pi}{2} - \alpha\right) = \sin\alpha$

38. $\sin\left(\frac{\pi}{2} - \alpha\right) = \cos\alpha$

39. $\sin(2\alpha) = 2\sin\alpha\cos\alpha$

40. $\cos(2\alpha) = 2\cos^2\alpha - 1$

9.2   Quadratic forms

Definition 9.2.1. By a *quadratic form* we mean a function which associates to every vector $\mathbf{x}$ from $\mathbb{R}^2$ the number $\mathbf{x}^T A \mathbf{x}$, where A is an 2×2 symmetric matrix.

Example 9.2.2. Find the quadratic form associated with the matrix

$$A = \begin{bmatrix} 7 & 5 \\ 5 & 2 \end{bmatrix}.$$

Solution.

$$\begin{bmatrix} x & y \end{bmatrix} \begin{bmatrix} 7 & 5 \\ 5 & 2 \end{bmatrix} \begin{bmatrix} x \\ y \end{bmatrix} = \begin{bmatrix} 7x+5y & 5x+2y \end{bmatrix} \begin{bmatrix} x \\ y \end{bmatrix} = 7x^2 + 2y^2 + 10xy$$

□

Definition 9.2.3. Let A be a symmetric 2×2 matrix.

(a) The quadratic form $\mathbf{x}^T A \mathbf{x}$ is called *positive definite* if $\mathbf{x}^T A \mathbf{x} > 0$ for all vectors $\mathbf{x}$ from $\mathbb{R}^2$ different from $\mathbf{0}$.

(b) The quadratic form $\mathbf{x}^T A \mathbf{x}$ is called *positive semidefinite* if $\mathbf{x}^T A \mathbf{x} \geq 0$ for all vectors $\mathbf{x}$ from $\mathbb{R}^2$.

(c) The quadratic form $\mathbf{x}^T A \mathbf{x}$ is called *negative definite* if $\mathbf{x}^T A \mathbf{x} < 0$ for all vectors $\mathbf{x}$ from $\mathbb{R}^2$ different from $\mathbf{0}$.

(d) The quadratic form $\mathbf{x}^T A \mathbf{x}$ is called *negative semidefinite* if $\mathbf{x}^T A \mathbf{x} \leq 0$ for all vectors $\mathbf{x}$ from $\mathbb{R}^2$.

(e) The quadratic form $\mathbf{x}^T A \mathbf{x}$ is called *indefinite* if $\mathbf{x}^T A \mathbf{x} > 0$ for a vector $\mathbf{x}$ from $\mathbb{R}^2$ and $\mathbf{y}^T A \mathbf{y} < 0$ for a vector $\mathbf{y}$ from $\mathbb{R}^2$.

Example 9.2.4. If $A = \begin{bmatrix} 1 & -1 \\ -1 & 2 \end{bmatrix}$, then for any $\mathbf{x} = \begin{bmatrix} x_1 \\ x_2 \end{bmatrix}$ we have

$$\mathbf{x}^T A \mathbf{x} = \begin{bmatrix} x_1 & x_2 \end{bmatrix} \begin{bmatrix} 1 & -1 \\ -1 & 2 \end{bmatrix} \begin{bmatrix} x_1 \\ x_2 \end{bmatrix}$$

$$= \begin{bmatrix} x_1 & x_2 \end{bmatrix} \begin{bmatrix} x_1 - x_2 \\ -x_1 + 2x_2 \end{bmatrix}$$

$$= x_1^2 - 2x_1 x_2 + 2x_2^2$$

$$= (x_1 - x_2)^2 + x_2^2.$$

Since

$$(x_1 - x_2)^2 + x_2^2 > 0$$

whenever at least one of the numbers x_1 and x_2 is different from 0, the quadratic form $\mathbf{x}^T A \mathbf{x}$ is positive definite.

To classify quadratic forms we will use the following result.

Theorem 9.2.5. *If A is an 2×2 symmetric matrix with eigenvalues λ_1, λ_2 and P is an orthogonal 2×2 matrix such that*

$$A = P \begin{bmatrix} \lambda_1 & 0 \\ 0 & \lambda_2 \end{bmatrix} P^T,$$

then

$$\mathbf{x}^T A \mathbf{x} = y_1^2 \lambda_1 + y_2^2 \lambda_2,$$

where

$$\begin{bmatrix} y_1 \\ y_2 \end{bmatrix} = P^T \mathbf{x}.$$

Proof. We have to show that, if A is a 2×2 symmetric matrix with eigenvalues λ_1, λ_2 and P is an orthogonal 2×2 matrix such that $A = P \begin{bmatrix} \lambda_1 & 0 \\ 0 & \lambda_2 \end{bmatrix} P^T$, then

$$\begin{bmatrix} x_1 & x_2 \end{bmatrix} A \begin{bmatrix} x_1 \\ x_2 \end{bmatrix} = y_1^2 \lambda_1 + y_2^2 \lambda_2,$$

where $\begin{bmatrix} y_1 \\ y_2 \end{bmatrix} = P^T \begin{bmatrix} x_1 \\ x_2 \end{bmatrix}$. First we note that

$$\begin{bmatrix} x_1 & x_2 \end{bmatrix} A \begin{bmatrix} x_1 \\ x_2 \end{bmatrix} = \begin{bmatrix} x_1 & x_2 \end{bmatrix} P \begin{bmatrix} \lambda_1 & 0 \\ 0 & \lambda_2 \end{bmatrix} P^T \begin{bmatrix} x_1 \\ x_2 \end{bmatrix}.$$

If

$$\begin{bmatrix} y_1 \\ y_2 \end{bmatrix} = P^T \begin{bmatrix} x_1 \\ x_2 \end{bmatrix},$$

then

$$\begin{bmatrix} y_1 & y_2 \end{bmatrix} = \begin{bmatrix} x_1 & x_2 \end{bmatrix} P$$

and consequently

$$\begin{bmatrix} x_1 & x_2 \end{bmatrix} P \begin{bmatrix} \lambda_1 & 0 \\ 0 & \lambda_2 \end{bmatrix} P^T \begin{bmatrix} x_1 \\ x_2 \end{bmatrix} = \begin{bmatrix} y_1 & y_2 \end{bmatrix} \begin{bmatrix} \lambda_1 & 0 \\ 0 & \lambda_2 \end{bmatrix} \begin{bmatrix} y_1 \\ y_2 \end{bmatrix} = y_1^2 \lambda_1 + y_2^2 \lambda_2.$$

$\square$

Theorem 9.2.6. *Let A be a symmetric 2×2 matrix with eigenvalues λ_1, λ_2.*

(a) *The quadratic form $\mathbf{x}^T A \mathbf{x}$ is positive definite if and only if $\lambda_1 > 0$ and $\lambda_2 > 0$.*

(b) *The quadratic form $\mathbf{x}^T A \mathbf{x}$ is positive semidefinite if and only if $\lambda_1 \geq 0$ and $\lambda_2 \geq 0$.*

(c) *The quadratic form $\mathbf{x}^T A \mathbf{x}$ is negative definite if and only if $\lambda_1 < 0$ and $\lambda_2 < 0$.*

(d) *The quadratic form $\mathbf{x}^T A \mathbf{x}$ is negative semidefinite if and only if $\lambda_1 \leq 0$ and $\lambda_2 \leq 0$.*

(e) *The quadratic form $\mathbf{x}^T A \mathbf{x}$ is indefinite if and only if one eigenvalue of A is strictly positive and one eigenvalue of A is strictly negative.*

Proof. We only prove that a quadratic form $\mathbf{x}^T A \mathbf{x}$ is positive definite if and only if the eigenvalues of the matrix A are strictly positive. The other proofs are similar.

We have to prove that, if A is a 2×2 symmetric matrix with eigenvalues λ_1 and λ_2, then $\lambda_1 > 0$ and $\lambda_2 > 0$ if and only if

$$\begin{bmatrix} x_1 & x_2 \end{bmatrix} A \begin{bmatrix} x_1 \\ x_2 \end{bmatrix} > 0$$

for all $\begin{bmatrix} x_1 \\ x_2 \end{bmatrix} \neq \begin{bmatrix} 0 \\ 0 \end{bmatrix}$. According to Theorem 9.2.5 there is an orthogonal 2×2 matrix P such that if

$$\begin{bmatrix} y_1 \\ y_2 \end{bmatrix} = P^T \begin{bmatrix} x_1 \\ x_2 \end{bmatrix},$$

then

$$\begin{bmatrix} x_1 & x_2 \end{bmatrix} A \begin{bmatrix} x_1 \\ x_2 \end{bmatrix} = y_1^2 \lambda_1 + y_2^2 \lambda_2.$$

If $\lambda_1 > 0$, $\lambda_2 > 0$, and $\begin{bmatrix} x_1 \\ x_2 \end{bmatrix} \neq \begin{bmatrix} 0 \\ 0 \end{bmatrix}$, then

$$\begin{bmatrix} y_1 \\ y_2 \end{bmatrix} = P^T \begin{bmatrix} x_1 \\ x_2 \end{bmatrix} \neq \begin{bmatrix} 0 \\ 0 \end{bmatrix}$$

and consequently

$$[x_1 \ x_2] A \begin{bmatrix} x_1 \\ x_2 \end{bmatrix} = y_1^2 \lambda_1 + y_2^2 \lambda_2 > 0.$$

Now assume that

$$[x_1 \ x_2] A \begin{bmatrix} x_1 \\ x_2 \end{bmatrix} > 0$$

for all $\begin{bmatrix} x_1 \\ x_2 \end{bmatrix} \neq \begin{bmatrix} 0 \\ 0 \end{bmatrix}$. If we take $\begin{bmatrix} x_1 \\ x_2 \end{bmatrix} = P \begin{bmatrix} 1 \\ 0 \end{bmatrix}$, then we have $\begin{bmatrix} x_1 \\ x_2 \end{bmatrix} \neq \begin{bmatrix} 0 \\ 0 \end{bmatrix}$ and consequently

$$[x_1 \ x_2] A \begin{bmatrix} x_1 \\ x_2 \end{bmatrix} = [1 \ 0] P^T P \begin{bmatrix} \lambda_1 & 0 \\ 0 & \lambda_2 \end{bmatrix} P^T P \begin{bmatrix} 1 \\ 0 \end{bmatrix}$$

$$= [1 \ 0] \begin{bmatrix} \lambda_1 & 0 \\ 0 & \lambda_2 \end{bmatrix} \begin{bmatrix} 1 \\ 0 \end{bmatrix} = \lambda_1 > 0.$$

Using a similar argument we can show that $\lambda_2 > 0$. $\qquad\square$

The general form of a quadratic form on $\mathbb{R}^2$ is

$$[x \ y] A \begin{bmatrix} x \\ y \end{bmatrix} = ax^2 + by^2 + cxy,$$

where a, b, c are arbitrary real numbers. The terms ax^2 and by^2 are called **quadratic terms** and cxy is called the **cross-product term**.

Note that, if A is a diagonal matrix, then the quadratic form $\mathbf{x}^T A \mathbf{x}$ has no cross-product terms. Since an orthogonal 2×2 matrix P corresponds to a change of base in $\mathbb{R}^2$, the representation

$$A = P \begin{bmatrix} \lambda_1 & 0 \\ 0 & \lambda_2 \end{bmatrix} P^T$$

in Theorem 9.2.5 can be used to find new variables in $\mathbb{R}^2$ for which the quadratic form has no cross-product terms.

Example 9.2.7. Classify the quadratic form $2x^2 + 17y^2 + 8xy$ and find a change of variables

$$\begin{bmatrix} x \\ y \end{bmatrix} = P \begin{bmatrix} x' \\ y' \end{bmatrix}$$

such the quadratic form expressed in these new variables has no cross-product term.

Solution. We have

$$2x^2 + 4y^2 + 8xy = [x \ y] \begin{bmatrix} 2 & 4 \\ 4 & 17 \end{bmatrix} \begin{bmatrix} x \\ y \end{bmatrix}$$

and

$$\begin{bmatrix} 2 & 4 \\ 4 & 17 \end{bmatrix} = P \begin{bmatrix} 1 & 0 \\ 0 & 18 \end{bmatrix} P^T,$$

where

$$P = \begin{bmatrix} \frac{4}{\sqrt{17}} & \frac{1}{\sqrt{17}} \\ -\frac{1}{\sqrt{17}} & \frac{4}{\sqrt{17}} \end{bmatrix}.$$

Since both eigenvalues are positive, the quadratic form is positive definite. In terms of the new variables x' and y' the quadratic form becomes

$$\begin{bmatrix} x' & y' \end{bmatrix} \begin{bmatrix} 1 & 0 \\ 0 & 18 \end{bmatrix} \begin{bmatrix} x' \\ y' \end{bmatrix} = (x')^2 + 18(y')^2.$$

$\square$

The Principal Axes Theorem

Theorem 9.2.8. *For every quadratic form $\mathbf{x}^T A \mathbf{x}$ on $\mathbb{R}^2$ there is an orthonormal basis $\{\mathbf{p}, \mathbf{p}^\perp\}$ of $\mathbb{R}^2$ such that the equation*

$$\mathbf{x}^T A \mathbf{x} = c$$

is equivalent to the equation

$$\lambda_1 y_1^2 + \lambda_2 y_2^2 = c,$$

where c is a real number, y_1, y_2 are the coordinates of $\mathbf{x}$ in the basis $\{\mathbf{p}, \mathbf{p}^\perp\}$, that is

$$\mathbf{x} = y_1 \mathbf{p} + y_2 \mathbf{p}^\perp,$$

and λ_1 and λ_2 are the eigenvalues of the symmetric matrix A.

Proof. By Theorem 3.3.12, the symmetric matrix A can be diagonalized, that is, there is an orthogonal matrix P with columns $\mathbf{p}_1$ and $\mathbf{p}_2$ and a diagonal matrix

$$D = \begin{bmatrix} \lambda_1 & 0 \\ 0 & \lambda_2 \end{bmatrix}$$

such that $A = PDP^{-1}$. Then

$$\mathbf{x}^T A \mathbf{x} - \mathbf{x}^T PDP^{-1}\mathbf{x} - (P^{-1}\mathbf{x})^T DP^{-1}\mathbf{x} = \mathbf{y}^T D\mathbf{y} = \lambda_1 y_1^2 + \lambda_2 y_2^2,$$

where $\mathbf{x} = P\mathbf{y}$. Note that we can always take

$$\begin{bmatrix} \mathbf{p}_1 & \mathbf{p}_2 \end{bmatrix} = \begin{bmatrix} \mathbf{p} & \mathbf{p}^\perp \end{bmatrix}.$$

$\square$

We note that, if $\mathbf{p} = \begin{bmatrix} q \\ r \end{bmatrix}$, then the function

$$R\left(\begin{bmatrix} a \\ b \end{bmatrix}\right) = q\begin{bmatrix} a \\ b \end{bmatrix} + r\begin{bmatrix} a \\ b \end{bmatrix}^{\perp} = q\begin{bmatrix} a \\ b \end{bmatrix} + r\begin{bmatrix} -b \\ a \end{bmatrix} = \begin{bmatrix} \mathbf{p} & \mathbf{p}^{\perp} \end{bmatrix}\begin{bmatrix} a \\ b \end{bmatrix}$$

defines a rotation such that $R\left(\begin{bmatrix} 1 \\ 0 \end{bmatrix}\right) = \mathbf{p}$ and $R\left(\begin{bmatrix} 0 \\ 1 \end{bmatrix}\right) = \mathbf{p}^{\perp}$.

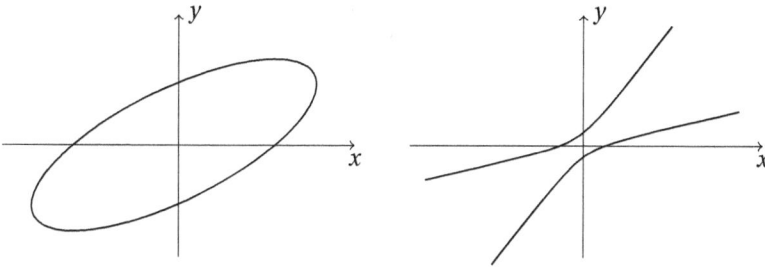

Figure 9.4: The graphs of $x^2 - 2xy + 2y^2 = 1$ and $x^2 - 4xy + 2y^2 = 1$.

The equation

$$ax^2 + by^2 + cxy = d,$$

where a, b, c, d are real numbers, with a, b, d different from 0, describes a curve in $\mathbb{R}^2$. For example, the graph of the equation $x^2 - 2xy + 2y^2 = 1$ is an ellipse and the graph of the equation $x^2 - 4xy + 2y^2 = 1$ is a hyperbola. It is not obvious why such similar equations produce curves that are very different. How can we tell without graphing these equations? The answer is easy if the equation does not have a cross-product term. The graph of any equation that can be written in the form

$$\frac{x^2}{a^2} + \frac{y^2}{b^2} = 1$$

for some $a > 0$ and $b > 0$ is always an ellipse and the graph of any equation that can be written in the form

$$\frac{x^2}{a^2} - \frac{y^2}{b^2} = 1$$

for some $a > 0$ and $b > 0$ is always a hyperbola. Since the shape of a curve in the plane does not depend on the choice of the coordinates we use, if we eliminate the cross-product term in the original equation, the form of the equation in the new variables will immediately tell us whether the graph is an ellipse or a hyperbola.

Example 9.2.9. Apply The Principal Axes Theorem to the equation

$$-13x^2 + 14\sqrt{3}xy + y^2 = 40$$

and classify the curve.

Solution. Since

$$-13x^2 + 14\sqrt{3}xy + y^2 = \begin{bmatrix} x & y \end{bmatrix} \begin{bmatrix} -13 & 7\sqrt{3} \\ 7\sqrt{3} & 1 \end{bmatrix} \begin{bmatrix} x \\ y \end{bmatrix}$$

$$= \begin{bmatrix} x & y \end{bmatrix} \begin{bmatrix} \frac{1}{2} & -\frac{\sqrt{3}}{2} \\ \frac{\sqrt{3}}{2} & \frac{1}{2} \end{bmatrix} \begin{bmatrix} 8 & 0 \\ 0 & -20 \end{bmatrix} \begin{bmatrix} \frac{1}{2} & \frac{\sqrt{3}}{2} \\ -\frac{\sqrt{3}}{2} & \frac{1}{2} \end{bmatrix} \begin{bmatrix} x \\ y \end{bmatrix},$$

the desired change of variables is

$$\begin{bmatrix} x \\ y \end{bmatrix} = \begin{bmatrix} \frac{1}{2} & \frac{\sqrt{3}}{2} \\ -\frac{\sqrt{3}}{2} & \frac{1}{2} \end{bmatrix}^{-1} \begin{bmatrix} x' \\ y' \end{bmatrix} = \begin{bmatrix} \frac{1}{2} & -\frac{\sqrt{3}}{2} \\ \frac{\sqrt{3}}{2} & \frac{1}{2} \end{bmatrix} \begin{bmatrix} x' \\ y' \end{bmatrix} = x' \begin{bmatrix} \frac{1}{2} \\ \frac{\sqrt{3}}{2} \end{bmatrix} + y' \begin{bmatrix} -\frac{\sqrt{3}}{2} \\ \frac{1}{2} \end{bmatrix}.$$

We note that

$$\begin{bmatrix} \frac{1}{2} \\ \frac{\sqrt{3}}{2} \end{bmatrix} = \begin{bmatrix} \cos\frac{\pi}{3} \\ \sin\frac{\pi}{3} \end{bmatrix} = \cos\frac{\pi}{3}\begin{bmatrix} 1 \\ 0 \end{bmatrix} + \sin\frac{\pi}{3}\begin{bmatrix} 0 \\ 1 \end{bmatrix},$$

and

$$\begin{bmatrix} -\frac{\sqrt{3}}{2} \\ \frac{1}{2} \end{bmatrix} = \begin{bmatrix} -\sin\frac{\pi}{3} \\ \cos\frac{\pi}{3} \end{bmatrix} = -\sin\frac{\pi}{3}\begin{bmatrix} 1 \\ 0 \end{bmatrix} + \cos\frac{\pi}{3}\begin{bmatrix} 0 \\ 1 \end{bmatrix} = \cos\frac{\pi}{3}\begin{bmatrix} 0 \\ 1 \end{bmatrix} + \sin\frac{\pi}{3}\begin{bmatrix} -1 \\ 0 \end{bmatrix},$$

so the new coordinate system is obtained by rotating the axes by $\frac{\pi}{3}$.

The equation in terms of the new variables x' and y' is

$$\begin{bmatrix} x' & y' \end{bmatrix} \begin{bmatrix} 8 & 0 \\ 0 & -20 \end{bmatrix} \begin{bmatrix} x' \\ y' \end{bmatrix} = 8(x')^2 - 20(y')^2 = 40$$

which can be written as

$$\frac{(x')^2}{5} - \frac{(y')^2}{2} = 1,$$

so the graph of the equation is a hyperbola. □

Example 9.2.10. Apply The Principal Axes Theorem to the equation

$$15x^2 + 2\sqrt{3}xy + 13y^2 = 48$$

and classify the curve.

Solution. Since

$$15x^2 + 2\sqrt{3}xy + 13y^2 = \begin{bmatrix} x & y \end{bmatrix} \begin{bmatrix} 15 & \sqrt{3} \\ \sqrt{3} & 13 \end{bmatrix} \begin{bmatrix} x \\ y \end{bmatrix}$$

$$= \begin{bmatrix} x & y \end{bmatrix} \begin{bmatrix} \frac{\sqrt{3}}{2} & -\frac{1}{2} \\ \frac{1}{2} & \frac{\sqrt{3}}{2} \end{bmatrix} \begin{bmatrix} 16 & 0 \\ 0 & 12 \end{bmatrix} \begin{bmatrix} \frac{\sqrt{3}}{2} & \frac{1}{2} \\ -\frac{1}{2} & \frac{\sqrt{3}}{2} \end{bmatrix} \begin{bmatrix} x \\ y \end{bmatrix},$$

the desired change of variables is

$$\begin{bmatrix} x \\ y \end{bmatrix} = \begin{bmatrix} \frac{\sqrt{3}}{2} & -\frac{1}{2} \\ \frac{1}{2} & \frac{\sqrt{3}}{2} \end{bmatrix} \begin{bmatrix} x' \\ y' \end{bmatrix} = x' \begin{bmatrix} \frac{\sqrt{3}}{2} \\ \frac{1}{2} \end{bmatrix} + y' \begin{bmatrix} -\frac{1}{2} \\ \frac{\sqrt{3}}{2} \end{bmatrix}.$$

We note that

$$\begin{bmatrix} \frac{\sqrt{3}}{2} \\ \frac{1}{2} \end{bmatrix} = \begin{bmatrix} \cos\frac{\pi}{6} \\ \sin\frac{\pi}{6} \end{bmatrix} = \cos\frac{\pi}{6} \begin{bmatrix} 1 \\ 0 \end{bmatrix} + \sin\frac{\pi}{6} \begin{bmatrix} 0 \\ 1 \end{bmatrix},$$

and

$$\begin{bmatrix} -\frac{1}{2} \\ \frac{\sqrt{3}}{2} \end{bmatrix} = \begin{bmatrix} -\sin\frac{\pi}{6} \\ \cos\frac{\pi}{6} \end{bmatrix} = -\sin\frac{\pi}{6} \begin{bmatrix} 1 \\ 0 \end{bmatrix} + \cos\frac{\pi}{6} \begin{bmatrix} 0 \\ 1 \end{bmatrix} = \cos\frac{\pi}{6} \begin{bmatrix} 0 \\ 1 \end{bmatrix} + \sin\frac{\pi}{6} \begin{bmatrix} -1 \\ 0 \end{bmatrix},$$

so the new coordinate system is obtained by rotating the axes by $\frac{\pi}{6}$.

The equation in terms of the new variables x' and y' is

$$\begin{bmatrix} x' & y' \end{bmatrix} \begin{bmatrix} 16 & 0 \\ 0 & 12 \end{bmatrix} \begin{bmatrix} x' \\ y' \end{bmatrix} = 16(x')^2 + 12(y')^2 = 48$$

or, equivalently,

$$\frac{(x')^2}{3} + \frac{(y')^2}{4} = 1,$$

so the graph of the equation is an ellipse.

$\square$

The Cholesky decomposition

Definition 9.2.11. An 2×2 symmetric matrix A is called **positive definite** if

$$\mathbf{x}^T A \mathbf{x} > 0$$

for all vectors $\mathbf{x}$ from $\mathbb{R}^2$ different from $\mathbf{0}$.

Note that the quadratic form $\mathbf{x}^T A \mathbf{x}$ is positive definite if and only if the matrix A is positive definite. The following theorem gives us a characterization of positive definite matrices. It can be used to easily produce examples of such matrices.

Theorem 9.2.12 (Cholesky decomposition). *Let A be a symmetric 2 × 2 matrix. The following conditions are equivalent:*

   (a) *A is positive definite;*

   (b) *There is a lower triangular 2 × 2 matrix M with strictly positive elements on the main diagonal such that*

$$A = M^T M.$$

Proof. First we show that (a) implies (b). Since

$$a_{11} = \begin{bmatrix} 1 & 0 \end{bmatrix} \begin{bmatrix} a_{11} & a_{21} \\ a_{21} & a_{22} \end{bmatrix} \begin{bmatrix} 1 \\ 0 \end{bmatrix} > 0$$

and

$$\begin{bmatrix} a_{11} & a_{21} \\ a_{21} & a_{22} \end{bmatrix} - \begin{bmatrix} \sqrt{a_{11}} \\ \frac{a_{21}}{\sqrt{a_{11}}} \end{bmatrix} \begin{bmatrix} \sqrt{a_{11}} & \frac{a_{21}}{\sqrt{a_{11}}} \end{bmatrix} = \begin{bmatrix} a_{11} & a_{21} \\ a_{21} & a_{22} \end{bmatrix} - \begin{bmatrix} a_{11} & a_{21} \\ a_{21} & \frac{a_{21}^2}{a_{11}} \end{bmatrix} = \begin{bmatrix} 0 & 0 \\ 0 & a_{22} - \frac{a_{21}^2}{a_{11}} \end{bmatrix},$$

we have

$$\begin{bmatrix} x_1 & x_2 \end{bmatrix} \begin{bmatrix} a_{11} & a_{21} \\ a_{21} & a_{22} \end{bmatrix} \begin{bmatrix} x_1 \\ x_2 \end{bmatrix} - \begin{bmatrix} x_1 & x_2 \end{bmatrix} \begin{bmatrix} \sqrt{a_{11}} \\ \frac{a_{21}}{\sqrt{a_{11}}} \end{bmatrix} \begin{bmatrix} \sqrt{a_{11}} & \frac{a_{21}}{\sqrt{a_{11}}} \end{bmatrix} \begin{bmatrix} x_1 \\ x_2 \end{bmatrix} = \begin{bmatrix} x_1 & x_2 \end{bmatrix} \begin{bmatrix} 0 & 0 \\ 0 & a_{22} - \frac{a_{21}^2}{a_{11}} \end{bmatrix} \begin{bmatrix} x_1 \\ x_2 \end{bmatrix}.$$

which can be written as

$$\begin{bmatrix} x_1 & x_2 \end{bmatrix} \begin{bmatrix} a_{11} & a_{21} \\ a_{21} & a_{22} \end{bmatrix} \begin{bmatrix} x_1 \\ x_2 \end{bmatrix} - \left(\sqrt{a_{11}} x_1 + \frac{a_{21}}{\sqrt{a_{11}}} x_2 \right)^2 = \left(a_{22} - \frac{a_{21}^2}{a_{11}} \right) x_2^2.$$

This shows that the matrix $\begin{bmatrix} a_{11} & a_{21} \\ a_{21} & a_{22} \end{bmatrix}$ is positive definite if and only if $a_{11} > 0$ and $a_{22} - \frac{a_{21}^2}{a_{11}} > 0$.

Now it is easy to verify that

$$\begin{bmatrix} a_{11} & a_{21} \\ a_{21} & a_{22} \end{bmatrix} = \begin{bmatrix} \sqrt{a_{11}} & 0 \\ \frac{a_{21}}{\sqrt{a_{11}}} & \sqrt{a_{22} - \frac{a_{21}^2}{a_{11}}} \end{bmatrix} \begin{bmatrix} \sqrt{a_{11}} & \frac{a_{21}}{\sqrt{a_{11}}} \\ 0 & \sqrt{a_{22} - \frac{a_{21}^2}{a_{11}}} \end{bmatrix}.$$

To prove that (b) implies (a) it suffices to observe that

$$\mathbf{x}^T A \mathbf{x} = \mathbf{x}^T M^T M \mathbf{x} = \| M \mathbf{x} \|^2$$

and that $\| M\mathbf{x} \| > 0$ if $\mathbf{x} \neq \mathbf{0}$, which is a consequence of the fact that the columns of the matrix M are linearly independent. $\qquad\square$

Definition 9.2.13. The decomposition of an 2×2 matrix A in the form

$$A = \begin{bmatrix} m_{11} & 0 \\ m_{21} & m_{22} \end{bmatrix} \begin{bmatrix} m_{11} & m_{21} \\ 0 & m_{22} \end{bmatrix},$$

where $m_{11} > 0$ and $m_{22} > 0$, is called the **Cholesky decomposition** of A.

Theorem 9.2.12 says that an 2×2 matrix is positive definite if and only if it has a Cholesky decomposition.

Example 9.2.14. Find the Cholesky decomposition of the matrix $A = \begin{bmatrix} 3 & 1 \\ 1 & 5 \end{bmatrix}$.

Proof. We follow the method of the proof of Theorem 9.2.12. If we take $\mathbf{v} = \begin{bmatrix} \sqrt{3} \\ \frac{1}{\sqrt{3}} \end{bmatrix}$,

then

$$\mathbf{v}\mathbf{v}^T = \begin{bmatrix} 3 & 1 \\ 1 & \frac{1}{3} \end{bmatrix} \quad \text{and} \quad A - \mathbf{v}\mathbf{v}^T = \begin{bmatrix} 0 & 0 \\ 0 & \frac{14}{3} \end{bmatrix}.$$

Consequently, the Cholesky decomposition of A is

$$A = \begin{bmatrix} \sqrt{3} & 0 \\ \frac{\sqrt{3}}{3} & \frac{\sqrt{42}}{3} \end{bmatrix} \begin{bmatrix} \sqrt{3} & \frac{\sqrt{3}}{3} \\ 0 & \frac{\sqrt{42}}{3} \end{bmatrix}.$$

$\square$

Example 9.2.15. Show that the matrix $A = \begin{bmatrix} 2 & 3 \\ 3 & 4 \end{bmatrix}$ is not positive definite.

Proof. We follow the method of the proof of Theorem 9.2.12. If we take $\mathbf{v} = \begin{bmatrix} \sqrt{2} \\ \frac{3}{\sqrt{2}} \end{bmatrix}$,

then

$$\mathbf{v}\mathbf{v}^T = \begin{bmatrix} 2 & 3 \\ 3 & \frac{9}{2} \end{bmatrix} \quad \text{and} \quad A - \mathbf{v}\mathbf{v}^T = \begin{bmatrix} 0 & 0 \\ 0 & -\frac{1}{2} \end{bmatrix}.$$

Consequently, the matrix A is not positive definite because $-\frac{1}{2} < 0$. $\square$

From the proof of Theorem 9.2.12 we get the following easy to use characterizations of 2×2 positive definite matrices.

Theorem 9.2.16. *A matrix* $\begin{bmatrix} a_{11} & a_{21} \\ a_{21} & a_{22} \end{bmatrix}$ *is positive definite if and only if*

$$a_{11} > 0 \quad and \quad \det \begin{bmatrix} a_{11} & a_{21} \\ a_{21} & a_{22} \end{bmatrix} > 0.$$

Proof. In the proof of Theorem 9.2.12 we show that the matrix $\begin{bmatrix} a_{11} & a_{21} \\ a_{21} & a_{22} \end{bmatrix}$ is positive definite if and only if $a_{11} > 0$ and $a_{22} - \frac{a_{21}^2}{a_{11}} > 0$. Our result is a consequence of the equality

$$a_{22} - \frac{a_{21}^2}{a_{11}} = \frac{1}{a_{11}} \det \begin{bmatrix} a_{11} & a_{21} \\ a_{21} & a_{22} \end{bmatrix}.$$

$\square$

Example 9.2.17. Show that the matrix $A = \begin{bmatrix} 3 & -1 \\ -1 & 5 \end{bmatrix}$ is positive definite.

Solution. Since $a_{11} = 3$ and $\det \begin{bmatrix} 3 & -1 \\ -1 & 5 \end{bmatrix} = 14$, the result follows from Theorem 9.2.16.

$\square$

Example 9.2.18. Show that the matrix $A = \begin{bmatrix} 4 & 7 \\ 7 & 5 \end{bmatrix}$ is not positive definite.

Solution. It suffices to note that $\det \begin{bmatrix} 4 & 7 \\ 7 & 5 \end{bmatrix} = -8$.

$\square$

The LU-decomposition of a positive definite matrix

Theorem 9.2.19. *An 2×2 symmetric matrix A is positive definite if and only if A has the LU-decomposition*

$$A = \begin{bmatrix} 1 & 0 \\ l_{21} & 1 \end{bmatrix} \begin{bmatrix} d_1 & l_{21} d_1 \\ 0 & d_2 \end{bmatrix}$$

such that $d_1 > 0$ and $d_2 > 0$.

Proof. Suppose that the matrix A is positive definite and has the Cholesky decomposition

$$A = \begin{bmatrix} m_{11} & 0 \\ m_{21} & m_{22} \end{bmatrix} \begin{bmatrix} m_{11} & m_{21} \\ 0 & m_{22} \end{bmatrix}.$$

Then the LU-decomposition of A is

$$A = \begin{bmatrix} 1 & 0 \\ m_{21} m_{11}^{-1} & 1 \end{bmatrix} \begin{bmatrix} m_{11}^2 & m_{21} m_{11} \\ 0 & m_{22}^2 \end{bmatrix}.$$

Now suppose that we have

$$A = \begin{bmatrix} 1 & 0 \\ l_{21} & 1 \end{bmatrix} \begin{bmatrix} d_1 & l_{21} d_1 \\ 0 & d_2 \end{bmatrix},$$

where $d_1 > 0$ and $d_2 > 0$. Then the Cholesky decomposition of A is

$$A = \begin{bmatrix} \sqrt{d_1} & 0 \\ l_{21} \sqrt{d_1} & \sqrt{d_2} \end{bmatrix} \begin{bmatrix} \sqrt{d_1} & l_{21} \sqrt{d_1} \\ 0 & \sqrt{d_2} \end{bmatrix},$$

which means that the matrix A is positive definite. □

The above result gives us a new method for calculating the Cholesky decomposition of a matrix.

Example 9.2.20. Using Theorem 9.2.19, find the Cholesky decomposition of the matrix

$$A = \begin{bmatrix} 3 & 2 \\ 2 & 7 \end{bmatrix}.$$

Solution. Since the LU-decomposition of the matrix A is

$$A = \begin{bmatrix} 1 & 0 \\ \frac{2}{3} & 1 \end{bmatrix} \begin{bmatrix} 3 & 2 \\ 0 & \frac{17}{3} \end{bmatrix},$$

the Cholesky decomposition of A is $A = MM^T$ where

$$M = \begin{bmatrix} \sqrt{3} & 0 \\ \frac{2}{\sqrt{3}} & \sqrt{\frac{17}{3}} \end{bmatrix}.$$

□

9.2.1 Exercises

Find the quadratic form associated with the given matrix.

1. $\begin{bmatrix} 2 & 1 \\ 1 & 7 \end{bmatrix}$

3. $\begin{bmatrix} -2 & -4 \\ -4 & 1 \end{bmatrix}$

2. $\begin{bmatrix} 7 & 9 \\ 9 & 3 \end{bmatrix}$

4. $\begin{bmatrix} -2 & -2 \\ -2 & 3 \end{bmatrix}$

Find the matrix associated with the given quadratic form.

5. $3x^2 + 14xy + 2y^2$

7. $x^2 + xy + 4y^2$

6. $9x^2 + 8xy + y^2$

8. $-x^2 - 3xy + y^2$

Classify the given quadratic form.

9. $\begin{bmatrix} x & y \end{bmatrix} \begin{bmatrix} -2 & -2 \\ -2 & 3 \end{bmatrix} \begin{bmatrix} x \\ y \end{bmatrix}$

11. $\begin{bmatrix} x & y \end{bmatrix} \begin{bmatrix} -2 & 1 \\ 1 & -2 \end{bmatrix} \begin{bmatrix} x \\ y \end{bmatrix}$

10. $\begin{bmatrix} x & y \end{bmatrix} \begin{bmatrix} 2 & -1 \\ -1 & 2 \end{bmatrix} \begin{bmatrix} x \\ y \end{bmatrix}$

12. $\begin{bmatrix} x & y \end{bmatrix} \begin{bmatrix} 0 & 4 \\ 4 & 15 \end{bmatrix} \begin{bmatrix} x \\ y \end{bmatrix}$

Find the Cholesky decomposition of the given matrix using the method from the proof of Theorem 9.2.12.

13. $A = \begin{bmatrix} 2 & -3 \\ -3 & 7 \end{bmatrix}$

14. $A = \begin{bmatrix} 1 & 1 \\ 1 & 5 \end{bmatrix}$

Determine if the given matrix is positive definite using Theorem 9.2.16.

15. $\begin{bmatrix} 3 & -3 \\ -3 & 5 \end{bmatrix}$

17. $\begin{bmatrix} 3 & 5 \\ 5 & 7 \end{bmatrix}$

16. $\begin{bmatrix} -2 & 4 \\ 4 & 9 \end{bmatrix}$

18. $\begin{bmatrix} 5 & -7 \\ -7 & 10 \end{bmatrix}$

Determine the Cholesky decomposition using the method from Example 9.2.20.

19. $A = \begin{bmatrix} 2 & 3 \\ 3 & 5 \end{bmatrix}$

20. $A = \begin{bmatrix} 4 & -5 \\ -5 & 7 \end{bmatrix}$

For the given matrix A complete the following:

(a) Find an orthogonal matrix $P = \begin{bmatrix} p_{11} & -p_{21} \\ p_{21} & p_{11} \end{bmatrix}$ satisfying the given conditions and such that the quadratic form $\begin{bmatrix} x' & y' \end{bmatrix} P^T AP \begin{bmatrix} x' \\ y' \end{bmatrix}$ has no cross-product term.

(b) Calculate $\begin{bmatrix} x' & y' \end{bmatrix} P^T AP \begin{bmatrix} x' \\ y' \end{bmatrix}$.

(c) Determine the rotation which rotates $\begin{bmatrix} 1 \\ 0 \end{bmatrix}$ to $\begin{bmatrix} p_{11} \\ p_{21} \end{bmatrix}$ and $\begin{bmatrix} 0 \\ 1 \end{bmatrix}$ to $\begin{bmatrix} -p_{21} \\ p_{11} \end{bmatrix}$ and the angle of that rotation.

21. $A = \begin{bmatrix} -3 & 3 \\ 3 & 5 \end{bmatrix}$, $p_{11} > 0$, $p_{21} > 0$.

25. $A = \begin{bmatrix} -3 & 3 \\ 3 & 5 \end{bmatrix}$, $p_{11} > 0$, $p_{21} < 0$.

22. $A = \begin{bmatrix} 8 & 2 \\ 2 & 11 \end{bmatrix}$, $p_{11} > 0$, $p_{21} > 0$.

26. $A = \begin{bmatrix} 8 & 2 \\ 2 & 11 \end{bmatrix}$, $p_{11} < 0$, $p_{21} > 0$.

23. $A = \begin{bmatrix} -3 & 3 \\ 3 & 5 \end{bmatrix}$, $p_{11} < 0$, $p_{21} > 0$.

27. $A = \begin{bmatrix} -3 & 3 \\ 3 & 5 \end{bmatrix}$, $p_{11} < 0$, $p_{21} < 0$.

24. $A = \begin{bmatrix} 8 & 2 \\ 2 & 11 \end{bmatrix}$, $p_{11} > 0$, $p_{21} < 0$.

28. $A = \begin{bmatrix} 8 & 2 \\ 2 & 11 \end{bmatrix}$, $p_{11} < 0$, $p_{21} < 0$.

29. Consider the equation $x^2 - 10xy + 27y^2 = \begin{bmatrix} x & y \end{bmatrix} \begin{bmatrix} 3 & -5 \\ -5 & 27 \end{bmatrix} \begin{bmatrix} x \\ y \end{bmatrix} = 14$.

    (a) Find an orthogonal matrix $P = \begin{bmatrix} p_{11} & -p_{21} \\ p_{21} & p_{11} \end{bmatrix}$ such that $p_{11} > 0$, $p_{21} > 0$,

    and such that the quadratic form $\begin{bmatrix} x' & y' \end{bmatrix} P^T \begin{bmatrix} 3 & -5 \\ -5 & 27 \end{bmatrix} P \begin{bmatrix} x' \\ y' \end{bmatrix}$ has no cross-

    product term, where $P \begin{bmatrix} x' \\ y' \end{bmatrix} = \begin{bmatrix} x \\ y \end{bmatrix}$.

    (b) Express the equation $\begin{bmatrix} x' & y' \end{bmatrix} P^T \begin{bmatrix} 3 & -5 \\ -5 & 27 \end{bmatrix} P \begin{bmatrix} x' \\ y' \end{bmatrix} = 14$ in the standard form,

    that is, $a(x')^2 + b(y')^2 = 1$.

    (c) Determine the rotation which rotates $\begin{bmatrix} 1 \\ 0 \end{bmatrix}$ to $\begin{bmatrix} p_{11} \\ p_{21} \end{bmatrix}$ and $\begin{bmatrix} 0 \\ 1 \end{bmatrix}$ to $\begin{bmatrix} -p_{21} \\ p_{11} \end{bmatrix}$ and

    the angle of that rotation.

30. Consider the equation $3x^2 + 16xy + 33y^2 = \begin{bmatrix} x & y \end{bmatrix} \begin{bmatrix} 3 & 8 \\ 8 & 33 \end{bmatrix} \begin{bmatrix} x \\ y \end{bmatrix} = 22$.

    (a) Find an orthogonal matrix $P = \begin{bmatrix} p_{11} & -p_{21} \\ p_{21} & p_{11} \end{bmatrix}$ such that $p_{11} > 0$, $p_{21} > 0$,

    and such that the quadratic form $\begin{bmatrix} x' & y' \end{bmatrix} P^T \begin{bmatrix} 3 & 8 \\ 8 & 33 \end{bmatrix} P \begin{bmatrix} x' \\ y' \end{bmatrix}$ has no cross-

    product term, where $P \begin{bmatrix} x' \\ y' \end{bmatrix} = \begin{bmatrix} x \\ y \end{bmatrix}$.

    (b) Express the equation $\begin{bmatrix} x' & y' \end{bmatrix} P^T \begin{bmatrix} 3 & 8 \\ 8 & 33 \end{bmatrix} P \begin{bmatrix} x' \\ y' \end{bmatrix} = 22$ in the standard form,

    that is, $a(x')^2 + b(y')^2 = 1$.

    (c) Determine the rotation which rotates $\begin{bmatrix} 1 \\ 0 \end{bmatrix}$ to $\begin{bmatrix} p_{11} \\ p_{21} \end{bmatrix}$ and $\begin{bmatrix} 0 \\ 1 \end{bmatrix}$ to $\begin{bmatrix} -p_{21} \\ p_{11} \end{bmatrix}$ and

    the angle of that rotation.

9.3 Rotations in $\mathbb{R}^3$

In Section 9.1 we were interested in rotations of vectors in $\mathbb{R}^2$ about the origin. The original vector $\mathbf{a}$ and the rotated vector $\mathbf{b}$ were elements of $\mathbb{R}^2$, which can be thought of as a vector plane in $\mathbb{R}^3$, namely the plane $\mathrm{Span}\left\{ \begin{bmatrix} 1 \\ 0 \\ 0 \end{bmatrix}, \begin{bmatrix} 0 \\ 1 \\ 0 \end{bmatrix} \right\}$. Now we would like to consider rotations about the origin in an arbitrary plane in $\mathbb{R}^3$.

When we described rotations in $\mathbb{R}^2$, it was clear what was meant by "counterclockwise rotation" and "clockwise rotation". When an arbitrary plane in $\mathbb{R}^3$ is considered, we use the "right-hand rule" to specify the direction of a rotation.

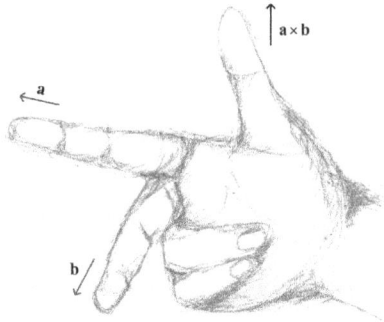

Figure 9.5: The right-han d rule.

It turns out that the vectors $\mathbf{a}$, $\mathbf{b}$, and $\mathbf{a} \times \mathbf{b}$ are like the fingers of your right hand with the index finger pointing in the direction of $\mathbf{a}$, the middle finger pointing in the direction of $\mathbf{b}$, and your thumb pointing in the direction of $\mathbf{a} \times \mathbf{b}$ (see Fig. 9.5). We will accept this fact for now and come back to it at the end of this chapter.

Rotations in a vector plane in $\mathbb{R}^3$

If a vector $\mathbf{a}$ in $\mathbb{R}^2$ is rotated counterclockwise about the origin by an angle α, the result is the vector $\mathbf{b} = \cos\alpha\, \mathbf{a} + \sin\alpha\, \mathbf{a}^\perp$. The fact that the vector $\mathbf{a}^\perp$ is orthogonal to $\mathbf{a}$ is important here, but note that there is another vector orthogonal to $\mathbf{a}$, namely $-\mathbf{a}^\perp$. From these two vectors we identify $\mathbf{a}^\perp$ as the vector obtained when $\mathbf{a}$ is rotated counterclockwise about the origin by the angle $\frac{\pi}{2}$. In this section we generalize this setup to an arbitrary vector plane in $\mathbb{R}^3$.

Recall that an arbitrary vector plane in $\mathbb{R}^3$ can be described by the equation $\mathbf{n}\cdot\mathbf{x} = 0$, where $\mathbf{n}$ is a vector different from the origin. Now consider a vector $\mathbf{a} \neq \mathbf{0}$ on that plane, that is $\mathbf{n}\cdot\mathbf{a} = 0$. We want to rotate $\mathbf{a}$ about the vector line $\mathrm{Span}\{\mathbf{n}\}$ oriented by the vector $\mathbf{n}$ in the plane $\mathbf{n}\cdot\mathbf{x} = 0$ and describe the resulting vector by a formula analogous to the one in $\mathbb{R}^2$, that is, $\mathbf{b} = \cos\alpha\, \mathbf{a} + \sin\alpha\, \mathbf{a}^\perp$. Since the perp operation is not available here, we need to find a replacement.

First suppose that $\mathbf{a} = s\mathbf{e} + t\mathbf{f}$ where $\{\mathbf{e}, \mathbf{f}\}$ is an orthonormal basis in the plane $\mathbf{n} \cdot \mathbf{x} = 0$, where $\mathbf{n} = \mathbf{e} \times \mathbf{f}$. It will be natural to define

$$\mathbf{a}_\mathbf{n}^{\perp} = -t\mathbf{e} + s\mathbf{f}.$$

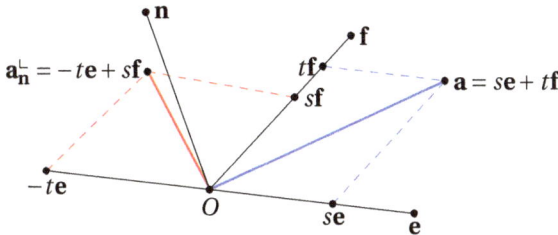

Figure 9.6: The vector $\mathbf{a}_\mathbf{n}^{\perp} = -t\mathbf{e} + s\mathbf{f}$.

The defined vector $\mathbf{a}_\mathbf{n}^{\perp}$ should be independent of the choice of an orthonormal basis in the plane $\mathbf{n} \cdot \mathbf{x} = 0$. Indeed, we have

$$-t\mathbf{e} + s\mathbf{f} = s\mathbf{f} - t\mathbf{e}$$
$$= s(\mathbf{e} \times \mathbf{f}) \times \mathbf{e} + t(\mathbf{e} \times \mathbf{f}) \times \mathbf{f}$$
$$= (\mathbf{e} \times \mathbf{f}) \times (s\mathbf{e} + t\mathbf{f})$$
$$= \mathbf{n} \times (s\mathbf{e} + t\mathbf{f})$$
$$= \mathbf{n} \times \mathbf{a},$$

where we use the identity

$$(\mathbf{x} \times \mathbf{y}) \times \mathbf{z} = (\mathbf{x} \cdot \mathbf{z})\mathbf{y} - (\mathbf{y} \cdot \mathbf{z})\mathbf{x}.$$

The vector $\mathbf{n} \times \mathbf{a}$ is clearly independent of the choice of an orthonormal basis in the plane $\mathbf{n} \cdot \mathbf{x} = 0$. Moreover, since $\mathbf{n} \times \mathbf{a} = -t\mathbf{e} + s\mathbf{f}$, it agrees with our intuition of what the vector $\mathbf{a}_\mathbf{n}^{\perp}$ should be.

Definition 9.3.1. For any vector $\mathbf{a}$ in $\mathbb{R}^3$ and any unit vector $\mathbf{n}$ in $\mathbb{R}^3$ we define

$$\mathbf{a}_\mathbf{n}^{\perp} = \mathbf{n} \times \mathbf{a}.$$

Note that, if $\mathbf{n} = \begin{bmatrix} 0 \\ 0 \\ 1 \end{bmatrix}$ and $\mathbf{a} = \begin{bmatrix} a_1 \\ a_2 \\ 0 \end{bmatrix}$, then $\mathbf{a}_\mathbf{n}^{\perp} = \begin{bmatrix} -a_2 \\ a_1 \\ 0 \end{bmatrix}$, which gives us the familiar perp operation in $\mathbb{R}^2$.

We are now ready to generalize the formula $\mathbf{b} = \cos\alpha\,\mathbf{a} + \sin\alpha\,\mathbf{a}^{\perp}$ to $\mathbb{R}^3$.

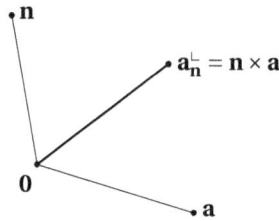

Figure 9.7: Vectors $\mathbf{n}$, $\mathbf{a}$, and $\mathbf{a_n^\perp}$.

Definition 9.3.2. Let $\mathbf{a}$ be a vector in the plane $\mathbf{n} \cdot \mathbf{x} = 0$ where $\|\mathbf{n}\| = 1$. By the *vector obtained by counterclockwise rotation by an angle α of $\mathbf{a}$ about the vector line* $\mathrm{Span}\{\mathbf{n}\}$ *oriented by the vector* $\mathbf{n}$ we mean the vector

$$\cos \alpha\, \mathbf{a} + \sin \alpha\, \mathbf{a_n^\perp} = \cos \alpha\, \mathbf{a} + \sin \alpha\, (\mathbf{n} \times \mathbf{a}).$$

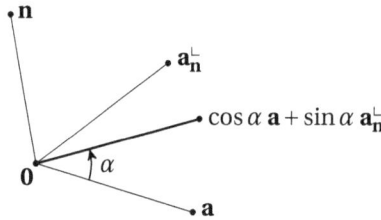

Figure 9.8: The vector $\cos \alpha\, \mathbf{a} + \sin \alpha\, \mathbf{a_n^\perp}$.

Note that for $\alpha = \frac{\pi}{2}$ we have

$$\cos \frac{\pi}{2}\, \mathbf{a} + \sin \frac{\pi}{2}\, \mathbf{a_n^\perp} = \mathbf{a_n^\perp},$$

so $\mathbf{a_n^\perp}$ is the vector obtained by rotating $\mathbf{a}$ in the plane $\mathbf{n} \cdot \mathbf{x} = 0$ about the vector line $\mathrm{Span}\{\mathbf{n}\}$ oriented by the vector $\mathbf{n}$ by the angle $\frac{\pi}{2}$.

If $\mathbf{n}$ is an arbitrary nonzero vector in $\mathbb{R}^3$, then the vector obtained by rotating $\mathbf{a}$ about the vector line $\mathrm{Span}\{\mathbf{n}\}$ oriented by the vector $\mathbf{n}$ by an angle α is

$$\cos \alpha\, \mathbf{a} + \sin \alpha\, \frac{1}{\|\mathbf{n}\|}(\mathbf{n} \times \mathbf{a}).$$

If p and q is any pair of numbers such that $p^2 + q^2 = 1$ and $\mathbf{n}$ a unit vector in $\mathbb{R}^3$, then for any vector $\mathbf{a}$ in the plane $\mathbf{x} \cdot \mathbf{n} = 0$ the vector

$$p\mathbf{a} + q(\mathbf{n} \times \mathbf{a})$$

is the counterclockwise rotation of **a** about the vector line Span{**n**} oriented by the vector **n** by an angle α where α is defined by $p = \cos \alpha$ and $q = \sin \alpha$.

Example 9.3.3. Let $\mathbf{n} = \begin{bmatrix} 1 \\ -1 \\ 1 \end{bmatrix}$ and $\mathbf{a} = \begin{bmatrix} 1 \\ 2 \\ 2 \end{bmatrix}$. Find the vector **b** obtained by rotating **a** counterclockwise by the angle $\alpha = \frac{\pi}{4}$ in the plane $\mathbf{n} \cdot \mathbf{x} = 0$ about the vector line Span{**n**} oriented by the vector **n**.

Solution. We know that

$$\mathbf{b} = \cos \frac{\pi}{4} \mathbf{a} + \sin \frac{\pi}{4} \mathbf{a}_\mathbf{n}^\perp = \frac{1}{\sqrt{2}} \mathbf{a} + \frac{1}{\sqrt{2}} \mathbf{a}_\mathbf{n}^\perp,$$

so we only need to find

$$\mathbf{a}_\mathbf{n}^\perp = \frac{1}{\|\mathbf{n}\|} (\mathbf{n} \times \mathbf{a}).$$

Since

$$\|\mathbf{n}\| = \sqrt{1^2 + (-1)^2 + 1^2} = \sqrt{3}$$

and

$$\mathbf{n} \times \mathbf{a} = \begin{bmatrix} 1 \\ -1 \\ 1 \end{bmatrix} \times \begin{bmatrix} 1 \\ 2 \\ 2 \end{bmatrix} = \begin{bmatrix} -4 \\ -1 \\ 3 \end{bmatrix},$$

we have

$$\mathbf{a}_\mathbf{n}^\perp = \frac{1}{\sqrt{3}} \begin{bmatrix} -4 \\ -1 \\ 3 \end{bmatrix}.$$

Consequently,

$$\mathbf{b} = \frac{1}{\sqrt{2}} \begin{bmatrix} 1 \\ 2 \\ 2 \end{bmatrix} + \frac{1}{\sqrt{2}} \frac{1}{\sqrt{3}} \begin{bmatrix} -4 \\ -1 \\ 3 \end{bmatrix}.$$

$\square$

Rotations in an arbitrary plane in $\mathbb{R}^3$

In the previous section we considered a vector plane $\mathbf{n} \cdot \mathbf{x} = 0$ and a vector **a** in that plane that is different from the origin. Now we would like to investigate the case when **a** is not in the plane $\mathbf{n} \cdot \mathbf{x} = 0$. In this case we will consider rotating **a** about the vector line Span{**n**} oriented by the vector **n**.

Let **a** and **n** be vectors in $\mathbb{R}^3$ different from the origin. To rotate **a** about the axis Span{**n**} oriented by the vector **n** we first decompose **a** into the component of **a** that is on the vector line Span{**n**} and the component of **a** that is perpendicular to that

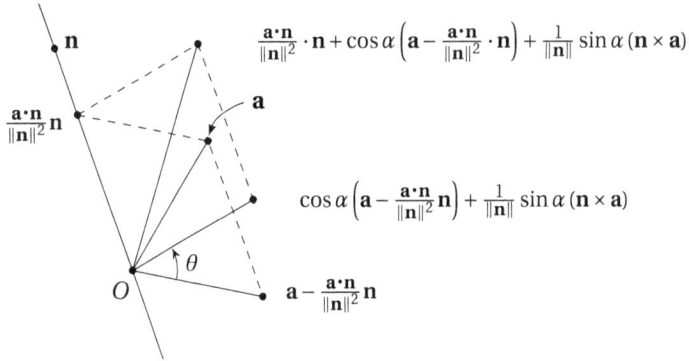

Figure 9.9: Rotating **a** about the vector line Span{**n**}.

line. Note that the component of **a** on Span{**n**} does not change when rotated about that line.

Recall that the projection of **a** onto the line Span{**n**} is $\dfrac{\mathbf{a}\cdot\mathbf{n}}{\|\mathbf{n}\|^2}\mathbf{n}$ (Theorem 4.2.10) and the projection of **a** onto the plane $\mathbf{x}\cdot\mathbf{n}=0$ is $\mathbf{a}-\dfrac{\mathbf{a}\cdot\mathbf{n}}{\|\mathbf{n}\|^2}\mathbf{n}$ (from the proof of Theorem 8.1.17). The rotation of $\mathbf{a}-\dfrac{\mathbf{a}\cdot\mathbf{n}}{\|\mathbf{n}\|^2}\mathbf{n}$ in the plane $\mathbf{x}\cdot\mathbf{n}=0$ by the angle α about the vector line Span{**n**} oriented by the vector **n** is

$$\cos\alpha\left(\mathbf{a}-\frac{\mathbf{a}\cdot\mathbf{n}}{\|\mathbf{n}\|^2}\mathbf{n}\right)+\frac{1}{\|\mathbf{n}\|}\sin\alpha\left(\mathbf{n}\times\left(\mathbf{a}-\frac{\mathbf{a}\cdot\mathbf{n}}{\|\mathbf{n}\|^2}\mathbf{n}\right)\right)$$

which reduces to

$$\cos\alpha\left(\mathbf{a}-\frac{\mathbf{a}\cdot\mathbf{n}}{\|\mathbf{n}\|^2}\mathbf{n}\right)+\frac{1}{\|\mathbf{n}\|}\sin\alpha\,(\mathbf{n}\times\mathbf{a}),$$

because $\mathbf{n}\times\mathbf{n}=0$. Consequently, the rotation of **a** about the axis Span{**n**} oriented by the vector **n** by angle α is

$$\frac{\mathbf{a}\cdot\mathbf{n}}{\|\mathbf{n}\|^2}\cdot\mathbf{n}+\cos\alpha\left(\mathbf{a}-\frac{\mathbf{a}\cdot\mathbf{n}}{\|\mathbf{n}\|^2}\cdot\mathbf{n}\right)+\frac{1}{\|\mathbf{n}\|}\sin\alpha\,(\mathbf{n}\times\mathbf{a}).$$

Example 9.3.4. Let

$$\mathbf{n}=\begin{bmatrix}1\\1\\-1\end{bmatrix}\quad\text{and}\quad\mathbf{a}=\begin{bmatrix}2\\-3\\5\end{bmatrix}$$

We would like to find the vector **b** obtained by rotating **a** about the vector line Span{**n**} oriented by the vector **n** by the angle $\alpha=\frac{\pi}{6}$.

Since

$$\|\mathbf{n}\| = \sqrt{3} \quad \text{and} \quad \mathbf{a} \cdot \mathbf{n} = \begin{bmatrix} 2 \\ -3 \\ 5 \end{bmatrix} \cdot \begin{bmatrix} 1 \\ 1 \\ -1 \end{bmatrix} = -6,$$

the projection of **a** onto the vector line Span{**n**} is

$$\frac{\mathbf{a} \cdot \mathbf{n}}{\|\mathbf{n}\|^2} \mathbf{n} = \frac{-6}{3} \begin{bmatrix} 1 \\ 1 \\ -1 \end{bmatrix} = \begin{bmatrix} -2 \\ -2 \\ 2 \end{bmatrix}$$

and the projection of **a** onto the plane $\mathbf{x} \cdot \mathbf{n} = 0$ is

$$\mathbf{a} - \frac{\mathbf{a} \cdot \mathbf{n}}{\|\mathbf{n}\|^2} \mathbf{n} = \begin{bmatrix} 2 \\ -3 \\ 5 \end{bmatrix} - \begin{bmatrix} -2 \\ -2 \\ 2 \end{bmatrix} = \begin{bmatrix} 4 \\ -1 \\ 3 \end{bmatrix}.$$

Since

$$\mathbf{n} \times \mathbf{a} = \begin{bmatrix} 1 \\ 1 \\ -1 \end{bmatrix} \times \begin{bmatrix} 2 \\ -3 \\ 5 \end{bmatrix} = \begin{bmatrix} 2 \\ -7 \\ -5 \end{bmatrix},$$

we have

$$\mathbf{b} = \frac{\mathbf{a} \cdot \mathbf{n}}{\|\mathbf{n}\|^2} \cdot \mathbf{n} + \cos\theta \left(\mathbf{a} - \frac{\mathbf{a} \cdot \mathbf{n}}{\|\mathbf{n}\|^2} \cdot \mathbf{n} \right) + \frac{1}{\|\mathbf{n}\|} \sin\theta \, (\mathbf{n} \times \mathbf{a})$$

$$= \begin{bmatrix} -2 \\ -2 \\ 2 \end{bmatrix} + \cos\frac{\pi}{6} \begin{bmatrix} 4 \\ -1 \\ 3 \end{bmatrix} + \frac{1}{\sqrt{3}} \sin\frac{\pi}{6} \begin{bmatrix} 2 \\ -7 \\ -5 \end{bmatrix}$$

$$= \begin{bmatrix} -2 \\ -2 \\ 2 \end{bmatrix} + \frac{\sqrt{3}}{2} \begin{bmatrix} 4 \\ -1 \\ 3 \end{bmatrix} + \frac{1}{\sqrt{3}} \frac{1}{2} \begin{bmatrix} 2 \\ -7 \\ -5 \end{bmatrix}$$

$$= \begin{bmatrix} -2 + \frac{7\sqrt{3}}{3} \\ -2 - \frac{5\sqrt{3}}{3} \\ 2 + \frac{2\sqrt{3}}{3} \end{bmatrix}.$$

9.3.1 Exercises

Find the rotation of the vector **a** about the vector line through **n** oriented by **n**.

1. $\mathbf{a} = \begin{bmatrix} 1 \\ 0 \\ 0 \end{bmatrix}, \mathbf{n} = \begin{bmatrix} 0 \\ 0 \\ 1 \end{bmatrix}$

2. $\mathbf{a} = \begin{bmatrix} 1 \\ 0 \\ 0 \end{bmatrix}, \mathbf{n} = \begin{bmatrix} 0 \\ 1 \\ 0 \end{bmatrix}$

3. $\mathbf{a} = \begin{bmatrix} 0 \\ 0 \\ 1 \end{bmatrix}, \mathbf{n} = \begin{bmatrix} 1 \\ 0 \\ 0 \end{bmatrix}$

5. $\mathbf{a} = \begin{bmatrix} 0 \\ 1 \\ 0 \end{bmatrix}, \mathbf{n} = \begin{bmatrix} 1 \\ 0 \\ 0 \end{bmatrix}$

4. $\mathbf{a} = \begin{bmatrix} 0 \\ 0 \\ 1 \end{bmatrix}, \mathbf{n} = \begin{bmatrix} 0 \\ 1 \\ 0 \end{bmatrix}$

6. $\mathbf{a} = \begin{bmatrix} 0 \\ 0 \\ 1 \end{bmatrix}, \mathbf{n} = \begin{bmatrix} 0 \\ 1 \\ 0 \end{bmatrix}$

7. Show that the rotation of the vector $\mathbf{a}$ about the vector line through $\begin{bmatrix} 1 \\ 0 \\ 0 \end{bmatrix}$ ori-

ented by $\begin{bmatrix} 1 \\ 0 \\ 0 \end{bmatrix}$ is $\begin{bmatrix} 1 & 0 & 0 \\ 0 & \cos\alpha & -\sin\alpha \\ 0 & \sin\alpha & \cos\alpha \end{bmatrix} \mathbf{a}$.

8. Show that the rotation of the vector $\mathbf{a}$ about the vector line through $\begin{bmatrix} 0 \\ 0 \\ 1 \end{bmatrix}$ ori-

ented by $\begin{bmatrix} 0 \\ 0 \\ 1 \end{bmatrix}$ is $\begin{bmatrix} \cos\alpha & -\sin\alpha & 0 \\ \sin\alpha & \cos\alpha & 0 \\ 0 & 0 & 1 \end{bmatrix} \mathbf{a}$.

9. Show that the rotation of the vector $\mathbf{a}$ about the vector line through $\begin{bmatrix} 0 \\ 1 \\ 0 \end{bmatrix}$ ori-

ented by $\begin{bmatrix} 0 \\ 1 \\ 0 \end{bmatrix}$ is $\begin{bmatrix} \cos\alpha & 0 & \sin\alpha \\ 0 & 1 & 0 \\ -\sin\alpha & 0 & \cos\alpha \end{bmatrix} \mathbf{a}$.

9.4   Cross product and the right-hand rule

Now we would like to offer a different intuitive interpretation of the cross product. The idea is to start with any two orthogonal vectors $\mathbf{n}$ and $\mathbf{a}$ of norm one and, by using only rotations, transform the space so that $\begin{bmatrix} 0 \\ 0 \\ 1 \end{bmatrix}$ becomes $\mathbf{n}$ and $\begin{bmatrix} 1 \\ 0 \\ 0 \end{bmatrix}$ becomes $\mathbf{a}$. Since $\begin{bmatrix} 0 \\ 0 \\ 1 \end{bmatrix} \times \begin{bmatrix} 1 \\ 0 \\ 0 \end{bmatrix} = \begin{bmatrix} 0 \\ 1 \\ 0 \end{bmatrix}$, the cross product $\mathbf{n} \times \mathbf{a}$ will be the new position of $\begin{bmatrix} 0 \\ 1 \\ 0 \end{bmatrix}$. this shows that the relative position of the vectors $\mathbf{a}$, $\mathbf{n} \times \mathbf{a}$, and $\mathbf{n}$ is the same as the relative position of the vectors $\begin{bmatrix} 1 \\ 0 \\ 0 \end{bmatrix}$, $\begin{bmatrix} 0 \\ 1 \\ 0 \end{bmatrix}$, and $\begin{bmatrix} 0 \\ 0 \\ 1 \end{bmatrix}$, which justifies the right-hand rule. It is important to note that rotations do not change the relative position of $\mathbf{n}$ and $\mathbf{a}$.

In order to obtain the intuitive interpretation of the cross product we proceed in the following way:

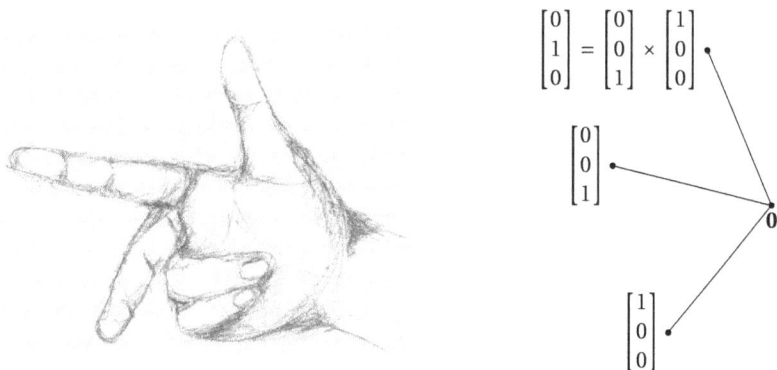

$$\begin{bmatrix}0\\1\\0\end{bmatrix} = \begin{bmatrix}0\\0\\1\end{bmatrix} \times \begin{bmatrix}1\\0\\0\end{bmatrix}$$

$$\begin{bmatrix}0\\0\\1\end{bmatrix}$$

$$0$$

$$\begin{bmatrix}1\\0\\0\end{bmatrix}$$

Figure 9.10: The right-hand rule.

If we have three orthonormal vectors **e**, **f**, and **g** in $\mathbb{R}^3$ which are in the same position as $\begin{bmatrix}1\\0\\0\end{bmatrix}$, $\begin{bmatrix}0\\1\\0\end{bmatrix}$, and $\begin{bmatrix}0\\0\\1\end{bmatrix}$ and, if the real numbers s and t satisfy $s^2 + t^2 = 1$, then the vectors $s\mathbf{e} + t\mathbf{f}$ and $-t\mathbf{e} + s\mathbf{f}$ and **g** are in the same position as $\begin{bmatrix}1\\0\\0\end{bmatrix}$, $\begin{bmatrix}0\\1\\0\end{bmatrix}$ and $\begin{bmatrix}0\\0\\1\end{bmatrix}$.

Let $\mathbf{n} = \begin{bmatrix}n_1\\n_2\\n_3\end{bmatrix}$ and $\mathbf{a} = \begin{bmatrix}a_1\\a_2\\a_3\end{bmatrix}$ be two orthogonal vectors of unit length, that is, $\mathbf{n} \cdot \mathbf{a} = 0$ and $\|\mathbf{n}\| = \|\mathbf{a}\| = 1$. We want to derive an expression for $\mathbf{n} \times \mathbf{a}$ from the intuitive equality $\begin{bmatrix}0\\0\\1\end{bmatrix} \times \begin{bmatrix}1\\0\\0\end{bmatrix} = \begin{bmatrix}0\\1\\0\end{bmatrix}$ by appropriately rotating vectors **n** and **a**. We proceed in three steps.

Step 1

The first step is the counterclockwise rotation about the z-axis until the vector $\begin{bmatrix}1\\0\\0\end{bmatrix}$ becomes a vector orthogonal to the vector $\mathbf{n} = \begin{bmatrix}n_1\\n_2\\n_3\end{bmatrix}$, for example the vector $\begin{bmatrix}\dfrac{n_2}{\sqrt{n_1^2+n_2^2}}\\ -\dfrac{n_1}{\sqrt{n_1^2+n_2^2}}\\ 0\end{bmatrix}$.

Since

$$\begin{bmatrix} \dfrac{n_2}{\sqrt{n_1^2+n_2^2}} \\ -\dfrac{n_1}{\sqrt{n_1^2+n_2^2}} \\ 0 \end{bmatrix} = \frac{n_2}{\sqrt{n_1^2+n_2^2}} \begin{bmatrix} 1 \\ 0 \\ 0 \end{bmatrix} - \frac{n_1}{\sqrt{n_1^2+n_2^2}} \begin{bmatrix} 0 \\ 1 \\ 0 \end{bmatrix}$$

and

$$\begin{bmatrix} \dfrac{n_1}{\sqrt{n_1^2+n_2^2}} \\ \dfrac{n_2}{\sqrt{n_1^2+n_2^2}} \\ 0 \end{bmatrix} = \frac{n_1}{\sqrt{n_1^2+n_2^2}} \begin{bmatrix} 1 \\ 0 \\ 0 \end{bmatrix} + \frac{n_2}{\sqrt{n_1^2+n_2^2}} \begin{bmatrix} 0 \\ 1 \\ 0 \end{bmatrix},$$

when we rotate the vector $\begin{bmatrix} \dfrac{n_2}{\sqrt{n_1^2+n_2^2}} \\ -\dfrac{n_1}{\sqrt{n_1^2+n_2^2}} \\ 0 \end{bmatrix}$ counterclockwise by $\frac{\pi}{2}$ about the z-axis ori-

ented by the vector $\begin{bmatrix} 0 \\ 0 \\ 1 \end{bmatrix}$, we obtain the vector $\begin{bmatrix} \dfrac{n_1}{\sqrt{n_1^2+n_2^2}} \\ \dfrac{n_2}{\sqrt{n_1^2+n_2^2}} \\ 0 \end{bmatrix}$.

Consequently if we define the angle α by

$$\frac{n_2}{\sqrt{n_1^2+n_2^2}} = \cos\alpha \quad \text{and} \quad -\frac{n_1}{\sqrt{n_1^2+n_2^2}} = \sin\alpha$$

when we rotate the vector $\begin{bmatrix} 1 \\ 0 \\ 0 \end{bmatrix}$ counterclockwise by α about the z-axis oriented by

the vector $\begin{bmatrix} 0 \\ 0 \\ 1 \end{bmatrix}$, we obtain the vector $\begin{bmatrix} \dfrac{n_2}{\sqrt{n_1^2+n_2^2}} \\ -\dfrac{n_1}{\sqrt{n_1^2+n_2^2}} \\ 0 \end{bmatrix}$ and when we rotate the vector $\begin{bmatrix} 0 \\ 1 \\ 0 \end{bmatrix}$

counterclockwise by α about the z-axis oriented by the vector $\begin{bmatrix} 0 \\ 0 \\ 1 \end{bmatrix}$, we obtain the

vector $\begin{bmatrix} \dfrac{n_1}{\sqrt{n_1^2+n_2^2}} \\ \dfrac{n_2}{\sqrt{n_1^2+n_2^2}} \\ 0 \end{bmatrix}$.

We summarize the effect of the rotation by the angle α in the following table.

Original position	After the first rotation
$\begin{bmatrix} 1 \\ 0 \\ 0 \end{bmatrix}$	$\begin{bmatrix} \dfrac{n_2}{\sqrt{n_1^2+n_2^2}} \\ -\dfrac{n_1}{\sqrt{n_1^2+n_2^2}} \\ 0 \end{bmatrix}$
$\begin{bmatrix} 0 \\ 1 \\ 0 \end{bmatrix}$	$\begin{bmatrix} \dfrac{n_1}{\sqrt{n_1^2+n_2^2}} \\ \dfrac{n_2}{\sqrt{n_1^2+n_2^2}} \\ 0 \end{bmatrix}$
$\begin{bmatrix} 0 \\ 0 \\ 1 \end{bmatrix}$	$\begin{bmatrix} 0 \\ 0 \\ 1 \end{bmatrix}$

Step 2

The second step is the counterclockwise rotation about the vector line through $\begin{bmatrix} \dfrac{n_2}{\sqrt{n_1^2+n_2^2}} \\ -\dfrac{n_1}{\sqrt{n_1^2+n_2^2}} \\ 0 \end{bmatrix}$

until $\begin{bmatrix} 0 \\ 0 \\ 1 \end{bmatrix}$ becomes $\mathbf{n} = \begin{bmatrix} n_1 \\ n_2 \\ n_3 \end{bmatrix}$. Note that this is possible since both $\begin{bmatrix} 0 \\ 0 \\ 1 \end{bmatrix}$ and $\mathbf{n} = \begin{bmatrix} n_1 \\ n_2 \\ n_3 \end{bmatrix}$

are orthogonal to $\begin{bmatrix} \dfrac{n_2}{\sqrt{n_1^2+n_2^2}} \\ -\dfrac{n_1}{\sqrt{n_1^2+n_2^2}} \\ 0 \end{bmatrix}$. Since

$$\mathbf{n} = \begin{bmatrix} n_1 \\ n_2 \\ n_3 \end{bmatrix} = n_3 \begin{bmatrix} 0 \\ 0 \\ 1 \end{bmatrix} + \sqrt{n_1^2+n_2^2} \begin{bmatrix} \dfrac{n_1}{\sqrt{n_1^2+n_2^2}} \\ \dfrac{n_2}{\sqrt{n_1^2+n_2^2}} \\ 0 \end{bmatrix}$$

and

$$\begin{bmatrix} \dfrac{n_1 n_3}{\sqrt{n_1^2+n_2^2}} \\ \dfrac{n_2 n_3}{\sqrt{n_1^2+n_2^2}} \\ -\sqrt{n_1^2+n_2^2} \end{bmatrix} = -\sqrt{n_1^2+n_2^2} \begin{bmatrix} 0 \\ 0 \\ 1 \end{bmatrix} + n_3 \begin{bmatrix} \dfrac{n_1}{\sqrt{n_1^2+n_2^2}} \\ \dfrac{n_2}{\sqrt{n_1^2+n_2^2}} \\ 0 \end{bmatrix},$$

when we rotate the vector $\mathbf{n} = \begin{bmatrix} n_1 \\ n_2 \\ n_3 \end{bmatrix}$ counterclockwise by $\frac{\pi}{2}$ about the vector

line through $\begin{bmatrix} \dfrac{n_2}{\sqrt{n_1^2+n_2^2}} \\ -\dfrac{n_1}{\sqrt{n_1^2+n_2^2}} \\ 0 \end{bmatrix}$ oriented by the vector $\begin{bmatrix} \dfrac{n_2}{\sqrt{n_1^2+n_2^2}} \\ -\dfrac{n_1}{\sqrt{n_1^2+n_2^2}} \\ 0 \end{bmatrix}$ we obtain

the vector $\begin{bmatrix} \dfrac{n_1 n_3}{\sqrt{n_1^2+n_2^2}} \\ \dfrac{n_2 n_3}{\sqrt{n_1^2+n_2^2}} \\ -\sqrt{n_1^2 + n_2^2} \end{bmatrix}$.

Consequently if we define the angle β by

$$n_3 = \cos\beta \quad \text{and} \quad \sqrt{n_1^2 + n_2^2} = \sin\beta$$

when we rotate the vector $\begin{bmatrix} 0 \\ 0 \\ 1 \end{bmatrix}$ counterclockwise by β about the vector line through

$\begin{bmatrix} \dfrac{n_2}{\sqrt{n_1^2+n_2^2}} \\ -\dfrac{n_1}{\sqrt{n_1^2+n_2^2}} \\ 0 \end{bmatrix}$ oriented by the vector $\begin{bmatrix} \dfrac{n_2}{\sqrt{n_1^2+n_2^2}} \\ -\dfrac{n_1}{\sqrt{n_1^2+n_2^2}} \\ 0 \end{bmatrix}$ we obtain the vector $\begin{bmatrix} n_1 \\ n_2 \\ n_3 \end{bmatrix}$ and when

we rotate the vector $\begin{bmatrix} \dfrac{n_1}{\sqrt{n_1^2+n_2^2}} \\ \dfrac{n_2}{\sqrt{n_1^2+n_2^2}} \\ 0 \end{bmatrix}$ counterclockwise by β about the vector line through

$\begin{bmatrix} \dfrac{n_2}{\sqrt{n_1^2+n_2^2}} \\ -\dfrac{n_1}{\sqrt{n_1^2+n_2^2}} \\ 0 \end{bmatrix}$ oriented by the vector $\begin{bmatrix} \dfrac{n_2}{\sqrt{n_1^2+n_2^2}} \\ -\dfrac{n_1}{\sqrt{n_1^2+n_2^2}} \\ 0 \end{bmatrix}$ we obtain the vector $\begin{bmatrix} \dfrac{n_1 n_3}{\sqrt{n_1^2+n_2^2}} \\ \dfrac{n_2 n_3}{\sqrt{n_1^2+n_2^2}} \\ -\sqrt{n_1^2 + n_2^2} \end{bmatrix}$.

We summarize the cumulative effect of both rotations by α and β in the following table.

Original position	After the first rotation	After the first two rotations
$\begin{bmatrix} 1 \\ 0 \\ 0 \end{bmatrix}$	$\begin{bmatrix} \dfrac{n_2}{\sqrt{n_1^2+n_2^2}} \\ -\dfrac{n_1}{\sqrt{n_1^2+n_2^2}} \\ 0 \end{bmatrix}$	$\begin{bmatrix} \dfrac{n_2}{\sqrt{n_1^2+n_2^2}} \\ -\dfrac{n_1}{\sqrt{n_1^2+n_2^2}} \\ 0 \end{bmatrix}$
$\begin{bmatrix} 0 \\ 1 \\ 0 \end{bmatrix}$	$\begin{bmatrix} \dfrac{n_1}{\sqrt{n_1^2+n_2^2}} \\ \dfrac{n_2}{\sqrt{n_1^2+n_2^2}} \\ 0 \end{bmatrix}$	$\begin{bmatrix} \dfrac{n_1 n_3}{\sqrt{n_1^2+n_2^2}} \\ \dfrac{n_2 n_3}{\sqrt{n_1^2+n_2^2}} \\ -\sqrt{n_1^2+n_2^2} \end{bmatrix}$
$\begin{bmatrix} 0 \\ 0 \\ 1 \end{bmatrix}$	$\begin{bmatrix} 0 \\ 0 \\ 1 \end{bmatrix}$	$\begin{bmatrix} n_1 \\ n_2 \\ n_3 \end{bmatrix}$

Step 3

In this step the vector line through $\mathbf{n}$ is the axis of rotation. Since $\mathbf{a}$ is in the vector plane $\mathbf{x} \cdot \mathbf{n} = 0$ and $\|\mathbf{a}\| = 1$, there are real numbers p and q such that $p^2 + q^2 = 1$ and

$$\mathbf{a} = p \begin{bmatrix} \dfrac{n_2}{\sqrt{n_1^2+n_2^2}} \\ -\dfrac{n_1}{\sqrt{n_1^2+n_2^2}} \\ 0 \end{bmatrix} + q \begin{bmatrix} \dfrac{n_1 n_3}{\sqrt{n_1^2+n_2^2}} \\ \dfrac{n_2 n_3}{\sqrt{n_1^2+n_2^2}} \\ -\sqrt{n_1^2+n_2^2} \end{bmatrix} .$$

Now, since

$$\mathbf{n} \times \begin{bmatrix} \dfrac{n_2}{\sqrt{n_1^2+n_2^2}} \\ -\dfrac{n_1}{\sqrt{n_1^2+n_2^2}} \\ 0 \end{bmatrix} = \begin{bmatrix} n_1 \\ n_2 \\ n_3 \end{bmatrix} \times \begin{bmatrix} \dfrac{n_2}{\sqrt{n_1^2+n_2^2}} \\ -\dfrac{n_1}{\sqrt{n_1^2+n_2^2}} \\ 0 \end{bmatrix} = \begin{bmatrix} \dfrac{n_1 n_3}{\sqrt{n_1^2+n_2^2}} \\ \dfrac{n_2 n_3}{\sqrt{n_1^2+n_2^2}} \\ -\sqrt{n_1^2+n_2^2} \end{bmatrix} \tag{9.6}$$

and

$$\mathbf{n} \times \begin{bmatrix} \dfrac{n_1 n_3}{\sqrt{n_1^2+n_2^2}} \\ \dfrac{n_2 n_3}{\sqrt{n_1^2+n_2^2}} \\ -\sqrt{n_1^2+n_2^2} \end{bmatrix} = \begin{bmatrix} n_1 \\ n_2 \\ n_3 \end{bmatrix} \times \begin{bmatrix} \dfrac{n_1 n_3}{\sqrt{n_1^2+n_2^2}} \\ \dfrac{n_2 n_3}{\sqrt{n_1^2+n_2^2}} \\ -\sqrt{n_1^2+n_2^2} \end{bmatrix} = \begin{bmatrix} \dfrac{-n_2}{\sqrt{n_1^2+n_2^2}} \\ \dfrac{n_1}{\sqrt{n_1^2+n_2^2}} \\ 0 \end{bmatrix} , \tag{9.7}$$

we get

$$\mathbf{n} \times \mathbf{a} = \mathbf{n} \times \left(p \begin{bmatrix} \dfrac{n_2}{\sqrt{n_1^2+n_2^2}} \\[2ex] -\dfrac{n_1}{\sqrt{n_1^2+n_2^2}} \\[2ex] 0 \end{bmatrix} + q \begin{bmatrix} \dfrac{n_1 n_3}{\sqrt{n_1^2+n_2^2}} \\[2ex] \dfrac{n_2 n_3}{\sqrt{n_1^2+n_2^2}} \\[2ex] -\sqrt{n_1^2 + n_2^2} \end{bmatrix} \right)$$

$$= p \left(\mathbf{n} \times \begin{bmatrix} \dfrac{n_2}{\sqrt{n_1^2+n_2^2}} \\[2ex] -\dfrac{n_1}{\sqrt{n_1^2+n_2^2}} \\[2ex] 0 \end{bmatrix} \right) + q \left(\mathbf{n} \times \begin{bmatrix} \dfrac{n_1 n_3}{\sqrt{n_1^2+n_2^2}} \\[2ex] \dfrac{n_2 n_3}{\sqrt{n_1^2+n_2^2}} \\[2ex] -\sqrt{n_1^2 + n_2^2} \end{bmatrix} \right)$$

$$= -q \begin{bmatrix} \dfrac{n_2}{\sqrt{n_1^2+n_2^2}} \\[2ex] -\dfrac{n_1}{\sqrt{n_1^2+n_2^2}} \\[2ex] 0 \end{bmatrix} + p \begin{bmatrix} \dfrac{n_1 n_3}{\sqrt{n_1^2+n_2^2}} \\[2ex] \dfrac{n_2 n_3}{\sqrt{n_1^2+n_2^2}} \\[2ex] -\sqrt{n_1^2 + n_2^2} \end{bmatrix} .$$

Note that $\mathbf{a}$ and $\mathbf{n} \times \mathbf{a}$ are both in the plane orthogonal to $\mathbf{n}$ and $\mathbf{n} \times \mathbf{a}$ can be obtained from $\mathbf{a}$ by counterclockwise rotation in that plane by $\frac{\pi}{2}$ about the vector line Span$\{\mathbf{n}\}$ oriented by the vector $\mathbf{n}$.

Consequently if we define the angle γ by

$$p = \cos\gamma \quad \text{and} \quad q = \sin\gamma$$

when we rotate the vector $\begin{bmatrix} \dfrac{n_2}{\sqrt{n_1^2+n_2^2}} \\[2ex] -\dfrac{n_1}{\sqrt{n_1^2+n_2^2}} \\[2ex] 0 \end{bmatrix}$ counterclockwise by γ about the vector line Span$\{\mathbf{n}\}$ oriented by the vector $\mathbf{n}$ we obtain the vector $\mathbf{a}$ and when we rotate the vector $\begin{bmatrix} \dfrac{n_1 n_3}{\sqrt{n_1^2+n_2^2}} \\[2ex] \dfrac{n_2 n_3}{\sqrt{n_1^2+n_2^2}} \\[2ex] -\sqrt{n_1^2 + n_2^2} \end{bmatrix}$ counterclockwise by γ about the vector line Span$\{\mathbf{n}\}$ oriented by the vector $\mathbf{n}$ we obtain the vector $\mathbf{n} \times \mathbf{a}$.

We summarize the cumulative effect of these three rotations (by α, β and γ) in the following table.

Original position	After the first rotation	After the first two rotations	After three rotations
$\begin{bmatrix} 1 \\ 0 \\ 0 \end{bmatrix}$	$\begin{bmatrix} \dfrac{n_2}{\sqrt{n_1^2+n_2^2}} \\ -\dfrac{n_1}{\sqrt{n_1^2+n_2^2}} \\ 0 \end{bmatrix}$	$\begin{bmatrix} \dfrac{n_2}{\sqrt{n_1^2+n_2^2}} \\ -\dfrac{n_1}{\sqrt{n_1^2+n_2^2}} \\ 0 \end{bmatrix}$	$\mathbf{a}$
$\begin{bmatrix} 0 \\ 1 \\ 0 \end{bmatrix}$	$\begin{bmatrix} \dfrac{n_1}{\sqrt{n_1^2+n_2^2}} \\ \dfrac{n_2}{\sqrt{n_1^2+n_2^2}} \\ 0 \end{bmatrix}$	$\begin{bmatrix} \dfrac{n_1 n_3}{\sqrt{n_1^2+n_2^2}} \\ \dfrac{n_2 n_3}{\sqrt{n_1^2+n_2^2}} \\ -\sqrt{n_1^2+n_2^2} \end{bmatrix}$	$\mathbf{n} \times \mathbf{a}$
$\begin{bmatrix} 0 \\ 0 \\ 1 \end{bmatrix}$	$\begin{bmatrix} 0 \\ 0 \\ 1 \end{bmatrix}$	$\mathbf{n}$	$\mathbf{n}$

Chapter 10

Problems in plane geometry

The classical plane geometry gives us an attractive opportunity to apply and understand linear algebra. In this chapter we use tools provided by linear algebra to solve nontrivial problems in plane geometry. The main purpose of the chapter is to give students a chance to practice newly acquired skills in the familiar context of geometry. We also hope that students will appreciate the elegance of algebraic solutions.

The applications are presented in the form of problems with complete solutions. With few exceptions, the tools used here were introduced in Chapters 3. The solutions are presented with fewer details than in the rest of the book. The reader is expected to work through the arguments and fill in the finer details.

In this chapter, following the standard notation in plane geometry, points in $\mathbb{R}^2$ will be denoted with capital letters $A, B, C, \ldots, X, Y, Z$ instead of $\mathbf{a}, \mathbf{b}, \mathbf{c}, \ldots, \mathbf{x}, \mathbf{y}, \mathbf{z}$. As before, we will identify points in $\mathbb{R}^2$ with vectors in $\mathbb{R}^2$. This will allow us to translate a purely geometric problem to a problem in linear algebra and then use the power of linear algebra to solve the problem in an elegant way.

10.1   Lines and circles

Let A and B be two distinct points in $\mathbb{R}^2$. The line segment connecting A and B will be denoted by AB.

Problem 10.1.1. *Let A, B, and C be distinct points in $\mathbb{R}^2$ such that the angle $\angle ACB$ is a right angle. Show that C is on the circle with diameter AB.*

Solution 1. The center of the circle is at $\dfrac{A+B}{2}$. It suffices to show that

$$\left\| A - \frac{A+B}{2} \right\| = \left\| C - \frac{A+B}{2} \right\|.$$

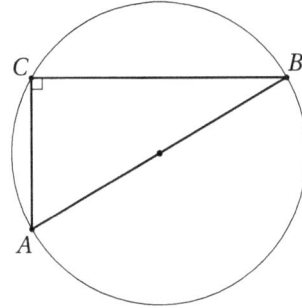

Figure 10.1: Problem 10.1.1.

Since the angle $\angle ACB$ is a right angle, we have $0 = (C - A) \cdot (C - B)$. Hence,

$$
\begin{aligned}
0 &= (C - A) \cdot (C - B) \\
&= \left(C - \frac{A+B}{2} + \frac{A+B}{2} - A \right) \cdot \left(C - \frac{A+B}{2} + \frac{A+B}{2} - B \right) \\
&= \left(C - \frac{A+B}{2} + \frac{B-A}{2} \right) \cdot \left(C - \frac{A+B}{2} - \frac{B-A}{2} \right) \\
&= \left\| C - \frac{A+B}{2} \right\|^2 - \left\| \frac{A-B}{2} \right\|^2 \\
&= \left\| C - \frac{A+B}{2} \right\|^2 - \left\| A - \frac{A+B}{2} \right\|^2,
\end{aligned}
$$

which gives us the desired equality.   □

It turns out that the algebraic part of this argument can be significantly simplified. Since the described property does not depend on the position of the triangle relative to the origin, we can choose its position to simplify calculations. In this case it is most convenient to assume that the middle of the line segment AB is at the origin. Then the solution becomes significantly simpler.

Solution 2. If we take the middle of the segment AB as the origin, then we have $B = -A$ and hence

$$
0 = (C - A) \cdot (C - B) = (C - A) \cdot (C + A) = C \cdot C - A \cdot A = \|C\|^2 - \|A\|^2,
$$

which gives us the desired equality.   □

As we can see, the algebraic part of the proof has been reduced to a single line. In the remaining examples in this chapter we will always try to find a position that gives us the simplest calculations. Sometimes it is not entirely obvious what that position is. In such a case we might have to try a couple of different positions before we discover the best one. On the other hand, in some examples it seems that there is no advantage in choosing a special position.

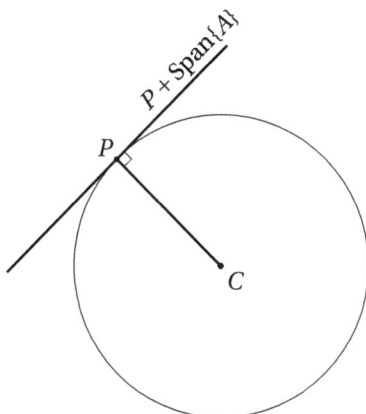

Figure 10.2: Problem 10.1.2.

Problem 10.1.2. *Let P be a point on a circle with the center at C. Show that the tangent to the circle at P is orthogonal to the line through the points P and C.*

Solution. Let A be a point such that $\|A\| = 1$. The common points of the circle and the line $P + \text{Span}\{A\}$, if such points exist, are given by the equation

$$\|P + tA - C\| = \|P - C\|,$$

or, equivalently,

$$(P + tA - C) \cdot (P + tA - C) = (P - C) \cdot (P - C).$$

This reduces to

$$t^2 + 2((P - C) \cdot A)t = 0,$$

since $\|A\| = 1$. This quadratic equation has exactly one solution if and only if

$$(P - C) \cdot A = 0.$$

□

Problem 10.1.3 (Chord Theorem). *If two chords in a circle intersect, then the product of the lengths of the two segments on one chord is equal to the product of the lengths of the two segments on the other chord.*

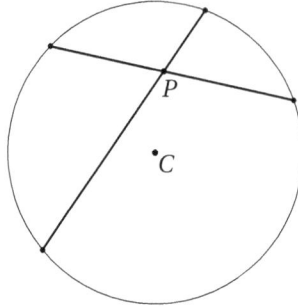

Figure 10.3: Problem 10.1.3.

Solution. Consider a circle with the center at C and the radius r. Let P be a point inside the circle, that is, $\|P-C\| < r$. Any line through P can be described as $P + \text{Span}\{A\}$ with $\|A\| = 1$. Note that the distance from P to a point $P + tA$ is $|t|$.

The intersection points of the line $P + \text{Span}\{A\}$ and the circle are given by the equation

$$\|P + tA - C\| = r,$$

which is equivalent to

$$t^2 + 2((P - C) \cdot A)t + \|P - C\|^2 - r^2 = 0.$$

Since

$$((P - C) \cdot A)^2 - \|P - C\|^2 + r^2 > 0,$$

the equation has two distinct roots, as expected. Moreover, the product of these roots is

$$-\|P - C\|^2 + r^2,$$

which is independent of A.   □

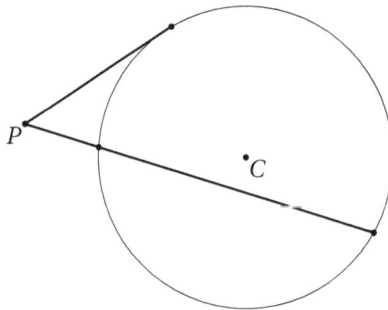

Figure 10.4: Problem 10.1.4.

> **Problem 10.1.4** (Tangent-Secant Theorem). *If a secant segment and tangent segment are drawn to a circle from the same external point, the product of the length of the secant segment and its external part equals the square of the length of the tangent segment.*

Solution. This theorem can be interpreted as a case of the Chord Theorem when the point P is outside the circle. It can proved by a modification of the argument presented above. We leave the proof as an exercise. $\qquad\square$

10.2 Triangles

In many solutions in this section we consider the triangle with vertices $C - C$, $A - C$, and $B - C$ instead of the triangle with vertices C, A, and B. This is done to simplify the calculations. Subtracting C from all points of the triangle results in translating the whole triangle without changing its size or shape. This allows us to solve the general problem by solving an "easier" problem.

> **Problem 10.2.1.** *Show that the three altitudes in a triangle intersect at a single point.*

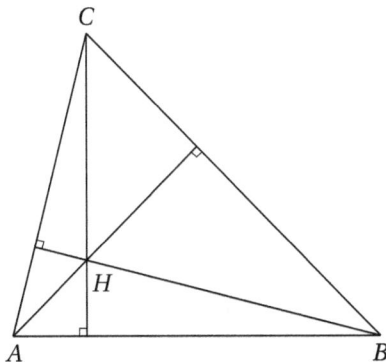

Figure 10.5: Problem 10.2.1.

Solution. Consider a triangle with vertices A, B, and C. To simplify the calculations, we assume that C is at the origin. The altitude from A is on the line $A + \text{Span}\{B^{\perp}\}$ and the altitude from B is on the line $B + \text{Span}\{A^{\perp}\}$. The intersection of these to lines can be found as the solution of the equation

$$A + tB^{\perp} = B + sA^{\perp}.$$

Note that this equation has a unique solution since $A^\perp$ and $B^\perp$ are linearly independent. Let's denote the intersection point by H. It suffices to prove that $H \cdot (A - B) = 0$. Indeed,

$$H \cdot (A - B) = H \cdot A - H \cdot B = (B + sA^\perp) \cdot A - (P + tB^\perp) \cdot B = B \cdot A - A \cdot B = 0.$$

$\square$

Problem 10.2.2. *Show that the three medians in a triangle intersect at a single point.*

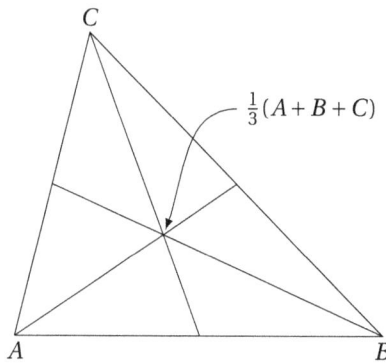

Figure 10.6: Medians in a triangle intersect at a single point.

Solution. The median from A is

$$A + t\left(\frac{B+C}{2} - A\right).$$

If we take $t = \frac{2}{3}$, we get the point $\frac{A+B+C}{3}$. This point is also on the medians from B and from C, since

$$\frac{A+B+C}{3} = B + \frac{2}{3}\left(\frac{A+C}{2} - B\right) = C + \frac{2}{3}\left(\frac{A+B}{2} - C\right).$$

$\square$

Problem 10.2.3. *Consider a triangle with vertices A, B, and C. Show that the bisectors intersect at a single point. The point of intersection is*

$$I = \frac{a}{a+b+c} A + \frac{b}{a+b+c} B + \frac{c}{a+b+c} C,$$

where $a = \|B - C\|$, $b = \|A - C\|$, and $c = \|A - B\|$.

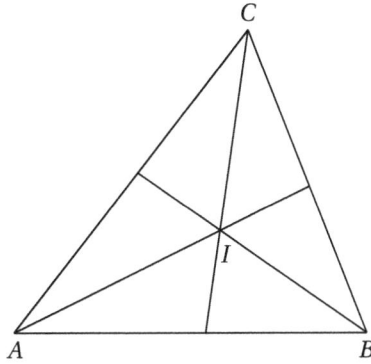

Figure 10.7: Bisectors in a triangle intersect at a single point.

Solution. We will show that the point is on the bisector from C. Observe that

$$I = C + \frac{a}{a+b+c}(A-C) + \frac{b}{a+b+c}(B-C)$$

$$= C + \frac{ab}{a+b+c}\frac{A-C}{b} + \frac{ab}{a+b+c}\frac{B-C}{a}$$

$$= C + \frac{ab}{a+b+c}\left(\frac{A-C}{\|A-C\|} + \frac{B-C}{\|B-C\|}\right),$$

from which it is clear that the point I is on the bisector from C. The same method can be used to check that I is on the other two bisectors. □

Problem 10.2.4. *Consider a triangle with vertices A, B, and C. Show that*

$$(A-C)\cdot(B-C) = \frac{a^2 + b^2 - c^2}{2},$$

where $a = \|B - C\|$, $b = \|A - C\|$ and $c = \|A - B\|$.

Solution. We have

$$\|A-B\|^2 = \|A-C+C-B\|^2$$

$$= \|A-C\|^2 + \|C-B\|^2 + 2(A-C)\cdot(C-B)$$

$$= \|A-C\|^2 + \|B-C\|^2 - 2(A-C)\cdot(B-C),$$

which gives us the desired equality. □

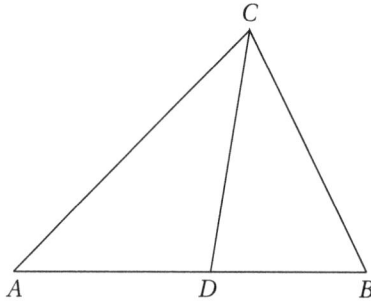

Figure 10.8: Problem 10.2.5.

Problem 10.2.5. *Consider a triangle with vertices A, B, and C. Denote the intersection of the bisector from the vertex C with the opposite side by D. Then*

$$\frac{\|A - D\|}{\|A - C\|} = \frac{\|B - D\|}{\|B - C\|}.$$

Solution. We place the vertex C at the origin. Now the bisector from the vertex C is on the line

$$\text{Span}\left\{\frac{A}{\|A\|} + \frac{B}{\|B\|}\right\}$$

and a point on the line through the points A and B is of the form

$$A + s(B - A).$$

Thus the intersection is given by the equation

$$t\left(\frac{A}{\|A\|} + \frac{B}{\|B\|}\right) = (1 - s)A + sB,$$

where t and s are real numbers to be determined. Since A and B are linearly independent, we must have

$$\frac{t}{\|A\|} = 1 - s \quad \text{and} \quad \frac{t}{\|B\|} = s.$$

Now, with $a = \|B\|$ and $b = \|A\|$, we obtain

$$s = \frac{b}{a + b},$$

and thus the intersection point is

$$\frac{a}{a + b}A + \frac{b}{a + b}B = D.$$

Now

$$D - A = \frac{a}{a+b}A + \frac{b}{a+b}B - A$$
$$= \frac{b}{a+b}B - \frac{b}{a+b}A = \frac{b}{a+b}(B - A)$$

and

$$D - B = \frac{a}{a+b}A + \frac{b}{a+b}B - B$$
$$= \frac{a}{a+b}A - \frac{a}{a+b}B = \frac{a}{a+b}(A - B).$$

Hence

$$\frac{\|D - A\|}{b} = \frac{\|A - B\|}{a+b} = \frac{\|D - B\|}{a}$$

or

$$\frac{\|D - A\|}{\|B\|} = \frac{\|D - B\|}{\|A\|}.$$

Finally, going back to the triangle with vertices A, B, and C, we get

$$\frac{\|A - D\|}{\|A - C\|} = \frac{\|B - D\|}{\|B' - C\|}.$$

□

Problem 10.2.6. *Consider a triangle with vertices A, B, and C. The radius r of the circle inscribed in the triangle is*

$$\frac{|\det[A - C \quad B - C]|}{a + b + c},$$

where $a = \|B - C\|$, $b = \|A - C\|$, and $c = \|A - B\|$.

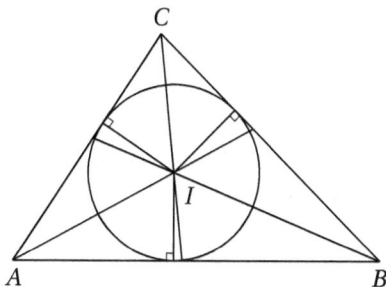

Figure 10.9: Inscribed circle.

Solution. We assume that C is at origin. From Problem 10.2.3 we know that the center of the circle is

$$I = \frac{a}{a+b+c}A + \frac{b}{a+b+c}B + \frac{c}{a+b+c}C = \frac{a}{a+b+c}A + \frac{b}{a+b+c}B.$$

The intersection point of the line through I perpendicular to the line through B and C with that line is given by the equation

$$\frac{a}{a+b+c}A + \frac{b}{a+b+c}B + tB^{\perp} = sB.$$

Hence

$$\frac{a}{a+b+c}A \cdot B^{\perp} + \frac{b}{a+b+c}B \cdot B^{\perp} + tB^{\perp} \cdot B^{\perp} = sB \cdot B^{\perp}$$

which simplifies to

$$\frac{a}{a+b+c}A \cdot B^{\perp} + t\|B\|^2 = 0.$$

Since $A \cdot B^{\perp} = \det\begin{bmatrix} B & A \end{bmatrix}$ and $\|B\|^2 = a^2$, we get

$$\frac{a}{a+b+c}\det\begin{bmatrix} B & A \end{bmatrix} + ta^2 = 0.$$

Hence

$$t = -\frac{\det\begin{bmatrix} B & A \end{bmatrix}}{a(a+b+c)} = \frac{\det\begin{bmatrix} A & B \end{bmatrix}}{a(a+b+c)}.$$

Now observe that the radius must be equal to $\|tB^{\perp}\|$ and thus

$$r = \|tB^{\perp}\| = \frac{|\det\begin{bmatrix} B & A \end{bmatrix}|}{a(a+b+c)}\|B^{\perp}\| = \frac{|\det\begin{bmatrix} A & B \end{bmatrix}|}{a+b+c},$$

because $\|B^{\perp}\| = \|B\| = a$. Switching back to A, B, and C we obtain

$$r = \frac{|\det\begin{bmatrix} A-C & B-C \end{bmatrix}|}{a+b+c}.$$

$\square$

Problem 10.2.7. *Consider a triangle with vertices A, B, and C. Show that the radius R of the circumcircle is given by*

$$R = \frac{abc}{2|\det\begin{bmatrix} A-C & B-C \end{bmatrix}|},$$

where $a = \|B-C\|$, $b = \|A-C\|$ and $c = \|A-B\|$.

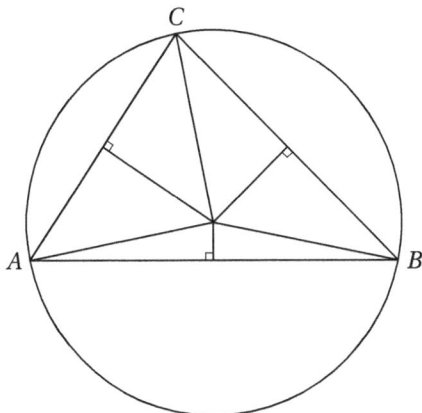

Figure 10.10: Problem 10.2.7.

Solution. We place the origin at C. The center of the circumcircle can be found from the equation

$$\frac{A}{2} + tA^{\perp} = \frac{B}{2} + sB^{\perp},$$

where t and s are real numbers to be determined. We multiply both sides by B and obtain

$$\frac{A \cdot B}{2} + tA^{\perp} \cdot B = \frac{\|B\|^2}{2} = \frac{a^2}{2}.$$

By solving for t we obtain

$$t = \frac{a^2 - A \cdot B}{2A^{\perp} \cdot B}. \tag{10.1}$$

Now observe that

$$R^2 = \left\| \frac{A}{2} + tA^{\perp} - A \right\|^2 = \left\| tA^{\perp} - \frac{A}{2} \right\|^2$$

$$= t^2 \|A^{\perp}\|^2 + \frac{\|A\|^2}{4} = t^2 \|A\|^2 + \frac{\|A\|^2}{4} = b^2 \left(t^2 + \frac{1}{4} \right).$$

From (10.1) we obtain

$$t^2 + \frac{1}{4} = \left(\frac{a^2 - A \cdot B}{2A^{\perp} \cdot B} \right)^2 + \frac{1}{4}$$

$$= \frac{a^4 - 2a^2 A \cdot B + (A \cdot B)^2 + (A^{\perp} \cdot B)^2}{4(A^{\perp} \cdot B)^2}$$

$$= \frac{a^4 - a^2(a^2 + b^2 - c^2) + a^2 b^2}{4(A^{\perp} \cdot B)^2} = \frac{a^2 c^2}{4(A^{\perp} \cdot B)^2}$$

$$= \frac{a^2 c^2}{4(\det [A \quad B])^2},$$

since
$$A \cdot B = \frac{a^2 + b^2 - c^2}{2},$$

as shown in Problem 10.2.4, and
$$(A \cdot B)^2 + (A^\perp \cdot B)^2 = \|A\|^2 \|B\|^2.$$

Consequently
$$R^2 = b^2 \left(t^2 + \frac{1}{4} \right) = \frac{a^2 b^2 c^2}{4 \det \begin{bmatrix} A & B \end{bmatrix}^2}$$

and thus
$$R = \frac{abc}{2|\det \begin{bmatrix} A & B \end{bmatrix}|}.$$

Recalling that $A = A - C$ and $B = B - C$ we obtain the desired equality
$$R = \frac{abc}{2|\det \begin{bmatrix} A - C & B - C \end{bmatrix}|}.$$

$\square$

Problem 10.2.8 (Nine-point circle). *Let H be the orthocenter of the triangle with vertices A, B, and C. Prove that the three points where the altitudes meet the opposite sides of the triangle and the midpoints of the line segments HA, HB, HC, AB, BC and CA are all on the same circle.*

Solution. We place the origin at H.

We have
$$\left\| \frac{C}{2} - \frac{A+B+C}{4} \right\|^2 = \left\| \frac{A+B}{2} - \frac{A+B+C}{4} \right\|^2 = \left\| \frac{A+B-C}{4} \right\|^2,$$

$$\left\| \frac{B}{2} - \frac{A+B+C}{4} \right\|^2 = \left\| \frac{A+C}{2} - \frac{A+B+C}{4} \right\|^2 = \left\| \frac{A+C-B}{4} \right\|^2$$

and
$$\left\| \frac{A}{2} - \frac{A+B+C}{4} \right\|^2 = \left\| \frac{B+C}{2} - \frac{A+B+C}{4} \right\|^2 = \left\| \frac{B+C-A}{4} \right\|^2.$$

Since
$$A \cdot (B - C) = B \cdot (A - C) = C \cdot (A - B) = 0,$$

by the Pythagorean Theorem, we have
$$\left\| \frac{A+B-C}{4} \right\|^2 = \frac{1}{16} \left(\|A\|^2 + \|B - C\|^2 \right)$$
$$= \frac{1}{16} \left(\|A\|^2 + \|C - B\|^2 \right)$$

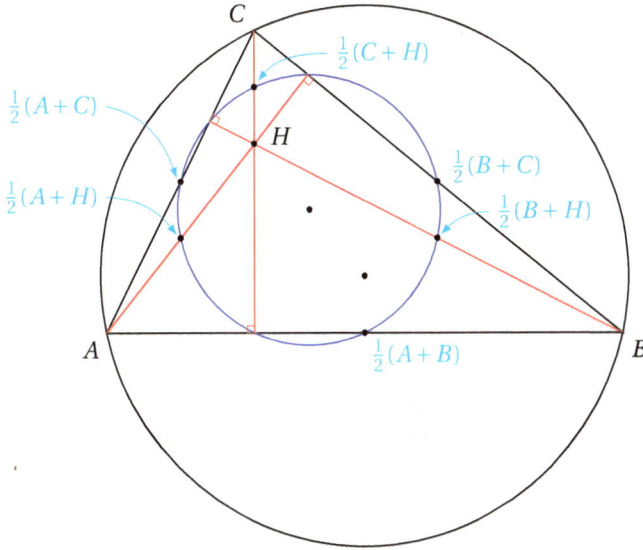

Figure 10.11: Problem 10.2.8: The nine-point circle.

$$= \left\| \frac{A + C - B}{4} \right\|^2$$

and

$$\left\| \frac{A + B - C}{4} \right\|^2 = \left\| \frac{B + A - C}{4} \right\|^2$$

$$= \frac{1}{16} \left(\|B\|^2 + \|A - C\|^2 \right)$$

$$= \frac{1}{16} \left(\|B\|^2 + \|C - A\|^2 \right)$$

$$= \left\| \frac{B + C - A}{4} \right\|^2.$$

Consequently,

$$\left\| \frac{A + B - C}{4} \right\| = \left\| \frac{A + C - B}{4} \right\| = \left\| \frac{B + C - A}{4} \right\|.$$

Therefore, the points

$$\frac{A}{2}, \frac{B}{2}, \frac{C}{2}, \frac{A + B}{2}, \frac{B + C}{2}, \text{ and } \frac{B + C}{2}$$

are on the circle centered at $\dfrac{A + B + C}{4}$ of radius $r = \dfrac{\|A + B - C\|}{4}$. Since,

$$\left\| \frac{B + C}{2} - \frac{A}{2} \right\| = 2r,$$

points $\dfrac{B+C}{2}$ and $\dfrac{A}{2}$ are on a diameter of that circle. If D is the point where the altitude from A meets the opposite side of the triangle, then we have

$$\left(\frac{B+C}{2}-D\right)\cdot\left(\frac{A}{2}-D\right)=0.$$

But this means that D is on the circle, by Problem 10.1.1. By a similar argument we can show that the points where the altitudes from B and C meet the opposite sides of the triangle are on the circle.   □

Problem 10.2.9. *Prove that the orthocenter H, the center N of the nine-point circle (see Problem 10.2.8), and the circumcenter Z are on the same line. Moreover, show that N is the midpoint of the segment with endpoints H and Z and the radius of the circumcircle is twice the radius of the nine-point circle.*

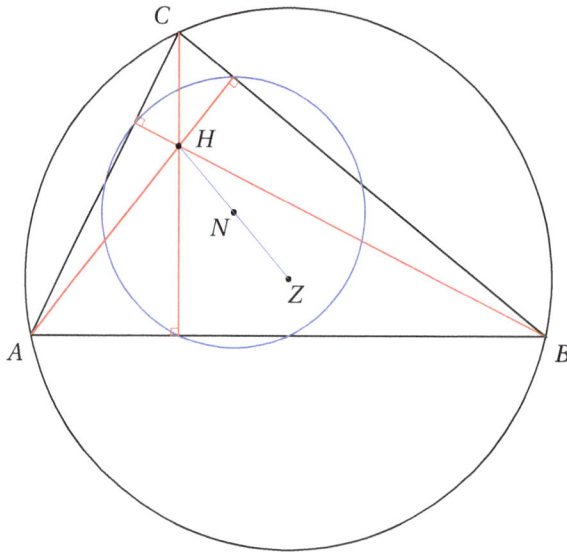

Figure 10.12: Problem 10.2.9.

Solution. We place H at the origin. Since

$$N=\frac{1}{4}(A+B+C),$$

the line which passes through H and N is

$$\text{Span}\{A+B+C\}.$$

By the hypothesis, we have $A \cdot (B - C) = 0$ and thus

$$\left(2N - \frac{1}{2}(B + C)\right) \cdot (B - C) = 0.$$

Similarly,

$$\left(2N - \frac{1}{2}(A + C)\right) \cdot (A - C) = 0$$

and

$$\left(2N - \frac{1}{2}(A + B)\right) \cdot (A - B) = 0,$$

which means that $2N = Z$ is the circumcenter. Now to finish the proof it suffices to observe that the square of the radius of the circumcircle is

$$\|Z - A\|^2 = \left\|\frac{1}{2}(A + B + C) - A\right\|^2 = 4\left\|\frac{1}{4}(B + C - A)\right\|^2.$$

$\square$

10.3 Geometry and trigonometry

In this section we assume basic knowledge of trigonometry. Properties of trigonometric functions used here were derived in Section 9.1.

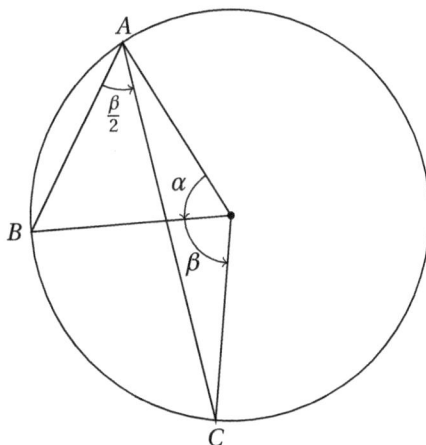

Figure 10.13: Problem 10.3.1

Problem 10.3.1. *Show that the inscribed angle equals half of the central angle.*

Solution. If we take the origin in the center of the circle, then

$$B = \cos\alpha\, A + \sin\alpha\, A^{\perp} \quad \text{and} \quad C = \cos(\alpha+\beta)A + \sin(\alpha+\beta)A^{\perp},$$

where $\alpha > 0$, $\beta > 0$, and $\alpha + \beta < 2\pi$. Then

$$B - A = (\cos\alpha - 1)A + \sin\alpha\, A^{\perp}$$
$$= -2\sin^2\frac{\alpha}{2}A + 2\sin\frac{\alpha}{2}\cos\frac{\alpha}{2}A^{\perp}$$
$$= 2\sin\frac{\alpha}{2}\left(\cos\left(\frac{\pi}{2}+\frac{\alpha}{2}\right)A + \sin\left(\frac{\pi}{2}+\frac{\alpha}{2}\right)A^{\perp}\right)$$

and

$$C - A = (\cos(\alpha+\beta) - 1)A + \sin(\alpha+\beta)A^{\perp}$$
$$= 2\sin\frac{\alpha+\beta}{2}\left(-\sin\frac{\alpha+\beta}{2}A + \cos\frac{\alpha+\beta}{2}A^{\perp}\right)$$
$$= 2\sin\frac{\alpha+\beta}{2}\left(\cos\left(\frac{\pi}{2}+\frac{\alpha}{2}+\frac{\beta}{2}\right)A + \sin\left(\frac{\pi}{2}+\frac{\alpha}{2}+\frac{\beta}{2}\right)A^{\perp}\right).$$

Note that $0 < \dfrac{\alpha}{2} < \pi$ and $0 < \dfrac{\alpha+\beta}{2} < \pi$. Hence $\sin\dfrac{\alpha}{2} > 0$ and $\sin\dfrac{\alpha+\beta}{2} > 0$ and

$$\angle(B-A, C-A) = \frac{\beta}{2}.$$

$\square$

Problem 10.3.2. *Show that in a triangle with vertices A, B, and C we have*

$$a = 2R\sin\alpha,$$

where $a = \|B - C\|$, R *is the radius of the circumcircle, and* $0 < \alpha = \angle(B-A, C-A) < \pi$.

Solution. If we take the origin in the center of the circle, then

$$C = \cos 2\alpha\, B + \sin 2\alpha\, B^{\perp},$$

by Problem 10.3.1. This yields

$$C \cdot B = \cos 2\alpha\, B \cdot B = R^2 \cos 2\alpha.$$

Now, since

$$a^2 = \|B - C\|^2$$
$$= \|B\|^2 + \|C\|^2 - 2B \cdot C$$

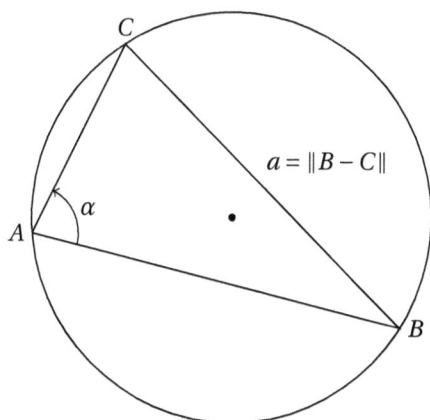

Figure 10.14: Problem 10.3.2

$$= 2R^2 - 2B \cdot C,$$

we have

$$a^2 = 2R^2 - 2R^2 \cos 2\alpha = 2R^2(1 - \cos 2\alpha) = 4R^2 (\sin \alpha)^2.$$

$\square$

Problem 10.3.3. *Show that, if*

$$\angle(B - A, C - A) = \angle(B - D, C - D),$$

then the points A, B, C, and D are on the same circle.

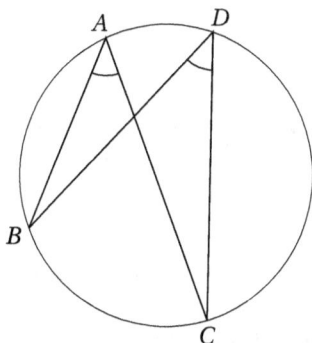

Figure 10.15: Problem 10.3.3.

Solution. Suppose that

$$\angle(B - A, C - A) = \angle(B - D, C - D) = \gamma.$$

Then

$$\frac{C - D}{\|C - D\|} = \cos\gamma \, \frac{B - D}{\|B - D\|} + \sin\gamma \, \frac{(B - D)^{\perp}}{\|B - D\|}. \tag{10.2}$$

Placing the origin at the center of the circumcircle of the triangle with vertices A, B, and C, we obtain

$$C = \cos 2\gamma \, B + \sin 2\gamma \, B^{\perp}. \tag{10.3}$$

Let $\|C - D\| = c$ and $\|B - D\| = b$. Using 10.2 and 10.3, we get

$$\cos 2\gamma \, \frac{B}{c} + \sin 2\gamma \, \frac{B^{\perp}}{c} - \cos\gamma \, \frac{B}{b} - \sin\gamma \, \frac{B^{\perp}}{b} = \frac{D}{c} - \cos\gamma \, \frac{D}{b} - \sin\gamma \, \frac{D^{\perp}}{b}$$

and, by calculating the square of the norm of both sides,

$$\|B\|^2 \left(\left(\frac{\cos 2\gamma}{c} - \frac{\cos\gamma}{b} \right)^2 + \left(\frac{\sin 2\gamma}{c} - \frac{\sin\gamma}{b} \right)^2 \right)$$

$$= \|D\|^2 \left(\left(\frac{1}{c} - \frac{\cos\gamma}{b} \right)^2 + \left(\frac{\sin\gamma}{b} \right)^2 \right).$$

Since

$$\left(\frac{\cos 2\gamma}{c} - \frac{\cos\gamma}{b} \right)^2 + \left(\frac{\sin 2\gamma}{c} - \frac{\sin\gamma}{b} \right)^2 = \frac{1}{c^2} - 2\frac{\cos\gamma}{bc} + \frac{1}{b^2}$$

$$= \left(\frac{1}{c} - \frac{\cos\gamma}{b} \right)^2 + \left(\frac{\sin\gamma}{b} \right)^2,$$

we conclude

$$\|D\| = \|B\|.$$

$\square$

10.4 Geometry problems from the International Mathematical Olympiads

In this section we present solutions to geometry problems from the International Mathematical Olympiads that were held in 2007, 2008, 2009, and 2010. We would like to point out that, while the solutions are not trivial, they are strightforward. They do not require any special tricks.

Problem 10.4.1 (IMO 2007 Vietnam). *In triangle ABC the bisector of angle BCA intersects the circumcircle again at R, the perpendicular bisector of BC at P, and the perpendicular bisector of AC at Q. The midpoint of BC is K and the midpoint of AC is L. Prove that the triangles RPK and RQL have the same area.*

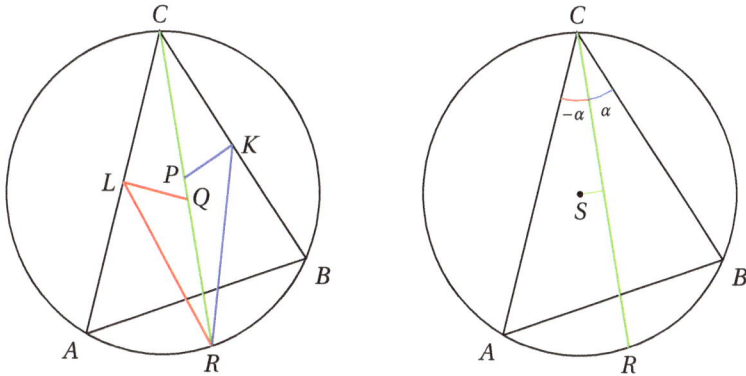

Figure 10.16: IMO 2007 Vietnam.

Solution. If we place the origin at C, then the circumcenter is

$$S = \frac{1}{2}R + sR^{\perp},$$

for some real number s.

Since CR is the bisector of the angle BCA, we have

$$B = t\left(\cos\alpha\, R + \sin\alpha\, R^{\perp}\right)$$

and

$$A = u\left(\cos(-\alpha)R + \sin(-\alpha)R^{\perp}\right),$$

for some t, u, and α. From

$$B - S = \left(t\cos\alpha - \frac{1}{2}\right)R + (t\sin\alpha - s)R^{\perp},$$

$$C - S = 0 - S = -\frac{1}{2}R - sR^{\perp},$$

and

$$\|B - S\|^2 = \|C - S\|^2,$$

we obtain

$$\left(t\cos\alpha - \frac{1}{2}\right)^2 + (t\sin\alpha - s)^2 = \left(\frac{1}{2}\right)^2 + s^2$$

and then

$$t^2 - t\cos\alpha - 2ts\sin\alpha = 0.$$

Since $t \neq 0$, we must have $t = \cos\alpha + 2s\sin\alpha$.

Now we note that point P is given by the equation

$$K + x\left(\cos\alpha R + \sin\alpha R^{\perp}\right)^{\perp} = yR.$$

Hence

$$\left(K + x\left(\cos\alpha R + \sin\alpha R^{\llcorner}\right)^{\llcorner}\right) \cdot \left(\cos\alpha R + \sin\alpha R^{\llcorner}\right) = yR \cdot \left(\cos\alpha R + \sin\alpha R^{\llcorner}\right),$$

which, in view of the equalities

$$\left(\cos\alpha R + \sin\alpha R^{\llcorner}\right)^{\llcorner} \cdot \left(\cos\alpha R + \sin\alpha R^{\llcorner}\right) = 0$$

and

$$\left(\cos\alpha R + \sin\alpha R^{\llcorner}\right) \cdot \left(\cos\alpha R + \sin\alpha R^{\llcorner}\right) = \|R\|^2,$$

simplifies to

$$\frac{1}{2}(\cos\alpha + 2s\sin\alpha) = y\cos\alpha$$

or

$$\frac{t}{2} = y\cos\alpha.$$

The area of the triangle RPK is

$$\frac{1}{2}\left|(K-R)\cdot(P-R)^{\llcorner}\right| = \frac{1}{2}\left|\left(\frac{1}{2}B - R\right)\cdot\left(\frac{t}{2\cos\alpha}R - R\right)^{\llcorner}\right|$$

$$= \frac{1}{2}\left|\left(\frac{t}{2}\left(\cos\alpha\, R + \sin\alpha\, R^{\llcorner}\right) - R\right)\cdot\left(\frac{t}{2\cos\alpha} - 1\right)R^{\llcorner}\right|$$

$$= \frac{1}{4}\left|t\sin\alpha\left(\frac{t}{2\cos\alpha} - 1\right)\right|\|R\|^2$$

$$= \frac{1}{4}\left|\frac{\sin\alpha}{\cos\alpha}\left(t^2 - 2t\cos\alpha\right)\right|\|R\|^2$$

$$= \frac{1}{4}\left|\frac{\sin\alpha}{\cos\alpha}\left(4s^2(\sin\alpha)^2 - (\cos\alpha)^2\right)\right|\|R\|^2.$$

Now we note that the value of this expression depends only on α, s, and R and does not change if we replace α with $-\alpha$. Thus we can conclude that the triangles RPK and RQL have the same area, because the area of the triangle RQL is obtained by replacing α with $-\alpha$ in the expression for the area of the triangle RPK.   □

Problem 10.4.2 (IMO 2008 Spain). *An acute-angled triangle ABC has ortho-center H. The circle passing through H with the center at the midpoint of BC intersects the line BC at A_1 and A_2. Similarly, the circle passing through H with the center at the midpoint of CA intersects the line CA at B_1 and B_2, and the circle passing through H with the center at the midpoint of AB intersects the line AB at C_1 and C_2. Show that A_1, A_2, B_1, B_2, C_1, C_2 lie on a circle.*

Solution. We place the origin at H. Since

$$A\cdot(B-C) = B\cdot(A-C) = C\cdot(A-B) = 0,$$

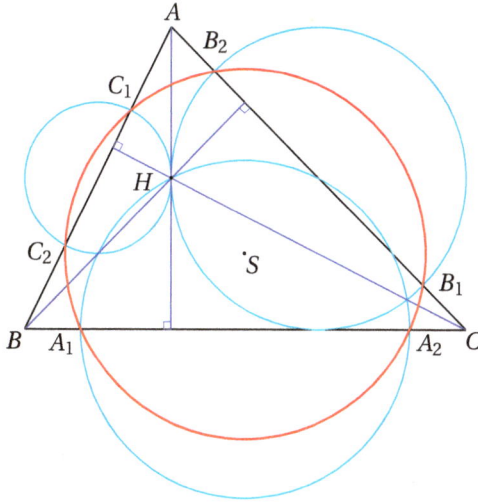

Figure 10.17: IMO 2008 Spain.

and, consequently,

$$A \cdot B = B \cdot C = A \cdot C, \tag{10.4}$$

it is easy to verify that the center of the circumcircle is $S = \frac{1}{2}(A + B + C)$.

Points A_1 and A_2 are on the line $B + \mathrm{Span}\{C - B\}$ and satisfy the equation

$$\left\| B + t(C - B) - \frac{B + C}{2} \right\|^2 = \left\| \frac{B + C}{2} \right\|^2,$$

which is equivalent to

$$t^2 \|B - C\|^2 - t\|B - C\|^2 - B \cdot C = 0.$$

Hence

$$t = \frac{1}{2}\left(1 \pm \frac{\|B + C\|}{\|B - C\|}\right).$$

Thus points A_1 and A_2 are

$$\frac{B + C}{2} \pm \frac{1}{2} \frac{\|B + C\|}{\|B - C\|}(C - B).$$

The square of the distance between the circumcenter S and the points A_1 and A_2 is

$$\left\| \frac{A + B + C}{2} - \left(\frac{B + C}{2} \pm \frac{1}{2} \frac{\|B + C\|}{\|B - C\|}(B - C) \right) \right\|^2 = \left\| \frac{A}{2} \pm \frac{1}{2} \frac{\|B + C\|}{\|B - C\|}(B - C) \right\|^2$$

$$= \frac{1}{4}\left(\|A\|^2 + \|B + C\|^2 \right)$$

Similarly, the square of the distance between the circumcenter and the points B_1 and B_2 is $\frac{1}{4}(\|B\|^2 + \|A+C\|^2)$ and the square of the distance between the circumcenter and the points C_1 and C_2 is $\frac{1}{4}(\|C\|^2 + \|A+B\|^2)$.

Now, from (10.4), we get

$$\|A\|^2 + \|B+C\|^2 = \|B\|^2 + \|A+C\|^2 = \|C\|^2 + \|A+B\|^2,$$

which proves that the points A_1, A_2, B_1, B_2, C_1, and C_2 lie on a circle centered at S. □

Problem 10.4.3 (IMO 2009 Germany). *Let ABC be a triangle with circumcenter S. The points P and Q are interior points on the sides CA and AB, respectively. Let K, L and M be the midpoints of the segments BP, CQ and PQ, respectively, and let Γ be the circle passing through K, L, and M. Suppose that the line PQ is tangent to the circle Γ. Prove that the line segments SP and SQ have the same length.*

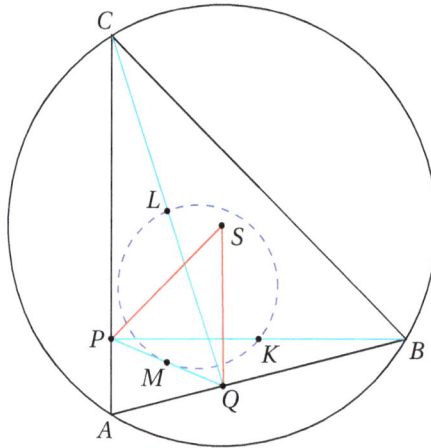

Figure 10.18: IMO 2009 Germany.

Solution. If we place the origin at A, we can express C as

$$C = \frac{c}{b}\left(uB + vB^\perp\right), \tag{10.5}$$

where $b = \|B\|$, $c = \|C\|$, $u = \cos\alpha$, $v = \sin\alpha$, and α is the angle between AB and AC. Moreover,

$$Q = sB, \quad \text{for some } 0 < s < 1,$$

$$P = t\frac{c}{b}\left(uB + vB^{\perp}\right), \text{ for some } 0 < t < 1,$$

$$K = \frac{B + P}{2},$$

$$L = \frac{C + Q}{2},$$

$$M = \frac{P + Q}{2}.$$

Let D be a point on the perpendicular bisector of the segment PQ, that is,

$$D = M + x(P - Q)^{\perp},$$

for some real number x. Note that D is the center of Γ and PQ is tangent to Γ if and only if

$$\|D - K\| = \|D - L\| = \|D - M\|.$$

The equality

$$\|D - K\| = \|D - M\|$$

is equivalent to

$$\left\|\frac{Q - B}{2} + x(P - Q)^{\perp}\right\|^2 = \left\|x(P - Q)^{\perp}\right\|^2.$$

When expressed in terms of B and $B^{\perp}$, the equality becomes

$$\left\|\frac{s - 1}{2}B + x\left(t\frac{c}{b}\left((u - s)B^{\perp} - vB\right)\right)\right\|^2 = \left\|x\left(t\frac{c}{b}\left((u - s)B^{\perp} - vB\right)\right)\right\|^2.$$

Hence

$$\frac{(s - 1)^2}{4}b^2 + 2\frac{s - 1}{2}B \cdot x\left(t\frac{c}{b}\left((u - s)B^{\perp} - vB\right)\right) = 0,$$

which simplifies to

$$\frac{(s - 1)^2}{4}b - xtc(s - 1)v = 0.$$

Solving for x we obtain

$$x = \frac{(s - 1)b}{4tcv}. \tag{10.6}$$

The equality

$$\|D - L\| = \|D - M\|$$

is equivalent to

$$\left\|\frac{P - C}{2} + x(P - Q)^{\perp}\right\|^2 = \left\|x(P - Q)^{\perp}\right\|^2,$$

which, when expressed in terms of B and $B^{\perp}$, becomes

$$\left\|\frac{t - 1}{2}\frac{c}{b}\left(uB + vB^{\perp}\right) + xt\frac{c}{b}\left((u - s)B^{\perp} - vB\right)\right\|^2 = \left\|xt\frac{c}{b}\left((u - s)B^{\perp} - vB\right)\right\|^2.$$

Consequently,

$$\frac{(t-1)^2}{4}c^2 - 2xscb\frac{t-1}{2}v = 0,$$

which gives us

$$x = \frac{(t-1)c}{4sbv}. \tag{10.7}$$

From (10.6) and (10.7) we obtain

$$t(t-1)c^2 = s(s-1)b^2. \tag{10.8}$$

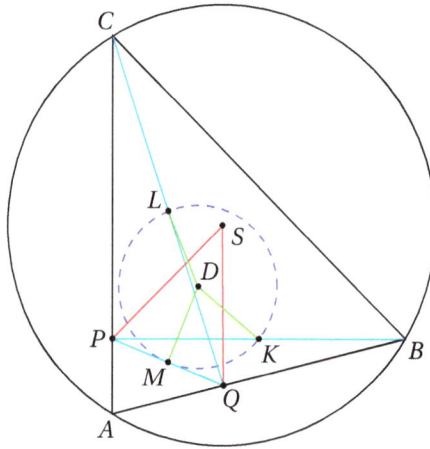

Figure 10.19: IMO 2009 Germany.

Now we show that (10.8) is equivalent to $\|S - P\| = \|S - Q\|$. Since the circumcenter S is at the intersection of the perpendicular bisectors of the segments AB and CA, we have

$$S = \frac{B}{2} + yB^\perp = \frac{C}{2} + zC^\perp,$$

for some real numbers y and z. Hence

$$\left(\frac{B}{2} + yB^\perp\right)\cdot C = \left(\frac{C}{2} + zC^\perp\right)\cdot C,$$

which, using (10.5), yields

$$\frac{1}{2}\frac{c}{b}ub^2 + y\frac{c}{b}vb^2 = \frac{c^2}{2}.$$

Solving for y we obtain

$$y = \frac{c - bu}{2bv}.$$

Consequently,

$$S = \frac{1}{2}B + \frac{c-bu}{2bv}B^{\perp},$$

which gives us

$$\|S - P\|^2 = \left(\left(\frac{1}{2} - t\frac{c}{b}u\right)^2 + \left(\frac{c-bu}{2bv} - t\frac{c}{b}v\right)^2\right)b^2$$

and

$$\|S - Q\|^2 = \left(\left(\frac{1}{2} - s\right)^2 + \left(\frac{c-bu}{2bv}\right)^2\right)b^2.$$

The equality

$$\|S - P\|^2 = \|S - Q\|^2$$

is thus equivalent to

$$\frac{1}{4} - s + s^2 + \left(\frac{c-bu}{2bv}\right)^2 = \frac{1}{4} - t\frac{c}{b}u + \left(t\frac{c}{b}u\right)^2 + \left(\frac{c-bu}{2bv}\right)^2 - 2t\frac{c-bu}{2bv}\frac{c}{b}v + \left(t\frac{c}{b}v\right)^2$$

or, after simplifying and using the fact that $u^2 + v^2 = 1$,

$$-s + s^2 = (-t + t^2)\frac{c^2}{b^2}.$$

But this is equivalent to (10.8), which completes the proof. □

Problem 10.4.4 (IMO 2010 Kazakhstan). *Let P be a point inside the triangle ABC. The lines AP, BP, and CP intersect the circumcircle Γ of triangle ABC again at the points K, L and M, respectively. The tangent to Γ at C intersects the line AB at S. Suppose that line segments SC and SP have the same length. Prove that the line segments MK and ML have the same length.*

Solution. If we place the origin at S, then

$$P = (\cos\beta)C + (\sin\beta)C^{\perp},$$

$$A = a\left((\cos\alpha)C + (\sin\alpha)C^{\perp}\right),$$

and

$$B = b\left((\cos\alpha)C + (\sin\alpha)C^{\perp}\right),$$

for some $a, b > 0$, where α is the angle between SC and SA and β is the angle between SC and SP.

From the Tangent-Secant Theorem (Problem 10.1.4) we have

$$\|C\|^2 = \|A\| \, \|B\| = a\|C\| b\|C\|.$$

Hence $ab = 1$ and

$$B = \frac{1}{a}\left((\cos\alpha)C + (\sin\alpha)C^{\perp}\right). \qquad (10.9)$$

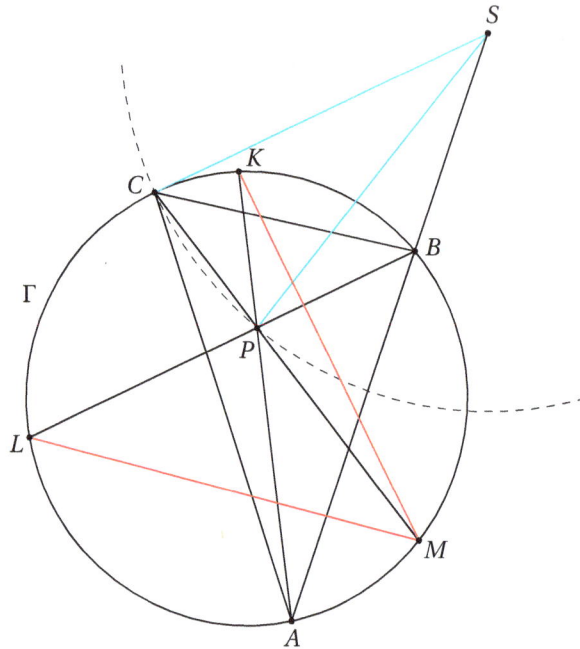

Figure 10.20: IMO 2010 Kazakhstan.

Note that

$$K = P + k(A - P), \quad L = P + l(B - P), \quad M = P + m(C - P),$$

for some $k, l, m < 0$. From the Chord Theorem (Problem 10.1.3) we have

$$\|C - P\| \, \|M - P\| = \|A - P\| \, \|K - P\| = \|B - P\| \, \|L - P\|.$$

Hence

$$m\|C - P\|^2 = k\|A - P\|^2 = l\|B - P\|^2$$

and

$$k = m\frac{\|C - P\|^2}{\|A - P\|^2} \quad \text{and} \quad l = m\frac{\|C - P\|^2}{\|B - P\|^2}.$$

Now

$$
\begin{aligned}
\|M - K\| &= \|m(C - P) - k(A - P)\| \\
&= \left\| m(C - P) - m\frac{\|C \ P\|^2}{\|A - P\|^2}(A - P) \right\| \\
&= |m| \left\| (C - P) - \frac{\|C - P\|^2}{\|A - P\|^2}(A - P) \right\| \\
&= |m| \left(\|C - P\|^2 + \frac{\|C - P\|^4}{\|A - P\|^2} - 2\frac{\|C - P\|^2}{\|A - P\|^2}(A - P) \cdot (C - P) \right)
\end{aligned}
$$

$$= |m| \|C - P\|^2 \left(1 + \frac{\|C - P\|^2}{\|A - P\|^2} - 2 \frac{1}{\|A - P\|^2} (A - P) \cdot (C - P) \right).$$

Similarly

$$\|M - L\| = |m| \|C - P\|^2 \left(1 + \frac{\|C - P\|^2}{\|B - P\|^2} - 2 \frac{1}{\|B - P\|^2} (B - P) \cdot (C - P) \right).$$

Therefore, to prove the equality

$$\|M - K\| = \|M - L\|,$$

it suffices to show that the value of

$$\frac{\|C - P\|^2}{\|A - P\|^2} - 2 \frac{1}{\|A - P\|^2} (A - P) \cdot (C - P) = \frac{\|(C - P)\|^2 - 2(A - P) \cdot (C - P)}{\|(A - P)\|^2}$$

does not change the value when we replace A by B.

We have

$$(A - P) \cdot (C - P) = ((a \cos \alpha - \cos \beta)(1 - \cos \beta) - (a \sin \alpha - \sin \beta) \sin \beta) \|C\|^2$$
$$= \left(a \cos(\alpha - \beta) + a \cos \alpha + (\cos \beta)^2 - \cos \beta + (\sin \beta)^2 \right) \|C\|^2$$
$$= ((a \cos(\alpha - \beta) + a \cos \alpha + 1 - \cos \beta)) \|C\|^2,$$

$$\|A - P\|^2 = \left((a \cos \alpha - \cos \beta)^2 + (a \sin \alpha - \sin \beta)^2 \right) \|C\|^2$$
$$= (a^2 + 1 - 2a \cos(\alpha - \beta)) \|C\|^2,$$

$$\|C - P\|^2 = \left((1 - \cos \beta)^2 + (\sin \beta)^2 \right) \|C\|^2$$
$$= (2 - 2 \cos \beta) \|C\|^2.$$

The above equalities yield

$$\|C - P\|^2 - 2(A - P) \cdot (C - P) = (a \cos(\alpha - \beta) + a \cos \alpha) \|C\|^2.$$

Similarly,

$$\|C - P\|^2 - 2(B - P) \cdot (C - P) = \left(\frac{1}{a} \cos(\alpha - \beta) + \frac{1}{a} \cos \alpha \right) \|C\|^2.$$

Consequently,

$$\frac{\|C - P\|^2 - 2(A - P) \cdot (C - P)}{\|A - P\|^2} = \frac{a \cos(\alpha - \beta) + a \cos \alpha}{a^2 + 1 - 2a \cos(\alpha - \beta)}$$
$$= \frac{\frac{1}{a} \cos(\alpha - \beta) + \frac{1}{a} \cos \alpha}{1 + \frac{1}{a^2} - 2 \frac{1}{a} \cos(\alpha - \beta)}$$
$$= \frac{\|C - P\|^2 - 2(B - P) \cdot (C - P)}{\|B - P\|^2},$$

which completes the proof. $\qquad \square$

Chapter 11

Problems for a computer algebra system

The following problems are intended to be solved with a computer algebra system like Maple, Mathematica, or Matlab. The purpose of these exercises is to give you an opportunity to gain some basic knowledge and some practice in using a computer algebra system to solve problems in linear algebra.

1. Determine the reduced row echelon form that the matrix $A = \begin{bmatrix} 2 & 3 & 5 & 1 \\ 4 & 7 & 9 & 2 \\ 8 & 15 & 17 & 4 \end{bmatrix}$.

2. Verify, using the reduced row echelon form, that the matrix $B = \begin{bmatrix} 3 & 1 & 2 \\ 8 & 3 & 4 \\ 5 & 7 & 2 \end{bmatrix}$

   is invertible and then solve the equation $BX = C$, where $C = \begin{bmatrix} 4 & 7 \\ 3 & 0 \\ 7 & 9 \end{bmatrix}$.

3. Calculate the inverse of the matrix $E = \begin{bmatrix} 2 & 1 & 5 & 2 \\ 5 & 8 & 7 & 4 \\ 1 & 5 & 7 & 3 \\ 4 & 3 & 3 & 2 \end{bmatrix}$.

4. Calculate the product $\begin{bmatrix} 1 & 4 & 0 \\ 0 & 1 & 0 \\ 0 & -3 & 1 \end{bmatrix} \begin{bmatrix} 1 & 0 & 5 \\ 0 & 1 & 2 \\ 0 & 0 & 1 \end{bmatrix} \begin{bmatrix} 2 & 1 & 3 \\ 1 & 1 & 4 \\ 7 & 2 & 5 \end{bmatrix}$.

5. Determine the projection of the vector $\mathbf{b} = \begin{bmatrix} 7 \\ 21 \\ 14 \end{bmatrix}$ on the vector plane Span$\{\mathbf{u}, \mathbf{v}\}$,

where $\mathbf{u} = \begin{bmatrix} 2 \\ 3 \\ 1 \end{bmatrix}$ and $\mathbf{v} = \begin{bmatrix} 4 \\ 1 \\ 3 \end{bmatrix}$.

6. Find numbers x and y that minimize

$$(2 - 2x - y)^2 + (1 - x - 3y)^2 + (1 - 5x - 2y)^2 = \left\| \mathbf{c} - F \begin{bmatrix} x \\ y \end{bmatrix} \right\|^2,$$

where $\mathbf{c} = \begin{bmatrix} 2 \\ 1 \\ 1 \end{bmatrix}$ and $F = \begin{bmatrix} 2 & 1 \\ 1 & 3 \\ 5 & 2 \end{bmatrix}$.

7. Calculate the determinant of the matrix $G = \begin{bmatrix} 7 & 2 & x \\ y & 2 & 5 \\ 4 & -3 & 8 \end{bmatrix}$.

8. Find the eigenvalues and eigenvectors of the matrix $H = \begin{bmatrix} 2 & 1 & 2 \\ 2 & 3 & 4 \\ 4 & 2 & 10 \end{bmatrix}$.

9. Find a symmetric matrix which has eigenvalues x, y, and z such that the vector

$\mathbf{p} = \begin{bmatrix} 3 \\ 1 \\ 4 \end{bmatrix}$ is an eigenvector corresponding to the eigenvalue x and the vector

$\mathbf{q} = \begin{bmatrix} 1 \\ -11 \\ 2 \end{bmatrix}$ is an eigenvector corresponding to the eigenvalue y.

10. Find the QR-decomposition of the matrix $K = \begin{bmatrix} 1 & 1 & 1 \\ 2 & -1 & 5 \\ 1 & 2 & -3 \end{bmatrix}$.

Chapter 12

Answers to selected exercises

Section 1.1

1 2

3 $\begin{bmatrix} 1 & 16 \end{bmatrix}$

5 $\begin{bmatrix} 12 \\ -18 \end{bmatrix}$

7 $\begin{bmatrix} 13 & 13 \\ 5 & -4 \end{bmatrix}$

9 $\begin{bmatrix} 13 & 1 \\ 65 & -4 \end{bmatrix}$

11 $\begin{bmatrix} 7p-2r & 7q-2s \\ 5p+3r & 5q+3s \end{bmatrix}$

13 $\begin{bmatrix} 7p+5q & -2p+3q \\ 7r+5s & -2r+3s \end{bmatrix}$

Section 1.2

1 $\begin{bmatrix} 4 & 5 \\ 3 & 8 \end{bmatrix}$

3 $\begin{bmatrix} 3 & 8 \\ 13 & 29 \end{bmatrix}$

5 $\begin{bmatrix} 19 & -35 \\ 5 & -9 \end{bmatrix}$

7 $\begin{bmatrix} 4 & 11 \\ 2 & 4 \end{bmatrix}$

9 $\begin{bmatrix} 8w+y & 8x+z \\ w & x \end{bmatrix}$

11 $\begin{bmatrix} -25w+15y & -25x+15z \\ -2w+y & -2x+z \end{bmatrix}$

13 $\begin{bmatrix} 1 & -8 \\ 0 & 1 \end{bmatrix}\begin{bmatrix} \frac{1}{9} & 0 \\ 0 & 1 \end{bmatrix} = \begin{bmatrix} \frac{1}{9} & -8 \\ 0 & 1 \end{bmatrix}$

15 $\begin{bmatrix} \frac{1}{3} & 0 \\ 0 & 1 \end{bmatrix}\begin{bmatrix} 1 & 0 \\ -7 & 1 \end{bmatrix}\begin{bmatrix} 1 & 1 \\ 0 & 1 \end{bmatrix}\begin{bmatrix} 1 & 0 \\ 0 & \frac{1}{2} \end{bmatrix} =$

$\begin{bmatrix} \frac{1}{3} & \frac{1}{6} \\ -7 & -3 \end{bmatrix}$

17 $\begin{bmatrix} 1 & -4 \\ 0 & 1 \end{bmatrix}$

19 $\begin{bmatrix} 1 & -\frac{2}{5} \\ 0 & \frac{1}{5} \end{bmatrix}$

21 $\begin{bmatrix} 1 & 0 \\ 0 & \frac{1}{2} \end{bmatrix}\begin{bmatrix} 1 & -2 \\ 0 & 1 \end{bmatrix}\begin{bmatrix} 1 & 0 \\ -5 & 1 \end{bmatrix} = \begin{bmatrix} 11 & -2 \\ -\frac{5}{2} & \frac{1}{2} \end{bmatrix}$

23 $\begin{bmatrix} 1 & 4 \\ 0 & 1 \end{bmatrix}\begin{bmatrix} \frac{1}{3} & 0 \\ 0 & 1 \end{bmatrix}\begin{bmatrix} 1 & 0 \\ 0 & \frac{1}{8} \end{bmatrix}\begin{bmatrix} 0 & 1 \\ 1 & 0 \end{bmatrix} = \begin{bmatrix} \frac{1}{2} & \frac{1}{3} \\ \frac{1}{8} & 0 \end{bmatrix}$

25 $\begin{bmatrix} 1 & 5 \\ 0 & 1 \end{bmatrix}\begin{bmatrix} 1 & 0 \\ 7 & 1 \end{bmatrix}$

27 $\begin{bmatrix} 1 & 0 \\ 5 & 1 \end{bmatrix}\begin{bmatrix} 1 & 0 \\ 0 & 2 \end{bmatrix}\begin{bmatrix} 1 & -2 \\ 0 & 1 \end{bmatrix}\begin{bmatrix} 3 & 0 \\ 0 & 1 \end{bmatrix}$

29 $\begin{bmatrix} 1 & 0 \\ 1 & 1 \end{bmatrix}\begin{bmatrix} 1 & 1 \\ 0 & 1 \end{bmatrix}\begin{bmatrix} 2 & 0 \\ 0 & 1 \end{bmatrix}\begin{bmatrix} 1 & 0 \\ 0 & 4 \end{bmatrix}$

459

31 $\begin{bmatrix} a & b \\ 0 & 0 \end{bmatrix}\begin{bmatrix} w & x \\ y & z \end{bmatrix} = \begin{bmatrix} aw+by & ax+bz \\ 0 & 0 \end{bmatrix} \neq \begin{bmatrix} 1 & 0 \\ 0 & 1 \end{bmatrix}$

33 If the matrix AB is invertible, then the matrix B is invertible since $((AB)^{-1}A)B = \begin{bmatrix} 1 & 0 \\ 0 & 1 \end{bmatrix}$.

35 If the matrix A is not invertible and $a \neq 0$ or $c \neq 0$, then there is an invertible matrix $\begin{bmatrix} \alpha & \beta \\ \gamma & \delta \end{bmatrix}$ (a product of elementary matrices) such that $\begin{bmatrix} \alpha & \beta \\ \gamma & \delta \end{bmatrix} A = \begin{bmatrix} 1 & k \\ 0 & 0 \end{bmatrix}$, where k is a real number. Hence $A = \begin{bmatrix} \alpha & \beta \\ \gamma & \delta \end{bmatrix}^{-1} \begin{bmatrix} 1 & k \\ 0 & 0 \end{bmatrix}$.

37 $a = 20$, $\begin{bmatrix} 1 & 0 \\ -5 & 1 \end{bmatrix}\begin{bmatrix} \frac{1}{2} & 0 \\ 0 & 1 \end{bmatrix} = \begin{bmatrix} \frac{1}{2} & 0 \\ -\frac{5}{2} & 1 \end{bmatrix}$

39 $a = \frac{14}{3}$, $\begin{bmatrix} 0 & 1 \\ 1 & 0 \end{bmatrix}\begin{bmatrix} 1 & -2 \\ 0 & 1 \end{bmatrix}\begin{bmatrix} 1 & 0 \\ 0 & \frac{1}{3} \end{bmatrix} = \begin{bmatrix} 0 & \frac{1}{3} \\ 1 & -\frac{2}{3} \end{bmatrix}$

41 $x = \frac{17}{7}$, $y = \frac{15}{7}$

43 $x = -\frac{u}{2} + \frac{3v}{2}$, $y = u - 2v$

45 $\begin{bmatrix} 1 & 0 \\ \frac{3}{5} & 1 \end{bmatrix}\begin{bmatrix} 5 & 7 \\ 0 & \frac{5}{19} \end{bmatrix}$

47 $\begin{bmatrix} 1 & 0 \\ \frac{5}{3} & 1 \end{bmatrix}\begin{bmatrix} 3 & 2 \\ 0 & -\frac{1}{3} \end{bmatrix}$

49 $\begin{bmatrix} 1 & 0 \\ \frac{q}{p} & 1 \end{bmatrix}\begin{bmatrix} p & p \\ 0 & p-q \end{bmatrix}$

51 $D^{-1}C^{-1}B^{-1}A^{-1}ABCD = D^{-1}C^{-1}B^{-1}BCD = D^{-1}C^{-1}CD = D^{-1}D = \begin{bmatrix} 1 & 0 \\ 0 & 1 \end{bmatrix}$

Section 1.3

1 7

3 −11

5 −bc

7 0

9 $\frac{1}{13}\begin{bmatrix} 5 & -2 \\ -1 & 3 \end{bmatrix}$

11 $\frac{1}{2}\begin{bmatrix} -4 & 2 \\ 3 & -1 \end{bmatrix}$

13 If $a \neq b$ and $a \neq -b$, then the inverse is $\frac{1}{a^2-b^2}\begin{bmatrix} a & -b \\ -b & a \end{bmatrix}$. If $a = b$ or $a = -b$, then the matrix is not invertible.

15 $\frac{1}{a^2+10}\begin{bmatrix} 5 & -a \\ a & 2 \end{bmatrix}$

17 If $3b - 2a \neq 0$, then the inverse is $\begin{bmatrix} \frac{b}{3b-2a} & -\frac{a}{3b-2a} \\ -\frac{2}{3b-2a} & \frac{3}{3b-2a} \end{bmatrix}$. If $3b - 2a = 0$, then the matrix is not invertible.

19 $x = -3$, $y = \frac{5}{2}$

21 $x = \frac{8}{31}$, $y = \frac{28}{31}$

23 $\det\begin{bmatrix} a_1 + ta_2 & b_1 + tb_2 \\ a_2 & b_2 \end{bmatrix} = (a_1 + ta_2)b_2 - (b_1 + tb_2)a_2 = a_1b_2 - b_1a_2 = \det\begin{bmatrix} a_1 & b_1 \\ a_2 & b_2 \end{bmatrix}$

25 $x = y = \frac{1}{5}$

27 $x = \frac{2}{47}$, $y = -\frac{7}{47}$

29 $x = \frac{a+5}{a^2-4a+10}$, $y = \frac{a-6}{a^2-4a+10}$

31 If $2a - 15 \neq 0$, then $x = \frac{-5s+at}{2a-15}$ and $y = \frac{2s-3t}{2a-15}$. If $2a - 15 = 0$ or $a = \frac{15}{2}$, then we have

$$\begin{cases} 2x + 5y = \frac{2s}{3} \\ 2x + 5y = \ t \end{cases}.$$

If $\frac{2s}{3} \neq t$, then there is no solution. If $\frac{2s}{3} = t$, then x is arbitrary and $y = \frac{t-2x}{5}$.

33 If $ab - 4 \neq 0$, then $x = \frac{-2s+at}{ab-4}$ and $y = \frac{bs-2t}{ab-4}$. If $ab - 4 = 0$ or $b = \frac{4}{a}$, then we have

$$\begin{cases} 2x + ay = \ s \\ 2x + ay = \frac{ta}{2} \end{cases}.$$

If $s \neq \frac{ta}{2}$, there is no solution. If $s = \frac{ta}{2}$, then x is arbitrary and $y = \frac{s-2x}{a}$.

Section 1.4

1 $\begin{bmatrix} 2 \\ -1 \end{bmatrix}$

3 $\begin{bmatrix} 1 \\ 4 \end{bmatrix}$

5 6 and 2

7 13 and 2

9 7 and 0

11 10 and −6

13 1 and $a - b$

15 3 and $15a + 3$

17 $\begin{bmatrix} 4 & 4 \\ -1 & -1 \end{bmatrix}$

19 $\frac{1}{7} \begin{bmatrix} 8s - t & -2s + 2t \\ 4s - 4t & -s + 8t \end{bmatrix}$

21 From $\det \begin{bmatrix} a-1 & b \\ 5 & 2 \end{bmatrix} = 0$ and $\det \begin{bmatrix} a-2 & b \\ 5 & 1 \end{bmatrix} = 0$ we get $a = 0$ and $b = -\frac{2}{5}$.

23 $A = PDP^{-1}$, where $P = \begin{bmatrix} 1 & 3 \\ 1 & -5 \end{bmatrix}$ and $D = \begin{bmatrix} 12 & 0 \\ 0 & 4 \end{bmatrix}$,

25 $A = PDP^{-1}$ where, $P = \begin{bmatrix} 1 & 4 \\ 3 & -1 \end{bmatrix}$ and $D = \begin{bmatrix} 14 & 0 \\ 0 & 1 \end{bmatrix}$.

27 $A = P_1 DP_1^{-1}$, where $P_1 = \begin{bmatrix} 1 & 1 \\ -1 & 8 \end{bmatrix}$ and $D = \begin{bmatrix} 1 & 0 \\ 0 & 10 \end{bmatrix}$.

$A = P_2 DP_2^{-1}$, where $P_2 = \begin{bmatrix} -1 & 2 \\ 1 & 16 \end{bmatrix}$ and $D = \begin{bmatrix} 1 & 0 \\ 0 & 10 \end{bmatrix}$.

29 Since $A = PDP^{-1}$, where $P = \begin{bmatrix} 2 & 1 \\ -1 & 4 \end{bmatrix}$ and $D = \begin{bmatrix} 2 & 0 \\ 0 & 11 \end{bmatrix}$, we have

$$A^n = PD^n P^{-1} = P \begin{bmatrix} 2^n & 0 \\ 0 & 11^n \end{bmatrix} P^{-1} = \frac{1}{9} \begin{bmatrix} 8 \cdot 2^n + 11^n & -2 \cdot 2^n + 2 \cdot 11^n \\ -4 \cdot 2^n + 4 \cdot 11^n & 2^n + 8 \cdot 11^n \end{bmatrix}.$$

31 Since $\begin{bmatrix} x_n \\ y_n \end{bmatrix} = A^n \begin{bmatrix} 1 \\ 1 \end{bmatrix}$, where $A = \begin{bmatrix} 3 & 7 \\ 1 & 9 \end{bmatrix}$, and $A = PDP^{-1}$, where $P = \begin{bmatrix} 1 & 7 \\ 1 & -1 \end{bmatrix}$ and $D = \begin{bmatrix} 10 & 0 \\ 0 & 2 \end{bmatrix}$, we have

$$A^n = PD^n P^{-1} = P \begin{bmatrix} 10^n & 0 \\ 0 & 2^n \end{bmatrix} P^{-1} = \frac{1}{8} \begin{bmatrix} 10^n + 7 \cdot 2^n & 7 \cdot 10^n - 7 \cdot 2^n \\ 10^n - 2^n & 7 \cdot 10^n + 2^n \end{bmatrix}.$$

Consequently $\begin{bmatrix} x_{33} \\ y_{33} \end{bmatrix} = A^{33} \begin{bmatrix} 1 \\ 1 \end{bmatrix} = \begin{bmatrix} 10^{33} \\ 10^{33} \end{bmatrix}$.

33 Since $\begin{bmatrix} x_n \\ y_n \end{bmatrix} = A^n \begin{bmatrix} 2 \\ 3 \end{bmatrix}$, where $A = \begin{bmatrix} \frac{3}{5} & \frac{1}{4} \\ \frac{2}{5} & \frac{3}{4} \end{bmatrix}$, and $A = PDP^{-1}$, where $P = \begin{bmatrix} 5 & 1 \\ 8 & -1 \end{bmatrix}$ and $D = \begin{bmatrix} 1 & 0 \\ 0 & \frac{7}{20} \end{bmatrix}$, we have $A^n = PD^n P^{-1} = P \begin{bmatrix} 1 & 0 \\ 0 & \left(\frac{7}{20}\right)^n \end{bmatrix} P^{-1}$. Consequently,

$$\lim_{n \to \infty} \begin{bmatrix} x_n \\ y_n \end{bmatrix} = \lim_{n \to \infty} A^n \begin{bmatrix} 2 \\ 3 \end{bmatrix} = P \begin{bmatrix} 1 & 0 \\ 0 & 0 \end{bmatrix} P^{-1} \begin{bmatrix} 2 \\ 3 \end{bmatrix} = \frac{1}{13} \begin{bmatrix} 5 & 5 \\ 8 & 8 \end{bmatrix} \begin{bmatrix} 2 \\ 3 \end{bmatrix} = \begin{bmatrix} \frac{25}{13} \\ \frac{40}{13} \end{bmatrix}.$$

Since $\begin{bmatrix} 2 \\ 3 \end{bmatrix} = \frac{5}{13} \begin{bmatrix} 5 \\ 8 \end{bmatrix} + \frac{1}{13} \begin{bmatrix} 1 \\ -1 \end{bmatrix}$, the solution is $\frac{5}{13} \begin{bmatrix} 5 \\ 8 \end{bmatrix} = \begin{bmatrix} \frac{25}{13} \\ \frac{40}{13} \end{bmatrix}$.

Section 2.1

1 $\begin{bmatrix} 10 & 0 & 1 & 13 \end{bmatrix}$

3 $\begin{bmatrix} 8 & 3 & -1 \\ 12 & 2 & -4 \end{bmatrix}$

5 $\begin{bmatrix} 15 & 10 & 5 \\ 5 & 20 & -5 \end{bmatrix}$

7 $\begin{bmatrix} 22 & -1 \\ 8 & 38 \\ 4 & -9 \end{bmatrix}$

9 $\begin{bmatrix} 2 & 5 \\ 4 & 20 \end{bmatrix}$

11 21

13 $\begin{bmatrix} 8 & 28 & 4 \\ 4 & 14 & 2 \\ 2 & 7 & 1 \\ 10 & 35 & 5 \end{bmatrix}$

15 $\begin{bmatrix} 2 & 1 & 2 \\ 1 & 1 & 2 \end{bmatrix}$

17 $\begin{bmatrix} 19 & 7 \\ 17 & 9 \end{bmatrix}$

19 $\begin{bmatrix} 15 & 20 & -5 & -5 \\ 3 & 4 & -1 & -1 \\ 6 & 8 & -2 & -2 \end{bmatrix}$

21 The first matrix has 4 columns and the second matrix has 3 rows.

23 From the assumptions we get $\begin{bmatrix} a_{11} & a_{12} \\ a_{21} & a_{22} \end{bmatrix} \begin{bmatrix} 1 & 0 \\ 0 & 1 \end{bmatrix} = \begin{bmatrix} b_{11} & b_{12} \\ b_{21} & b_{22} \end{bmatrix} \begin{bmatrix} 1 & 0 \\ 0 & 1 \end{bmatrix}$ and thus

$$\begin{bmatrix} a_{11} & a_{12} \\ a_{21} & a_{22} \end{bmatrix} = \begin{bmatrix} a_{11} & a_{12} \\ a_{21} & a_{22} \end{bmatrix} \begin{bmatrix} 1 & 0 \\ 0 & 1 \end{bmatrix} = \begin{bmatrix} b_{11} & b_{12} \\ b_{21} & b_{22} \end{bmatrix} \begin{bmatrix} 1 & 0 \\ 0 & 1 \end{bmatrix} = \begin{bmatrix} b_{11} & b_{12} \\ b_{21} & b_{22} \end{bmatrix}.$$

25 From the assumptions we get

$$\begin{bmatrix} a_{11} & a_{12} & a_{13} & a_{14} \end{bmatrix} \begin{bmatrix} 1 & 0 & 0 & 0 \\ 0 & 1 & 0 & 0 \\ 0 & 0 & 1 & 0 \\ 0 & 0 & 0 & 1 \end{bmatrix} = \begin{bmatrix} b_{11} & b_{12} & b_{13} & b_{14} \end{bmatrix} \begin{bmatrix} 1 & 0 & 0 & 0 \\ 0 & 1 & 0 & 0 \\ 0 & 0 & 1 & 0 \\ 0 & 0 & 0 & 1 \end{bmatrix}.$$

Consequently,

$$[a_{11} \ a_{12} \ a_{13} \ a_{14}] = [a_{11} \ a_{12} \ a_{13} \ a_{14}] \begin{bmatrix} 1 & 0 & 0 & 0 \\ 0 & 1 & 0 & 0 \\ 0 & 0 & 1 & 0 \\ 0 & 0 & 0 & 1 \end{bmatrix}$$

$$= [b_{11} \ b_{12} \ b_{13} \ b_{14}] \begin{bmatrix} 1 & 0 & 0 & 0 \\ 0 & 1 & 0 & 0 \\ 0 & 0 & 1 & 0 \\ 0 & 0 & 0 & 1 \end{bmatrix} = [b_{11} \ b_{12} \ b_{13} \ b_{14}]$$

27 $A^T = \begin{bmatrix} 1 \\ 2 \\ -3 \end{bmatrix}$
  29 $A^T = \begin{bmatrix} 1 & 3 \\ 2 & 4 \\ 5 & 2 \end{bmatrix}$

31

$$\left(\begin{bmatrix} a_1 & a_2 \\ a_3 & a_4 \end{bmatrix} \begin{bmatrix} b_1 & b_2 \\ b_3 & b_4 \end{bmatrix} \right)^T = \left(\begin{bmatrix} [a_1 \ a_2] \begin{bmatrix} b_1 \\ b_3 \end{bmatrix} & [a_1 \ a_2] \begin{bmatrix} b_2 \\ b_4 \end{bmatrix} \\ [a_3 \ a_4] \begin{bmatrix} b_1 \\ b_3 \end{bmatrix} & [a_3 \ a_4] \begin{bmatrix} b_2 \\ b_4 \end{bmatrix} \end{bmatrix} \right)^T .$$

$$= \begin{bmatrix} [a_1 \ a_2] \begin{bmatrix} b_1 \\ b_3 \end{bmatrix} & [a_3 \ a_4] \begin{bmatrix} b_1 \\ b_3 \end{bmatrix} \\ [a_1 \ a_2] \begin{bmatrix} b_2 \\ b_4 \end{bmatrix} & [a_3 \ a_4] \begin{bmatrix} b_2 \\ b_4 \end{bmatrix} \end{bmatrix}$$

$$= \begin{bmatrix} [b_1 \ b_3] \begin{bmatrix} a_1 \\ a_2 \end{bmatrix} & [b_1 \ b_3] \begin{bmatrix} a_3 \\ a_4 \end{bmatrix} \\ [b_2 \ b_4] \begin{bmatrix} a_1 \\ a_2 \end{bmatrix} & [b_2 \ b_4] \begin{bmatrix} a_3 \\ a_4 \end{bmatrix} \end{bmatrix}$$

$$= \begin{bmatrix} b_1 & b_3 \\ b_2 & b_4 \end{bmatrix} \begin{bmatrix} a_1 & a_3 \\ a_2 & a_4 \end{bmatrix}$$

$$= \begin{bmatrix} b_1 & b_2 \\ b_3 & b_4 \end{bmatrix}^T \begin{bmatrix} a_1 & a_2 \\ a_3 & a_4 \end{bmatrix}^T$$

33 $\det A = \det \begin{bmatrix} a_{11} & a_{12} \\ a_{21} & a_{22} \end{bmatrix} = a_{11}a_{22} - a_{12}a_{21} = \det \begin{bmatrix} a_{11} & a_{21} \\ a_{12} & a_{22} \end{bmatrix} = \det A^T$

35 $(AA^T)^T = (A^T)^T A^T = AA^T$, because $(A^T)^T = A$.

Section 2.2

1 $\begin{cases} 2x + y = 3 \\ 3x + 2y = 4 \end{cases}$
  5 $\begin{bmatrix} 1 & 0 & 0 \\ 0 & 1 & 0 \end{bmatrix}$

3 $\begin{bmatrix} 1 & \frac{7}{2} \\ 0 & 0 \end{bmatrix}$
  7 $\begin{bmatrix} 1 & 0 & \frac{1}{2} \\ 0 & 1 & -\frac{1}{2} \\ 0 & 0 & 0 \end{bmatrix}$

9 $\begin{bmatrix} 1 & 0 & 2 \\ 0 & 1 & -1 \\ 0 & 0 & 0 \\ 0 & 0 & 0 \end{bmatrix}$

17 $\begin{bmatrix} 1 & 0 & 1 & 0 & -p-3q \\ 0 & 1 & -2 & 0 & 3p \\ 0 & 0 & 0 & 1 & 2q \end{bmatrix}$

11 $\begin{bmatrix} 1 & 0 & 0 & -\frac{1}{3} \\ 0 & 1 & 0 & \frac{2}{3} \\ 0 & 0 & 1 & \frac{1}{3} \end{bmatrix}$

19 $\begin{bmatrix} 1 & 0 & 0 & \frac{3p}{2}+\frac{q}{2}-\frac{5r}{2} \\ 0 & 1 & 0 & -\frac{p}{2}+\frac{q}{2}+\frac{r}{2} \\ 0 & 0 & 1 & -\frac{p}{2}-\frac{q}{2}+\frac{3r}{2} \end{bmatrix}$

13 $\begin{bmatrix} 1 & 0 & 0 & -2 & 5 \\ 0 & 1 & 0 & -3 & 3 \\ 0 & 0 & 1 & 1 & \frac{2}{3} \end{bmatrix}$

15 $\begin{bmatrix} 1 & 0 & -2 & -p+2q \\ 0 & 1 & 3 & p-q \\ 0 & 0 & 0 & 0 \end{bmatrix}$

21 $\begin{bmatrix} 1 & a \\ 0 & 0 \\ 0 & 0 \\ 0 & 0 \end{bmatrix}, \begin{bmatrix} 0 & 1 \\ 0 & 0 \\ 0 & 0 \\ 0 & 0 \end{bmatrix}$

23 $\begin{bmatrix} 1 & 0 & a \\ 0 & 1 & b \\ 0 & 0 & 0 \end{bmatrix}, \begin{bmatrix} 1 & a & 0 \\ 0 & 0 & 1 \\ 0 & 0 & 0 \end{bmatrix}, \begin{bmatrix} 0 & 1 & 0 \\ 0 & 0 & 1 \\ 0 & 0 & 0 \end{bmatrix}$

25 $\begin{bmatrix} 1 & 0 & 0 & a \\ 0 & 1 & 0 & b \\ 0 & 0 & 1 & c \end{bmatrix}, \begin{bmatrix} 1 & 0 & a & 0 \\ 0 & 1 & b & 0 \\ 0 & 0 & 0 & 1 \end{bmatrix}, \begin{bmatrix} 1 & a & 0 & 0 \\ 0 & 0 & 1 & 0 \\ 0 & 0 & 0 & 1 \end{bmatrix}, \begin{bmatrix} 0 & 1 & 0 & 0 \\ 0 & 0 & 1 & 0 \\ 0 & 0 & 0 & 1 \end{bmatrix}$

27 $x = -1$ and $y = 2$

31 $x = \frac{9}{4}$, $y = \frac{5}{4}$, $z = -\frac{7}{4}$

29 $x = -2 + 3z$ and $y = 5 - 5z$

33 No solutions.

35 If $r = \frac{p}{4} + \frac{q}{4}$, then $x = \frac{3p}{8} - \frac{q}{8} - \frac{1}{2}z$ and $y = -\frac{p}{8} + \frac{3q}{8} - \frac{1}{2}z$. If $r \neq \frac{p}{4} + \frac{q}{4}$, then there are no solutions.

37 If $r = p + 2q$, then $x = \frac{p}{2} + \frac{q}{2} - 2z$, $y = \frac{p}{2} - \frac{q}{2} - z$, and z is arbitrary. If $r \neq p + 2q$, there are no solutions.

39 If $r = p + q$ and $s = 2p + q$, then $x = p - q + y$, $z = -p + 2q - 3y$, and y is arbitrary. If $r \neq p + q$ or $s \neq 2p + q$, then there are no solutions.

41 If $r = p - 2q$, then $x = 2p - 5q + 4z - 3w$, $y = -p + 3q - 3z + w$, z and w are arbitrary. If $r \neq p - 2q$, then there are no solutions.

Section 2.3

1 $\begin{bmatrix} a_1 & b_1 & c_1 & d_1 \\ a_2 + 4a_3 & b_2 + 4b_3 & c_2 + 4c_3 & d_2 + 4d_3 \\ a_3 & b_3 & c_3 & d_3 \end{bmatrix}$

5 $\begin{bmatrix} 0 \\ 1 \\ 0 \\ 0 \\ 0 \end{bmatrix}$

3 $\begin{bmatrix} a_1 & b_1 & c_1 & d_1 \\ a_2 - a_1 & b_2 - b_1 & c_2 - c_1 & d_2 - d_1 \\ a_3 + 7a_1 & b_3 + 7b_1 & c_3 + 7c_1 & d_3 + 7d_1 \end{bmatrix}$

7 $\begin{bmatrix} a_1 + 3a_2 & b_1 + 3b_2 \\ a_2 & b_2 \\ a_3 & b_3 \\ a_4 + 8a_2 & b_4 + 8b_2 \end{bmatrix}$

9 $A = \begin{bmatrix} 10 & 2 & -3 \\ 5 & 1 & -2 \\ 24 & 5 & -9 \end{bmatrix}$

15 $P = \begin{bmatrix} 0 & 1 & 0 \\ 0 & 0 & 1 \\ 1 & 0 & 0 \end{bmatrix}$

11 $P = \begin{bmatrix} 5 & 0 & 0 \\ 0 & 4 & 0 \\ 0 & 0 & 2 \end{bmatrix}$

17 $P = \begin{bmatrix} 0 & j & 1 & 0 \\ 0 & 1 & 0 & 0 \\ 0 & k & 0 & 0 \\ 0 & m & 0 & 1 \end{bmatrix}$

13 $P = \begin{bmatrix} 0 & 0 & 1 & 0 \\ 0 & 1 & 0 & 0 \\ 1 & 0 & 0 & 0 \\ 0 & 0 & 0 & 1 \end{bmatrix}$

19 $P = \begin{bmatrix} 1 & 0 & 0 & 3 & 0 \\ 0 & 1 & 0 & 1 & 0 \\ 0 & 0 & 1 & 5 & 0 \\ 0 & 0 & 0 & 1 & 0 \\ 0 & 0 & 0 & 7 & 1 \end{bmatrix}$

21 Since $\begin{bmatrix} 1 & 1 & 0 \\ 0 & 1 & 0 \\ 0 & 0 & 1 \end{bmatrix} \begin{bmatrix} 1 & 0 & 0 \\ 0 & \frac{1}{8} & 0 \\ 0 & 0 & 1 \end{bmatrix} \begin{bmatrix} 1 & 0 & 0 \\ -11 & 1 & 0 \\ 4 & 0 & 1 \end{bmatrix} \begin{bmatrix} 1 & 0 & 0 \\ 0 & 0 & 1 \\ 0 & 1 & 0 \end{bmatrix} \begin{bmatrix} -4 & 4 \\ 11 & -3 \\ 1 & -1 \end{bmatrix} = \begin{bmatrix} 1 & 0 \\ 0 & 1 \\ 0 & 0 \end{bmatrix}$,

we have $P = \begin{bmatrix} 1 & 1 & 0 \\ 0 & 1 & 0 \\ 0 & 0 & 1 \end{bmatrix} \begin{bmatrix} 1 & 0 & 0 \\ 0 & \frac{1}{8} & 0 \\ 0 & 0 & 1 \end{bmatrix} \begin{bmatrix} 1 & 0 & 0 \\ -11 & 1 & 0 \\ 4 & 0 & 1 \end{bmatrix} \begin{bmatrix} 1 & 0 & 0 \\ 0 & 0 & 1 \\ 0 & 1 & 0 \end{bmatrix} = \frac{1}{8} \begin{bmatrix} 0 & 1 & -3 \\ 0 & 1 & -11 \\ 8 & 0 & 32 \end{bmatrix}$

23 Since $\begin{bmatrix} 1 & -1 & 0 \\ 0 & 1 & 0 \\ 0 & 2 & 1 \end{bmatrix} \begin{bmatrix} 1 & 0 & 0 \\ 0 & \frac{1}{2} & 0 \\ 0 & 0 & 1 \end{bmatrix} \begin{bmatrix} 1 & 0 & 0 \\ 1 & 1 & 0 \\ -1 & 0 & 1 \end{bmatrix} \begin{bmatrix} 1 & 1 & 3 \\ -1 & 1 & 1 \\ 1 & -1 & -1 \end{bmatrix} = \begin{bmatrix} 1 & 0 & 1 \\ 0 & 1 & 2 \\ 0 & 0 & 0 \end{bmatrix}$,

we have $P = \begin{bmatrix} 1 & -1 & 0 \\ 0 & 1 & 0 \\ 0 & 2 & 1 \end{bmatrix} \begin{bmatrix} 1 & 0 & 0 \\ 0 & \frac{1}{2} & 0 \\ 0 & 0 & 1 \end{bmatrix} \begin{bmatrix} 1 & 0 & 0 \\ 1 & 1 & 0 \\ -1 & 0 & 1 \end{bmatrix} = \frac{1}{2} \begin{bmatrix} 1 & -1 & 0 \\ 1 & 1 & 0 \\ 0 & 2 & 2 \end{bmatrix}$

25 $A = \begin{bmatrix} 0 & 0 & 1 \\ 0 & 2 & 0 \\ 1 & 0 & 0 \end{bmatrix}$, $A^{-1} = \begin{bmatrix} 0 & 0 & 1 \\ 0 & 1 & 0 \\ 1 & 0 & 0 \end{bmatrix} \begin{bmatrix} 1 & 0 & 0 \\ 0 & \frac{1}{2} & 0 \\ 0 & 0 & 1 \end{bmatrix} = \begin{bmatrix} 0 & 0 & 1 \\ 0 & \frac{1}{2} & 0 \\ 1 & 0 & 0 \end{bmatrix}$,

$\begin{bmatrix} 0 & 0 & 1 \\ 0 & 2 & 0 \\ 1 & 0 & 0 \end{bmatrix} \begin{bmatrix} 0 & 0 & 1 \\ 0 & \frac{1}{2} & 0 \\ 1 & 0 & 0 \end{bmatrix} = \begin{bmatrix} 1 & 0 & 0 \\ 0 & 1 & 0 \\ 0 & 0 & 1 \end{bmatrix}$

27 $A = \begin{bmatrix} 1 & 4 & 0 \\ 0 & 7 & 2 \\ 0 & 1 & 0 \end{bmatrix}$, $A^{-1} = \begin{bmatrix} 1 & 0 & 0 \\ 0 & 0 & 1 \\ 0 & 1 & 0 \end{bmatrix} \begin{bmatrix} 1 & 0 & 0 \\ 0 & \frac{1}{2} & 0 \\ 0 & 0 & 1 \end{bmatrix} \begin{bmatrix} 1 & 0 & -4 \\ 0 & 1 & -7 \\ 0 & 0 & 1 \end{bmatrix} = \begin{bmatrix} 1 & 0 & -4 \\ 0 & 0 & 1 \\ 0 & \frac{1}{2} & -\frac{7}{2} \end{bmatrix}$, $\begin{bmatrix} 1 & 4 & 0 \\ 0 & 7 & 2 \\ 0 & 1 & 0 \end{bmatrix} \begin{bmatrix} 1 & 0 & -4 \\ 0 & 0 & 1 \\ 0 & \frac{1}{2} & -\frac{7}{2} \end{bmatrix} =$

$\begin{bmatrix} 1 & 0 & 0 \\ 0 & 1 & 0 \\ 0 & 0 & 1 \end{bmatrix}$

29 $A = \begin{bmatrix} 0 & 0 & 1 & 0 \\ 0 & 3 & 0 & 0 \\ 1 & 0 & 0 & 0 \\ 0 & 0 & 0 & 1 \end{bmatrix}$, $A^{-1} = \begin{bmatrix} 0 & 0 & 1 & 0 \\ 0 & 1 & 0 & 0 \\ 1 & 0 & 0 & 0 \\ 0 & 0 & 0 & 1 \end{bmatrix} \begin{bmatrix} 1 & 0 & 0 & 0 \\ 0 & \frac{1}{3} & 0 & 0 \\ 0 & 0 & 1 & 0 \\ 0 & 0 & 0 & 1 \end{bmatrix} = \begin{bmatrix} 0 & 0 & 1 & 0 \\ 0 & \frac{1}{3} & 0 & 0 \\ 1 & 0 & 0 & 0 \\ 0 & 0 & 0 & 1 \end{bmatrix}$,

$\begin{bmatrix} 0 & 0 & 1 & 0 \\ 0 & 3 & 0 & 0 \\ 1 & 0 & 0 & 0 \\ 0 & 0 & 0 & 1 \end{bmatrix} \begin{bmatrix} 0 & 0 & 1 & 0 \\ 0 & \frac{1}{3} & 0 & 0 \\ 1 & 0 & 0 & 0 \\ 0 & 0 & 0 & 1 \end{bmatrix} = \begin{bmatrix} 1 & 0 & 0 & 0 \\ 0 & 1 & 0 & 0 \\ 0 & 0 & 1 & 0 \\ 0 & 0 & 0 & 1 \end{bmatrix}$

31 $A = \begin{bmatrix} 1 & 0 & 0 & 3 \\ 0 & 1 & 0 & 5 \\ 0 & 0 & 0 & 1 \\ 0 & 0 & 1 & 2 \end{bmatrix}$, $A^{-1} = \begin{bmatrix} 1 & 0 & 0 & -3 \\ 0 & 1 & 0 & -5 \\ 0 & 0 & 1 & -2 \\ 0 & 0 & 0 & 1 \end{bmatrix} \begin{bmatrix} 1 & 0 & 0 & 0 \\ 0 & 1 & 0 & 0 \\ 0 & 0 & 0 & 1 \\ 0 & 0 & 1 & 0 \end{bmatrix} = \begin{bmatrix} 1 & 0 & -3 & 0 \\ 0 & 1 & -5 & 0 \\ 0 & 0 & -2 & 1 \\ 0 & 0 & 1 & 0 \end{bmatrix}$,

$\begin{bmatrix} 1 & 0 & 0 & 3 \\ 0 & 1 & 0 & 5 \\ 0 & 0 & 0 & 1 \\ 0 & 0 & 1 & 2 \end{bmatrix} \begin{bmatrix} 1 & 0 & -3 & 0 \\ 0 & 1 & -5 & 0 \\ 0 & 0 & -2 & 1 \\ 0 & 0 & 1 & 0 \end{bmatrix} = \begin{bmatrix} 1 & 0 & 0 & 0 \\ 0 & 1 & 0 & 0 \\ 0 & 0 & 1 & 0 \\ 0 & 0 & 0 & 1 \end{bmatrix}$

33 $\frac{1}{5}\begin{bmatrix} 5 & 3 & 2 \\ 0 & -1 & 1 \end{bmatrix}$

35 $\frac{1}{3}\begin{bmatrix} -1 & -2 \\ 1 & 2 \\ 2 & 1 \end{bmatrix}$

37 Since $\begin{bmatrix} 2 & 1 & 1 & 1 & 0 & 0 \\ 1 & 3 & 1 & 0 & 1 & 0 \\ -1 & 1 & 1 & 0 & 0 & 1 \end{bmatrix} \sim \begin{bmatrix} 1 & 0 & 0 & \frac{1}{3} & 0 & -\frac{1}{3} \\ 0 & 1 & 0 & -\frac{1}{3} & \frac{1}{2} & -\frac{1}{6} \\ 0 & 0 & 1 & \frac{2}{3} & -\frac{1}{2} & \frac{5}{6} \end{bmatrix}$,

we have $\begin{bmatrix} 2 & 1 & 1 \\ 1 & 3 & 1 \\ -1 & 1 & 1 \end{bmatrix}^{-1} = \begin{bmatrix} \frac{1}{3} & 0 & -\frac{1}{3} \\ -\frac{1}{3} & \frac{1}{2} & -\frac{1}{6} \\ \frac{2}{3} & -\frac{1}{2} & \frac{5}{6} \end{bmatrix}$.

39 $A^{-1} = \frac{1}{2}\begin{bmatrix} 3 & -1 & -1 \\ -2 & 0 & 2 \\ -2 & 2 & 0 \end{bmatrix}$

41 $A^{-1} = \frac{1}{8}\begin{bmatrix} -1 & -2 & 3 & 2 \\ 2 & 4 & 2 & -12 \\ 3 & -2 & -1 & 2 \\ -2 & 4 & -2 & 4 \end{bmatrix}$

43 The matrix A is not invertible because $A \sim \begin{bmatrix} 1 & 0 & 0 & 1 \\ 0 & 1 & 0 & 1 \\ 0 & 0 & 1 & -1 \\ 0 & 0 & 0 & 0 \end{bmatrix}$.

45 $A^{-1} = \begin{bmatrix} -1 & 0 & 1 \\ 3 & -1 & -1 \\ -1 & 1 & 0 \end{bmatrix}$, $A = \begin{bmatrix} 1 & 0 & 1 \\ 0 & 1 & 1 \\ 0 & 0 & 1 \end{bmatrix} \begin{bmatrix} 1 & 0 & 0 \\ 0 & 1 & 0 \\ 0 & 1 & 1 \end{bmatrix} \begin{bmatrix} 1 & 0 & 0 \\ 0 & 0 & 1 \\ 0 & 1 & 0 \end{bmatrix} \begin{bmatrix} 1 & 0 & 0 \\ -3 & 1 & 0 \\ 1 & 0 & 1 \end{bmatrix} \begin{bmatrix} -1 & 0 & 0 \\ 0 & 1 & 0 \\ 0 & 0 & 1 \end{bmatrix}$, because

$\begin{bmatrix} 1 & 0 & 1 \\ 0 & 1 & 1 \\ 0 & 0 & 1 \end{bmatrix} \begin{bmatrix} 1 & 0 & 0 \\ 0 & 1 & 0 \\ 0 & 1 & 1 \end{bmatrix} \begin{bmatrix} 1 & 0 & 0 \\ 0 & 0 & 1 \\ 0 & 1 & 0 \end{bmatrix} \begin{bmatrix} 1 & 0 & 0 \\ -3 & 1 & 0 \\ 1 & 0 & 1 \end{bmatrix} \begin{bmatrix} -1 & 0 & 0 \\ 0 & 1 & 0 \\ 0 & 0 & 1 \end{bmatrix} A^{-1} = \begin{bmatrix} 1 & 0 & 0 \\ 0 & 1 & 0 \\ 0 & 0 & 1 \end{bmatrix}$. Note that we can

solve the exercise as in Example 2.3.18

47

$A^{-1} = \begin{bmatrix} 1 & 0 & 6 \\ 0 & 1 & -4 \\ 0 & 0 & 1 \end{bmatrix} \begin{bmatrix} 1 & 0 & 0 \\ 0 & 1 & 0 \\ 0 & 0 & -\frac{1}{3} \end{bmatrix} \begin{bmatrix} 1 & -2 & 0 \\ 0 & 1 & 0 \\ 0 & 0 & 1 \end{bmatrix} \begin{bmatrix} 1 & 0 & 0 \\ 0 & -1 & 0 \\ 0 & 0 & 1 \end{bmatrix} \begin{bmatrix} 1 & 0 & 0 \\ 0 & 0 & 1 \\ 0 & 1 & 0 \end{bmatrix} \begin{bmatrix} 1 & 0 & 0 \\ -2 & 1 & 0 \\ -3 & 0 & 1 \end{bmatrix} \begin{bmatrix} 0 & 1 & 0 \\ 1 & 0 & 0 \\ 0 & 0 & 1 \end{bmatrix}$

$= \begin{bmatrix} -2 & -1 & 2 \\ \frac{4}{3} & \frac{1}{3} & -1 \\ -\frac{1}{3} & \frac{2}{3} & 0 \end{bmatrix}$

49 $A = \begin{bmatrix} 1 & 0 & 0 \\ 1 & 1 & 0 \\ 1 & 1 & 1 \end{bmatrix} \begin{bmatrix} 1 & 0 & 1 \\ 0 & 1 & 0 \\ 0 & 0 & -1 \end{bmatrix}$

51 $A = \begin{bmatrix} 1 & 0 & 0 \\ \frac{1}{2} & 1 & 0 \\ \frac{1}{2} & \frac{1}{3} & 1 \end{bmatrix} \begin{bmatrix} 2 & 1 & 1 \\ 0 & \frac{3}{2} & \frac{1}{2} \\ 0 & 0 & \frac{4}{3} \end{bmatrix}$

Section 3.1

1 $\mathbf{d} = \mathbf{b} - \mathbf{a} + \mathbf{c} = \begin{bmatrix} 1 \\ 2 \end{bmatrix}$.

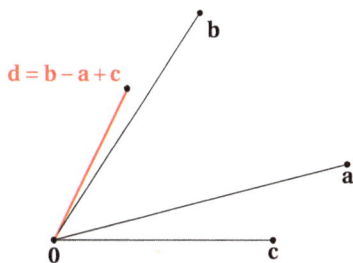

3 Here is an example of two vectors **a** and **b** and the vector $2\mathbf{a} + \frac{1}{3}\mathbf{b}$.

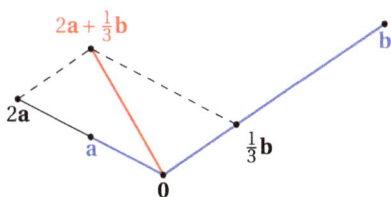

Here is another example of two vectors **a** and **b** and the vector $2\mathbf{a} + \frac{1}{3}\mathbf{b}$.

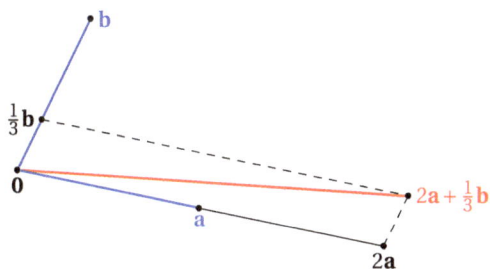

5 Here is an example of two vectors **a** and **b** and the vector $-\frac{4}{3}\mathbf{a} + 2\mathbf{b}$.

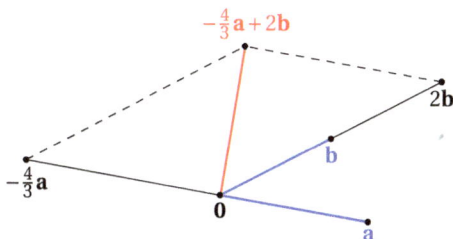

Here is another example of two vectors **a** and **b** and the vector $-\frac{4}{3}\mathbf{a} + 2\mathbf{b}$.

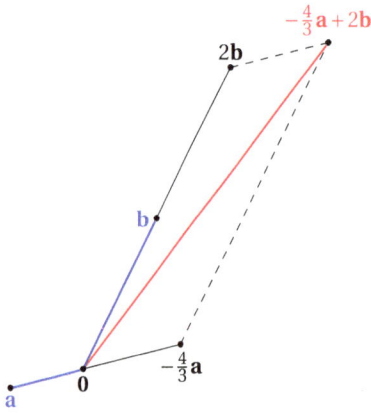

7 Here is an example of two vectors **a** and **b** and the vector $-0.5\mathbf{a} - 0.75\mathbf{b}$.

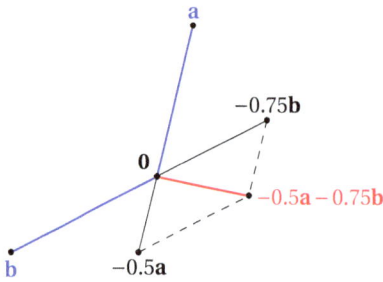

Here is another example of two vectors **a** and **b** and the vector $-0.5\mathbf{a} - 0.75\mathbf{b}$.

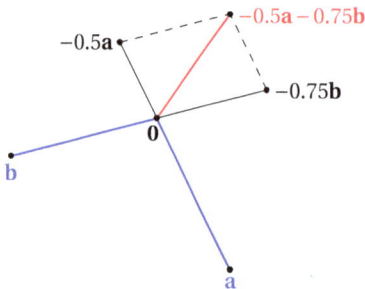

9 Here is an example of three vectors **a**, **b**, **c**, and the vector $\frac{1}{2}\mathbf{a} + \mathbf{b} - \frac{5}{3}\mathbf{c}$.

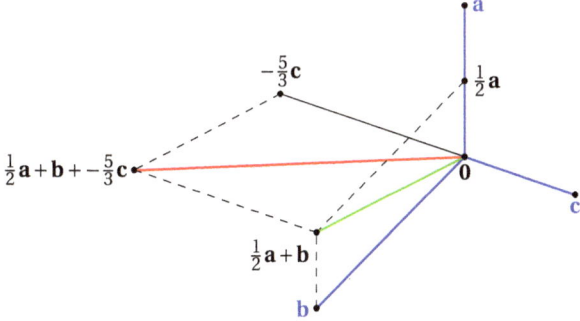

Here is another example of three vectors **a**, **b**, **c**, and the vector $\frac{1}{2}\mathbf{a}+\mathbf{b}-\frac{5}{3}\mathbf{c}$.

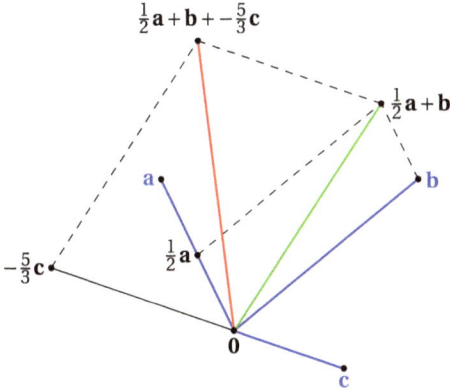

11 Here is an example of three vectors **a**, **b**, **c**, and the vector $-\frac{3}{4}\mathbf{a}+\frac{1}{3}\mathbf{b}+3\mathbf{c}$.

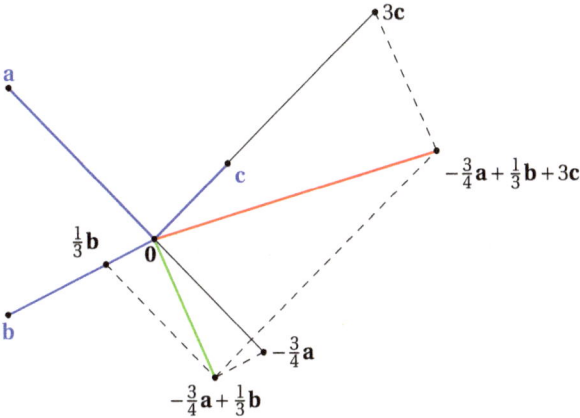

Here is another example of three vectors **a**, **b**, **c**, and the vector $-\frac{3}{4}\mathbf{a}+\frac{1}{3}\mathbf{b}+3\mathbf{c}$.

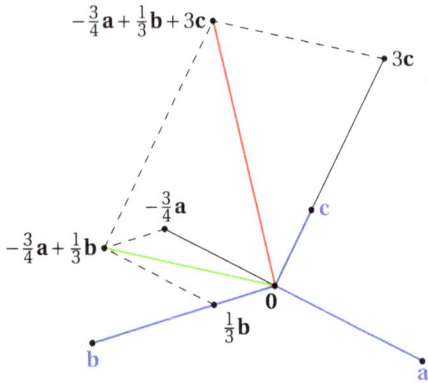

13 We have $\mathbf{u} = -\frac{5}{7}\mathbf{v}$.

15 The equality $\begin{bmatrix} 5 \\ -5 \end{bmatrix} = c \begin{bmatrix} 7 \\ 7 \end{bmatrix}$ is not possible.

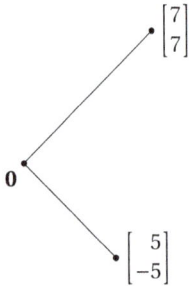

17 We must have $\det \begin{bmatrix} \mathbf{u} & \mathbf{v} \end{bmatrix} = 0$ which gives us $a = -\frac{21}{2}$.

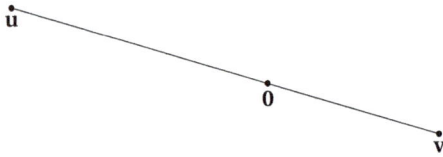

19 $a = 2$ or $a = 5$.

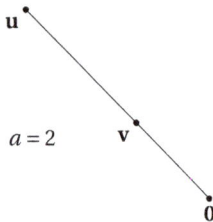

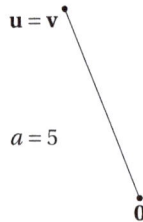

21 $a = 1$ or $a = -9$.

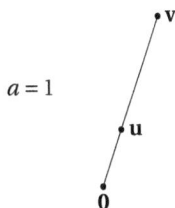

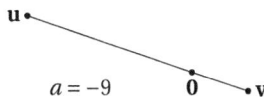

23 $a = 2$ or $a = 7$.

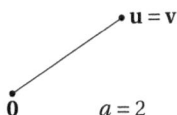

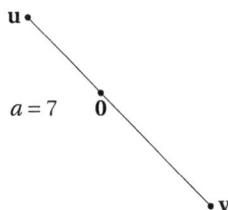

25 The vectors $\begin{bmatrix} 5 \\ 2 \end{bmatrix}$ and $\begin{bmatrix} -1 \\ 4 \end{bmatrix}$ are linearly independent.

27 The vectors $\begin{bmatrix} a \\ b \end{bmatrix}$ and $\begin{bmatrix} c \\ a \end{bmatrix}$ are linearly independent because $\det\left(\begin{bmatrix} a \\ b \end{bmatrix} \quad \begin{bmatrix} c \\ a \end{bmatrix}\right) = a^2 - bc >$ 0.

29 Since $\begin{bmatrix} 1 \\ 2 \end{bmatrix} = 2\begin{bmatrix} 1 \\ 1 \end{bmatrix} - \begin{bmatrix} 1 \\ 0 \end{bmatrix}$, the coordinates are 2 and -1.

31 Since $\begin{bmatrix} 1 \\ -2 \end{bmatrix} = \frac{1}{3}\begin{bmatrix} 3 \\ 1 \end{bmatrix} - \frac{7}{3}\begin{bmatrix} 0 \\ 1 \end{bmatrix}$, the coordinates are $\frac{1}{3}$ and $-\frac{7}{3}$.

33 The coordinates of $\begin{bmatrix} 1 \\ 0 \end{bmatrix}$ with respect to the basis $\left\{\begin{bmatrix} 1 \\ 1 \end{bmatrix}, \begin{bmatrix} 0 \\ 1 \end{bmatrix}\right\}$ are 1 and -1. The coordinates of $\begin{bmatrix} 1 \\ 0 \end{bmatrix}$ with respect to the basis $\left\{\begin{bmatrix} 1 \\ 1 \end{bmatrix}, \begin{bmatrix} 1 \\ 3 \end{bmatrix}\right\}$ are $\frac{3}{2}$ and $-\frac{1}{2}$.

Section 3.2

1 5

3 $\sqrt{34}$

5 $\frac{\sqrt{2}}{\sqrt{a}}$

7 $\sqrt{13}$

9 $\sqrt{2}|a|$

11 11

13 0

15 $a = -2$, see Figure 12.1

17 $a = -1$ or a=1, see Figure 12.2

19 $a = 1$ or $a = 4$, see Figure 12.3

21 See Figure 12.4

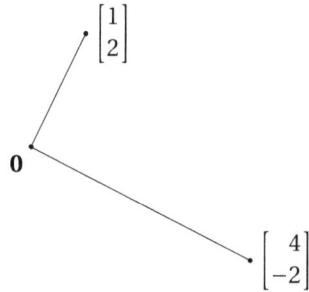

Figure 12.1: Solution to Exercise 15.

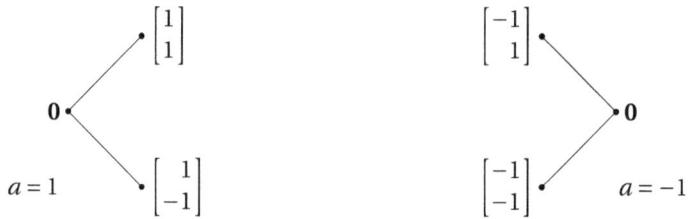

Figure 12.2: Solutions to Exercise 17.

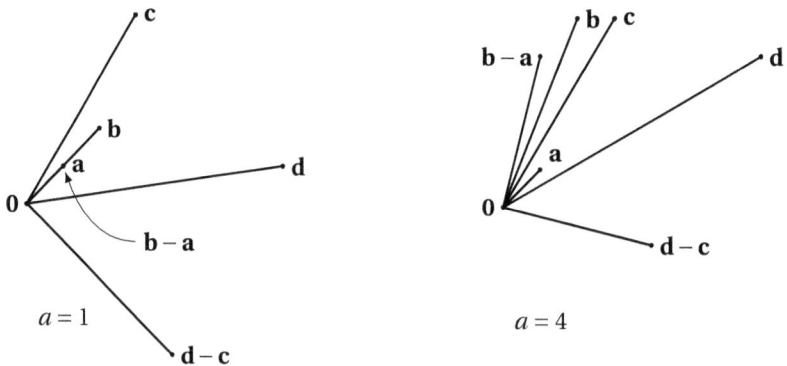

Figure 12.3: Solutions to Exercise 19.

23 $\mathbf{p} = \dfrac{\mathbf{b} \cdot \mathbf{u}}{\|\mathbf{u}\|^2} \mathbf{u} = \dfrac{5}{10} \begin{bmatrix} 1 \\ 3 \end{bmatrix} = \begin{bmatrix} \frac{1}{2} \\ \frac{3}{2} \end{bmatrix}$

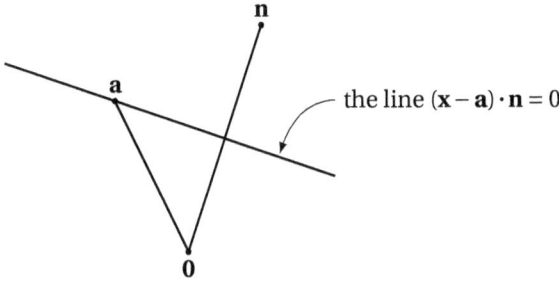

Figure 12.4: Solution to Exercise 21.

25 $\mathbf{p} = \frac{\mathbf{b} \cdot \mathbf{u}}{\|\mathbf{u}\|^2} \mathbf{u} = \frac{x+y}{2} \begin{bmatrix} 1 \\ 1 \end{bmatrix} = \begin{bmatrix} \frac{x+y}{2} \\ \frac{x+y}{2} \end{bmatrix}$

27 The projection matrix on Span{$\mathbf{u}$} is $\frac{1}{2} \begin{bmatrix} 1 & 1 \\ 1 & 1 \end{bmatrix}$ and the projection of $\begin{bmatrix} 1 \\ 0 \end{bmatrix}$ on Span{$\mathbf{u}$} is

$\begin{bmatrix} \frac{1}{2} \\ \frac{1}{2} \end{bmatrix}$.

29 The projection matrix on Span{$\mathbf{u}$} is $\frac{1}{5} \begin{bmatrix} 4 & -2 \\ -2 & 1 \end{bmatrix}$ and the projection of $\begin{bmatrix} x \\ y \end{bmatrix}$ on Span{$\mathbf{u}$}

is $\begin{bmatrix} \frac{4x-2y}{5} \\ \frac{-2x+y}{5} \end{bmatrix}$.

31 $\begin{bmatrix} -1 \\ 2 \end{bmatrix}$ 33 $\begin{bmatrix} y \\ x \end{bmatrix}$ 35 $A = \frac{1}{5} \begin{bmatrix} 4 & -3 \\ -3 & -4 \end{bmatrix}$

37 The vector $\begin{bmatrix} x \\ y \end{bmatrix}$ satisfies the equation $3x - 4y = 0$ if and only if the vector $\begin{bmatrix} x \\ y \end{bmatrix}$ is in

Span$\left\{ \begin{bmatrix} 3 \\ -4 \end{bmatrix}^{\perp} \right\} =$ Span$\left\{ \begin{bmatrix} 4 \\ 3 \end{bmatrix} \right\}$.

39 $x - y = 0$ 43 $\frac{1}{\sqrt{2}}$

41 $x + 3y = 0$ 45 $\frac{13}{2}$

47 Let $\mathbf{a} = \begin{bmatrix} a_1 \\ a_2 \end{bmatrix}$ and $\mathbf{b} = \begin{bmatrix} b_1 \\ b_2 \end{bmatrix}$. Then $\mathbf{b} \cdot \mathbf{a}^{\perp} = \begin{bmatrix} b_1 \\ b_2 \end{bmatrix} \cdot \begin{bmatrix} -a_2 \\ a_1 \end{bmatrix} = -b_1 a_2 + b_2 a_1 = \det \begin{bmatrix} \mathbf{a} & \mathbf{b} \end{bmatrix}$.

49 $\det \begin{bmatrix} \mathbf{a}^{\perp} & \mathbf{b}^{\perp} \end{bmatrix} = \mathbf{b}^{\perp} \cdot (\mathbf{a}^{\perp})^{\perp} = -\mathbf{a} \cdot \mathbf{b}^{\perp} = -\det \begin{bmatrix} \mathbf{b} & \mathbf{a} \end{bmatrix} = \det \begin{bmatrix} \mathbf{a} & \mathbf{b} \end{bmatrix}$

51 There are three possibilities:

(a) The vectors $\begin{bmatrix} a_1 \\ b_1 \end{bmatrix}$ and $\begin{bmatrix} a_2 \\ b_2 \end{bmatrix}$ are linearly independent,

(b) The vectors $\begin{bmatrix} a_1 \\ b_1 \end{bmatrix}$ and $\begin{bmatrix} a_2 \\ b_2 \end{bmatrix}$ are linearly dependent and at least one of them is

different from the zero vector $\begin{bmatrix} 0 \\ 0 \end{bmatrix}$,

(c) $\begin{bmatrix} a_1 \\ b_1 \end{bmatrix} = \begin{bmatrix} a_2 \\ b_2 \end{bmatrix} = \begin{bmatrix} 0 \\ 0 \end{bmatrix}$.

In the first case we have $\det \begin{bmatrix} a_1 & b_1 \\ a_2 & b_2 \end{bmatrix} \neq 0$ and thus the equation $\begin{bmatrix} a_1 & b_1 \\ a_2 & b_2 \end{bmatrix} \begin{bmatrix} x \\ y \end{bmatrix} = \begin{bmatrix} 0 \\ 0 \end{bmatrix}$ has

a unique solution $\begin{bmatrix} x \\ y \end{bmatrix} = \begin{bmatrix} 0 \\ 0 \end{bmatrix}$, by Cramer's Rule (Theorem 1.3.10). So in this case we

have $\mathbf{S} = \left\{ \begin{bmatrix} 0 \\ 0 \end{bmatrix} \right\}$.

Now we assume that the vectors $\begin{bmatrix} a_1 \\ b_1 \end{bmatrix}$ and $\begin{bmatrix} a_2 \\ b_2 \end{bmatrix}$ are linearly dependent and at least one

of them is different from the zero vector. Suppose that $\begin{bmatrix} a_1 \\ b_1 \end{bmatrix} \neq \begin{bmatrix} 0 \\ 0 \end{bmatrix}$. Then there is a

number c such that

$$\begin{bmatrix} a_2 \\ b_2 \end{bmatrix} = c \begin{bmatrix} a_1 \\ b_1 \end{bmatrix}. \tag{12.1}$$

The equation $\begin{bmatrix} a_1 & b_1 \\ a_2 & b_2 \end{bmatrix} \begin{bmatrix} x \\ y \end{bmatrix} = \begin{bmatrix} 0 \\ 0 \end{bmatrix}$ is equivalent to the system

$$\begin{cases} a_1 x + b_1 y = 0 \\ a_2 x + b_2 y = 0 \end{cases},$$

which can be written as

$$\begin{cases} a_1 x + b_1 y = 0 \\ c a_1 x + c a_2 y = 0 \end{cases},$$

by (12.1). But this means that the equation $\begin{bmatrix} a_1 & b_1 \\ a_2 & b_2 \end{bmatrix} \begin{bmatrix} x \\ y \end{bmatrix} = \begin{bmatrix} 0 \\ 0 \end{bmatrix}$ is equivalent to

$$a_1 x + b_1 y = 0.$$

From Theorem 3.2.22 the general solution of this equation is

$$\begin{bmatrix} x \\ y \end{bmatrix} = t \begin{bmatrix} -b_1 \\ a_1 \end{bmatrix},$$

where t is an arbitrary real number. In other words, $\begin{bmatrix} x \\ y \end{bmatrix}$ is a solution of the equation

(51) if an only if $\begin{bmatrix} x \\ y \end{bmatrix}$ is in $\mathrm{Span}\left\{ \begin{bmatrix} -b_1 \\ a_1 \end{bmatrix} \right\}$, so $\mathbf{S}$ is a vector line in this case. The case when

the vector $\begin{bmatrix} a_2 \\ b_2 \end{bmatrix} \neq \begin{bmatrix} 0 \\ 0 \end{bmatrix}$ is handled in a similar way.

Finally we note that, if $\begin{bmatrix} a_1 \\ b_1 \end{bmatrix} = \begin{bmatrix} a_2 \\ b_2 \end{bmatrix} = \begin{bmatrix} 0 \\ 0 \end{bmatrix}$, then every vector $\begin{bmatrix} x \\ y \end{bmatrix}$ in $\mathbb{R}^2$ satisfies the

equation

$$\begin{bmatrix} a_1 & b_1 \\ a_2 & b_2 \end{bmatrix} \begin{bmatrix} x \\ y \end{bmatrix} = \begin{bmatrix} 0 & 0 \\ 0 & 0 \end{bmatrix} \begin{bmatrix} x \\ y \end{bmatrix} = \begin{bmatrix} 0 \\ 0 \end{bmatrix},$$

so $\mathbf{S} = \mathbb{R}^2$ in this case.

53    $\dfrac{\|\mathbf{p}\|}{\|\mathbf{a}\|} = \dfrac{\left\| \frac{\mathbf{a} \cdot \mathbf{b}}{\|\mathbf{b}\|^2} \mathbf{b} \right\|}{\|\mathbf{a}\|} = \dfrac{|\mathbf{a} \cdot \mathbf{b}|}{\|\mathbf{a}\| \|\mathbf{b}\|}$

55 Note that, if the vectors **a** and **b** are linearly independent, then $\mathbf{b}\cdot\mathbf{a}^{\perp} \neq 0$ and $\mathbf{a}\cdot\mathbf{b}^{\perp} \neq 0$, by Exercise 48, so the expressions in $x = \dfrac{\mathbf{c}\cdot\mathbf{b}^{\perp}}{\mathbf{a}\cdot\mathbf{b}^{\perp}}$ and $y = \dfrac{\mathbf{c}\cdot\mathbf{a}^{\perp}}{\mathbf{b}\cdot\mathbf{a}^{\perp}}$ make sense.

The equation has a solution because the set $\{\mathbf{a},\mathbf{b}\}$ is a basis of $\mathbb{R}^2$.

Now, if numbers x and y satisfy the equation $x\mathbf{a} + y\mathbf{b} = \mathbf{c}$, then $x\mathbf{a}\cdot\mathbf{b}^{\perp} + y\mathbf{b}\cdot\mathbf{b}^{\perp} = \mathbf{c}\cdot\mathbf{b}^{\perp}$ and $x\mathbf{a}\cdot\mathbf{a}^{\perp} + y\mathbf{b}\cdot\mathbf{a}^{\perp} = \mathbf{c}\cdot\mathbf{a}^{\perp}$. Since $\mathbf{a}\cdot\mathbf{a}^{\perp} = 0$, $\mathbf{b}\cdot\mathbf{b}^{\perp} = 0$, $\mathbf{b}\cdot\mathbf{a}^{\perp} \neq 0$, and $\mathbf{a}\cdot\mathbf{b}^{\perp} \neq 0$, we must have $x = \dfrac{\mathbf{c}\cdot\mathbf{b}^{\perp}}{\mathbf{a}\cdot\mathbf{b}^{\perp}}$ and $y = \dfrac{\mathbf{c}\cdot\mathbf{a}^{\perp}}{\mathbf{b}\cdot\mathbf{a}^{\perp}}$.

57 $A = 2\dfrac{1}{\|\mathbf{u}\|^2}(\mathbf{u}\mathbf{u}^T) - \begin{bmatrix} 1 & 0 \\ 0 & 1 \end{bmatrix}$ and we have $A^T = A$.

Section 3.3

1 We have $\begin{bmatrix} 3 \\ 2 \end{bmatrix} \cdot \begin{bmatrix} -2 \\ 3 \end{bmatrix} = 0$.

3 $P = \dfrac{1}{\sqrt{5}}\begin{bmatrix} 1 & 2 \\ -2 & 1 \end{bmatrix}, D = \begin{bmatrix} 4 & 0 \\ 0 & 9 \end{bmatrix}$

5 $P = \dfrac{1}{\sqrt{5}}\begin{bmatrix} -1 & 2 \\ 2 & 1 \end{bmatrix}, D = \begin{bmatrix} -7 & 0 \\ 0 & 13 \end{bmatrix}$

7 $P = \dfrac{1}{\sqrt{26}}\begin{bmatrix} 5 & 1 \\ 1 & -5 \end{bmatrix}, D = \begin{bmatrix} 28 & 0 \\ 0 & 2 \end{bmatrix}$

9 $P = \dfrac{1}{\sqrt{2}}\begin{bmatrix} -1 & 1 \\ 1 & 1 \end{bmatrix}, D = \begin{bmatrix} 2 & 0 \\ 0 & 2a+2 \end{bmatrix}$

11 $A = \dfrac{7}{2}\begin{bmatrix} 1 \\ 1 \end{bmatrix}\begin{bmatrix} 1 & 1 \end{bmatrix} + \dfrac{3}{2}\begin{bmatrix} 1 \\ -1 \end{bmatrix}\begin{bmatrix} 1 & -1 \end{bmatrix}$

13 $A = \dfrac{4}{5}\begin{bmatrix} 1 \\ -2 \end{bmatrix}\begin{bmatrix} 1 & -2 \end{bmatrix} + \dfrac{9}{5}\begin{bmatrix} 2 \\ 1 \end{bmatrix}\begin{bmatrix} 2 & 1 \end{bmatrix}$

15 $A = \begin{bmatrix} \frac{1}{2} & \frac{3}{2} \\ \frac{3}{2} & \frac{1}{2} \end{bmatrix}$

17 $A = \dfrac{1}{5}\begin{bmatrix} 4\alpha+\beta & 2\alpha-2\beta \\ 2\alpha-2\beta & \alpha+4\beta \end{bmatrix}$

19 $A = PDP^T$ where $P = \dfrac{1}{\sqrt{2}}\begin{bmatrix} 1 & 1 \\ 1 & -1 \end{bmatrix}$ and $D = \begin{bmatrix} 1 & 0 \\ 0 & 0 \end{bmatrix}$.

21 $A = PDP^T$ where $P = \dfrac{1}{\sqrt{10}}\begin{bmatrix} 1 & 3 \\ 3 & -1 \end{bmatrix}$ and $D = \begin{bmatrix} 1 & 0 \\ 0 & 0 \end{bmatrix}$.

23 $A = \begin{bmatrix} \frac{15}{17} & \frac{8}{17} \\ \frac{8}{17} & -\frac{15}{17} \end{bmatrix} = PDP^T$ where $P = \dfrac{1}{\sqrt{17}}\begin{bmatrix} 4 & -1 \\ 1 & 4 \end{bmatrix}$ and $D = \begin{bmatrix} 1 & 0 \\ 0 & -1 \end{bmatrix}$.

25 The vector $\begin{bmatrix} a \\ b \end{bmatrix}$ is an eigenvector corresponding to the eigenvalue 1 and the vector $\begin{bmatrix} -b \\ a \end{bmatrix}$ is an eigenvector corresponding to the eigenvalue 0.

27 Let $A = \begin{bmatrix} a & b \\ b & c \end{bmatrix}$, $\mathbf{u} = \begin{bmatrix} u_1 \\ u_2 \end{bmatrix}$ and $\mathbf{v} = \begin{bmatrix} v_1 \\ v_2 \end{bmatrix}$. The desired equality is a consequence of the calculations:

$$(A\mathbf{u})\cdot\mathbf{v} = \left(\begin{bmatrix} a & b \\ b & c \end{bmatrix}\begin{bmatrix} u_1 \\ u_2 \end{bmatrix}\right)\cdot\begin{bmatrix} v_1 \\ v_2 \end{bmatrix} = (au_1 + bu_2)v_1 + (bu_1 + cu_2)v_2$$

$$\mathbf{u}\cdot(A\mathbf{v}) = \begin{bmatrix} u_1 \\ u_2 \end{bmatrix}\cdot\left(\begin{bmatrix} a & b \\ b & c \end{bmatrix}\begin{bmatrix} v_1 \\ v_2 \end{bmatrix}\right) = u_1(av_1 + bv_2) + u_2(bv_1 + cv_2)$$

29 $A = \begin{bmatrix} \frac{1}{\sqrt{2}} & -\frac{1}{\sqrt{2}} \\ \frac{1}{\sqrt{2}} & \frac{1}{\sqrt{2}} \end{bmatrix}\begin{bmatrix} \sqrt{2} & \frac{3}{\sqrt{2}} \\ 0 & \frac{1}{\sqrt{2}} \end{bmatrix}$

31 $A = \frac{1}{a^2+b^2} \begin{bmatrix} a \\ b \end{bmatrix} \begin{bmatrix} a & b \end{bmatrix} + \frac{1}{a^2+b^2} \begin{bmatrix} -b \\ a \end{bmatrix} \begin{bmatrix} -b & a \end{bmatrix}$ where $\begin{bmatrix} a \\ b \end{bmatrix} \neq \begin{bmatrix} 0 \\ 0 \end{bmatrix}$.

Section 4.1

---

1 $\begin{bmatrix} 1 \\ 2 \\ -1 \end{bmatrix} = -\frac{1}{3} \begin{bmatrix} -3 \\ -6 \\ 3 \end{bmatrix}$

11 There is no c such that $\begin{bmatrix} 1 \\ -3 \\ 1 \end{bmatrix} = c \begin{bmatrix} -1 \\ 3 \\ 2 \end{bmatrix}$.

3 There is no c such that $\begin{bmatrix} 1 \\ 2 \\ 3 \end{bmatrix} = c \begin{bmatrix} 1 \\ 0 \\ -1 \end{bmatrix}$.

13 $\begin{bmatrix} 1 \\ 1 \\ 1 \end{bmatrix} = -2 \begin{bmatrix} 1 \\ -1 \\ 1 \end{bmatrix} + \begin{bmatrix} 3 \\ -1 \\ 3 \end{bmatrix}$

5 $\begin{bmatrix} 3 \\ 2 \\ 1 \end{bmatrix} = \frac{1}{4} \begin{bmatrix} 12 \\ 4 \\ 8 \end{bmatrix}$

15 $\begin{bmatrix} 1 \\ 2 \\ 1 \end{bmatrix} = \begin{bmatrix} 2 \\ 2 \\ 3 \end{bmatrix} - \begin{bmatrix} 1 \\ 0 \\ 2 \end{bmatrix}$

7 $\begin{bmatrix} 1 \\ -3 \\ 1 \end{bmatrix} = - \begin{bmatrix} -1 \\ 3 \\ -1 \end{bmatrix}$

17 $\mathbf{x} = \mathbf{u} + 2\mathbf{v}$

9 There is no c such that $\begin{bmatrix} 3 \\ 1 \\ 2 \end{bmatrix} = c \begin{bmatrix} 3 \\ 1 \\ 3 \end{bmatrix}$.

19 $\mathbf{x}$ is not in Span $\{\mathbf{u}, \mathbf{v}\}$.

21 $\mathbf{x} = \frac{4}{3}\mathbf{u} + \frac{5}{3}\mathbf{v}$

23 We have $\begin{bmatrix} 3 \\ 5 \\ -1 \end{bmatrix} = \begin{bmatrix} 1 \\ 1 \\ 1 \end{bmatrix} + 2 \begin{bmatrix} 1 \\ 2 \\ -1 \end{bmatrix}$ and $\begin{bmatrix} 3 \\ 4 \\ 1 \end{bmatrix} = 2 \begin{bmatrix} 1 \\ 1 \\ 1 \end{bmatrix} + \begin{bmatrix} 1 \\ 2 \\ -1 \end{bmatrix}$ and the vectors $\begin{bmatrix} 3 \\ 5 \\ -1 \end{bmatrix}$ and $\begin{bmatrix} 3 \\ 4 \\ 1 \end{bmatrix}$ are linearly independent.

25 We have $\begin{bmatrix} 5 \\ 7 \\ 1 \end{bmatrix} = 3 \begin{bmatrix} 1 \\ 1 \\ 1 \end{bmatrix} + 2 \begin{bmatrix} 1 \\ 2 \\ -1 \end{bmatrix}$ and $\begin{bmatrix} 2 \\ 1 \\ 4 \end{bmatrix} = 3 \begin{bmatrix} 1 \\ 1 \\ 1 \end{bmatrix} - \begin{bmatrix} 1 \\ 2 \\ -1 \end{bmatrix}$ and the vectors $\begin{bmatrix} 5 \\ 7 \\ 1 \end{bmatrix}$ and $\begin{bmatrix} 2 \\ 1 \\ 4 \end{bmatrix}$ are linearly independent.

27 The transition matrix is $\begin{bmatrix} 1 & 2 \\ 2 & 1 \end{bmatrix}$ and, because $\begin{bmatrix} 1 & 2 \\ 2 & 1 \end{bmatrix} \begin{bmatrix} 5 \\ 2 \end{bmatrix} = \begin{bmatrix} 9 \\ 12 \end{bmatrix}$, we have $\mathbf{w} = 9 \begin{bmatrix} 1 \\ 1 \\ 1 \end{bmatrix} + 12 \begin{bmatrix} 1 \\ 2 \\ -1 \end{bmatrix}$.

29 The transition matrix is $\begin{bmatrix} 3 & 3 \\ 2 & -1 \end{bmatrix}$ and, because $\begin{bmatrix} 3 & 3 \\ 2 & -1 \end{bmatrix} \begin{bmatrix} a \\ b \end{bmatrix} = \begin{bmatrix} 3a+3b \\ 2a-b \end{bmatrix}$, we have $\mathbf{w} = (3a+3b) \begin{bmatrix} 1 \\ 1 \\ 1 \end{bmatrix} + (2a-b) \begin{bmatrix} 1 \\ 2 \\ -1 \end{bmatrix}$.

31 We have $\begin{bmatrix} \mathbf{u} & \mathbf{v} \end{bmatrix} = \begin{bmatrix} \mathbf{a} & \mathbf{b} \end{bmatrix} \begin{bmatrix} 2 & 3 \\ 1 & 5 \end{bmatrix}$ and $\begin{bmatrix} \mathbf{u} & \mathbf{v} \end{bmatrix} \begin{bmatrix} 2 & 3 \\ 1 & 5 \end{bmatrix}^{-1} = \begin{bmatrix} \mathbf{a} & \mathbf{b} \end{bmatrix}$. This means that the transition matrix from the basis $\{\mathbf{u}, \mathbf{v}\}$ to the basis $\{\mathbf{a}, \mathbf{b}\}$ is $\begin{bmatrix} 2 & 3 \\ 1 & 5 \end{bmatrix}$ and the transition matrix from the basis $\{\mathbf{a}, \mathbf{b}\}$ to the basis $\{\mathbf{u}, \mathbf{v}\}$ is $\begin{bmatrix} 2 & 3 \\ 1 & 5 \end{bmatrix}^{-1}$.

33 We have $[\mathbf{u}\ \mathbf{v}] = [\mathbf{a}\ \mathbf{b}]\begin{bmatrix} 3 & 0 \\ 2 & 1 \end{bmatrix}$ and $[\mathbf{u}\ \mathbf{v}]\begin{bmatrix} 3 & 0 \\ 2 & 1 \end{bmatrix}^{-1} = [\mathbf{a}\ \mathbf{b}]$. This means that the tran-

sition matrix from the basis $\{\mathbf{u}, \mathbf{v}\}$ to the basis $\{\mathbf{a}, \mathbf{b}\}$ is $\begin{bmatrix} 3 & 0 \\ 2 & 1 \end{bmatrix}$ and the transition matrix

from the basis $\{\mathbf{a}, \mathbf{b}\}$ to the basis $\{\mathbf{u}, \mathbf{v}\}$ is $\begin{bmatrix} 3 & 0 \\ 2 & 1 \end{bmatrix}^{-1}$.

35 We have $[\mathbf{u}\ \mathbf{v}]\begin{bmatrix} 2 & 1 \\ 1 & 2 \end{bmatrix}^{-1} = [\mathbf{a}\ \mathbf{b}]$ or $[\mathbf{u}\ \mathbf{v}]\left(\frac{1}{3}\begin{bmatrix} 2 & -1 \\ -1 & 2 \end{bmatrix}\right) = [\mathbf{a}\ \mathbf{b}]$. This means that the

transition matrix from the basis $\{\mathbf{a}, \mathbf{b}\}$ to the basis $\{\mathbf{u}, \mathbf{v}\}$ is $\frac{1}{3}\begin{bmatrix} 2 & -1 \\ -1 & 2 \end{bmatrix}$ and we have $\mathbf{w} =$

$[\mathbf{a}\ \mathbf{b}]\begin{bmatrix} 1 \\ 1 \end{bmatrix} = [\mathbf{u}\ \mathbf{v}]\left(\frac{1}{3}\begin{bmatrix} 2 & -1 \\ -1 & 2 \end{bmatrix}\right)\begin{bmatrix} 1 \\ 1 \end{bmatrix} = \frac{1}{3}\mathbf{u} + \frac{1}{3}\mathbf{v}$.

37 We have $[\mathbf{u}\ \mathbf{v}]\begin{bmatrix} 3 & 1 \\ 2 & 5 \end{bmatrix}^{-1} = [\mathbf{a}\ \mathbf{b}]$ or $[\mathbf{u}\ \mathbf{v}]\left(\frac{1}{13}\begin{bmatrix} 5 & -1 \\ -2 & 3 \end{bmatrix}\right) = [\mathbf{a}\ \mathbf{b}]$. This means that the

transition matrix from the basis $\{\mathbf{a}, \mathbf{b}\}$ to the basis $\{\mathbf{u}, \mathbf{v}\}$ is $\frac{1}{13}\begin{bmatrix} 5 & -1 \\ -2 & 3 \end{bmatrix}$ and we have

$\mathbf{w} = [\mathbf{a}\ \mathbf{b}]\begin{bmatrix} 4 \\ 3 \end{bmatrix} = [\mathbf{u}\ \mathbf{v}]\left(\frac{1}{13}\begin{bmatrix} 5 & -1 \\ -2 & 3 \end{bmatrix}\right)\begin{bmatrix} 4 \\ 3 \end{bmatrix} = \frac{17}{13}\mathbf{u} + \frac{1}{13}\mathbf{v}$.

39 Let $\mathbf{u} = p\mathbf{a} + q\mathbf{b}$. One of the numbers p or q must be different from 0. If $p \neq 0$, then $\{\mathbf{b}, \mathbf{u}\}$
is a basis of the vector plane Span$\{\mathbf{a}, \mathbf{b}\}$. Indeed, if $x\mathbf{b} + y\mathbf{u} = \mathbf{0}$, then $x\mathbf{b} + y(p\mathbf{a} + q\mathbf{b}) = \mathbf{0}$
or $yp\mathbf{a} + (x + yq)\mathbf{b} = \mathbf{0}$. Thus $x = y = 0$, because $p \neq 0$. Consequently, the vectors $\mathbf{b}$ and
$\mathbf{u}$ are linearly independent. This means that the set $\{\mathbf{b}, \mathbf{u}\}$ is a basis of the vector plane
Span$\{\mathbf{a}, \mathbf{b}\}$.

If $q \neq 0$, then $\{\mathbf{a}, \mathbf{u}\}$ is a basis of the vector plane Span$\{\mathbf{a}, \mathbf{b}\}$.

Section 4.2

---

1 25

3 $\sqrt{30}$

5 3

7 Span $\left\{\begin{bmatrix} 1 \\ -2 \\ 0 \end{bmatrix}, \begin{bmatrix} 0 \\ 2 \\ 1 \end{bmatrix}\right\}$

9 $\frac{1}{27}\begin{bmatrix} 5 \\ 1 \\ 1 \end{bmatrix}$

11 $\frac{7}{9}\begin{bmatrix} 1 \\ 2 \\ 2 \end{bmatrix}$

13 $x = \frac{3}{11}$

15 $\frac{1}{5}\begin{bmatrix} 1 & 0 & 2 \\ 0 & 0 & 0 \\ 2 & 0 & 4 \end{bmatrix}$

17 $\frac{1}{38}\begin{bmatrix} 9 & -6 & 15 \\ -6 & 4 & -10 \\ 15 & -10 & 25 \end{bmatrix}$

19 $\begin{bmatrix} 0 \\ 0 \\ 0 \end{bmatrix}$

21 $\frac{1}{9}\begin{bmatrix} x + 2y + 2z \\ 2x + 4y + 4z \\ 2x + 4y + 4z \end{bmatrix}$

23 $\left\{\begin{bmatrix} 1 \\ 1 \\ -2 \end{bmatrix}, \begin{bmatrix} 3 \\ 1 \\ 2 \end{bmatrix}\right\}$ and $\left\{\begin{bmatrix} 1 \\ 0 \\ 2 \end{bmatrix}, \begin{bmatrix} 8 \\ 5 \\ -4 \end{bmatrix}\right\}$

25 $\left\{\begin{bmatrix} 2 \\ 1 \\ 5 \end{bmatrix}, \begin{bmatrix} 19 \\ 12 \\ -10 \end{bmatrix}\right\}$ and $\left\{\begin{bmatrix} 3 \\ 2 \\ -4 \end{bmatrix}, \begin{bmatrix} 94 \\ 53 \\ 97 \end{bmatrix}\right\}$

27 Span $\left\{ \begin{bmatrix} 1 \\ -2 \\ 0 \end{bmatrix}, \begin{bmatrix} 4 \\ 2 \\ 5 \end{bmatrix} \right\}$

29 Span $\left\{ \begin{bmatrix} \frac{1}{\sqrt{2}} \\ \frac{1}{\sqrt{2}} \\ 0 \end{bmatrix}, \begin{bmatrix} -\frac{1}{\sqrt{3}} \\ \frac{1}{\sqrt{3}} \\ \frac{1}{\sqrt{3}} \end{bmatrix} \right\}$

31 $\frac{1}{2} \begin{bmatrix} -1 \\ 1 \\ 2 \end{bmatrix}$

33 $\frac{1}{27} \begin{bmatrix} -1 \\ -7 \\ 25 \end{bmatrix}$

35 $\frac{2}{7} \begin{bmatrix} 2 \\ 1 \\ 4 \end{bmatrix}$

37 $\frac{1}{3} \begin{bmatrix} 8 \\ 7 \\ 1 \end{bmatrix}$

39 $\frac{\sqrt{2}}{2}$

41 $\sqrt{2}$

43 $\frac{1}{2} \begin{bmatrix} 1 & 0 & -1 \\ 0 & 2 & 0 \\ -1 & 0 & 1 \end{bmatrix}$

45 $\frac{1}{3} \begin{bmatrix} 2 & 1 & 1 \\ 1 & 2 & -1 \\ 1 & -1 & 2 \end{bmatrix}$

47 $\frac{1}{11} \begin{bmatrix} 10 & 3 & -1 \\ 3 & 2 & 3 \\ -1 & 3 & 10 \end{bmatrix}$

49 $\frac{1}{5} \begin{bmatrix} 1 & 2 & 0 \\ 2 & 4 & 0 \\ 0 & 0 & 5 \end{bmatrix}$

51 $x = 5$ and $y = 8$

53 $x = -\frac{1}{2}$ and $y = \frac{1}{2}$

55 $x = -\frac{1}{2} + t, y = t$. The vectors $\begin{bmatrix} -1 \\ -2 \\ 1 \end{bmatrix}$ and $\begin{bmatrix} 1 \\ 2 \\ -1 \end{bmatrix}$ are linearly dependent.

57 $x = 0$ and $y = \frac{3}{2}$

59 $x = -\frac{3}{4}$ and $y = -\frac{1}{4}$

61 $\frac{1}{2} \begin{bmatrix} 2 \\ 1 \\ -1 \end{bmatrix}$

63 $\frac{2}{3} \begin{bmatrix} 1 \\ 1 \\ 2 \end{bmatrix}$

65 $x = \frac{1}{3}, y = \frac{4}{3}$

67 $\frac{1}{9} \begin{bmatrix} 5 & 4 & 2 \\ 4 & 5 & -2 \\ 2 & -2 & 8 \end{bmatrix}$

69

$$\mathbf{p} = A(A^T A)^{-1} A^T \mathbf{b} = \begin{bmatrix} \mathbf{u} & \mathbf{v} \end{bmatrix} \begin{bmatrix} \mathbf{u} \cdot \mathbf{u} & \mathbf{u} \cdot \mathbf{v} \\ \mathbf{u} \cdot \mathbf{v} & \mathbf{v} \cdot \mathbf{v} \end{bmatrix}^{-1} \begin{bmatrix} \mathbf{b} \cdot \mathbf{u} \\ \mathbf{b} \cdot \mathbf{v} \end{bmatrix} = \begin{bmatrix} \mathbf{u} & \mathbf{v} \end{bmatrix} \begin{bmatrix} \mathbf{u} \cdot \mathbf{u} & 0 \\ 0 & \mathbf{v} \cdot \mathbf{v} \end{bmatrix}^{-1} \begin{bmatrix} \mathbf{b} \cdot \mathbf{u} \\ \mathbf{b} \cdot \mathbf{v} \end{bmatrix}$$

$$= \begin{bmatrix} \mathbf{u} & \mathbf{v} \end{bmatrix} \begin{bmatrix} \frac{1}{\mathbf{u} \cdot \mathbf{u}} & 0 \\ 0 & \frac{1}{\mathbf{v} \cdot \mathbf{v}} \end{bmatrix} \begin{bmatrix} \mathbf{b} \cdot \mathbf{u} \\ \mathbf{b} \cdot \mathbf{v} \end{bmatrix} = \begin{bmatrix} \mathbf{u} & \mathbf{v} \end{bmatrix} \begin{bmatrix} \frac{\mathbf{b} \cdot \mathbf{u}}{\mathbf{u} \cdot \mathbf{u}} \\ \frac{\mathbf{b} \cdot \mathbf{v}}{\mathbf{v} \cdot \mathbf{v}} \end{bmatrix} = \frac{\mathbf{b} \cdot \mathbf{u}}{\mathbf{u} \cdot \mathbf{u}} \mathbf{u} + \frac{\mathbf{b} \cdot \mathbf{v}}{\mathbf{v} \cdot \mathbf{v}} \mathbf{v}$$

71 $\frac{17}{2} - \frac{27}{14} x$

73 $\frac{7}{13} - \frac{11}{13} x$

75 $\begin{bmatrix} \frac{2}{\sqrt{5}} & \frac{2}{3\sqrt{5}} \\ \frac{1}{\sqrt{5}} & -\frac{4}{3\sqrt{5}} \\ 0 & \frac{\sqrt{5}}{3} \end{bmatrix} \begin{bmatrix} \sqrt{5} & -\frac{1}{\sqrt{5}} \\ 0 & \frac{3}{\sqrt{5}} \end{bmatrix}$

77 $\begin{bmatrix} \frac{1}{\sqrt{2}} & \frac{1}{\sqrt{38}} \\ \frac{1}{\sqrt{2}} & -\frac{1}{\sqrt{38}} \\ 0 & \frac{6}{\sqrt{38}} \end{bmatrix} \begin{bmatrix} \sqrt{2} & \frac{1}{\sqrt{2}} \\ 0 & \frac{\sqrt{38}}{2} \end{bmatrix}$

Section 5.1

1 $\begin{bmatrix} -32 \\ 17 \\ 18 \end{bmatrix}$

3 $\begin{bmatrix} 11 \\ -1 \\ -7 \end{bmatrix}$

5 $\text{Span} \left\{ \begin{bmatrix} 2 \\ -9 \\ -4 \end{bmatrix} \right\}$

7 $\text{Span} \left\{ \begin{bmatrix} -3 \\ -3 \\ 3 \end{bmatrix} \right\} = \text{Span} \left\{ \begin{bmatrix} 1 \\ 1 \\ -1 \end{bmatrix} \right\}$

9 $-x - 2y + 3z = 0$

11 $-14x + 23y + z = 0$

13 $\det A = 18$

15 $\det A = -1$

17 $\det \begin{bmatrix} \mathbf{a} & \mathbf{a} & \mathbf{b} \end{bmatrix} = \det \begin{bmatrix} \mathbf{a} & \mathbf{b} & \mathbf{a} \end{bmatrix} = 0$, because $\mathbf{a} \cdot (\mathbf{b} \times \mathbf{a}) = \mathbf{a} \cdot (\mathbf{a} \times \mathbf{b}) = 0$, and $\det \begin{bmatrix} \mathbf{b} & \mathbf{a} & \mathbf{a} \end{bmatrix} = 0$, because $\mathbf{a} \times \mathbf{a} = \mathbf{0}$.

19 $\det \begin{bmatrix} \mathbf{a} + \mathbf{d} & \mathbf{b} & \mathbf{c} \end{bmatrix} = (\mathbf{a} + \mathbf{d}) \cdot (\mathbf{b} \times \mathbf{c}) = \mathbf{a} \cdot (\mathbf{b} \times \mathbf{c}) + \mathbf{d} \cdot (\mathbf{b} \times \mathbf{c}) = \det \begin{bmatrix} \mathbf{a} & \mathbf{b} & \mathbf{c} \end{bmatrix} + \det \begin{bmatrix} \mathbf{d} & \mathbf{b} & \mathbf{c} \end{bmatrix}$

21 $\det \begin{bmatrix} \mathbf{a} + s\mathbf{b} & \mathbf{b} & \mathbf{c} + t\mathbf{b} \end{bmatrix} = (\mathbf{a} + s\mathbf{b}) \cdot (\mathbf{b} \times (\mathbf{c} + t\mathbf{b})) = \mathbf{a} \cdot (\mathbf{b} \times \mathbf{c}) + s\mathbf{b} \cdot (\mathbf{b} \times \mathbf{c}) + t\mathbf{a} \cdot (\mathbf{b} \times \mathbf{b}) + st\mathbf{b} \cdot (\mathbf{b} \times \mathbf{b}) = \mathbf{a} \cdot (\mathbf{b} \times \mathbf{c}) = \det \begin{bmatrix} \mathbf{a} & \mathbf{b} & \mathbf{c} \end{bmatrix}$

25

$$\det \begin{bmatrix} a_1 + sa_2 & b_1 + sb_2 & c_1 + sc_2 \\ a_2 & b_2 & c_2 \\ a_3 + ta_2 & b_3 + tb_2 & c_3 + tc_2 \end{bmatrix} = \det \begin{bmatrix} a_1 + sa_2 & a_2 & a_3 + ta_2 \\ b_1 + sb_2 & b_2 & b_3 + tb_2 \\ c_1 + sc_2 & c_2 & c_3 + tc_2 \end{bmatrix}$$

$$= \det \begin{bmatrix} a_1 & a_2 & a_3 \\ b_1 & b_2 & b_3 \\ c_1 & c_2 & c_3 \end{bmatrix} = \det \begin{bmatrix} a_1 & b_1 & c_1 \\ a_2 & b_2 & c_2 \\ a_3 & b_3 & c_3 \end{bmatrix}$$

27 The first and the third column of the matrix $\begin{bmatrix} 2 & 1 & 2 \\ 2 & 4 & 2 \\ 7 & 5 & 7 \end{bmatrix}$ are equal.

29 33 (The result is a consequence of the equality $\det \begin{bmatrix} \mathbf{a} + 9\mathbf{b} & \mathbf{b} & \mathbf{c} \end{bmatrix} = \det \begin{bmatrix} \mathbf{a} & \mathbf{b} & \mathbf{c} \end{bmatrix}$.)

31 99 (The result is a consequence of the equality $\det \begin{bmatrix} 3\mathbf{a} + 5\mathbf{b} & \mathbf{b} & \mathbf{c} \end{bmatrix} = 3\det \begin{bmatrix} \mathbf{a} & \mathbf{b} & \mathbf{c} \end{bmatrix}$.)

33 Use Theorem 5.1.16 and $\det \begin{bmatrix} 1 & s & 0 \\ 0 & 1 & 0 \\ 0 & t & 1 \end{bmatrix} = 1$.

Section 5.2

1 $\begin{bmatrix} -1 & 1 & -1 \\ 1 & 1 & -3 \\ -1 & -3 & 5 \end{bmatrix}$

3 $\begin{bmatrix} 5 & -6 & -2 \\ 0 & -2 & 1 \\ -10 & 13 & 1 \end{bmatrix}$

5 $-\frac{1}{4} \begin{bmatrix} 4 & -4 & 0 \\ -9 & 4 & 1 \\ 1 & 0 & -1 \end{bmatrix}$

7 $\begin{bmatrix} 9 & -4 & -1 \\ -7 & 17 & -2 \\ -4 & -1 & 6 \end{bmatrix}$

9 $\begin{bmatrix} -1 & 1 & 1 \\ 1 & -1 & 0 \\ 1 & 0 & -1 \end{bmatrix}$

11 $4\det\begin{bmatrix} 1 & 5 \\ 1 & 3 \end{bmatrix} - 3\det\begin{bmatrix} 2 & 5 \\ 1 & 3 \end{bmatrix} + 2\det\begin{bmatrix} 2 & 1 \\ 1 & 1 \end{bmatrix} = -9$

13 $-3\det\begin{bmatrix} 2 & 5 \\ 1 & 3 \end{bmatrix} + 1\det\begin{bmatrix} 4 & 2 \\ 1 & 3 \end{bmatrix} - 1\det\begin{bmatrix} 4 & 2 \\ 2 & 5 \end{bmatrix} = -9$

15 $-\det\begin{bmatrix} 1 & 1 \\ 3 & 1 \end{bmatrix} - 1\det\begin{bmatrix} 4 & 1 \\ 2 & 3 \end{bmatrix} = -8$

19 $7\det\begin{bmatrix} 4 & 1 \\ 1 & 2 \end{bmatrix} = 49$

21 $x = 2$, $y = -\frac{1}{5}$, $z = -\frac{7}{5}$

17 $-3\det\begin{bmatrix} 2 & 1 \\ 3 & 1 \end{bmatrix} = 3$

23 $x = -\frac{2}{5}$, $y = \frac{4}{5}$, $z = -\frac{1}{5}$

Section 5.3

1 Linearly independent because $\begin{bmatrix} 3 & 1 & 3 \\ 3 & 3 & 1 \\ 1 & 3 & 3 \end{bmatrix} \sim \begin{bmatrix} 1 & 0 & 0 \\ 0 & 1 & 0 \\ 0 & 0 & 1 \end{bmatrix}$.

3 Linearly dependent because $\begin{bmatrix} 1 & 3 & 1 \\ 3 & 2 & 2 \\ 2 & -1 & 1 \end{bmatrix} \sim \begin{bmatrix} 1 & 0 & \frac{4}{7} \\ 0 & 1 & \frac{1}{7} \\ 0 & 0 & 0 \end{bmatrix}$.

5 Linearly independent because $\begin{bmatrix} 1 & 1 & 1 \\ 1 & 1 & 0 \\ 1 & 0 & 0 \end{bmatrix} \sim \begin{bmatrix} 1 & 0 & 0 \\ 0 & 1 & 0 \\ 0 & 0 & 1 \end{bmatrix}$.

7 Linearly dependent because $\begin{bmatrix} 2 & 1 & 1 \\ 1 & 3 & 4 \\ 9 & 7 & 8 \end{bmatrix} \sim \begin{bmatrix} 1 & 0 & -\frac{1}{5} \\ 0 & 1 & \frac{7}{5} \\ 0 & 0 & 0 \end{bmatrix}$.

9 $\begin{bmatrix} \mathbf{a} & \mathbf{b} & \mathbf{c} \end{bmatrix} \sim \begin{bmatrix} 1 & 0 & p \\ 0 & 1 & q \\ 0 & 0 & 0 \end{bmatrix}$, $p \neq 0$, or $\begin{bmatrix} \mathbf{a} & \mathbf{b} & \mathbf{c} \end{bmatrix} \sim \begin{bmatrix} 0 & 1 & 0 \\ 0 & 0 & 1 \\ 0 & 0 & 0 \end{bmatrix}$

11 $x - y + z = 0$. The vectors are linearly dependent.

13 $4x + y - z = 0$. The vectors are linearly dependent.

15 $\mathbf{c} = x\mathbf{a} + y\mathbf{b} = \begin{bmatrix} x + y \\ 2x + y \\ 3x + y \end{bmatrix}$

19 $a = 3$, $\begin{bmatrix} 3 \\ 3 \\ 1 \end{bmatrix} = -\begin{bmatrix} 1 \\ -1 \\ 1 \end{bmatrix} + 2\begin{bmatrix} 2 \\ 1 \\ 1 \end{bmatrix}$

21 Any $a \neq 2$.

23 Any $a \neq 1$.

17 $a = 2$, $\begin{bmatrix} 2 \\ 1 \\ 1 \end{bmatrix} = \begin{bmatrix} 1 \\ 1 \\ 0 \end{bmatrix} + \begin{bmatrix} 1 \\ 0 \\ 1 \end{bmatrix}$

25 $\mathbf{x} = 3\mathbf{a} - \mathbf{b} + \frac{1}{2}\mathbf{c}$

27 $\mathbf{x} = \frac{3}{4}\mathbf{a} - \frac{1}{4}\mathbf{b} - \frac{1}{4}\mathbf{c}$

29 $a = 2$ or $a = 5$.

31 $a = -11$ and $\begin{bmatrix} x \\ y \\ z \end{bmatrix} = t \begin{bmatrix} 5 \\ -1 \\ -7 \end{bmatrix}$.

33 Because the implication (a) implies (b) is immediate, we only have to show the other implication. If $\alpha\mathbf{u} + \beta\mathbf{v}$ is an arbitrary vector in Span$\{\mathbf{u}, \mathbf{v}\}$, then we have

$$\alpha\mathbf{u} + \beta\mathbf{v} = \alpha(p\mathbf{a} + q\mathbf{b} + r\mathbf{c}) + \beta(x\mathbf{a} + y\mathbf{b} + z\mathbf{c}) = (p\alpha + x\beta)\mathbf{a} + (q\alpha + y\beta)\mathbf{b} + (r\alpha + z\beta)\mathbf{c},$$

which means that $\alpha\mathbf{u} + \beta\mathbf{v}$ is in Span$\{\mathbf{a}, \mathbf{b}, \mathbf{c}\}$. We show in the same way that an arbitrary element in Span$\{\mathbf{a}, \mathbf{b}, \mathbf{c}\}$ is in Span$\{\mathbf{u}, \mathbf{v}\}$ which completes the proof.

35 $\det \begin{bmatrix} \mathbf{a} & \mathbf{b} & \mathbf{a} \times \mathbf{b} \end{bmatrix} = (\mathbf{a} \times \mathbf{b}) \cdot (\mathbf{a} \times \mathbf{b}) = \|\mathbf{a} \times \mathbf{b}\|^2$

37 $\det \begin{bmatrix} \mathbf{a} & \mathbf{b} & \mathbf{a} \times \mathbf{b} \end{bmatrix} = \|\mathbf{a} \times \mathbf{b}\|^2 > 0$

39 We have $(\mathbf{c} - (r\mathbf{a} + s\mathbf{b})) \cdot \mathbf{a} = 0$ and $(\mathbf{c} - (r\mathbf{a} + s\mathbf{b})) \cdot \mathbf{b} = 0$. This shows that $r\mathbf{a} + s\mathbf{b}$ is the projection of $\mathbf{c}$ on the vector plane Span$\{\mathbf{a}, \mathbf{b}\}$. We also have $\mathbf{c} - t(\mathbf{a} \times \mathbf{b}) \cdot (\mathbf{a} \times \mathbf{b}) = 0$. This shows that $t(\mathbf{a} \times \mathbf{b})$ is the projection of $\mathbf{c}$ on the vector line Span$\{\mathbf{a} \times \mathbf{b}\}$.

41 $\|\mathbf{c} - \mathbf{p}\| = \left\| \mathbf{c} - \left(\mathbf{c} - \frac{\mathbf{c} \cdot (\mathbf{a} \times \mathbf{b})}{\|\mathbf{a} \times \mathbf{b}\|^2} (\mathbf{a} \times \mathbf{b}) \right) \right\| = \frac{|\mathbf{c} \cdot (\mathbf{a} \times \mathbf{b})|}{\|\mathbf{a} \times \mathbf{b}\|} = \frac{\left| \det \begin{bmatrix} \mathbf{a} & \mathbf{b} & \mathbf{c} \end{bmatrix} \right|}{\|\mathbf{a} \times \mathbf{b}\|}$

43 $A = \frac{1}{2} \|\mathbf{b}\| \frac{\|\mathbf{a} \times \mathbf{b}\|}{\|\mathbf{b}\|} = \frac{1}{2} \|\mathbf{a} \times \mathbf{b}\|$

45 Let $\mathbf{u} = p\mathbf{a} + q\mathbf{b} + r\mathbf{c}$. One of the numbers p, q, r must be different from 0. Suppose $p \neq 0$. We show now that $\{\mathbf{b}, \mathbf{c}, \mathbf{u}\}$ is a basis in $\mathbb{R}^3$. If $x\mathbf{b} + y\mathbf{c} + z\mathbf{u} = \mathbf{0}$, then $x\mathbf{b} + y\mathbf{c} + z(p\mathbf{a} + q\mathbf{b} + r\mathbf{c}) = \mathbf{0}$ or $zp\mathbf{a} + (x + zq)\mathbf{b} + (y + zr)\mathbf{c} = \mathbf{0}$. Hence $x = y = z = 0$, because $p \neq 0$. Consequently the vectors $\mathbf{b}, \mathbf{c}, \mathbf{u}$ are linearly independent, which means that the set $\{\mathbf{b}, \mathbf{c}, \mathbf{u}\}$ is a basis in $\mathbb{R}^3$.

If $q \neq 0$ we obtain that $\{\mathbf{a}, \mathbf{c}, \mathbf{u}\}$ is a basis in $\mathbb{R}^3$ and if $r \neq 0$ we obtain that $\{\mathbf{a}, \mathbf{b}, \mathbf{u}\}$ is a basis in $\mathbb{R}^3$.

Section 5.4

1 2
3 1
5 2
7 1

9 3

11 rank$A = 2$

13 rank$A = 3$

Chapter 6

1 $\sqrt{7}$ and $\sqrt{5}$

3 $\sqrt{50}$ and $\sqrt{10}$

5 $\left\{ A\begin{bmatrix} 1 \\ 1 \end{bmatrix}, A\begin{bmatrix} 1 \\ -1 \end{bmatrix} \right\} = \left\{ \begin{bmatrix} 1 \\ 1 \\ -3 \end{bmatrix}, \begin{bmatrix} 1 \\ 2 \\ 1 \end{bmatrix} \right\}$

7 $\left\{ A\begin{bmatrix} 1 \\ 2 \end{bmatrix}, A\begin{bmatrix} 2 \\ -1 \end{bmatrix} \right\} = \left\{ \begin{bmatrix} 4 \\ 5 \\ -3 \end{bmatrix}, \begin{bmatrix} 3 \\ 0 \\ 4 \end{bmatrix} \right\}$

9 $A = \sigma_1 \mathbf{u}_1 \mathbf{v}_1^T + \sigma_2 \mathbf{u}_2 \mathbf{v}_2^T$, where $\mathbf{u}_1 = \frac{1}{\sqrt{3}} \begin{bmatrix} 1 \\ 1 \\ 1 \end{bmatrix}$, $\mathbf{u}_2 = \frac{1}{\sqrt{2}} \begin{bmatrix} 1 \\ 0 \\ -1 \end{bmatrix}$, $\mathbf{v}_1 = \frac{1}{\sqrt{5}} \begin{bmatrix} 1 \\ 2 \end{bmatrix}$, $\mathbf{v}_2 = \frac{1}{\sqrt{5}} \begin{bmatrix} 2 \\ -1 \end{bmatrix}$,

and the singular values are $\sqrt{15}$ and $\sigma_2 = \sqrt{10}$.

11   $A = \sigma_1 \mathbf{u}_1 \mathbf{v}_1^T + \sigma_2 \mathbf{u}_2 \mathbf{v}_2^T$, where $\mathbf{u}_1 = \frac{1}{\sqrt{6}} \begin{bmatrix} 1 \\ -1 \\ 2 \end{bmatrix}$, $\mathbf{u}_2 = \frac{1}{\sqrt{2}} \begin{bmatrix} -1 \\ -1 \\ 0 \end{bmatrix}$, $\mathbf{v}_1 = \frac{1}{\sqrt{10}} \begin{bmatrix} 3 \\ -1 \end{bmatrix}$ and $\mathbf{v}_2 =$

$\frac{1}{\sqrt{10}} \begin{bmatrix} 1 \\ 3 \end{bmatrix}$, and the singular values are $\sigma_1 = \sqrt{15}$ and $\sigma_2 = \sqrt{5}$.

13   $A = \begin{bmatrix} \mathbf{u}_1 & \mathbf{u}_2 & \mathbf{u}_3 \end{bmatrix} \begin{bmatrix} \sqrt{15} & 0 \\ 0 & \sqrt{10} \\ 0 & 0 \end{bmatrix} \begin{bmatrix} \mathbf{v}_1^T \\ \mathbf{v}_2^T \end{bmatrix}$, $\mathbf{u}_3 = \frac{1}{\sqrt{6}} \begin{bmatrix} 1 \\ -2 \\ 1 \end{bmatrix}$

15   $A = \begin{bmatrix} \mathbf{u}_1 & \mathbf{u}_2 & \mathbf{u}_3 \end{bmatrix} \begin{bmatrix} \sqrt{15} & 0 \\ 0 & \sqrt{10} \\ 0 & 0 \end{bmatrix} \begin{bmatrix} \mathbf{v}_1^T \\ \mathbf{v}_2^T \end{bmatrix}$, $\mathbf{u}_3 = \frac{1}{\sqrt{3}} \begin{bmatrix} 1 \\ -1 \\ -1 \end{bmatrix}$

17   If $A = \sigma_1 \mathbf{u}_1 \mathbf{v}_1^T + \sigma_2 \mathbf{u}_2 \mathbf{v}_2^T$, then $A^T = \sigma_1 (\mathbf{v}_1^T)^T \mathbf{u}_1^T + \sigma_2 (\mathbf{v}_2^T)^T \mathbf{u}_2^T = \sigma_1 \mathbf{v}_1 \mathbf{u}_1^T + \sigma_2 \mathbf{v}_2 \mathbf{u}_2^T$.

19   If $A = \sigma_1 \mathbf{u}_1 \mathbf{v}_1^T + \sigma_2 \mathbf{u}_2 \mathbf{v}_2^T$, then $A^T A = \sigma_1^2 \mathbf{v}_1 \mathbf{v}_1^T + \sigma_2^2 \mathbf{v}_2 \mathbf{v}_2^T$, by exercise 18. Hence $A^T A(\mathbf{v}_1) = (\sigma_1^2 \mathbf{v}_1 \mathbf{v}_1^T + \sigma_2^2 \mathbf{v}_2 \mathbf{v}_2^T)(\mathbf{v}_1) = \sigma_1^2 \mathbf{v}_1$ and $A^T A(\mathbf{v}_2) = (\sigma_1^2 \mathbf{v}_1 \mathbf{v}_1^T + \sigma_2^2 \mathbf{v}_2 \mathbf{v}_2^T)(\mathbf{v}_2) = \sigma_2^2 \mathbf{v}_2$. Therefore σ_1 and σ_2 are the singular values of the matrix A.

Section 7.1

1   3 is an eigenvalue because $\det \begin{bmatrix} 7-3 & 4 & 9 \\ 1 & 5-3 & 1 \\ 1 & 2 & 4-3 \end{bmatrix} = \det \begin{bmatrix} 4 & 4 & 9 \\ 1 & 2 & 1 \\ 1 & 2 & 1 \end{bmatrix} = 0.$

3   1 is an eigenvalue because $\det \begin{bmatrix} 2-1 & 1 & 1 \\ 1 & 2-1 & 1 \\ 8 & 3 & 3-1 \end{bmatrix} = \det \begin{bmatrix} 1 & 1 & 1 \\ 1 & 1 & 1 \\ 8 & 3 & 2 \end{bmatrix} = 0.$

5   1 is an eigenvalue because $\det \begin{bmatrix} 3-1 & 4 & 1 \\ 5 & 7-1 & 9 \\ 2 & 4 & 2-1 \end{bmatrix} = \det \begin{bmatrix} 2 & 4 & 1 \\ 5 & 6 & 9 \\ 2 & 4 & 1 \end{bmatrix} = 0.$

7   3 is an eigenvalue because $\det \begin{bmatrix} 7-3 & 8 & 4 \\ 2 & 9-3 & 3 \\ 2 & 4 & 5-3 \end{bmatrix} = \det \begin{bmatrix} 4 & 8 & 4 \\ 2 & 6 & 3 \\ 2 & 4 & 2 \end{bmatrix} = 2\det \begin{bmatrix} 2 & 4 & 2 \\ 2 & 6 & 3 \\ 2 & 4 & 2 \end{bmatrix} =$
  0.

9   1 is an eigenvalue because $\det \begin{bmatrix} 5-1 & 4 & 4 \\ 4 & 5-1 & 4 \\ 4 & 1 & 8-1 \end{bmatrix} = \det \begin{bmatrix} 4 & 4 & 4 \\ 4 & 4 & 4 \\ 4 & 1 & 7 \end{bmatrix} = 0$ and 4 is an eigen-

  value because $\det \begin{bmatrix} 5-4 & 4 & 4 \\ 4 & 5-4 & 4 \\ 4 & 1 & 8-4 \end{bmatrix} = \det \begin{bmatrix} 1 & 4 & 4 \\ 4 & 1 & 4 \\ 4 & 1 & 4 \end{bmatrix} = 0$

11   Span $\left\{ \begin{bmatrix} 1 \\ 1 \\ 1 \end{bmatrix} \right\}$

13   Span $\left\{ \begin{bmatrix} 1 \\ 1 \\ 1 \end{bmatrix} \right\}$

15   Span $\left\{ \begin{bmatrix} -5 \\ 0 \\ 1 \end{bmatrix}, \begin{bmatrix} -2 \\ 1 \\ 0 \end{bmatrix} \right\}$

17   Span $\left\{ \begin{bmatrix} 1 \\ 1 \\ -4 \end{bmatrix} \right\}$

19 0,2,3

21 $1, \frac{5-\sqrt{5}}{2}, \frac{5+\sqrt{5}}{2}$

23 $P = \begin{bmatrix} 0 & 1 & 0 \\ 2 & -1 & 0 \\ 1 & 1 & 1 \end{bmatrix}, D = \begin{bmatrix} 2 & 0 & 0 \\ 0 & 1 & 0 \\ 0 & 0 & 0 \end{bmatrix}$

25 $P = \begin{bmatrix} -1 & -2 & 2 \\ 0 & 1 & 1 \\ 1 & 0 & 1 \end{bmatrix}, D = \begin{bmatrix} 1 & 0 & 0 \\ 0 & 1 & 0 \\ 0 & 0 & 6 \end{bmatrix}$

27 $P = \begin{bmatrix} -2 & -2 & 1 \\ 0 & 1 & 2 \\ 1 & 0 & 2 \end{bmatrix}, D = \begin{bmatrix} 3 & 0 & 0 \\ 0 & 3 & 0 \\ 0 & 0 & 12 \end{bmatrix}$

29 The eigenvalues are 2 (double) and 3. The eigenspace corresponding to the eigenvalue 2 is Span $\left\{ \begin{bmatrix} 0 \\ 0 \\ 1 \end{bmatrix} \right\}$ and the eigenspace corresponding to the eigenvalue 3 is Span $\left\{ \begin{bmatrix} 1 \\ 1 \\ 2 \end{bmatrix} \right\}$.

Consequently it is not possible to diagonalize the matrix.

Section 7.2

---

1 $P = \begin{bmatrix} \frac{1}{\sqrt{2}} & \frac{1}{\sqrt{3}} & \frac{1}{\sqrt{6}} \\ -\frac{1}{\sqrt{2}} & \frac{1}{\sqrt{3}} & \frac{1}{\sqrt{6}} \\ 0 & -\frac{1}{\sqrt{3}} & \frac{2}{\sqrt{6}} \end{bmatrix}, D = \begin{bmatrix} -1 & 0 & 0 \\ 0 & 1 & 0 \\ 0 & 0 & 7 \end{bmatrix}$

3 $P = \begin{bmatrix} \frac{1}{\sqrt{2}} & \frac{1}{\sqrt{3}} & \frac{1}{\sqrt{6}} \\ 0 & -\frac{1}{\sqrt{3}} & \frac{2}{\sqrt{6}} \\ -\frac{1}{\sqrt{2}} & \frac{1}{\sqrt{3}} & \frac{1}{\sqrt{6}} \end{bmatrix}, D = \begin{bmatrix} 12 & 0 & 0 \\ 0 & 18 & 0 \\ 0 & 0 & 6 \end{bmatrix}$

5 $P = \begin{bmatrix} -\frac{2}{\sqrt{5}} & \frac{1}{30} & \frac{1}{\sqrt{6}} \\ 0 & -\frac{5}{30} & \frac{1}{\sqrt{6}} \\ \frac{1}{\sqrt{5}} & \frac{2}{30} & \frac{2}{\sqrt{6}} \end{bmatrix}, D = \begin{bmatrix} 1 & 0 & 0 \\ 0 & 1 & 0 \\ 0 & 0 & 13 \end{bmatrix}$

7 $P = \begin{bmatrix} \frac{3}{\sqrt{11}} & \frac{1}{\sqrt{10}} & \frac{3}{\sqrt{110}} \\ \frac{1}{\sqrt{11}} & -\frac{3}{\sqrt{10}} & \frac{1}{\sqrt{110}} \\ \frac{1}{\sqrt{11}} & 0 & -\frac{10}{\sqrt{110}} \end{bmatrix}, D = \begin{bmatrix} 33 & 0 & 0 \\ 0 & 0 & 0 \\ 0 & 0 & 0 \end{bmatrix}$

9 $P = \begin{bmatrix} \frac{2}{\sqrt{6}} & \frac{1}{\sqrt{5}} & \frac{2}{\sqrt{30}} \\ \frac{1}{\sqrt{6}} & -\frac{2}{\sqrt{5}} & \frac{1}{\sqrt{30}} \\ \frac{1}{\sqrt{6}} & 0 & -\frac{5}{\sqrt{30}} \end{bmatrix}, D = \begin{bmatrix} 10 & 0 & 0 \\ 0 & 10 & 0 \\ 0 & 0 & -20 \end{bmatrix}$

11 $A = \frac{12}{2} \begin{bmatrix} 1 \\ 0 \\ -1 \end{bmatrix} \begin{bmatrix} 1 & 0 & -1 \end{bmatrix} + \frac{18}{3} \begin{bmatrix} 1 \\ -1 \\ 1 \end{bmatrix} \begin{bmatrix} 1 & -1 & 1 \end{bmatrix} + \frac{6}{6} \begin{bmatrix} 1 \\ 2 \\ 1 \end{bmatrix} \begin{bmatrix} 1 & 2 & 1 \end{bmatrix}$

13 $A = \frac{10}{6} \begin{bmatrix} 2 \\ 1 \\ 1 \end{bmatrix} \begin{bmatrix} 2 & 1 & 1 \end{bmatrix} + \frac{10}{5} \begin{bmatrix} 1 \\ -2 \\ 0 \end{bmatrix} \begin{bmatrix} 1 & -2 & 0 \end{bmatrix} - \frac{20}{30} \begin{bmatrix} 2 \\ 1 \\ -5 \end{bmatrix} \begin{bmatrix} 2 & 1 & -5 \end{bmatrix}$

15 $A = \begin{bmatrix} 27 & -13 & -5 \\ -13 & 30 & -8 \\ -5 & -8 & 22 \end{bmatrix}$

17 $A = \begin{bmatrix} \frac{41}{2} & \frac{25}{2} & 0 \\ \frac{25}{2} & \frac{41}{2} & 0 \\ 0 & 0 & 33 \end{bmatrix}$

19 $\begin{bmatrix} 1 & 2 & 0 \\ 1 & 0 & 1 \\ 0 & 1 & 1 \end{bmatrix} = \begin{bmatrix} \frac{1}{\sqrt{2}} & \frac{1}{\sqrt{3}} & -\frac{1}{\sqrt{6}} \\ \frac{1}{\sqrt{2}} & -\frac{1}{\sqrt{3}} & \frac{1}{\sqrt{6}} \\ 0 & \frac{1}{\sqrt{3}} & \frac{2}{\sqrt{6}} \end{bmatrix} \begin{bmatrix} \sqrt{2} & \sqrt{2} & \frac{1}{\sqrt{2}} \\ 0 & \sqrt{3} & 0 \\ 0 & 0 & \frac{3}{\sqrt{6}} \end{bmatrix}$

21 Let $A = \begin{bmatrix} a & b & c \\ b & d & e \\ c & e & f \end{bmatrix}$. Then

$$\det \begin{bmatrix} a-\lambda & b & c \\ b & d-\lambda & e \\ c & e & f-\lambda \end{bmatrix} = (a-\lambda)(d-\lambda)(f-\lambda)+2bce-(a-\lambda)e^2-(d-\lambda)c^2-(f-\lambda)b^2.$$

Let
$$\Phi(\lambda) = (a-\lambda)(d-\lambda)(f-\lambda)+2bce-(a-\lambda)e^2-(d-\lambda)c^2-(f-\lambda)b^2.$$
If $\Phi(\lambda) = -(\lambda-\alpha)^3$, then $\Phi'(\lambda) = -3(\lambda-\alpha)^2$ and $\Phi''(\lambda) = -6(\lambda-\alpha)$. Since $\Phi''(\lambda) = 2(a+d+f)-6\lambda$, we have

$$2(a+d+f)-6\lambda = -6(\lambda-\alpha)$$

and consequently $\alpha = \frac{1}{3}(a+d+f)$. Now we have

$$\Phi'(\lambda) = -(a-\lambda)(d-\lambda)-(a-\lambda)(f-\lambda)-(a-\lambda)(d-\lambda)+e^2+c^2+b^2$$
$$= -3\lambda^2+2(a+d+f)\lambda-df-af-ad+e^2+c^2+b^2$$

and, since $\alpha = \frac{1}{3}(a+d+f)$,

$$\Phi'(\alpha) = -\frac{1}{3}(a+d+f)^2+\frac{2}{3}(a+d+f)^2-df-af-ad+e^2+c^2+b^2$$
$$= \frac{1}{3}(a+d+f)^2-df-af-ad+e^2+c^2+b^2$$
$$= \frac{1}{6}(a-d)^2+(d-f)^2+(f-a)^2)+e^2+c^2+b^2.$$

But $\Phi'(\alpha) = 0$, so we must have $a = d = f = \alpha$ and $e = c = b = 0$.

Section 8.1

1 $2x+y=1$

3 $3x+2y=11$

5 $\left(\begin{bmatrix} x \\ y \end{bmatrix} - \frac{1}{2}\begin{bmatrix} 3 \\ 1 \end{bmatrix}\right) \cdot \begin{bmatrix} 3 \\ 1 \end{bmatrix} = 0$

13 $\det \begin{bmatrix} 1 & 1 & 1 \\ x_1 & a_1 & b_1 \\ x_2 & a_2 & b_2 \end{bmatrix} = -\det \begin{bmatrix} x_1-a_1 & b_1-a_1 \\ x_2-a_2 & b_2-a_2 \end{bmatrix} = 0$

7 $\frac{2}{5}\begin{bmatrix} 3 \\ 4 \end{bmatrix}$

9 $\frac{1}{2}\begin{bmatrix} -3 \\ 1 \end{bmatrix}$

11 $\frac{5}{2}$

Section 8.2

1 $\frac{1}{3}\begin{bmatrix} 1 \\ 1 \\ 4 \end{bmatrix}$

3 $x + y - 4z = 6$

5 $x - y - z = 1$

7 $2x + 3y + 2z = 9$

9 $\left(\begin{bmatrix} x \\ y \\ z \end{bmatrix} - \frac{2}{11}\begin{bmatrix} 3 \\ -1 \\ 1 \end{bmatrix} \right) \cdot \begin{bmatrix} 3 \\ -1 \\ 1 \end{bmatrix} = 0$

11 $\begin{bmatrix} 0 \\ 0 \\ 2 \end{bmatrix} + \mathrm{Span}\left\{ \begin{bmatrix} 1 \\ 0 \\ -3 \end{bmatrix}, \begin{bmatrix} 0 \\ 1 \\ 1 \end{bmatrix} \right\}$

13 $\frac{1}{26}\begin{bmatrix} 17 \\ 23 \\ 12 \end{bmatrix}$

15 $\sqrt{\frac{117}{13}}$

17 $\frac{1}{3}\begin{bmatrix} 2 \\ 4 \\ 5 \end{bmatrix}$

19 $3x - y + z = 3$

21 $x + y + z = 1$

23 $\frac{\sqrt{2}}{2}$

25 $\frac{\sqrt{3}}{2}$

27 $\frac{2}{3}$

29 2

31 The conditions $(\mathbf{p} - \mathbf{a}) \cdot (\mathbf{u} \times \mathbf{v}) = 0$, $(\mathbf{q} - \mathbf{p}) \cdot \mathbf{u} = 0$, and $(\mathbf{q} - \mathbf{p}) \cdot \mathbf{v} = 0$ are equivalent to the fact that $\mathbf{p}$ is in the plane $\mathbf{a} + \mathrm{Span}\{\mathbf{u}, \mathbf{v}\}$ and there is a real number t such that $\mathbf{p} = \mathbf{q} + t(\mathbf{u} \times \mathbf{v})$.

Section 9.1

1 By Theorem 9.1.4 we have $ps - qt = 0$ and $pt + qs = 1$. Hence

$$p = p(pt + qs) = psq + tp^2 = tq^2 + tp^2 = t(q^2 + p^2) = t$$

and

$$q = q(pt + qs) = sq^2 + tpq = sq^2 + sp^2 = s(q^2 + p^2) = s.$$

3 By Theorem 9.1.4 we have $ps - qt = -1$ and $pt + qs = 0$. Hence

$$-p = p(ps - qt) = sp^2 - ptq = tp^2 + sq^2 = s(q^2 + p^2) = s$$

and

$$-q = qsp - tq^2 = -tp^2 - tq^2 = -t(p^2 + q^2) = -t.$$

5 Using Theorem 9.1.4 we obtain

$$\mathbf{c} = (p^2 - q^2)\mathbf{a} + 2pq\mathbf{a}^{\perp} = (p^2 + p^2 - p^2 - q^2)\mathbf{a} + 2pq\mathbf{a}^{\perp}$$
$$= (2p^2 - (p^2 + q^2))\mathbf{a} + 2pq\mathbf{a}^{\perp} = (2p^2 - 1)\mathbf{a} + 2pq\mathbf{a}^{\perp}.$$

7 If $\mathbf{a} = \begin{bmatrix} a_1 \\ a_2 \end{bmatrix}$, then $\mathbf{b} = p\mathbf{a} + q\mathbf{a}^{\perp} = p\begin{bmatrix} a_1 \\ a_2 \end{bmatrix} + q\begin{bmatrix} -a_2 \\ a_1 \end{bmatrix} = \begin{bmatrix} pa_1 - qa_2 \\ pa_2 + qa_1 \end{bmatrix} = \begin{bmatrix} p & -q \\ q & p \end{bmatrix}\begin{bmatrix} a_1 \\ a_2 \end{bmatrix}$.

9 $L(M(\mathbf{x})) = p\mathbf{x}^{\perp} + q(\mathbf{x}^{\perp})^{\perp} = -q\mathbf{x} + p\mathbf{x}^{\perp}$

11 Since $(4p^3 - 3p)\mathbf{a} + (3q - 4q^3)\mathbf{a}^{\perp} = -\mathbf{a}$, we have $4p^3 - 3p = -1$ and $3q - 4q^3 = 0$.

13 Since $(2p^2 - 1)\mathbf{a} + 2pq\mathbf{a}^\perp = \frac{1}{2}\mathbf{a} + \frac{\sqrt{3}}{2}\mathbf{a}^\perp$, we have $2p^2 - 1 = \frac{1}{2}$ and $2pq = \frac{\sqrt{3}}{2}$. Consequently, $p = \frac{\sqrt{3}}{2}$ and $q = \frac{1}{2}$.

15 This is an immediate consequence of Exercise 11.

17 Since $(2p^2 - 1)\mathbf{x} + 2pq\mathbf{x}^\perp = -\mathbf{x}$, we have $2p^2 - 1 = -1$ and $2pq = 0$. Consequently, $p = 0$ and $q = 1$.

19 $\|\mathbf{b} - \mathbf{a}\| = \|\mathbf{a}\|$

21 This is a direct consequence of Theorem 9.1.6.

25 $\cos\frac{7\pi}{12} = \cos(\frac{\pi}{3} + \frac{\pi}{4}) = \frac{1}{2}\frac{\sqrt{2}}{2} - \frac{\sqrt{3}}{2}\frac{\sqrt{2}}{2}$

27 $\cos\frac{\pi}{12} = \cos(\frac{\pi}{3} - \frac{\pi}{4}) = \frac{1}{2}\frac{\sqrt{2}}{2} + \frac{\sqrt{3}}{2}\frac{\sqrt{2}}{2}$

29 $\cos\frac{5\pi}{12} = \cos(\frac{\pi}{4} + \frac{\pi}{6}) = \frac{\sqrt{3}}{2}\frac{\sqrt{2}}{2} - \frac{1}{2}\frac{\sqrt{2}}{2}$

31 $\cos(\theta + \pi) = \cos\theta\cos\pi - \sin\theta\sin\pi = -\cos\theta$

33 This is a direct consequence of Theorem 9.1.6.

35 This is a direct consequence of Theorem 9.1.6.

37 This is a direct consequence of Theorem 9.1.6.

39 This is a direct consequence of Theorem 9.1.6.

Section 9.2

1 $2x^2 + 2xy + 5y^2$

3 $-2x^2 - 8xy + y^2$

5 $\begin{bmatrix} 3 & 7 \\ 7 & 2 \end{bmatrix}$

7 $\begin{bmatrix} 1 & \frac{1}{2} \\ \frac{1}{2} & 4 \end{bmatrix}$

9 Positive semidefinite.

11 Negative definite.

13 $\begin{bmatrix} \sqrt{2} & 0 \\ -\frac{3\sqrt{2}}{2} & \frac{\sqrt{10}}{2} \end{bmatrix} \begin{bmatrix} \sqrt{2} & -\frac{3\sqrt{2}}{2} \\ 0 & \frac{\sqrt{10}}{2} \end{bmatrix}$

15 The matrix is positive definite.

17 The matrix is not positive definite.

19 $\begin{bmatrix} \sqrt{2} & 0 \\ \frac{3\sqrt{2}}{2} & \frac{\sqrt{2}}{2} \end{bmatrix} \begin{bmatrix} \sqrt{2} & \frac{3\sqrt{2}}{2} \\ 0 & \frac{\sqrt{2}}{2} \end{bmatrix}$

21 (a) Since $\begin{bmatrix} -3 & 3 \\ 3 & 5 \end{bmatrix} = \begin{bmatrix} 1 & -3 \\ 3 & 1 \end{bmatrix} \begin{bmatrix} 6 & 0 \\ 0 & -4 \end{bmatrix} \begin{bmatrix} 1 & -3 \\ 3 & 1 \end{bmatrix}^{-1} = \begin{bmatrix} \frac{1}{\sqrt{10}} & -\frac{3}{\sqrt{10}} \\ \frac{3}{\sqrt{10}} & \frac{1}{\sqrt{10}} \end{bmatrix} \begin{bmatrix} 6 & 0 \\ 0 & -4 \end{bmatrix} \begin{bmatrix} \frac{1}{\sqrt{10}} & -\frac{3}{\sqrt{10}} \\ \frac{3}{\sqrt{10}} & \frac{1}{\sqrt{10}} \end{bmatrix}^T$,

we can take $P = \begin{bmatrix} \frac{1}{\sqrt{10}} & -\frac{3}{\sqrt{10}} \\ \frac{3}{\sqrt{10}} & \frac{1}{\sqrt{10}} \end{bmatrix}$.

(b) $\begin{bmatrix} x' & y' \end{bmatrix} P^T \begin{bmatrix} -3 & 3 \\ 3 & 5 \end{bmatrix} P \begin{bmatrix} x' \\ y' \end{bmatrix} = 6(x')^2 - 4(y')^2$

(c) Since $R\left(\begin{bmatrix} a \\ b \end{bmatrix}\right) = \frac{1}{\sqrt{10}}\begin{bmatrix} a \\ b \end{bmatrix} + \frac{3}{\sqrt{10}}\begin{bmatrix} -b \\ a \end{bmatrix}$, we have $\begin{bmatrix} \frac{1}{\sqrt{10}} \\ \frac{3}{\sqrt{10}} \end{bmatrix} = \frac{1}{\sqrt{10}}\begin{bmatrix} 1 \\ 0 \end{bmatrix} + \frac{3}{\sqrt{10}}\begin{bmatrix} 0 \\ 1 \end{bmatrix}$ and $\begin{bmatrix} -\frac{3}{\sqrt{10}} \\ \frac{1}{\sqrt{10}} \end{bmatrix} =$

$\frac{1}{\sqrt{10}}\begin{bmatrix} 0 \\ 1 \end{bmatrix} + \frac{3}{\sqrt{10}}\begin{bmatrix} -1 \\ 0 \end{bmatrix}$. Now, because $\cos^{-1}\frac{1}{\sqrt{10}} \approx 71°$, the angle of the rotation is approximately $71°$.

23 (a) Since $\begin{bmatrix} -3 & 3 \\ 3 & 5 \end{bmatrix} = \begin{bmatrix} 1 & -3 \\ 3 & 1 \end{bmatrix} \begin{bmatrix} 6 & 0 \\ 0 & -4 \end{bmatrix} \begin{bmatrix} 1 & -3 \\ 3 & 1 \end{bmatrix}^{-1} = \begin{bmatrix} -\frac{3}{\sqrt{10}} & -\frac{1}{\sqrt{10}} \\ \frac{1}{\sqrt{10}} & -\frac{3}{\sqrt{10}} \end{bmatrix} \begin{bmatrix} -4 & 0 \\ 0 & 6 \end{bmatrix} \begin{bmatrix} -\frac{3}{\sqrt{10}} & -\frac{1}{\sqrt{10}} \\ \frac{1}{\sqrt{10}} & -\frac{3}{\sqrt{10}} \end{bmatrix}^T$,

we can take $P = \begin{bmatrix} -\frac{3}{\sqrt{10}} & -\frac{1}{\sqrt{10}} \\ \frac{1}{\sqrt{10}} & -\frac{3}{\sqrt{10}} \end{bmatrix}$.

(b) $\begin{bmatrix} x' & y' \end{bmatrix} P^T \begin{bmatrix} -3 & 3 \\ 3 & 5 \end{bmatrix} P \begin{bmatrix} x' \\ y' \end{bmatrix} = -4(x')^2 + 6(y')^2$

(c) The rotation is $R\left(\begin{bmatrix} a \\ b \end{bmatrix}\right) = -\frac{3}{\sqrt{10}} \begin{bmatrix} a \\ b \end{bmatrix} + \frac{1}{\sqrt{10}} \begin{bmatrix} -b \\ a \end{bmatrix}$ and, because $\cos^{-1} \frac{3}{\sqrt{10}} \approx 19°$, the angle of the rotation is approximately $180° - 19° = 161°$.

25 (a) Since $\begin{bmatrix} -3 & 3 \\ 3 & 5 \end{bmatrix} = \begin{bmatrix} 1 & -3 \\ 3 & 1 \end{bmatrix} \begin{bmatrix} 6 & 0 \\ 0 & -4 \end{bmatrix} \begin{bmatrix} 1 & -3 \\ 3 & 1 \end{bmatrix}^{-1} = \begin{bmatrix} \frac{3}{\sqrt{10}} & \frac{1}{\sqrt{10}} \\ -\frac{1}{\sqrt{10}} & \frac{3}{\sqrt{10}} \end{bmatrix} \begin{bmatrix} -4 & 0 \\ 0 & 6 \end{bmatrix} \begin{bmatrix} \frac{3}{\sqrt{10}} & \frac{1}{\sqrt{10}} \\ -\frac{1}{\sqrt{10}} & \frac{3}{\sqrt{10}} \end{bmatrix}^T$,

we can take $P = \begin{bmatrix} \frac{3}{\sqrt{10}} & \frac{1}{\sqrt{10}} \\ -\frac{1}{\sqrt{10}} & \frac{3}{\sqrt{10}} \end{bmatrix}$.

(b) $\begin{bmatrix} x' & y' \end{bmatrix} P^T \begin{bmatrix} -3 & 3 \\ 3 & 5 \end{bmatrix} P \begin{bmatrix} x' \\ y' \end{bmatrix} = -4(x')^2 + 6(y')^2$

(c) The rotation is $R\left(\begin{bmatrix} a \\ b \end{bmatrix}\right) = \frac{3}{\sqrt{10}} \begin{bmatrix} a \\ b \end{bmatrix} - \frac{1}{\sqrt{10}} \begin{bmatrix} -b \\ a \end{bmatrix}$ and, because $\cos^{-1} \frac{3}{\sqrt{10}} \approx 19°$, the angle of the rotation is approximately $360° - 19° = 341°$.

27 (a) Since $\begin{bmatrix} -3 & 3 \\ 3 & 5 \end{bmatrix} = \begin{bmatrix} 1 & -3 \\ 3 & 1 \end{bmatrix} \begin{bmatrix} 6 & 0 \\ 0 & -4 \end{bmatrix} \begin{bmatrix} 1 & -3 \\ 3 & 1 \end{bmatrix}^{-1} = \begin{bmatrix} -\frac{1}{\sqrt{10}} & \frac{3}{\sqrt{10}} \\ -\frac{3}{\sqrt{10}} & -\frac{1}{\sqrt{10}} \end{bmatrix} \begin{bmatrix} 6 & 0 \\ 0 & -4 \end{bmatrix} \begin{bmatrix} -\frac{1}{\sqrt{10}} & \frac{3}{\sqrt{10}} \\ -\frac{3}{\sqrt{10}} & -\frac{1}{\sqrt{10}} \end{bmatrix}^T$,

we can take $P = \begin{bmatrix} -\frac{1}{\sqrt{10}} & \frac{3}{\sqrt{10}} \\ -\frac{3}{\sqrt{10}} & -\frac{1}{\sqrt{10}} \end{bmatrix}$.

(b) $\begin{bmatrix} x' & y' \end{bmatrix} P^T \begin{bmatrix} -3 & 3 \\ 3 & 5 \end{bmatrix} P \begin{bmatrix} x' \\ y' \end{bmatrix} = 6(x')^2 - 4(y')^2$

(c) The rotation is $R\left(\begin{bmatrix} a \\ b \end{bmatrix}\right) = -\frac{1}{\sqrt{10}} \begin{bmatrix} a \\ b \end{bmatrix} - \frac{3}{\sqrt{10}} \begin{bmatrix} -b \\ a \end{bmatrix}$ and, because $\cos^{-1} \frac{1}{\sqrt{10}} \approx 71°$, the angle of the rotation is approximately $180° + 71° = 251°$.

29 (a) Since

$$\begin{bmatrix} x & y \end{bmatrix} \begin{bmatrix} 3 & -5 \\ -5 & 27 \end{bmatrix} \begin{bmatrix} x \\ y \end{bmatrix} = \begin{bmatrix} x & y \end{bmatrix} \begin{bmatrix} 5 & -1 \\ 1 & 5 \end{bmatrix} \begin{bmatrix} 2 & 0 \\ 0 & 28 \end{bmatrix} \begin{bmatrix} 5 & -1 \\ 1 & 5 \end{bmatrix}^{-1} \begin{bmatrix} x \\ y \end{bmatrix}$$

$$= \begin{bmatrix} x & y \end{bmatrix} \begin{bmatrix} \frac{5}{\sqrt{26}} & -\frac{1}{\sqrt{26}} \\ \frac{1}{\sqrt{26}} & \frac{5}{\sqrt{26}} \end{bmatrix} \begin{bmatrix} 2 & 0 \\ 0 & 28 \end{bmatrix} \begin{bmatrix} \frac{5}{\sqrt{26}} & -\frac{1}{\sqrt{26}} \\ \frac{1}{\sqrt{26}} & \frac{5}{\sqrt{26}} \end{bmatrix}^T \begin{bmatrix} x \\ y \end{bmatrix},$$

we can take $P = \begin{bmatrix} \frac{5}{\sqrt{26}} & -\frac{1}{\sqrt{26}} \\ \frac{1}{\sqrt{26}} & \frac{5}{\sqrt{26}} \end{bmatrix}$.

(b) Since $\begin{bmatrix} x \\ y \end{bmatrix} = \begin{bmatrix} \frac{5}{\sqrt{26}} & -\frac{1}{\sqrt{26}} \\ \frac{1}{\sqrt{26}} & \frac{5}{\sqrt{26}} \end{bmatrix} \begin{bmatrix} x' \\ y' \end{bmatrix}$, the equation $\begin{bmatrix} x' & y' \end{bmatrix} P^T \begin{bmatrix} 3 & -5 \\ -5 & 27 \end{bmatrix} P \begin{bmatrix} x' \\ y' \end{bmatrix} = 14$ be-

comes $\frac{(x')^2}{7} + \frac{(y')^2}{\frac{1}{2}} = 1$.

(c) The rotation is $R\left(\begin{bmatrix} a \\ b \end{bmatrix}\right) = \frac{5}{\sqrt{26}}\begin{bmatrix} a \\ b \end{bmatrix} + \frac{1}{\sqrt{26}}\begin{bmatrix} -b \\ a \end{bmatrix}$ and the angle of rotation is $\cos^{-1}\frac{5}{\sqrt{26}} \approx 11°$.

Section 9.3

1 $\cos\alpha\begin{bmatrix} 1 \\ 0 \\ 0 \end{bmatrix} + \sin\alpha\left(\begin{bmatrix} 0 \\ 0 \\ 1 \end{bmatrix} \times \begin{bmatrix} 1 \\ 0 \\ 0 \end{bmatrix}\right) = \begin{bmatrix} \cos\alpha \\ \sin\alpha \\ 0 \end{bmatrix}$

3 $\cos\alpha\begin{bmatrix} 0 \\ 0 \\ 1 \end{bmatrix} + \sin\alpha\left(\begin{bmatrix} 1 \\ 0 \\ 0 \end{bmatrix} \times \begin{bmatrix} 0 \\ 0 \\ 1 \end{bmatrix}\right) = \begin{bmatrix} 0 \\ -\sin\alpha \\ \cos\alpha \end{bmatrix}$

5 $\cos\alpha\begin{bmatrix} 0 \\ 1 \\ 0 \end{bmatrix} + \sin\alpha\left(\begin{bmatrix} 1 \\ 0 \\ 0 \end{bmatrix} \times \begin{bmatrix} 0 \\ 1 \\ 0 \end{bmatrix}\right) = \begin{bmatrix} 0 \\ \cos\alpha \\ \sin\alpha \end{bmatrix}$

7 If $\mathbf{a} = \begin{bmatrix} a_1 \\ a_2 \\ a_3 \end{bmatrix}$ and $\mathbf{n} = \begin{bmatrix} 1 \\ 0 \\ 0 \end{bmatrix}$, then

$$\frac{\mathbf{a}\cdot\mathbf{n}}{\|\mathbf{n}\|^2}\cdot\mathbf{n} + \cos\alpha\left(\mathbf{a} - \frac{\mathbf{a}\cdot\mathbf{n}}{\|\mathbf{n}\|^2}\cdot\mathbf{n}\right) + \frac{1}{\|\mathbf{n}\|}\sin\alpha\,(\mathbf{n}\times\mathbf{a}) = \begin{bmatrix} a_1 \\ a_2\cos\alpha - a_3\sin\alpha \\ a_2\sin\alpha + a_3\cos\alpha \end{bmatrix}.$$

9 If $\mathbf{a} = \begin{bmatrix} a_1 \\ a_2 \\ a_3 \end{bmatrix}$ and $\mathbf{n} = \begin{bmatrix} 0 \\ 1 \\ 0 \end{bmatrix}$, then

$$\frac{\mathbf{a}\cdot\mathbf{n}}{\|\mathbf{n}\|^2}\cdot\mathbf{n} + \cos\alpha\left(\mathbf{a} - \frac{\mathbf{a}\cdot\mathbf{n}}{\|\mathbf{n}\|^2}\cdot\mathbf{n}\right) + \frac{1}{\|\mathbf{n}\|}\sin\alpha\,(\mathbf{n}\times\mathbf{a}) = \begin{bmatrix} a_1\cos\alpha + a_3\sin\alpha \\ a_2 \\ -a_1\sin\alpha + a_3\cos\alpha \end{bmatrix}.$$

Chapter 11

The solutions for the problems for a computer algebra system presented here are written in the Maple code. It is necessary to include
> *with(LinearAlgebra)* :
at the beginning of your Maple document.

1   > *A* := *Matrix*([[2, 3, 5, 1], [4, 7, 9, 2], [8, 15, 17, 4]]);

   > *ReducedRowEchelonForm(A)*;

2   > *B* := *Matrix*([[3, 2, 1], [8, 3, 4], [5, 7, 2]]);

   > *ReducedRowEchelonForm(B)*;

   > *C* := *Matrix*([[4, 7], [3, 0], [7, 9]]);

   > *LinearSolve(B, C)*;

3   > *E* := *Matrix*([[2, 1, 5, 2], [3, 8, 7, 4], [1, 5, 7, 3], [4, 3, 3, 2]]);

> *MatrixInverse(E);*

4   > *Matrix([[1, 4, 0], [0, 1, 0], [0, −3, 1]]).Matrix([[1, 0, 5], [0, 1, 2], [0, 0, 1]]).*
       Matrix([[2, 1, 3], [1, 1, 4], [7, 2, 5]]);

5   > *u := Vector([2, 3, 1]);*

      > *v := Vector([4, 1, 3]);*

      > *P := ProjectionMatrix({u, v});*

      > *b := Vector([7, 21, 14]);*

      > *P.b;*

6   > *F := Matrix([[2, 1], [1, 3], [5, 2]]);*

      > *c := Vector([2, 1, 1]);*

      > *LeastSquares(F, c);*

7   > *G := Matrix([[7, 2, x], [y, 2, 5], [4, −3, 8]]);*

      > *Determinant(G);*

8   > *H := Matrix([[2, 1, 2], [2, 3, 4], [4, 2, 10]]);*

      > *Eigenvectors(H);*

9   > *p := Vector([3, 1, 4]);*

      > *q := Vector([1, −11, 2]);*

      > *r := CrossProduct(p, q);*

      > $A1 := \dfrac{x}{(Norm(p, 2))^2} OuterProductMatrix(p, p);$

      > $A2 := \dfrac{y}{(Norm(q, 2))^2} OuterProductMatrix(q, q);$

      > $A3 := \dfrac{z}{(Norm(r, 2))^2} OuterProductMatrix(r, r);$

      > *A1 + A2 + A3;*

10   > *K := Matrix([[1, 1, 1], [2, −1, 5], [1, 2, −3]]);*

      > *QRDecomposition(K);*

Bibliography

[1] D. Atanasiu, Linjär Algebra och Geometri, Göteborgs Universitet, 1994.

[2] J.S.R. Chisholm, Vectors in three-dimensional space, Cambridge University Press, 1978.

[3] J. Dieudonné, Algèbre linéaire et géométrie élémentaire, Hermann, 1964.

[4] T.W. Körner, Vectors Pure and Applied, Cambridge University Press, 2013.

[5] S. Lang, Introduction to Linear Algebra, Springer, 2nd edition, 1997.

[6] R. Larson, Elementary Linear Algebra, Cengage, 8th edition, 2017.

[7] D. Lay, S. Lay, and J. McDonald, Linear Algebra and Its Applications, 5th Edition, Pearson, 2016.

[8] L. Spence, A. Insel, and S. Friedberg, Elementary Linear Algebra, Pearson, 2nd edition, 2007.

Index

$m \times n$ matrix, 55

addition of matrices, 57
adjoint matrix, 254
area of a triangle, 158
associativity of matrix multiplication, 61
augmented matrix, 76

back substitution, 25, 94
basic variable, 93
basis, 135, 140, 181, 190, 270
best approximation, 149, 205, 211

Cartesian coordinates, 132, 179
change of basis, 193
characteristic polynomial, 315
Cholesky decomposition, 409
Chord Theorem, 431
column of a matrix, 55
column space of a matrix, 288
consumption matrix, 37
coordinates of a vector, 141, 190
coordinates of a vectors, 272
Cramer's Rule, 32, 261
cross product, 236
cross-product term of a quadratic form, 403

demand vector, 37
determinant, 28, 244
diagonal matrix, 39, 319
diagonalizable matrix, 46, 319
dimension of a vector subspace, 284
distance between two points, 144, 201
distance from a point to a line, 156, 363, 371
distance from a point to a plane, 379
dot product, 145, 200

eigenspace, 318
eigenvalue, 44, 314

eigenvector, 45, 317
elementary matrix, 14, 101
elementary operations, 71
elementary row operations, 76
entry of a matrix, 55
equality of matrices, 56
equation of a plane, 374
equation of a vector plane, 240
equivalent matrices, 98

forward substitution, 25
free variable, 93

Gauss-Jordan form, 79
Gaussian elimination, 78
Gram-Schimdt process, 349

identity matrix, 62
indefinite quadratic form, 400
inverse matrix, 10, 106
invertible matrix, 10, 106

leading one, 79
leading term, 79
least squares, 216
least-squares line, 223
Leontief model, 35
line, 355, 370
linear combination, 141, 190, 270
linearly dependence, 275
linearly dependent vectors, 135, 182, 264
linearly independent vectors, 139, 185, 269
lower triangular matrix, 23
LU-decomposition, 23, 120
LU-factorization, 23, 120

matrix, 55
multiplication of matrices by numbers, 58

negative definite quadratic form, 400
negative semidefinite quadratic form,
            400
nine-point circle, 440
norm of a vector, 143, 200
nullity of a matrix, 312
nullspace of a matrix, 312

observed values, 223
origin, 132, 179
orthogonal basis, 156, 208, 275
orthogonal diagonalization, 167
orthogonal matrix, 165, 331
orthogonal vectors, 202
orthogonally diagonalizable matrix, 333
orthonormal basis, 208, 333
outer product expansion, 297
output vector, 37

Parallelogram law, 201
parameter values, 223
perp operation, 153
pivot column, 80
pivot position, 80
pivot variable, 93
plane through a point orthogonal to a
            line, 372
plane through three points, 381
positive definite matrix, 407
positive definite quadratic form, 400
positive semidefinite quadratic form, 400
predicted values, 223
Principal Axes Theorem, 404
product of matrices, 2, 59, 60
production vector, 37
projection, 150, 206, 214
projection matrix, 153, 207, 215
projection of a point on a line, 363, 371
projection of a point on a plane, 377

QR factorization, 173, 225, 348
quadratic form, 400
quadratic terms of a quadratic form, 403

rank of a matrix, 288
rank theorem for 3×3 matrices, 285
rank-nullity theorem, 312
reduced row echelon form, 79
reflection, 152
regression line, 223

residuals, 223
right-hand rule, 414
rotation, 391, 414
row interchange, 76
row of a matrix, 55
row replacement, 76
row scaling, 76
row space of a matrix, 288

scalar multiplication, 58
simple matrix, 104
singular value decomposition, 301
singular values, 292
span, 134, 186
spectral decomposition, 169, 346
square matrix, 55
sum of matrices, 57
symmetric matrix, 68

Tangent-Secant Theorem, 433
transition matrix, 194
transpose of a matrix, 66

unit matrix, 62
unit vector, 144, 200
upper triangular matrix, 23

vector line, 134, 181
vector plane, 186
vector subspace, 134, 181, 186, 271
volume of a tetrahedron, 282

zero matrix, 57